MATERIALS RESEARCH SOCIETY SYMPOSIA PROCEEDINGS

MATERIALS RESEARCH SOCIETY SYMPOSIA PROCEEDINGS

MATERIALS RESEARCH SOCIETY SYMPOSIA PROCEEDINGS

Better Ceramics Through Chemistry II

Symposium held April 15-19, 1986, Palo Alto, California, U.S.A.

EDITORS:

C. Jeffrey Brinker
Sandia National Laboratories, Albuquerque, New Mexico, U.S.A.

David E. Clark
University of Florida, Gainesville, Florida, U.S.A.

Donald R. Ulrich
Air Force Office of Scientific Research, Washington D.C., U.S.A.

MRS | MATERIALS RESEARCH SOCIETY
Pittsburgh, Pennsylvania

This work was supported by the Air Force Office of Scientific Research, Air Force Systems Command, USAF, under Grant Number AFOSR-86-0108.

Published by:

Materials Research Society
9800 McKnight Road, Suite 327
Pittsburgh, Pennsylvania 15237
telephone (412) 367-3003

Library of Congress Cataloging in Publication Data

Better ceramics through chemistry.

(Materials Research Society symposia proceedings, ISSN 0272-9172 ; v. 73)
Proceedings of the Second Materials Research Society Symposium on Better Ceramics Through Chemistry.
Includes bibliographies and index.
1. Ceramics—Congresses. I. Brinker, C. Jeffrey.   II. Clark, David E.   III. Ulrich, Donald R.   IV. Materials Research Society.   V. Materials Research Society Symposium on Better Ceramics Through Chemistry (2nd : 1986 : Palo Alto, Calif.)
TP786.B484   1986   666   86-17969
ISBN 0-931837-39-1

Manufactured in the United States of America

Manufactured by Publishers Choice Book Mfg. Co.
Mars, Pennsylvania 16046

Contents

PART XI: NEW INITIATIVES/NOVEL MATERIALS

Preface

This volume contains the proceedings of the second Materials Research Society Symposium on Better Ceramics Through Chemistry which was held April 15-18, 1986 in Palo Alto, California. This symposium, which was wholly funded by the Air Force Office of Scientific Research, brought together over 250 international researchers from government laboratories, universities, and industry with diverse backgrounds ranging from polymer physics and chemistry to ceramic engineering. As the symposium title suggests, the central theme of the symposium was to improve upon the current state of the art in ceramic materials by exploring synthetic chemical routes, e.g., solution processing and polymer pyrolysis, as alternatives to conventional processing of natural minerals mined from the earth.

The general topic area, chemical synthesis of ceramics from molecular precursors, is one of the most rapidly evolving fields of material science. Based on the number of submitted papers, the interest in this subject has roughly doubled since 1984 when the first symposium on Better Ceramics Through Chemistry was held in Albuquerque, NM (Vol. 32 in MRS Symposia Proceedings). This growth largely reflects the recent involvement of synthetic, inorganic chemists and physicists. These two groups of scientists have made excellent contributions in the areas of molecular precursors, non-oxide synthesis, organically modified ceramics, nuclear magnetic resonance spectroscopy, small angle scattering and the physics of random systems. It should be clear from the contents of this volume that a multidisciplinary approach is essential to making further advances in ceramic synthesis and processing.

The symposium addressed the synthesis, structure, and applications of ceramic materials derived from molecular precursors. The program was organized to follow the evolution of structure from its inception in solution through gelation (or precipitation) drying, heating, and consolidation. Finally, applications of chemically-derived ceramics were addressed and comparisons were made between the properties and structures of conventional and chemically-derived ceramic materials.

101 papers were presented either orally or as posters. 99 of the presented papers appear here. All papers were subjected to a peer review after which authors had an opportunity to respond to its reviewer's comments. Because a primary goal of the editors was rapid publication, the revised papers were accepted without further revision. Thus, the published papers reflect the views and standards of the authors not necessarily those of the editors.

June 12, 1986   C. Jeffrey Brinker
Albuquerque, NM   David E. Clark
                                 Donald R. Ulrich

PART I

Solution Chemistry and
Synthesis I: Gels

A MOLECULAR BUILDING-BLOCK APPROACH TO THE SYNTHESIS OF CERAMIC MATERIALS

W. G. KLEMPERER, V. V. MAINZ, AND D. M. MILLAR
University of Illinois, School of Chemical Sciences and Materials Research
Laboratory, Urbana, IL 61801

ABSTRACT

Ceramic materials are generally prepared from structurally simple
starting materials, with the consequence that structural properties are
difficult to control on a molecular level. This difficulty might be addressed by following the approach taken in polymer chemistry in which molecular building blocks are first prepared and then polymerized in a subsequent step. In the present case, the polysilicic acid esters $[Si_2O]-$
$(OCH_3)_6$, $[Si_3O_2](OCH_3)_8$, and $[Si_8O_{12}](OCH_3)_8$ are prepared and then polymerized by hydrolysis/condensation. Silicon-29 solution nuclear magnetic
resonance (NMR) spectroscopy is used to estimate the extent to which the
molecular frameworks of these monomers are retained during the course of
hydrolysis and condensation.

INTRODUCTION

The impact of chemistry on materials science has perhaps been most
dramatic in polymer science, where small structural modifications on the
molecular scale are readily seen to affect the bulk properties of materials. Consider, for example, the polyester "EKONOL" [1], 1, and the polymer 2 that differs only by the introduction of CH_2 groups [2]. Since this
group introduces flexibility into the relatively rigid structure of 1, its
effect on physical properties is considerable: whereas "EKONOL" is suitable for continuous use at temperatures over 300°C, polymer 2 melts at
205-210°C. The ability to observe the effect of molecular structure on
physical properties derives from an ability to prepare polymer 1 from monomer 3 and polymer 2 from monomer 4. The reaction conditions employed for
these condensation polymerizations are sufficiently mild to insure that
the monomers' molecular structures are retained during the polymerization
process.

This paper is concerned with an effort to extend the molecular building-block approach used in polymer science to the synthesis of ceramic materials, with the ultimate objective of altering the properties of inorganic solids by varying the structure and composition of their synthetic building-blocks. Specifically, four different monomers will be considered, the silicic acid esters $Si(OCH_3)_4$ (tetramethyl orthosilicate, 5), $[Si_2O](OCH_3)_6$ (hexamethyl disilicate, 6), $[Si_3O_2](OCH_3)_8$ (octamethyl trisilicate, 7), and $[Si_8O_{12}](OCH_3)_8$ (octamethoxy octasilsesquioxane, 8).

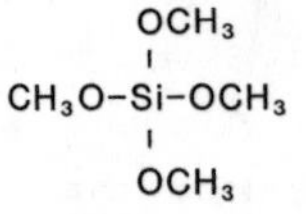

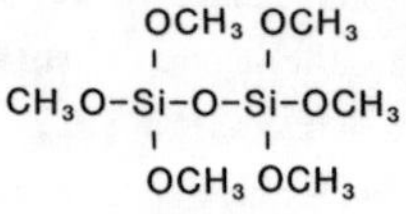

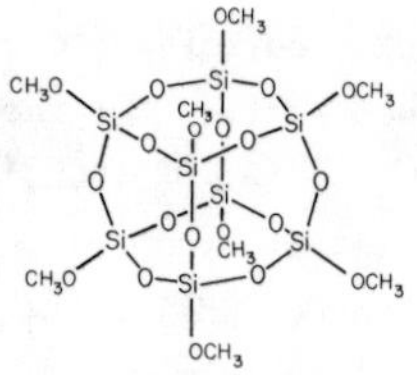

Polysilicic acid esters were selected for examination since numerous studies have shown that monosilicic acid esters like 5 can be polymerized by hydrolysis/condensation under chemically mild, low temperature conditions (see Scheme I). The resulting sols have a number of remarkable properties. Hydrolysis of $Si(OCH_2CH_3)_4$, for example, can yield solutions from which fine, glass-like fibers can be drawn [3], and larger objects such as optical lenses up to 6 cm in diameter have even been obtained using sufficiently long processing times [4].

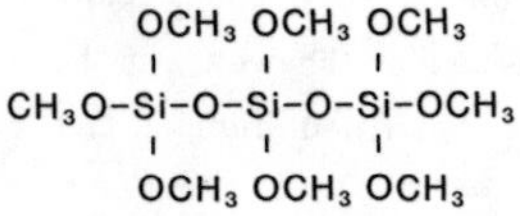

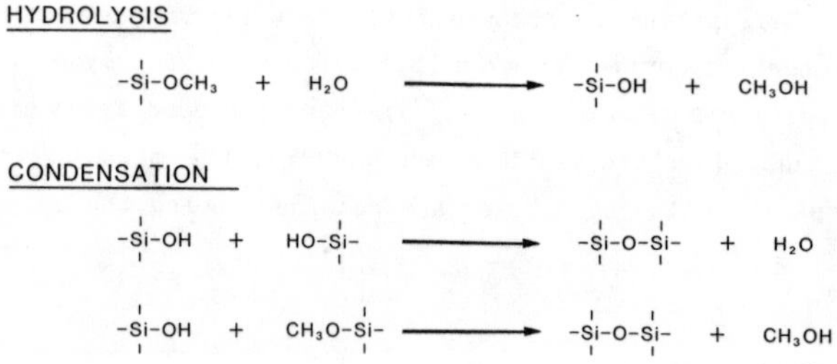

SCHEME I

POLYSILICATE CHAIN MONOMERS

The two simplest possible polysilicate chain monomers **6** and **7** are easily prepared according to equations (1) and (2), respectively, and pro-

$$Cl_3SiOSiCl_3 \xrightarrow[\text{(ii) } CH_3OH]{\text{(i) } HC(OCH_3)_3} (CH_3O)_3SiOSi(OCH_3) \tag{1}$$

$$Cl_3SiOSiCl_2OSiCl_3 \xrightarrow[\text{(ii) } CH_3OH]{\text{(i) } HC(OCH_3)_3} (CH_3O)_3SiOSi(OCH_3)_2OSi(OCH_3)_3 \tag{2}$$

vide a logical starting point for determining whether siloxane monomers can retain their structural integrity during hydrolysis and subsequent condensation. As illustrated in Scheme II for the disiloxane monomer **6**, hydrolysis can proceed without siloxane degradation if water attacks Si-OCH$_3$ bonds only. Hydrolysis of Si-OSi bonds, however, can cleave the disiloxane monomer into two fragments. In a similar fashion, condensation reactions of the type shown in Scheme II between two silicic acids yield no fragmentation if no Si-OSi bonds are cleaved. Condensations between a silicic acid and a silicic acid ester have two analogous options, only one of which yields disiloxane fragmentation. Note that Scheme II does not include alchoholysis reactions that can lead to esterification of a silicic acid, transesterification of an ester, or cleavage of a siloxane linkage. Although not discussed here, alchoholysis can be studied using deuterium-labeled methanol, CD$_3$OH [5].

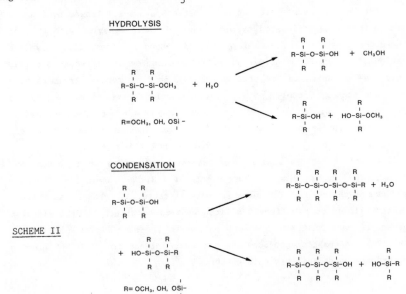

SCHEME II

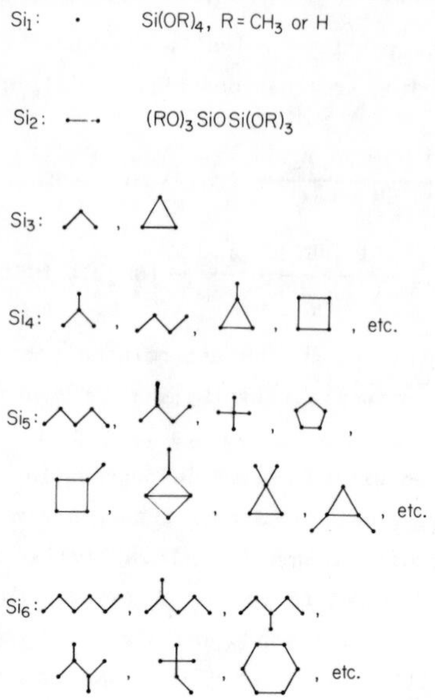

Si_1: • $Si(OR)_4$, R = CH_3 or H

Si_2: ⟶ $(RO)_3 SiOSi(OR)_3$

Si_3: ⋀ , △

Si_4: ⋏ , ⋏⋏ , △ , □ , etc.

Si_5: ⋀⋀ , ⋏⋏ , ✝ , ⬠ ,

□ , ◇ , ⋈ , △ , etc.

Si_6: ⋀⋀⋀ , ⋀⋀⋀ , ⋏⋀ ,

⋏⋏ , ✝ , ⬡ , etc.

SCHEME III

The actual hydrolysis and condensation pathways followed by 6 and 7 can in theory be delineated by comparison of their behavior with that of the monosilicate ester, $Si(OCH_3)_4$ (5). As shown in Scheme III, polymerization of 5 can in principal yield a large number of different species, only some of which can be obtained from 6 or 7 if monomer fragmentation does not occur. Clearly, no Si_1 species can be derived from 6 or 7, no Si_n species where n is odd can be derived from 6, etc., unless siloxane monomer units are fragmented. A physical technique is fortunately available that can provide some information concerning the course of polysilicic acid ester hydrolysis/condensation, namely, ^{29}Si NMR spectroscopy. Marsmann [6] has examined the dependence of ^{29}Si NMR chemical shifts upon structural environment for numerous polysilicates, and shown that silicon centers linked to n different silicon centers by direct Si-O-Si linkages generally have ^{29}Si NMR chemical shifts in the same region (see Scheme IV). This information provides the basis for interpreting ^{29}Si NMR spectra of hydrolysed polysilicic acid esters.

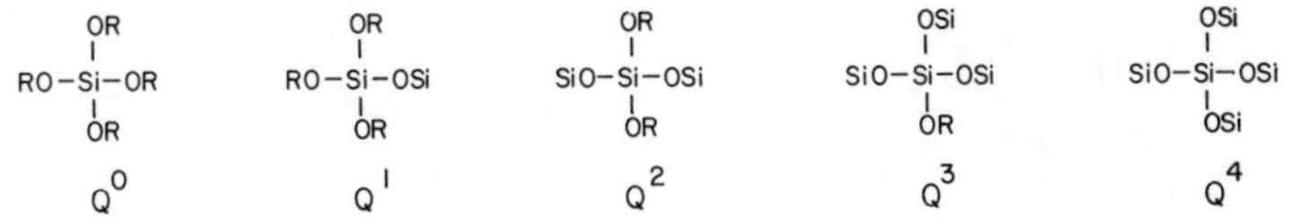

$$
\begin{array}{ccccc}
\text{OR} & \text{OR} & \text{OR} & \text{OSi} & \text{OSi} \\
| & | & | & | & | \\
\text{RO-Si-OR} & \text{RO-Si-OSi} & \text{SiO-Si-OSi} & \text{SiO-Si-OSi} & \text{SiO-Si-OSi} \\
| & | & | & | & | \\
\text{OR} & \text{OR} & \text{OR} & \text{OR} & \text{OSi} \\
Q^0 & Q^1 & Q^2 & Q^3 & Q^4
\end{array}
$$

R = H or CH_3

SCHEME IV

The ^{29}Si NMR spectrum of methanolic $(CH_3O)_3SiOSi(OCH_3)_3$, **6**, measured 3 hours after addition of 15 equivalents of water (see Figure 1a), shows that hydrolysis of both $Si-OCH_3$ and $Si-OSi$ bonds occurs, but that $Si-OCH_3$ hydrolysis is more rapid, as evidenced by the relative intensities of the $(CH_3O)_3SiO\underline{Si}(OCH_3)_2(OH)$ and $(CH_3O)_3\underline{Si}(OH)$ resonances. The spectrum shown in Figure 1b, measured after 9 hours reaction time, shows that extensive condensation has taken place, and that the trisilicate ester $(CH_3O)_3SiOSi-(OCH_3)_2OSi(OCH_3)_3$ is one of the condensation products, again providing unambiguous evidence of monomer degradation.

In light of the results just mentioned, it is somewhat surprising to note that hydrolysis of the trisilicate ester **7** yields no detectable amount of $(CH_3O)_3SiOH$, indicating that the rate of $Si-OSi$ hydrolysis is much slower than the rate of $Si-OCH_3$ hydrolysis (see Figure 2a). The ^{29}Si NMR spectrum shown in Figure 2b was measured in the presence of an acid catalyst, and still displays no $(CH_3O)_3SiOH$ resonance. Unfortunately, the presence or absence of disilicate species resulting from degradation is difficult to establish since the Q^2 region is dominated by resonances arising from $(CH_3O)_3SiOSi(OCH_3)_2OSi(OCH_3)_3$ and its hydrolysis products. It is tempting to attribute the hydrolytic stability of **7** relative to **6** to steric factors, but there is insufficient data available to provide clear support for such an effect.

POLYSILICATE CAGE MONOMER

The cubic monomer **8** is prepared by the route given in equations (3) and (4) [7]. Whereas the flexible chain monomers **6** and **7** offer the

$$[Si_8O_{12}]H_8 \xrightarrow{Cl_2} [Si_8O_{12}]Cl_8 \tag{3}$$

$$[Si_8O_{12}]Cl_8 \xrightarrow{HC(OCH_3)_3} [Si_8O_{12}](OCH_3)_8 \tag{4}$$

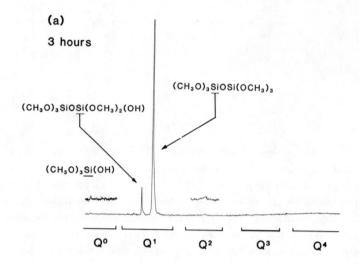

(a)

3 hours

$(CH_3O)_3SiOSi(OCH_3)_3$

$(CH_3O)_3SiOSi(OCH_3)_2(OH)$

$(CH_3O)_3\underline{Si}(OH)$

Q⁰   Q¹   Q²   Q³   Q⁴

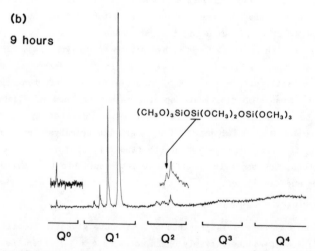

(b)

9 hours

$(CH_3O)_3SiOSi(OCH_3)_2OSi(OCH_3)_3$

Q⁰   Q¹   Q²   Q³   Q⁴

<u>Figure 1.</u> $^{29}Si\{^1H\}$ 59.6-MHz FT NMR spectra of 1.2 M $[Si_2O](OCH_3)_6$ in CH_3OH containing 15 equivalents H_2O, measured with gated decoupling (a) 3 hours and (b) 9 hours after hydrolysis. In each spectrum 356 transients were acquired using a 4 second recycle time. The chemical shift range shown is -75 to -115 ppm relative to TMS. The solution contained 0.015 M $Cr(acac)_3$ to reduce relaxation times; no hydrolysis/condensation catalysts were employed.

9

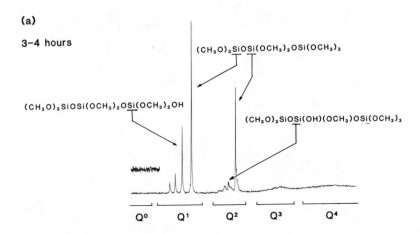

(a)

3-4 hours

$(CH_3O)_3SiOSi(OCH_3)_2OSi(OCH_3)_3$

$(CH_3O)_3SiOSi(OCH_3)_2OSi(OCH_3)_2OH$

$(CH_3O)_3SiOSi(OH)(OCH_3)OSi(OCH_3)_3$

Q^0 Q^1 Q^2 Q^3 Q^4

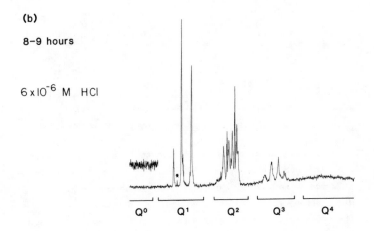

(b)

8-9 hours

6×10^{-6} M HCl

Q^0 Q^1 Q^2 Q^3 Q^4

Figure 2. $^{29}Si\{^1H\}$ 59.6-MHz FT NMR spectra of 1.2 M $[Si_3O_2](OCH_3)_8$ in CH_3OH/H_2O solution measured with gated decoupling (a) 3-4 hours after un-catalyzed hydrolysis with 20 equivalents of H_2O, and (b) 8-9 hours after acid-catalyzed hydrolysis with 2 equivalents of H_2O and 6×10^{-6} M in HCl. In each spectrum 356 transients were acquired using a 4 second recycle time. The chemical shift range shown is -75 to -115 ppm relative to TMS. The solutions contained 0.015 M $Cr(acac)_3$ to reduce relaxation times. The small resonance in spectrum (b) marked with an asterisk is assigned to a cyclic trisiloxane.

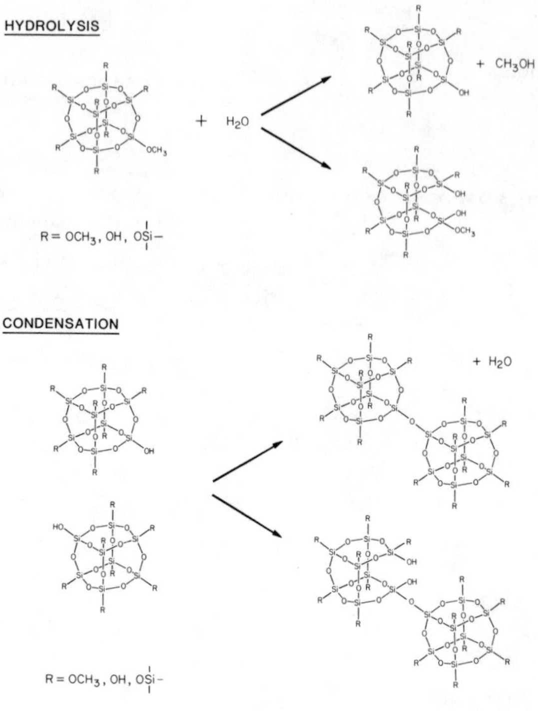

HYDROLYSIS

$R = OCH_3, OH, OSi-$

CONDENSATION

$R = OCH_3, OH, OSi-$

SCHEME V

possibility of generating strong gels through their built-in crosslinking capabilities, the cubic cage monomer **8** could potentially exert an even greater influence on gel structure due to its structural rigidity. This rigidity might affect local structure and bulk properties in much the same way that the rigidity of monomer **3** affects the properties of polymer **1**.

As illustrated in Scheme V, the cubic $[Si_8O_{12}](OCH_3)_8$ monomer contains only Q^3 silicon centers, all of which are symmetry-equivalent, with the consequence that cage degradation can be monitored by the appearance of Q^2 silicon centers and cage condensation can be monitored by the appearance of Q^4 silicon centers. The ^{29}Si NMR spectrum shown in Figure 3a, measured 8 hours after addition of 8 equivalents of water in CH_3CN solution, shows the resonances expected when $[Si_8O_{12}](OCH_3)_8$ is hydrolysed to $[Si_8O_{12}](OCH_3)_7(OH)$ without degradation of its cubic Si_8O_{12} core: (a) no

(a)

8 hrs

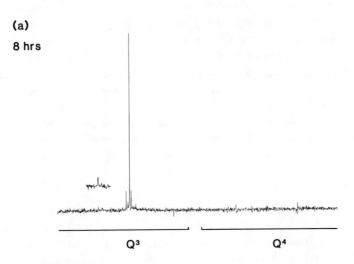

Q³   Q⁴

(b)

27 hrs

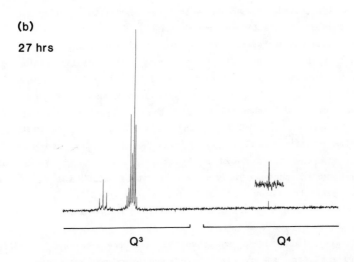

Q³   Q⁴

Figure 3. $^{29}Si\{^1H\}$ 59.6-MHz FT NMR spectra of 0.1 M $[Si_8O_{12}](OCH_3)_8$ in CH_3CN/H_2O solution measured with gated decoupling (a) 8 hrs and (b) 27 hours after hydrolysis with 8 equivalents of H_2O. In each spectrum 356 transients were acquired using (a) a 4 second and (b) a 30 second recycle time. The chemical shift range shown ; is -99.5 to -115 ppm relative to TMS. The solution contained 0.015 M $Cr(acac)_3$ to reduce relaxation times. No hydrolysis/condensation catalysts were employed.

Q^1, Q^2, or Q^4 resonances are observed; (b) a single SiOH resonance is observed in the Q^3 region; and (c) four SiOCH$_3$ resonances are observed in the Q^3 region, including a large resonance arising from $[Si_8O_{12}](OCH_3)_8$ plus three smaller resonances, with relative intensities 3:1:3, arising from $[Si_8O_{12}](OCH_3)_7(OH)$ [7]. After 27 hours have passed, the spectrum shown in Figure 3b is obtained. Here, no Q^2 resonances are observed but a single resonance appears in the Q^4 region, indicating that condensation has occurred without measurable amounts of cube degradation. Taken together, these results show that hydrolysis and condensation of 8 proceed with retention, not inversion, of configuration at silicon.

CONCLUSIONS

The principal conclusion to be drawn from the results outlined above is that certain polysilic acid esters, namely $[Si_3O_2](OCH_3)_8$ and $[Si_8O_{12}]$-$(OCH_3)_8$, can be hydrolysed and condensed without measurable degradation of their polysilicate frameworks during the initial stages of hydrolysis/condensation. It must be emphasized, however, that retention of the molecular building-block structure during the initial stages of hydrolysis and condensation provides no guarantee that degradation will not be extensive during subsequent gelation and drying stages.

Although the emphasis of the present study has been on the retention of molecular building-block structures, many of the results obtained give insights into other aspects of silicic acid ester hydrolysis and condensation. Consider, for example, the ^{29}Si NMR spectra shown in Figure 2. Comparison of the intensities of the resonances arising from $(CH_3O)_3SiOSi(OCH_3)(OH)OSi(OCH_3)_3$ and $(CH_3O)_3SiOSi(OCH_3)_2OSi(OCH_3)_2(OH)$ shows the relative hydrolysis rates of internal (Q^2) and end-group (Q^1) Si-OCH$_3$ groups. Comparison of condensation products in the absence of a catalyst and the presence of an acid catalyst can also be made. For example, cyclic trisiloxane is not observed when $[Si_3O_2](OCH_3)_8$ is hydrolyzed without using a catalyst, but is observed in the presence of an acid catalyst (see Figure 2b and Figure 2 caption).

Finally, it should be mentioned that gels prepared from polysilicic acid esters offer the possibility of producing new materials even if structural degradation of the monomers is extensive, simply because the amount of hydrolysis and condensation needed to obtain complete reaction to SiO$_2$ depends upon the chemical composition of the monomer. Consider,

for example, the complete hydrolysis/condensation of $Si(OCH_3)_4$ (5) and
$[Si_8O_{12}](OCH_3)_8$ (8) given in equations (5) and (6). Formation of a given

$$8\ Si(OCH_3)_4 + 16\ H_2O \longrightarrow 8\ SiO_2 + 32\ CH_3OH \tag{5}$$

$$[Si_8O_{12}](OCH_3)_8 + 4\ H_2O \longrightarrow 8\ SiO_2 + 8\ CH_3OH \tag{6}$$

quantity of SiO_2 from 8 instead of 5 requires only a fourth as much hydrolysis and condensation.

ACKNOWLEDGEMENTS

This research was supported by the U. S. Department of Energy, Division of Material Science, under Contract DE-ACO2-76ERO1198. We are grateful to Professors David Payne, Evan Melhado, and Joan Dawson for assistance and advice.

REFERENCES

1. J. Economy, B. E. Novak, S. G. Cottis, Am. Chem. Soc. Div. Polym. Chem. Preprints 11, 332 (1970).

2. J. G. Cook, J. T. Dickson, A. R. Lowe, J. R. Whinfield, U. S. Patent No. 2 471 023 (24 May 1949).

3. M. Ebelmen, Ann. de Chimie et de Phys. 16, 129 (1846); S. Sakka and K. Kamiya, in Proc. Int. Symp. Factors in Densification and Sintering of Oxide and Nonoxide Ceramics, edited by S. Somiya and S. Saito (Tokyo Institute of Technology, Tokyo, 1978), p. 101.

4. M. Ebelmen, Comptes Rend. de L'Acad. des Sciences 25, 854 (1847).

5. The ^{29}Si chemical shifts of CH_3O- and CD_3O-substituted silicons differ significantly: V. V. Mainz and D. M. Millar (private communication).

6. H. C. Marsmann and M. Vonegehr, in Soluble Silicates, edited by J. S. Falcone, Jr. (American Chemical Society, Washington, 1982), p. 73; H. C. Marsmann, E. Meyer, M. Vongehr, E. F. Weber, Makromol. Chem. 184, 1817 (1983).

7. V. W. Day, W. G. Klemperer, V. V. Mainz, D. M. Millar, J. Am. Chem. Soc. 107, 8262 (1985).

A SOLID STATE MULTINUCLEAR MAGNETIC RESONANCE
STUDY OF THE SOL-GEL PROCESS USING
POLYSILICATE PRECURSORS

W.G. KLEMPERER, V.V. MAINZ AND D.M. MILLAR
School of Chemical Sciences and Materials Research Laboratory,
University of Illinois, Urbana, IL 61801

ABSTRACT

A solid state multinuclear NMR study of the sol-gel process was performed using the molecular building blocks tetramethoxysilane, hexamethoxydisiloxane, octamethoxytrisiloxane and octamethoxyoctasilsesquioxane as
precursor monomers. Water content, solvent content, and hydrolysis/con-
densation processes were monitored using ^{17}O, ^{13}C, and ^{29}Si FT, FTMAS and
CPMAS NMR techniques.

INTRODUCTION

The mechanism by which silicon alkoxides hydrolyze to gels has been
investigated on the molecular level by several groups using multinuclear
solution nuclear magnetic resonance (NMR) spectroscopy. The hydrolysis
and condensation reactions responsible for gelation (Scheme I) can be fol-

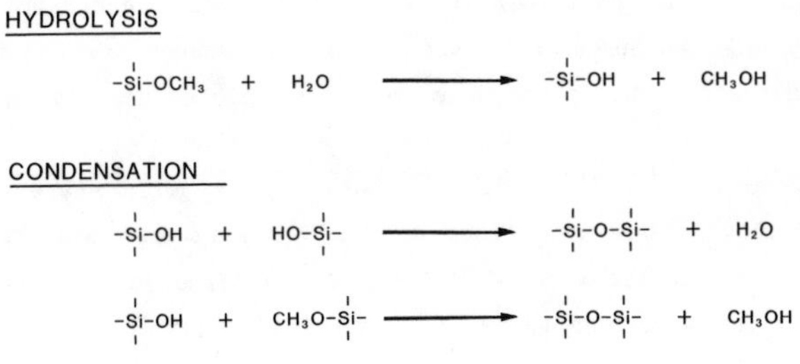

HYDROLYSIS

$$-Si-OCH_3 \ + \ H_2O \ \longrightarrow \ -Si-OH \ + \ CH_3OH$$

CONDENSATION

$$-Si-OH \ + \ HO-Si- \ \longrightarrow \ -Si-O-Si- \ + \ H_2O$$

$$-Si-OH \ + \ CH_3O-Si- \ \longrightarrow \ -Si-O-Si- \ + \ CH_3OH$$

SCHEME I

lowed in solution by the appearance and disappearance of signals arising
from different low molecular weight oligomers. For example, a wealth of
information can be obtained from ^{29}Si NMR spectra because the chemical
shifts are very sensitive to the local environment about silicon and the
narrow lines allow the resolution and assignment of the many species
formed in solution [1-5]. In a similar fashion, the water content in gels
can be observed through ^{17}O NMR by using ^{17}O-enriched water, and the or-

ganic solvent content can be observed through ^{13}C NMR. Finally, ^{1}H NMR can be used to monitor early stages of the hydrolysis/condensation process [6].

Unfortunately, less information is obtainable from multinuclear solution NMR studies as gelation proceeds. The resonances due to large oligomers broaden and eventually disappear as their molecular weights increase and they become less mobile, so that only resonances due to smaller, more mobile species are observed. As a result, there is virtually no chemical information available regarding the later stages of the gelation process. The solidified gel should be well-suited for study, however, if routine Fourier transform (FT) NMR experiments are supplemented by recently-developed magic angle spinning (MAS) solid state NMR techniques [7]. Three FT NMR techniques are in theory suitable for examining silica gels. The first is routine FT NMR which requires no MAS. For example, ^{17}O NMR data for water can be acquired on non-spinning solid samples because the mobility of the water molecules in these gels is relatively large. The acquisition of ^{17}O NMR data using this technique has been demonstrated using hydrated solid proteins [8]. The second method which could be applied to the study of gels is the FTMAS NMR experiment. In ^{13}C NMR spectroscopy, this technique is sensitive to mobile or loosely bound molecules such as adsorbates on a substrate [9]. The ^{29}Si FTMAS experiment might provide similar information regarding silicon-containing materials in the sample. This can be contrasted with data obtained from the third method, the cross polarization (CP) MAS NMR experiment, which in ^{29}Si and ^{13}C NMR enhances the intensity of resonances arising from nonmobile silicon and carbon atoms, respectively, located near hydrogen atoms [7].

In this paper, we illustrate the utility of solid state NMR techniques by following the hydrolyses of four polysilicate precursor molecules: 1 tetramethoxysilane $Si(OCH_3)_4$, 2 hexamethoxydisiloxane $(Si_2O)(OCH_3)_6$, 3 octamethoxytrisiloxane $(Si_3O_2)(OCH_3)_8$, and 4 octamethoxyoctasilsesquioxane $(Si_8O_{12})(OCH_3)_8$. The principal objective of this work is to compare and contrast the results of the solution ^{29}Si NMR studies discussed in the previous paper with the results of solid state NMR studies of the same systems following gelation. Specifically, can solid state NMR techniques show whether the structural integrity of 3 and 4, maintained in solution during the initial stages of hydrolysis and condensation, is preserved during the gelation and drying stages? A more general

OCH₃ structures:

```
        OCH₃                      OCH₃ OCH₃
         |                          |    |
CH₃O-Si-OCH₃              CH₃O-Si-O-Si-OCH₃
         |                          |    |
        OCH₃                      OCH₃ OCH₃

         1                              2
```

```
        OCH₃ OCH₃ OCH₃
         |    |    |
CH₃O-Si-O-Si-O-Si-OCH₃
         |    |    |
        OCH₃ OCH₃ OCH₃

              3
```

4

objective, however, is a determination of the scope and limitations of solid state NMR techniques as applied to sol-gel processes. Can water content be determined? Can the concentration of organic solvents, such as methanol, be monitored in the gel? Can meaningful data on the hydrolysis/condensation process be obtained in the solid state?

^{17}O NMR: MEASUREMENT OF WATER CONTENT

The change in water content during the hydrolysis/condensation of polysilicates can be monitored by ^{17}O FTNMR. In a typical experiment, the silicate ester 1 is treated with 10 equivalents of 6% ^{17}O enriched water at 25°C. Initially, loss of intensity of the water signal is observed as 1 hydrolyzes, followed by growth of this signal as condensation becomes the predominant process in the solution (see Scheme I). Upon gelation (after approximately 24 hours), the intensity of the water signal becomes constant. If the gel is then dried at 100°C, the intensity of the water signal decreases steadily with time, until after 12 hours at 110°C no signal is observed. Note that no ^{17}O signals for any Si-O groups are observed, since these groups are relatively immobile compared to water.

^{13}C NMR: METHANOL AND METHOXIDE CONTENT

The methanol content of gels can be monitored by ^{13}C FTMAS and CPMAS NMR. Spinning at >3 KHz puts several constraints upon the investigation of this system since centrifugation of the sample in the spinner can result in phase separation. With "fresh" gels (sols that have just gelled), solvent loss from the gel can often occur under these conditions, making

18

quantitative measurements of methanol content in the gel difficult to perform.   This necessitates working with gels that have been dried for a
short time and taking data very quickly.   Using this technique, spectra
can be observed for soft, friable gels as well as for harder materials.

In the reaction of 1 with four equivalents of water, only methanol is
observed in the ^{13}C FTMAS spectrum.   It has the relatively sharp line
characteristic of liquid-like molecules, indicating the substantial mobility of some of the methanol molecules in the gel [9].   In comparison, Figures 1a and 1b show the CPMAS spectrum of a fresh gel and a gel which has
been dried for 24 hours at 40°C.   Two signals are observed, the peak at
52.5 ppm arising from $SiOCH_3$ groups and the peak at 50.4 ppm arising from
methanol. Methanol is the predominant species in the fresh gel and the
loss of this methanol upon drying is evident in Figure 1b.   Drying this
gel at 60°C for 24 hours causes a further drop in the methanol content
observed by both FTMAS and CPMAS NMR.   Continued drying of the gel at 80°C
causes a decrease in both the $SiOCH_3$ groups and methanol at a relatively
constant rate until signals from both species can no longer be observed
under the conditions of the CPMAS experiment.

We have performed experiments similar to those described above using
octamethoxytrisiloxane 3.   In principle, two resonances due to the two
different $SiOCH_3$ groups should be observed in the ^{13}C NMR spectrum of pure

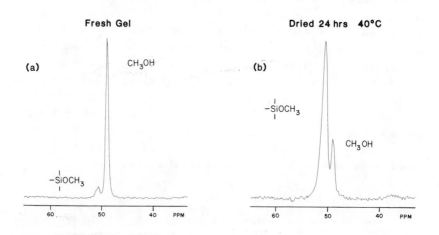

Figure 1. ^{13}C CPMAS NMR spectra obtained at 37.9 MHz of gels prepared
from 2.9M $Si(OCH_3)_4$ in CH_3OH hydrolyzed with 4 equivalents of H_2O a) gelled sealed at 25°C for 24 hours and b) gelled sealed at 25°C for 24 hours
and dried open at 40°C for 24 hours (8000 scans, 15 ms contact times, 2-s
repetitions, 25°C).   No acid or base catalysts were employed.

3. However, these two different $SiOCH_3$ environments are not resolvable in the solid state FTMAS or CPMAS NMR spectra. Hydrolysis experiments reveal no appreciable spectral differences compared with that of tetramethoxysilane 1.

^{29}Si NMR: SILICATE HYDROLYSIS AND CONDENSATION

The local environment about silicon centers in silicates, conventionally described using the Q^n nomenclature shown in Figure 2, has been found to give rise to characteristic ^{29}Si chemical shifts. Several research groups [10,11] have used these correlations to establish the kinds of environments present in silica gel by ^{29}Si CPMAS NMR spectroscopy.

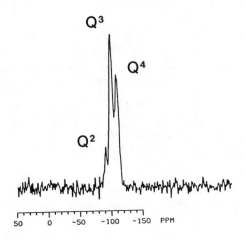

Figure 2. A typical ^{29}Si CPMAS spectrum of commercial silica gel obtained by conventional methods [10]. Q^n nomenclature is indicated below, in which a Q^n silicon is linked to n other silicons by Si-O-Si linkages.

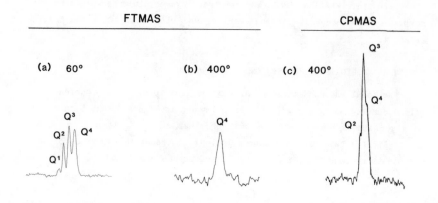

<u>Figure 3</u>. $^{29}Si\{^1H\}$ FTMAS NMR spectra obtained at 71.5 MHz of gels pre-
pared from 2.9 M $Si(OCH_3)_4$ in CH_3OH hydrolyzed with 4 equivalents of H_2O
a) gelled sealed at 40°C for 24 hours and dried for 8 hours at 60°C, b)
gelled sealed at 25°C for 40 hours and dried 6 days at 25°C, then heated
to 400°C by 50°C increments every 12 hours (300 scans, 30-s repetitions
25°C.), and c) ^{29}Si CPMAS NMR obtained at 71.5 MHz of the same sample described in (b) above (20,000 scans, 29-ms contact time, 2-s repetitions
25°C).   The chemical shift range shown is 0 to -200 ppm relative to TMS.
No acid or base catalysts were employed.

Figure 3a shows the ^{29}Si FT MAS NMR spectrum of the gel obtained by
reacting 1 with four equivalents of water for 24 hours at 25°C followed by
drying for 48 hours at 60°C.   The gel contains no observable Q^0, but does
contain Q^1, Q^2, Q^3 and Q^4 silicon centers.   Drying this gel at 400°C (see
Figure 3 caption for conditions) gives a material whose ^{29}Si FT MAS spectrum is shown in Figure 3b. The spectrum shows ^{29}Si intensity only in the
Q^4 region.   This does not indicate that the system is fully condensed,
only that the Q^1-Q^3 resonance intensities have decreased with respect to
the Q^4 resonance, and may be present but hidden under the observed signal.
The ^{29}Si CPMAS spectrum of the gel dried at 400°C (Figure 3c) clearly indicates the presence of not only Q^3 species but also the Q^2 silicon species. Therefore, even at 400°C there are still appreciable amounts of
uncondensed silicon centers in this hard, glass-like material.

Similar hydrolysis experiments were performed using the polysilicate precursors **2** and **3**. Both of the gels obtained from these materials had gross physical characteristics very different from the gels obtained from **1**. They were harder after gelling, much less friable, had far smoother surfaces, and had a lesser tendency to cloud upon drying than gels obtained from **1**. The ^{29}Si solid state MAS spectra of the gels obtained from

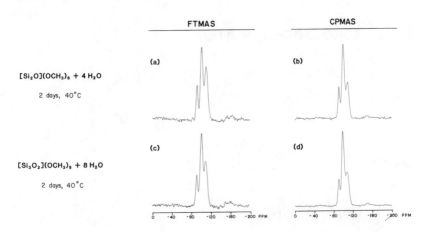

Figure 4. a) ^{29}Si{^{1}H} FTMAS NMR spectrum of gel prepared from 1.8 M [Si$_2$O](OCH$_3$)$_6$ in CH$_3$OH hydrolyzed with 4 equivalents H$_2$O and gelled sealed 2 days at 40°C. b) ^{29}Si CPMAS NMR spectrum of gel prepared in (a) above. c) ^{29}Si{^{1}H} FTMAS NMR spectrum of gel prepared from 1.8 M [Si$_3$O$_2$](OCH$_3$)$_8$ in CH$_3$OH hydrolyzed with 8 equivalents H$_2$O and gelled sealed 2 days at 40°C. d) ^{29}Si CPMAS NMR spectrum of gel prepared in c) above, all spectra were obtained at 59.6 MHz (FTMAS: 160 scans, 30-s repetitions, 25°C, CPMAS: 800 scans; 6-ms contact times, 10-s repetitions, 25°C). No acid or base catalysts were employed.

2 and **3** are shown in Figure 4. These spectra are essentially identical in appearance to spectra of the tetramethoxysilane system, with the CP MAS spectra for both systems showing the expected enhancement of Q^2 and Q^3 versus Q^4 species. Clearly, the gross features of the spectra give no evidence of the macroscopic differences observed in the solid materials.

The ^{29}Si FTMAS NMR spectrum of hydrolyzed **4** is shown in Figure 5. This spectrum shows the presence of Q^2 as well as Q^3 and Q^4 silicon centers. Since the cube structure of **4** contains only Q^3 silicons, simple hydrolysis/condensation with retention of the cube structure will convert Q^3 silicon exclusively to Q^4 silicon [5]. Q^0, Q^1, and Q^2 silicon centers will therefore never be observed unless the cube structure is degraded at

22

some stage of the gelation process. The presence of Q^2 centers unambiguously implies cleavage of the cube structure. This result contrasts with solution ^{29}Si spectroscopic studies made in the very early stages of hydrolysis/condensation, where no degradation was observed.

Note that extreme caution must be exercised when interpreting the relative intensities of ^{29}Si CPMAS data. The intensity of a ^{29}Si CPMAS

(a) $[Si_3O_2](OCH_3)_8 + 8H_2O$

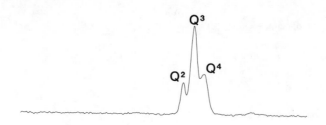

(b) $[Si_8O_{12}](OCH_3)_8 + 8H_2O$

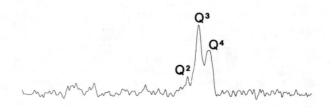

Figure 5. a) $^{29}Si\{^1H\}$ FTMAS NMR spectrum obtained at 59.6 MHz of gel prepared from 1.8 M $[Si_3O_2](OCH_3)_8$ in CH_3OH hydrolyzed with 8 equivalents, H_2O and gelled sealed 2 days at 40°C (160 scans, 30-s repetitions, 25°C). b) $^{29}Si\{^1H\}$ FTMAS NMR spectrum obtained at 71.5 MHz of gel prepared from 0.7 M $[Si_8O_{12}](OCH_3)_8$ in CH_3CN hydrolyzed with 8 equivalents H_2O gelled 2 days at 40°C. The chemical shift range shown is 50 to -50 ppm relative to TMS. (2156 scans, 56.38-ms contact time, 30-s repetitions, 25°C). No acid or base catalysts were employed.

resonance is related not only to the number of ^{29}Si nuclei of a given type present in the sample, but is also related to the proton environment of that type of silicon nuclei. For example, Figure 6 shows the effect of using formamide as a cosolvent in the tetramethoxysilane system. The ^{29}Si

CPMAS spectrum of a sample with formamide (Figure 6d) shows a much greater enhancement of the Q^3 silicon nuclei than that observed in a spectrum from a similar sample without formamide (Figure 6b). This does not necessarily reflect the presence of more Q^3 silicon nuclei in this sample, but may be related to the presence of a greater number of protons in the vicinity of these ^{29}Si nuclei.

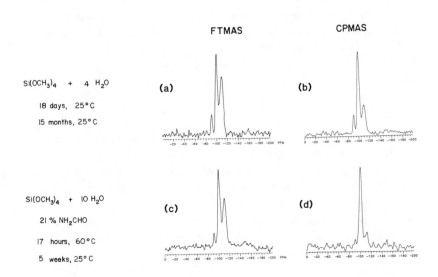

Figure 6. a) $^{29}Si\{^1H\}$ FTMAS NMR spectrum of a gel prepared from 2.9 M $Si(OCH_3)_4$ in CH_3OH hydrolyzed with 4 equivalents of H_2O and gelled sealed 18 days at 25°C and dried open 11 months at 25°C (2000 scans, 30-S repetitions, 25°C). b) ^{29}Si CPMAS NMR spectrum of gel prepared in (a) above (160 scans, 6-ms contact times, 10-s repetitions, 25°C). c) $^{29}Si\{^1H\}$ FTMAS NMR spectrum of gel prepared from 1.7 M $Si(OCH_3)_4$ in 50:50 V/V $CH_3OH:NH_2CHO$ hydrolyzed with 10 equivalents H_2O, 0.53 M HNO_3, gelled sealed 16 hours at 60°C and dried open for 4 weeks at 25°C (1200 scans, 30-s repetitions, 25°C). d) ^{29}Si CPMAS spectrum of gel prepared in (c) above (240 scans, 12-ms contact times, 15-s repetitions, 25°C).

CONCLUSION

The studies reviewed above show the dangers inherent in extrapolating the results of solution NMR studies to gelled samples. According to ^{29}Si NMR spectroscopy, hydrolysis/condensation chemistry continues to occur in the gel and may take a different path than that followed in solution.

In addition, ^{17}O NMR proves to be a convenient method of measuring the change in water content in sol-gel systems during processing, and ^{13}C NMR shows great promise for monitoring hydrolysis as well as the presence of organic additives in gels as a function of time.

It must be emphasized that the results presented here are preliminary in nature and therefore have only qualitative significance. Once the dependence of solid state NMR spectra on instrument parameters and processing conditions has been better defined, solid state NMR techniques should be capable of yielding quantitative information regarding dynamics, through relaxation time measurements [12], and structure, through the detailed analysis of chemical shift and lineshape data [13]. In this fashion, the effects of processing variables such as solvent, acid/base content, time of gelation, etc., on the gel nanostructure might be clarified.

ACKNOWLEDGEMENTS

This research was supported by the United States Department of Energy, Division of Material Science, under Contract DE-AC02-76ER01198.

We are grateful to Kristy Hendrich at the University of Illinois School of Chemical Sciences Molecular Spectroscopy Laboratory for technical assistance in measuring NMR spectra. We are also extremely grateful to Drs. Bruce Hawkins, Charles Bronnimann, and James Frye at the Colorado State University Regional NMR Center for teaching us how to measure and interpret solid state FT NMR spectra.

REFERENCES

1. R. K. Harris, C. T. G. Knight and D. N. Smith, J. Chem. Soc. Chem. Comm. 726 (1980).

2. H. C. Marsmann, E. Meyer, M. Vongehr and E. F. Weber, Makromol. Chem. 184, 1817 (1983).

3. I. Artaki, M. Bradley, T. W. Zerda, and J. Jonas, J. Phys. Chem. 89, 4399 (1985).

4. G. Orcel and L. Hench, J. Noncryst. Solids 79, 177 (1986)

5. W. G. Klemperer, V. V. Mainz, and D. M. Millar, this volume.

6. R. A. Assink and B. D. Kay, in Better Ceramics Through Chemistry, edited by C. J. Brinker, D. R. Ulrich, and D. E. Clark (Elsevier Science Publishers, New York, 1984), p. 301.

7. C. A. Fyfe, Solid State NMR for Chemists (C.F.C. Press, Ontario, 1983).

8. Y. Tricot and W. Niederberger, Biophys. Chem. 9, 195 (1979).

9. G. E. Maciel, J. F. Haw, I.-S. Chuang, B. L. Hawkins, T. A. Early, D. R. McKay and Leon Petrakis, J. Am. Chem. Soc. 105, 5529 (1983).

10. G. E. Maciel and D. W. Sindorf, J. Am. Chem. Soc. 102, 7606 (1980).

11. E. T. Lippmaa, A. V. Samosan, V. V. Brei and Y. I. Gorlov, Dokl. Akad. Nauk SSSR 259, 403 (1981).

12. ^{17}O NMR: see reference 8; ^{13}C NMR: see reference 7, pp. 324-329; ^{29}Si NMR: G. C. Levy, J. D. Cargioli, P. C. Juliano and T. D. Mitchell, J. Am. Chem. Soc. 95, 3445 (1973).

13. a) J. V. Smith and C. S. Blackwell, Nature 303, 223 (1983); b) E. Dupree and R. F. Pettifer, Nature 308, 523 (1984).

STUDIES OF THE INITIAL STEPS IN SOL-GEL PROCESSING OF Si(OR)$_4$:
^{29}Si NMR OF ALKOXYSILANE AND ALKOXYSILOXANE SOLUTIONS

CAROL A. BALFE[†] AND SHERYL L. MARTINEZ
Sandia National Laboratories, Albuquerque, NM 87185

ABSTRACT

Silicon-29 NMR has been used to investigate the initial steps of hydrolysis and condensation of tetramethoxysilane (TMOS), hexamethoxy-disiloxane, and octamethoxytrisiloxane in both basic and acidic methanol solutions. Comparisons of the spectra obtained from the various starting solutions and literature reports of mono-, di-, and tri-silicic acids in aqueous solution were used to identify and assign chemical shifts to condensation intermediates in the TMOS system. The observation of cyclic intermediates in the acid catalysis of TMOS in methanol and the chemical shifts of several products of its hydrolysis and condensation are reported.

INTRODUCTION

Currently, there is strong interest in the mechanistic details of the various stages of the hydrolysis and condensation reactions in alcohol solution of tetraalkoxysilanes, Si(OR)$_4$, the basis of the "sol-gel" process for producing glasses at low temperatures relative to conventional melt-processes. Similar systems, both acid and base catalyzed, were studied during the early years of the silicone industry, and hydrolysis rate constants and condensation products have been reported for such silicone precursors as R$_x$Si(OR')$_{4-x}$ where R = CH$_3$, C$_2$H$_5$ and R' = C$_2$H$_5$ (x=0,1,2,3), or R' = CH$_3$, n-C$_3$H$_7$, i-C$_3$H$_7$, n-C$_4$H$_9$ (x=1). [1-5] Typically, the distribution of condensation products is strongly dependent on the concentration of water relative to SiOR units. Very low water concentrations (1 H$_2$O/3 SiOR) lead to low molecular weight, linear polymers and small cyclics (R$_2$SiO)$_n$ (n=3-6). Higher water to SiOR ratios lead to progressively more complex products, dominated by bicyclic polymers and 8-membered cage structures of the general formula, (SiO$_{1.5}$)$_8$.

In later work, ^{29}Si NMR studies of aqueous solutions of mono-, di-, and tri-silicic acids in acid solution were reported by Engelhardt and coworkers. [6,7] The first intermediate observed was cyclotrisilicic acid, followed at later times by formation of linear polymers and higher order rings. These workers also observed an equilibrium distribution of condensation products at long reaction times which was independent of the nature of the starting material. [7] Recently Artaki, et al. [12], reported the chemical shifts of the monomeric and dimeric hydrolysis and condensation products of TMOS in methanol under acidic conditions.

Our goal in this work was to probe the initial hydrolysis and condensation steps of the tetramethoxy ester of silicic acid, TMOS, in methanol under both acidic and basic conditions using ^{29}Si NMR. We have followed the hydrolysis reactions of the dimeric and linear trimeric methoxy esters of silicic acid in order to assign the chemical shifts and to examine the reactivity of possible condensation intermediates. We report here the observation of cyclic intermediates in the acid catalysis of TMOS in

*This work performed at Sandia National Laboratories supported by the U.S. Department of Energy under contract number DE-AC04-76DP00789.

†Current address: Raychem Corporation, Menlo Park, CA 94025

methanol, as well as the chemical shifts of these and several linear products of its hydrolysis and condensation.

EXPERIMENTAL

TMOS obtained from Alfa was treated by bubbling dry NH_3 through, followed by distillation under Ar from which the middle fraction was used. Hexamethoxydisiloxane was prepared according to the method of Klemperer and coworkers. [8] Two higher boiling fractions were obtained from the distillation of the dimer during the purification step and were identified by GC/MS and ^{29}Si NMR as linear trimer and a mixture of linear trimer and cyclic tetramer. Deuterated methanol was obtained from Merck and used as received.

^{29}Si NMR experiments were performed using a GE 361 spectrometer at an observation frequency of 71.727 MHz in the quadrature detection mode, employing a 45° observation pulse of 17 μs and a gated decoupling sequence to suppress the nuclear Overhauser effect. 100 transients were acquired using a recycle time of 3.704 s. The probe temperature was 23°C and was actively regulated using the spectrometer's variable temperature system. Broadband decoupling of protons was accomplished in the heteronuclear mode using 2 watts of forward power and 800 Hz of frequency modulation.

Chromium (III) acetylacetonate ($Cr(acac)_3$, 0.015 M) was added as a paramagnetic relaxing agent. Control experiments in the absence of $Cr(acac)_3$ were done to verify that the reaction path and rate was not affected by its presence. A solution of $[CH_3(CH_3O)_2Si]_2O$ in benzene with 0.015 M $Cr(acac)_3$ in a sealed 5 mm tube was inserted coaxially in the 12 mm sample tube and served as a chemical shift and integration standard. Solutions were made acidic (pH ≤ 2) by addition of 1 M HCl; they were made basic (pH ~9) by addition of 1 M KOH. The "pH" was measured and is reported for reference purposes only, since there is not a direct correspondence of pH values measured in methanol with standard aqueous pH values.

RESULTS

The notation used to describe the silicic acid ester structures is that of Engelhardt. [9] The basic structural unit is described as a "Q" unit, the symbol commonly used to represent a silicon bonded by 4 oxygen atoms. The number of Q-units attached to the one under consideration is indicated by a superscript. The notation does not distinguish between SiOR, SiOH, or SiO^- groups. Thus, $Q^1Q^2Q^1$ designates a trimer which could be $(CH_3O)_3Si$-O-$Si(CH_3O)_2$-O-$Si(OCH_3)_3$ or $(HO)_3Si$-O-$Si(CH_3O)_2$-O-$Si(OH)_3$ or combinations thereof.

The chemical shifts of the starting materials, TMOS, (monomer, Q^0) hexamethoxydisiloxane (dimer, Q_2^1), and octamethoxytrisiloxane (trimer, $Q^1Q^2Q^1$), and octamethoxycyclotetrasiloxane (cyclic tetramer, Q_4^2) are listed in Table I. These shifts are in close agreement with those reported by Marsmann. [10] The small differences we observe are presumably due to solvent effects.

Acid solutions of the mono-, di-, and tri-silicic acid methyl esters exhibited complex solution spectra. In order to assign the resonances which appeared during acid hydrolysis and condensation of the TMOS solutions, comparisons were made with those of the dimer and trimer, as well as with a mixture of trimer and cyclic tetramer. Figure 1 shows a comparison of the Q^1 region (-80 to -88 ppm) for monomer, dimer, and trimer. Figure 2 shows a similar comparison of the Q^2 region (-90 to -95 ppm). The chemical shifts of the observed resonances are tabulated in Table II. The resonances listed represent a single peak in an envelope of several resonances. In most

Table I
Chemical Shifts of Starting Materials in CH_3OH
(Assignments of Marsmann [10] are given in parentheses.)

Compound	"Q" Notation	Chemical Shifts (δ ppm relative to TMS, $Si(CH_3)_4$)		
		Q^0	Q^1	Q^2
$Si(OCH_3)_4$ (TMOS)	Q^0	-78.4 (-78.59)		
$[(CH_3O)_3Si]_2O$ (Dimer)	$(Q^1)_2$	-	-86.0 (-85.82)	-
$(CH_3O)_8Si_3O_2$ (Trimer)	$Q^1Q^2Q^1$	-	-85.9 (-86.00)	-93.7 (-93.64)
$[(CH_3))_2SiO]_4$ (Cyclic Tetramer)*	$(Q^2)_4$	-	-	-92.9 (-93.10)

* Cyclic tetramer assignment is made from the spectrum of a mixture of trimer and cyclic tetramer.

cases, several closely spaced resonances (~0.1 ppm apart) appeared in a given chemical shift region as illustrated in Figures 1 and 2. The values listed in Table II are meant to identify the regions and typically represent the first major peak to appear in each of the regions during the course of hydrolysis. Assignments are based on a comparison with literature reports and with the hydrolysis patterns of the various compounds reported here.

In basic solutions, the predominant spectral change observed over a period of 3 to 10 hours was the loss in intensity of the resonances due to starting material. Although hydrolysis proceeded rapidly (no TMOS was present after 4 hours, ~20% of dimer and ~30% of trimer were present after 7 hours), the subsequent condensation reactions were so fast relative to hydrolysis that gelation occurred in each case within 1 hour of addition of base. Hence, spectra exhibited a loss in intensity of the resonance due to starting material, some broadening of resonances due to viscosity effects on relaxation time, and few resonances due to hydrolysis of condensation products (See Figure 3).

The small resonance at -85.94 ppm in Figure 3a appeared within 15 min for samples of TMOS and can be assigned to the dimeric condensation product, $[(CH_3O)_3Si]_2O$ [10]. The peak at -82.69 ppm in the dimer solutions (Figure 3b) is attributed to a hydrolysis product, probably an end group of a linear chain of the general formula, $-OSi(OCH_3)(OH)_2$. No resonances due to hydrolysis or condensation intermediates were observed for the trimer. Since the basic solutions gelled during the course of the NMR measurements, the total integrated intensities of the solution spectra steadily decreased. Solid state NMR measurements would be required to probe the structures of the solid phase intermediates.

DISCUSSION

The results described above comprise the first systematic comparison study by ^{29}Si NMR of the hydrolysis and condensation reactions of the methoxy esters of mono-, di-, and tri-silicic acid in methanol solution

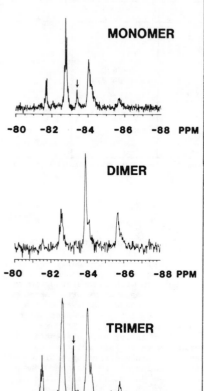

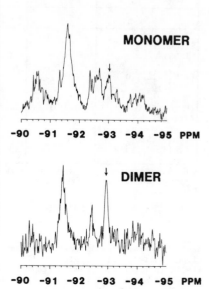

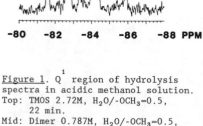

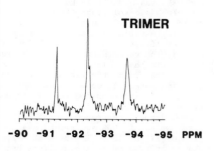

Figure 1. Q[1] region of hydrolysis
spectra in acidic methanol solution.
Top: TMOS 2.72M, $H_2O/-OCH_3=0.5$,
    22 min.
Mid: Dimer 0.787M, $H_2O/-OCH_3=0.5$,
    115 min.
Bot: Trimer 0.55M, $H_2O/-OCH_3=1$,
    7 min.
Arrows point to cyclic trimer
peaks.

Figure 2. Q[2] region of hydrolysis
spectra in acidic methanol solution.
Top: TMOS 2.72M, $H_2O/-OCH_3=0.5$,
    329 min.
Mid: Dimer 0.77M, $H_2O/-OCH_3=0.5$,
    115 min.
Bot: Trimer 0.55M, $H_2O/-OCH_3=1$,
    7 min.
Arrows point to peaks proposed to be
due to cyclic tetramer.

Table II
Chemical Shifts of Products of Acid Hydrolysis
of TMOS, Dimer, and Trimer in CH_3OH.
(in ppm relative to TMS)

Chemical Shift Region	TMOS	Dimer	Trimer	Assignment
Q^0	-76.1 -74.5 -73.3 -72.3			$(CH_3O)_3\underline{Si}OH$ $(CH_3O)_2\underline{Si}(OH)_2$ $(CH_3O)\underline{Si}(OH)_3$ $\underline{Si}(OH)_4$
Q^1	-85.7 -84.0 -83.4 -82.8 -81.8	-85.7 -83.9 -82.7 -81.7	-84.1 -83.3 -82.7 -81.6	$(CH_3O)_3\underline{Si}$-O-Si $(CH_3O)_2(OH)\underline{Si}$-O-Si cyclic trimer $[(CH_3O)_2SiO]_3$ $(CH_3O)(OH)_2\underline{Si}$-O-Si $(OH)_3\underline{Si}$-O-Si
Q^2	-92.8 -92.4 -91.3 -90.5	-93.9 -93.0 -92.4 -91.4 -90.5	-93.7 -92.3 -91.3 -90.5	Si-O-$(CH_3O)_2\underline{Si}$-O-Si cyclic tetramer $[(CH_3O)_2SiO]_4$ See Discussion See Discussion Si-O-$(OH)_2\underline{Si}$-O-Si

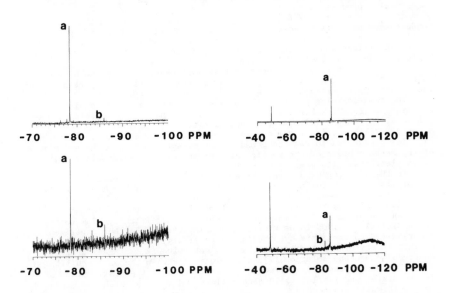

Figure 3a. TMOS 2.72M, $H_2O/-OCH_3=1$, pH≈9. Top: after 13 min; peak a, -78.41 ppm; b, -85.94 ppm. Bottom: after 103 min; a, -78.41 ppm; b, -85.94 ppm.

Figure 3b. Dimer 0.77M, $H_2O/-OCH_3=0.5$ pH≈9. Top: after 7 min; a, -85.99 ppm Bottom: after 490 min; a, -85.99 ppm; b, -82.69 ppm. The peak at -48.71 ppm is the $[CH_3(CH_3O)_2Si]_2O$ standard under conditions similar to those used in sol-gel preparation of glasses (i.e., under conditions of either acid or base catalysis).

In acidic methanol solutions, resonances observed in the Q^0 and Q^1 regions during the course of hydrolysis (Table II) can be assigned in accordance with values reported in recent literature. [7,11,12] The four hydrolysis products of TMOS appear downfield of the initial resonance at -78.4 ppm. The peaks at -76.1, -74.5, -73.3, and -72.3 ppm are assigned to the first, second, third and final hydrolysis products respectively. Similarly, in the Q^1 region, the resonances near -84.0, -82.7, and -81.7 ppm are assigned to the successive hydrolysis products of an end group on a linear chain.

These assignments do not account for the resonance observed at -83.3 to -83.4 ppm in solutions starting with monomer or linear trimer. The absence of a peak in this region during hydrolysis of the dimer supports the assignment of this chemical shift to that of a cyclic trimer species. Although the chemical shift of the methoxyester of cyclotrisilicic acid has not been reported, those of the ethoxyester [10] and of cyclotrisilicates [6,7] are known. In each case, the chemical shift of the cyclotrisilicic acid species is observed at ~9 ppm downfield from the corresponding Q^2 of the linear trimer, as is observed in this work. In further support of this assignment, the cyclotrisilicate has been observed by Engelhardt et al [8] in aqueous acid studies.

The assignment of the chemical shifts observed in the Q^2 region is more complex. The resonance at -93.7 to -93.8 ppm is due to a Si middle group with two methoxyesters, in close agreement with the chemical shift of the starting linear trimer and with the reports of Marsmann [10] (see Table I). In a similar manner, the resonance at -90.4 to -90.5 ppm may be due to Si-O-Si(OH)$_2$-O-Si middle groups, again based on the chemical shifts reported by Marsmann. [10] The fact that this resonance appears in the condensation products of monomer, dimer, and trimer suggests that in all cases, linear condensation products are formed.

The resonance at -92.8 to -92.9 ppm which appears only in the condensation products of monomer and dimer can be assigned to that of a cyclic tetramer, based on the chemical shift observed in the spectrum of the mixture of trimer and cyclic tetramer and the reports of Marsmann [10] (see Table I). The absence of this resonance in the early condensation products of the linear trimer lends support to this assignment, and to the hypothesis that hydrolytic decomposition of the trimer is kinetically unimportant under the conditions reported here.

The two remaining peaks in the Q^2 region cannot be assigned with certainty without further data. One of these resonances is certainly due to a Si-O-Si(OH)(OCH$_3$)-O-Si middle group. The other is possibly due to a Q^2 of a higher cyclic species (cyclic hexamer or octamer) or alternatively to bicyclo- precursors to cage compounds. Experiments to follow the hydrolysis of the methoxyesters of the cyclic tetramer and cyclic trimer are required to complete these assignments.

In the work reported here, there was no detectable evidence of hydrolysis of the siloxane bond itself during either acid or base hydrolysis of the dimer or linear trimer, as evidenced by the lack of resonances in the Q^0 region during hydrolysis. It is probable then that the decomposition of condensation products to form monomeric units is sufficiently slow during the hydrolysis of TMOS to be kinetically unimportant. Further experiments are required to test this hypothesis: specifically, it is desirable to monitor intermediates for longer times to determine if, as Engelhardt observed in aqueous solution [7], all solutions attain a common equilibrium distribution of products. Likewise, determination of the distribution of intermediates under mildly acidic conditions (pH 2 to 7) which are conducive to gel formation is needed.

The appearance of the cyclic trimer soon after the addition of acid
(~30 minutes) to solutions of TMOS and the appearance of the cyclic tetramer
at later times (~2 hours) is, again, in agreement with the findings of
Engelhardt for aqueous systems. Work is in progress to ascertain the
relative concentrations of the cyclic intermediates as a function of time.

ACKNOWLEDGMENTS

The authors gratefully acknowledge the assistance of James Satterlee of
the University of New Mexico for assistance with the use of the NMR
instrument, which instrument was partially funded by NSF grant number CHE-
8201374. They also thank Walter Klemperer, Vera Mainz, and Dean Millar of
the University of Illinois for sharing details of the synthesis of
hexamethoxydisiloxane prior to publication, and their Sandia colleagues,
Philip Rodacy for the GC/MS, and Roger Assink and Bruce Bunker, for helpful
discussions.

REFERENCES

1. M. G. Voronkov, L. A. Zhagata, Zh. Obschei Khimii, 37, 2764 (1976).
2. H. J. Fletcher, M. J. Hunter, J. Amer. Chem. Soc., 71, 2922 (1949).
3. M. M. Sprung, F. O. Guenther, J. Amer. Chem. Soc., 77, 4173 (1955).
4. M. M. Sprung, F. O. Guenther, J. Amer. Chem. Soc., 77, 6054 (1955).
5. R. Aelion, A. Loebbel, F. Eirich, J. Amer. Chem. Soc., 72, 5705 (1950).
6. V. G. Engelhardt, W. Altenburg, D. Hoebbel and W. Wieker, Z. Anorg,
    Allg. Chem., 428, 43 (1977).
7. D. Hoebbel, G. Garzo, G. Engelhardt, and A. Till, Z. Anorg. Allg.
    Chem., 450, 5 (1979).
8. W. G. Klemperer, V. V. Mainz, D. Millar, U. of Illinois Chem. Dept.,
    Manuscript in preparation.
9. G. Engelhardt, H. Jancke, D. Hoebbel, and W. Wieker, Z. Chem., 14, 109
    (1974).
10. H. C. Marsmann, E. Meyer, M. Vongehr, E. F. Weber, Makromol. Chem.,
    184, 1817 (1983).
11. V. G. Engelhardt, W. Altenburg, D. Hoebbel, and W. Wieker, Z. Anorg.
    Allg. Chem., 437, 249 (1977).
12. I. Artaki, M. Bradley, T. W. Zerda, and J. Jonas, J. Phys. Chem., 89,
    4399 (1985).

THE ROLE OF CHEMICAL ADDITIVES IN SOL-GEL PROCESSING

L. L. HENCH, G. ORCEL AND J. L. NOGUES
Advanced Materials Research Center, College of Engineering, University of
Florida, One Progress Blvd., #14, Alachua, FL 32615

ABSTRACT

The effect of various concentrations of formamide with and without
acid catalysis on TMOS derived silica sol and gel structures and physical
properties is described using a quantitative structural model. The model
is based upon ^{29}Si NMR, SAXS, Raman and FTIR spectroscopy, and an acid
solubility test. Changes in chemical reactions during drying due to
formamide are presented using FTIR, DSC and TGA data.

INTRODUCTION

Sol-Gel processing offers the potential of rapidly producing large,
homogeneous, near net shape silica and silicate optics such as lenses,
mirrors, filters, waveguides, laser hosts, and non-linear optic matrices.
However, in order to realize this potential it is essential to avoid
cracking of gel derived devices during processing. Of the six sol-gel
process steps, 1) Mixing and Sol Formation, 2) Casting, 3) Gelation, 4)
Aging, 5) Drying, and 6) Densification, fracture occurs most frequently
during drying. However, densification can result in fracture as well.

In order to avoid cracking during drying it is necessary to minimize
drying stresses. Zarzycki's review [1] of the many factors that lead to
drying stresses shows that it is necessary to control pore size distributions and the rate of evaporation of the pore liquor. Use of critical
point drying solves the drying problems resulting in low density aerogels.
Previous articles have shown that use of a chemical additive, such as
formamide ($CHONH_2$), can modify the pore distribution of dried xerogels and
thereby influence the drying behavior [9]. It has been suggested [2] that
the role of formamide is related to its effect on the hydrolysis and condensation rate of sol formation, as measured by ^{29}Si NMR [2]. Goals of
this paper are: 1) produce a structural model of gel formation, with and
without various concentrations of formamide, based upon the NMR data; 2)
compare the structural model with Guinier radii and fractal dimensions
obtained from small angle X-ray scattering (SAXS) studies of the same sol
formulas; 3) compare the model with Fourier transform IR liquid cell reaction data; 4) characterize the dried gel structures using N_2 adsorption
isotherms for quantitative pore distribution analysis, and laser Raman
scattering and acidic Mo solubility tests for the solid phase analysis; 5)
relate physical properties of the gels such as density and microhardness
to the model and structural characteristics of the gels; and 6) determine
the pore surface reactions during drying using FTIR and TGA and relate
them to the model and gel structure and additive chemistry.

SOL FORMATION AND GELATION

Details of the procedure to form the sol is published elsewhere [2]
as is the method for obtaining the ^{29}Si NMR data. Eight sols prepared
from TMOS [$Si(OCH_3)_4$] were studied:

SW55: methanol solvent, no formamide, and pH 5.5 DI water.
SFX: X vol % formamide in the methanol solvent, pH 5.5 DI water
with X = 10, 20, 25, 30, 40, 50.
SF23: 25 vol % formamide in the methanol solvent, pH 3 HCl water.

The relative volume percent of constituents to yield a silica sol with a Si concentration of 1.6 M and R ratio (# mole H_2O/# mole TMOS) = 10 is given in the summary Figs. 5-8. Rate constants of hydrolysis of TMOS to $(MeO)_3SiOH$ (k_H) and condensation of $(MeO)_3SiOH$ to $(MeO)_3SiOSi(MeO)_3$ (k_C) are given in Figs. 5-8. The rate constants were calculated using the time dependence of formation of hydrolyzed TMOS monomers, dimers, and higher ordered polymers (see ref. 2 for details).

The same procedure was used for the four sols studied in the SAXS experiments. Details of the SAXS study, conducted at the National Small Angle X-ray Scattering Laboratory at Oak Ridge National Laboratory, are given elsewhere in this Conference Proceedings [3]. Both Guinier radii and fractal dimensions of the sols were calculated [12,13] and are included in Figs. 5-8.

Confirmation of the relative rates of hydrolysis of the four sols were obtained using a Fourier transform IR spectrometer (Nicolet 20SX) and a cylindrical internal reflection (Circle) cell (Barnes Analytical). Details of the FTIR kinetics experiments will appear elsewhere [5]. Thirty-two scans were accumulated over a minute period while the sol was reacting in the circle cell at 23°C. The scans were repeated on 20 min to 1 hr schedules up to t/tg = 1.3 depending on length of gelation time (tg). Figure 1 compares the IR spectra of the sols at one-half of the time of gelation (t/tg = 0.5).

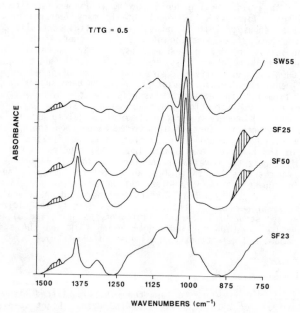

Fig. 1. FTIR spectra of various SiO_2 sols for a reduced time t/tg = 0.5 at 23°C.

In these experiments tg is defined as the time required for a sol to become rigid when tilted through 90°. The length of tg, at 23°C, for the sols decreased from 33 hr for the sol without formamide to 5 hr with 50 vol. % formamide (F) (Table 1). Most of the decrease in tg was achieved with only 10 vol. % F present, e.g., 8 hr. At 50°C, gelation time is still decreased with formamide present but is reduced to 3X instead of 6X at 23°C.

Returning to Fig. 1, note the two cross hatched IR absorbtion bands at 1448 cm^{-1} and 825 cm^{-1}. The 1448 cm^{-1} band is assigned to the C-H vibrational mode of CH_3OH [6]. Although there is some overlap of 1465 cm^{-1} C-H mode of $Si(OCH_3)_4$, there is a quite distinct increase in intensity due to release of CH_3OH in the hydrolysis reaction shown in Fig. 3.

Concurrently, there is a marked decrease in the 825 cm^{-1} band characteristic of the vibrational modes in $Si(OCH_3)_4$ as hydrolysis proceeds. Figure 1 shows that the sols containing formamide still contain substantial quantities of unhydrolyzed TMOS monomer at t/tg = 0.5. This result corresponds to that found in the NMR study, Figs. 5-8.

Discussion of the time dependent changes of the IR bands from 950 cm^{-1} to 1200 cm^{-1} which contain mixed vibrational modes Si-O-Si, Si-O-C, C-N, etc. will be reported elsewhere [5].

GEL CHARACTERISTICS

The pore network of the gels with and without formamide was characterized using N_2 adsorption isotherms (Quantachrome, Autosorb 6) using a procedure described previously [7]. It is important to note that all samples were given the following post gelation treatment prior to determining the pore analyses: Gelation occurred at 50°C at the times shown in Table I, the gels were maintained at 50°C in their liquor (aging) for the remainder of 24 hr, followed by another 24 hr aging at 60°C. Drying occurred during 1 day at 60°C and 1 day at 110°C. Following drying all samples for N_2 adsorption analyses were heated to 200°C in air and 250°C under vacuum immediately prior to analysis. Results are summarized in Table I. The gel without formamide (SW55) developed a very high surface area (750 m^2/g) and a very broad pore distribution with a very small average pore radius (PR = 14 Å). With only 10 vol. % formamide the SA dropped by more than one-half to 330 m^2/g and the PR increased by 5X to 74 Å. Increasing the formamide content continued to decrease SA and increase PR up to 30 vol. % formamide. At larger vol. % formamide the SA increased and PR decreased slightly. A comparison of the shape of the pore distributions of some of these gels is shown in refs. [7], [8] and [9]. The gels with formamide have a much more narrow distribution of pore sizes.

Raman spectroscopy was used to assess the structural scale of the solid network of the gels. Details are given in ref. [9]. Raman intensity increased up to t/tg = 2 after which it remained constant. The gels containing formamide showed a faster increase in intensity and a final value which was 50% larger than gels without formamide.

The scale of the gel solid network was deduced as follows [10]: The Raman intensity (I_R) is inversely proportional to the silica depolymerization rate constant, k_D. Thus, I_R can be related to sol particle size (d) as: $I_R = A\,d^{1/b}$, where A is a function independent of d. Using empirical values of 1/b = 3.48 and A = 6.8 (see ref.10) the time dependent change in sol particle size can be calculated.

Table I. Effect of Formamide on SiO_2 Gel Characteristics.

Gel#	SW55	SF10	SF20	SF25	SF30	SF40	SF50
Formamide Vol.%	0	10	20	25	30	40	50
tg at 23°C	33:53	8:13	7:05	6:33	6:19	6:00	5:43
tg at 50°C	3:00	1:07	1:05	1:11	1:11	1:10	1:05
$SA(m^2/g)$	750	330	341	359	249	284	292
PR (A)	14	74	69	76	130	122	110
HV at 60°C	51	23	14	5.1	2.7	1.3	0.8
d at 60°C	1.482	1.473	1.465	1.413	1.407	1.330	1.285
HV at 110°C	66	28	17	11	5.8	3.2	1.5
d at 110°C	1.505	1.224*	1.098*	1.418	1.399	1.352	1.319
WL at 100°C	8.9	10.9	10.4	-	12.1	14.1	9.5
WL at 200°C	13.1	30.1	44.0	-	41.2	57.7	56.5
WL at 250°C	14.8	32.9	45.5	-	44.4	59.9	59.2

tg in hr; SA=Specific Surface Area; PR=Average Pore Radius;
HV=Vickers Microhardness; d=Bulk Density in g/cm^3; WL=Weight Loss in %.
*Measurements done on small samples.

Sol SW55, without formamide, produced an effective particle diameter at t/tg = 0.5 of 15 Å. Sol SF25, had 16 Å particles, and SF50, 17 Å particles at t/tg = 0.5. At t/tg = 1.0 SW55 had 20 Å particles; SF25 had 22 Å particles, and SF50, 28 Å particles. Thus, the major difference between the sols was the rate of growth of the particles. The faster growth rate of the sol with formamide led to a structure after aging for t/tg = 1.0 of 35 Å particles compared with 26 Å for the gel without formamide.

There are two important differences between the Raman data and SAXS data: 1) The size of the solid network at t/tg = 1.0 determined by Raman spectroscopy is about one half to one-third that of the Guinier radii; 2) The differences in size between the sols increase as time progresses towards gelation in the Raman data whereas it is the reverse for the SAXS data. Both the Guinier radii and the fractal dimension of the gels are approximately equivalent at t/tg = 1.0 whereas the formamide gels are 40% larger in dimension.

Physical properties of the dried gels also show very large differences depending on formamide content. Table I summarizes the gel microhardness (HV = Vickers microhardness number, Leco Micromet) and gel density as a function of volume percent formamide. The microhardness was measured with a 10 g load and 200X or 1,000X magnification, five measurements per data point. Gel density was determined using a mercury pyncnometer, repeating each measurement five times per sample. Values for both HV and density in Table I are approximately ±2%.

The microhardness of the gel without formamide is quite high, 51 at 60°C and 66 at 110°C. Addition of formamide progressively decreases the hardness to as low as 0.8 at 60°C for 50 vol. % formamide. This is more than a sixty fold decrease in hardness.

There is also a decrease in gel density as formamide is added to the sol. However, the effect is very much less than is observed for micro-hardness. The density of the SW55 xerogel is quite high, 1.482 g/cc even after only 60°C drying. Fifty volume percent formamide reduces the 60°C dried gel density to 1.285 g/cc.

DRYING REACTIONS

Although addition of formamide can be used primarily to control gel pore size and structural scale, this chemical additive will also affect the chemical reactions during the last stage of drying. Recently FTIR has been used to understand the temperature evolution of TMOS derived silica thin films [6]. It was found that removal of water from the pores and the gel surface is easier when the gels are prepared in basic conditions and have a larger pore structure. Gels prepared using TMOS in methanolic sol-utions under either acidic or basic conditions resulted in the oxidation of residual methanol in the pores. In the temperature range from 170°C to 220°C, methanol oxidation occurs in the presence of silanol and siloxane bonds and forms formaldehyde, formic acid, and formates. These species were still present in the gel at 300°C. Monolithicity can be maintained only if these molecules are removed without developing excessive stresses.

Differential scanning calorimetry (DSC) was used to study the thermal drying reactions of the gels with and without formamide. The DuPont 1090 system was used with a 10°C/min heating rate. Figure 2 shows that the gels with formamide have a much larger broad exothermic peak between 250-350°C, indicating that organic molecules are being oxidized. Figure 2 also shows that gel SW55 (without formamide) has a single dehydration endothermic peak at about 125°C. In contrast the gels with formamide have double endothermic peaks centered around 110 and 210°C, the former being the dehydration peak and the latter corresponding to the decomposition of formamide.

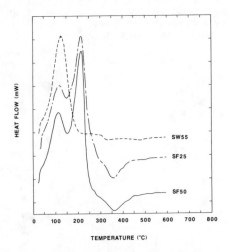

Fig. 2. DSC curves of SiO_2 gels prepared with different amounts of formamide.

The DSC data indicate that the effect of formamide is to block the water molecules in the gel pores from chemisorbing on the silanol bonds on the gel surface. Thus the water is present as free water in the gel pores and evaporates at the B.P. of water. In contrast, without formamide the pore water is chemisorbed to the silanol groups in the pore network via hydrogen bonding and therefore requires a higher temperature to desorb. The very small mean pore radius contributes to the higher desorbtion temperature as well.

These differences in pore liquor chemistry strongly affects the temperature dependence of weight loss during drying. Table I compares the weight loss, measured by DuPont 1090 TGA, for the gels with various volume percentages of formamide. There is relatively little difference between the gels at 100°C; they all lose about the percentage of weight by evaporation of water. However, at 200°C and 250°C there is a very much larger loss in weight for the gels containing formamide because of the higher decomposition temperature for this additive. It is this delayed loss of pore liquor that can aid in control of the drying rate of the gels. However, as shown in Table I, the strength of the formamide gels, i.e., microhardness is very much lower, at least at 110°C. Consequently, the thermal schedule for aging and drying of these gels below 200°C must be sufficient to increase strength to avoid cracking during formamide decomposition. Even then removal of the reaction products formed between 250-350°C can cause sufficient stress to fracture the gel.

STRUCTURAL MODEL

In order to integrate the many experiments described above it is essential to use a structural model. We have chosen a 2D graphical model using symmetrical molecules for simplicity. Figures 3 and 4 establish the basis for the model and its scale. A 6 Å unhydrolyzed TMOS molecule [$Si(OCH_3)_4$] is represented as a small unfilled circle in Figs. 5 to 8. A hydrolyzed or partially hydrolyzed TMOS molecule is represented as a filled circle in Figs. 5-8. Dimers are designated by connecting two hydrolyzed monomers together and higher ordered Si-O-Si-O-Si polymers are designated by a connection of lines. A dimer is 6 to 8 Å. A linear chain of four Si-O-Si polymer units is 12 Å (Fig. 4).

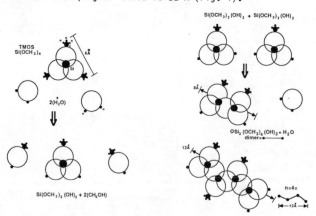

Fig. 3 and 4. Schematic representation of the structural units comprising a SiO_2 gel.

As reaction time (t) proceeds towards gelation time (tg) in increments of t/tg = 0.1, 0.3, 0.5, 1.0, it is assumed that hydrolysis has preceded formation of dimers, and dimers have preceded formation of higher ordered polymers.

A randomly generated array of 100 TMOS monomers is the basis for the model shown in Figs. 5-8. All sol formulations assume the same arrangement of 100 monomeric units at t = 0. NMR measurements [2] of the time dependence of formation of the hydrolyzed monomeric units (M), dimers (D) and higher order polymers (HP) provide the data for creating the time dependent model. For this first approximation, no distinction has been made between the various types of hydrolyzed monomers (M1-4) or dimers (D1-4) although this information is available from the NMR study [2]. A second order model is underway which will include the additional monomer and dimer information [11].

The Si atoms present in higher ordered polymers (HP) have been assigned in the model using the following assumptions: 1) HP are formed from dimers first, if they are present; 2) HP incorporate hydrolyzed monomers before adding unhydrolyzed monomers; 3) HP growth occurs by adding nearest neighbor monomers; 4) Size and shape of HP are approximated to be consistent with the Guiner radius and fractal dimension; 5) HP growth occurs so as to result in gelation at t/tg = 1; 6) Gelation requires at least one interconnection of HP units across the 100 unit array.

Figures 5 to 8 show the time dependence of gel formation using the model. The time increment, gel code #, gel formulation, Guinier radius, fractal dimension, and gel physical properties are given in each figure.

Gels SW55, no formamide, and SF50, with 50 vol.% formamide represent the extremes. At t/tg = 0.1, only four monomers are hydrolyzed in the SF50 sol and 1 dimer is formed. By t/tg = 0.3, there is still only 1 dimer and 1 trimer present. Even at t/tg = 0.5 only 8 monomers have formed siloxane bonds. However, the hydrolyzed monomers that are formed quickly polymerize into a network at gelation with structural units of 30 Å.

Without formamide, hydrolysis is rapid. Consequently, when condensation occurs a large number of small silica polymer units are formed of 20 Å radii. As time progresses the remaining monomers join the previously formed polymer units so that when gelation occurs there are many more branches for interconnection between the primary aggregate and neighbors, forming a much stronger gel network.

Sols SF25 and SF23 compare the effects of formamide alone (SF25) and formamide with acid catalysis (SF23). The network formed with 25% formamide is very similar to that described above for 50%, for the same reasons. Addition of acid increases the rate of hydrolysis by 3X which produces a smaller scale network when gelation occurs with more interparticle interconnects. This makes the gel stronger. The network and resulting porosity is more uniform however than the gel without formamide. Consequently, the combination of formamide and acid provide a very useful means for modifying silica gel processing, structure, and resultant properties with predictable results.

42

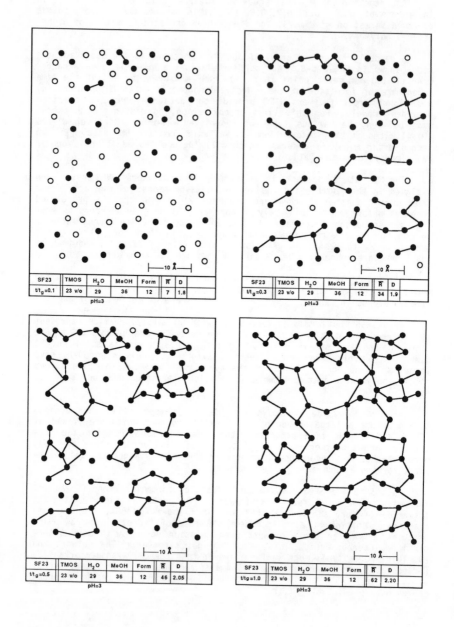

SF23	TMOS	H₂O	MeOH	Form	R̄	D	
t/t_g=0.1	23 v/o	29	36	12	7	1.8	

pH=3

SF23	TMOS	H₂O	MeOH	Form	R̄	D	
t/t_g=0.3	23 v/o	29	36	12	34	1.9	

pH=3

SF23	TMOS	H₂O	MeOH	Form	R̄	D	
t/t_g=0.5	23 v/o	29	36	12	46	2.05	

pH=3

SF23	TMOS	H₂O	MeOH	Form	R̄	D	
t/tg=1.0	23 v/o	29	36	12	62	2.20	

pH=3

Fig. 5. Time dependent change in the model of a gel structure of a SF23 sol.

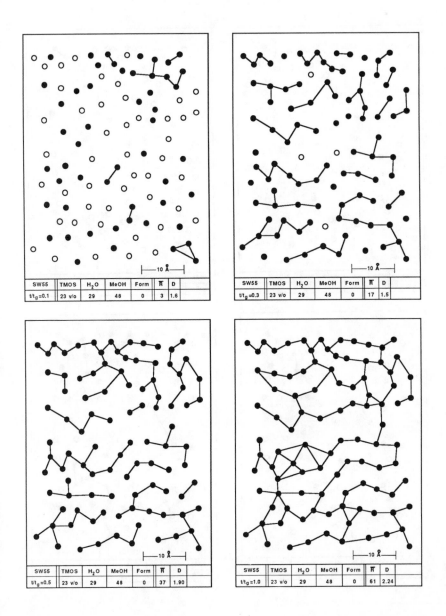

SW55	TMOS	H$_2$O	MeOH	Form	$\bar{R}$	D	
t/t$_g$ =0.1	23 v/o	29	48	0	3	1.6	

SW55	TMOS	H$_2$O	MeOH	Form	$\bar{R}$	D	
t/t$_g$ =0.3	23 v/o	29	48	0	17	1.5	

SW55	TMOS	H$_2$O	MeOH	Form	$\bar{R}$	D	
t/t$_g$ =0.5	23 v/o	29	48	0	37	1.90	

SW55	TMOS	H$_2$O	MeOH	Form	$\bar{R}$	D	
t/t$_g$ =1.0	23 v/o	29	48	0	61	2.24	

Fig. 6. Time dependent change in the model of a gel structure of a SW55 sol.

44

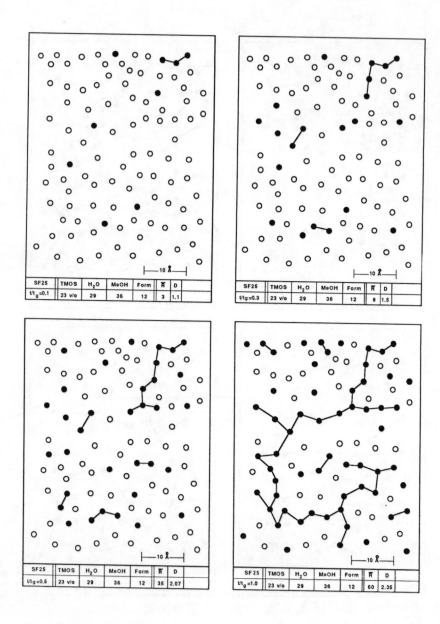

SF25	TMOS	H₂O	MeOH	Form	R̄	D	
t/tg=0.1	23 v/o	29	36	12	3	1.1	

SF25	TMOS	H₂O	MeOH	Form	R̄	D	
t/tg=0.3	23 v/o	29	36	12	8	1.5	

SF25	TMOS	H₂O	MeOH	Form	R̄	D	
t/tg=0.5	23 v/o	29	36	12	35	2.07	

SF25	TMOS	H₂O	MeOH	Form	R̄	D	
t/tg=1.0	23 v/o	29	36	12	60	2.35	

Fig. 7. Time dependent change in the model of a gel structure of a SF25 sol.

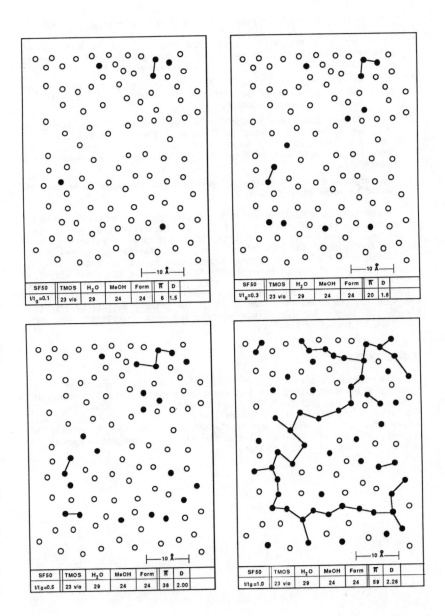

Fig. 8. Time dependent change in the model of a gel structure of a SF50 sol.

SUMMARY

The silica gel characterization studies yield the following facts:

1) Raman and acid solubility tests show that the gel is composed of primary silica polymer units of 20 Å for gels without formamide. With formamide the units increase to a maximum of 30 Å.

2) SAXS shows that the primary silica polymeric units make up an aggregate of 60 Å radius and a fractal dimension of 2.3, and both are independent of formamide content.

3) The ^{29}Si NMR based growth model shows that the gel without formamide forms an aggregate with many small pores (14 Å) within the aggregate and more interconnecting branches between the aggregates. With formamide the aggregate develops large intra and interaggregate pores (70-140 Å).

4) The large pores in the formamide gels and few aggregate interconnects result in lower strengths.

These structural differences are a consequence of formamide markedly decreasing the rate of hydrolysis of TMOS and increasing the rate of condensation. These changes in reaction rates produce a very large difference in structure at the time of gelation. With formamide, the gel forms at tg with only half of the monomer polymerized in the gel network. Consequently, there are large pores in the formamide gels at tg and very few interconnects between the primary silica polymer units. The aggregate structure is basically the same, however with and without formamide, hence the same fractal dimension. The scale of the primary aggregate structure however is larger for the formamide gels and most importantly the pore structure is increased ten-fold.

ACKNOWLEDGEMENTS

The authors gratefully acknowledge the financial support of AFOSR contract #F49620-83-C-0072 and the research assistance of I. Artaki, M. Bradley, T. Zerda, and Prof. J. Jonas of the University of Illinois, J. Phallipou of the University of Montpellier, J.S. Lin of Oak Ridge National Laboratory, and M. Wilson and Professor Gould of the University of Florida, and especially D.R. Ulrich for his continued encouragement to pursue interinstitutional research.

REFERENCES

1. J. Zarzycki, in Ultrastructure Processing of Ceramics, Glasses and Composites, edited by L.L. Hench and D.R. Ulrich (Wiley, New York, 1984) pp 27-42.
2. Gerard Orcel and Larry Hench, J. Non-Cryst. Solids 79, 177-194 (1986).
3. G. Orcel, R.W. Gould, and L.L. Hench, presented at the 1986 Spring Meeting, Palo Alto, CA 1986 (these proceedings).
4. C.F. Brinker, K.D. Keefer, D.W. Schaefer, R.A. Assink, R.D. Kay and C.S. Ashley, J. Non-Cryst. Solids 63, 45 (1984).
5. J.L. Nogues, G. Orcel and L.L. Hench to be submitted.
6. G. Orcel, J. Phalippou and L.L. Hench, submitted to J. Non-Cryst. Solids.
7. S. Wallace and L.L. Hench in Science of Ceramic Chemical Processing, edited by L.L. Hench and D.R. Ulrich (J. Wiley, New York, 1986) pp 148-155.
8. L.L. Hench, in Science of Ceramic Chemical Processing, edited by L.L. Hench and D.R. Ulrich (J. Wiley, New York, 1986) pp 52-64.
9. I. Artaki, M. Bradley, T.W. Zerda, J. Jonas, G. Orcel and L. Hench, in Science of Ceramic Chemical Processing, edited by L.L. Hench and D.R. Ulrich (J. Wiley, New York, 1986) pp 73-80.
10. L.L. Hench and G. Orcel, presented at the Third International Workshop on Glasses and Class-Ceramics from Gels, Montpellier, France, 1985 (in press).
11. G. Orcel, Ph.D. Thesis, University of Florida, 1986.
12. D.W. Schaefer and K.D. Keefer, in Better Ceramics Through Chemistry, edited by C.J. Brinker, D.E. Clark and D.R. Ulrich (North Holland, New York, 1984) pp 1-14.
13. K.D. Keefer, in Better Ceramics Through Chemistry, edited by C.J. Brinker, D.E. Clark and D.R. Ulrich (North Holland, New York, 1984) pp 15-24.

SOLUTION CHEMISTRY OF SILICATE AND BORATE MATERIALS*

B. C. BUNKER, Sandia National Laboratories, Albuquerque, NM 87185

ABSTRACT

Reactions which create and destroy Si-O-Si, Si-O-B, and B-O-B bonds occur in leached layers on borosilicate glasses during glass dissolution. Solid state NMR, Raman, and TEM techniques have been used to determine the distribution of species formed in leached layers as a function of solution pH and the mechanisms of glass-water reactions. The chemical principles which govern the reactivity of silicate and borate bonds in leached layers appear to be similar to those which govern the formation of silicate and borate gels via sol-gel techniques.

INTRODUCTION

The initial stage in preparing sol-gel glasses is the hydrolysis and condensation of dissolved solution species such as tetraethylorthosilicate (TEOS) to form a hydrosilicate gel. Hydrosilicate gels can also be produced from solids via selective leaching of silicate glasses. Since the glass → gel transition is almost the reverse of the sol → gel transition, many of the reversible reactions which control the growth and structure of the resulting hydrosilicate networks are the same. We have used solid state NMR and Raman spectroscopies to study which molecular species are produced in hydrosilicate gels during leaching of sodium borosilicate glasses. We have also examined the resulting network structures using transmission electron microscopy. We find that the observed structural rearrangements which occur as a result of polymerization and depolymerization in leached layers agree with classical models [1] for the reactions of Si-O-Si, Si-O-B, and B-O-B with water. Applying the same models to sol-gel systems has enabled us to predict trends in reactivity in the polymerization of both silicate and borate networks.

STRUCTURAL REARRANGEMENTS IN LEACHED LAYERS

Three sodium borosilicate glass compositions of various Na:B ratios were leached in aqueous solutions having various pH's. Compositions are referred to by listing mole%'s in the order $Na_2O \cdot B_2O_3 \cdot SiO_2$ (e.g., $30 \cdot 10 \cdot 60 = Na_2O \cdot 10B_2O_3 \cdot 60SiO_2$). Raman measurements were taken on intact leached layers in solution. NMR measurements were obtained on samples of leached glass powders. Details of experimental procedures, results, and

*This work performed at Sandia National Laboratories supported by the U.S. Department of Energy under contract number DE-AC04-76DP00789.

conclusions concerning leaching mechanisms will be published elsewhere.
[2,3] In this section, results are highlighted which pertain to silica
polymerization in leached layers and the hydrolysis of Si-O-B bonds.

When sodium borosilicate glasses dissolve in acid, almost all Na and B
is selectively leached to produce a hydrosilicate surface layer. Raman
spectra [4] of glasses before and after leaching (Fig. 1) indicate that:
1) Different glass compositions have substantially different silicate
network structures prior to leaching. Based on Raman intensities, the
30·10·60 glass contains almost one non-bridging oxygen (nbo) (see 1100 cm^{-1}
band) per Si, while the 10·30·60 glass has a negligible number of nbo's.
2) After leaching, the structure of the silicate network is substantially
different. For the 30·10·60 glass, nbo's have been converted into silanol
groups (at 980 cm^{-1}) via ion exchange. The silanol concentration estimated
from Raman intensities (SiOH:Si = 0.2) is much lower than the initial nbo
concentration (nbo:Si = 1) because of extensive repolymerization of silanols
to form Si-O-Si bonds. 3) The structure of hydrosilicate phases produced is
identical, regardless of the structure of the starting glass. The initial
glass structure is completely broken down and reconstructed by reactions
controlled primarily by the solution chemistry. At pH 1, hydrosilicate
leached layers contain high concentrations of ring structures containing 4
SiO$_4$ tetrahedra (490 cm^{-1} band) and a high degree of polymerization (SiOH:Si
= 0.2). In glass leached in neutral or basic solutions (Fig. 2), even fewer
SiOH are present (SiOH:Si = .05), and larger silicate rings (at 430 cm^{-1})
predominate.

Solid state ^{17}O NMR spectra (Fig. 3) of sodium borosilicate glasses
dissolved in H$_2$^{17}O show that water reacts extensively with the silicate
network during leaching. Labeled oxygen is incorporated into nbo sites and
Si-O-Si, Si-O-B, and B-O-B bridging sites. The distribution of ^{17}O depends
on both the solution pH and the glass composition. For 30·10·60, most ^{17}O
occupies nbo sites when the glass is dissolved at pH 12, while at pH 1, all
^{17}O is present in Si-O-Si bonds or as molecular water. For the glasses
leached at pH 1, the concentration of ^{17}O in Si-O-Si sites is higher for the
30·10·60 glass (which has a high nbo content prior to leaching) than for the
10·30·60 glass which initially contains few nbo's, but high concentrations
of Si-O-B and B-O-B bonds.

Transmission electron microscopy (TEM) and BET gas adsorption were used
to characterize how the molecular structures described above relate to
macroscopic gel structures. For samples leached at pH 1, the gel consists
of chains comprised of 10-15 A particles (Fig. 4a). The gels are highly
porous, having specific surface areas of up to 600 M^2/gm with average pore
diameters of 20-35 A. Samples leached at neutral or high pH appear to

consist of aggregates of 150-200 Å diameter colloidal silica particles
(Fig. 4b).

REACTION MECHANISMS

The formation of hydrosilicate phases during dissolution of sodium
borosilicate glasses is rationalized on the basis of the formation of

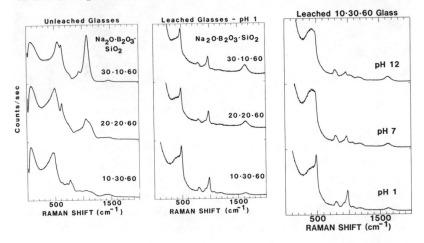

Figure 1. Raman spectra of unleached and leached sodium borosilicate glasses
vs. glass composition, leaching at pH 2, 70°C
Figure 2. Raman spectra of leached glass vs. solution pH

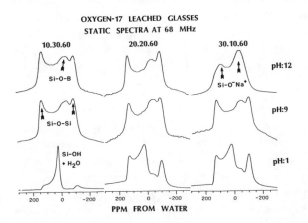

Figure 3. ^{17}O NMR spectra of borosilicate glasses leached at 70°C in $H_2{}^{17}O$

52

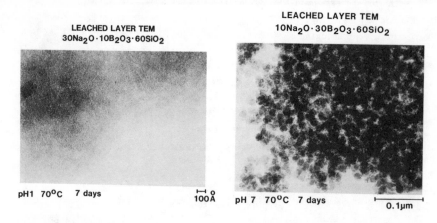

LEACHED LAYER TEM
$30Na_2O \cdot 10B_2O_3 \cdot 60SiO_2$

LEACHED LAYER TEM
$10Na_2O \cdot 30B_2O_3 \cdot 60SiO_2$

pH1 70°C 7 days |———| $\overset{0}{100Å}$ pH 7 70°C 7 days |———| 0.1µm

Figure 4. Transmission electron micrographs of glass leached at (a) pH 1 and
(b) pH 12

silanol groups via ion exchange (on nbo's) or hydrolysis of Si-O-B bonds,
followed by a combination of silanol polymerization and hydrolytic
depolymerization. Previous results [1] suggest that the
polymerization/depolymerization reactions involve nucleophilic attack by
anionic species above pH 2, and electrophilic attack by cationic species
below pH 2. Since the pH in leached layers is above pH 2 for all leaching
conditions described here (even when the solution pH is 1), only the
nucleophilic reactions will be discussed to describe the ^{17}O NMR results.

Above pH 12, the predominant reaction is nucleophilic attack of OH^- on
nbo sites to form an unstable five-coordinate intermediate:

$$\geqslant Si - O^-Na^+ + {}^{17}OH^- \rightarrow \left[\geqslant Si \overset{\diagup {}^{17}OH}{\diagdown O^-Na+} \right]^- \rightarrow \geqslant Si - {}^{17}O^-Na^+ + OH^- \quad (1)$$

If the intermediate decomposes via expulsion of the nbo (as shown), ^{17}O is
incorporated into the glass without changing the local structure. If an
oxygen in a Si-O-Si bond is expelled, the result is hydrolytic
depolymerization of the silicate network. Depolymerization is favored on
nbo sites because the network distortions required for the formation of the
five-coordinate intermediate become easier as the site is attached to the
network by fewer bridging oxygens. Therefore, chains are easiest to
depolymerize, followed by rings, and totally polymerized silica (all
bridging oxygens).

Below pH 10, ion exchange on nbo sites leads to the formation of silanol groups. The ^{17}O NMR results show that the silanols react with each other to form bridging oxygens via condensation reactions (presumed to be nucleophilic):

$$\geqslant Si - O^- + Si(OH)_4 \;\rightarrow\; [\geqslant Si\text{-}O\text{-}Si(OH)_4]^- \;\rightarrow\; \geqslant Si\text{-}O\text{-}Si(OH)_3 + OH^- \qquad (2)$$

Since silanols on highly polymerized sites are more acidic than those on silicic acid [1], and since silicic acid can readily rearrange during nucleophilic attack to form the five-coordinate intermediate, the preferred condensation reaction is between highly polymerized silica and silicic acid. Combining the predicted trends in reactivity for condensation with trends described above for depolymerization, it is clear why hydrosilicates containing a random distribution of silanols (as in glass immediately after leaching) are unstable, and reorganize into highly polymerized silica containing few (if any) silanols.

For B rich glasses, ^{17}O and ^{11}B NMR data suggest that hydrolysis of Si-O-B and B-O-B bonds involves electrophilic attack of protons on the bridging oxygen in acids and nucleophilic attack of OH$^-$ on neutral trigonal B sites in bases as shown below:

$$[\geqslant Si\text{-}O\text{-}B\lessgtr\,]^- \;+\; H^+ \;\rightarrow\; \geqslant Si \overset{\displaystyle H}{\underset{}{\,^{\nearrow}O^{\nwarrow}}} B\lessgtr \;\rightarrow\; \geqslant Si - OH + B\lessgtr \qquad (3)$$

$$\geqslant Si\text{-}O\text{-}B\lessgtr \;+\; ^{17}OH^- \;\rightarrow\; \left[\geqslant Si\text{-}O\text{-}B\lessgtr_{^{17}OH}\right]^- \;\rightarrow\; \geqslant Si - O^- + \overset{\diagup}{\underset{^{17}OH}{B}} \qquad (4)$$

Although silanol groups are created by the hydrolytic release of B, none of the silanols created during dissolution in $H_2\,^{17}O$ are predicted to be labeled, so little ^{17}O is incorporated into bridging O sites by their subsequent polymerization (see Fig. 3, 10·30·60, pH 1).

IMPLICATIONS FOR GEL FORMATION

Silicate Gels

Gel formation in leached layers on glass approximates the polymerization conditions found in the sol-gel process when hydrolysis is complete, i.e. when the only reactive monomer present is $Si(OH)_4$. Near pH 2, the ^{29}Si NMR data of Englehardt, et al. [4], show that polymerization of $Si(OH)_4$ proceeds from monomer to dimer to cyclic trimer to cyclic tetramer, and finally to higher order rings. The rings are the basic units which interconnect to form gels.

Ring structures are favored over polymeric chain structures because 1) the end groups of short chains such as the linear trimer are more acidic than those of $Si(OH)_4$ and 2) the chain ends are held in close proximity to each other. Since the concentrations of ionized forms of $Si(OH)_4$ required for further chain extension are low, the linear trimer reacts with itself to form the cyclic trimer before it forms the linear tetramer. Conversion of the cyclic trimer to the cyclic tetramer requires temporary ring rupture via hydrolysis, and subsequent condensation with more monomer, followed by ring closure. Therefore, the kinetics of formation of higher order rings are slower than the kinetics of formation of the cyclic trimer. Molecular orbital calculations [6] suggest that the cyclic trimer contains strained Si-O bonds which are expected to be more susceptible to hydrolytic bond rupture reactions than normal Si-O bonds. Therefore, the conversion from the cyclic trimer to the cyclic tetramer is expected to be faster than the conversion of the relatively unstrained cyclic tetramer into higher order rings.

For the hydrosilicate layers on glass leached in acid, the Raman spectra suggest that conversion of the cyclic tetramer to higher order rings is slow, and that the cyclic tetramer forms the basic unit which polymerizes to form the gel network via condensation with residual silicic acid in solution. The dimensions of the cyclic tetramer and other small silicate rings and cages (such as the cubic octamer) are approximately the same size as the fine-scale (10 Å) particles observed via TEM which interconnect to form the network structure of the acid leached gels. Therefore, it appears that in acid, gel structures are controlled by the structures of isolated rings which form in early stages of the polymerization process. The linear polymeric silicate chains reported for acid-catalyzed sol-gel glasses [7] appear to be unstable relative to small ring structures for the case of complete alkoxide hydrolysis.

In highly basic solutions, NMR studies by Harris and Knight [8] show that both short chain and ring structures are present in equilibrium. Extensive polymerization is inhibited because all silicate species are negatively charged at high pH, repelling each other, and because so much hydroxide is present that depolymerization reactions are favored. Below pH 11, it is probable that polymerization rapidly forms small rings (as in acids) which serve as nuclei for particle growth. Particle growth is promoted in basic solutions (in contrast to retarded growth in acids) because both polymerization and hydrolytic depolymerization reactions proceed much faster due to the much higher concentrations of nucleophilic species in solution. The growth of silica particles continues until the 100-200 A particles observed in leached layers via TEM are formed.

Borate Gels

Mechanisms describing polymerization in borate gels are the reverse of the hydrolysis reactions used to describe boron dissolution from glass. Borate gels are formed by reactions involving two monomeric species: neutral, trigonal-planar boric acid and the tetrahedral borate anion. Both species must be present for gelation to occur. The relative concentrations of the two species are controlled by the solution pH:

$$B(OH)_3 + H_2O \leftrightarrow B(OH)_4^- + H^+, \ pK_a = 9 \qquad (5)$$

Above pH 11, the dominant species in solution is $B(OH)_4^-$. The borate anion will not polymerize with other borate anions because: 1) both species are anionic and repel each other, and 2) both species are coordinatively saturated (surrounded by 4 oxygens) and thus cannot polymerize with an oxygen donated by another borate anion. Below pH 4, where the dominant solution species is $B(OH)_3$, polymerization is inhibited because the hydroxyl groups on boric acid are not basic enough to have a large affinity for another boric acid molecule to form a bridging oxygen. Between pH 4 and 11, polymerization is promoted, e.g.,:

$$2B(OH)_3 + B(OH)_4^- \leftrightarrow B_3O_3(OH)_4^- + 3H_2O, \ K = 110 \qquad (6)$$

The anionic ring trimer thus formed can polymerize with other coordinatively unsaturated trigonal boron species eventually to form a gel network. As in silicate gels, the formation of small rings appears to precede gelation. The above predictions agree with the observations of Brinker [9], who has successfully synthesized and characterized borate gels.

Acknowledgements

I would like to thank my Sandia colleagues, D. R. Tallant and K. L. Higgins, for Raman spectra and interpretations, T. J. Headley for TEM results, M. S. Harrington for preparing the glasses, D. L. Lamppa for conducting leaching experiments, and C. A. Balfe, T. A. Michalske, K. D. Keefer, and C. J. Brinker for useful discussions. I would also like to thank G. L. Turner and R. J. Kirkpatrick at the University of Illinois for NMR data and interpretations.

56

References

1. R. K. Iler, The Chemistry of Silica, John Wiley & Sons, New York, 1979.

2. B. C. Bunker, G. W. Arnold, D. E. Day, and P. J. Bray, "The Effect of Molecular Structure on Borosilicate Leaching," submitted J. Non. Cryst. Solids.

3. B. C. Bunker, D. R. Tallant, T. J. Headley, R. J. Kirkpatrick, and G. L. Turner, "Molecular Structures in Leached Borosilicate Glass," in preparation.

4. D. R. Tallant, B. C. Bunker, C. J. Brinker, and C. A. Balfe, "Raman Spectra of Rings in Silicate Materials," this proceedings.

5. Von G. Engelhardt, W. Altenburg, D. Hoebbel, and W. Wieker, "Untersuchungen zur Kondensation der Monokieselsaure," Z. Anorg. Allg. Chem., 428, 43 (1977).

6. F. L. Galeener, "Planar Rings in Vitreous Silica," J. Non. Cryst. Solids, 49, 53 (1982).

7. C. J. Brinker, K. D. Keefer, D. W. Schaeffer, and C. S. Ashley, "Sol-Gel Transition in Simple Silicates," J. Non. Cryst. Solids, 48, 47 (1982).

8. R. K. Harris and C. T. G. Knight, "Silicon-29 Nuclear Magnetic Resonance Studies of Aqueous Silicate Solutions, Part 6," J. Chem. Soc. Farad. 2, 79, 1539, (1983).

9. C. J. Brinker, K. J. Ward, K. D. Keefer, E. Holupka, P. J. Bray, and R. K. Pearson, "Synthesis and Structure of Borate Based Aerogels,"

POLYMER GROWTH AND GELATION IN BORATE-BASED SYSTEMS†

C. J. BRINKER*, K. J. WARD*, K. D. KEEFER*, E. HOLUPKA** AND P. J. BRAY**
*Sandia National Laboratories, Albuquerque, NM 87185
**Brown University, Providence, RI

ABSTRACT

In situ FTIR, NMR and SAXS were used to investigate the synthesis and molecular structure of $xLi_2O \cdot (1-x)B_2O_3$ gels derived from tri-n-butyl borate (TBB) and lithium methoxide. In solution the fraction of tetrahedrally coordinated borons (N_4) increases linearly with x, and a critical value of N_4 must be exceeded in order to form gels. The primary criterion affecting gel formation is the kinetic stability of borate bonds toward molecules which are able to undergo dissociative chemisorption.

INTRODUCTION - Solution Chemistry of Borates

Although silica gels exist in nature and synthetic silica gels were reviewed in the literature as early as 1928 [1], the first synthesis of a borate-based gel was reported in 1984 [2]. This historical comment is intended to reflect the highly restrictive conditions under which polymer growth and gelation occur in borate systems as compared to silicate systems. The differences in solution chemistry of borates compared to silicates results from the fact that in the majority of cases the boron atom is trigonal coplanar with sp^2 hybridization. In this configuration boron is electron deficient in the sense of the Lewis octet theory. Much of the solution behavior of borates which we report here is related to the resulting electrophilicity of the trigonal boron atom making it susceptible to attack by water (hydrolysis) or alcohol (alcoholysis) according to the S_N2 mechanism:

$$ \tag{1} $$

†This work performed at Sandia National Laboratories supported by the U.S. Department of Energy under contract number DE-AC04-76DP00789.

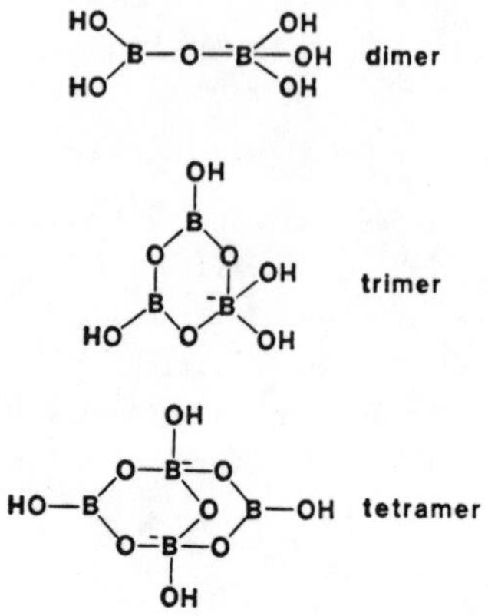

Figure 1. Aqueous borate species: dimer, trimer, and tetramer proposed by Edwards and Ross [3].

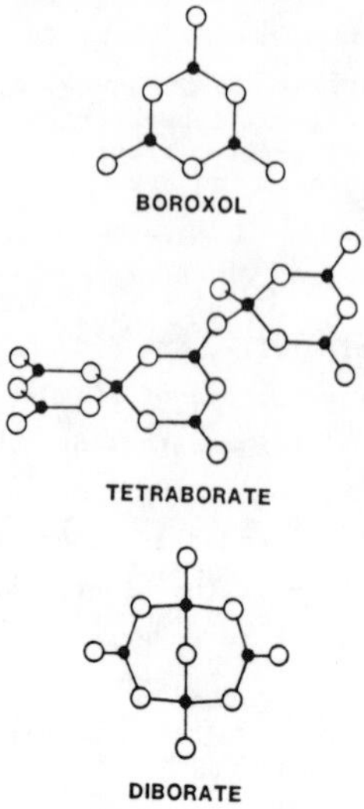

Figure 2. Examples of borate units postulated to exist in alkali borate glasses according to Krogh-Moe [4].

Like silicon, boron can assume a tetrahedral configuration with sp^3 hybridization and a negative charge:

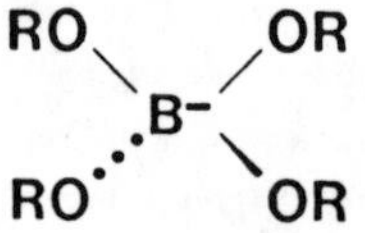

$$\begin{matrix} RO & & OR \\ & \diagdown \!\! B^- \!\! \diagup & \\ RO^{\cdots} & & OR \end{matrix}$$ (2)

But because boron has no available low energy d orbitals, tetrahedrally coordinated boron cannot form the five coordinate sp^3d transition state required for the S_N2 mechanism (Eq. 1). Therefore compounds involving tetrahedral borons are kinetically stable towards hydrolysis and alcoholysis.

A third difference between borates and silicates is the tendency for polyborates to form structures composed of cyclic trimers both in solution and in crystalline and amorphous alkali borates and alkali borate hydrates. The tendency toward ring formation and the stability of tetrahedrally coordinated borons with respect to hydrolysis form the basis of "rules" proposed by Edwards and Ross [3] for the formation of aqueous polyborates. Postulated structures for the hydrated polyborate dimer, trimer, and tetramer are shown in Figure 1. These authors suggest that longer chain polyanions are formed in solution by repeated dehydration or fusion of rings at tetrahedral boron atoms, although in aqueous systems there are no reports of gel formation.

Borate structures similar to those shown in Figure 1 are also proposed [4] to exist in crystalline alkali-borate hydrates (Figure 2). In this case the alkali ion (rather than the proton) charge compensates the borate polyanion. ^{11}B NMR [5,6], has shown that for binary alkali borate glasses and crystals ($xR_2O \cdot (1-x)\ B_2O_3$) the percentage of tetrahedrally coordinated borons (N_4) varies with x according to $N_4 = x/(1-x)$ for $x \leq 0.33$. Therefore two further structural differences between borate and silicate systems are: 1) borate glasses and gels are not continuous random networks, instead, they are are composed of well-defined structural units giving rise to intermediate range order; 2) the addition of alkali oxide to a borate glass causes the connectivity of the borate network to increase (the boron anomaly) due to an increase in bridging oxygen linkages (B-O-B) with increasing BO_4^- tetrahedra [4].

This paper describes the formation and molecular structure of xLi_2O $(1-x)B_2O_3$ gels derived from tri-n-butyl borate (TBB) and lithium methoxide in mixed alcohol/water or ether/water solvents. In situ techniques, including NMR, FTIR, and SAXS are used to obtain structural and chemical information concerning polymer growth and gelation. The primary criterion affecting gel formation is the kinetic stability of the borate-network toward dissociative reactions such as hydrolysis and alcoholysis.

EXPERIMENTAL

Using methods developed by Moore and co-workers [7] as a starting point, gel formation in the lithia-borate system was systematically investigated. Lithium methoxide ($LiOCH_3$) dissolved in methanol was diluted with 2-methoxyethanol or THF. This solution was added to requisite volumes of TBB to obtain molar compositions $xLi_2O \cdot (1-x)\ B_2O_3$, where x varied from 0 to 0.4. Neglecting the solvents (alcohol or THF), the compositions investigated are represented by six tie lines in the H_2O, B_2O_3, Li_2O system (Fig. 3). Initial solutions were prepared with H_2O/alkoxide ratios (y) of 0 to 0.4. Polymer growth was observed only after exposure of these initial solutions to 100% relative humidity (RH) at 25°C for "aging" periods of up to several days. Figure 3 defines the approximate gel forming regions in the methoxyethanol and THF systems.

Experimental details of the FTIR, NMR and SAXS experiments are published elsewhere [8].

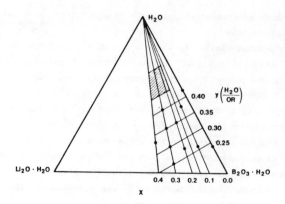

<u>Figure 3</u>. Investigated partial ternary $Li_2O \cdot H_2O$, $B_2O_3 \cdot H_2O$, H_2O system neglecting solvents. Tie lines represent constant $\bar{x}$ at varying degrees of hydrolysis (y). The crosshatched area represents the approximate gel forming region in methoxyethanol, the crosshatched plus dotted region represents the gel forming region in THF. The region $x > 0.4$ has not been investigated.

RESULTS AND DISCUSSION

Gelation in this organoborate system involves 1) the partial hydrolysis of the borate precursor (TBB) 2) condensation to form small primary units, and 3) linkage of units to form macromolecular networks. Both a minimum extent of hydrolysis and a minimum fraction of tetrahedrally coordinated borons must be exceeded in order to obtain gels. In this section we summarize our results concerning the 3 stages of gel formation itemized above. A more detailed report will be published elsewhere.

Boron Coordination

Figure 4 shows the fraction of tetrahedrally coordinated borons (N_4 determined by ^{11}B NMR) as a function of x including all investigated values of y (0 to 0.4). N_4 varies as $N_4 = (0.43)x + 0.104$ regardless of the initial y and remains constant during subsequent aging steps in 100% RH (increasing y). FTIR spectra of the pure solution components and mixtures of components [9] identify lithium methoxide as the solution component which causes boron to assume tetrahedral coordination:

$$LiOR' + \begin{array}{c} RO \quad OR \\ \backslash / \\ B \\ | \\ OR \end{array} \rightleftharpoons Li^+ \left[\begin{array}{c} OR \\ | \\ RO-B \cdots OR \\ | \\ OR' \end{array} \right]^- \qquad (3)$$

This process occurs by nucleophilic attack of the methoxide oxygen on boron. FTIR and high resolution ^{11}B NMR show that H_2O, butanol, and methanol do not produce 4-coordinated borons, suggesting that the basicity of the oxygen largely determines whether or not boron will increase its coordination. (For example, carboxylic acids can stabilize tetrahedral boron, whereas aliphatic alcohols do not. [10])

If all the added LiOMe resulted in the production of tetrahedral boron, N_4 would vary with x as $\frac{x}{1-x}$. The linear dependence of N_4 on x (Fig. 4) suggests that an equilibrium is established in which some of the LiOMe remains unreacted. FTIR spectral comparisons show evidence for unreacted LiOMe in hydrolyzed alkali borate systems. Both ^{11}B NMR and FTIR show that the non-zero intercept in Figure 4 is a result of ~10% tetrahedral boron in the "pure" tri-n-butyl borate reagent.

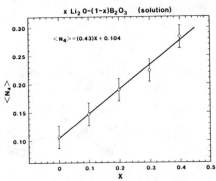

Figure 4. Fraction of 4-coordinated borons (N_4) as a function of x including all H_2O/OR ratios.

Hydrolysis and Growth

SAXS of borates prepared in 2-methoxyethanol indicates that, for $y < 0.4$, growth (if any) of polyborates it is limited to species < 10 Å in radius. Measurable growth is observed only during aging in 100% RH where y exceeds 0.4 (see Fig. 5). High resolution ^{11}B NMR and FTIR spectroscopy indicate that the addition of water to unhydrolyzed solutions ($y = 0$) initially causes a change in the chemical environment of 3-coordinated boron species. N_4 and initially the environment of 4-coordinated boron species remains unchanged. This behavior supports the nucleophilic hydrolysis mechanism indicated in Eq. 1. Due to the lack of sp^3d hybridization, hydrolysis (according to Eq. 1) occurs exclusively on 3-coordinated boron species with concerted loss of alcohol (N_4 does not increase) as confirmed by high resolution ^{11}B NMR.

62

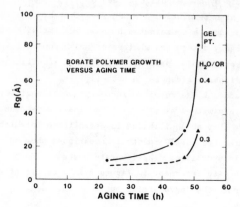

<u>Figure 5</u>.   Borate polymer growth versus aging time in 100% RH for initial
          H_2O/OR (y) ratios of 0.4 and 0.3. Gel point for initial y = 0.4
          sample indicated.

The condensation process also involves the nucleophilic attack of
oxygen on electrophilic, trigonal boron atoms:

$$ \tag{4} $$

Thus, we expect that condensation between two 4-coordinated species is
forbidden, and in reactions between 4- and 3-coordinated species, the
oxygen residing on the 4-coordinated boron is incorporated in the B-O-B
bond.

Due to the proposed [3] stability of the cyclic trimer, (Figure 1) we
expect that the initial stage of growth involves the condensation of
partially hydrolyzed 3-coordinated species with unhydrolyzed 4-coordinated
species to form primary units composed of 6-membered rings, for example:

$$ \tag{5} $$

which on average requires y ≥ 0.3. The concentration and distribution of
primary species (Fig. 2) would depend on the extent of hydrolysis, the

fraction of tetrahedral borons (N_4), and the relative stabilities of the various possible primary units. All of these units are of such small size that they would not be detected in the SAXS experiments (Figure 5).

Additional hydrolysis of trigonal boron is required in order to form extended polymers. For example:

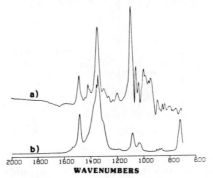

$$(6)$$

This establishes $y = 0.4$ as the minimum average value required for polymer growth. Using SAXS, we observed no growth for $y < 0.4$. For an initial $y = 0.4$ measurable growth occurred only after aging in 100% RH (Figure 5).

To examine the structures of growing polymers, FTIR subtraction spectra were derived. Either the solvent spectrum was subtracted from the

Figure 6. a) FTIR Solution spectrum (x = 0.3) after 24 hours of aging (solvents subtracted)
b) FTIR spectrum of trimethoxyboroxine

solution spectrum after a specific aging time or the spectra of solutions aged for two different times were subtracted. In the latter case, the difference spectrum represents product species. Figure 6 compares the difference spectrum for x = 0.3 obtained after 24 hrs of aging (solution spectrum - solvent spectrum) to the spectrum of trimethoxyboroxine. The similarities in the ~1300-1500 cm^{-1} region (trigonal B-O assym. stretch [11]) in Figs. 6a and b suggest that after 24 hours of aging the environment of the trigonal borons in solution is similar to those in trimethoxyboroxine:

$$(7)$$

64

Dissimilarities in the ~850-1100 cm^{-1} region (tetrahedral B-O assym. stretch [11]) reflect that N_4 = 0.22 for the lithium borate solution (x = 0.3), whereas N_4 = 0 for trimethoxyboroxine. As shown by the difference spectrum, 120-72 hours of aging (Fig. 7c), additional hydrolysis causes the environment of the 3-coordinated borons (1300-1500 cm^{-1}) to become more complex. The comparison made in Fig. 7 suggests that during the later stages of aging the products formed by continued hydrolysis and condensation are similar in structure to partially hydrolyzed lithium tetraborate.

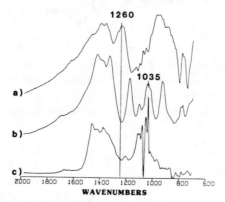

Figure 7. FTIR spectra of a) anhydrous crystalline lithium tetraborate, b) partially hydrolyzed lithium tetraborate, and c) difference spectrum 120-72 hours of aging.

Criteria for Gel Formation

Figure 3 defines the approximate gel forming regions in the Li_2O, B_2O_3, H_2O ternary for x ≤ 0.4. Apart from a critical concentration of H_2O, Figure 3 also suggests a minimum concentration of Li_2O which must be exceeded in order to obtain gels. Because N_4 varies linearly with x (Fig. 4), this limitation in gel formation presumably arises due to a critical requirement for 4-coordinated borons. For example in 2-methoxyethanol no polymer growth is observed for x ≤ 0.1, [N_4 ≤ 0.14] and only very weak gels are formed when x = 0.2[N_4 = 0.19]. In THF, gels are observed for x ≥ 0.15[N_4 ≥ 0.17]. These experimental observations can be explained on the basis of the stability of B-O-B bonds towards dissociative reactions of the type:

$$ROH + \ \text{B—O—B} \rightleftharpoons \text{B—O—B (intermediate)} \rightleftharpoons \text{B} + \text{B} \quad (8)$$

Because boron cannot utilize sp^{3}d hybridization, this nucleophilic mechanism is not possible for 4-coordinated borons. Thus, the kinetic stability of B-O-B bonds is expected to decrease in the following order [$\equiv$B — O — B$\equiv$] > [$\equiv$B — O — ·B$=$] > [$=$B — O — B$=$] . This is consistent with hydrolysis mechanisms proposed by Bunker to explain the aqueous corrosion of alkali borosilicate glasses [12].

For all systems investigated $N_4 \leq 0.27$. Thus there are no possible structural models which both exclude all linkages between two trigonal borons: $=$B — O — B$=$ and fully incorporate boron in an "infinite" network. Since B-O-B bonds exist both within primary structural units (Figs. 1 and 2) and between primary units, the N_4 criterion may reflect the kinetic stability of either of these types of borate bonds. For vitreous B_2O_3 (composed completely of trigonal borons residing in boroxol units) Krogh-Moe [4] determined that $=$B — O — B$=$.bonds which link together units are more susceptible to hydrolysis than $=$B — O — B$=$ bonds within units. Thus in the solution environment employed in gel processing we should expect that gel formation will be limited by the kinetic stability of borate bonds which link together primary structural units.

In order to test this hypothesis and determine an appropriate structural model for solution-derived lithium borate networks (gels), we performed FTIR investigations of crystalline alkali borates and partially hydrated crystalline alkali-borates (with N_4 = 0.25 to 0.50). A prominent spectral feature (Figs. 8a and b) of the anhydrous tetra- and triborate compounds is the ~1260 cm^{-1} vibration due to oxygens bridging between trigonal borons contained in separate primary units [4] (Fig. 9a). Anhydrous lithium diborate is composed exclusively of diborate units linked by oxygens bridging between 3- and 4-coordinated borons and exhibits no 1260 cm^{-1} vibration (Fig. 8c). Partial hydrolysis of the tetra- (and tri-

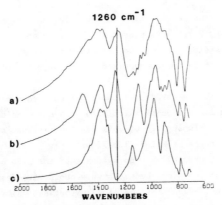

Figure 8.   FTIR spectra of anhydrous crystalline lithia borates:
a) tetraborate b) triborate and c) diborate.

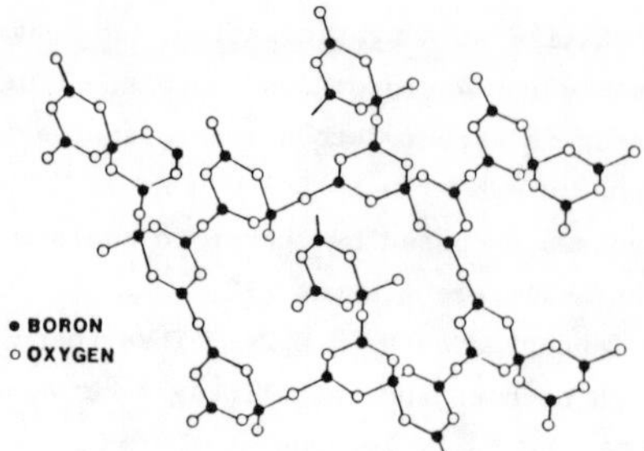

Figure 9. a) A portion of the crystalline tetraborate network showing linkages between primary units:
$\equiv B - O - B\equiv$ and $=B - O - B\equiv$.

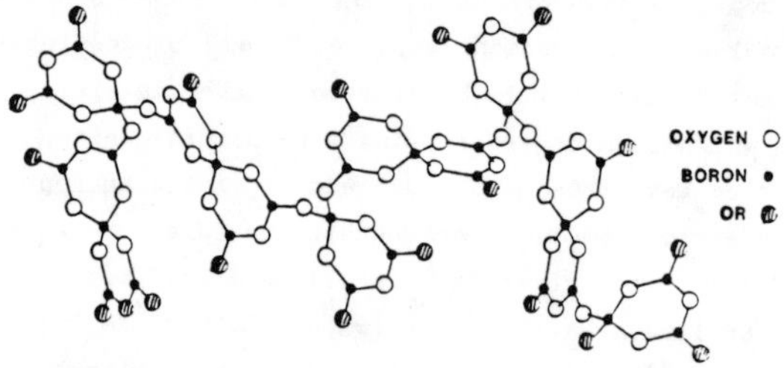

b) Tetraborate network in which there is no $=B - O - B\equiv$ bonding between units.

borates) causes a dramatic reduction in the relative intensity of the 1260 cm^{-1} vibration (Fig. 7b). This proves that oxygens which bridge between trigonal borons residing in separate primary units (e.g., tetraborates or triborates) are unstable in aqueous (and presumably alcoholic) environments.

Thus, any reasonable structural model proposed to explain borate gel networks formed in mixed alcohol/water solutions must exclude $=B - O - B\equiv$ bonds between units. In support of this idea, difference spectra obtained during the latter stages of aging (120-72 hrs, Fig. 7c), where we measure considerable polymer growth, show no evidence of the 1260 cm^{-1} band. Because condensation reactions between two 4-coordinated borons are forbidden (Eq. 4) we deduce that the linkages between units responsible for polymer growth and gelation are oxygens bridging 3- and 4-coordinated borons. From the greatly increased relative intensity of the 1035 cm^{-1}

band (Fig. 7c) during the latter stages of aging, we tentatively assign this vibration to the assymetric B-O stretch of $\equiv$B$-$O$-$B$=$ linkages between units. This vibration is also prominent in the spectrum of partially hydrated (partially depolymerized) lithium tetraborate (Fig. 7b).

The similar shapes of the envelopes of vibrations associated with trigonal (1300-1500 cm^{-1}) and tetrahedral (850-1100 cm^{-1}) borons in partially hydrated lithium tetraborate (Fig. 9b) and the 120-72 hour product species (Fig. 9c) allow us to conclude that the polyborate species formed during the latter stages of aging are structurally similar to partially hydrolyzed lithium tetraborate. The strong bands at about 930 and 1175 cm^{-1}, present only in the partially hydrolyzed lithium tetraborate spectrum, are assigned to B-O-H out-of-plane and in-plane bending, respectively, on the basis of D_2O and $H_2^{18}O$ hydrolyses. Since under the present hydrolysis conditions (aging at 25°C and 100% RH) the concentration of B-OH appears to be very low in the borate gel, we conclude that hydrolysis is rate limiting with respect to polymer growth and gelation. This makes isolation and identification of the hydrolysis products difficult in these systems.

To arrive at a structural model for lithium borate gels (x = 0.3) based on a tetraborate primary unit, we exclude all $=$B $-$ O $-$ B$=$ linkages between units and terminate non-bonded sites with alkoxides (Fig. 9b). Perhaps, coincidentally, this tetraborate borate network represents the lowest average mole fraction of tetrahedral borons (N_4 = 0.25) in which a continuous network can be constructed which excludes $=$B $-$ O $-$ B$=$ between units and includes all borons. In methoxyethanol, stiff gels were formed for x = 0.3 (N_4 = 0.23), whereas very weak gels were formed when x = 0.2 (N_4 = 0.19). This indicates that for x = 0.2 less of the total boron in solution is incorporated in the tetraborate network as required by the lower average N_4.

Replacing 2-methoxyethanol with the cyclic ether, THF,

$$\begin{array}{c} H_2C \text{———} CH_2 \\ \diagup \qquad\qquad \diagdown \\ H_2C \qquad\qquad\quad CH_2 \\ \diagdown \qquad\quad \diagup \\ O \end{array} \qquad (9)$$

which due to the non-labile protons does not react according to Eq. 8, broadens the gel forming region to include x $\geq$ 0.15 (Fig. 3). This reduced requirement for N_4 reflects the reduced concentration of alcohols which can undergo dissociative reactions with B-O-B linkages (Eq. 8). Future investigations will address whether or not gels formed in the system x = 0.15 are structurally different from those formed when x $\geq$ 0.2.

CONCLUSIONS

Gel formation in organoboron systems involves the partial hydrolysis of the borate precursor, condensation of monomers to form primary units and linkage of primary units to form extended networks. The hydrolysis and condensation processes as well as the kinetic stability of the borate network are all influenced by the electrophilic behavior of the trigonal boron atom. In solution environments containing molecules which can dissociatively react with trigonal boron (H_2O, ROH, etc.) a critical fraction of 4-coordinated borons must be exceeded in order to obtain gels. The gel network appears to be composed of tetraborate units crosslinked between 3- and 4-coordinated borons, exclusively.

ACKNOWLEDGEMENTS

The authors thank B. C. Bunker for many informative discussions. The technical assistance of C. S. Ashley is greatly appreciated.

REFERENCES

1. K. Wolf and M. Praetorius, Metallbourse 18, 453 (1928).
2. N. Tohge, G.S. Moore and J.D. Mackenzie, J. Non. Cryst. Solids 63, 95 (1984).
3. J.O. Edwards and V. Ross, J. Inorg. Nucl. Chem. 15, 329 (1960).
4. J. Krogh-Moe, Phys. Chem. Glasses 6, 46 (1965).
5. A.H. Silver and P.J. Bray, J. Chem. Phys. 29, 984 (1958).
6. S.E. Svanson, E. Forslind and J. Krogh-Moe, J. Phys. Chem. 66, 174 (1962).
7. M.C. Weinberg, G.F. Neilson, G.L. Smith, B. Dunn, G.S Moore and J.D. Mackenzie, J. Mat. Sci. 20 1501 (1985).
8. C.J. Brinker, K.J. Ward, K.D. Keefer, E. Holupka, P.J. Bray and R.K. Pearson, Aerogels, edited by J. Fricke (Springer Verlag, Berlin, 1986).
9. D. M. Haaland,R. G. Easterling, and D. A. Vopicko, J. Appl. Spectroscopy, 39, 73 (1985).
10. K. Kustin and Richard Pizer, J. Am. Chem. Soc. 91-2, 317 (1969).
11. S.D. Ross, The Infrared Spectra of Minerals, edited by V.C. Farmer (The Mineralogical Society, London, 1974).
12. B.C. Bunker, this proceedings.

PART II

Solution Chemistry and Synthesis II: Powders

NUCLEATION AND GROWTH OF UNIFORM m-ZRO$_2$

A. BLEIER[*] AND R. M. CANNON[**]
[*] Oak Ridge National Laboratory, P. O. Box X, Oak Ridge, TN 37831
[**] Lawrence Berkeley Laboratory, Hearst Mining Building, University of California, Berkeley, CA 94720

ABSTRACT

Hydrothermal treatment of zirconyl salt solutions produces uniform m-ZrO$_2$ powder on the order of 80 nm. This powder is porous, has a 3-nm crystallite size, and exhibits an unusually high degree of crystallographic alignment within particles. The generation of this powder occurs via a complex process involving nucleation, growth, and controlled agglomeration of primary particles. Particle formation, crystallographic alignment and particulate uniformity are explained in terms of solution reactions and colloidal behavior.

INTRODUCTION

Controlled hydrothermal treatment of zirconyl salt solutions produces uniform monoclinic zirconia powder that is 50-% porous, has a specific surface area as high as 180 m^2 g^{-1}, and is polycrystalline with a 3-nm crystallite size, but exhibits an unusually high degree of crystallographic, intraparticle alignment.[1] Figure 1 contains transmission electron micrographs of typical powder produced using 0.2 M ZrO(NO$_3$)$_2$; similar powder has been obtained from ZrOCl$_2$.[2,3] Nucleation and growth of these particles was controlled by slowly aging the solution at 98 °C for 70 h. Under these conditions, sufficient hydrolysis of zirconyl species occurs[2] to generate complex species of significantly less solubility than that of the starting salt. Table I lists typical properties of the powder while Table II indicates that some reagents impact the final product differently.

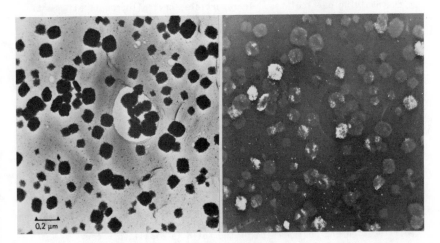

0.2 μm

Figure 1: (a) Light field and (b) dark field TEM pictures of typical m-ZrO$_2$ powder produced at 98 °C using ZrO(NO$_3$)$_2$. Bright areas in (b) emphasize alignment of crystallites comprising particles.

Table I. Properties of Typical Powder, 98 °C, 72 h[1].

Property	Characterization	Analytical Methods
Composition	m-ZrO$_2$	X-ray Diffraction, EDAX
% v/v H$_2$O	50	TGA
Surface Area, m^2 g^{-1}	>100	BET
Particle Size, nm	10 - 100[a]	TEM, SEM, Light Scattering
Isoelectric Point	6.5 - 7.0	Microelectrophoresis

[a] Increases with concentration in the range 0.002 to 0.2 M Zr.

Table II. Reagents for Studies of 0.2, 0.02, and 0.002 M Zr.

Anion	Salt	Product	Relative Acidity[a]
NO$_3$$^-$	ZrO(NO$_3$)$_2$·2H$_2$O	Powder	1.0
Cl$^-$	ZrOCl$_2$·8H$_2$O	Powder	1.1
ClO$_4$$^-$	ZrO(ClO$_4$)$_2$·8H$_2$O	Powder (0.2 M)	1.7
SO$_4$$^{2-}$	ZrOSO$_4$·H$_2$SO$_4$·3H$_2$O	Gel	0.6
Cl$^-$	ZrCl$_4$	Powder (0.02 M)	0.8

[a] Normalized to the number of equivalents of acid per mole ZrO(NO$_3$)$_2$.

The synthesis process for m-ZrO$_2$ circumvents the need for calcination since amorphous gelatinous material is not an isolated phase, intermediate between the initial solution and the desired oxide powder. Traditional considerations[4,5] of solution hydrolysis, solute complexation, and nucleation and growth only partly explain the process. The roles of colloidal stability and, therefore, agglomeration of precipitated matter must be taken into account. Consequently, uniform particulates are adequately explained only by considering the specific effects of synthesis conditions, e.g. pH, metal concentration, ionic strength, and temperature, on the evolving particulates. Intraparticle, crystallographic alignment can be also understood in terms of colloidal interactions and dipole-dipole forces and inherent material properties such as crystal structure, solvation, and molar volume.

Previous parts of this work, being prepared for publication, focus on details of the synthesis and characterization of uniform m-ZrO$_2$[1], its processing properties, and pertinent sintering behavior[6,7]. Following a summary of the synthesis, this paper describes a model developed to understand the generation and evolution of the powder. General quantitative models for the hydrous metal oxide surface and an appropriate incorporation of colloidal considerations improve the elucidation of specific mechanisms that determine the nature of the precipitated phases. The fundamental principles that seem to govern growth of the uniform particles in Figure 1 necessarily include controlled agglomeration of 3-nm crystallites of m-ZrO$_2$.

MATERIALS AND METHODS[1,6-9]

Aqueous solutions of ZrO(NO$_3$)$_2$ at desired concentrations were aged for extended times at temperatures between 90 and 116 °C. Most studies were conducted at 98 °C. The resultant sols were cleaned via repeated centrifugation and redispersion of the solid in distilled water.

Techniques used to study the powder and its generation include Tyndall scattering to determine initial precipitation conditions, electron microscopy to evaluate particle size and microstructure, x-ray diffraction to examine dried powder compacts, electron diffraction to identify crystallinity for individual particles, BET surface area analysis and in situ photon correlation spectroscopy, dynamic light scattering, and Stokes' law to estimate particle size. Microelectrophoresis and Henry's equation[10] were used to evaluate zeta potential. Potentiometric titration yielded information on surface charge characteristics of selected powders and equivalent weight of initial zirconium salts; the second quantity corresponds to the mass of salt that neutralizes one mole of OH⁻ ions.[11] Finally, atomic absorption was used to measure the amount of zirconium remaining in solution and thermal analysis was used to investigate selected powders.

POWDER SYNTHESIS[1,3,9]

Aging solutions of $ZrO(NO_3)_2$ in the range 0.002 to 0.2 M at or near 98 °C produced a colloidally stable, white precipitate. Similar powders are produced from solutions of some other Zr-containing salts, whereas gels result from solutions of zirconyl sulfate; see Table II. Microscopic observations on powder derived from 0.2 M Zr indicate a spheroidal particle size of 80 nm, whereas photon correlation and sedimentation data indicated respective values of 99.5 and 96 nm.

The effects of heating time, concentration of zirconyl salt, aging temperature, and anion were determined. An increase in either of the first two quantities increases the final particle size. A slight reduction in x-ray line broadening with increased heating time was also noted[7]. Particles obtained from 0.02 M Zr seem to be less spherical, as well as finer (Table I), than those formed from 0.2 M solutions. Precipitation of $m-ZrO_2$ during the initial 72 h is very sensitive to temperature in the range 90 to 98 °C, with 94 to 96 °C being a particularly important range when the initial concentration is 0.2 M Zr. Yield is virtually 100 % above 96 °C, noticeably less that this value below 94 °C, and nil at 90 °C; aging between 115 and 120 °C for 24 h produces uniform $m-ZrO_2$ powder that is structurally similar by TEM analysis to the product obtained at 98 °C after 72 h. Finally, though similar powder may be obtained with other salts, Table II indicates that the specific hydrolysis route may depend on the acid-base character of the anion and that the final product is also sensitive to the anion; the table also notes the similar reactivity of zirconyl nitrate and chloride.

COLLOIDAL CHARACTERIZATION AND CERAMIC PROCESSING[1,3,6-9]

Colloidal Characterization. The powder described in Table I (a) does not significantly agglomerate at pH <4.5 and >9.5; (b) forms small but somewhat stable agglomerates in the ranges pH 4.5 to 6.5 and 8.3 to 9.5; and (c) coagulates and rapidly settles within the range pH 6.5 to 8.3. Electrophoretic and acid-base behavior agrees well with that of other zirconium oxides.[12]

Potentiometric titration, used to probe porosity evident in Figure 1, demonstrated that equilibration between particles and an aqueous environment is slowly attained, but if the time of "equilibration" between particles and solution is sufficiently long, on the order of 1 h, most of the internal surface reacts with its environment.

Ceramic Processing. Green pieces, produced by sedimentation, fired in air at 1250 °C for one hour, and examined by SEM of fracture surfaces, demonstrated[7] that the interior of fired pieces is nearly free of pores. This feature derives, in part, from efficient packing in prefired pieces.[1]

GENERAL THEORETICAL CONSIDERATIONS

Nucleation and growth apparently must be separated to prepare particles of nearly monodisperse size distribution and of uniform compositional and shape characteristics. According to LaMer's theory[13], homogeneous nucleation of monodisperse sols requires the nucleation rate to increase monotonically to a critical supersaturation range, at which point the nucleation products are stable and instantly reduce the degree of super-saturation and after which point, growth of these nuclei ensures that super-saturation can not reach the critical range again. Moreover, solubility decreases as growth continues owing to the reduced specific surface area. If crystals nucleate, their growth is often governed by the rate at which solute in the proper orientation reaches their surfaces.

Though primary particles grow by adsorption of solute species, an equally important role is ascribed to aggregation. Specific experimental conditions determine whether the respective roles of these two processes are of similar magnitude or one dominates, and so dictate which of the vastly different possible morphologies develop. Finally, aging, e.g. by Ostwald ripening, and chemical and structural transformations from meta-stable phases to more thermodynamically stable ones complete the general list of phenomena comprising the generation of uniform powders.

The powder characteristics for specific materials depend on mechanisms and relative rates of competing reactions and processes, corresponding to the experimental conditions. Initial concentration of zirconyl salt, pH, and temperature are critical parameters determining whether uniform $m\text{-}ZrO_2$ or amorphous, gelatinous precipitate is obtained. Rate of hydrolysis leading to the polymerization of solute species is, moreover viewed by Clearfield[14], as the most important consideration regarding whether crystal-lization or gelation predominates for precipitates derived from zirconyl salts. Additional solution species, i.e. other than those of the metal and the various components of water, also influence the nature of the precipitate; a detrimental effect of sulfate species on the generation of uniform $m\text{-}ZrO_2$ was reported by Beckhart[15], agreeing with the concepts of Clearfield[14] and Blumenthal[16] and supporting the conclusions of Matijević et al.[17]

Consideration of hydrolysis and polymerization in acidic, zirconyl solutions[14,18], fundamental colloid behavior[4,19,20], and emerging perceptions of intermolecular and surface forces[21] suggests the following mechanism for generation of the uniform $m\text{-}ZrO_2$.

Hydrolysis, Complexation, and Precipitation. Upon dissolution, zirconyl salts liberate cationic ZrO^{2+} species that rapidly hydrolyze, generating hydroxylated sites, and ultimately lead to stoichiometrically monodisperse, soluble tetramers of the form, $[Zr(OH)_2 \cdot 4H_2O]_4^{8+}$; these tetramers have a square, planar array of Zr atoms and overall diameter and length respectively, of 0.898 and 0.582 nm.[18] Upon subjection to heat and under certain conditions, these complexes react further to produce their own dimers and higher order assemblies that may constitute two-dimensional sheets. Either the individual tetramers or their two-dimensional networks associate into a three-dimensional structure, via condensation of water between hydroxylated sites on adjacent tetramers or layers. Under proper conditions a fluorite or derivative crystal structure obtains directly. Nucleation appears favorable when the cluster or aggregation number of this quasicrystalline, 3-dimensional assembly approaches 12 to 24 tetramers, a range supported elsewhere[14]. Aging promotes crystal growth and removal of local defects; though intermediate structures may resemble cubic ZrO_2, $m\text{-}ZrO_2$ eventually forms and attains a 3-nm crystallite size.

Based on this scenario, the charge density of the tetrameric cation is estimated to be 44 μC cm^{-2}; a similar value is anticipated for the next generation's clusters or agglomerates. Interestingly, the maximum charge density of $m\text{-}ZrO_2$, based on available potentiometric titration data[12], is 40 μC cm^{-2}, supporting the scheme just given.

The neutralization of solutions containing zirconyl salts is, of course, related to the process postulated here. If base is added to acidic solutions, gelation occurs rather than formation of crystalline particles. Available data[2,14] suggest that 0.2 M Zr is the solubility at pH 1.15 and that the total concentration of zirconium is reduced by at least 60 % for an increase in pH of 0.1, whereas it is reduced to 0.001 % of its initial value for an increase of 1.0 pH-unit. Thus, syntheses based on neutralization evidently constitute a chemical shock and, thereby, induce very rapid polymerization of tetramers, generating polymeric structures that ultimately precipitate as gels, exhibiting high entropy, partial hydration, and entrapped counterions. The requisite extensive reconstruction and purification to crystallize such three-dimensional polymers takes long times or high temperatures. Thus, this sequence of events critically contrasts that of the process described first and seems fundamentally incapable of directly leading to the formation of powders, particularly those that are colloidally stable or crystalline.

These differences arise from specific reactions involving the cationic tetramer. Crystal growth requires their parallel, but staggered alignment. Moreover, at surfaces where contact is edge-to-edge, bonding entails dissociation of H_2O to establish hydroxyl or oxide bridges, and face-to-face bonding entails concurrent condensation of H_2O. In contrast, a tetramer corner can attach to a face-center without either dissociation or condensation; thus, such attachment could be more rapid, albeit less energetically favorable. Bonding of additional tetrameric cations to the misoriented one provides a basis for forming a three-dimensional, amorphous network. Apparently, achieving crystalline primary particles requires supersaturation sufficiently low that individual tetramers can align and bond properly during growth. Acidic solutions may also facilitate crystal formation because firstly, the positive particle surface charge inhibits approach of other tetramers and secondly, the reduction in surface hydroxyl groups lessens the amount of H_2O-condensation needed.

Formation of Uniform Secondary Particulates. Though some previous work[14,18] describes the nucleation and growth of $m-ZrO_2$ from solutions similar to those examined here and in related research[1,3,9], the origins of uniform secondary structures, such as those in Figure 1, have not been described for zirconia. Using the treatments of Hogg et al.[20] for interactions between electrical double layers and Israelachvili[21] for van der Waals forces, the colloidal stability of 3-nm $m-ZrO_2$ and subsequent aggregates of these primary particles was investigated. Theoretical examination, of the three synthesis conditions, 0.2, 0.02, and 0.002 M Zr, at respective pH-values of 0.5, 1.5, and 2.5,[1,8,9] demonstrates with the aid of Table III that colloidal, primary 3-nm crystallites are: (a) quite unstable when produced by the first set of conditions, having a barrier to agglomeration of ~0.3 kT, (b) more stable under the second set but still likely to agglomerate significantly, since the predicted barrier is ~7 kT, and (c) very stable under the third set of conditions which correspond to a barrier of 24 kT. These conclusions agree well with experimental observations suggesting secondary particle formation, as stated earlier and described in detail elsewhere[1]. Similarly, the experimental finding[1] that these sets of conditions ultimately produce respective particulates of diameter 80, 20, and <10 nm is also supported by these calculations, revealing the stability to homocoagulation of the porous particulates in Figure 1. Finally, Table III predicts that heterocoagulation between large particulates and primary particles occurs for the higher initial Zr-concentrations. This table and indirect data[1,7] support the conclusion that secondary particles principally form by heterocoagulation involving attachment of individual small crystallites, a process usefully perceived as analogous to precipitation of macromolecules.

The mechanism leading to the observed crystallographic alignment within large particles (Figure 1b) is not fully understood, but several factors are envisaged to induce a tendency for primary particles to align during the

Table 3. Colloidal Stability of m-ZrO$_2$ Particles at 20°C[9]

[Zr], M	pH	Diameter, nm	Barrier, kT	Agglomerate Growth
0.16	0.49	3	0.3	Highly Favorable
		80	14	Borderline
		Mixed	0.8	Highly Favorable
0.016	1.48	3	7	Highly Favorable
		20	49	Negligible
		Mixed	13	Borderline
0.0016	2.48	3	24	Negligible
		10	80	Negligible
		Mixed	37	Negligible

successful collisions causing coagulation. Longer range forces, and likely more potent ones, arise from a nonuniform surface charge distribution around the crystallite surface; the resulting dipole or multipole forces would operate over separations approximating the particle size. Particle faceting would lead to torques and forces favoring the closer center-to-center approach attained with aligned facets.

Finally, significant effects of hydrated Stern layers[19,21] emerge from detailed analysis[3,11,22] of the interactions among particulates during synthesis. Under the synthesis conditions, the Stern layer helps to stabilize growing agglomerates consisting of 3-nm m-ZrO$_2$ and to retard the loss of the crystallites by homocoagulation. This retardation of coagulation enables agglomerate structures to grow slowly and more uniformly. Crystallite alignment demonstrated in Figure 1 may also ultimately rely on the existence of Stern layers and their impact on the final stages of the heterocoagulation process that leads to growth of secondary particulates; such layers would slow the approach and allow time for rotation or alignment into favorable configurations.

In summary, producing uniform, secondary particulates comprised of primary, crystalline particles evidently depends upon several factors. Though a critical supersaturation range is required to induce copious nucleation, it must keep growth from solution appropriately slow to allow alignment of solute and dissociation or condensation of H$_2$O during attachment. This slow growth at higher Zr-concentrations is apparently slower than that controlled by diffusion and permits agglomeration to ensue before the crystallites become large, i.e. at ~3 nm, curtailing further crystallite growth. A relatively narrow secondary particle size distribution ultimately evolves owing to the increased colloidal stability of large, porous particulates. In very dilute solution, only stable primary crystalline particles are expected because of their greater colloidal stability (Table III).

CLOSING REMARKS

Elements of the model described for the synthesis of uniform m-ZrO$_2$ may not be unique but the principles may also apply to other oxide materials, particularly those that transform to polycrystalline ceramic powders upon calcination at elevated temperature. Key factors are: (a) generation of discrete solute species, in this case [Zr(OH)$_2$·4H$_2$O]$_4^{8+}$, (b) slow growth of these entities toward controlled nucleation and growth of precipitate, and (c) uniform agglomeration of primary particles.

A few citations that support the general logic underpinning the sequence of events proffered here are noteworthy at this point. Firstly,

Matijević[5] has demonstrated that solute species of discrete stoichiometry may lead to crystalline particles, whereas metals that exhibit broad speciation in solution under synthesis conditions produce amorphous particles; assignment of a specific tetrameric species in the synthesis scheme for m-ZrO_2 is consistent with the generic principle developed by Matijević after studying numerous chemical systems. Secondly, traditional theory for the generation of monodisperse particles[5,13] certainly includes controlled growth, both prior to and subsequent to nucleation, as noted earlier. Distinguishing features of the synthesis described here with those that produce gels underscore this point. Thirdly, few reports appear in the literature describing processes in which controlled agglomeration of primary particles leads to uniform suprastructures and none seem to describe synthesis of zirconia. However, among the reports of controlled generation of secondary particles involving other compositions, those of de Bruyn and coworkers[23] for FeOOH are most pertinent to this study. Though oriented coagulation is considered by these researchers, they favor an alternative route that is based on long-term stability of an amorphous "goethite-like" precipitate. To the contrary for m-ZrO_2, Bleier and Angelini[9] reported that only crystalline intermediates could be detected in early stages of nucleation. However, transformation of amorphous precipitate to goethite[23], accompanied by densification, apparently does occur prior to formation of secondary particles. Finally, since homo- and heterocoagulation calculations for m-ZrO_2 support a model involving controlled agglomeration to produce uniform secondary structures, a similar treatment for goethite may explain monodispersity in that system and also in the very recently noted[5] one for monodisperse CeO_2. Similarly, secondary structure within low density particles of monodisperse amorphous TiO_2, formed by alkoxide hydrolysis[24] suggests that they also form by controlled coagulation of primary particles.

ACKNOWLEDGEMENTS

The authors gratefully acknowledge Drs. P. Angelini and P. Sklad for numerous discussions and Ms. F. Stooksbury for help in assembling the manuscript. This research was sponsored by the Division of Materials Science, U.S. Department of Energy, under Contract No. DE-AC05-84OR21400 with Martin Marietta Energy Systems, Inc. Teledyne, Wah Chang Albany, Albany, OR kindly donated high-purity zirconyl chloride and nitrate.

REFERENCES

1. A. Bleier and R. M. Cannon, "Synthesis and Characterization of Uniform Zirconia", In preparation for the Am. Ceram. Soc.; Am. Ceram. Soc. Bull. 61 (3), 336 (1982).
2. A. Bleier, "Synthesis of ZrO_2-Based Powders Using Solution Techniques", Presented at the MRS Annual Meeting, Boston, MA, 1982, Paper No. N3.2; A. N. Ermakov, I. N. Marov, and V. K. Balyaeva, Russ. J. Inorg. Chem. 8, 845 (1963); H. Bilinski, M. Branica, and L. G. Sillen, Acta Chem. Scand. 20, 853 (1966); E. Matijević, K. G. Mathai, and M. Kerker, J. Phys. Chem. 66, 1799 (1962); Stability Constants of Metal Ion Complexes, Spec. Publ. Nos. 17 and 25, edited by L. G. Sillen and A. E. Martell (Chem. Soc., London, 1964 and 1971), pp. 45-46 and 18, respectively.
3. A. Bleier, in Ultrastructure Processing of Ceramics, Glasses, and Composites, edited by L. L. Hench and D. R. Ulrich (John Wiley & Sons Inc., New York, 1984), p. 391.
4. J. Th. G. Overbeek, in Colloid Science, edited by H. R. Kruyt (Elsevier Publishing Co., Amsterdam, 1952), pp. 63-68.

78

5. E. Matijević, Langmuir $\underline{2}$, 12 (1986); Acc. Chem. Res. $\underline{14}$, 22 (1981); Pure Appl. Chem. $\underline{50}$, 1193 (1978).
6. E. Rozier, "Effect of Grain Size on Sintering Behavior, Phase Transformation, and Toughness in Unstabilized Zirconia", B.Sc. Thesis, Massachusetts Institute of Technology, Cambridge, MA, 1982, 57 p.
7. R. M. Cannon and A. Bleier, "Sintering of Ultrafine ZrO_2 Powder", In preparation for the Am. Ceram. Soc.; Am. Ceram. Soc. Bull. $\underline{61}$ (8), 811 (1982).
8. W. C. Hasz and A. Bleier, Am. Ceram. Soc. Bull. $\underline{62}$ (3), 376 (1983).
9. (a) A. Bleier and P. Angelini, in Abstracts: 59th Colloid and Surface Science Symposium (Am. Chem. Soc., Potsdam, NY, 1985), Paper No. 346; (b) S. Spooner, P. Angelini, P. F. Becher, A. Bleier, W. D. Bond, and J. Brynestad, in Extended Abstracts: American Ceramic Society, 87th Annual Meeting (Am. Ceram. Soc., Columbus, OH, 1985), p. 3.
10. D. C. Henry, Proc. Roy. Soc. A133, 106 (1931).
11. (a) A. Bleier, in Advances in Materials Characterization, Material Science Research, Vol. 15, edited by D. R. Rossington, R. A. Condrate, and R. L. Snyder (Plenum Pr., New York, 1983), p. 499; (b) W. C. Hasz and A. Bleier, in Advances in Materials Characterization II, Materials Science Research, Vol. 19, edited by R. L. Snyder, R. A. Condrate, Sr., and P. F. Johnson (Plenum Pr. New York, 1985), p. 189.
12. F. S. Mandel and H. G. Spencer, J. Colloid Interface Sci. $\underline{77}$, 57 (1980); A. E. Regazzoni, M. A. Blesa, and A. J. G. Marato, J. Colloid Interface Sci. $\underline{91}$, 560 (1983); S. K. Milonjic, Z. E. Ilic, and M. M. Kopecni, Colloids and Surfaces $\underline{6}$, 167 (1983).
13. (a) V. K. La Mer and M. D. Barnes, J. Colloid Sci. $\underline{1}$, 71 (1946); (b) V. K. La Mer and A. S. Kenyon, ibidem. $\underline{2}$, 257 (1947); (c) V. K. La Mer and R. H. Dinegar, J. Am. Chem. $\underline{72}$, 4847 (1950).
14. A. Clearfield, Rev. Pure Appl. Chem. $\underline{14}$, 91 (1964).
15. G. H. Beckhart, "Zirconium Dioxide Synthesis", UROP Report, Massachusetts Institute of Technology, Cambridge, MA, May 1981, 20 p.
16. W. B. Blumenthal, The Chemical Behavior of Zirconium (D. Van Nostrand Co., Inc., Princeton, 1958), p. 240.
17. E. Matijević, A. Watanabe, and M. Kerker, Kolloid Z. Z. Polym. $\underline{235}$, 1200 (1969).
18. J. R. Fryer, J. L. Hutchinson, and R. Paterson, J. Colloid Interface Sci. $\underline{34}$, 238 (1970).
19. (a) P. C. Hiemenz, Principles of Colloid and Surface Chemistry, 2nd ed. (Marcel Dekker, Inc, New York, 1986), 815 p.; (b) R. D. Vold and M. J. Vold, Colloid and Interface Chemistry (Addison-Wesley, Reading, MA, 1983), 694 p.; (c) O. Stern, Z. Elektrochem. $\underline{30}$, 508 (1924).
20. R. Hogg, T. W. Healy, and D. W. Fuerstenau, Trans. Faraday Soc. $\underline{62}$, 1638 (1966).
21. J. N. Israelachvili, Intermolecular and Surface Forces (Academic Pr. London, 1985), 296 p.
22. R. O. James and G. A. Parks, in Surface and Colloid Science, Vol. 12 edited by E. Matijević (Plenum Pr., New York, 1982), p. 119.
23. (a) J. H. A. van der Woude, P. L. de Bruyn, and J. Pieters, Colloid and Surfaces 9, 173 (1984); (b) J. H. A. van der Woude and P. L. de Bruyn, Colloids and Surfaces $\underline{12}$, 179 (1984).
24. (a) L. H. Edelson, K. Gaugler, and A. M. Glaeser, Am. Ceram. Soc. Bull. $\underline{65}$, 504 (1986); (b) D. G. Pickles and E. Lilley, J. Am. Ceram. Soc. $\underline{68}$ (9), C-222 (1985); (c) E. Lilley and D. G. Pickles, in Extended Abstracts: American Ceramic Society, 88th Annual Meeting (Am. Ceram. Soc., Columbus, OH, 1986), p. 60; Am. Ceram. Soc. Bull. $\underline{65}$, 502 (1986).

PRECIPITATION OF MONODISPERSE CERAMIC PARTICLES;
THEORETICAL MODELS

PAUL CALVERT
University of Sussex, Brighton, BN1 9QJ, England.
and Ceramics Processing Research Laboratory,
Massachusetts Institute of Technology, Cambridge, MA 02139

ABSTRACT

Monodisperse ceramic powders with sizes in the range of 0.1-μm have been prepared by a number of groups. Generally a dilute solution of a metal compound is hydrolysed to produce amorphous, spherical particles which may be crystallized by subsequent heat treatments. Emulsion polymerization can produce similarly monodisperse latices if the polymerization conditions are carefully controlled. This paper discusses models of the precipitation process which are based on crystallization from aqueous solutions but are believed to be more generally applicable. From simulations it is concluded that monodisperse particles are a most unlikely outcome of a batch precipitation process. Surface active compounds could play an important role in controlling particle size. Inhomogenous mixing during the early stages of reaction should also be important.

INTRODUCTION

There is a great deal of interest in improving the performance of ceramics in electronic and structural applications by careful control of the size, size distribution and packing of the powders prior to sintering. In principle it should be possible to produce ceramic green states comprising single-size, sub-micron particles packed with crystalline regularity. Such highly ordered green bodies would be expected sinter readily to high density, fine-grained ceramics. In an effort to produce narrow distribution powders there has been much study of the processes for the precipitation of oxide ceramic powders from solutions of metal alkoxides (1).

One other route to new and improved ceramic materials would be to imitate natural composite materials such as bones, shells and teeth where the ceramic phase is precipitated into a pre-existing matrix of polymeric gel. In shells, mineral contents above 90 wt% are achieved with great control of particle size and orientation. In the mineralization process the polymer gel may play the passive role of controlling diffusion of the precipitating species and of preventing agglomeration of the particles. Alternatively, there may be some specific surface interaction between the polymer and the particles leading to controlled nucleation. Such precipitation in polymeric matrices could be used to produce ceramic green bodies or as a method of producing composite materials. We at Sussex are studying the precipitation of organic and inorganic crystals into polymers as a route to highly ordered composite materials.

Work on the production and packing of monodisperse ceramic particles has been encouraged by studies on polymer latices produced by emulsion polymerization. Careful control of the polymerization conditions allows the production of sub-micron spheres with a very narrow particle size distribution (2). Packing studies of these particles has demonstrated the possibility of producing colloidal crystals with interesting properties (3, 4).

In this paper I briefly discuss methods for the production of monodisperse ceramics and for polymer latices and then outline some simple models for crystallization processes which help to elucidate the requirements for the production of monodisperse particles.

MONODISPERSE OXIDE PARTICLES

Examples of the preparation of monodisperse ceramic particles have been described, including transition metal oxides (5), silica (6), zirconia (1) and titania (1). In each case the system is characterised by being dilute such that the volume of particles is about 1% of that of the suspending solvent. For such processes to become commerically viable it is probably necessary that concentrations be increased or that efficient continuous flow processes be developed. Two factors may be of significance in limiting the particle concentration. One is the role of concentration in controlling nucleation rates and particle growth kinetics. The second process is that of particle agglomeration which is more likely to occur at high particle densities and will lead to a broadening of the size distribution.

The observations on monodisperse particle production from silicon, titanium and zirconium alkoxides can be summarized in terms of three reactions:

$$\geq M\text{-}OR \ + \ H_2O \rightleftharpoons \ \geq M\text{-}OH \ + \ ROH \qquad (1)$$
$$\geq M\text{-}OH \ + \ \geq M\text{-}OR \rightleftharpoons \ \geq M\text{-}O\text{-}M\leq \ + \ ROH \qquad (2)$$
$$\geq M\text{-}OH \ + \ \geq M\text{-}OH \rightleftharpoons \ \geq M\text{-}O\text{-}M\leq \ + \ H_2O \qquad (3)$$

The alkoxide is hydrolysed (reaction 1) to form a hydroxyl which may then either condense with another hydroxyl or with an alkoxide (reactions 2 and 3). One important set of questions concerns whether these reactions occur at random or whether there is an enhanced reaction rate of species where the metal centre has already reacted once. The overall reaction kinetics will be determined by the balance of the reactions rates for the three reactions where either hydrolysis or condensation may be rate-determining.

For silicon tetra-alkoxides monodisperse particles are generally formed from basic solution (6). Base enhances the rate of condensation with respect to hydrolysis of silicon alkoxides and seems to favour the formation of less soluble highly cross-linked forms. In titanium and zirconium the condensation rates are already fast compared to hydrolysis (7). In each case the formation of particles is very sensitive to the substituents on the alkoxide. It has also been found for titanium that small amounts of triethylamine are necessary for formation of monodisperse particles (8).

As a first approximation, a discussion of the particle size distribution need not involve the details of the reaction kinetics. The reaction can be viewed as producing precipitating species which nucleate and then grow as crystalline or amorphous particles. The precipitation behaviour is then governed by the phase diagram for the system and the transport properties.

MONODISPERSE PARTICLES FROM EMULSION POLYMERIZATION

There is an enormous literature on emulsion polymerization stretching back almost 40 years. Primary concerns have been the analysis of reaction rate data and the control of molecular weight distributions. The system comprises an emulsion of monomer with an excess of surfactant present as micelles. An aqueous initiator produces free radicals which enter the micelles and start polymerization. As the polymerization proceeds monomer diffuses into the micelles which become polymer particles. Under ideal conditions (Smith-Ewart Case II) each micelle has either one radical or none as radical-radical recombination rapidly occurs when a second radical enters an active particle. Under these circumstances the rate of increase of particle size is governed only by the polymerization rate and not by the particle radius (d(Volume)/d(time) is independent of radius). Hence the particle size distribution tends to narrow as the particles grow. However, as long as micelles are present new particles will also nucleate.

A number of techniques have been developed for the production of monodisperse particles. In principle, polymerizations with low emulsifier concentrations will lead to a limit to the production of new particles and so give a narrow size distribution. The problem is that agglomeration tends to occur when insufficient emulsifier is present. This situation can be controlled to some extent by the use of a mixture of anionic and non-ionic emulsifiers. Another alternative is to seed the polymerization with very fine particles but it is again necessary to work with low emulsifier levels which limits the stability of the emulsion to low particle concentrations and small particles. A typical latex particle size distribution extends from 60-100nm at 80nm mean whereas, with seeding, the range can be reduced to +/-5nm. Ugelstad et al. (9) have described techniques for producing large (up to 100μm) monodisperse particles by techniques involving swelling particles with a low molecular weight liquid and a monomer in a two-stage process. This process can give dispersions of 1-2%.

In summary, the production of monodisperse latex particles is a complex process where it is difficult to achieve stability at high particle concentrations and that the relationship between growth rate and particle radius is important in determining the final size distribution.

PRECIPITATION REACTIONS

Two models can be used to describe particle precipitation: agglomeration of small particles or growth by molecular addition. In the first case we regard the particle as composed of many smaller particles with intervening pores, such a structure would normally appear as very irregular (10). The alternative is to regard the precipitation as one where surface energy is minimized during the addition process such that pores do not form and the final particle is uniformly dense. Crystal growth and coacervation from liquid or glassy mixtures are examples of molecular addition. Since the alkoxide particles generally form as amorphous spheres they can probably be regarded as the coacervation of polymeric species from solution. The controlling processes will then be nucleation, which is governed by the surface energies, and growth which will depend on the interfacial reactions kinetics while the particles are small but will subsequently become controlled by diffusion of the soluble species to the particle surface.

GROWTH SIMULATIONS

As a preliminary to looking for special effects resulting from the interaction of polymers with precipitating systems we have developed a computer programme to simulate crystal nucleation and growth in an unstirred liquid. Using this model we can investigate the response of the particle size distribution to changes in the solubility, diffusion rate and surface energy of the precipitate. These calculations also enable us to investigate the changes in crystal size distribution that should occur on going from a normal liquid medium to a polymeric solvent. The computations are based on published data for the concentration dependence of nucleation and growth rates. This dependence is then coupled with a growth model which converts from an interfacial to a diffusion-controlled growth mechanism as the crystal becomes sufficiently large. So far two materials have been modelled, barium sulphate and potassium sulphate, both growing from aqueous solution. These were chosen because nucleation and growth data are available and because they represent a relatively insoluble and a relatively soluble system (10^{-5}M and 0.63 M respectively).

Matejevic (5) has discussed the requirements for producing monodisperse particles and has pointed out that the essential requirement is a brief burst of nucleation followed by an extended growth period during which no new particles form. Thus, if we wish to produce crystals of a

particular size, for instance 1/2 micron, we need to produce, within a
short time, an abundance of seeds in the size range 0-100nm from which can
grow the required size crystals. If we consider a system in which the
solute is being added at a constant rate the simulations show that the
dissolved concentration rises steadily until nucleation occurs then drops
to a more or less steady state as the precipitate grows without further
nucleation (Figure 1). Table 1 shows the effect of solute addition rate on
the initial size distribution of each of the sulphates in water. If we are
looking for an initial crystal size of 100nm or less, which can then be
grown up to 1/2μm powder, rapid addition of solute is required. The solute
may not necessarily be added directly but may be produced by some in-situ
reaction as in the alkoxide precipitation.

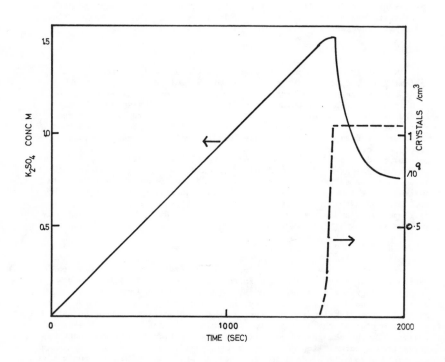

DEVELOPMENT OF CONCENTRATION AND NUMBER OF CRYSTALS
FOR K_2SO_4 ADDED AT 10^{-3} MOL/L/S

FIGURE 1

TABLE 1
CRYSTAL SIZE AT PEAK NUCLEATION RATE

ADDITION RATE MOL/L/S	BARIUM SULPHATE		POTASSIUM SULPHATE	
	RADIUS μm	NO. CRYSTALS /ML	RADIUS μm	NO. CRYSTALS /ML
10^{-6}	9.1	$9.5 \cdot 10^3$		
10^{-5}	3.5	$2.2 \cdot 10^5$	29	$6.0 \cdot 10^3$
10^{-4}	1.4	$4.4 \cdot 10^6$	5.5	$2.0 \cdot 10^6$
10^{-3}	0.6	$1.5 \cdot 10^8$	1.0	$5.0 \cdot 10^8$
10^{-2}	0.2	$3.0 \cdot 10^9$	0.13	$2.0 \cdot 10^{11}$
10^{-1}	0.09	$6.3 \cdot 10^9$	0.025	$6.0 \cdot 10^{13}$

In Table 2 is shown the effect of decreasing the diffusion rate in the medium, for instance by going to a solution in a liquid polymer or increasing the molecular weight of the precipitationg species. Under these circumstances small crystals are readily formed.

TABLE 2
EFFECT OF DIFFUSION RATE ON PARTICLE SIZE
ADDITION RATE 0.001 MOL/L/S

DIFFUSION COEFFICIENT cm^2/s	CRYSTAL RADIUS, μm	
	BARIUM SULPHATE	POTASSIUM SULPHATE
10^{-5}	0.60	1.00
10^{-6}	0.29	0.10
10^{-7}	0.11	0.022
10^{-8}	0.04	

The particle size is very sensitive to surface energy, Table 3 shows that a 25% reduction in surface energy of barium sulphate produces a 60-fold reduction in the initial crystal size.

SILICA, TITANIA AND ZIRCONIA

In the alkoxide precipitation the precipitating species is a partly hydrolysed oligomer for which the rate of production is the equivalent to the addition rate of solute in the crystallization models. The nucleation process will not be strictly like that of the crystals since it is a coacervation of an amorphous precipitate but the kinetic scheme is similar. The particle growth will depend on interface kinetics and diffusion in a similar way to crystallization. To obtain a sharp burst of nucleation it will be necessary to achieve a rapid production rate for the insoluble species. This will be enhanced by a lower solubility.

We also want to aid the nucleation kinetics, for instance by use of impurities which could reduce the energy of the particle-liquid interface. Further we would expect to benefit by reducing the diffusion rates within the medium either by increasing the viscosity of the reaction solution or increasing the molecular weight of the precipitating species. Clearly many of the changes which may be made will modify several aspects of the process and for this reason models of this type should be a great aid to interpretation of experimental results.

On the basis of the model calculations we see that only by achieving very high addition rates is it possible to get the initial burst of nucleation necessary to produce monodisperse particles. In a batch system this could only be obtained locally during the addition of reagents. Subsequent mixing would distribute the nuclei throughout the solution and allow them to grow. Factors which reduce the surface energy without

affecting solubility should improve the chance of obtaining the rapid
nucleation.

TABLE 3
EFFECT OF SURFACE ENERGY ON PARTICLE SIZE
ADDITION RATE 0.001 MOL/L/S

SURFACE ENERGY $mJ.M^{-2}$	PARTICLE RADIUS μm	NO. CRYSTALS /ML
BARIUM SULPHATE		
100	0.01	2.10^{12}
115	0.1	2.10^{10}
130	0.6	$1.5.10^{8}$
145	1.7	5.10^{6}
160	4.3	5.10^{5}
POTASSIUM SULPHATE		
27	0.025	$1.4.10^{10}$
30	0.44	$2.2.10^{9}$
33	0.90	$4.6.10^{8}$
36	1.10	$3.3.10^{8}$
39	3.15	$2.8.10^{7}$

ACKNOWLEDGEMENTS

The work on precipitation in polymers is supported by the Venture
Research Unit of the British Petroleum Company PLC and that on ceramic
particles by the M.I.T. Ceramics Processing Consortium.

REFERENCES

1) B.Fegley and E.A.Barringer, Mat.Res.Soc.Symp.Proc. 32, 187 (1984).
2) G.Lichti, R.G.Gilbert and D.H.Napper in Emulsion Polymerization, edited
   by I.Piirma (Academic Press,New York, 1982), pp.94-145.
3) S.Hachisu and K.Takano, Adv. Colloid Interface Sci. 16, 233 (1982).
4) P.Pieranski, Contemp. Phys. 24, 25 (1983).
5) E.Matejevic, Acc.Chem.Res. 14, 22 (1981).
6) W.Stover, A.Fink and E.Bohn, J.Colloid Interface Sci. 26, 62 (1968).
7) M.Prassas and L.L.Hench in Ultrastructures Processing of Ceramics,
   Glasses and Composites edited by L.L.Hench and D.R.Ulrich, (Wiley,
   New York, 1984), p. 100.
8) P.Nahass, G.H.Wiseman, C.A.Sobon and H.K.Bowen (to be published).
9) J.Ugelstad, H.R.Mfutakamba, P.C.Mork, T.Ellingsen, A.Berge, R.Schmid,
   L.Holm, A.Jorgedal, F.K.Hamsen and K.Nustad, J.Polymer Sci. Symposia,
   72, 225 (1985).
10) H.E.Stanley, F.Family and H.Gould, J.Polymer Sci Symposia, 73, 19
   (1985).

EFFECT OF IN SITU POLYMER ADSORPTION ON THE SYNTHESIS OF MONODISPERSE TiO₂ POWDERS

J. H. Jean and T. A. Ring
Department of Materials Science and Engineering
Massachusetts Institute of Technology
Cambridge, MA 02139

ABSTRACT

Monodisperse and spherical titania powders were prepared by the hydrolysis of titanium tetraethoxide in ethanol with hydroxy-propyl cellulose(HPC) present during particle precipitation. The mean particle size decreased as the concentration of HPC increased. The critical concentration of HPC for the in situ stabilization of titania powder suspensions was determined by a photographic method, and found to be independent of the molecular weight of HPC within the molecular range of 60,000 to 300,000. The amount of HPC adsorbed on the particle surface, determined by a colorimetric method using a UV-visible spectrophotometer, decreased with the increasing molecular weight of HPC. Suspension stability was confirmed by theoretical calculations based on experimental results.

INTRODUCTION

To produce monodisperse ceramic powders precipitated from a generating reaction, agglomeration during particle formation must be prevented. The stability of a reacting suspension can be obtained by electrostatic or steric means. For a system stabilized using the electrostatic method, the pH of the reacting suspension must be different from that at the isoelectric point of the precipitated powders as in the generation of monodisperse silica powders by the hydrolysis of tetraethylorthosilicate (TEOS) with NH₃[1]. However, agglomerated TiO₂ powders were always observed using electrolytic additives[2], because the reaction kinetics were completely different. A sterically stabilizing dispersant, hydroxy-propyl cellulose (HPC), was therefore added during particle precipitation. This investigation includes studies of the synthesis and packing of monodisperse TiO₂ powders, measurement of the HPC adsorption isotherms using the colorimetric method, and the theoretical calculations of suspension stability in the presence of HPC.

EXPERIMENTAL

Monodisperse and amorphous TiO₂ powders were prepared by the controlled hydrolysis of titanium tetraethoxide* in 200-proof anhydrous ethanol with a dispersant, hydroxy-propyl cellulose (HPC)**. Two molecular weights of HPC, G-type (MW 300,000) and E-type (MW 60,000), were used in this research. The overall hydrolysis and condensation processes can be expressed as

$$Ti(OC_2H_5)_4 + 4H_2O \xrightarrow[C_2H_5OH]{HPC} TiO_2(s) + 4C_2H_5OH$$

These two reactants were dissolved separately in equal volumes of ethanol and HPC was added in the ethoxide vessels; the resulting two solutions were injected into a tee mixing section using a pair of syringes operated by a single plunger. To remove the heterogeneous sites, the solutions were fitered through 0.22 μm Millipore filters before mixing. Precipitation was conducted in a glove box under a dry nitrogen atmosphere at 25° C. When the powders

---

*Alfa Products, Danvers MA

** Hercules Inc, Wilmington DE

precipitated, the solution became cloudy; the time needed for observation of the turbid solution increased with a decrease in concentration of either reactant and with an increase in HPC concentration. The characteristics of powders prepared by this method have been described in a previous paper[3]. Powder size was determined by the direct measurement of transmission electron microscopy (TEM) photos using an image analyzer; 200~300 particles were measured for each sample. The fraction of singlets on the top, fracture, and bottom surfaces of green compacts were determined from the direct measurement of SEM photos. From the percentage of singlets observed on these surfaces versus HPC concentrations, the critical HPC concentration for the <u>in situ</u> stabilization of these powder suspensions was determined.

The amount of HPC adsorbed, in gm/gm titania powders (Γ), can be calculated by performing a mass balance using the following equation

$$\Gamma = (C_0 - C_s)/W_t \qquad (1)$$

where C_0 is the initial HPC concentration, C_s is the equilibrium HPC concentration in the supernatant solution after the powder is removed, and W_t is the weight of titania powders. Because the <u>in situ</u> adsorption of HPC onto titania powders is too small to be easily detected using a colorimetric method[4], agglomerated titania powders of 0.5 μm diameter prepared in the absence of HPC were used as absorbant. A typical experiment to determine the amount of HPC adsorbed onto amorphous titania powders was conducted as follows. HPC was dissolved in pure ethanol using a magnetic stirrer at 40° C. Ten milliliters of an HPC solution of known concentration were introduced into a 15 cc plastic centrifuge tube which contained 0.4 gm amorphous titania powders. To prevent the agglomerated powders from settling, an end-over-end mixer was used to mix the powders and the HPC solution at 25°C. The supernatant solutions were removed from the mixer after 24 hours using a 0.22 μm filter. The polymer concentration left in the supernatant solution was determined by a colorimetric method using a phenol-HPC complex in concentrated sulfuric acid, which has adsorption peaks at 336 and 488 nm. The calibration curves of absorbance versus HPC concentrations were constructed from the analysis of absorbance spectra of a series of solutions with known HPC concentrations. Ten samples of each solution were measured and an average calculated. The two peaks at the wavelengths of 336 and 488 nm increased linearly with increasing HPC concentration when the HPC concentration was less than 0.855×10^{-3} gm/cc.

RESULTS AND DISCUSSION

Figure 1 shows a typical TEM photo for amorphous, spherical and monodisperse titania powders prepared with 0.2M $Ti(OC_2H_5)_4$/0.8M H_2O and

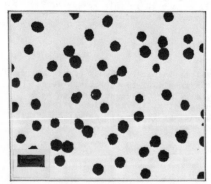

Figure 1. TEM micrograph of a powder prepared with 0.2M titanium ethoxide/ 0.8M H_2O, and 1.14×10^{-3} gm/cc HPC-E, an hour after induction (Bar = 1 μm).

1.14×10^{-3} gm/cc HPC-E in pure ethanol. The mean size determined using an image analyzer is 0.36 μm, and the width of the size distribution, defined as the ratio between standard deviation and mean size, is 0.1. These particles are unlike the agglomerated powders prepared without HPC at this concentration shown in Figure 2.

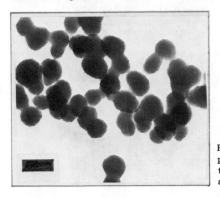

Figure 2. TEM photo of reaction product for the system with 0.2M titanium ethoxide and 0.8M H_2O, an hour after induction (Bar = 1 μm).

Typical SEM micrographs of samples from the top, fracture, and bottom surfaces of sediments prepared by gravitational settling from solutions with various concentrations of HPC-G are shown in Figures 3A-C, respectively. Examination of these micrographs reveals that the particles are agglomerated and poorly packed in the fracture and bottom surfaces as the concentration of HPC-G is reduced. The particle packing becomes denser and more uniform as the HPC-G concentration is increased. In addition, the mean particle size, determined from direct measurement in the SEM photos, decreases as HPC-G

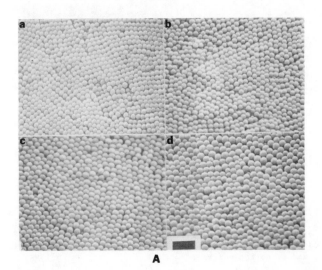

A

88

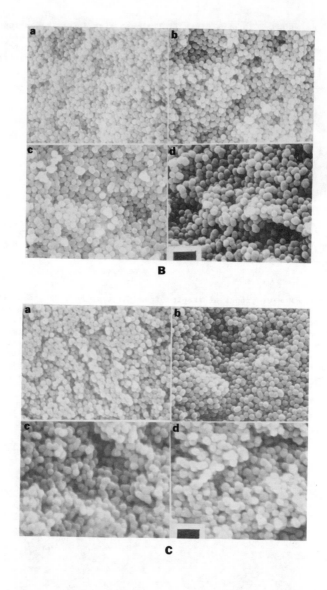

Figure 3. SEM photos of sedimented powder compacts formed by gravitational settling for the system with 0.15M titanium ethoxide/0.8M H_2O, and various concentrations of HPC-G: (a) 1.71×10^{-3} gm/cc, (b) 0.342×10^{-3} gm/cc, (c) 0.244×10^{-3} gm/cc, (d) 0.17×10^{-3} gm/cc – A, top surface, B, fracture surface, and C, bottom surface.

concentration is increased. Figure 4 shows the effect of various
concentrations of HPC-G on the fraction of singlets on the top, fracture, and
bottom surfaces of settled compacts. A critical concentration was found to be
0.342×10^{-3} gm/cc HPC-G for the <u>in situ</u> stabilization of titania powders
prepared from the solution with 0.15M titanium ethoxide and 0.8M water. The
same critical concentration of HPC-E was also found at these reactant
concentrations[3], indicating that the critical HPC concentration for <u>in situ</u>
stabilization of titania powders is independent of the 60,000 to 300,000
molecular weight of HPC.

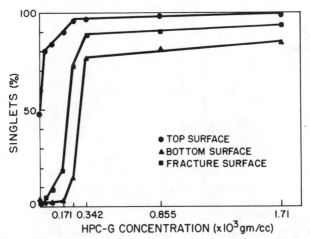

Figure 4. Percentage of singlets on top, fracture, and bottom surfaces of
green compacts prepared by gravitational settling versus HPC-G concentrations
for the system with 0.15M titanium ethoxide and 0.8M H_2O.

The isotherms for the adsorption of HPC onto the agglomerated titania
powders are shown in Figure 5. For both isotherms, the initial slopes do not
generally show a very high affinity of HPC for titania powders. No agreement
with Langmuir[5] or Simha-Frisch-Eirich (SFE) equations[6] was investigated. An
adsorption plateau was observed at 1.8×10^{-3} gm HPC/gm TiO_2 for HPC-G. No
plateau was observed at high HPC-E concentration, although the slope of the
adsorption isotherm curve decreased with increasing HPC-E concentration. The
plateau for HPC-E is likely to occur at higher concentrations since HPC-E has
higher solubility than HPC-G in ethanol. The maximum adsorption concentrations
are several times greater than that calculated by assuming that a monolayer of
HPC polymer is adsorbed on the external powder surface.

An attempt to measure the amount of HPC adsorbed onto titania powders
during precipitation using the colorimetric method was unsuccessful due to the
small amount of HPC adsorbed onto titania powders. Therefore, the amount of
HPC adsorbed onto titania powders has to be calculated from a comparison of
the weight of the powders precipitated and the adsorption isotherms determined
using agglomerated powders prepared in the absence of HPC. Surprisingly, only
3.22% of the initial HPC concentration corresponding to 0.117×10^{-4} gm/cc HPC
is adsorbed onto titania powders during precipitation, with initial HPC
concentrations of 0.342×10^{-3} gm/cc for 0.15M titanium ethoxide and 0.8M water.
It is estimated that large amounts of HPC (i.e., 0.8×10^{-3} gm/cc) are needed if
HPC molecules are irreversibly anchored by a chemical bond with the titania
surface. This is almost two orders of magnitude larger than the experimental
values of HPC adsorbed onto the particles, which suggests that the HPC

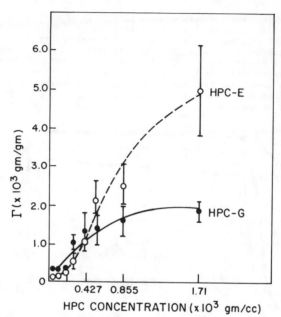

Figure 5. Amount of HPC adsorbed onto amorphous titania powders versus
initial HPC concentrations.

molecules are reversibly adsorbed and are not incorporated into the titania
powder structure during particle formation. If the adsorption of HPC molecules
onto the titania powders is irreversible, particle growth will cause a uniform
decrease of the effective HPC content in solution. At a certain stage during
the particle growth process, agglomeration will take place because most of the
HPC molecules lost to the surface would be buried in the particle rather than
available for steric stabilization on the particle surface. This would lead to
a deficiency in the steric barrier of HPC and cause agglomeration. However,
monodisperse rather than agglomerated titania powders were always observed
when the system contained sufficient HPC concentration (i.e., 0.342×10^{-3}
gm/cc) for 0.15M titanium ethoxide and 0.8M water. On the other hand, if
irreversible adsorption of HPC molecules on titania powder surface occurs, an
inverse relation between particle size and HPC concentration as given in
Figure 3A-C was not observed. Based on these analyses, the adsorption of HPC
onto titania powder surfaces during precipitation is reversible, and most of
the adsorbed HPC stays on the external particle surfaces.

Some basic requirements for the reversible adsorption of HPC during
particle formation include a fast adsorption-desorption, corresponding to a
very low activation energy for the adsorption. This generally means a physical
interaction between the HPC molecules and titania powders. Fast adsorption-
desorption compared with the powder precipitation process prevents the HPC
molecules from being incorporated into the particle structure and prevents
particle agglomeration throughout growth. The low activation energy between
hydrogen bonds allows the process of adsorption-desorption of HPC to be fast
enough to redistribute the HPC molecules on the particle surface.

If it is assumed that the critical HPC concentration corresponds to
monolayer adsorption on the external particle surface, the thickness of the
polymer layers can be estimated from the particle number densities, particle
size, and the concentration of HPC adsorbed on the particle surface.

Experimental results show the particle number density, 6.0×10^{10} #/cc, and mean size, 0.48 μm, for a solution with a critical HPC concentration (0.342×10^{-3} gm/cc) and 0.15M titanium ethoxide/0.8M water. The total surface area of particles in one milliliter of solution is calculated to be $4.34 \times 10^{18} Å^2$. Assuming that the HPC molecules on the particle surface are spherical coils, the calculated diameter is 222 Å for HPC-E and 497 Å for HPC-G, which agrees with the polymer sizes measured by GPC and light scattering[2]. Therefore, the adsorbed layer thickness for these precipitated powders was estimated to be 222 Å for HPC-E and 497 Å for HPC-G.

The change in free energy as a result of the overlap of the adsorbed polymers on the colliding particles is the sum of the free energy change on mixing, the excluded volume effect, the surface energy change on mixing, and an elastic energy of repulsion. The following equation was derived by Fischer[7] to calculate this free energy change

$$V_S = 4/3 \cdot \pi RTBC_i^2 (\delta - H/2)^2 (3r + 2\delta + H/2) \qquad (2)$$

where B is the second virial coefficient for HPC in ethanol, C_i is the concentration of HPC in the adsorbed layer (gm/cc), δ is the adsorbed layer thickness, r is the mean particle size, and H is the separation distance. The second virial coefficient of HPC is 7.7×10^{-4} [8], as determined by light scattering. The concentrations of HPC polymer in the adsorbed layer, C_i, are obtained from the results of the thickness of the adsorbed layer and the amount of HPC polymer adsorbed per particle. It was found that C_i is equal to 1.269×10^{-2} gm/cc for HPC-E and 4.65×10^{-3} gm/cc for HPC-G.

The repulsive energy, V_S, arising from the interaction of two 0.48 μm particles with the adsorbed HPC-E layer thickness, 222 Å, is shown in Figure 6. As a result of r>>H, the interaction energy between two 0.48 μm particles with

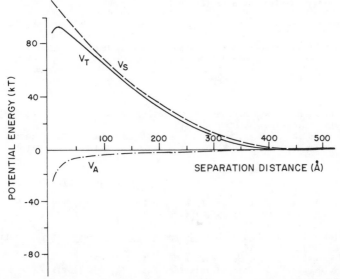

Figure 6. Potential energy versus particle separation distance. Curve V_S is the steric repulsion developed from the interpenetration of 222 Å of HPC-E on titania powder surface, curve V_A is the attractive force arising from the interaction of Van der Waals force, and curve V_T is the sum of V_S and V_A. An activation free energy for particle agglomertion of 94 kT is present.

the adsorbed polymer layer resulting from the Van der Waals force derived by Vold is[9]

$$V_A = -1/12 \cdot \{[A_{22}^{1/2} - A_{33}^{1/2}]^2 [(r+\delta)/H] + [A_{33}^{1/2} - A_{11}^{1/2}]^2 [r/(H+2\delta)]$$

$$+ 4r[A_{22}^{1/2} - A_{33}^{1/2}][A_{33}^{1/2} - A_{11}^{1/2}][r+\delta]/[(H+\delta)(2r+\delta)]\} \qquad (3)$$

where A_{11}, A_{22} and A_{33} are the Hamaker constants of titania powder, ethanol and HPC, respectively, A_{11} is 4.2×10^{-22} J^{10}, A_{22} is 3.563×10^{-20} J^{11}, and A_{33} is 1.6×10^{-20} J^{12}. The interaction energy (V_A) and total potential energy $(V_T = V_S + V_A)$ are also given in Figure 6 as a function of separation distance between two particles. The sum of V_A and V_S are also given in Figure 6. V_A dominates when the two particles are at very close appproach, and as a consequence, the net interaction curve predicts a large energy barrier for particle agglomeration. As noted in Figure 6, the activation free energy is 94 kT for particle agglomeration for the suspension with a critical HPC-E concentration present during particle formation. A similar treatment was performed for the particles with the thickness of the adsorbed HPC-G layer, 497 Å, and an activation energy of 70 kT was found. Those two activation energies are much larger than the 15 kT[13] needed for a stable suspension, therefore, experimental results indicate that suspensions with a critical HPC concentration are stable.

CONCLUSIONS

The effect of HPC on the _in situ_ stabilization of TiO_2 powders is significant. The critical concentration of HPC is 0.342×10^{-3} gm/cc for 0.15M titanium ethoxide and 0.8M water, and is independent of the molecular weight of HPC in the range of 60,000 to 300,000. The stability of suspensions with critical HPC concentrations is confirmed by theoretical calculations of the activation energy barrier for aggregation.

ACKNOWLEDGEMENT

This material is based upon work supported by the National Science Foundation under Grant No. MEA-8310530, and by IBM.

REFERENCES

1. W. Stober, A. Fink, and E. Bohn, J. Colloid Interface Sci., _26_, 62 (1968).
2. J. H. Jean, Ph.D. Thesis, M. I. T., (1986).
3. J. H. Jean and T. A. Ring, submitted to the J. Am. Ceram. Soc., 1986.
4. M. Dubois, K. E. Gilles, J. K. Hamilton and P. A. Smith, Anal. Chem., _28_, 350 (1956).
5. T. Langmuir, Science, _88_, 450 (1938).
6. R. Simha, H. L. Frisch and F. R. Eirich, J. Phys. Chem., _57_, 584 (1952).
7. E. W. Fischer, Kolloid-Z, _160_, 120 (1958).
8. M. G. Wirick and M. H. Waldman, J. Appl. Polym. Sci., _44_, 579 (1970).
9. M. J. Vold, J. Colloid Sci., _16_, 1 (1961).
10. E. A. Barringer, Ph.D. thesis, M. I. T., 1983.
11. P. C. Hiemenz, _Principles of Colloid and Surface Chemistry_, Marcel Dekker, New York, 1977.
12. R. J. Samules, J. Polym. Sci. Part A-2, _7_, 1197 (1969).
13. J. Th. G. Overbeek, J. Colloid Interface Sci., _58_, 408 (1977).

SYNTHESIS OF SUBMICRON, NARROW SIZE DISTRIBUTION SPHERICAL ZINCITE

ROBERT H. HEISTAND II AND YEE-HO CHIA
The Dow Chemical Company, New England Laboratory, P. O. Box 400, Wayland, MA 01778

ABSTRACT

Zincite has been produced by the controlled hydrolysis of an alkylzinc alkoxide (ethylzinc-_t_-butoxide) resulting in $\sim 0.2 \mu$m spherical particles with a narrow size distribution consisting of 150 Å crystallites. The surface area is 30 m^2/g. Variation of the concentration of the water drastically affects the particle and crystallite sizes. The results of a systematic study of the hydrolysis parameters are reported here.

INTRODUCTION

The recognition that progress in advanced ceramic materials will result from improved microstructures which are in turn dependent on the raw ceramic powders has led to the quest for the "ideal" powder. Recent studies on varistor materials have been directed to this end.[1,2] Since the varistor device is based on a grain boundary phenomenon, a more uniform, smaller grained microstructure would produce a higher alpha, more volume efficient device.[1-7] Two chemical routes for zinc oxide relating to varistor powders have been published. One is a sol-gel process involving hydrous zinc oxides, and the other is the calcination of oxalate precursors.[2] We have been investigating an alternative route to precipitate zincite directly.[8] Ideally for device fabrication, the raw zinc oxide powder should be submicron, spheroidal, non-agglomerated, and have a narrow size distribution and moderately low surface area. This would result in a dense sintered microstructure of uniform, micron-size grains.

A survey of the zinc chemical literature produced a promising candidate for a non-aqueous hydrolysis reaction. Alkylzinc alkoxides are quantitatively formed by the reaction of dialkylzinc with alcohols.[9,10] The alkylzinc alkoxides form tetrameric complexes that are soluble and well characterized (Figure 1.).[11,12] It is interesting to speculate that the cubic zinc-oxygen core would enhance the probability of crystallization during hydrolysis. Ethylzinc-_t_-butoxide was chosen as the initial complex for the reaction. It is readily soluble in hydrocarbon solvents such as hexane or toluene and can be hydrolyzed in a binary system of toluene/ethanol. By controlling the hydrolysis parameters, crystalline zinc oxide in the form of zincite can be produced to have the following characteristics: 0.15 to 0.70 μm size, spherical to bipyramidal morphology, a narrow to wide size distribution, in an agglomerated or dispersed state, and surface areas of 6 to 70 m^2/g. The results of the effect of the water concentrations, reaction times, aging and washing solutions on the powder properties will be presented here.

EXPERIMENTAL

Ethylzinc-_t_-butoxide was synthesized from diethylzinc and _t_-butanol according to Matsui et al[11] in toluene. Diethylzinc was purchased from Aldrich Chemical Co. as a 15% solution in toluene. A standard reaction consists of rapidly admixing a solution made of 30 ml of toluene and 3.25 g (19.4 mmol) ethylzinc-_t_-butoxide with a solution made of 67 ml absolute ethanol and 6.0 ml (330 mmol) distilled water. The reaction is performed at room temperature in a well stirred vessel under an inert dry argon

94

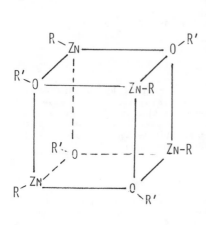

Fig. 1. Molecular geometry of alkylzinc alkoxides.

Fig. 2. SEM Micrograph of the reference zincite powder.

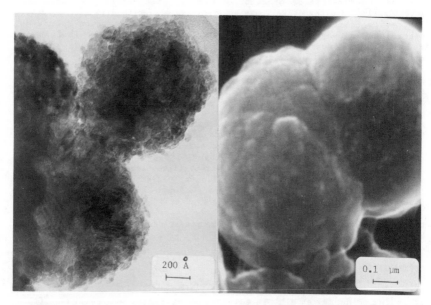

Fig. 3. TEM Micrograph of a typical zincite powder.

Fig. 4. SEM Micrograph of a powder from the 2.8 water equivalents hydrolysis.

atmosphere. The solution turns turbid within 60 seconds. The residence time in the reactor is three hours, followed by centrifugation at 2310 G for 10 minutes. The supernatant is discarded and the resulting pellet is redispersed in $\sim$ 50 ml of absolute ethanol by ultrasonication at 80 watts for 3-5 minutes. This washing procedure is repeated one more time. Variations to this procedure discussed in the text are less water, longer residence times and washing with water instead of ethanol. One other parameter that was investigated was the aging of the sample as a washed, but solvent containing, cake resulting from the centrifugation. For surface area (single point B.E.T.) and x-ray powder diffraction analysis, the zincite was dried at 90-110°C under vacuum for 20 hours. Electron microscopy was performed on a Jeol 100-CX Analytical Electron Microscope.

RESULTS AND DISCUSSION

Concentration of Water

The reference reaction produces spheroidal zincite shown in Figure 2. X-ray powder diffraction reveals only the hexagonal zincite phase. The particle size by scanning electron microscopy (SEM) is 0.18 µm, however, x-ray line broadening analysis reveals a crystallite size of 150Å. This is consistent with a surface area of 30 m^2/g and the typical transmission electron micrograph in Figure 3. Obviously the zincite particles are porous polycrystalline spheres. Ideally for microstructure development during sintering, the particles should be non-porous and have a lower surface area.

To facilitate mixing on a large scale, a longer induction time would be beneficial. One way of lengthening the induction time is to reduce the concentration of water. The results are found in Table I and the extreme case is pictured in Figure 4. Halving the concentration of water has little effect, but reduction to a quarter and a sixth lengthened the induction time. The particle size does increase, but at the expense of a highly agglomerated powder with smaller crystallites and an even higher surface area.

TABLE I

EFFECT OF WATER CONCENTRATIONS ON ZnO

POWDER CHARACTERISTICS

AMOUNT OF H_2O USED (ML)	$[H_2O]$/ $[ETZN^TBU]$	PARTICLE SIZE (Å)	CRYSTALLITE SIZE (Å)	BET EQUIV. SPHERICAL DIA. (Å)	BET SURFACE AREA (M^2/G)
6	17.0	1800 - 2000	150	370	30
3	8.5	2000 - 2500	150	340	32
1.5	4.3	$\sim$ 5000	100	170	66
1	2.8	$\sim$ 7500	60	160	68

Aging

A possible way to lower the surface area of a powder is to age the sample.[13] The solvent containing cake resulting from centrifugation was allowed to age in a closed vessel for 19 days. The reaction scale was five fold the standard reaction, but all other conditions remained the same. A sample was dried and analyzed on the first, second, fifth, ninth and nineteenth day. The results are listed in Table II. No effect is noted until after the fifth day. The SEM micrographs for the unaged and nineteenth day samples are shown in Figure 5. The crystallite size increased by nearly a factor of three and the surface area decreased

TABLE II

EFFECT OF AGING OF "WET" CAKES ON ZnO

POWDER CHARACTERISTICS

AGING TIME (DAY)	PARTICLE SIZE & SHAPE ($\AA$)	CRYSTALLITE SIZE ($\AA$)	BET EQUIV. SPHERICAL DIA. ($\AA$)	BET SURFACE AREA (m^2/g)
0	1400$\AA$ AND SPHEROIDAL	~150	308	36
1	1400$\AA$ AND SPHEROIDAL	~150	358	31
2	1400$\AA$ AND SPHEROIDAL	~150	326	34
5	SPHEROIDAL AND ELLIPTICAL	150—160	300	37
9	IRREGULAR	290—440	844	13
19	IRREGULAR (SOME WITH HEXAGONAL SHAPE)	280—460	844	13

Fig. 5. SEM Micrographs of (A) unaged and (B) nineteen day zincite.

accordingly. In the micrograph, the hexagonal morphology of zincite is beginning to show. As expected the 150Å crystallites are highly reactive and even small quantities of ethanol are sufficient to allow recrystallization to occur. This is consistent with the reactivity reported for hydrous zinc oxide.

Residence Time and Aqueous Washing

In light of the aging studies, the residence time was increased to 16 hours. The scale and all other conditions were the same as the aging study. Little change occurs with longer residence time. The micrograph in Figure 6A shows some deterioration of the spherical morphology but the powder characteristics listed in Table III (ethanol wash) show only a slight change in surface area and crystallite size compared to the unaged sample in Table II. However, if the colloidal zinc oxide is washed with water instead of

TABLE III

EFFECT OF WASHING SOLUTIONS ON
ZnO POWDER CHARACTERISTICS

WASHING SOLUTION	NUMBER OF WASHINGS	PARTICLE SIZE & SHAPE (Å)	CRYSTALLITE SIZE (Å)	BET EQUIV. SPHE. DIA. (Å)	BET SURFACE AREA (M^2/G)
ETHANOL	2	~ 1500 SPHERICAL & IRREGULAR	190 - 220	378	29
ETHANOL	3	NO CHANGE	190 - 260	378	29
WATER	2	BIPYRAMIDAL	300 - 550	1097	10
WATER	3	NO CHANGE	320 - 490	1097	10

absolute ethanol, a dramatic recrystallization occurs. This results in bipyramidal particles depicted in Figure 6B. The surface area is 10 m^2/g and the crystallite size is 300-500Å. Adding an additional ethanol or aqueous wash does little to change the crystallite size. This is tabulated in Table III for comparison. The same results were obtained for a three hour residence time reaction. The aqueous wash is a very effective and efficient treatment to produce lower surface area zincite while not drastically affecting the agglomeration or overall particle size. This may be due to the ultrasonication step inducing a pseudohydrothermal recrystallization.

Conclusion

The hydrolysis of ethylzinc-t-butoxide produces submicron polycrystalline zincite. The particle size, shape, state of agglomeration, size distribution, crystallite size and surface area are highly affected by the concentration of water during hydrolysis and washing. Spheroidal, 0.2 μm, narrow size distribution, 150Å polycrystalline, 30 m^2/g, unagglomerated zincite is produced under one set of conditions. Changing the washing solution from absolute ethanol to water produces ~ 0.2 μm bipyramidal zincite particles consisting of ~450Å crystallites and having a surface area of 10 m^2/g. The results of the studies on water concentration, aging, and washing suggest a hydrolysis mechanism where the zinc oxide crystal growth is highly dependent on the amount of water present during or after precipitation.

98

Fig. 6. SEM Micrographs of 16 hour residence time (A) ethanol washed
or (B) water washed zinc oxide.

ACKNOWLEDGEMENT

The authors wish to thank Evelyn Ekmejian for the surface area analyses.

REFERENCES

1. R. J. Lauf and W. D. Bond, Cer. Bull., 63, 278-281 (1984).
2. R. G. Dosch and K. M. Kimball, Sandia National Labs Report, SAND
    85-0195, 1985.
3. M. Matsuoka, Jap. J. Appl. Phys., 10, 736-746 (1971).
4. H. R. Philipp and L. M. Levison, J. Appl. Phys., 48, 1621-1627 (1977).
5. P. R. Emtage, J. Appl. Phys., 50, 6833-6837 (1979).
6. J. Wong, J. Appl. Phys., 46, 1653-1659 (1975).
7. H. R. Philipp, G. D. Mahan, L. M. Levinson, Oak Ridge National Laboratory
    Sub/84-17457/1, 1984.
8. R. H. Heistand II, Y. Oguri, H. Okamara, W. C. Moffatt, B. Novich,
    E. A. Barringer, H.. K. Bowen, Proceedings of the Second Conference on
    Ultrastructure Processing of Ceramics, Glasses, and Composites,
    (John Wiley & Sons, New York, 1985) in print.
9. G. Allen, J. M. Bruce, D. W. Farren, F. G. Hutchinson, J. Chem. Soc.
    B 1966, 799-803
10. J. M. Bruce, B. C. Cutsforth, D. W. Farren, F. M. Rabagliati, D. R.
    Reed, J. Chem. Soc., B 1966, 1020-1024.
11. Y. Matsui, K.Kamiya, M. Nishikawa, Y. Tomiie, Bull. Chem. Soc. Jap. 39
    1828 (1966).
12. H. M. M. Shearer and C. B. Spencer, Chem. Comm., 1966 194.
13. B. Basak, D. R. Glasson, S. A. A. Jayaweera, Particle Growth in
    Suspensions, edited by A. L. Smith, (Academic Press, London, 1973)
    pp. 109-112.

PREPARATION OF CERAMIC POWDERS FROM EMULSIONS

MUFIT AKINC and KERRY RICHARDSON
Department of Materials Science & Engineering
Iowa State University, Ames, Iowa 50011

ABSTRACT

Yttrium oxide powders were prepared from water-in-oil type emulsions by loading the yttrium ions into the aqueous phase. Emulsions were characterized with respect to droplet size and distribution, emulsion type, and time and temperature stability. Precursor powders were obtained from the emulsions by evaporation of the aqueous phase in a hot oil bath. Powder characteristics, such as size, shape, composition, and sinterability as a function of procedural variables, were determined. The new technique appears to be practical and may be economically feasible.

INTRODUCTION

The role of powder characteristics on the processing and on the properties of the final ceramic material has long been recognized [1,2]. Recently more emphasis has been given to research on the synthesis of powders to produce particles with a control over stoichiometry, impurity levels, particle size and shape [3]. A variety of techniques is available for preparation of ceramic powders. Among these are chemical precipitation in aqueous or organic solutions, gas-phase reactions, spray roasting, cryochemical processing, hydrothermal processing, and others [4]. The aim of each technique is to produce well-behaved powders. It has been shown that small, uniform, unagglomerated powders lead to dense and uniform microstructures at relatively low sintering temperatures [5,6].

Formation of small, uniform particles for a number of systems has been demonstrated by Matijević and his coworkers [7]. Generally, however, these homogeneous precipitation procedures involve aging times too long, temperatures, too elevated, and solutions too dilute for them to be a viable route for ceramic powder synthesis. In a few cases, formation of monosize powders were achieved by hydrolysis of metal alkoxide solutions in alcoholic solutions [8]. However, a method to produce small uniform and unagglomerated ceramic powders that would be applicable to a variety of single and multicomponent systems and that would be economically feasible is yet to be developed.

The primary goal of this research was to study the formation of fine, unagglomerated yttria powders by the use of the emulsion technique. More specifically, the preparation, characterization, and evaporation of water-in-oil type emulsions to produce yttria powders was investigated.

TECHNIQUE

Among the various techniques, precipitation of precursor particles from solution is still the most widely used technique both in the laboratory and in industry. The aqueous route is the oldest and still the most popular technique because of its simplicity, safety, low cost, and traditional familiarity to industry. However, conventional aqueous precipitation often produces nonreproducible results, probably because of the technique's extreme sensitivity to subtle variations in conditions and procedures. Problems of flocculation and subsequent formation of larger, irregular agglomerates can be eliminated if isolated particles are produced within small liquid droplets. Upon removal of the solvent, a powder consisting of solid particles may be obtained. Small liquid droplets may be produced either in a gas (aerosol) or in a liquid (emulsion) phase. Spray roasting and spray drying are based on the former technology; the emulsion process is the topic of the present study.

EMULSIONS

Emulsion technology is well known and widely used in chemical industries. However, the majority of emulsions are of the oil-in-water (o/w) type, in which droplets of an organic phase are dispersed in an aqueous phase [9]. The type of emulsions of interest in this study are the ones wherein aqueous droplets are dispersed in an organic phase (w/o type emulsions).

One of the most critical steps in the preparation of a given emulsion is the choice of emulsifying agents [10]. Hydrophile-Lipophile Balance (HLB) values of emulsifying the agents are commonly used to determine whether a particular emulsifying agent favors an o/w or w/o emulsion formation. HLB values for surface-active compounds can be obtained from the following empirical formula [9]:

$$HLB = \sum_i Mi -0.475n + 7$$

where Mi corresponds to the hydrophilic group number for the ith group, n is the number of CH_2 groups in the compound, and -0.475 is the value of the CH_2 group. Emulsifying agents that favor a w/o type emulsion have an HLB value around seven. Higher HLB values favor o/w type emulsions [11].

One of the important characteristics of an emulsion is the droplet size. The size of the droplets is a function of the nature of the phases, the quantity and type of surfactant, and the processing methods. Higher concentrations of surfactant and a higher rate of mixing produces small droplet size emulsions.

Among other factors, emulsion stability depends greatly on the strength of the interfacial film produced by the emulsifying agent. Higher emulsifying agent concentration leads to a close-packed arrangement of these molecules on the surface of the droplets and strengthens the interfacial films [10]. Stability of emulsions may be improved, especially in w/o type emulsions, by steric effects. Stability may also be improved if the density difference between dispersed phase and continuous phase is small and by increasing the viscosity of the continuous phase [12]. Furthermore, presence of solid particles is believed to increase the stability of the emulsion, especially if the solid particles are in the internal phase [13,14].

Emulsion drying of an aqueous solution of salts such as sulphates in hot kerosene has been mentioned in the literature [4,15,16], but no detailed information was available concerning the characteristics of the powders produced.

The literature review indicates that w/o type emulsions having a broad range of internal phase ratios and containing solid particles can be prepared and are quite stable under normal conditions. Furthermore, experimental evidence exists in the literature that the solvent in the droplets can be evaporated to produce solid particles dispersed in the continuous organic phase [16].

EXPERIMENTAL

Materials

In this study a wide variety of materials was tried. These include benzene, toluene, kerosene, and mineral oil for the continuous organic phase; distilled water and aqueous $Y(NO_3)_3$ for the dispersed phase; and Arlacel 83, Span 60, Span 80, Tween 80, Tween 85 (all from ICI Americas, Inc.), Aerosol-OT (Fisher), and Pluronic L62 (BASF) as emulsifying agent. Benzene and toluene were reagent grade and were used without further purification; kerosene and mineral oil (B.P ∿ 360°C) were purchased locally. Kerosene was filtered before use. Emulsifiers were supplied by the manufacturers and were used without further treatment. $Y(NO_3)_3$ solutions were prepared by dissolving 99.9% pure Y_2O_3 (Molycorp #B-261) in excess HNO_3 to obtain 0.44M Y^{+3} stock solution.

Procedure

Emulsions were prepared by dissolving the emulsifying agent in the organic liquid at room temperature, followed by addition of the aqueous phase dropwise while stirring. A fraction of the aqueous phase was varied between 10% and 75% by volume; concentration of emulsifying agent was varied between 1% and 10% by volume. Temperature stability of the emulsion was studied from room temperature up to 100°C. Some of the emulsions were stored under normal conditions to check their stability with time. Emulsion type was determined by measuring the electrical resistance of the emulsions. High resistance was indicative of continuous organic phase, while low resistance indicated that the aqueous solution was in the continuous phase.

Droplet size distribution of the emulsions was studied both by optical microscopy (Zeiss, Ultraphot II) and by employing a particle size analyzer (Microtrac Small Particle Size Analyzer). The roles of emulsifier concentration, internal volume, phase content, and method of mixing on the droplet size were studied.

Formation of solid particles was accomplished by dropwise addition of the emulsion to a hot organic liquid that had the same composition as the continuous phase of the emulsion. During this process the temperature and the atmosphere were varied. Temperature was varied between 200° and 240°C in the case of mineral oil and 80 and 180°C in the case of kerosene. Evaporation was carried out in air, in nitrogen and under reduced pressure (∿1 torr).

Precursor particles were recovered by filtration, which was followed by washing the precipitate with toluene to remove excess oil phase. In some cases, solid particles were recovered by centrifugation. Precursor particles were further dried overnight at 110°C and calcined at 850°C to obtain oxide particles.

Powders were characterized both in precursor and oxide state with respect to size, shape, and degree of agglomeration. Powders were pressed into compacts without binders and sintered at 1750°C for two hours in vacuum. Another set of compacts was prepared from a commercial* yttria powder and sintered together with the powders obtained by the emulsion evaporation technique for comparison. Densities of sintered compacts were determined by the liquid displacement method.

---

*Molycorp Lot B-261.

RESULTS AND DISCUSSION

Emulsions

A series of emulsions was prepared to select a working system for powder preparation studies. Preliminary experiments revealed that a number of organic, aqueous, and emulsifier phases having a wide range of compositions produce stable w/o type emulsions. During the course of study, it was observed that anionic surfactants such as Aerosol-OT produce w/o emulsions with the toluene/water system at room temperature; these emulsions became unstable at elevated temperatures or in the presence of ions in the aqueous phase. On the other hand, nonionic surfactants such as Pluronic L62 or Arlacel 83 produced w/o emulsions using a variety of organic/aqueous systems that were stable in a wide range of temperatures as well as in the presence of ionic salts. Therefore, remaining experiments were carried out using nonionic surfactants. Emulsion characteristics of a toluene/aqueous $Y(NO_3)_3$/Span 60 system as a function of emulsifier concentration and aqueous phase volume ratio are summarized in Table I. It is clear from the table that w/o type, stable emulsions were formed for a broad range of emulsifier concentrations and aqueous phase volume ratios. Figure 1 shows the optical micrographs of emulsions shown in Table I. In series A the aqueous phase was kept constant and emulsifier concentration was varied; in series B the aqueous phase to emulsifier concentration was kept constant while aqueous phase content was changed. It appears that as the emulsifier concentration increases the droplet size decreases (Fig. 1a, b, c), and as the aqueous phase content increases the droplets become larger (Fig. 1d, e, f). When the aqueous phase content is more than 50% the droplet becomes very irregular in shape and distribution. When these emulsions were subjected to ultrasonic disruption, droplet size was reduced to submicron range, but considerable agglomeration of the droplets, especially in high aqueous phase contents, was observed. For emulsion evaporation studies, high boiling point organic phases using various types of emulsifying agents were investigated. Table II shows the types of oils and emulsifiers used along with the characteristics of the emulsions prepared.

In each case the aqueous phase was a 0.44M Y^{+3} solution. Volume ratios of each phase were 70/20/10 for the organic/aqueous/emulsifier system, and emulsions were prepared at room temperature with mechanical stirring.

Table 2 shows that all three organic liquids produce w/o type emulsions with Span 85, Span 80, and Arlacel 83; Tween 85 and Tween 80 produce o/w or w/o type emulsions depending on the organic phase. These results confirm the

Table I. Emulsions with toluene/0.44MY(NO_3)_3/Span 60 system.

Emulsion Designation	Composition v/o Toluene	Aqueous	Span 60	Characteristics
A1	74.0	25.0	1.0	w/o, stable, 4-40 μm
A2	73.5	24.5	2.0	w/o, stable, 4-32 μm
A3	72.0	24.0	4.0	w/o, stable, 1-20 μm
A4	70.5	23.5	6.0	w/o, stable, 1-20 μm
A5	69.0	23.0	8.0	w/o, stable, 1-10 μm
B1	74.0	25.0	1.0	w/o, stable, 4-40 μm
B2	59.0	39.4	1.6	w/o, stable, 8-70 μm
B3	49.0	49.0	2.0	w/o, stable, 8-130 μm
B4	39.0	58.4	2.6	w/o, not stable
B5	24.0	73.0	3.0	w/o, not stable

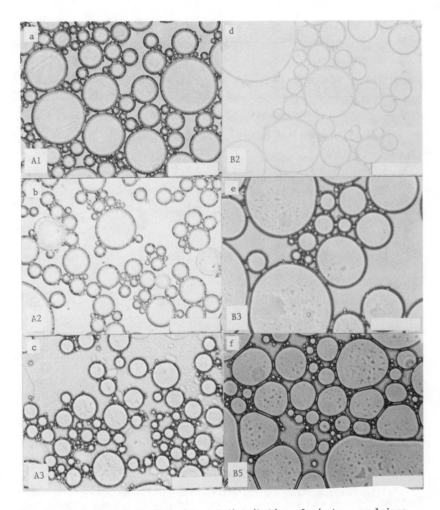

Fig. 1.   Variation of droplet size and distribution of w/o type emulsions.
Compositions of emulsions are given in Table 1.

predictions that the lower HLB values favor w/o type emulsions, while higher
HLB values generally yield o/w type emulsions.   Among the organic phases
studied, kerosene appears to form less stable emulsions than the others.
However, this condition was circumvented when the emulsions were treated with
a sonic disruptor.   This treatment not only reduced the size of the droplets,
but also rendered the emulsions very stable.

   Droplet size distribution of 70% mineral oil, 25% aqueous 0.44M $Y(NO_3)_3$,
and 5% Arlacel 83 is given in Fig. 2.   The two charts shown in the figure
compare the effect of mixing method on the droplet size distribution.   Figure 2a shows droplet distribution for an emulsion that was prepared by stirring; Fig. 2b shows same emulsion after sonic treatment.   Comparison of the
two charts indicates that after ultrasonic treatment the distribution shifts
to a smaller size and becomes narrower, having a peak at about 1 μm.   Similar
results were obtained for the kerosene/$Y(NO_3)_3$/Tween 85 system as well.   The

Table II. Emulsions with higher boiling point organic phases.

| Emulsifier | Organic Phases | | |
	Mineral Oil	Paraffin Oil	Kerosene
SPAN 85 HLB = 1.8	w/o separates in one hour $\emptyset$ = 5 - 45 µm	w/o viscous and stable $\emptyset$ = 5 - 40 µm	w/o separates in 15 minutes $\emptyset$ = 1 - 15 µm
SPAN 80 HLB = 4.3	w/o stable one day $\emptyset$ = 2 - 30 µm	w/o stable $\emptyset$ = 1 - 10 µm	w/o separates in $\emptyset$ = 1 - 50 µm
ARLACEL HLB = 3.7	w/o stable one hour $\emptyset$ = 1 - 15 µm	w/o stable $\emptyset$ = 1 - 15 µm	w/o separates quickly *
TWEEN 85 HLB = 11	o/w very viscous *	o/w very viscous *	w/o stable for 8 hours $\emptyset$ = 1 - 5 µm
TWEEN 80 HLB = 15	w/o very viscous $\emptyset$ = 5 - 40 µm	o/w viscous *	o/w viscous *

*
$\emptyset$ = not applicable.

size distributions remained relatively constant over a period of time, so that the stability of the emulsions was satisfactory.

Heating of the emulsions from room temperature to 100° C were determined by recording the resistivity of the emulsion as a function of temperature. All of the emulsions prepared by nonionic surfactants in this study have exhibited an excellent stability up to boiling. These findings agree well with the literature [17].

Emulsion Evaporation and Characterization of the Powders

Following preparation and characterization of the emulsions, mineral oil/$Y(NO_3)_3$/Arlacel 83 and Kerosene/$Y(NO_3)_3$/Arlacel 83 systems evaporated at different temperatures and atmospheres. The precursor particles were brown to black in color, indicating an organic residue. Scanning electron micrographs of oxide powders derived from mineral oil and kerosene are given in Fig. 3. Particles are generally spherical in shape with considerable size variation ranging from 0.4 µm to 2.5 µm. In addition, few large agglomerates are formed (Fig. 3a). Electron micrographs also reveal that the particles are not fully dense. Some of the spheres are hollow and some are sponge-like (Fig. 3b). Kerosene-derived powders appear to be smaller and more uniform and appear to consist of dense spheres (Fig. 3c and 3d). The particle size distribution of yttria powders obtained from mineral oil/$Y(NO_3)_3$/ Arlacel 83

system at various evaporation temperatures is plotted in Fig. 4. At 200° C, the distribution is bimodal. Increasing temperature not only eliminates the agglomerates but also shifts the peak to smaller sizes. However, no significant improvement is observed above 230°C. Mean particle size for the temperature range of 230 to 250°C is around 1.5 µm. Variation of median particle size as a function of temperature in various atmospheres is given in Fig. 5.

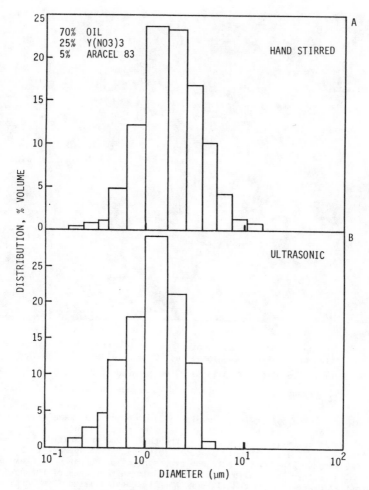

Fig. 2. Droplet size distribution of w/o type emulsions. a) Hand stirred;
b) Ultrasonically treated for 5 minutes.

As shown for air atmosphere in Fig. 4, the median particle size of the powders
decreases with increasing evaporation temperature for nitrogen and a vacuum
atmosphere as well. However, the influence of temperature on the median
particle is not very drastic. A more important observation is that the particle distribution becomes more uniform and narrow as the temperature is
increased. It was assumed that the optimal temperature for this system was
240°C and the effect of emulsifier concentration on the previous system was
investigated by evaporating emulsions in a nitrogen atmosphere. As the emulsifier concentration was increased, the distributions were shifted to smaller
sizes (Fig. 6). The major change in size occurred between 1% and 4% Arlacel 83 concentrations; above this, particle size distribution was narrow and
uniform. The distribution peak for emulsifier concentrations of 6% to 10% is
around one micron. These results suggest that spherical, micron-size oxide
particles can be produced relatively easily. One point, however, remains

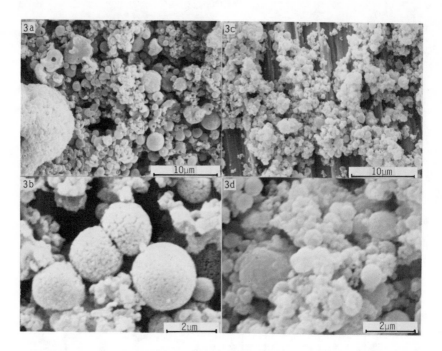

Fig. 3. Scanning electron micrographs of yttria powders derived from evaporation of w/o type emulsions. a) 70% mineral oil, 25% aqueous
0.44 Y^{+3} solution, 5% Arlacel 83 evaporated at 240°C; b) 73% kerosene,
25% aqueous 0.44 Y^{+3} solution and 2% Arlacel 83 evaporated at 180° C.

unexplained: The droplets in the emulsion were roughly the same size as the
oxide particles. It was expected that the oxide particles derived from these
droplets would be smaller; this was not the case. One explanation might be
that the particles are extremely porous. This assumption appears to be
justified because, in fact, electron micrographs of precursor particles appear
to be hollow rather than solid, and the specific surface area of the powders
was about an order of magnitude larger than the one estimated from particle
size data.

Complete chemical analysis of the precursors has yet to be done. TGA/
DTA analysis of the precursors showed three characteristic exothermic peaks.
The first peak appears at 290°, the second at 390°, and the last one at 470°C.
Control experiments performed by mixing of raw Y_2O_3 powder, mineral oil, and
Arlacel 83 have shown that the first two peaks are associated with combustion
of mineral oil and the last one with Arlacel 83 decomposition. Thus yttrium
is either chemically or physically associated with organics of the emulsion
medium. Absence of any endothermic peak may imply that the precursors are
organometallic compound of yttrium, rather than having a simple, inorganic
composition.

Powder compacts derived from emulsion evaporation achieved a sintered
density of 95.5% theoretical, while control compacts derived from raw Y_2O_3
sintered to only a 86.5% theoretical under identical conditions. Furthermore,
closer examination of the pellets showed that the emulsion derived specimens
were translucent with some relatively large defect pockets. It is presumed

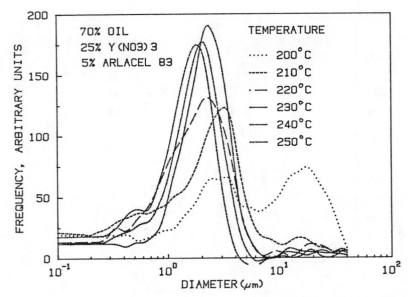

Fig. 4. Particle size distribution of yttria powders obtained from evaporation of mineral oil/Y(NO$_3$)$_3$/Arlacel 83 emulsions at various temperatures.

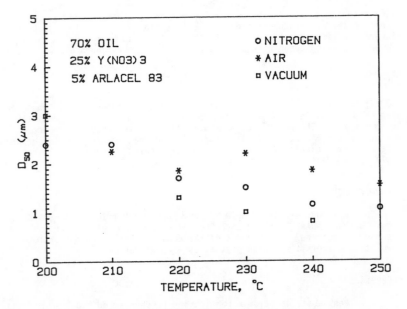

Fig. 5. Variation of median particle size, D$_{50}$, of yttria powders obtained from mineral oil/Y(NO$_3$)$_3$/Arlacel 83 in various atmospheres.

108

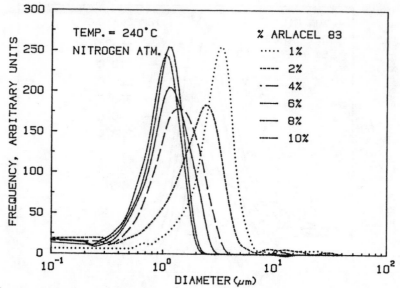

Fig. 6. Particle size distribution of yttria powders as a function of emulsifier concentration. In these experiments, mineral oil-to-aqueous
phase ratio was kept constant at 2.8, and evaporation temperature
was 240°C.

that these large defects are due to incomplete calcination and may eventually
be eliminated by careful control over calcination and sintering. Even the
sintered density of a theoretical 95.5% equals or surpasses most of the
powder processing techniques that were studied in our laboratory [18,19].

SUMMARY

It was shown in this study that small (1 to 2 μm) spherical, unagglomerated oxide particles may be obtained by evaporation of w/o type emulsions in
hot oil baths. The new method offers an alternative route to prepare a
variety of ceramic powders. It is anticipated that the method will be practical and economically feasible because it is based on well-established
emulsion and distillation technologies.

ACKNOWLEDGMENT

This work was supported by U.S. Army Research Office, Durham, N.C.,
under Contract #DAAG29-85-K-0086. Seminal discussions with A. Crowson are
greatly appreciated. The support of the Engineering Research Institute of
Iowa State University is also gratefully acknowledged.

REFERENCES

1. J.L. Pentecost, in Ceramic Fabrication Processes, edited by F.F. Wang
   (Academic Press, New York, 1976), pp. 1-14.

2. D.W. Johnson, Jr. and P.K. Gallagher, in Ceramic Processing Before Firing, edited by G.Y. Onoda and L.L. Hench (John Wiley, New York, 1978), pp. 125-139.

3. W.D. Kinery, in Ceramic Powders, edited by P. Vincenzini (Elsevier, Amsterdam, 1983), pp. 3-18.

4. D.W. Johnson, Jr., Am. Ceram. Soc. Bull. 60 (2), 221-224 (1981).

5. W.H. Rhodes, J. Am. Ceram. Soc. 64, 19-23 (1981).

6. E.A. Barringer, R. Brook, H.K. Bowen in Sintering and Heterogeneous Catalysis, edited by G.C. Kuczynski, A.E. Miller, G.A. Sargent (Plenum Press, New York, 1984), pp. 1-21.

7. Egon Matijević, in Ultrastructure Processing of Ceramics, Glasses, and Composites, edited by L.L. Hench and D.R. Ulrich (John Wiley & Sons, New York, 1984).

8. Bruce Fegley, Jr. and E.A. Barringer, in Better Ceramics Through Chemistry, edited by C.J. Brinker, D.E. Clark, and D.R. Ulrich (North-Holland, New York, 1984), pp. 187-198.

9. Paul Becher, Emulsions: Theory and Practice, 2nd ed. (Reinhold, New York, 1965).

10. T.F. Tadros and Brian Vincent, in Encyclopedia of Emulsion Technology, edited by Paul Becher (Marcel Dekker, New York, 1983), pp. 129-209.

11. A.L. Smith, Theory and Practice of Emulsion Technology, (Academic Press, New York, 1976).

12. S. Marti, T. Nervo, J. Perierd, G. Rieos, Coll. Polym. Sci. 253 (3), 220-224 (1975).

13. G.M. Glagoleva, P.N. Kraglyakov, A.F. Kovetskii, Izv. Sib. Otd. Akad. Nauk SSSR Ser. Khim. Mauk 1972 (6), 9-12 (Chem. Abs. 83:115605e).

14. H. Bennet, J.L. Bishop, M.F. Wulfinshoff, Practical Emulsions II, (Chemical Publishing Co., New York, 1968), pp. 106-107.

15. A.L. Stuijts, in Proc. Int. Conf. Ferrites, edited by Yittoshino, S. Lida, M. Sugamoto (University Park Press, Baltimore, 1971), pp. 108-113.

16. P. Reynen, H. Bastius, M. Fiedler in Ceramic Powders, edited by P. Vincenzini (Elsevier, Amsterdam, 1983), pp. 499-504.

17. K. Shinoda and H. Kunieda, in Encyclopedia of Emulsion Technology, edited by Paul Becher (Marcel Dekker, New York, 1983), pp. 337-367.

18. M.D. Rasmussen, M. Akinc, D. Milius, M.G. McTaggart, Am. Ceram. Soc. Bull. 64 (2), 314-318 (1985).

19. M.D. Rasmussen, M. Akinc, M.F. Berard, Ceramics International 10 (3), 99-104 (1984).

SPINEL FORMATION FROM MAGNESIUM ALUMINIUM DOUBLE ALKOXIDES

KENNETH JONES, THOMAS J. DAVIES*, HAROLD G. EMBLEM, AND PETER PARKES
Dept. of Chemistry and *Dept. of Metallurgy, University of Manchester
Institute of Science and Technology, Manchester M60 1QD, England.

ABSTRACT

Double alkoxides of the general formula $Mg[Al(OR)_4]_2$ where R is iso-Pr
or sec-Bu were prepared by reacting aluminium and magnesium metals together
with the alcohol ROH and purified by vacuum distillation. They were
characterised by IR, proton and ^{27}Al NMR spectroscopy and MW determination.
A magnesium aluminium double alkoxide was also prepared by treating the
commercially available aluminium alkoxide 'Aliso B' [a mixed aluminium
(iso-propoxide)(sec-butoxide)] with magnesium and iso-propanol. Treatment
of magnesium aluminium double alkoxides with water and an alkanolamine
(preferably triethanolamine) gives a rigid coherent gel. Viscosity
measurements and ^{27}Al NMR spectroscopy suggest that the double alkoxide
does not break down to its constituents during hydrolysis. The air-dried
gel was shown by XRD to convert quantitatively to spinel on firing to
1500°C. The resistance of the double alkoxide moiety to hydrolysis
explains the ease of conversion to spinel on firing. The gel has been used
to bind alumina and magnesia grain. Gels suitable for binding refractory
grain were obtained only when the alkanolamine content corresponded to one
alkanolamine group per metal atom. Electron micrographic and XRD studies
showed that in fired refractory pieces, the bonding phase was spinel.

1 INTRODUCTION

Spinels are generally prepared [1] by fusion of a mixture of the two
component oxides. This is usually a high temperature reaction. Only
spinel itself $(MgO.Al_2O_3)$ has been used [2] pure as a special refractory.
$MgO-Al_2O_3$ mixtures, which form spinel on heating, may be used to line
induction-melting furnaces. The present work describes methods for the
preparation of magnesium-aluminium double alkoxides having the oxide
stoichiometry of spinel $(MgO.Al_2O_3)$, the characterisation of these
alkoxides, their conversion to the mineral spinel, and their use in binding
magnesia and alumina grain where they form spinel as the grain binding
phase.

2 MAGNESIUM-ALUMINIUM DOUBLE ALKOXIDES

When a magnesium aluminium double alkoxide is to be used for the
preparation of spinel or for the binding of refractory grain, the organic
content should be as low as practical. The iso-propoxide and sec-butoxide
are convenient being liquid at ambient temperature and miscible with most
common organic solvents.

2.1 Preparation of magnesium aluminium double alkoxides

Magnesium aluminium double alkoxides were first prepared [3] by
refluxing together the two independently prepared component alkoxides. It
was known [4] later that separate preparation of the magnesium alkoxide is
not necessary, the double alkoxide being formed by treatment of magnesium

with an alcohol in the presence of the corresponding aluminium alkoxide. In the present work, double alkoxides of the general formula $Mg[Al(OR)_4]_2$ where R is iso-Pr or sec-Bu have been prepared [5,6] by treating the alcohol ROH with aluminium and magnesium metals together and the product purified by vacuum distillation. The reaction was vigorous and proceeded rapidly.

$$Mg + 2Al + 8iso-PrOH \longrightarrow Mg[Al(Oiso-Pr)_4]_2 + 4H_2 \qquad (1)$$

The double alkoxides were characterised by IR, proton and ^{27}Al NMR spectroscopy and MW determination. A magnesium aluminium double alkoxide was also prepared [7] by treating the commercially available [8] aluminium alkoxide 'Aliso B' [a mixed aluminium (iso-propoxide)(sec-butoxide)] with magnesium and iso-propanol. The product double alkoxide 'Mag Aliso B' was purified by vacuum distillation.

2.2 Hydrolysis and gelation of magnesium aluminium double alkoxides

Rigid coherent gels suitable for binding alumina or mangesia grain mixes were formed [5,6,7] when the magnesium aluminium double alkoxide was first mixed with triethanolamine then hydrolysed with a water-iso-propanol mixture. Gelation was dependent on a critical concentration of triethanolamine with only a 1:1 stoichiometry of alkanolamine group per metal atom giving a rigid coherent gel. Viscosity measurements and ^{27}Al NMR spectra of the gelling system indicate that the double alkoxide does not break down to the separate metal components during hydrolysis. XRD showed that the air-dried gel converted quantitatively to spinel on firing for 1 hour at 1500°C. The resistance of the double alkoxide moiety to hydrolysis explains the easy conversion of the gel to spinel on firing and the formation of a spinel bonding phase in fired refractory pieces obtained from alumina or magnesia grain.

3 EXPERIMENTAL

3.1 Materials and methods

Magnesium metal was used as ribbon or as powder. Aluminium metal was used as foil and as turnings. 'Aliso B' was used as received [8]. Other reagents were of 'laboratory grade' and were used as received.

Proton NMR spectra were recorded on a Perkin Elmer R12 spectrometer at 90 MHz or at 220 MHz on a Perkin Elmer R34 spectrometer. ^{27}Al NMR spectra were run at 20.85 MHz on a Bruker WP80 spectrometer. IR spectra were recorded on a Perkin Elmer 598 instrument as capillary films between KBr plates. Double alkoxide products were handled in a dry box under nitrogen. Viscosities were measured using a Brookfield Synchro-Lectric Viscometer (Model LVT). Molecular weights were determined ebullioscopically [9]. XRD measurements were obtained from a Philips PW1729 x-ray generator and Debye-Scherrer powder camera (diam = 12.5 cm) using CuK$_\alpha$ radiation; d-spacings and relative intensities being recorded on a Joyce-Loebel recording microdensitometer.

3.2 Preparation of magnesium aluminium double alkoxides

3.2.1 Magnesium aluminium iso-propoxide

Magnesium ribbon (11.5 g), aluminium turnings (20.0 g) and aluminium foil (3.0 g) were placed in a 2000 cm³ flask containing 1000 cm³ iso-propanol with a trace of iodine as catalyst. Reaction was complete after refluxing for 12 hours. The product was filtered (Schlenk) under reduced pressure. After excess solvent was removed under vacuum using a rotary evaporator, a clear viscous product remained which crystallised on standing overnight. Analysis gave (required %) 4.0 Mg(4.5), 10.7 Al(9.8), 51.0 C(52.3), 9.9 H(10.2), and 9.3 O(8.0). Ebullioscopic MW determination in iso-propanol solution gave an association of 1.05. The IR spectrum was 2620w, 1455m, 1365m, 1185sh, 1165vs, 1125vs, 1030vs, 970vs, 950sh, 850sh, 835sh, 685vs, 620s, 540m, 455m. The proton NMR agreed with the reported [4] spectrum. The ^{27}Al NMR spectrum of a solution in CDCl$_3$ showed a single resonance (61 ppm, W½ = 800 Hz) typical of aluminium in a tetrahedral environment.

3.2.2 Magnesium aluminium sec-butoxide

The procedure used was similar to that described above, using magnesium ribbon (2.5 g) and aluminium foil (5.4 g) in sec-butanol (250 cm^3). The product distilled at 145°C/0.1 mmHg, yield 51 g, 77%. Analysis gave (required %) 3.6 Mg(3.7), 7.4 Al(8.1), 57.1 C(58.0), 11.0 H(10.9), and 20.9 O(19.3). The IR spectrum was 2950vs, 2920vs, 2870vs, 2840sh, 2670w, 1450s, 1360vs, 1265m, 1170vs, 1140vs, 1110vs, 1050vs, 1030vs, 990vs, 960m, 920vs, 890sh, 830s, 795sh, 780m, 690s, 580s, 540sh, 495w, 450m, 380w. The proton NMR spectrum [6] is consistent with the formula Mg[Al(Osec-Bu)$_4$]$_2$.

3.2.3 'Mag Aliso B'

'Aliso B' (200 g) was mixed with iso-propanol (200 cm^3) and magnesium powder (15 g) added then refluxed for ca. 56 hours, during which time mercury(II) chloride (0.5 g) catalyst was added at 8 hour intervals. The double alkoxide was purified by vacuum distillation (140°C/0.1 mmHg) to give a yield of 76%. The product is liquid at ambient temperature and is miscible with or soluble in common organic solvents. 'Mag Aliso B' may be predominantly magnesium aluminium iso-propoxide as a result of the exchange reaction:

$$\text{)AlOsecBu} + \text{iso-PrOH} \rightleftharpoons \text{)AlOisoPr} + \text{secBuOH} \tag{2}$$

which may occur during preparation between the solvent iso-propanol and the sec-butoxy groups present in 'Aliso B'.

3.3 Hydrolysis and gelation procedure and viscosity determination

The required volumes of the double alkoxide and alkanolamine (monoethanolamine or triethanolamine) were measured by syringe into a glass tube, stoppered, and shaken thoroughly until the heat evolved had dissipated and a single mobile liquid phase resulted. After equilibration for at least 1 hour, the required volume of iso-propanol-water mixture was syringed into the tube, noting the time, then mixing vigorously, and the gel-time recorded. The procedure was repeated throughout the range of variables studied, and the quantities of reagents used in the formation of rigid coherent gels from 'Mag Aliso B' are listed in Table I.

For viscosity determinations, gellable compositions were also prepared as in Table I. The time of addition of the iso-propanol-water mixture was noted, then the spinning disc of the Brookfield viscometer was immersed in the mixture and the viscosity recorded against time as the composition gelled. The log (viscosity)/time plot for a double alkoxide gelling composition {comprising 10.0 cm^3 Mg[Al(Osec-Bu)$_4$]$_2$, 2.0 cm^3 triethanol-amine, 3.7 cm^3 water, and 7.4 cm^3 iso-propanol} which had similar gelling

Table I   Formation of rigid coherent gels from 'Mag Aliso B'

'Mag Aliso B' (cm³)	triethanolamine (cm³)	water (cm³)	iso-propanol (cm³)	Gel time (min)
5	1	1.9	2.1	1.5
5	1	1.85	2.15	5
5	1	1.8	2.2	20
5	1	1.75	2.25	480

Table II   XRD Data for 'Mag Aliso B' fired gels and spinel

Fired 'Mag Aliso B' gel 2θ	d-spacing	relative intensity	spinel [10] d-spacing	relative intensity
19	4.62	32	4.67	35
26.5	3.36	5		
36.75	2.86	30	2.86	40
37	2.43	100	2.44	100
45	2.01	57	2.02	65
55.75	1.65	9	1.65	10
59.45	1.55	43	1.55	45
62.75	1.48	2		
65.25	1.43	69	1.43	55
68.5	1.37	2	1.37	4
69.5	1.35	2		
74	1.28	9	1.28	4
77.25	1.23	9	1.23	8
82.74	1.16	6	1.17	6
85.5	1.13	2	1.13	2

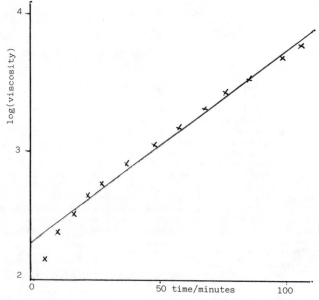

Figure I   log(viscosity)/time plot for $Mg[Al(OsecBu)_4]_2$ gelling composition

characteristics to 'Mag Aliso B' compositions, is shown in Figure I.

3.4 Formation of spinel and binding of refractory grain

The gels obtained from the hydrolysis of double alkoxides were allowed to dry in air then fired at 1500°C for 1 hour. X-ray powder diffraction patterns of the fired products were recorded and compared with data reported for spinel [10], the results showing complete conversion to spinel, the comparison for the fired product obtained from 'Mag Aliso B' given in Table II being typical. Refractory shapes were prepared from alpha-alumina grain mixes and magnesia grain mixes using the binder gels obtained from 'Mag Aliso B'. Details of these preparations are given in references [6] and [7]. After firing, XRD and electron micrographic studies including scanning diffraction techniques confirmed that the bonding phase was spinel.

4 RESULTS AND DISCUSSION

^{27}Al NMR spectroscopy has been used to investigate species of high symmetry and is particularly successful with those of cubic symmetry. Thus for similar ligands, aluminium species with tetrahedral co-ordination give signals to low field of species in an octahedral environment. $Al(OH_2)_6^{3+}$ is usually taken as the standard reference at zero ppm, with tetrahedral $Al(OH)_4^-$ appearing at +80 ppm [11]. The ^{27}Al NMR signals reported here for magnesium aluminium double alkoxides appear downfield of $Al(OH_2)_6^{3+}$ (zero), and are given the positive sign convention. Also, the spectra recorded during gelation of magnesium aluminium double alkoxide systems are similar to those of the parent alkoxide. These results are consistent with the bimetallic unit of the double metal alkoxide remaining intact during the hydrolysis and gelation.

The log(viscosity)/time plots for the double alkoxide gelling systems have a linear portion (see Figure I). This indicates [6] that the double metal alkoxides gel by a random polymerisation and do not break down into separate constituents on hydrolysis. The survival intact of the bimetallic unit during hydrolysis and gelation of magnesium aluminiun double alkoxides is an important factor contributing to the easy conversion of the gel to spinel on firing. It also explains why spinel is the bonding phase in fired refractory shapes prepared from alumina or magnesia grain bonded with a gel derived from the double magnesium aluminium alkoxide 'Mag Aliso B'.

Acknowledgement

Financial support from Zirconal Processes Ltd., Dewsbury, West Yorkshire, UK, is gratefully acknowledged.

References

1.   P.B. Moore in McGraw-Hill Encylopedia of Science and Technology, 3rd ed. (McGraw-Hill Book Company, New York, 1982) Vol. 12, pp. 903-904.

2.   J.D. Gilchrist, Fuels and Refractories. (Pergamon Press, Oxford, 1963) pp. 134-135.

3.   H. Meerwein and T. Bersin, Ann. Chim. 476, 113 (1929).

116

4. R.C. Mehrotra, S. Goel, A.B. Goel, R.B. King and K.C. Nainan, Inorg. Chim. Acta 29, 131 (1978).

5. H.G. Emblem, K. Jones and P. Parkes, British Patent No. 2 128 604A (19 October 1982).

6. P. Parkes, PhD thesis, University of Manchester, 1984.

7. H.G. Emblem, K. Jones and P. Parkes, European Patent Appliation No. 0 063 034 (7 April 1982).

8. Manchem Ltd., Ashton New Road, Manchester M11 4AT, UK;Data Sheet 'Aliso and Aliso B'.

9. C. Heitler, Analyst 83, 223 (1958).

10. ASTM PD File, Data Card 21-1152.

11. E.G. Derouane, J.B. Naggy, Z. Gabelica and N. Blom, Zeolites 2, 229 (1983).

SYNTHESIS OF OXIDE CERAMIC POWDERS BY AQUEOUS COPRECIPITATION

J. R. MOYER, A. R. PRUNIER, JR., N. N. HUGHES, AND R. C. WINTERTON
Central Research Inorganic Materials and Catalysis Laboratory
The Dow Chemical Company, Midland, MI 48674

ABSTRACT

Aqueous precipitation in a computer-controlled continuous crystallizer has been applied to the preparation of cordierite, mullite, alumina, and zirconia precursor powders. The particles are dense, spherical, and x-ray amorphous. The powders have a particle size range of 0.2-5 μm and a surface area of 1-10 m^2/g The properties of the powders and of some green and fired pieces are described.

INTRODUCTION

The fine, homogeneous powders obtained from alkoxides are outstanding ceramic precursor materials. Because the various components are mixed at the atomic level before precipitation, the densification and crystallization occur at remarkably low temperatures. However, alkoxide-derived powders are very expensive. This is illustrated for cordierite in Table 1. The cost of its silica content varies enormously depending upon the source. The relative cost of aqueous metal salts versus their alkoxides is even more dramatic.

Table I
Relative Costs Per Lb. SiO_2

Compound	$/lb.	%SiO_2	$/lb. SiO_2	$/lb. Cordierite
Tetraethylorthosilicate	1.66	28.8	5.75	2.96
40% Aq. Colloidal Silica	0.572	40.0	1.43	0.73
Sodium Silicate (aq)	0.157	28.7	0.547	0.28

This work was undertaken to develop a process for making ceramic powders using aqueous solutions. The initial objective was a cordierite powder which gives fully dense pieces with excellent properties. Achieving this would require precise control over stoichiometry and the size of particles. A second objective was greenware with at least 55% of the theoretical density (t.d.).

Experimental

Aluminum nitrate (60% $Al(NO_3)_3 \cdot 9H_2O$) and magnesium nitrate (66.7% $Mg(NO_3)_2 \cdot 6H_2O$) were purchased from Minerals Research and Development Corporation. Aluminum chloride (50% $AlCl_3 \cdot 6H_2O$) was obtained from Upjohn Fine Chemicals; zirconyl nitrate (37.8% $ZrO(NO_3)_2$) from Magnesium Elek- tron. The colloidal silica was the AS grade of Dupont's LUDOX® (40% SiO_2). The aqueous sodium silicate was The PQ Corporation's Type N (28.7% SiO_2).

The precipitations were carried out at 75°C in a 5 gallon continuous crystallizer which was monitored and controlled by a Dow-built laboratory microcomputer system. Gaseous ammonia was used for pH control.

The products were mixtures of hydroxides, amorphous by x-ray. Analyses of the products (by neutron activation) showed that the desired stoichiometry was maintained to ±2% (relative). Small quantities of precipitate were settled by centrifugation; larger quantities were allowed to settle in 55 gallon drums. In both cases, the supernatant liquid was replaced with water and the solids were reslurried. This sequence was repeated until the conductivity of the supernatant liquid reached 500 microohms. Small samples were pan-dried at 120°C. Larger quantities were spray-dried at 250°C.

CORDIERITE

Particle size

Figures 1a and 1b show scanning electron micrographs of a cordierite powder prepared by aqueous coprecipitation. The average particle size is 2.8±1.3 μm, but it is clearly an agglomeration of much smaller particles. The surface area of this material, 130 m^2/g by single point BET nitrogen adsorption, corresponds to a crystallite diameter of 0.018 μm.

Figure 1. Scanning electron micrographs of cordierite precursor powder dried at 120°C.

Calcination

On heating to 500°C, the powder loses 35% of its weight and gains surface area. Figure 2 shows a calcination curve for a cordierite precursor powder prepared from aqueous Mg and Al nitrates and either aqueous sodium silicate or an aqueous suspension of colloidal silica. The material made using aqueous sodium silicate contained about 3500 ppm sodium. Its high sodium content makes its physical properties unacceptable. Switching to an aqueous suspension of colloidal silica (Dupont's Ludox®) lowered the Na content to <200 ppm. The properties of this material are described below.

Cordierite processing

The surface area of the Ludox®-derived powder has a critical influence upon the green and fired densities. Powders with low surface area (1-2 m^2/g) can be processed to yield greenware of 55-60% the t.d. but are unsinterable. Powders with very high surface area (>50 m^2/g) can be sintered to >97% t.d. but give greenware with only 31-34% t.d. Mercury porosimetry data show that the high surface area of these agglomerated powders consists of fine (0.02-0.2 μm) internal porosity. Calculations based on the porosimetry data indicate that even the smallest agglomerates (0.2 μm) would contain about 40% internal porosity.

The processing of powders calcined to various surface areas has shown that at least 30 m^2/g is needed to achieve high fired densities. Such powders yield greenware having 35-40% t.d. With a slight chemical modification of the powder, we have achieved 55-60% dense greenware which fires to >98% t.d. with only limited degradation of physical properties. The polished and etched surface shown in Figure 3 reveals a microstructure in which the grains are submicron in size.

Figure 4 compares x-ray diffraction patterns (using Cu K$_\alpha$ radiation) of a precipitated powder with a sol-gel product prepared in this laboratory. After two hours at 1200°C the sol-gel powder is transformed to cordierite whereas the coprecipitated powder contains spinel, cristobalite and amorphous phases. However, further heat treatment for two hours at 1400°C makes the two products nearly indistinguishable.

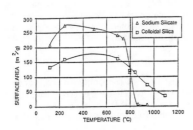

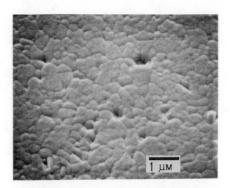

Figure 2. The effect of temperature upon the surface area of cordierite precursor powder.

Figure 3. Scanning electron micrograph of polished and etched surface of 97% dense cordierite body.

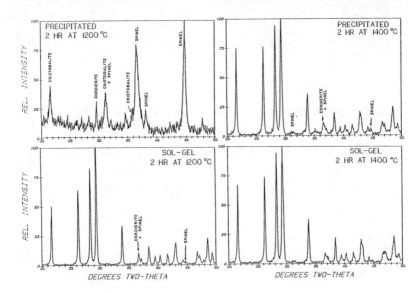

Figure 4. X-ray diffraction patterns of cordierite powders prepared by coprecipitation from aqueous solutions (upper) and from alkoxides (lower) and heated for 2 h at 1200° (left) and 1400° (right).

Properties of cordierite pieces

A cordierite precursor powder (14.09% MgO, 35.34% Al_2O_3, 50.49% SiO_2 and 120 ppm Na_2O by neutron activation analysis) with specific surface area 31 m^2/g (single point BET) was ball-milled in distilled water using high alumina jar and media plus a small amount of dispersant (Daxad 32, W. R. Grace & Co.). After milling for 20 hours, a polymer binder formulation* was added. The slip obtained after another 18 hours of milling was used "as is", without any filtering, de-airing or sedimentation. The mean particle size of the porous agglomerates after milling was 1-2 μm (Micromeritics 5000ET x-ray sedigraph) with an approximation correction for agglomerate porosity.

* Mostly Gelvatol 20-30, Monsanto Corp., and Methocel K35LV, the Dow Chemical Co.

Greenware disks ~80 mm in diameter were cast from the slip using a vacuum filtration assembly with #42 Whatman filter paper to catch the particles. After binder burnout, the disks were 33% of the t.d. of the precursor powder (2.85 g/cc). The disks were fired in air, heating 3 hours from 1070 to 1250°C to promote densification and another 2 hours at 1420°C to develop the cordierite phase. A careful correlation between apparent density and microstructural porosity showed the fired disks to be 96.7-97.1% of t.d. (2.545 g/cc). (Densities >99% t.d. have been achieved for small pieces of this material if the slip is filtered and/or sedimented to eliminate the larger agglomerates.) Specimens for determining the coefficient of thermal expansion (C.T.E.), dielectric properties, elastic moduli, and flexural strength were machined from these fired disks.

The C.T.E from 25 to 650°C was determined using a computer automated differential dilatometer (Theta Dilatronic II) and an NBS fused silica reference bar. Values for both the true (or instantaneous) and the average C.T.E. are listed in Table II. These values are comparable to literature values for pure cordierites [2] and are lower than those obtained from commercial glass-ceramic or alkali-fluxed materials.

Dielectric properties for this cordierite were measured by an independent testing laboratory (A. P. Sheppard Microwave and Electronic Consultants, Atlanta GA). Microwave (9.375 GHz) properties were determined by a transmission line method (ASTM D 2520-81) on a specimen machined to 2.26 x 1.25 x 0.996 cm; 1 MHz properties were determined on a disk (0.32 cm x 2.54 cm dia.) using an A-C loss method (ASTM D 150-81 and D 2149-79) and referenced against a fused silica standard. The results, listed in Table III, show that these materials have a significantly lower dielectric constant at 1 MHz (4.00) than has been reported previously for relatively pure cordierites (e.g., 4.8-5.0 [1,2]). The microwave properties of these relatively dense pieces are comparable to those observed in more porous commercial products.

Table II	Table III

Thermal Expansion of Cordierite			The Dielectric Properties of Cordierite			
Temp. (°C)	True C.T.E. (10^{-6} °C^{-1})	Ave. C.T.E. ($25°$-T, 10^{-6}°C^{-1})	Freq. (MHz)	Temp (°C)	Dielectric Constant	Loss Tangent
50	0.3	0.09	1	26	4.00	<0.01
100	0.7	0.3	9375	26	4.42	<0.0008
200	1.3	0.7	9375	300	4.42	<0.0008
300	1.7	1.0				
400	2.0	1.2				
500	2.2_5	1.4				
600	2.4	1.5_5				
650	2.5	1.6				

Dynamic elastic moduli were determined using an electrostatically driven resonance tester (James Instrument Inc., E-meter with electrostatic test bench). Calibration was performed using an NBS elasticity standard. Both Young's modulus, 132 GPa via longitudinal resonance, and shear modulus, 50.3 GPa via torsional resonance, are very close to literature values for pure cordierites [2]. From these two moduli Poissons's ratio (0.312) and the bulk modulus (117 GPa) were calculated.

Flexural strengths were determined on cordierite bars machined and tested according to MIL-STD-1942(MR). Machined specimens were 4 x 3 mm in cross section, from 45 to 50 mm long. They were broken in 4 point mode using a 40 mm span. The data from 32 such bars are plotted in Figure 5 using Weibull analysis scaling. The mean strength is 120 MPa and the Weibull coefficient is 20. The limited fractographic examination completed to date points to large (80-100 μm) spherical voids and inclusions as the fracture origins. The strength values are about half those obtained from glass-ceramics but are comparable to strengths of cordierites derived from oxides [1] It is felt that improved strength can be achieved through refinements in the slip processing/casting procedure.

MULLITE

Particle Size

Figure 6 shows a scanning electron micrograph of a mullite precursor powder after drying at 120°C. The mean diameter is 1.5±0.8 μm. The surface area is 2.8 m^2/g, which corrresponds to a particle diameter of 0.6 μm.

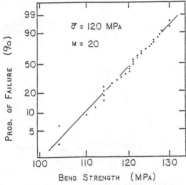

Figure 5. 4 point bend strengths of cordierite

Figure 6. Scanning electron micrograph of mullite precursor powder

Calcination

The effect of heating upon the surface area is shown in Figure 7. At 500°C the surface area increased to 225 m^2/g and the weight loss plateaued at 35%. At 1250°C the surface area was 3.4 m^2/g. X-ray diffraction patterns show that the solid is amorphous after 4 h at 1150°C but is mullite at 1250°C. (Figure 8). Phase development is actually complete at 1200°C. This is characteristic of well mixed oxides prepared by a variety of techniques: "flash hydrolysis" of a mixture of $SiCl_4$ and $AlCl_3$ [3], hydrolyzing a mixture of alkoxides [4], hydrolyzing tetraethylorthosilicate (TEOS) on gelatinous $Al(OH)_3$ [5], and pyrolyzing a mixture of silicon tetraacetate and aqueous $Al(NO_3)_3$ [6]. Work on the processing and fabricating of pieces from this mullite precursor powder is in progress.

ZIRCONIA

Aqueous zirconyl nitrate was precipitated in a slightly acidic solution to give a hydrous oxide with a BET surface area of 1.6 m^2/g. A scanning electron micrograph is shown in Figure 9a. The particles appear to have a diameter of 4 μm. The surface area corresponds to a particle size of about 0.7 μm. The powder is amorphous by x-ray diffraction. Phase development begins at 400°C. After 2 h at 450°C the powder is tetragonal ZrO_2 with no trace of the monoclinic phase. At 650° it is mostly monoclinic.

ALUMINA

Figure 9b shows a scanning electron micrograph of hydrated alumina after drying at 120°C. The particles have an average diameter of 2.2±0.5 μm and a BET surface area of 10 m^2/g. The powder is amorphous by x-ray diffraction. We have made no attempt to convert it to α-alumina or to process it.

122

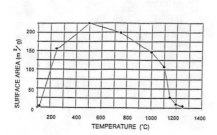

Figure 7. The effect of temperature upon the surface area of mullite precursor powder

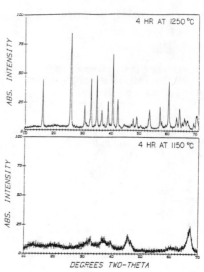

Figure 8. X-ray diffraction patterns of mullite precursor powder after heating for 4 h at 1150° (bottom) and 1250° (top).

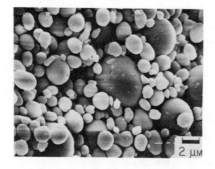

Figure 9. Scanning electron micrographs of precursor powders of zirconia (left) and alumina (right), dried at 120°C.

REFERENCES

1. D. Lewis and J. R. Spann, Proc. Symp. Electromagn. Windows 15, 165 (1980)
2. B. H. Mussler and M. W. Shafer, Amer. Cer. Soc. Bull 63(5), 705 (1984)
3. J. D. Crofts and W. W. Marshall, Trans. Brit. Ceram. Soc. 66(3), 121 (1967)
4. K. S. Mazdiyasni and L. M. Brown, J. Am. Ceram. Soc. 55(11), 54 (1972)
5. B. L. Metcalfe and J. H. Sant, Trans. Brit. Ceram. Soc. 74(6), 193 (1975)
6. G. Y. Meng and R. A. Huggins, Mat. Res. Bull. 18, 581 (1983)

PRECIPITATION OF PZT AND PLZT POWDERS USING A CONTINUOUS REACTOR

R. W. SCHWARTZ, D. J. EICHORST AND D. A. PAYNE
Department of Ceramic Engineering and Materials Research Laboratory
University of Illinois at Urbana-Champaign, Urbana, IL 61801

ABSTRACT

PZT and PLZT powders were prepared from nitrate and chloride precursors in a continuous constant volume precipitator. After precipitation, the powders were dried by a variety of methods, including: spray-drying, freeze-drying (in liquid nitrogen), and centrifugal freeze-drying. Spray-dried powders were found to have a spherical morphology, and to be solid. The particle size was in the micron range. Powders dried by nitrogen freeze-drying were characterized by an open morphology of agglomerated platelets. For centrifugally freeze-dried powders, particle size analyses were found to fit a population balance model, giving crystallite, cluster, and agglomerate population densities and growth rates.

INTRODUCTION

Production of high quality ceramic products, such as transparent electro-optic materials, requires better control of chemical composition and ceramic microstructure than is possible by conventional ceramic processing methods. For example, in the preparation of multicomponent ceramics (e.g., PLZT) by conventional mixed oxide routes, heterogeneities in chemical distribution can degrade optimum performance characteristics. Therefore, so as to prepare materials of uniform composition on a microscopic scale, and to obtain desired properties, chemical solution liquid-mix methods are increasingly used [1]. In the present study, precipitations of PZT (lead zirconium titanate) and PLZT (lanthanum modified PZT) from an ammonium hydroxide solution were carried out to prepare compositions near the rhombohedral-tetragonal morphotropic phase boundary. The materials are of interest because of their piezoelectric and electro-optic properties.

Characteristics of chemically prepared powders depend critically upon precipitation conditions. Therefore, a continuous, constant volume reactor was designed and constructed for the study of experimental variables. For example, the primary effects and secondary interactions of concentration, temperature, pH, and residence time in the reactor, on particle size distributions, morphology, and surface area characteristics were investigated by statistically designed experiments based on Plackett-Burman analysis [2]. Details of the investigation are reported elsewhere [3].

In addition, the use of a constant volume precipitator allows for the interpretation of particle size data in terms of a population balance approach [4,5]. Population densities of crystallites, clusters, and agglomerates in a given size range can be evaluated by analysing the logarithm of the population density as a function of particle size. Graphical interpretation gives the initial particle density and growth rate for each type of species. From consideration of clustering and agglomeration mechanisms, the method of analysis takes the form:

$$n_i = n_i^0 \exp\left[-1/r_i\tau\right] \qquad (1)$$

where i may be x (crystallites), c (clusters), or a (agglomerates); and n_i, the average population density; n_i^0, the initial "nuclei" density; 1, the particle size; r_i, the growth rate; and τ, the residence time.

124

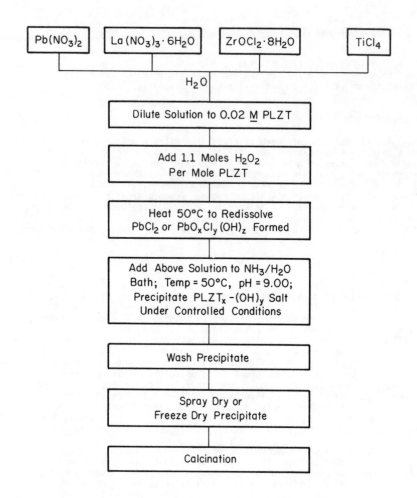

Fig. 1. Flow diagram for the preparation of PLZT powders.

EXPERIMENTAL

PLZT powders were synthesized according to the chemical formula $Pb_{1-x}La_x(Zr_yTi_{1-y})_{1-x/4}O_3$. For example, PLZT 9065 corresponds to x = 0.090 and y = 0.650; and PZT 53:47 corresponds to y = 0.530. The powders were prepared from nitrate and chloride precursors, according to the procedure outlined in Fig. 1 [6]. Figure 2 illustrates the equipment used, which consisted of a reaction vessel and powder trap kept at constant temperature, and an overflow vessel for constant volume conditions. The system volume of sixteen liters allowed for good control of precipitation parameters, especially temperature and pH. Precipitation was accomplished by spray atomizing the stock solution into the reaction vessel under controlled and reproducible conditions. Ammonium hydroxide additions were used to induce precipitation and maintain the desired pH.

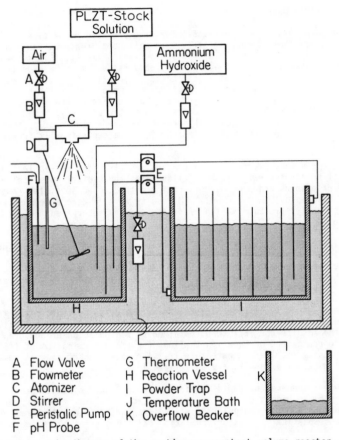

Fig. 2. Schematic diagram of the continuous constant volume reactor.

A Flow Valve	G Thermometer
B Flowmeter	H Reaction Vessel
C Atomizer	I Powder Trap
D Stirrer	J Temperature Bath
E Peristalic Pump	K Overflow Beaker
F pH Probe	

After precipitation the powders were washed repeatedly with deionized water to remove undesirable adsorbed anion species. Powders were then dried by a variety of independent methods, including: (i) spray-drying (Buchi 190 Laboratory Spray-Drier) of a 60:40 isopropanol:water suspension of the precipitate; (ii) freeze-drying into a liquid nitrogen bath; and (iii) centrifugal freeze-drying (Virtis Bio-Dryer). The drying methods were evaluated for the subsequent handling of fine powders.

Heat-treatments were carried out in an electric furnace using a double crucible method, with 90:10 lead zirconate:lead oxide atmosphere powder [7]. Dried powders were calcined at temperatures varying from 400 to 700°C, and were subsequently analysed by X-ray diffraction.

Particle size and morphology were determined on an ISI DS-130 SEM. Samples were prepared by dispersing small quantities of powder onto a specimen mount containing double sided tape. Compositional information was determined by Energy Dispersive Spectroscopy (EDS). A Micromeretics Sedigraph (5000 E) was also used for the determination of particle size distributions.

RESULTS AND DISCUSSION

The purpose of the investigation was the preparation of fine uniform powders for enhanced sintering activity. While the precipitation process gave good control of particle size distributions while still in suspension, the drying methods tended to agglomerate powders, as is to be expected. Figure 3 compares data for as-precipitated (AP) PLZT 6060, with spray-dried (SD), and centrifugally freeze-dried (CFD) powder of the same batch, after re-suspension using an ultrasonic probe. Each drying method increased the average particle size and range, compared with the as-precipitated condition. The figure illustrates that spray-dried powders were of a finer particle size and narrower size distribution than centrifugally freeze-dried powders.

The morphology of spray-dried powders was also more desirable from a practical point-of-view. Figure 4 illustrates the typical spherical morphology of powders produced by this drying method. The particles ranged in size between 0.10 and 3.0 µm. SEM results were in good agreement with Sedigraph analysis. Examination of fracturegraphs determined the particles to be solid.

Dried powders were determined to be amorphous by X-ray diffraction, and crystallized on heat-treatment to the perovskite phase by 600-700°C. The total weight loss was 16 percent, as the dried hydroxide tranformed to the oxide phase. EDS and SEM analyses verified the synthesized powders were of uniform composition.

Powders that were frozen in suspension and dried by centrifugal freeze drying were found to follow the population balance model. Figure 5 illustrates the results for PLZT 6060, obtained by manipulation of Sedigraph data. Rearrangement of Equation 1 allows for the determination of initial nuclei density (n_i^0) and growth rate (r_i) for each type of species, as shown in the figure. The results are summarized in Table 1.

Powders prepared by rapid freeze-drying in liquid nitrogen were characterized by an open structure of agglomerated platelets. The morphology was inappropriate for manipulation of Sedigraph data (i.e., in terms of equivalent spherical size) for population balance studies. The low bulk

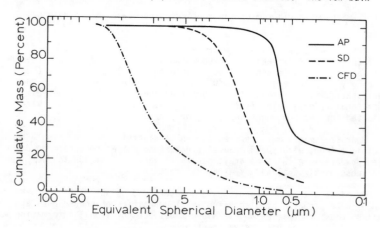

Fig. 3. Sedigraph Analysis for as-precipitated (AP), spray-dried (SD), and centrifugally freeze-dried (CFD) PLZT 6060.

127

Fig. 4. SEM photomicrograph of spray-dried PZT 53:47.

densities of powders produced by this drying method were not well suited for
pressing operations in their present form. Exchange of the solvent used in
precipitation (water) with an organic solvent (as used in spray-drying)
could conceivably result in a more suitable morphology. Differences in
morphology observed for powders dried by various methods indicates that
surface chemistry could have a significant effect in drying as well as
during precipitation.

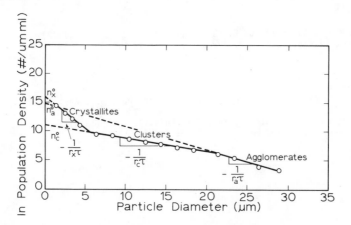

Fig. 5. Population balance analysis for PLZT 6060 (CFD).

Table 1. Typical Initial Nuclei Densities and Growth Rates for PLZT 6060
Powder Calculated from Population Balance Data

	Initial Nuclei Density n_i^0 (#/μm-ml)	Growth Rate r_i (nm/sec)
Crystallites (x)	$5.6 \cdot 10^6$	5.1
Clusters (c)	$8.2 \cdot 10^2$	19.3
Agglomerates (a)	$9.4 \cdot 10^5$	13.3

SUMMARY

Free flowing powders of spherical morphology were produced by aqueous
precipitation followed by spray-drying. The spherical morphology was
atypical for multicomponent systems prepared by aqueous precipitation,
followed by simple drying methods; but is highly desirable from a practical
point of view. The use of a continuous constant volume precipitation system
gave good control of the precipitation variables. Furthermore, results of
the population balance analysis indicated that both clustering and agglomeration mechanisms were applicable for centrifugally freeze-dried powders.
Work in progress indicates that closer control of particle size distribution
is possible for spherical particles produced by this inexpensive aqueous
precipitation method.

ACKNOWLEDGEMENTS

The work was supported by the U. S. Department of Energy, Division of
Materials Sciences, under contract DE-AC02-76ER01198. The use of the Center
for Electron Microscopy at the University of Illinois at Urbana-Champaign is
also gratefully acknowledged. We also acknowledge the technical assistance
of L. M. Falter, W. G. Fahrenholtz, and R. F. Falkner.

REFERENCES

1. G. H. Haertling and C. E. Land, Ferroelectrics 3, 269 (1972).
2. R. L. Plackett and J. P. Burman, Biometrika 33, 305 (1948).
3. D. J. Eichorst, M.S. Thesis, University of Illinois at Urbana-Champaign
   (1986).
4. M. A. Larson and A. D. Randolph, Chem. Eng. Prog. Symp. Ser. 65, 1
   (1969).
5. L. E. Burkhart, R. C. Hoyt, and T. Oolman, Mat. Sci. Res. 13, 23
   (1980).
6. M. Murata and K. Wakino, Mat. Res. Bull. 11, 323 (1976).
7. G. S. Snow, J. Am. Cer. Soc. 56, 91 (1973).

SYNTHESIS AND SINTERING COMPARISON OF CORDIERITE POWDERS

J.C.BERNIER, J.L. REHSPRINGER, S.VILMINOT and P.POIX
Departement Science des Matériaux, U.A.440 du C.N.R.S. E.N.S.C.S. 1 rue
Blaise Pascal 67008 Strasbourg Cedex France.

ABSTRACT

Cordierite and cordierite based glass ceramics are promising materials for electronic packaging. Preparations by organic and inorganic precursors along sol-gel processing are reported. Compared with conventional methods, they show a drastic decrease of the sintering temperatures. A process able to giving better than 95% theoretical density ceramic below 1000°C, with good electronic properties is described.

INTRODUCTION

TABLE I
Physical properties of some dielectric materials

Ceramic	Dielect. constant	tg δ 10^{-4}	Thermal cond. (W.cm-2)	Expansion coeff.(10^{-6} K^{-1})
Al_2O_3	9.5	1	0.3	6
Al_2O_3 96	9	20	0.2	6.4
Cordierite	5	20	0.02-0.1	2.2

Consideration of the preceeding table shows that cordierite is a promising material for electronic packaging.

This material with chemical formula $2MgO.2Al_2O_3.5SiO_2$ is usually obtained by high temperature treatment (T>1300°C) of the mixed oxides.The crystallization scheme reveals the intermediate formation of $MgAl_2O_4$ and SiO_2 which react at higher temperature to give rise to α-cordierite (1).

Crystallization from the glass obtained by quenching the melted oxides mixture (T>1600°C) does not proceed in the same way. The first step corresponds to the formation of μ-cordierite, also called stuffed quartz. In a second step, μ-cordierite transforms to β-cordierite. Whereas μ-cordierite is a low temperature metastable modification of cordierite, the α and β forms are both high temperature stable modifications. α(orthorhombic) differs from β(hexagonal) by the presence of disorder in the MO_4 (M=Si or Al) tetrahedra occupancy.

Sol-gel methods usually allow a decrease of the sintering temperatures by yielding powders with homogeneity at the molecular scale. Such a method has been developed for cordierite based ceramics.

SYNTHESIS OF THE POWDER

For a Mg-containing multicomponent oxide system such as cordierite, the use of only alkoxide precursors is not possible due to the insolubility of magnesium alkoxides. If alkoxides have to be used to form the network of the gel, other soluble metal salts can be used as source for the other components. In this paper, we present the results obtained using two alkoxide (Al and Si) precursors, with Mg being introduced as acetate.

Silicon ethoxide, TEOS, and aluminum s-butoxide are both dissolved in propanol-2. A solution of magnesium acetate in acetic acid is prepared and both solutions are mixed with vigorous stirring giving rise to a stable

130

solution. Hydrolysis and polycondensation are performed using hydrazine
hydrate, $N_2H_4 \cdot H_2O$. The gelation proceeds very rapidly and allows one to
obtain a very rigid gel. Thermolysis up to $250^{\circ}C$ eliminates most of the
solvents, as shown on the T.G.A. curve (Figure 1). An oxidizing washing
using hydrogen peroxide, H_2O_2, is done in order to eliminate the residual
organics. If this washing is not done, small carbon particles appear
during sintering. The powder is then air fired at $700^{\circ}C$ for two hours. The
resulting cordierite precursor is a white powder. The process is shown
schematically in Figure 2 (2).

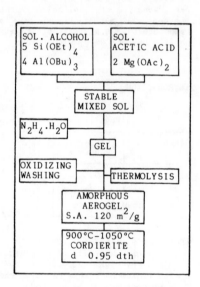

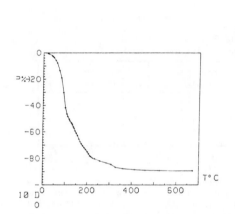

Figure 1: T.G.A. curve of the gel   Figure 2: Flow chart for process.

If anhydrous magnesium acetate is used, the mixing of the alcoholic
(TEOS and $Al(OBu)_3$) and acetic $(Mg(OAc)_2)$ solutions gives a clear
solution. If hydrated magnesium acetate is used, the mixing is immediately
accompanied by gelation due to hydrolysis of aluminum alkoxide. Addition
of hydrazine hydrate dissolves the emulsion and gelation occurs slowly.
It may be considered that hydrazine hydrate plays the role of a
stabilising agent by chelation. Such a result was already observed by
Debsikdar (3) on the formation of a soluble chelate between aluminum
butoxide and acetylacetone and for the stabilization of titanium
isopropoxide by addition of triethanolamine(4).

Substitution of hydrazine hydrate by ammonia as hydrolysis agent has
also been tested. The resulting gel is more transparent and the same
thermal treatment has been applied followed by H_2O_2 washing.

POWDER CHARACTERIZATION

The powders obtained after thermal treatment at $700^{\circ}C$ have been
characterized by granulometry, surface area measurements and S.E.M.
observations.

During the gelation process, many parameters can be varied such as
temperature, dilution, hydrolysis rate, pH, etc. In some cases, a

relationship between the experimental conditions and the powder characteristics is difficult to see.

According to our experiments, the same powder characteristics are observed for samples issued from both routes previously described.

The hydrolysis rate, i.e. the rate at which hydrazine hydrate or ammonia is added into the solution, has a noticeable influence on the grain size of the final powder. Decreasing the hydrolysis rate leads to agglomerates with decreasing mean size (Table I). A decrease of particle size is also observed upon dilution.

B.E.T. measurements show that the powders exhibit specific areas from 200 to 300 m^2xg^{-1} depending on the experimental procedure.

TABLE II
Characteristics of the powder and related ceramics

Parameter		Powder characteristics		Sintering		X-Rays after sint.
		Mean size(μ)	Spec.area	T shr.	d/dth	
Dilution	L	10-30	180-230	800	0.9-0.95	μ and β
	H	0.5-3	220-290	800	0.9-0.98	μ and M
Hydrol. rate	L	0.5-10	240-290	800	0.95	μ and β
	H	5-30	180-230	770	0.9-1.05	μ,S,Q
Temper.	RT	10-30	180-240	800	0.9-0.95	μ and β
	50°C	3-15	200-250	800	0.9-0.98	β,S,Q

L=low, H=high, μ= μcordierite, β= βcordierite, M=mullite, S=MgAl$_2$O$_4$ Q=quartz. d/dth is calculated with dth=2.512g.cm^{-3}(β-cordierite) μ-cordierite has a d value of 2.587g.cm^{-3}.

Scanning electron micrographs of the powder obtained after air firing at 250°C show nearly spherical particles around 1 μm in diameter. These spheres are agglomerates of smaller grains, 0.01-0.03 μm , as shown on Figure 3.

 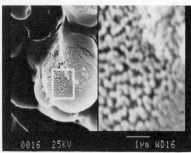

Figure 3: S.E.M. micrographs on gel after air-firing at 250°C.

132

SINTERING RESULTS AND CRYSTALLIZATION SCHEME

Whereas sintering from the mixture of oxides starts at temperatures higher than $1300^{\circ}C$, a strong decrease is observed in the case of sol-gel obtained powders as shown on figure 4.

The final densifications for powders with analogous mean size distribution issued from both preparation routes are in the same order of magnitude.

If the granulometric distribution is centered at higher values (10 to 30 μm), the dilatometric curve (Figure 4-b) reveals that sintering is not completely achieved in the range $800-900^{\circ}C$.

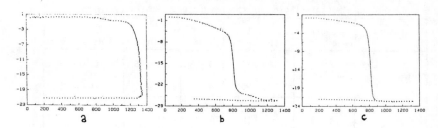

Figure 4: Dilatometric curves on cordierite by a, ceramic route, b.c, sol-gel process.

The final density, according to the preparation conditions can vary from 90 to 98% of the theoretical value.

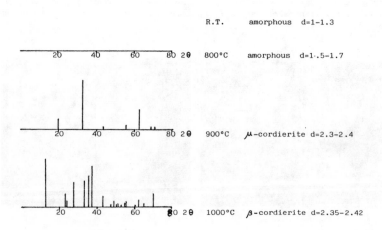

Figure 5: Evolution of X-ray patterns during sintering.

Crystallization has been followed by X-Ray diffraction using a furnace attached on the diffractometer. The amorphous precursor has a green density between 1 and 1.3 $g.cm^{-3}$ (40 to 50% of the final value) depending on the preparation used. Sintering begins in the amorphous state at a temperature near 800°C. μ-cordierite appears in the range 840-900°C, β-cordierite is obtained between 950 and 1050°C. The development of the diffraction patterns during sintering is shown on figure 5.

DIELECTRIC MEASUREMENTS

Dielectric measurements have been performed on pellets sintered at the desired temperature for a frequency of 1 MHz. The measured dielectric constant is around = 4.5 with tg = 0.002. The variation of c/c versus temperature (Figure 6) is below 120 ppm/K.

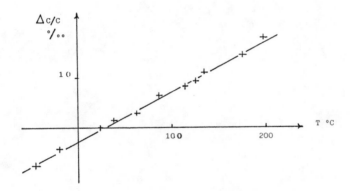

Figure 6: $\Delta C/C = f(T)$ curve for β-cordierite by sol-gel process.

SUMMARY AND DISCUSSION

The methods previously described allow the elaboration of powders which sinter at much lower temperatures than using the classical ceramic way.

Due to the complexity of this system, control of the gelation process is quite complicated. The three elements precipitate at pH values which do not overlap. Both alkoxide precursors have different hydrolysis rates. Aluminum alkoxide is very sensitive to water but TEOS is not. This difference in hydrolysis rates can lead to inhomogeneities during the gelification process, as illustrated by comparison of the results obtained from both hydrolysis agents. The amount of a second phase after sintering is always higher in the case of ammonia than hydrazine hydrate. Complexation of aluminum hydrolysis species by hydrazine hydrate has been proposed to explain the better homogeneity of the gel.

An evolution of the powder characteristics according to the gelation conditions has been emphasized. However, the densification and purity of the material obtained after sintering is not directly related to the morphology of the initial powder. As seen in Table I, smaller grain sizes can lead to increasing proportions of a second phase.

In summary, a process has been developed allowing the obtention of pure β-cordierite at 1000°C with a final densification to 96% of the theoretical value.

ACKNOWLEDGMENT.

To CRICERAM and XERAM of PECHINEY group for their partially support .

REFERENCES

(1)   A.F.   BESSONOV   and   E.V.   BESSONOVA,   Izv.Akad.Nauk   SSSR,
Neorg.Mater.,20(1),(1984),92-96.

(2) J.C. BERNIER, P. POIX, J.L. REHSPRINGER, G. VILMIN French Patent n=85-
10873

(3) J.C. DEBSIKDAR, J.Mater.Sci.,20,(1985),4454-4458.

(4) A. STIEGELSCHNITT, K. WEISSKOPF and J. PAISE Science of Ceramics 12
Ceramurgia 12 (1984).

SYNTHESIS AND PROCESSING OF THE SOL-GEL
DERIVED β-QUARTZ LITHIUM ALUMINUM SILICATES

J. COVINO*, F. G. A. DE LAAT**, AND R. A. WELSBIE**
* Engineering Sciences Division, Research Department, Naval Weapons Center, China Lake, CA 93555-6001
** Litton Guidance and Control Systems, 5500 Canoga Avenue, Woodland Hills, CA 91365

ABSTRACT

Lithium Aluminum Silicate (LAS) glass-ceramic compositions with and without phosphorous have been synthesized by Sol-Gel techniques. Resulting LAS-type powders are herein designated as NZ and NZP. X-Ray analysis, thermogravimetric analysis (TGA), particle size measurements, and thermal dilatometric shrinkage measurements have been performed on these samples. The NZ and NZP powders in calcined form, as well as commercially-available LAS glass-ceramic produce x-ray diffraction pattern very similar to the pattern of Virgilite $Li_xAl_xSi_{3-x}O_6$ (x=0.5-1.0). There is little difference between powders with and without phosphorous in the diffuse reflectance spectra (DRS). Preliminary results show that the material can be easily processed into glass ceramics.

INTRODUCTION

Since the production of Cer-vit ceased, there has been an increasing dependence on Schott's Zerodur and Cornings ULE (type 7971) glass for applications requiring ultra-low thermal expansivity. There are applications where neither ULE or Zerodur are acceptable. For example, ULE has a high helium permeability and cannot be used for ultra-precision measurement equipment[1-2]. Because Zerodur has some small instability on thermal cycling near -23 and -177°C, which are in the operational temperature range of ultraprecision measurement equipment, it cannot be used for all ultra precision measurement equipment[3]. Furthermore, the few remaining commercial β-quartz LAS glass-ceramics are not of a reproducibly acceptable quality, nor are they easily available. Thus, there is a need of new oxide glass ceramics with ultra-low expansion coefficients and low helium permeabilities.

In this paper, Sol-Gel syntheses of β-quartz lithium aluminum silicates, which are similar in composition to the nomical composition of a commercially-available LAS glass-ceramic system, will be discussed. Details of some material properties and processing conditions will be presented. Among some of the characterization tools which have been employed are X-ray spectroscopy, TGA, DRS, scanning electron microscopy (SEM), particle size analyzer, and dilatometer thermal analyses.

EXPERIMENTAL

The Sol-Gel synthetic technique has been used to synthesize nitric acid pH-adjusted LAS-like materials, designated as NZP and NZ, with phosphorous pentoxide and without, respectively. The compositions for NZ and NZP were chosen to be close to the nominal composition of a commercially-available LAS glass-ceramic system. Table 1 summarizes these compositions.

The Sol-Gel technique involves the formation of sol via a hydrolysis reaction of the metal alkoxides at relatively low temperatures (30-200°C). In the sol, a partially polymerized material is formed. This is followed by a slow gelation. The porous gel thus obtained is subsequently dried and heated to

136

form a monolithic oxide glass[4-5]. Figure 1 illustrates the step by step process from solution to processed powder used in this work. Sol-Gel produced NZ and NZP powders were calcined at 600 and 800°C in order to study material properties at the elevated temperatures.

Table 1. Compositions of Nominal LAS Glass-Ceramic as Compared to Synthetic Sol-Gel-Derived NZ and NZP.

Constituents	Nominal LAS	NZ	NZP
Silicon dioxide (SiO_2)	55.50	61.35	56.10
Aluminum oxide (Al_2O_3)	25.30	27.97	25.60
Lithium oxide (Li_2O)	3.70	4.08	3.70
Titanium dioxide (TiO_2)	2.30	2.30	2.30
Magnesium oxide (MgO)	1.00	1.00	1.00
Zirconium dioxide (ZrO_2)	1.90	1.90	1.90
Zinc oxide (ZnO)	1.40	1.40	1.40
Phosphorous pentoxide (P_2O_5)	7.90	----	8.00
Miscellaneous oxides (arsenic, iron, potassium, calcium, and sodium)	1.01	----	----

FIG. 1. Composite Photograph of Sol-Gel Process Stages for the Manufacture of Glass-ceramics. (a) Upper left: The SOL, i.e., the mixture of precursors, solvents, and nitric acid prior to hydrolysis. (b) upper center: The GEL in wet bulk form right after completion of hydrolysis. (c) Upper right: The wet GEL broken up after initial air-drying in the fumehood. (d) Lower left: Dehydrated GEL in granulated form after exposure to the IR heat lamp. (e) Lower left center: First POWDER stage after dehydrated GEL has been ballmilled. (f) Lower right center: Brown-colored POWDER after first calcining stage (up to 400°C). (g) Lower right: White, fine POWDER after final calcining stage (up to 600°C).

Sample Characterization

X-Ray Analysis. Powder data was collected using a Phillips diffractometer with a θ-compensating slit and diffracted beam monochromator scintillator (CuKα = 1.5418Å, kα = 1.5405Å, Kα$_2$ = 1.5444Å) with pulse height discrimination and copper sources.

Thermogravimetric Analysis (TGA). This was performed on a DuPont thermogravimetric analyzer using a quartz bucket. The heating rate used was 2°C/min to a maximum temperature of 1000°C in air. The balance accuracy was ±0.01 mg and sample sizes ranged from 20 to 30 mg.

Diffuse Reflectance Analysis (DRA). A Nicolet 7199 Fourier Transform Infrared spectrometer with a Harrick diffuse reflectance cell was used for diffuse reflectance measurements.

Particle Size Determination. Three methods were used for particle size determination. The first method employed was SEM in order to measure particle size distribution. Scanning electron micrographs were taken on an Amray 1400 with 40Å lateral resolution. The sieving of particles was a second method. Particles between 50-260 microns were separated and weight to determine percentage of particles in the 50-260 micron range. A centrifuging technique was used to measure particles less than 50 microns in size. In this analysis, a Horiba Capa-500 particle analyzer was used. The method works on a centrifugal technique using a powder/water sample. The change in optical absorption is measured as the sample is centrifuged.

Dilatometric Thermal Analyses. Dilatometric measurements were performed on NZ and NZP powders in order to study ease of processing and processing parameters. A dual push-rod quartz dilatometer was used on powder samples. A load of 20 grams was applied from room temperature to 1200°C. In this measurement, the powders are first pressed to about 40-50% of theoretical density to form a pellet. This pellet is then placed in a dual push-rod quartz dilatometer in which sample shrinkage is measured under a fixed 20-gram load as a function of temperature.

Solidification and Transformation Experiments. A variety of casting experiments on NZP and NZ powders have been performed. Several cast samples are illustrated in Figure 2.

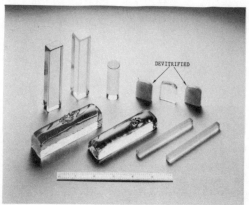

FIG. 2. A Variety of cast and Devitrified NZ Glass Samples.

After numerous experiments with NZ material, a devitrification temperature schedule has been established. The best cycle appears to be: heating at a rate slightly below 3°C/min to the nucleation temperature of 735°C, and holding it there for several hours; again, heating at a rate slightly above 2°C/min to the crystallization temperature of 800°C, then holding it there about 16 times longer than at the nucleation temperature. NZ materials produced by this cycle are designated as NZ-AH. The discoloration of the devitrified samples is probably due to some residual carbon still remaining in the glass ceramic.

RESULTS AND DISCUSSION

Thermogravimetric analyses of NZ and NZP powders were performed. Figure 3 shows the percent weight loss vs. temperature curves for both NZ and NZP from room temperature at 1000°C. As can be seen in the figure, both NZ and NZP lose most of their weight by 300°C. This weight loss is primarily due to the removal of physically bound water and some residual organic matter. By 800°C, both NZ and NZP have lost all the chemically bonded water.

138

Figure 4 represents typical X-ray powder patterns for both NZ and NZP before and after calcining at 800°C for 6 days. Samples before calcining are poorly crystalline while samples after calcining have a powder pattern which can be indexed as the virgilite structure. This structure is a stuffed disordered β-quartz structure. The virgilite structure ($Li_xAl_xSi_{3-x}O_6$ where x=0.5-1.0) has a hexagonal unit cell with a=5.132(1)Å, c=5.454(1)Å (6). Currently, the commercially-available β-quartz LAS glass-ceramic material has the identical virgilite powder pattern as the NZ and NZP powders. X-Ray powder patterns of the TGA residue for both NZ and NZP could also be indexed as the virgilite structure.

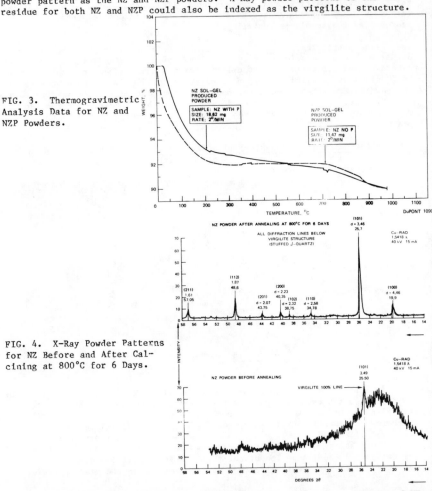

FIG. 3. Thermogravimetric Analysis Data for NZ and NZP Powders.

FIG. 4. X-Ray Powder Patterns for NZ Before and After Calcining at 800°C for 6 Days.

Diffuse reflectance spectra in the infrared were taken on a variety of NZ and NZP powders. The objective was to determine if added phosphorous made any differences to the Sol-Gel derived NZ and NZP samples at room temperature, 600, and 800°C. Figures 5 and 6 summarize the temperature dependence diffuse reflectance spectra of NZ and NZP, respectively. The NZP sample shows an increased retention for H_2O over NZ. This was determined by comparing the relative decrease of the strong broad band from ≈2800-3800 cm^{-1} $\nu(OH)$ in NZ and NZP. Also both samples lose water as a function of annealing temperature.

In all spectra, the strong broad band at ≈1000 cm^{-1} is the Si-O stretch. The Al-O stretch is seen at ≈700 cm^{-1} and the band at 400-500 cm^{-1} is the Si-O-Si bend, skeletal vibrations, and/or other M-O stretches.

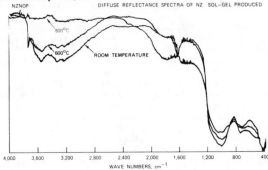

FIG. 5. Temperature Dependence Diffuse Reflectance Spectra of NZ.

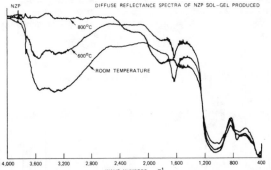

FIG. 6. Temperature Dependence Diffuse Reflectance Spectra of NZP.

In the region between 1400 cm^{-1}-400 cm^{-1}, there are also some minor differences between NZ and NZP. The Si-O stretch shifts slightly to higher frequency at about 1230 cm^{-1}. In phosphate zeolites, the Si-O vibration has been shown to shift to higher frequency by the introduction of phosphorous[7]. Some of these low frequency differences between the NZ and NZP spectra may be artifacts subtraction when looking at spectra at the edge of strong bands.

Other important differences are seen in the water and OH regions of the spectra. The sharp band at 3744 cm^{-1} is υ(OH) from free Si-OH, the strong broad band from 2800-3800 cm^{-1} is due to υ(OH), associated OH, and the 1630 cm^{-1} is the bending vibration of H_2O. The 1630 cm^{-1} band indicates H_2O molecules or water of crystallization. There is little, if any, evidence for Al-OH except the weak band at 3710 cm^{-1}, and the free Al-OH band at 3795 cm^{-1} is not observed. There is also no evidence for P-OH at 3655 cm^{-1}, as observed in SiO_2[8]. The three bands on the broad associated OH band are difficult to interpret except by comparison with zeolites. The band at ≈3560 cm^{-1} may correspond with the 3540 cm^{-1} band in zeolites for lattice OH, and the band at ≈3240 cm^{-1} may correspond with the 3200 cm^{-1} zeolite band due to H_2O in the channels. There is no evidence for water coordinated to metals due to the lack of sharp bands in the 650-800 cm^{-1} region.

At 600°C no structural differences are apparent, but associated H_2O has decreased relative to free OH and both NZ and NZP have lost H_2O. The 1630 cm^{-1} band remains indicating that H_2O is not just physically absorbed. At 800°C, again little structural differences between NZ and NZP are seen, but both show increased band structure on the Si-O and Al-O vibrations. The

υ(Si-O) show bands at 1070, 990, and 920 cm^{-1}, while the υ(Al-O) region shows a shift to higher frequency and a relatively sharp band appearing at 750 cm^{-1}. The 750 cm^{-1} band may correspond with the 758 cm^{-1} band in alumino-silicates due to replacement of silicon by aluminum in the silica tetrahedra[9].

Particle size determination on NZ and NZP powders were performed employing three different methods. The particle size distribution from 50-260 microns particles was determined by sieving, while for 50 microns or smaller the particle size distribution was determined utilizing the Horiba Capa-500 particle analyzer. About 50% (cumulative mass percentage) of the particles were larger than 50 microns for the NZP powder and 72% of the particles for the NZ powder. The distribution tables and graphs obtained with the Capa-500 analyzer are given in Figures 7 and 8 for NZ and NZP, respectively. The average particle size in the 0-50 micron range for the NZ powder was 14.17 microns (Figure 7) and for the NZP powder 13.46 microns (Figure 8). Both figures also show the particle frequencies (in percentages) of a given size range from 2.5-50.0 microns in intervals of 2.5 microns. The scanning electron micrographs (Figures 9 and 10) for NZ and NZP powders, respectively, confirm that the samples have a variety of particle sizes and shapes. These powders are too coarse for subsequent processing such as ceramic injection molding (CIM) or hot isostatic pressing (HIP). It is expected that some preprocessing steps are necessary.

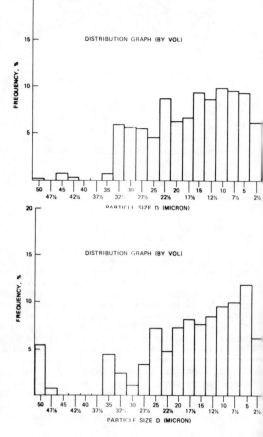

FIG. 7. Particle Size Distribution by Volume Obtained with the Capa-500 Analyzer From -300 Mesh (50 Microns) and Smaller of the NZ Sol-Gel-Produced Powder.

DISTRIBUTION TABLE (BY VOL)

D (MICRON)	F (%)	R (%)
50.0 – 47½	0.2	0.2
47½ – 45.0	0.0	0.2
45.0 – 42½	1.5	1.7
42½ – 40.0	0.6	2.3
40.0 – 37½	0.0	2.3
37½ – 35.0	0.0	2.3
35.0 – 32½	1.7	4.0
32½ – 30.0	5.9	9.9
30.0 – 27½	5.6	15.5
27½ – 25.0	5.5	21.0
25.0 – 22½	4.5	25.5
22½ – 20.0	8.7	34.2
20.0 – 17½	6.2	40.4
17½ – 15.0	6.6	47.0
15.0 – 12½	9.1	56.1
12½ – 10.0	8.8	64.9
10.0 – 7½	9.9	74.8
7½ – 5.0	9.6	84.4
5.0 – 2½	9.4	9.38
2½ – 0.0	6.2	100.0

$D_{average}$ = 14.17 MICRON

FIG. 8. Particle Size Distribution by Volume Obtained with the Capa-500 Analyzer From -300 Mesh (50 Microns) and Smaller of the NZP Sol-Gel-Produced Powder.

DISTRIBUTION TABLE (BY VOL)

D (MICRON)	F (%)	R (%)
50.0 – 47½	5.4	5.4
47½ – 45.0	0.8	6.1
45.0 – 42½	0.0	6.1
42½ – 40.0	0.0	6.1
40.0 – 37½	0.0	6.1
35.0 – 32½	4.5	10.6
32½ – 30.0	2.4	13.0
30.0 – 27½	1.1	14.1
27½ – 25.0	3.3	17.4
25.0 – 22½	7.3	24.7
22½ – 20.0	4.8	29.5
20.0 – 17½	7.3	36.8
17½ – 15.0	8.4	45.2
15.0 – 12½	7.8	53.0
12½ – 10.0	8.7	61.7
10.0 – 7½	9.7	71.4
7½ – 5.0	10.0	81.4
5.0 – 2½	12.2	93.6
2½ – 0.0	6.4	100.0

$D_{average}$ = 13.46 MICRON

In order to investigate the processing conditions for NZ and NZP powders, dilatometric thermal analyses were performed. Figure 11 depicts the dilatometric thermal analyses of both the NZ and NZP powders, as obtained with a dual push-rod quartz dilatometer to 1200°C. In the plot of $\Delta L/L_o$ vs. temperature, three distinct regions can be seen. From 0–700°C both NZ and NZP display a small linear shrinkage, probably due to water loss. Between 750–900°C a sharp shrinkage is taking place in both samples. As confirmed by X-ray and TGA results, this is the region in which crystallization takes place. In the NZP sample, the region between 800–900°C shows as a change in slope. This is probably an artifact in the data, and therefore, has no real significance. In the region between 900–1200°C both samples have stopped shrinking. The total shrinkage is 19.4% for the NZ and 23.7% for the NZP powder. The shrinkage along with a well-defined region (700–900°C) indicates that both powder formulations can be densified rather easily and should not create insurmountable problems during sintering.

(a) Magnification 20 KV x 90.   (b) Magnification 30 KV x 1200.
FIG. 9.   Electron Micrographs of the NZ Sol-Gel Produced Powder.

(a) Magnification 30 KV x 220.   (b) Magnification 30 KV x 1400.
FIG. 10.   Electron Micrographs of the NZP Sol-Gel Produced Powder.

FIG. 11.   Dilatometric Thermal Analysis of NZ and NZP Sol-Gel-Produced Powder.
L_o = 21.7 mm.
Load: 20 g.

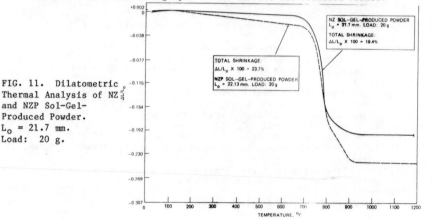

Preliminary casting experiments have been performed on NZ and NZP powders. Results illustrate that both NZ and NZP can be cast quite readily to produce a β-quartz lithium aluminum-silicate glass ceramic. Results of casting experiments show that the most promising candidate was NZ-AH (produced by heating NZ at 2.93°C/min to 735°C, hold for 6 hours; heating at 2.17°C/min to 800°C, hold for 96 hours; cool at 8°C/min). This glass ceramic material exhibited an 82% crystallinity with crystallite size averaging 400Å. The NZ-AH material had a predominantly Virgilite (or β-eucryptite since the two x-ray powder patterns are very similar) and a 3-5% Alpha-Spodumene impurity[10].

CONCLUSIONS

Sol-Gel synthesis of β-quartz lithium aluminum silicates have been performed and the material properties have been determined. Both NZ and NZP samples crystallize with the virgilite structure which is a stuffed disordered β-quartz structure. It should be noted that most commercially-available β-quartz LAS glass-ceramics also crystallize as the virgilite structure. The Sol-Gel process produces NZ and NZP powders with varying particles sizes. Although, these particles are not of ideal morphology for processing, preliminary data suggests that the materials can be processed without too much difficulty.

In this work, the role of phosphorous in lithium aluminum silicates has been investigated. Only minor differences are seen between NZ and NZP powders in the DRS. These differences become even less pronounced as the samples are annealed at elevated temperatures.

Solidification experiments have been successful. NZ material can be crystallized to 82% crystallinity having crystallite size averaging 400Å. These results are in agreement with the currently most popular, commercially-available β-quartz LAS glass-ceramic material.

ACKNOWLEDGEMENT

The authors would like to thank Melvin P. Nadler (Naval Weapons Center and Gary Messing (Pennsylvania State University) for their contributions. The authors would also like to thank the Naval Air Systems Command for the support of this research.

REFERENCES

1. S. F. Jacobs, S. C. Johnston, G. A. Hansen, Appl. Optics 23 (17), 3014-3017, (1984).
2. J. E. Shelby, J. Amer. Ceram. Soc. 55 (4) 195-197 (1972).
3. J. J. Shaffer, H. E. Bennett, Appl. Optics 23 (17), 2852-2853 (1984).
4. D. Levy, R. Reisfeld and D. Avnir, Chem. Phys. Letters 109 (6), 593 (1984).
5. Glasses and Glass Ceramics From Gels, Proceedings of the International Workshop on Glasses and Glass Ceramics From Gels, ed., V. Gottairdi (North Holland Publishing Company, Amsterdam, 1982).
6. B. French, Am. Mineral. 63, 461 (1978).
7. E. H. Glanigen, R. W. Grace, Adv. Chem. Ser. 101 76 (1971).
8. L. H. Little, Infrared Spectra of Absorbed Species (Academic Press, 1966).
9. A. V. Kiselev, V. I. Lygin, Infrared Spectra of Surface Compounds (John Wiley and Sons, Inc., New York, 1975).
10. Deere, et. al., Rock Forming Minerals, Vol. 2A, p. 530 (1978) (C. Roob, JCPDS grant/in/aid, 1980).

PART III

Characterization of
Chemically Derived Ceramics

PROLONGED CHAOTIC OSCILLATIONS DURING THE GEL/XEROGEL
TRANSITION IN SILICON TETRAMETHOXIDE POLYMERIZATION
AS DETECTED BY PYRENE EXCIMERIZATION.

VERED R. KAUFMAN AND DAVID AVNIR
Department of Organic Chemistry, The Hebrew University of Jerusalem,
Jerusalem 91904, Israel.

ABSTRACT

When the polymerization of $Si(OCH_3)_4$ is carried out in the presence
of surface active agents, prolonged oscillations (over 1000 hrs) at the
gel/xerogel transition are observed. The oscillations are of large
amplitude, they are slow (several hrs/period), and they exhibit a chaotic
behaviour. The probe by which these oscillations are observed is emission
from excited state monomeric and excimeric pyrene. It is suggested
tentatively that the driving forces for this oscillation are the structural
relaxation of the secondary polymeric gel structure and the dispersion of
adsorbed pyrene to thermodynamically favored adsorption sites. Relevant
models could be those of oscillatory sol/gel phase transitions and of
oscillatory polymerization reactions. We are unaware of previous
observations of oscillations in sol/gel systems.

1. BACKGROUND

1.1 Oscillating Chemical Reactions.

(It is impossible to review here this very wide field for the reader
who is unfamiliar with chemical oscillations. The intention of this
section is to provide the interested reader with some key ideas, necessary
vocabulary and suggested reading).

Oscillating chemical reactions are one of the most fascinating
phenomena in chemistry. Contrary to the monotonous progress of most
chemical reactions, one rarely encounters the unusual situation in which
one or most of the reactive chemical components are consumed and formed in
periodic cycles. Much theoretical and experimental effort has been
invested in the oscillating chemical reactions [1-6], especially in the
Belousov-Zhabotinskii reaction and its derivatives [2,5]. These reactions
fall into the more general frame of scientific interest in the evolution
of spatial and temporal order. Several general approaches have been
developed for the description and analysis of such phenomena. The main
ones: a) Haken's synergetics approach [7]; b) Prigogine's irreversible
thermodynamics and concept of dissipative structures [8]; c) the detailed
analysis of networks of rate equations, coupled to each other and in many
cases to mass transport terms. The latter approach is the one which so
far has been used most widely. A number of requirements and characteristic
features of the oscillating reactions have emerged from these studies:

(a) Oscillations occur only in a non-equilibrium situation.

(b) The rate equations are non-linear. This non-linearity can originate,
    for instance, in catalytic or inhibitory feed-back loops.

(c) The system must be open. This condition is found in CSTR (continuous
    stirred tank reactor), which is open in the full sense of the word.
    However, "open" need not mean feeding from an external source;

quasi-open conditions are sufficient, and these are found in situations where the solution of reactants has a large enough supply of starting materials and has the capacity to absorb or to dispose products.

(d) For many years it was believed that it is impossible to obtain oscillating reactions in homogeneous solutions. This is no longer the case - it has been shown conclusively that heterogeneity is not necessary.

(e) An important recent development in this field has been the recognition that oscillations need not show up in a regularly pattern [9]. Increasingly more experimental cases are reported in which a chaotic pattern of oscillations is observed [10]. Whereas initially such chaotic behaviour was attributed to experimental fluctuations, many theoretical studies indeed indicate that networks of non-linear rate equations may result in chaotic periodicity [11]. An interesting observation made is that after sufficiently long time, ordered patterns emerge out of the chaotic system [12]. Our preliminary indications suggest that the oscillator discovered in our laboratory is chaotic.

1.2 Pyrene Excimerization During the Sol/Gel Transition in SiO_2.

The oscillations were observed during the follow up of structural changes that occur in silica gel formed by the polymerization:

$$nSi(OCH_3)_4 \xrightarrow[\text{H}^+ \text{ or } OH^-]{H_2O} (SiO_2)_n + 4nCH_3OH \qquad (1)$$

The analytical probe used in that study has been the emission from excited state pyrene (Py) or more specifically the changes in the E/(E+M) ratio of intensities of fluorescence of Py* (Step 4, M) and of the excimer (PyPy)* (Step 5, E)

$$Py \xrightarrow{h\nu_1} Py* \qquad (2)$$

$$Py* + Py \longrightarrow (PyPy)* \qquad (3)$$

$$Py* \longrightarrow Py + h\nu_2 \qquad (M) \quad (4)$$

$$(PyPy)* \longrightarrow Py + Py + h\nu_3 \qquad (E) \quad (5)$$

Fig. 1 shows typical Py monomer and excimer emissions.

The use of Py photophysics for the study of the monomer-polymer-sol-gel-xerogel transitions in the polymerization (1), has been described in detail elsewhere [13-15], and only the main points will be repeated here.

Large variations in the E/(E+M) ratio were observed along the polymerization-desiccation stages. The variations were in many instances between the almost extreme possible ratio values of ∿0.8 and zero. One common pattern found is a gradual increase in excimer intensity to a maximum value, followed by decrease to zero excimer (left side of Fig. 3). The time scale involved is ∿200 hrs, i.e., without any correlation to the time scale of the photochemical events.

The interpretation of these changes [13-15] has been linked to the knowledge accumulated on the behaviour of Py in heterogeneous environments on the one hand [16-18], and on the sol/gel transition in metal alkoxides

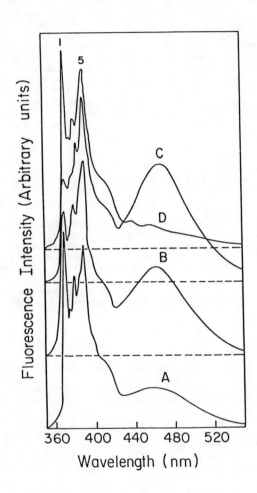

Fig. 1. Changes in relative emission intensities of monomeric
(360-400 nm) and excimeric (420-520 nm) pyrene at various
stages of the sol/gel/xerogel transition in Si(OCH₃)₄
polymerization. A: Shortly after preparation of the reactive solution. B and C: growth of the excimer emission
intensity (∿100 and 180 hrs). D: Final situation in the
absence of a surface active agent (∿250 hrs). Isolated
pyrene molecules are trapped in the silica xerogel.

on the other hand [19,20]. The main parameter probed by Py in the silica
system is the evolution of surface irregularity and the continuous decrease
in average pore size, a conclusion reached by studying the effects of
surface fractal dimension [21,22] and pore size of various SiO_2 materials on
Py excimerization [18].   It was found that these structural changes
continue much beyond gelation point. Furthermore, the geometrical parameter probed by Py is of molecular scales only, and on such scale lengths,
the gelation point emerged as a point of no special characteristics.
Most of the structural changes occur during the post-gelation transition
to xerogel, and it is during this time that we observe the oscillations
reported below. Changes in the course of the reaction, exerted by changing
the pH, the water/silane ratio or the substituents on the silicon, were
detected by this [13-15] and by other [20,23] fluorescent probes.

2. EXPERIMENTAL DETAILS

Chemicals: Pyrene (Aldrich, 99%) was purified by crystallization
from ethanol. Tetramethyl orthosilicate was purchased from Aldrich.
Methanol was spectroscopic grade (Merck). Water was triple distilled.

Samples preparation: A typical starting solution was composed of
46.4 ml methanol, 36.2 ml TMOS , 17.4 ml water, 0.022 gr ($1 \cdot 10^{-3}$ M) Py
and $3.5 \cdot 10^{-5}$ M Span 80, (sorbitol mono-oleate, a non-ionic surface active
agent). From this solution were taken 3 ml samples and placed in small
bottles covered with aluminium foil, perforated with one (~ 0.5 mm) hole.

Observation of Oscillations: Fluorescence measurements were carried
out on an LS-5 Perkin-Elmer Luminescence Spectrometer in front-face
geometry.

3. THE PHENOMENON

Under usual circumstances, the gel/xerogel transition proceeds
monotonously until a final equilibrium xerogel structure is formed. The
Py probe shows that this stage is reached after several hundreds of hours:
the excimer emission drops to a minimum value (close to zero), and stays
like that from then on. However, when a small amount of a surface active
agent (SAA) is added to the starting mixture, a completely different
pattern of behaviour is observed: After ~ 150 hrs, during which the usual
pattern of behaviour is observed, chaotic oscillations with a period of
several hrs appear and persist for >1000 hrs, i.e., the system does not
reach equilibrium for many weeks ! (Figs. 2, 3). We are unaware of
previous reports on oscillations in sol/gel processes, although oscillations
and chaotic kinetic profiles, have been predicted theoretically in such
systems (see below).

The following observations were made:

(a) The Surface Active Agent (SAA).

Various SAA's can be used with roughly the same effects. These include long chain esters and ethers like Triton X-100, Tween 60 and 80,
Span 80, Brij 96 and Myrj 45, as well as quaternary ammonium salts like
cetyltrimethylammonium bromide. Very little SAA is needed in order to
observe the oscillation. The oscillations in Figs. 2, 3 are obtained
with as little as $4 \cdot 10^{-6}$ M Triton X-100. On the other hand, too much
SAA ($>1 \cdot 10^{-3}$ M) quenches the oscillation. It should be noted that the
optimal concentration range (10^{-5} - 10^{-4} M) is much lower than the

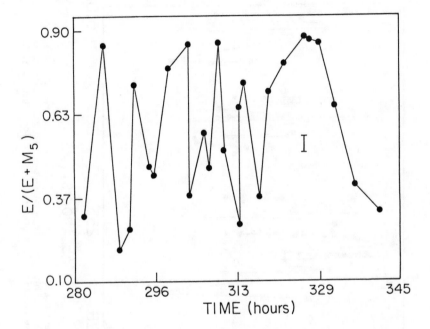

Fig. 2. A section of the oscillatory behaviour (280-340 hrs) under the conditions described in Section 2. E is the emission intensity of the excimer at its maximum. M_5 is the emission intensity of peak 5 of the monomer (see Fig 1). Oscillations start after ~150 hrs as shown in Fig 3. Notice that the amplitude covers most of its maximal possible range (0 - 1.0). The vertical bar in this and in the next Figure indicate the maximal fluctuation in the experimental points.

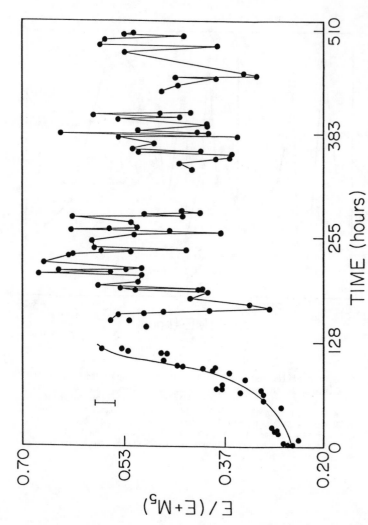

Fig. 3. The superimposed results from six samples prepared at 4 hrs interval from each other. Sinchronicity between the samples is evident at the early stages of the reaction. See previous caption for other details.

concentration of the fluorescent probe (10^{-3} M, see Discussion). Preliminary results with another fluorescent probe, 1,3-di(1-pyrenyl)propane [17], reveal oscillations without the presence of an SAA.

(b) Fluctuations in Measurements.

The amplitude of the oscillations is much larger than the fluctuations of measurements. The vertical bar in Figs. 2, 3 indicates the maximal size of the fluctuations.

(c) Continuous Measurements.

The average period is slow. Consequently we could not perform continuous measurements and so the line which connects the points (Figs. 2, 3) serves only the graphic clarity of the presentation of data. However, single oscillations were measured in a semi-continuous way (one measurement every ten minutes) to ascertain the authenticity of the oscillations. One such oscillation is shown in Fig. 4. Prolonged continuous fluorescent measurements are in preparation.

(d) Synchronicity Within the Same Glass Sample.

The xerogel is not always a monolithic block. It is brittle and can be easily fractured into several large grains. These grains were used to determine the synchronicity of oscillations within one sample. Five grains, ∿3 mm in diameter, were taken for this experiment, and the results are collected in Table 1. It is seen that (a) the oscillations are synchronous (in the solid state!); and (b) the system drifts slightly out of synchronicity with time.

(e) Synchronicity Between Different Samples.

The following experiment was carried out: six samples were prepared at intervals of 4 hrs between one sample preparation and the next, covering a cycle of 24 hrs. The idea was to test for inter sample synchronicity, and if such exists, then measurement of the six samples at a given time, will give, in fact, information for the whole 24 hrs cycle that preceeded the time of measurement. The results are shown in Fig. 3. It is clear that synchronicity exists through the non-oscillatory part at the early stage of the gel/xerogel transition, but no clear cut conclusion can be reached beyond ∿150 hrs.

(f) Excitation Spectra.

Excitation spectra taken at the excimer emission maximum (λ_{ex}=450 nm) overlap with the excitation spectra taken at the monomeric region (λ_{ex}=360 nm) throughout the pre-oscillatory and oscillatory modes. We recall that spectral shifts between these two excitations is an indication of the existence of ground state Py aggregates which cannot disperse to a more favorable thermodynamic distribution [16-18]. Such a situation is observed at the final xerogel stage in the absence of an SAA [13-15]. On the other hand, the overlapping excitation spectra in the oscillating case indicate that free diffusion is at least partially allowed. For typical excitation spectra see ref. 13-15.

152

Table 1.   Synchronicity of the Oscillations

Time (hrs)	Standard deviation of intensity readings between five grains (%)
336	1.0
360	4.0
384	3.0
408	3.5
432	5.7
456	6.3
504	7.0

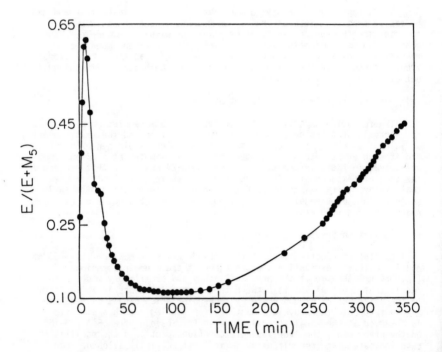

Fig. 4.   Semi-continuous recording of one of the cycles. See Fig. 2 for other details.

DISCUSSION

The key ingredient which allows observation of the oscillations is the SAA. However, it should be noted that the chemical structure of the SAA is a secondary parameter, and that only minute quantities are needed: oscillations are observed even at Py:SAA ratio of 1000:1. It seems therefore that the SAA does not participate as a chemically reactive component in the network of processes and reactions leading to the oscillations, but rather as an agent which alters the physical behaviour of the system. A well known role of emulsifiers in polymerization processes is to homogenize the system [20,24]. Thus, one cannot exclude the possibility that oscillations occur in the absence of an SAA on locally microscopic scales but that these oscillations cannot be observed macroscopically because they are not synchronous. In such a picture, the major role of the SAA is in homogenizing the system, thus allowing the synchronicity. We find the synchronicity throughout the solid, one of the remarkable features of this oscillator. Some preliminary support for these ideas comes from the fact that when 1,3-di(1-pyrenyl)propane is used as a fluorescent probe, chaotic oscillations are observed even in the absence of an SAA [23]; apparently, the fluorescent probe itself acts as an emulsifier.

The question is then, what oscillates ? The small quantity of SAA seems to indicate that the oscillations are linked to gel features which are far beyond molecular scales so that on molar basis there is sufficient SAA to affect these larger features. One possibility is [25] that we are observing oscillations in the secondary gel structure which is composed of supra-molecular, perhaps non-porous, building blocks. The fluorescent probe, as we have learned from previous studies [13-15], detects crowdedness of the environment and is capable of diffusing slowly from crowded to more open pores. At this early stage we can only tentatively speculate that the oscillations originate from the interplay between the continuous diffusion of the probe to the thermodynamically more stable situation of dispersed Py molecules (large pores; minimal excimer emission [18]) and the rearrangement of the secondary structure, a process which carries the Py molecules back into narrow crowded pores (high excimer emission). The driving force for the secondary structure rearrangement could be the energy released by the relaxation of the non-equilibrium structure "frozen-in" at the gel point, as suggested recently by Brinker et al. [26].

A relevant question here is whether the formation of the Si-O-Si bond is reversible. Basically the answer must be negative: the tetrahedral siloxane is stable to the neutral hydrolysis (or methanolysis) conditions of the residual water molecules. We also believe that the amount of angle-distorted metastable Si-O-Si bonds which can re-open, is not large. However, reversibility in this system is possible if one takes into consideration bonds with solvent molecules and interparticulate hydrogen bonding [27,28]. It is possible for residual water molecules to facilitate bond formation and bond re-opening as follows:

$$\equiv Si\text{-}OH \cdots \underset{H}{O}\text{-}H \cdots \underset{H}{O}\text{-}Si \equiv \; \rightleftharpoons \; 2 \equiv Si\text{-}OH + H_2O \qquad (6)$$

and even in the absence of water:

$$\equiv Si\text{-}OH \cdots \underset{H}{O}\text{-}Si \equiv \; \rightleftharpoons \; 2\equiv Si\text{-}OH \qquad (7)$$

It seems to us therefore, that if these reversible bond formations are

154

taken into account, then theories of reversible sol/gel phase transitions [27,28] might be applicable. It is therefore very interesting to notice that Klonowski has predicted the possibility of oscillations in reversible cross linked sol/gel systems [25,29]. Pisman and Kuchanov have studied theoretically kinetic instabilities which arise during gelation in branched condensation and chain polymerization [30]. The applicability of these ideas to our system, is under investigation.

Next we refer briefly to the chaotic nature of the oscillations. The problematics in overcoming the barrier of skepticism that chaotic behaviour is nothing more than experimental fluctuations, has been nicely described by Epstein et al. [5]. However, as mentioned in Section 1.1, it became well established in recent years [4,5,9,10] from theory, computations and experiments in oscillating reactions, that deterministic chaos is an authentic phenomenon, i.e., that what looks like a random process is in fact a complex temporal structure described mathematically by, e.g., strange attractors. The non-continuous way of data recording in the present study, precludes at the moment any possibility to model the apparently chaotic periodicity. We mention again that measurement fluctuations are much smaller than the amplitude changes (Figs. 2, 3) and that the amplitude covers most of the possible range (0 - 1.0).

Another possible approach for analysing the oscillator we found is from the point of view of oscillating polymerization reactions, since basically, the synthesis of the silica gel is a polymerization reaction. Studies in this field concentrated mainly on CSTR conditions, with special attention in several cases to polymerizations in the presence of an emulsifier [31]. Most of the work on oscillating polymerization is theoretical or tries to model the few experimental observations made [32,33]. In these experimental observations, an oscillatory product output from polymerization reactors were recorded [31,34,35]. The number of cycles in these oscillations is usually quite limited (e.g., 2-4 [35]). An interesting effect in such reactions has been proposed by Knorr et al. [36]: these authors demonstrated mathematically that viscosity effects are sufficient to induce multiple steady states in an isothermal CSTR. This so called "gel effect" [32] is a feed-back loop in which increase in viscosity, decreases the termination rate in (radical) polymerizations. Again, the applicability of these polymerization studies to our oscillator is under study.

Finally, we would like to mention that there is also an on-going activity in our laboratory in the evolution of spatial dissipative structure in chemically reactive liquid interfaces [37].

Acknowledgments: Helpful discussions with W. Klonowski, R.M. Noyes and M.L. Kagan are gratefully acknowledged. Klonowski drew our attention to the possible role of the secondary gel structure in the oscillations. This study is supported by the NCRD, Israel and the KFA, Julich, West Germany, by the Volkswagen Foundation (under "Synergetics") and by the F. Haber Research Center for Molecular Dynamics, Jerusalem.

Comments from colleagues on this report are most welcome.

REFERENCES

1. S. Kai and S.C. Müller, Sci. Form, 1, 9 (1985).
2. R.M. Noyes and R.J. Field, Ann. Rev. Phys. Chem., 25, 95 (1974).
3. R.M. Noyes, Ber. Bunesges. Phys. Chem. 84, 295 (1980), and articles in that issue.
4. L.M. Pismen, Chem. Eng. Sci., 35, 1950 (1980).
5. I.R. Epstein, K. Kustin, P. De Kepper and M. Orban, Sci. Am. 248, 96 (March 1983).
6. G. Nicolis and J. Portnow, Chem. Rev. 73, 365 (1973).
7. H. Haken, "Synergetics, an Introduction", 3rd ed., Springer Verlag, Berlin, 1983.
8. G. Nicolis and I. Prigogine , "Self Organization in Non-Equilibrium Systems", Wiley, N.Y. (1977).
9. V. Hlavacek and P. Van Rompay, Chem. Eng. Sci., 36, 1587 (1981).
10. E.g., J.L. Hudson, M. Hart and D. Marinko, J. Chem. Phys. 71, 1601 (1979); C. Vidal, J.C. Roux, S. Bachelart and A. Rossi, Ann. N.Y. Acad. Sci., 357, 377 (1980).
11. "Chaos and Order in Nature", H. Haken, Ed., Springer Verlag, Berlin, 1984.
12. "Evolution of Order and Chaos", H. Haken, Ed., Springer Verlag, Berlin, 1982.
13. V.R. Kaufman, D. Levy and D. Avnir, J. Non Cryst. Solids, 1986, in press.
14. V.R. Kaufman and D. Avnir, Proc. Int. Conf. Unconventional Solids, H. Scher, Ed., Plenum Press, 1986.
15. V.R. Kaufman and D. Avnir, Langmuir, submitted, 1986.
16. E.g., P. de Mayo, L.V. Natarajan and W.R. Ware, Chem. Phys. Lett., 107, 187 (1984); P. Levitz, H. Van Damme and P. Keravis, J. Phys. Chem., 88, 2228 (1984); D. Oelkrug and M. Radjaipur, Z. Phys. Chem., 123, 163 (1980); K. Chandrasekaran and J.K. Thomas, J. Colloid Interface Sci., 100, 116 (1984).
17. D. Avnir, R. Busse, M. Ottolenghi, E. Wellner and K. Zachariasse, J. Phys. Chem., 89, 3521 (1985).
18. E. Wellner, M. Ottolenghi, D. Avnir and D. Huppert, Langmuir, submitted, 1986.
19. C.J. Brinker and G.W.Scherer, J. Non-Cryst. Solids, 70, 301 (1985) and references cited therein.
20. D. Levy and D. Avnir in ref. 14; D. Avnir, V.R. Kaufman and R. Reisfeld, J. Non Cryst. Solids, 74, 395 (1985); D. Avnir, D. Levy and R. Reisfeld, J. Phys. Chem., 88, 5956 (1984); D. Levy, R. Reisfeld and D. Avnir, Chem. Phys. Lett., 109, 593 (1984).

21. D. Farin, A. Volpert and D. Avnir, J. Am. Chem. Soc., 107, 3368, 5319 (1985) and previous publications.
22. D. Avnir, this volume.
23. To be published.
24. J.K. Lissent, "Emulsions and Emulsion Technology", Dekker, N.Y., 1974.
25. W. Klonowski, J. Mat. Res., submitted, 1985.
26. C.J. Brinker, E.P. Roth, G.W. Scherer and D.R. Tallant, J. Non-Cryst. Solids, 71, 171 (1985).
27. C.K. Hu, Ann. Rep. Inst. Phys. Acad. Sinica, 14, 7 (1984).
28. A. Coniglio, H.E. Stanley and W. Klein, Phys. Rev. Lett., 42, 578 (1979).
29. W. Klonowski, J. Appl. Phys., 58, 2883 (1985).
30. L.M. Pismen and S.I. Kuchanov, Vysokomol. Soed. (Polymer Sci. USSR), 13, 689, 2035 (1971)
31. C. Kiparssides, J.F. MacGregor and A.E. Hamielec, J. Appl. Polym. Sci., 23, 401 (1979).
32. V.A. Kirilov and W.H. Ray, Chem. Eng. Sci., 33, 1499 (1978); J.W. Hammer, T.A. Akramov and W.H. Ray, ibid., 36, 1897 (1981); R. Jaisinghani and W.H. Ray, ibid., 32, 811 (1977).

156

33. C. Hyver, J. Chem. Phys. 83, 850 (1985).
34. H. Gerrens and K. Kuchner, Br. Polym. J., 2, 18 (1970); G. Ley and H. Gerrens, Macromol. Chem. 175, 563 (1974).
35. J.J. Owen, C.T. Steele, P.T. Parker and E.W. Carrier, Ind. Eng. Chem., 39, 110 (1947); B. Jacobi, Angew. Chem., 64, 539 (1952).
36. R.S. Knorr and K.F. O'Driscoll, J. Appl. Polym. Sci., 14, 2683 (1970).
37. D. Avnir and M. Kagan, Nature, 307, 717 (1984).

HIGH FIELD ^{1}H NMR STUDIES OF SOL-GEL KINETICS*

BRUCE D. KAY AND ROGER A. ASSINK
Sandia National Laboratories, Albuquerque, NM 87185, USA

ABSTRACT

High resolution ^{1}H NMR spectroscopy at high magnetic fields is employed to study the reaction kinetics of the $Si(OCH_3)_4:CH_3OH:H_2O$ sol-gel system. Both the overall extent of reaction as a function of time and the equilibrium distribution of species are measured. In acid catalyzed solution, condensation is the rate limiting step while in base catalyzed solution, hydrolysis becomes rate limiting. A kinetic model in which the rate of hydrolysis is assumed to be independent of the adjacent functional groups is presented. This model correctly predicts the distribution of product species during the initial stages of the sol-gel reaction.

INTRODUCTION

In a previous publication [1], we reported on the sol-gel transition of acid and base catalyzed tetraethylorthosilicate (TEOS). A simple kinetic model was used to predict the overall extent of reaction. We have extended these results to the tetramethylorthosilicate (TMOS) system. The kinetic model was expanded to include the concentration of 15 distinct silicon species based on their next to nearest neighbor groups ($-CH_3$, -H or -Si). The predictions of this model are compared to experiment for solutions which had come to equilibrium with limiting amounts of water.

EXPERIMENTAL

The standard sol-gel solution was 2.24 molar TMOS in methanol. To this solution 1.64 mmole/l HCl and various mole ratios of water were added. The base catalyzed second stage solution contained a 1.64 mmole/l excess of NaOH. The solutions were allowed to react and their spectra recorded at 25°C.

^{1}H NMR spectra were recorded at 360 MHz on a General Electric spectrometer at the University of New Mexico and at 200 MHz on a Chemagnetics spectrometer at Sandia National Laboratories. Four pulses were recorded with a 4 s delay between pulses. All resonances were referenced to tetramethylsilane (TMS) via the methanol methyl group, whose chemical shift is reported as 3.47 ppm relative to TMS [2]. Spectral simulations were performed using the NMRCAP program on a General Electric 1280 data station.

RESULTS AND DISCUSSION

Acid vs Base Catalysis

Figure 1 shows the proton spectrum and shift assignments for a typical acid catalyzed sol-gel solution. By comparing the integral of the set of resonances corresponding to methyl groups bonded to a silicon to

---

*This work performed at Sandia National Laboratories supported by the U. S. Department of Energy under contract number DE-AC04-76DP00789.

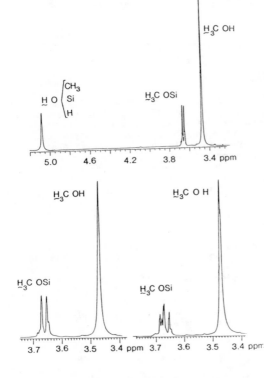

Fig. 1. ^{1}H NMR spectra of the initial hydrolysis for a) the entire proton spectrum at short time (3 min), and expansions of the methyl groups for b) short (3 min), and c) long (90 min) times.

the integral of all methyl groups, the overall extent of hydrolysis is monitored. In cases where the amount of water is limited, this analysis also permits an indirect measure of the extent of condensation as shown by the two possible functional reactions:

Hydrolysis

$$-Si \, OEt- + H_2O \longrightarrow -Si-OH + EtOH$$

Condensation

$$2 \, -Si-OH \longrightarrow -Si-O-Si- + H_2O.$$

Figure 2 shows the overall extent of hydrolysis for the two stage acid and base catalyzed TMOS solutions. The first stage contained a 1:1 H_2O:TMOS mole ratio while the second stages contained 4:1 H_2O:TMOS mole ratios. The overall extent of hydrolysis proceeds very rapidly in the first stage acid catalyzed solution until it reaches a point where all of the initial water

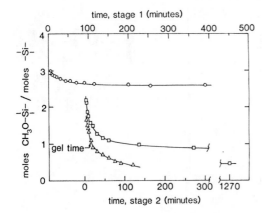

Fig. 2. Moles SiOCH$_3$/mole Si versus time for first stage (O), second stage acid catalyzed (□) and second stage base catalyzed (Δ) solutions. The base catalyzed solution gels relatively early with respect to hydrolysis compared to the acid catalyzed solution.

has reacted by hydrolysis. Because this point is reached by the time of the first measurement (3 min), the hydrolysis rate constant cannot be determined from this data. This behavior indicates that condensation is rate limiting in the acid catalyzed first stage. Further proof for this conclusion is presented below. For the base catalyzed second stage, gelation occurs rapidly (15 ± 2 min), while the acid catalyzed second stage displays long gelation times (115 ± 20 hr). These results parallel our previous findings for TEOS [1,3].

Hydrolysis vs Condensation

To provide further evidence that acid catalyzed hydrolysis occurs more rapidly than condensation, the overall extent of reaction for solutions having H$_2$O:TMOS ratios ranging from 1/2 to 3 were measured as a function of time. The results are shown in Figure 3. For H$_2$O:TMOS ratios of 2 and less, the reaction proceeds rapidly to the stoichiometry corresponding to complete consumption of the initial water assuming no condensation. For an H$_2$O:TMOS ratio of 3, the hydrolysis and condensation rates begin to compete. For large H$_2$O:TMOS ratios the rate of approach to complete consumption of the initial H$_2$O by hydrolysis decreases because the alkoxide concentration undergoes an appreciable decrease. Under these same conditions, the rate of condensation increases dramatically because it scales quadratically with the concentration of silanols formed.

Next to Nearest Neighbor Kinetic Model

In our previous study of TEOS [1] we presented a simple kinetic model based solely on functional group reactivity. This model accurately described the overall kinetics of hydrolysis and condensation. As an extension of this simple model we consider the 15 distinguishable local chemical environments based on a silicon atom's next to nearest neighbors. Figure 4 displays these 15 species in matrix form. To completely describe the kinetics of the 15 species requires 10 hydrolysis and 55 condensation rate constants. Again, we assume that all the hydrolysis and condensation rate constants depend only on the functional group reactivity (ie, the rate

160

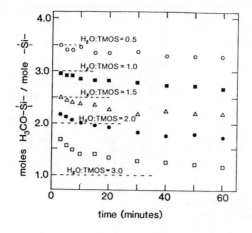

Fig. 3. Moles $SiOCH_3$/mole Si versus time for 0.5 (O), 1.0 (■), 1.5 (Δ), 2.0 (●) and 3.0 (□) moles of H_2O. The dotted lines correspond to the ratio expected if only hydrolysis took place.

of hydrolysis of species (400) is just four times the rate of hydrolysis of species (130)). However, we now retain the 15 distinct silicon species of the next to nearest neighbor model and numerically integrate the coupled rate equations describing their temporal evolution. This model incorporates only the forward rate constants and thus implicitly assumes that the equilibrium strongly favors the product species. The validity of this

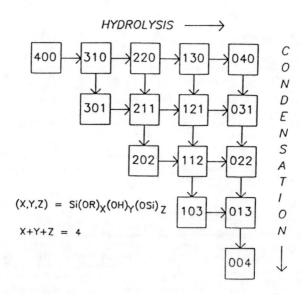

Fig. 4. Possible chemical speciation at the next to nearest neighbor level displayed in matrix form.

assumption is supported by the kinetic data in Figure 3. A detailed
description of the kinetic model will be presented elsewhere [4].

In order to allow a convenient comparison with experiment, we modelled
systems which had come to equilibrium with a H_2O:TMOS ratio of 2:1 or less
(for ratios above 2:1 the only species present would be (004) after
hydrolysis and condensation had both gone to completion). The species
present in such systems all lie along the diagonal of the sol-gel matrix.
The results of the numerical integration are displayed graphically in Figure
5; a comparison with experimental results is presented in the next section.

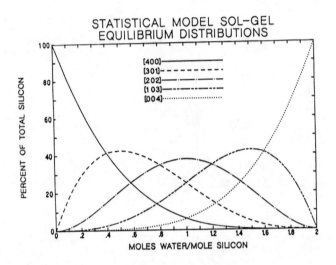

STATISTICAL MODEL SOL-GEL
EQUILIBRIUM DISTRIBUTIONS

Fig. 5. A numerical simulation of the equilibium distribution of species
of the next to nearest neighbor model. The rate of hydrolysis is
assumed to depend only on the functional group reactivity.

In principle, the kinetic model can be extended to next to next nearest
neighbors which, for example, is necessary to distinguish between a silicon
which is part of a dimer versus a silicon which is simply an end group of a
larger polymer. This extension is impractical, however, as 1365 species
exist, and 364 hydrolysis rate constants and 66,430 condensation rate constants are necessary to describe the kinetics. These numbers ignore
topological features such as rings or cage structures as well as
stereochemical effects.

Equilibrium Distribution of Species

To test the next to nearest neighbor model, solutions containing
various H_2O:TMOS ratios were prepared and allowed to react for 3 months.
The spectra of these solutions were readily resolved into four components
corresponding to the (400), (301), (202) and (103) silicon species of the
sol-gel matrix (Fig. 4). The (004) species is not observable by ^{1}H NMR.
The spectra of representative solutions and their computer simulations are
shown in Figure 6.

162

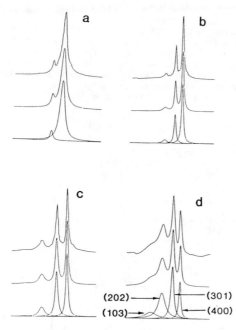

Fig. 6. Expansion of the methoxyl resonances for nominal H_2O:TMOS mole ratios of a) 0.0, b) 0.25, c) 0.50 and d) 1.00 . The upper traces are the experimental spectra, the center traces are the simulated spectra and the lower traces are the individual components of the simulated spectra.

(202) → (301)
(103) → (400)

Note that the spectrum of the solution containing no additional H_2O already contains a small amount of (301) species in addition to the (400) (TMOS) species. This excess H_2O may have been present in the TMOS, CH_3OH or may have been scavenged from the atmosphere or from the vessel walls since no extraordinary attempts to eliminate H_2O were made. Thus, this same amount of H_2O is assumed to be present in all of the solutions. A comparison of the experimentally determined distribution of species versus the theoretical distribution is shown in Table I. For a H_2O:TMOS of 1 or less, the agreement is very good. For ratios above 1 the solution gels, thereby prohibiting spectral analysis.

CONCLUSION

The study of the overall extent of reaction versus time for solutions with varius amounts of H_2O indicates that in acid catalyzed solutions, condensation is rate limiting. The agreement between the observed equilibrium distribution of product species and the next-to-nearest neighbor model indicates that under acid catalyzed conditions, the rate of hydrolysis can be approximated reasonably well by considering only the concentration of the various functional groups. Additional experiments measuring the time evolution of silicon species are being performed to determine if the condensation reactions can also be described using simple functional group kinetics.

Table I. The experimental and theoretical equilibrium speciation for various H_2O:TMOS mole ratios.

H_2O:TMOS Mole Ratio nominal	actual	Species	Experimental percent (±3)	Theoretical percent
0.000	0.063	400	89.1	88.5
		301	10.9	10.9
		202	–	0.6
		103	–	0.0
		004	–	0.0
0.125	0.188	400	63.0	67.7
		301	24.4	27.7
		202	4.6	4.3
		103	–	0.3
		004	–	0.0
0.25	0.313	400	52.8	50.9
		301	33.3	37.5
		202	13.9	10.3
		103	–	1.3
		004	–	0.0
0.50	0.563	400	31.4	26.7
		301	38.4	42.0
		202	24.3	24.5
		103	5.4	6.3
		004	(0.5)*	0.5
0.75	0.813	400	13.6	12.3
		301	34.4	34.2
		202	35.8	35.1
		103	13.6	15.8
		004	(2.6)*	2.6
1.00	1.063	400	7.6	4.7
		301	20.3	21.7
		202	36.0	37.4
		103	28.2	28.3
		004	(7.9)*	7.9

* Actual (004) species is not detected by ^{1}H NMR and is equated to the theoretical value for normalization purposes.

164

ACKNOWLEDGEMENTS

The NMR spectrometer at the University of New Mexico is partially
supported by NSF Grant CHE 8201374. The authors express their sincere
thanks to Dr. J. Satterlee (UNM) for assistance in operating the
spectrometer. The technical assistance of J. A. Ramos and D. A. Schneider
is gratefully acknowledged.

REFERENCES

1. R. A. Assink and B. D. Kay, Mat. Res. Soc. Symp. Proc. 32, 301 (1984).

2. N. S. Bhacca, L. D. Johnson, and J. N. Shoolery, NMR Spectra Catalog
    (Varian Associates, Palo Alto, CA, 1962).

3. C. J. Brinker, K. D. Keefer, D. W. Schaefer, R. A. Assink, B. D. Kay,
    and C. S. Ashley, J. Non-Crystalline Solids 63, 45 (1984).

4. B. D. Kay and R. A. Assink, manuscript in preparation.

SILICON -29 SOLIDS NMR-MAS CHARACTERIZATION OF NON-OXIDE POWDERS AND FIBERS

KENNETH E. INKROTT*, STEPHEN M. WHARRY** AND DAN J. O'DONNELL[+]
*Catalyst Resources Inc., Pasadena, TX 77507; **Phillips Petroleum Company, Research and Development, Bartlesville, Oklahoma 74004; [+]Chemagnetics Inc., Ft. Collins, CO 80524

ABSTRACT

Powder x-ray diffraction has been the most common method for rapid structural analysis and identification of crystalline ceramic raw materials. However, new structural ceramic raw materials often have line-broadened or featureless XRD patterns due to structural strain or small crystallite size. Characterization of various silicon carbide and silicon nitride powders and ceramic fibers by Si-29 NMR-MAS spectroscopy has revealed structural details and differences in these materials previously indistinguishable by XRD or other routine methods. Unique variables intrinsic to a given ceramic synthesis process are reflected in the NMR-MAS spectra of the resulting products.

INTRODUCTION

The recent commercialization of NMR-MAS hardware and software has provided an analytical tool complementary to XRD for characterizing solid inorganic materials (1). NMR spectroscopy gives information concerning chemical bonding and phase influences on certain isotopic nuclei (including Si-29) regardless of degree of crystallinity. While XRD is a simple and quick technique for identification of crystalline solids, it does not give discrete diffraction peaks for amorphous materials, and in practice, can grossly misrepresent the phase distributions present in a multicomponent solid containing substances of varying degrees of crystallinity or crystallite size.

The use of silicon carbide in advanced structural applications will require starting powders which are submicron in size in order to obtain fine-grained microstructures. In fact, many of the alternative methods being researched for fine SiC synthesis yield products which exhibit broad diffraction peaks due to small crystallite size. The resulting XRD patterns bear little resemblance to the known patterns for various SiC polymorphs. Because starting powder characteristics contribute to final sintered component physical properties, it is important to be able to fully differentiate the chemical nature of a specific powder feedstock intrinsic to it as a result of unique synthetic variables.

Hartman, et. al., (2), have recently reported and rationalized the Si-29 and C-13 solids NMR of highly crystalline α- and β-SiC. Pure cubic β-SiC has only one Si-29 NMR peak at -18.3 ppm from tetramethylsilane (TMS), while the pure 6H modification of hexagonal α-SiC has three peaks of equal area at -13.9, -20.2 and -24.5 ppm. However, all of the other approximately 170 α-SiC polymorphs (3) should also exhibit three peaks with approximately similar chemical shifts because of nearly identical structural environments within the immediate coordination surroundings of the three unique silicon sites (2).

Amorphous or poorly crystalline SiC by definition does not have such structural regularity, and NMR-MAS spectra of samples from various suppliers and processes differ substantially as reported below. In addition, NMR-MAS spectroscopy can distinguish Si-N silicon chemical shifts from Si-C. Preliminary analysis of crystalline and amorphous Si_3N_4 are reported below. One viable application of solids NMR in this system would be the study of structural evolution of ceramic fibers from polymeric SiNCO species.

EXPERIMENTAL

A model of WP 100/200 SY NMR-MAS spectrometer (IBM Instruments Inc., Danbury, CT 06810), operating at 39.76 MHz, was used for all the Si-29 NMR data. Relaxation time for Si atoms in β-SiC is 129 s, while the Tl's for the chemical shifts at -13.9, -20.2 and -24.5 ppm for α-SiC are 307, 313 and 362 s, respectively. All SiC NMR spectra were therefore obtained with at least 1500 s delay times between pulsed data acquisition to avoid saturation of the more slowly relaxing species. Si_3N_4 NMR were obtained with 500 s delays.

XRD data were obtained from a Rotaflex RU-200B rotating anode x-ray source (Rigaku Corporation, Tokyo, Japan). Annealing studies utilized a graphite furnace (Model 1000A, Astro Industries Inc., Santa Barbara, CA 93101).

Sample sources are as follows: Figure 1, Samples 1 and 2 (H. C. Starck, 280 Park Avenue, New York, NY 10017), Sample 3 (Ibiden Company Ltd., distributed by Mitsui & Company Inc., New York, NY 10166-0130), Sample 4 (Advanced Refractory Technologies Inc., Buffalo, NY 14207); Figure 2, SiC-40 (Los Alamos National Laboratory, Los Alamos, NM 87545); Figure 3, Sample 9 (Polycarbosilane PX, Nippon Carbon Company Ltd., distributed by Dow Corning, Midland, MI 48640), Samples 10-12 (SiC fiber type NLP-101, Nippon Carbon Company Ltd., distributed by Dow Corning), Sample 13 (SiC whisker type SC-10, ARCO Silag Operation, Greer, SC 29651), Sample 14 (SiC whisker type 3-4CD, Los Alamos National Laboratory); Figure 4, 60 Wt. % Cab-O-Sil fumed silica (Cabot Corporation, Tuscola, IL 61953) mixed with 40 Wt. % carbon black (Phillips Chemical Company, Bartlesville, OK 74004) and heated for 3 hours at the indicated temperature under argon purge; Figure 5, Samples 19 and 20 (Cerac, Milwaukee, WI 53233), Sample 21 (Si_3N_4-11C, Los Alamos National Laboratory).

RESULTS AND DISCUSSION

Interpretation of the Si-29 NMR-MAS spectra of various β-SiC powders is not straightforward. Figure 1 compares the chemical shifts in α-SiC (Sample 1) with three commercial β-SiC powders (Samples 2-4).

167

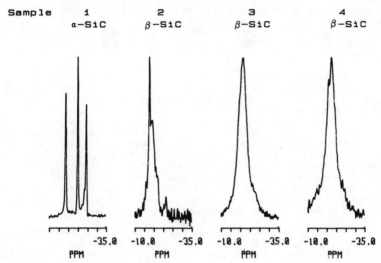

Fig. 1. Silicon-29 NMR-MAS spectra of commercial silicon carbides.

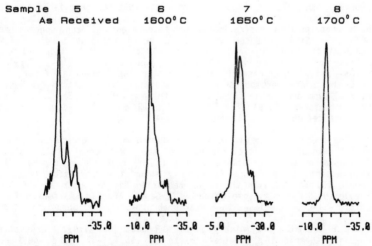

Fig.2. Silicon-29 NMR-MAS spectra of an RF plasma-synthesized SiC powder, annealed for 1 hr under helium at the indicated temperatures.

Sample 2 contains some α-SiC impurity, but the main discrepancy arises from the main chemical shift at -16.4 ppm in Sample 2, which is only a shoulder on the main chemical shift peak at -18.3 ppm in Samples 3 and 4. Analysis of the variable temperature studies in Figures 2-4, along with the data for SiC whiskers and fibers, points out how dependent the silicon chemical environment is on the process used for making the SiC material.

Synthesis of SiC from SiH_4 and CH_4 in a high temperature, radio frequency generated argon plasma (4) yields Sample 5 (Fig. 2). This material has a major Si-29 NMR chemical shift at -16.4 ppm, with minor upfield peaks which coincide with two upfield peaks of α-SiC (Sample 1, Fig. 1) at -20.2 and -24.6 ppm. Annealing this material at various temperatures under an atmosphere of helium causes little change below 1500°C. At 1600°C, the expected β-SiC peak at -18.3 ppm is growing into the spectra (Sample 6), obscuring the minor peak at -20.2 ppm. At 1650°C (Sample 7), the original anomalous chemical shift peak at -16.4 ppm has drastically attenuated while the expected β-SiC peak has grown in intensity. Finally, at 1700°C (Sample 8), all peaks have disappeared except that for β-SiC at -18.3 ppm. The anomalous peak at -16.4 ppm is now just a small shoulder indistinguishable at the scale shown.

The absence or presence of the major peak at -16.4 and minor peaks at -20.2 and -24.5 ppm is not merely a disorder phenomenon, since XRD of the annealed materials in Figure 2 gives much sharper line widths than the as-prepared material. Sample 3 (Fig. 1) contains small amounts of the 2H modification, assigned mainly by a weak XRD peak at 33.6 2θ. This weak peak also appears in the plasma material after annealing, which narrows the broad, overlapping XRD peak at 35.6 2θ (common to both 2H and β-SIC) to the extent that it no longer obscures the lower angle 2H diffraction peak.

SiC whiskers and fibers display Si-29 NMR characteristics similar to powders. Samples 9-12 (Fig. 3) demonstrate that the genesis of SiC from a polycarbosilane precursor occurs via an oxidized continuous fiber (5,6) which contains the entire range of tetravalent silicon species from $-SiC_4-$ to $-SiO_4-$. Based on previous model compound studies (5,7), NMR chemical shifts in Sample 10 (Fig. 3) at 0, -30, -70 and -110 ppm indicate the presence of substantial amounts of $-C_3SiO-$, $-C_2SiO_2-$, $-CSiO_3-$ and $-SiO_4-$, respectively. Heating the commercial fiber under inert atmosphere decomposes it to SiC similar to a plasma-synthesized material (compare Sample 12, Fig. 3, with Sample 5, Fig. 2). In contrast, the SiC whiskers (8,9) (Fig. 3, Samples 13 and 14) have the expected -SiC chemical shift at -18.3 ppm.

The presence of metallic species during whisker growth can promote β-SiC without substantial formation of the low-temperature 2H modification. Carbothermal reduction of the C/SiO_2 controls with metal catalysts absent as in Figure 4, clearly shows the transition from SiO_2 (Sample 15) to SiC with the major resulting SiC NMR peak at -16.4 ppm (Sample 18). There is some evidence in Sample 16 (Fig. 4) for an intermediate -Si-C-O- species with a weak peak at -30 to -50 ppm.

Figure 5 presents preliminary Si-29 NMR data for Si_3N_4. Crystalline α-Si_3N_4 (Sample 19, chemical shifts at -46.4, -47.3 and -48.1 ppm) can be distinguished from β-Si_3N_4 (Sample 20, chemical shift at -48.4 ppm), but resolution for the plasma-synthesized Si_3N_4 material (Sample 21) is not as good as that seen in the SiC material.

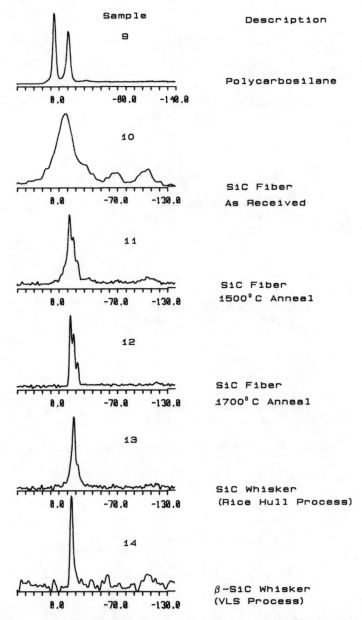

Fig. 3. Silicon-29 NMR-MAS spectra of commercial SiC fiber precursor, SiC fiber (as-received and treated at the indicated temperature under argon), and SiC whiskers.

170

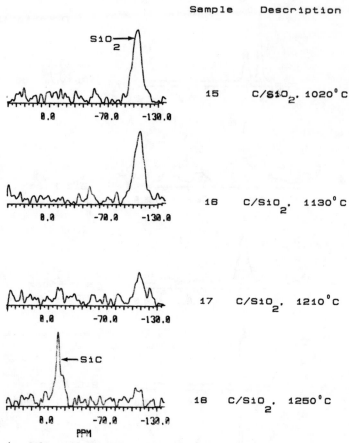

Fig. 4. Silicon-29 NMR-MAS spectra of C/SiO_2 , heated under argon.

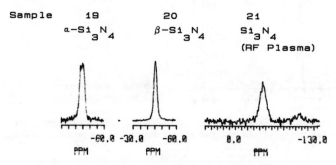

Fig. 5. Silicon-29 NMR-MAS spectra of Si_3N_4 powders.

ACKNOWLEDGMENTS

We thank G. J. Vogt and J. Katz of Los Alamos National Laboratory for samples of plasma-synthesized SiC powder and VLS SiC whiskers, and Phillips Petroleum Company for permission to publish this work.

REFERENCES

1. E. Oldfield and R. J. Kirkpatrick, Science 227 (4694), 1537 (1985).
2. G. R. Finlay, J. S. Hartman, M. F. Richardson and B. L. Williams, J. Chem. Soc., Chem. Commun. 3 159 (1985).
3. J. Schlichtung, et. al., in Gmelin Handbook of Inorganic Chemistry, 8th ed., Si-Silicon, Supplement Vol. B2 (Springer-Verlag, New York, 1984).
4. G. J. Vogt, C. M. Hollabaugh, D. E. Hull, L. R. Newkirk and J. J. Petrovic, Mat. Res. Soc. Symp. Proc. 30, 283 (1984).
5. S. Yajima, K. Okamura and Y. Hasegawa, J. Mater. Sci. 18 (12), 3633 (1983).
6. G. Simon and A. R. Bunsell, J. Mater. Sci. 19 (11) 3649 (1984).
7. G. E. Maciel, D. W. Sindorf and V. J. Bartuska, J. Chromatography 205 (2) 438 (1981).
8. R. L. Beatty and F. H. Wyman, G.B. Patent No. 2 122 982 (25 January 1984).
9. J. V. Milewski, F. D. Gac, J. J. Petrovic and S. R. Skaggs, J. Mater. Sci. 20 (4), 1160 (1985).

KINETIC INVESTIGATIONS OF ALKOXYSILANE SOL-GEL PROCESSING

K.A. Hardman-Rhyne, T.D. Coyle and E.P. Lewis
Ceramics Division, National Bureau of Standards, Gaithersburg, MD 20899
and S. Spooner, Oak Ridge National Laboratory, P.O. Box X, Oak Ridge, TN
37830

ABSTRACT

Effective control of sol-gel glass processing requires a detailed
understanding of the kinetic and mechanistic aspects of the process. We
have investigated structural evolution at the molecular level in the
tetramethoxysilane (TMOS) hydrolysis reaction by various spectroscopic
techniques. Development of structure at the macromolecular level and
evolution of particle/network dimensionality have been studied by
small-angle x-ray scattering (SAXS) in both hydrogenated and fully
deuterated reactions for silica macromolecular structural development.

INTRODUCTION

Alkoxide-based sol-gel processing is finding increasing use as a
route to multicomponent ceramic powders and glasses for high technology
applications. Successful and cost-effective design of sol-gel processes
will require basic understanding of the molecular structure of sol-gel
precursors and of the kinetic evolution of structure from solution to the
ultimate ceramic product. Our investigations of the molecular structure
of species in alkoxide systems include the kinetic aspects of sol-gel
reactions and the relation of sol-gel process conditions to
microstructure of product powders and glasses with the objective of
defining key process variables and guide lines for the development of
practical sol-gel processing.

Although extensive literature exist relating the chemistry and
structural aspects involved in silica based alkoxide studies [1,2,3],
these systems are very complicated and still not fully understood. The
basic chemistry of silica gel or powder particle formation includes both
hydrolysis and condensation steps [2,3] such as

$$\text{Hydrolysis:} \quad \text{Si(OR)}_4 + \text{H}_2\text{O} \xrightarrow{\text{H}^+ \text{ or OH}^-} \text{HO-Si(OR)}_3 + \text{ROH}$$

$$\text{Oligomerization:} \quad \text{(RO)}_3\text{Si} - \text{OH} + \text{HO} -\text{Si(OR)}_3 \xrightarrow{-\text{H}_2\text{O}} \text{(RO)}_3\text{Si-O-Si(OR)}_3$$

and Growth: rings, chains or cross-linking of -Si-O-Si- networks.

The degree of branching and thus the final structure is very dependent on
the solvents, R groups, pH, H_2O to Si(OR)_4 ratio, impurities such as Cl-
and many other factors [4]. Even in the very early stages of hydrolysis
the mechanisms and resulting structure formations are affected.

Small angle x-ray scattering techniques offer considerable
information in determining structures on the 1 to 100 nm scale which is
the significant range for macromolecular development and small particles
[4,5]. While atomic structure can be obtained from analyzing Bragg
peaks, submicrostructure features such as particle size, the degree of
polymeric branching, surface area of dense colloids and extent of the

polymeric species can be obtained in the small angle region (less than 5 degrees in scattering angle, ε). In this paper the data are presented as a function of scattering vector Q which is $2\pi\varepsilon/\lambda$ in the small angle limit and λ is the x-ray wavelength. Therefore, SAXS measurements were obtained for determination of particle size and the degree of molecular branching. This knowledge, along with previous chemistry information [2,6] can suggest mechanisms for the silica polymerization process.

We would like to extend these structural studies to larger sizes so that the entire solution to gel to solid process could be monitored. This can be accomplished with the use of small angle neutron scattering (SANS) techniques [7] which are sensitive to sizes up to 5 μm. However hydrogen scatters incoherently which greatly reduces the signal to noise ratio so deuterated materials would be necessary. The kinetics and the mechanisms of the structure development could be affected by the deuteration. Both neutral and acidic (0.001M HCl or DCl) solutions were measured. Previous work [2] has shown that the hydrolysis TMOS in neutral solution is not as fast as solutions with an acid catalyst. An isotope effect may indicate which of the many sensitive parameters predominates in the early stages of structure formation.

EXPERIMENT

The molar ratios of the solutions prepared are given in Table 1. The H_2O or 0.001M HCl solution were added to TMOS and CH_3OH. The hydrolysis reaction with D_2O or 0.001M DCl was with fully deuterated TMOS and CD_3OD. The deuterated TMOS was produced by reacting $SiCl_4$ with an excess of CD_3OD:

$$SiCl_4 \quad + \quad 4CD_3OD \quad \xrightarrow{\quad O(CH_2CH_3)_2 \quad} \quad Si(OCD_3)_4 \quad + \quad 4DCl.$$

The $Si(OCD_3)_4$ was treated further to reduce the Cl^- content below 0.1%. The products and several hydrolysis reactions were characterized with IR, NMR, CIR-FTIR, and gas chromatography techniques.

Table I. Compositional Molar Ratios

	TMOS	CH_3OH	H_2O	Gel Time
HCl	1	5.63	3.9	37 hours
DCl	1	5.42	4.3	23 hours
H_2O	1	5.63	4.8	26 hours
D_2O	1	5.42	4.3	16 hours

The SAXS measurements were collected at the Oak Ridge National Laboratory. The 10-m SAXS camera utilizes a rotating-anode x-ray source, pyrolytic graphite crystal monochromatization of the incident beam, pinhole collimation, and a two-dimensional position-sensitive proportional counter. The x-ray source was operated at 50 kV and 80 mA with an incident wavelength of 1.542Å(CuKα). The samples were 1mm thick and transmission measurements were between 0.2 and 0.36. Run times varied from 15 minutes to 1 hour and were taken at room temperature (~23C). All raw data were corrected for background and detector sensitivity. This corrected intensity data were circularly averaged (no anisotropic

scattering was observed) and analyzed as a function of the scattering
vector, Q.

RESULTS AND DISCUSSION

There are two major areas, Guinier and power law, in the small angle
x-ray scattering curve that contain useful information relating to the
structure development of the polymerized silica. Detailed explanations of
these scattering areas and how the Guinier radius, R_G, and the power law
exponent, D, relate to the mechanisms can be found elsewhere [4,5]. The
initial measurements are emphasized in that the greatest change in both
the extent and type of polymer branching are expected to occur. The data
are discussed in terms of reduced time, t', which is the time from the
beginning of the hydrolysis reaction divided by the time required for the
solution to gel. All gel times are given in Table I and the reduced times
of the sample are listed in Table II.

The Guinier radius relates to the size of the particles in a dilute
colloidal solution or the correlation length in systems such as this one
where there are overlapping or interacting chains of molecules. The R_G
values were calculated from the small Q regime, $Q < 0.6$ nm^{-1} and
qualitatively measure the breadth of silica macromolecular branching.
Generally the R_G values were between 5.0 and 6.5 nm for all samples
analyzed except in the D_2O case where the R_G range was 3.3 to 4.0 nm. The
Q range analyzed for these materials did fit the Guinier criteria of
R_G *Q < 2 however only 4 to 6 data points were used in the fit. Also the
lowest Q value used was 0.15 nm^{-1} and smaller Q values would have been
desirable. These R_G values are significantly larger than those reported
by Yamane et.al. [2] where R_G values ranged from 0.5 to 2.0 nm. However
our analysis including the Q range consideration would not account for
this large difference.

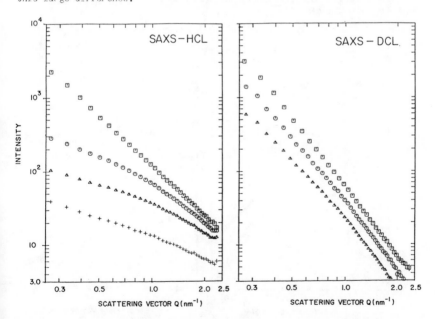

Figure 1. (left) SAXS-HCl t'=0.10(t), 0.16(Δ), 0.22(0), 0.31(□);
(right) SAXS-DCl t'=0.05(Δ), 0.10(0), 0.39(□).

The power law regime is sensitive to the type of silica polymerization or the degree of cross-linked chains that might exist in the solution. This structure can be associated with various mechanisms that can be theoretically calculated to give a power law exponent [4,8,9] which can be obtained from the SAXS data as follows:

$$I(Q) = AQ^{-D}.$$

The results are given in Table II along with the Q range used to obtain the various D values. A comparison of the HCl and DCl catalyzed reactions can be seen in Figure 1. The HCl catalyzed reaction (Fig. 1) has unusually low D values until t' = 0.31 where D is 2.25 indicating a slower development of the polymer network and a more linear chain like structure such as a randomly branched ideal polymer [9]. The DCl catalyzed reaction (Fig. 1) shows significant scattering at low Q values very early in the reaction t' = 0.05 and a very well established, highly cross-linked polymeric structure with D around 2.9. Both the structure and the rate of structural development appear quite different for HCl and DCl catalyzed reactions although the compositions were not very different. Surprisingly, the gel time for the HCl solution was much longer than that of the DCl solution.

Again, the gel time for the D_2O hydrolyzed reaction was much shorter than that of the HCL solution but other D_2O reactions with similar compositions gelled around 40 hours rather than the 16 hours found in this experiment. The D_2O hydrolyzed reaction (see Fig. 2) were analyzed at slightly higher Q values and had D values around 1.78. This value is consistent with a D value one would expect to see if the reaction of the silica macromolecules were controlled by a multiparticle diffusion limited aggregation mechanism [8]. The H_2O hydrolyzed reaction (see Fig. 2) suggests strong cross-linked polymer network with D values around 2.9 much like the DCl case.

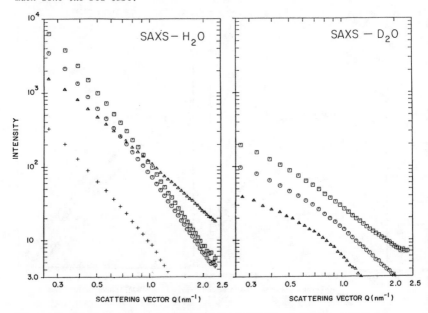

Figure 2. (left) SAXS-H_2O t'=0.053(t), 0.196(Δ), 0.278(0), 0.90( );
          (right) SAXS-D O t'=0.055(Δ), 0.134(0), 0.364( ).

Table II. Power Law (D) Results

	Reduced time (t')	Power Law (D)	Q range (nm^{-1})
D_2O	0.055	1.75	0.7 to 2.1
	0.134	1.72	0.7 to 2.2
	0.364	1.84	0.7 to 2.2
H_2O	0.058	2.73	0.3 to 1.1
	0.196	2.08	0.3 to 2.35
	0.278	2.95	0.3 to 1.1
	0.900	3.10	0.3 to 1.0
DCl	0.05	2.57	0.3 to 1.1
	0.10	2.88	0.3 to 2.3
	0.39	3.10	0.3 to 2.3
HCl	0.10	0.80	0.2 to 1.0
	0.16	0.76	0.25 to 0.95
	0.22	1.09	0.3 to 0.95
	0.31	2.25	0.25 to 2.4

CONCLUSIONS

In the D_2O hydrolyzed reaction case the short gel time, smaller R_G values and a power law exponent D of around 1.8 indicate a slightly different mechanism more sensitive to small particles aggregation than polymerized silica network formation. The other reactions appear more polymeric in nature with strong cross-linking in the DCl and H_2O case and much weaker cross-linking in the HCl case. The interpretations for the scattering data will be more straight forward when the chemical data from gas chromatograph, NMR and IR data are completed. This will add greatly in understanding the elementary reactions so that specific models exist for the chemical processes. Further studies are underway to monitor more advance stages in the gel formation of these systems and to study thicker samples of the deuterated materials with small angle neutron scattering.

REFERENCES

1. R.K. Iler, The Chemistry of Silica (John Wiley, New York, 1979), pp. 172-311.

2. Masayuki Yamane, Satoru Inoue and Atsuo Yasumori, J. of Non-Cryst. Sol. 63, 13 (1984).

3. H. Schmidt, H. Scholze and A. Kaiser, J. of Non-Cryst. Sol. 63, 1 (1984).

4. K.D. Keefer, Mat. Res. Soc. Symp. Proc. 32, 15 (1984).

5. Dale W. Schaefer and Keith D. Keefer, Mat. Res. Soc. Symp. Proc. 32, 1 (1984).

6. C.J. Brinker, K.D. Keefer, D.W. Schaefer, R.A. Assink, B.D. Kay and C.S. Ashley, J. of Non-Cryst. Sol. 63, 45 (1984).

178

7. K.A. Hardman-Rhyne and N.F. Berk, <u>Materials Characterization for Systems Performance and Reliability</u>, ed. James W. McCauley and Volker Weiss, Plenum Publishing Corp., New York, 257, (1986).

8. M. Kolb and R. Jullien, J. Physique Lett. <u>45</u>, L977 (1984).

9. Dale W. Schaefer and Keith D. Keefer, to be published in <u>Ultrastructure Properties of Materials</u>, ed. D. Ulrich.

APPLICATION OF SANS TO CERAMIC CHARACTERIZATION

K.G. FRASE*, K.A. HARDMAN-RHYNE AND N.F. BERK
Institute for Materials Science and Engineering,
National Bureau of Standards, Gaithersburg MD 20899

ABSTRACT

Traditionally, small angle neutron scattering (SANS) has been used to study dilute concentrations of defects 1 - 100 nm in size. Recent extensions of the scattering theory have allowed the expansion of the technique to include larger sizes through the use of multiple
scattering. With multiple small angle neutron scattering, defects (pores, microcracks, precipitates) up to 10 μm in size can be studied. SANS is inherently a non-destructive, bulk probe of
microstructure, with wide applications in the characterization of materials.

A number of studies of ceramic materials using multiple and traditional (single particle
diffraction) small angle neutron scattering will be discussed. The emphasis will be on the
strength of the technique in the characterization of materials. Particular examples will include:
the assessment of pore size distributions in spinel compacts as a function of sintering and
agglomeration, the characterization of primary and secondary particle sizes in precipitated aggregates, and the determination of microporosity in MDF cements.

INTRODUCTION

SANS has traditionally been used to study small voids and defects in solids of the 1 - 100
nm range. These single particle scattering measurements have yielded information on creep
cavitation flaws in metals [1,2] and porosity in glasses [3], as well as microcracking in ceramics
[4]. By adapting methodology from polymer science, SANS can also be used to characterize the
initial stages of sol-gel reactions [5]. (Scattering can arise from any interface in the bulk of the
material across which there is a change in chemical composition, and hence a change in ρ, the
scattering length density. Therefore voids, precipitates, grain boundary phases, and microcracks,
if present at more than one volume percent, may contribute to the observed scattering.) Such
single particle scattering experiments require thin (<3mm) samples with a dilute (<2%) concentration of the voids or flaws of interest, in order to minimize multiple scattering effects.

In the past few years, an extension of the small angle scattering theory, which incorporates
multiple scattering effects, has been developed [6,7]. This extension has allowed an expansion
of the sizes of microstructural features which can be seen experimentally with SANS up to several μm [8]. By using multiple scattering data, such features as agglomeration and green-state
porosity can be observed non-destructively [9]. By combining single particle diffraction and
multiple scattering, the whole range of 0.001 - 10 μm can be probed.

Neutron scattering is inherently a bulk technique, and the energy of the thermal neutron
beam is low, so that the experiment is non-destructive. The neutrons interact with the nuclei of
the atoms with a coherent neutron scattering length, b, which depends in an unsystematic way
on the atomic weight of the species. In particular, low molecular weight materials and isotopic

* Current address: I.B.M. T.J.Watson Research Center, P.O.Box 218, Yorktown Heights, NY
10598.

species can be studied readily with no loss in information. The SANS results are not as affected by surface inhomogeneities as are comparable Xray data.

The anomalous dependence of b on isotopic weight can be used to highlight particular features. For example, the b for hydrogen is -0.374(10^{-12} cm) while the b for deuterium is 0.667 (10^{-12} cm)[10]. Thus it is possible to partially deuterate solvents, for example, to exactly match the ρ of a matrix so that only air-filled voids, or closed porosity, are "visible" to the neutrons as a separate phase. This effect has been widely used by polymer scientists, and has also been used to advantage in the analysis of MDF cements [11,12]. The applicability to sol-gel samples is obvious, and is discussed in [13].

EXPERIMENTAL TECHNIQUES AND ANALYSIS

The SANS instrument at the NBS reactor [14] has the ability to change the incident neutron wavelength, λ, from 0.5 to 1.0 nm by changing the speed of a rotating helical channel velocity selector. This feature is central to the multiple scattering experiments described here, as the multiple scattering is wavelength dependent, and the functional form of that dependence is a critical component in the data analysis. The wavelength used and the distance from the sample to the position-sensitive, two-dimensional area detector determine the range of wave vector, Q, ($=4\pi\sin\varepsilon/\lambda$, where ε is the scattering angle) assessible in the experiment. These factors, along with the sizes of the cadmium collimating apertures in the evacuated flight path, fix the minimum Q resolution.

There is of course a trade-off between neutron flux at the sample and the precision of the collimation and the data. The installation of a cold source at the NBS reactor will improve data collection at higher wavelengths by shifting the peak in the flux vs wavelength distribution to longer wavelengths. The availability of wavelengths longer than 1.0 nm with reasonable fluxes will also permit measurements of larger size scattering centers using multiple scattering SANS.

In the single particle diffraction experiments, a centered beam stop is used on the detector to protect it from the unscattered direct beam, which limits the mimimum usable Q data. In the multiple scattering experiments, no beam stop is used, as the incident beam is almost completely scattered, yielding data down to very small Q, which is essential to the analysis. In experiments with high overall intensity on the detector, thin cadmium foils are used to attenuate the incident beam.

In both the single particle and beam broadening (multiple scattering) cases the scattering tends to be isotropic, and the data are circularly averaged in most cases. Other averaging methods are possible for anisotropic scattering. The data are thus reduced to a set of intensity vs. scattering vector (Q) values, which are corrected for instrumental and background effects.

Three distinct methods of analysis are used, depending on the Q range and the nature of the scattering. In the single particle diffraction case, the scattering at low Q for spherical scatterers is given by:

$$I \propto (\rho)^2 V_p^2 \exp\{ - R_g^2 Q^2/3\} \qquad (1)$$

where ρ is the difference in neutron scattering length densities between the scatterer and the matrix, often referred to as the "material contrast", V_p is the volume of the particle and R_g is the scatterer's radius of gyration. For non-spherical scatterers such as microcracks, other relationships for the radius of gyration are needed, but the Q^2 dependence is retained. Notice that there is no wavelength dependence to the single particle scattering.

The low Q region of the multiple scattering data reflects the wavelength dependent "beam broadening" effects. The degree of scattering can be characterized [6] by the dimensionless "scattering power", Z, which is the ratio of the sample thickness to the scattering mean free path. Z also depends on the material contrast (ρ), the scatterer radius (R), the volume fraction of scatterers (ϕ), and the neutron wavelength (λ). Single particle scattering can thus be seen as the lower limit to this function, where the combination these variables is sufficiently small to make the probability of multiple scattering events negligible, and thus minimize the observed wavelength dependence [6].

In a sample in which multiple scattering occurs, the scattering reflects a rough λ^2 dependence, as can be seen in Fig. 1. Similar effects of the sample thickness on multiple scattering can be seen [15]. It is important to note that both multiple scattering and Guinier scattering (single particle diffraction) show a similar Q^2 dependence at low Q (see Fig. 2). It is primarily the wavelength dependence which allows them to be differentiated.

The shape of the multiple scattering at each wavelength is characterized by its scattering radius of curvature $(r_c[Q])$ at $Q=0$, which is related to the full-width at half-maximum of the scattering by: $q(FWHM) = 2.335\, r_c[Q]$. After subtracting out the effect of the unscattered beam, the values of $r_c[Q]$ for the sample as a function of wavelength are fit empirically and modelled numerically to determine the concentration and size of the scatterers. Readers interested in the details of the theory and analytical method are referred to [6,7,8].

In the large Q regime for both single particle and multiple scattering SANS [6,7], theory predicts that the scattering at large Q is described by Porod's law:

$$I \propto \frac{2\pi(\rho)^2 S}{V\ Q^4} \tag{2}$$

where V is the sample volume seen by the neutron beam and S is the total surface area of scatterers in V. It is important, to collect data at high enough Q to pick up a true Porod (Q^{-4}) dependence in the scattering, and to accurately correct the data for background effects. The Porod intensities can be put on an absolute intensity scale by using a secondary intensity standard, so that quantitative surface area values can be obtained [16]. The combination of small Q and large Q methods provides more information than either method alone. Depending on the underlying model assumptions and information from other techniques, it is possible to characterize particle size distributions and make inferences about morphology.

REVIEW OF EXPERIMENTAL RESULTS FROM CERAMIC MATERIALS

A number of experiments will be summarized which demonstrate the capabilities of SANS as a non-destructive analytical tool in the study of precipitated powders, powder compacts, and controlled porosity materials.

Powders

The initial study of a ceramic powder by multiple scattering in combination with single particle diffraction used spherical α-alumina particles and showed that particle size distributions could be modelled by comparing the high and low Q scattering [15] . Moreover these data were compared with laser light scattering which confirmed the model lognormal particle size distribution determined by SANS.

A study of precipitated ("chemically derived") powders using SANS was recently undertaken to address the question of the "raspberry" morphology observed by Heistand et al [17] in their ZnO. Powder samples of ZnO (provided by Yee-Ho Chia at Dow Chemical) and AlOOH (provided by Ralph Nelson at DuPont) were analysed using both single particle diffraction and multiple scattering SANS [18].

The AlOOH material, although composed of very fine crystallites, determined by single particle diffraction to be approximately 7 nm in size, formed large agglomerates. These large agglomerates concentrate the scattering into very small angles and the samples were not thick enough for multiple scattering to be observed. The irregular and large scale ($>10\mu m$) structure of the material was confirmed by inspection in the SEM.

The ZnO material, however, showed considerable beam broadening. By collecting data at both small and large Q, it was possible to optimize both the beam broadening and the Porod data. Previous studies of the ZnO powder by the workers at Dow [19] had suggested that the raspberries were 0.1 to 0.2 μm in size. Their Xray line broadening work suggested a crystallite size of $\simeq$ 15 nm, but there was little reason to believe that the crystallites were densely packed in the raspberries.

The SANS analysis also showed two levels of structure. The Porod data, collected in the short instrumental configuration, at 0.6 nm and using a 2.0 mm thick sample, gave a radius of 17 nm and a scattering interface surface area of 36 m^2/g. Dow's gas adsorption (BET) work had also indicated a surface area of 36 m^2/g. The multiple scattering data, collected with a 5mm thick sample cell and the long configuration, gave a radius of 0.1 to 0.2 μm. The observation of such disparate scatterer sizes in the different Q regimes appears to be consistent with the assignment of a raspberry morphology.

This powder study shows the strength of SANS to access well-defined size distributions in a bulk powder, and underscores the importance of sample preparation, experimental design, and using complementary techniques, in order to maximize the useful information obtained.

Compacted Ceramic Materials

Three studies of ceramic compacts using SANS serve to highlight the variety of experiments possible with the technique. The most straightforward of these is the study in 1983 of Fe and W inclusions in fully dense MgO-doped Si_3N_4 [20] . Care was taken to assure that no porosity or microcracks were present to contribute to the scattering, and a thin sample was used so that only the single particle scattering from the inclusions was observed. The results were confirmed by TEM experiments, which showed spherical particles identified as Fe. The W inclusions, as observed in the TEM, were several μm in size and could not be seen with single particle diffraction. Guinier and Porod analysis suggested a distribution in inclusion radii from 24 to 16 nm. Microscopy is a useful complementary technique, as it provides information about spatial distributions and can sometimes, as in this case, be used to identify the scatterers.

The second study, by E.D.Case and C.J.Glinka [4], also combined another experimental technique, in this case elasticity measurements, with SANS to characterize the microcracking in sintered yttrium chromite. $YCrO_3$ undergoes microcracking at temperatures above $1100^\circ C$; this microcracking can be annealed out at lower temperatures without significantly changing the underlying porosity or microstructure. The scattering from these underlying features can thus be subtracted out, and the scattering from the microcracking itself analysed. By combining SANS with elasticity measurements, they determined values for the average crack radius, the crack opening displacement, and crack number density. This study used thin samples and single particle diffraction in the SANS experiments.

The third study used multiple scattering effects as well as single particle diffraction and characterized pore size distributions in spinel compacts in a sintering experiment [9,21]. Spinel powder (particle size 0.38 μm as determined by SANS) was compacted into disks $\simeq$ 6 mm thick with and without the incorporation of 20 vol% hard agglomerates (nominally 5 μm in size). These compacts were analysed using SANS in their green state and after firings at $1000^\circ C$ (12 hr), $1300^\circ C$ (12 hr) and $1500^\circ C$ (3 hr). The green state porosities and surface areas were also measured by mercury porosimetry (MP) and BET.

Because of the large amount of multiple scattering, particularly in the green state samples, it was necessary to collect data at relatively large Q for the Porod fit. The analysis of all the data is not yet complete, but the preliminary results suggest that the agglomerates impede both initial compaction and densification.

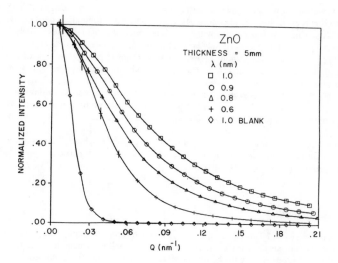

Figure 1. Normalized intensity vs. scattering vector, **Q**, for ZnO powder in 5 mm thick cell at six wavelengths and the direct beam (blank) at 1.0 nm.

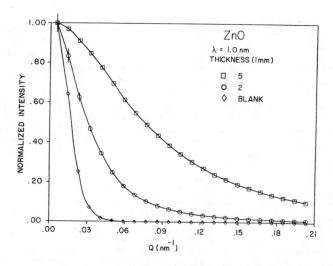

Figure 2. Normalized intensity vs. scattering vector, **Q**, for ZnO powder, in 2 mm and 5 mm thick cells, and the direct beam (blank) at 1.0 nm, showing the Q^2 dependence of the scattering at low **Q**.

SANS is very sensitive to multiple scattering from porosity, as can be seen in Fig. 3. Some multiple scattering is even observed in compacts sintered to 97-98% theoretical density. During the 1000°C bisque-firing, in the material without agglomerates, the smaller pores shrink; the 1500°C material (97% dense) showed only large (1.5 μm) sized residual porosity.

In the green state compact without agglomerates, two sizes of pores were indicated by the data: one at about 0.88 μm, obtained from small Q measurements, and the other at 0.4 μm, obtained from large Q measurements. The surface area from SANS was comparable to that from BET ($\simeq$ 4.0 m^2 /g) and the total porosity value (44%) was consistent with the MP results. The correlation between SANS and MP porosity levels suggests that the porosity is all open. The pore size as determined by porosimetry was 0.2 μm, and the MP surface area was accordingly much too large.

We have interpreted these results as indicative of the classic "ink-bottle" type of porosity, or in this case, empty packing sites of volume comparable to the powder particles, connected by throats at the tetrahedral junctions of the particles.

Controlled Porosity Solids

The microporosity in 10 μm silica spheres used in liquid chromatography was characterized by Glinka et al [3] using single particle diffraction methods. Their work demonstrated the usefullness of H/D isotope mixing to change the effective material contrast in porous materials. In their case, when the solvent contrast matched that of the matrix, no scattering was observed, demonstrating the connectivity and accessibility of essentially all of the porosity. They point out the applicability of such methods to the study of adsorbed layers and catalytic surfaces on porous substrates.

Similarly, Pearson et al [11,12] at Harwell have used H/D substitutions to identify the source of scattering in a single particle diffraction study of a variety of cements, including the so-called Macro-Defect-Free (MDF) cement. Very thin samples (<1mm) were used and only the microporosity was studied. The scatterers were identified as pores, probably tetrahedral in shape at the intersection of roughly spherical gel particles. Regardless of the water/cement ratio, conventional portland cement showed micropores at 5 and 10 nm, while the MDF cements (also based on portland cement) showed a shifted pore size distribution at 2.5 and 7 nm, although TEM studies could discern no difference in the gel structure or microporosity between the two types of cement. Due to the nature of the experiment, no information about larger microstructural features could be obtained.

A preliminary attempt at characterizing MDF cements using both single particle diffraction and multiple scattering SANS reproduced the 2% microporosity result observed by the Harwell/ICI group. Three samples were studied, containing 0%, 0.5% and 3% by weight of polyvinyl alcohol as a rheological aid to the shear mixing of the MDF samples. The cement matrix was based on a commercial calcium aluminate cement, and the samples were prepared by Emmanuel Cooper at I.B.M. [18].

All three samples showed reasonable Guinier behavior and a scatterer radius of 3 nm. The Porod region indicated about 2% scatterers of 2 nm for the sample without polymer, and 2% scatterers of 1.25 nm in size for the two samples with polymer. Pearson's conclusion about the MDF shear mixing affecting the microporosity seems to be correct. This microporosity is surely in the amorphous gel phase.

There was a small, but discernable multiple scattering component to all three data sets. Unfortunately, SEM work showed that the scale of the multiphase regions in the sample-- islands of various hydrated and anhydrous calcium aluminate phases in the gel matrix -- was on the order of 1 μm. Therefore, the multiple scattering could well be arising from the phase boundaries as well as from the porosity. Careful control of the microstructure and hydration state of future samples, and perhaps the use of H/D matrix matching, may yield clearly interpretable multiple scattering data.

Spinel Sintering

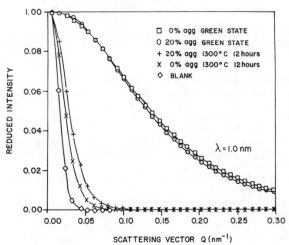

Figure 3. Normalized scattering intensity vs scattering vector, **Q**, for four spinel compacts, loose spinel powder, and the direct beam, at 1.0 nm.

CONCLUSIONS

The combination of single particle diffraction and multiple small angle neutron scattering allows the non-destructive evaluation of a wide variety of ceramic materials and properties. Particularly in the area of chemically derived ceramics there are many features which SANS can probe, in a quantitative manner. Some of these applications include the primary and secondary structure in precipitated aggregates, the growth of precipitates in solution and the polymerization of sol-gel suspensions. The combination of single particle diffraction and multiple scattering allows the extension of such studies into the densification stages -- how does a raspberry powder morphology affect the pore size evolution during sintering, for example? In sintered materials, the ability of SANS to assess bulk closed porosity in a quantitative manner is nearly unique.

The use of complementary techniques along with the SANS and careful experimental design can increase the amount of useful information obtained. Complementary techniques can identify the scatterers, check for spatial distributions and help distinguish open and closed features. The careful use of sample preparation, neutron wavelength, contrast matching and experimental configuration can optimize the SANS data. As higher wavelengths and new facilities (like a sintering furnace in the SANS sample chamber) become available, wider and wider applications of the technique will be realized.

Acknowledgements
One of us (K.G.F.) acknowledges the support of a National Research Council post-doctoral fellowship at the National Bureau of Standards.

186

REFERENCES:

1. J.R. Weertman, in <u>Non-destructive Evaluations: Microstructural Characterization and Reliability Strategies</u> , edited by O.Buck and S.M.Wolf (AIME, New York, 1981), pp. 147-168.
2. R.A.Page and J.Lankford, Comm. of the Amer. Ceram. Soc. C-146 (1983).
3. C.J.Glinka, L.C.Sander, S.A.Wise, M.L.Hunnicutt and C.H.Lochmuller, Anal.Chem. 57, 2079-2084 (1985).
4. E.D.Case and C.J.Glinka, J.Mat.Sci. 19, 2962 (1984).
5. K.A.Berglund, K.D.Keefer and R.G.Dosch, this conf.
6. N.F.Berk and K.A.Hardman-Rhyne, J.Appl.Cryst. 18, 467-472 (1985).
7. N.F.Berk and K.A.Hardman-Rhyne, Physica B, (1986), in press.
8. K.A.Hardman-Rhyne, Proc. of the 31st Sagamore Research Conference, August 1984, (Plenum Publ.,New York, 1986) in press.
9. K.A.Hardman-Rhyne, K.G.Frase and N.F.Berk, Physica B, (1986), in press.
10. G.Kostorz, in <u>Treatise on Materials Science and Technology</u> , Vol.15, edited by G.Kostorz (Acad.Press, New York, 1979), pp. 5-8.
11. D.Pearson, A.Allen, C.G.Windsor, N.McN.Alford and D.D.Double, J.Mat.Sci. 18, 430 (1983).
12. D.Pearson and A.J.Allen, J.Mat.Sci.20, 303 (1985).
13. K.A.Hardman-Rhyne, T.D.Coyle and E.P.Lewis, this conf.
14. C.J.Glinka, AIP Conference Proc. 89, Neutron Scattering (1981), p.395.
15. K.A.Hardman-Rhyne and N.F.Berk, J.Appl.Cryst. 18 , 473-479 (1985).
16. C.J.Glinka, NBS SANS manual (internal publication).
17. R.Heistand, II., MIT-CPRL report #Q3, 18-20, (1984).
18. K.G.Frase, to be published.
19. Yee-Ho Chia (private communication)
20. K.A.Hardman-Rhyne, N.F.Berk and E.R.Fuller,Jr.,J.Research (N.B.S.) 89, 17 (1984).
21. K.G.Frase and K.A.Hardman-Rhyne, in preparation.

CHARACTERIZATION OF THE SOL-GEL TRANSITION OF ALUMINA SOLS PREPARED FROM ALUMINUM ALCOXIDES VIA ^{27}AL NMR

William L. Olson and Lorenz J. Bauer, Signal Research Center, 50 E. Algonquin Rd., Des Plaines, IL 60017.

ABSTRACT

^{27}Al NMR spectra were obtained on a series of alumina sols prepared by the hydrolysis of aluminum sec-butoxide. Subtle but distinct differences were observed in the solution ^{27}Al NMR spectra of sols which varied in appearance from being very milky to completely transparent. No changes were observed in the ^{27}Al spectra of sols which had been aged. The adddition of sufficient quantities of acid or base to gel the sol precipitated dramatic changes in the ^{27}Al spectra. Aluminum-27 NMR was found to be a highly useful tool for probing the sol-gel transformation of this system.

INTRODUCTION

In order to address many of the pressing questions concerning the processing of sol-gel materials [1], techniques must be developed for characterizing these materials at the molecular level during processing. Of particular interest are techniques that allow the gelation process to be followed in situ in order to directly measure the effects that additives such as drying control chemical additives (DCCA's) [2], alcohols, salts, acids and bases have on the gelation chemistry of the sol.

Multinuclear NMR spectroscopy is a powerful technique for the characterization of sol-gel materials. ^{29}Si NMR has proven to be extremely useful for elucidating the hydrolysis/polymerization chemistry of Si(OMe)$_4$ [2-3]. ^{27}Al NMR has been used for the solution characterization of oligomeric aluminum alkoxide complexes [4] and the aqueous polymerization products of aluminum salts such as aluminum trichloride [5-12]. ^{27}Al MASNMR is widely used for the characterization of microporous materials such as zeolites and molecular sieves [13-15].

Widely used as precursors for the preparation of reactive alumina powders, aluminum alcoxides are an important starting material for the preparation of a variety of ceramic materials. This paper describes the use of ^{27}Al NMR spectroscopy for the characterization of the sol-gel transition of alumina sols treated with acids or bases. The alumina sol was prepared by the hydrolysis of aluminum sec-butoxide [16-20]. This characterization technique was found to provide useful insights into the gelation chemistry of this ceramic precursor.

EXPERIMENTAL

The alumina sol was prepared as described by Yoldas [16]. In a typical preparation, aluminum sec-butoxide (244g) was poured with stirring into 750 mL of distilled water. After equilibrating the mixture for thirty minutes, concentrated HNO$_3$ (3mL) was added and the reaction heated for 24 hours at 100C (acid/Al mole ratio = 0.05). Periodically the reactor was shaken to aid in dissolving the gelatinous deposits which formed on the reactor walls. The reaction mixture was then cooled to room temperature and the sol poured into a separatory funnel. The reaction mixture separated into two transparent layers. The aqueous bottom layer which contained the alumina sol was collected for study.

The concentration of aluminum in the sol determined by atomic absorption spectroscopy (AAS) was 1.3 M. The sol consisted of 8 percent by weight aluminum hydroxide. As prepared, the sol contained a large amount (12-15%) of sec-butanol. The gelation studies reported herein used a form of the sol which contained 12.7 mass percent sec-butanol. Gelation experiments were also conducted on alcohol-free alumina sols. The alcohol was removed by rotary evaporation of the sol to half volume followed by redilution with distilled water. This procedure was repeated twice to insure that all of the alcohol had been removed via its azeotrope. No significant amount of aluminum was detected in the distillate (AAS).

The ^{27}Al NMR measurements reported herein were made at 78 MHz on a Nicolet NT-300 spectrometer. A few measurements were also made at 130 MHz on a 11.7 tesla spectrometer. Although the spectra were comparable, a small dependance of the linewidth and position of the sol resonances on the field strength was observed possibly arising from quadrupolar interactions [21]. All chemical shifts are reported relative to an aqueous 0.1M aluminum chloride solution. Since the intensities and linewidths of the resonances depend on the pulse width and power levels, extreme care was taken to maintain consistency between the experiments and to limit saturation effects. Typical experimental parameters included the accumulation of 80 transients with a delay time of 1 second between pulses, a sweep width of 15,151Hz, a pulse width of 33 microseconds (90 degrees) and an acquisition delay of 3 microseconds.

The NMR spectra were obtained on static (i.e., nonspinning) samples in 20mm tubes. In the gelation studies, the first spectra were recorded 2 minutes following addition of the gelling agent. Relative intensities of the aluminum species were determined by computer simulation of the spectra. The best fit to the observed lineshapes was obtained using composite lines made up of 80% Lorentzian and 20% Gaussian character.

RESULTS

General Features: A typical spectrum of an alumina sol is presented in Figure 1. Simulation of this spectrum reveals that the peak at 8 ppm is too broad to be fitted with a single line. The asymmetric nature of the peak and its extreme width are better approximated by using a combination of two peaks, a relatively narrow (90Hz) line at 8 ppm and a broad (2200Hz) resonance with its center near 0 ppm. The relative intensities of the broad and narrow (8ppm) resonances, although dependent on the sol preparation procedure, were normally 40 and 60 percent respectively. Occasionally a third narrow (50 Hz) peak at 1.5 ppm arising from $[Al(H_2O)_6]^{3+}$ was observed. This resonance, whether present as a shoulder on the 8ppm line or well resolved, rarely accounted for more than 10 percent of the total spectral intensity.

From their chemical shifts, we can confidently assign these resonances as arising from octahedral aluminum species in the sol [5]. No peaks near 60 ppm, indicative of tetrahedrally-coordinated aluminum atoms, were resolved. The broad resonance near 0ppm, is tentatively assigned to aluminum atoms in the high molecular weight polymers that occupy sites in the polymer where the relative symmetry of the aluminum site is poorly defined (highly anisotropic). The relatively narrow resonance at 5ppm presumably arises from the more symmetric (octahedral) aluminum sites in the sol polymers that possess lower electric field gradients at the aluminum center.

When portions of a transparent sol were stored at 7 C and room temperature and analyzed periodically by ^{27}Al NMR, no significant differences/changes were observed in the spectra of the sol over a three month period. The relative linewidths and positions of the ^{27}Al sol resonances remained remarkably constant and gave no evidence of any time dependent (i.e., aging) behavior. Removal of the alcohol from the sol also produced no apparent change in these measured parameters.

The ^{27}Al NMR spectra of the alcoxide-derived sols, were also remarkably invariant to temperature. Spectra obtained at 70C were found to give good agreement with spectra obtained at room temperature. This is an important observation since it provides information concerning the fluxional character of the sol-gel polymers. Unlike aluminum alcoxide complexes which often give markedly narrower ^{27}Al NMR lines at elevated temperatures (due to rapid ligand reorientation and a more symmetric time-averaged environment for the aluminum nuclei) [4], the linewidths of the aluminum sol resonances remained constant. This result implies that either the aluminum atoms reside in sites that are already rapidly fluxional (i.e., averaged) at room temperature or that the coordination sphere of the aluminum atoms within the sol-gel polymers is fairly rigid (where raising the temperature does little to change the time averaged environment of the observed Al species). Although we suspect that the majority of aluminum species belong to the latter category, we have no evidence to disprove the former mechanism.

While the ^{27}Al spectra of individual sols were remarkably constant, subtle differences were observed between the spectra of Al sols prepared using different experimental conditions. The primary difference between the spectra involved the relative linewidth of the 5 ppm resonance and the intensity of $[Al(H_2O)_6]^{3+}$ peak. In general, sols exhibiting a milky appearance (a common property of sols prepared at lower temperatures) gave spectra with broader Al resonances than their clear counterparts with the intensity of the baseline resonance increasing at the expense of the 5ppm resonance. While these differences were very apparent at the qualitative level, exact quantification of these features proved to be more difficult. A small change in the phasing of a spectrum was found to produce significant shifts in the computer simulated positions and relative intensities of the broad and narrow octahedral resonances. Often these changes were on the order of 5-10 percent of the calculated values. Thus, although useful as a qualitative tool for probing the structure of alumina sols, it was not possible to obtain an unambiguous differentiation between different batches of alumina sols.

GELATION WITH ACID

^{27}Al NMR spectra of an alumina sol obtained at periodic intervals following addition of a small quantity of 1.8M HNO_3 (acid/Al=0.08) are displayed in Figure 2. The principal change in the spectra with time is the gradual disappearance of the 8ppm resonance with a corresponding increase in the intensity of the baseline resonance at 0 ppm (see Figure 3). A third peak at 1.5 ppm which prior to addition of the HNO_3 was evident only as a subtle shoulder on the 8ppm resonance, is now clearly resolved in the spectra. The relative intensities of the three components were determined by computer deconvolution of the spectra (see Figure 4). The relative intensities of the resonances as a function of gelation time are plotted in Figure 5. This plot clearly shows the reciprocal relationship of the 8 and 0 ppm resonance intensities during the gelling process while the intensity of the sharp resonance at 1.5ppm remains

190

constant. The gradual decrease in intensity of the 8ppm resonance is consistent with the gelation of the free (isotropic) alumina polymers to form a more amorphous (anisotropic) polymer network. The markedly restricted molecular motion of the polymers coupled with the increase in anisotropy of the sol during gelation both contribute to the pronounced line broadening that is observed during the sol-gel transition.

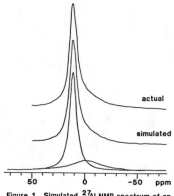

Figure 1. Simulated ^{27}Al NMR spectrum of an alumina sol

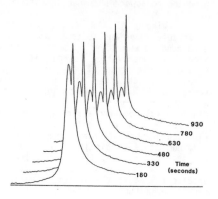

Figure 2. ^{27}Al NMR Spectra of an aluminum sec-butoxide derived sol taken at intervals of 150 seconds following addition of HNO$_3$

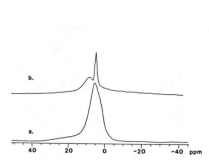

Figure 3. ^{27}Al NMR resonances of an Aluminum Alcosol a.) before and b.) 14 minutes after addition of nitric acid

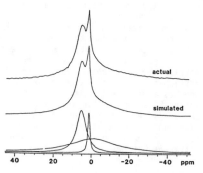

Figure 4. Deconvoluted ^{27}Al NMR Spectrum of an acid-treated Alumina Sol.

^{27}Al NMR spectra of the resulting transparent gel taken 24 hours after addition of the acid revealed that the peak at 1.5 ppm had dramatically sharpened (see Figure 6). This peak presumably arises from $[Al(H_2O)_6]^{3+}$ that is dissolved in the interstitial water contained in the gel. From intensity comparisons with stock solutions of aqueous aluminum trichloride [6-7,11], it was estimated that the concentration of aluminum hexahydrate in the gel was 0.03M. This corresponds to approximately three percent of all the aluminum in the sample.

Similiar results were obtained when HCl was used as the gelling agent. The time dependent features of the ^{27}Al NMR spectra were very

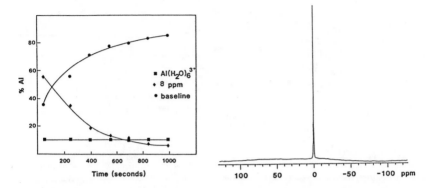

Figure 5. Relative Intensities of Al sol resonances after addition of acid

Figure 6. ^{27}Al NMR Spectrum of an HCl-treated Alumina Sol taken 24 hours after gelation.

comparable to the HNO$_3$ modified sol. Both acids produced gels that were clear in appearance approximately 25 minutes after their addition to the sol with comparable amounts of interstitial $[Al(H_2O)_6]^{3+}$ incorporated into the final gel structure.

A similiar series of gelation experiments were conducted on rotary evaporated alumina sols that did not contain any sec-butanol. The removal of the alcohol was found to profoundly change the overall gelation kinetics of the sol. In contrast to previous results where addition of the acidic gelling agent precipitated rapid changes in the ^{27}Al NMR spectrum of the sol, addition of the HCl to the alcohol-free sol gave no pronounced change in the spectrum for 40 minutes. Figure 7 shows a series of spectra of the acid-treated sol taken 0.03, 0.66, and 5 hours following addition of the HCl. Although the formation of the gel is obviously much slower, many similiarities are evident between the spectra of the original and rotary evaporated alumina sols in intermediate stages of gelation. The appearance of the spectra at 5 hours is very comparable to the sec-butanol modified gel spectra obtained after 14 minutes.

GELATION WITH BASE

The gelation experiment was carried out by mixing 0.33M NH$_4$OH (1mL) with 15mL of alumina sol that contained sec-butanol (base/Al=0.02). Addition of the base was made rapidly via syringe since contact of the base with the sol caused the formation of local precipitates. The sample was shaken vigorously after addition in order to evenly distribute the precipitates before placing the sample in the spectrometer for study.

A series of spectra obtained on a base treated sol is presented in Figure 8. Unlike the acid-treated sols that contained sec-butanol, there was no rapid decrease in the intensity of the 8 ppm line upon addition of the ammonium hydroxide solution. The main change in the spectrum consisted of the gradual dissappearance of the 8ppm resonance. Unlike the acid-treated sols, there was no peak at 1.5ppm that could be assigned to monomeric octahedral aluminum species, consistent with the chemistry of this complex [22-23]. The sol gelled four hours after addition of the ammonium hydroxide. ^{27}Al NMR spectra of the gel 24 hours later exhibited a

192

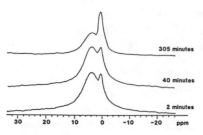

Figure 7. ^{27}Al NMR Spectra of an alcohol-free sol treated with HCl

305 minutes

40 minutes

2 minutes

30 20 10 0 -10 -20 ppm

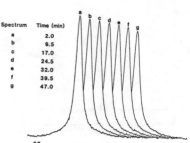

Spectrum	Time (min)
a	2.0
b	9.5
c	17.0
d	24.5
e	32.0
f	39.5
g	47.0

Figure 8. ^{27}Al NMR Spectra of a base-treated Aluminum Alcoxide derived Sol.

broad, poorly resolved peak at 10 ppm.

DISCUSSION:

Overall, ^{27}Al NMR spectroscopy is an excellent method for examining the gelation chemistry of Al sols. The high natural abundance (100%), excellent sensitivity, and relatively short relaxation times of the nucleus permit the rapid acquisition of data with good signal to noise levels. This technique is particularly well adapted for studying the gelation kinetics of Al containing sols. By examining the dependance of the gelation rate on such factors as the identity and concentration of alcohol and gelling agent, this technique provides an in situ method for quantitatively determining the impact of these variables on the gelation characteristics of these systems.

Although following the gelation process by ^{27}Al NMR is fairly straightforward, it is difficult to determine how much of the Al in the sample is actually observed during the experiment. Since aluminum-27 is a quadrupolar nucleus (spin 5/2), small changes in the local environment (i.e., ligand arrangement) surrounding the nucleus can produce dramatic changes in the linewidth and intensity of the resonance (due to electric field gradients). If the aluminum atom is sitting in a highly symmetric environment, the transitions between the spin states will be degenerate and give rise to a single sharp line (or group of lines) as is found for $[Al(H_2O)_6]^{3+}$. If the symmetry of the complex is lower however, the degeneracy of the spin states will be removed and fewer transitions will be observed. In a situation where only the $+\frac{1}{2} - -\frac{1}{2}$ transition is sampled, the maximum intensity of the resonance can only be 33% of that for a highly symmetric complex [24-25].

Although it is difficult to quantify how much aluminum is giving rise to the NMR signal, the experiments described above provide an excellent, qualitative description of the gelation process. It is also noteworthy that the spectrum of the aluminum alcoxide sol presented in Figure 1, corresponds closely to that reported recently by Roy and coworkers [26] for the solid state ^{27}Al NMR spectrum of an alumina gel prepared by the hydrolysis of aluminum sec-butoxide. ^{27}Al MASNMR spectra obtained on gels prepared using this precursor exhibited a single broad resonance near 7 ppm. The close similiarity between the spectra of the sol and gel forms of materials prepared from aluminum sec-butoxide is compelling evidence for the promise this method holds for the characterization of many "amorphous" materials of interest in ceramics.

CONCLUSIONS

^{27}Al NMR spectroscopy was found to be a powerful tool for examining the gelation chemistry of alumina sols. The relative intensities of the resonances and their corresponding linewidths are easily measureable and provide an excellent method for examining the gelation kinetics of aluminum containing sols. Coupled with viscometry, this technique holds considerable promise for examining the complex gelation chemistry of aluminum containing sols in a systematic manner [27].

REFERENCES

1. D. R. Uhlmann, B. J. J. Zelinski and G. E. Wnek, Better Ceramics Through Chemistry; Mat. Res. Soc. Symp. Proc. Vol. 32, Edited by C. J. Brinker (Elsevier, NY, 1984), pp. 59-70.
2. I. Artaki, T. W. Zerda, and J. Jonas, Mat. Lett. 3, 493 (1985).
3. G. Orcel and L. Hench, J. Non-Cryst. Solids 63, 177 (1986).
4. O. Kriz, B. Casensky, A. Lycka, J. Fusek, and S. Hermanek, J. Magn. Res. 60, 375 (1984).
5. J. W. Akitt and A. Farthing, J. Chem. Soc. Dalton 1981, 1624.
6. J. Y. Bottero, J. M. Cases, P. Rubini, and F. Flessinger, C. R. Acad. Sc. Paris Serie D 284, 1033 (1977).
7. J. Y. Bottero, J. M. Cases, F. Flessinger, and J. E. Poirier, J. Phys. Chem. 84, 2933 (1980).
8. J. W. Akitt, and A. Farthing, J. Magn. Res. 32, 345 (1978).
9. J. W. Akitt, N. N. Greenwood, B. L. Khandelwal, and G. D. Lester, J. Chem. Soc. Dalton 1972, 604.
10. J. W. Akitt and A. Farthing, J. Chem. Soc. Dalton 1981, 1606.
11. J. W. Akitt and A. Farthing, J. Chem. Soc. Dalton 1981, 1617.
12. J. W. Akitt, N. N. Greenwood and G. D. Lester, J. Chem. Soc. (A) 1969, 803.
13. D. Muller, E. Jahn, B. Fahlke, G. Ladwig, and U. Haubenreisser, Zeolites 5, 53 (1985).
14. D. Freude and H. J. Behrens, Crystals Res. Tech. 16, K36 (1981).
15. C. A. Fyfe, J. M. Thomas, J. Klinowski, and G. C. Gobbi, Angew. Chem. Int. Ed. Engl. 22, 259 (1983).
16. B. E. Yoldas, Amer. Ceram. Soc. Bull. 54, 289 (1975).
17. B. E. Yoldas, J. Mater. Sci. 10, 1856 (1975).
18. B. E. Yoldas, Amer. Ceram. Soc. Bull. 59, 286 (1975).
19. B. E. Yoldas, Amer. Ceram. Soc. Bull. 59, 479 (1980).
20. B. E. Yoldas, J. Amer. Ceram. Soc. 65, 387 (1982).
21. W. L. Olson and L. B. Welsh, to be published.
22. E. Grunwald and D. W. Fong, J. Phys. Chem. 73, 650 (1969).
23. D. W. Fong and E. Grunwald, J. Am. Chem. Soc. 91, 2413 (1969).
24. D. Fenzke, D. Freude, T. Frohlich and J. Haase, Chem. Phys. Lett. 111, 171 (1984).
25. E. Fukushima and S. B. W. Roeder, Experimental Pulse NMR A Nuts and Bolts Approach, 1st ed. (Addison-Wesley Publishing Co., Reading, MA, 1981), pp. 157-161; pp. 106-112.
26. S. Komarneni, R. Roy, C. A. Fyfe, and G. J. Kennedy, J. Am. Ceram. Soc. 68, C243 (1985).
27. W. L. Olson and L. J. Bauer, to be published.

MICROSTRUCTURAL STUDIES OF TRANSPARENT SILICA GELS AND AEROGELS

PARAM H. TEWARI, ARLON J. HUNT, K.D. LOFFTUS AND J.G. LIEBER,
Applied Science Division, Lawrence Berkeley Laboratory, University of California,
Berkeley, CA 94720, USA

ABSTRACT

Transmission electron microscopy, spectroscopy, light scattering and surface area measurements have been used to characterize the microstructural details of transparent silica aerogels. Light scattering intensity measurements have also been used to differentiate acid and base catalyzed hydrolysis and condensation reaction kinetics of $Si(OC_2H_5)_4$. A plot of logarithm of the light scattering intensity vs. logarithm of time during gelation gives distinctly different slopes for acid and base catalyzed systems (6 for acid and 2 for base). These slopes for the two systems can be explained on the basis of the point of zero charge of silica and the resulting interactions of the uncharged and charged particles in the systems.

The diameter of the scatterers calculated from light scattering data using Rayleigh theory for independent scattering is much larger than the particle diameter measured from TEM. The particle size initially observed by TEM before beam damage is in the range of 2-3 nm. However, due to sintering in the electron beam, the particle size grows to 5 nm or larger. The particle sizes calculated from surface area measurements are in agreement with TEM data. The transparency of the aerogel is determined by the heterogeneity introduced by the pore sizes which, depending on the mode of preparation, range from 20-60 nm (macro-pores).

INTRODUCTION

Aerogels are low density porous materials made by supercritical drying of gels [1-2]. We have been developing an industrially viable new process for making transparent silica aerogels for window glazings because of their excellent thermal insulation properties [3-5]. Transparent silica aerogels have been produced by base catalyzed hydrolysis of tetramethylorthosilicate, TMOS [6-9]. Since TMOS is highly toxic, a less toxic and cheaper material, tetraethylorthosilicate, TEOS, is more desirable for making the process industrially acceptable. Base catalysis of TEOS has not produced transparent aerogels until recently [10-14]. Acid catalyzed TEOS hydrolysis produces transparent silica gels but exhibit significant shrinkage during aging [14-16].

Acid and base catalysis of TEOS is reported to differ in many characteristics [17]. It is postulated that acid catalysis of TEOS produces gels through a polymeric precursor, and base catalysis produces particles that link to form gels [17-19]. There are numerous studies that provide evidence of particle formation in the base catalysis of metal alkoxides. However, the evidence of a polymeric precursor for silica gels has been indirect at best [19-20].

We report here some microstructural properties of these gels and aerogels by transmission electron microscopy, light scattering, transmission spectra and surface area studies that are helpful in elucidating gel formation mechanisms.

EXPERIMENTAL

Hydrolysis and condensation of TEOS, $Si(OC_2H_5)_4$ was performed using acid and base catalysis. The molar ratio of H_2O: TEOS, catalyst concentration, temperature and solution pH were varied over a wide range of conditions to optimize the properties of the alcogel/aerogel [11].

Supercritical drying of the alcogels was achieved by a process developed in our laboratory [10]. The angular distribution of light scattering intensity of the sol-gel system was measured from the initial fluid state through the gelation and aging process, by a technique described earlier [21]. Helium-neon (633 nm) and helium cadmium (442 nm) laser sources were used for the measurements. These measurements were helpful in optimizing a formulation with a minimum in the scattering intensity for the gels.

Transmission electron microscopy on supercritically dried aerogels was performed using a Zeiss 10A TEM operating at 60-80 kV. Samples for TEM studies were prepared by several techniques. In one approach, the aerogel sample was infiltrated with Medcast 812 resin (Ted Pella Inc.) for 42 hours and then cured at $60^{\circ}C$ for 48 hours. Thin sectioning of such samples   with glass knives was not practical due to the inherent hardness of the sample. Thin sections of 50-70 nm thickness were successfully cut using a diamond knife and a Reichart Ultracut E ultramicrotome. The sections were mounted on carbon stabilized, formvar coated 400 mesh copper grids. However, this technique did not provide ultimate resolution of the aerogel structure due to limitations in section thickness and from the resin itself.

In another approach, aerogel was ground to a fine dust and applied to a carbon stabilized, formvar coated copper grid. To enhance adhesion of the powder to the formvar grids, the grids were first heated to approximately $50^{\circ}C$ and a cotton swab saturated with ethylene dichloride was waved over the warmed grids. The powdered aerogel was applied to the grids while still warm. The grids were stabilized by vacuum carbon deposition. This preparation technique gave reasonable microstructural details, but introduced the possibility of damage to the structure caused by grinding.

The most successful approach used flakes of silica aerogel. The flakes were separated by a sharp knife from the outer periphery of the containers used in supercritical drying of aerogel. The flakes were mounted on carbon stabilized, formvar coated 400 mesh copper grids and vacuum carbon coated. This preparation provides the finest resolution of the aerogel ultrastructure although sintering of the aerogel subunits into larger sizes could be seen during exposure in the electron beam. The inherent thermally insulating properties undoubtedly promoted such sintering. Minimal exposure times were used when photographing the specimens to minimize sintering and the resultant specimen drift.

Transmission spectra were obtained using a Perkin Elmer Lambda 9 UV/VIS/NIR spectrophotometer. The aerogel samples were heated in a furnace and stored in a dessicator for optical studies. Surface area was measured by the BET nitrogen adsorption method.

RESULTS AND DISCUSSION

Optical Transmission Studies

Figure 1 shows the optical transmission spectra of aerogels obtained from base catalysis of TEOS and supercritically dried by two methods: 1) by CO_2 at 40°C and 2) by C_2H_5OH at high temperature and pressure (270°C, 1800 psi). It is clear that

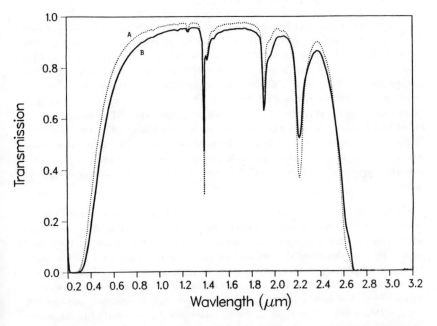

1. Optical transmission of base catalyzed TEOS aerogels supercritically dried by (A) CO_2 at 40°C and (B) by high temperature and high pressure supercritical drying by C_2H_5OH. (both samples 16 mm thick and heated in oven at 200°C to remove moisture, after supercritical drying).

CO_2 supercritically dried samples have higher optical transmission. Similarly, a CO_2 dried TMOS aerogel sample showed improved transmission [12]. The peaks between 1400 and 2000 nm are due to adsorbed H_2O and between 2200 and 2800 nm are combinations of O-H and Si-O fundamentals. The rising part of the curve between 200 and 700 nm is due to scattering.

Rubin and Lampert [5] analyzed this part of the curve for Rayleigh scattering and fitted the data to the expression:

$$\ln I = \ln I_o - K \left(\frac{1}{\lambda}\right)^4 \qquad [1]$$

where I is the transmitted intensity at wavelength λ, and K is a constant determined by fitting the data. Assuming independent Rayleigh scattering from silica spheres and deriving the total number of scatterers per unit area from sample thickness and density of the aerogel, they obtained a characteristic size of the scatterer (12.4 nm). They also showed that this value of the scatterer was consistent with the particle size obtained from their TEM work (10.4 nm).

Accepting similar validity of Rayleigh scattering conditions, we fitted our data to the above expression [Equation 1]. The least square fit of the data was good with a correlation coefficient of 0.9999. The characteristic scatterer diameter derived from curve fitting based on the index of refraction of silica is 8-12 nm, depending on the aerogel clarity. This value is significantly larger than the particle sizes determined by electron microscopy discussed later. It seems that the validity of Rayleigh scattering for SiO_2 particles in the aerogels is questionable. The match between TEM and optically derived particle sizes obtained by Rubin and Lampert [5] was likely due to sintering of the SiO_2 particles in the electron beam. The primary cause of scattering in aerogel is attributed to inhomogeneities in the average density of the scattering system [22]. This is verified by our TEM and light scattering studies discussed later.

Figure 2 is a transmission electron micrograph of a flake of base catalyzed TMOS aerogel. These flake preparations enable the best resolution of the aerogel ultrastructure since they frequently contain exceedingly thin areas on their outer edges. In these thin areas the two dimensional structure of the gel is visible, unlike the microtomed samples that contain many layers of pores and particles. The linking of the particles into chains and cross-linking of the chains can be seen in the figure. From this micrograph it can be determined that the smallest particles sizes are less than 5 nm. A variation of particle sizes can also be observed in this micrograph. The variation occurs because some areas of the micrograph were exposed to the electron beams for longer times. This exposure causes the particles to sinter to larger sizes. This sintering can be easily observed as the sample is exposed to the beam. Some sintered particles are as large as 25 nm. Similar results are obtained for other TEOS aerogels using thin section TEM studies. The particle sizes are 3-4 nm for both acid and base catalyzed aerogels, although there are some differences in the pore sizes due to differences in density of aerogels and their method of preparation. Nevertheless, the particles sizes are similar in all aerogels studied.

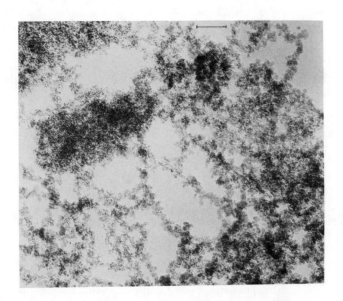

2.   Transmission electron micrograph of a base catalyzed TMOS aerogel after carbon coating. The bar corresponds to 100 nm.

Transmission electron micrographs also show that aerogels contain a distribution of pore sizes. Identifiable larger pores (macropores) accounted for about 11% of the volume in the examined samples. It is these macropores that seem to account for the intensity of the light scattering and hence on the clarity of the aerogels. It will be shown later that minimizing the presence of these macropores give aerogels of lower light scattering with better clarity.

Surface Area of Aerogels

The measured surface areas for the acid and base catalyzed aerogels, whether dried by CO_2 supercritical drying or by high temperature C_2H_5OH supercritical drying, are between 850-960 m^2/g. The particle size calculated from the surface area, assuming independent mono size spheres, is 3.0±0.1 nm. This value is consistent with the sizes determined by TEM measurements. Thus, particle sizes obtained from surface area, and TEM studies do not show any dependence on the preparation process. However, hazier samples gave larger scatterer diameter as calculated by Rayleigh theory, because of larger heterogeneity in macropores (also seen in TEM studies).

Differences Between Acid and Base Catalyzed Aerogels as Observed by Light Scattering Studies of Alcogels

There is a distinct difference in the texture and shrinkage between the acid and base catalyzed TEOS alcogels. The acid catalyzed TEOS alcogels are quite stiff in touch and shrink significantly during aging, whereas the base catalyzed TEOS alcogels are soft in touch and shrink little compared to the acid catalyzed alcogels. These differences persist after they are dried to aerogels. Acid catalyzed aerogels are more brittle and stiff than the base catalyzed aerogels that are more resilient to flexing. However, these are the qualitative physical differences. The observed differences from light scattering studies are discussed below.

During gelation, light scattering intensity of the alcosol increases with time until it reaches a limiting value for both the acid and base catalyzed systems. The scattering intensity as well as its rate of increase with time depends on the conditions of the initial fluid mixtures. On filtering the reactant mixture through a 0.2 μm millipore filter, the value of the intensity as well as its rate of increase is reduced significantly (by a factor of 2 to 4). Also, filtering minimizes variations in the rate of increase of intensity from run to run. A further decrease in the absolute value of the light scattering intensity is obtained by steam cleaning the scattering cell. This decrease in the intensity is caused by the removal of the suspended impurities which act as scattering centers and centers for nucleation sites for the sol-gel system. Once these impurities are removed, homogeneous nucleation dominates, resulting in less scattering. Therefore, all measurements for light scattering rates were made in steam cleaned scattering cells containing samples that were carefully filtered.

The characteristic difference in the light scattering intensity with time (nomalized to gelation time) for the acid and base catalyzed TEOS system is shown in figure 3. The time axis is nomalized to gelation time to accommodate vastly differing times in the same graph. It is apparent that changes in the light scattering intensity for the acid and base catalyzed systems are different. The increase in light scattering intensity ceases well before full gelation occurs in the base catalyzed system (Figure 3-B), whereas in the acid catalyzed system, scattering intensity continues to increase much after gelation (Figure 3-A). This difference may be explained by the nature of the linkages in the two cases. In the base catalyzed system, SiO_2 particles are highly charged and have a high repulsive barrier. However, when this barrier is overcome, the particles come close together and strong linkages occur, with little changes in the linkages that may produce heterogeneity in the pore structures. Thus, there is no change in the light scattering once the linkages have occurred. On the contrary, the acid catalyzed system has a pH of ≈ 1.6 which is close to the point of zero charge of SiO_2. At this pH, the attachments that occur are very weak. During gelation rearrangements of microgels occur giving new macropores that produce heterogeneity and an increase in the light scattering intensity. The low repulsive barrier allows slow rearrangement of the linked chains in the microgels leading to shrinkage and increase of light scattering during aging.

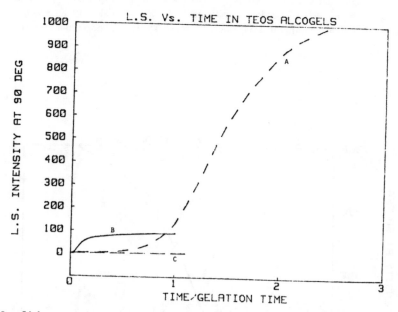

3. Light scattering intensity measured at 90° for different alcogels during gelation.

(A) Acid catalyzed TEOS, H_2O: TEOS = 5.9:1 Gelation time 1.5 hrs: SiO_2 4 volume percent.

(B) Base catalyzed TEOS, H_2O: TEOS = 25:1 Gelation time 5 hrs: SiO_2 2 volume percent.

(C) Acid catalyzed TEOS, H_2O:TEOS 2.3:1 Gelation time 120 hrs: SiO_2 12 volume percent.

The high density alcogel (Figure 3-C), on the other hand, shows very low light intensity and very little increase in light scattering intensity with time. This is because the increase in the TEOS content gives rise to increased nuclei population which reduces the number of macropores and ultimately the pore heterogeneity.

202

Thus the high density system has a very homogeneous microstructure with little heterogeneity during the entire gelation process. The resultant homogeneity of the microstructure produces insignificant rise in the light scattering intensity during gelation.

This observation of almost no increase in the light scattering intensity during gelation for the high density silica alcogels occurs for both with and without fluoride ions added to the system. The increase in the light scattering intensity is less than 10^{-3} of that of the low density silica gels. Also, the absolute value of the initial light scattering intensity for the high density gels is 10^3 smaller than that observed for the low density alcogels. These observations along with the TEM studies confirm that the light scattering intensity is associated with the heterogeneity of the macropores in the alcogels of the low density gels. As the silica density increases, the macropores almost disappear leaving only micropores with a homogeneous structure.

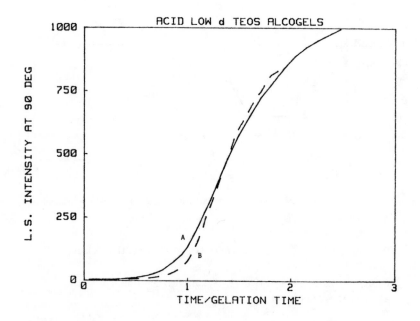

4. Light scattering intensity for acid catalyzed TEOS with two H_2O: TEOS ratios.
    (A) H_2O : TEOS = 6:1, Gelation time 1.5 hours.
    (B) H_2O : TEOS = 2:1, Gelation time 20 hours.

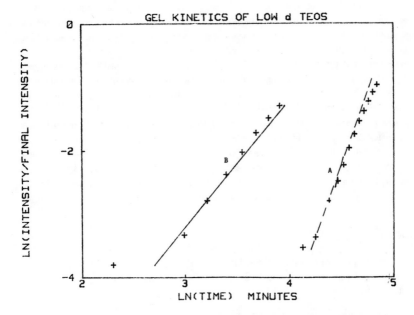

5. Log light scattering vs. log time for acid and base catalyzed TEOS.
(A) acid, (B) base.

In Figure 4, the light scattering intensity data for two acid catalyzed systems with H_2O: TEOS ratios of 2:1 and 6:1 are given. Although there is a significant difference in the gelation times of the two systems, the shape of the curves remains essentially the same for the two systems. (The sample with 6:1 H_2O: TEOS ratio gels in 1.5 hours, compared to 20 hours for a H_2O: TEOS ratio of 2:1). A small difference in the shape of the two curves is seen mainly in the early stages of the gelation process.

When the log of the scattering intensity is plotted against the log of time during gelation for the acid and base catalyzed TEOS alcogels, a linear graph is obtained for the early part of the gelation curve (Figure 5). The slopes of these graphs are 5.8 for the acid catalyzed and 2.2 for the base catalyzed gels respectively. If it is assumed that the gel formation is due to particle precursors (growth occurs by accretion on a sphere), a slope of six should be obtained because scattering intensity is proportional to the square of the volume. On the other hand, for a linear (polymeric) growth a slope of two is predicted [23]. The experimental data for the acid catalyzed system gives a slope of 5.8 and for the base catalyzed system 2.2, respectively. This is contrary to our concept of a linear polymeric growth in the acid and particle growth in the base catalyzed gelation of TEOS. Nevertheless, a slope of six and two for the acid and base catalyzed gelation can be rationalized on colloidal principles. Acid catalyzed TEOS systems are near the point of zero charge of silica. As such, there is very small repulsive barrier between the condensing silica system. Therefore, linking or growth in such systems will be on all sides, i.e., three dimensional or volumetric growth that will give a slope of six for log-log plot. However, in the base catalyzed case, the particles are highly charged (pH = 9). In this system agglomeration or particle attachment occurs preferentially at the ends, to surmount the charge barrier [24-25], and hence a slope of two. Thus, these results give a crude picture of the physical processes occurring during the gelation process. For a more quantitative explanation, kinetics of the initial processes need to be studied by a more sensitive light scattering technique, which can follow the very early events in the gelation process. Atomic Resolution TEM work on supercritically dried microgels and aerogels needs to be done to study their early attachments. However, these studies need to be done on a cold stage to minimize sintering damage to the microstructure.

ACKNOWLEDGEMENTS

The authors are extremely grateful to Dr. Paul Berdahl for his keen interest in the work and for helpful discussions and to Dr. Michael Rubin for the use of his spectrophotometer.

This work was supported by the Assistant Secretary for Conservation and Renewable Energy, Office of Solar Heat Technologies, Passive and Hybrid Solar Division of the U.S. Department of Energy under Contract No. DE-AC03-76SF00098.

REFERENCES

1. S.S. Kistler, Nature, 127, 741, (1931).

2. S.S. Kistler, J. Phys. Chem. 36, 52, (1932).

3. R. Caps and J. Fricke, Solar Energy 36, 361 (1986).

4. P. Scheuerpflug, R. Caps, D. Buttner, and J. Fricke, Int. J. Heat Mass Transfer, vol. 28 number 12, 2299 (1985).

5. M. Rubin and C. Lampert: Solar Energy Materials 7, 393 (1983).

6. S.S. Henning and L. Svenson, Physic. Scripta 23, 697, (1981).

7. G. Poelz, R. Riethmulle, Nuclear Instruments and Methods, 195, 491, (1982).

8. G.A. Nicholaon and S.J. Teichner: Bull. Soc. Chim. 5, 1900, 1909, (1968).

9. S.A. Teichner, G.A. Nicholaon, M.A. Vicarini and G.E.E. Gardes, Adv. Colloid and Interface Sc. 5, 245 (1976).

10. P.H. Tewari, A.J. Hunt, and K.D. Lofftus, Materials Letters 3, 362, 1985.

11. P.H. Tewari, A.J. Hunt and K.D. Lofftus, 2nd International Conference Proceedings on Ultrastructure Processing of Ceramics, Glasses, and Composites, Feb. 28, 1985.

12. P.H. Tewari, A.J. Hunt and K.D. Lofftus, Proceedings of the 1st International Symp. on Aerogels, Wurzburg, Germany, Sept. 23-25 (1985).

13. P.H. Tewari, A.J. Hunt, K.D. Lofftus, Proceedings of the 1st International Symp. on Aerogels, Wurzburg, Germany, Sept. 23-25 (1985).

14. R. Russo and A.J. Hunt, accepted for publication in Journal of Non-Crystalline Solids (1985).

15. W.J. Schmitt: M.S. Thesis, Dept. of Chem. Eng., University of Wisconsin, 1982.

16. W.J. Schmitt: Annual Meeting of AICHE, New Orleans, 1981.

17. R.K. Iler, Chemistry of Silica, John Wiley, New York, 1979 (page 225 and references therein).

18. B.E. Yolds, MRS Symposium Proc. Vol.24, 1984, Elsevier Publishing Co., page 291.

19. S. Saka and K. Kamiya, J. Non-Crystalline Solids, 48, 31, (1982).

20. C.J. Brinker K.D. Keefer, D.W. Schaefer and C.S. Ashley, J. Noncrystalline Solids, 48, 47, 1982.

21. A.J. Hunt: Ultra Structure Processing of Ceramics, Glasses and Composites, Edited by L.L. Hench and D.R. Ulrich, John Wiley & Sons, NY 1984.

22. A.J. Hunt and P. Berdahl, Materials Research Soc. Proceedings 32, 275, 1984, North Holland Press, New York 1984.

23. Paul Berdahl, private communication.

24. R.K. Iler, Chemistry of Silica, p. 226, John Wiley, New York, 1979.

25. A.G. Rees, J. Phys. Chem. 55, 1340 (1951).

STEREO-TEM IMAGING OF SOL-GEL GLASS SURFACES

George C. Ruben* and Merrill W. Shafer**
*Dept. of Biological Sciences, Dartmouth College, Hanover, NH 03755
**IBM T. J. Watson Research Center, Box 218, Yorktown Heights, NY 10598

ABSTRACT

Sol-gel top and cross-fractured surfaces have been Pt-C replicated (9.7-11.3 Å thick) and carbon film backed (100 -145 Å thick) and separated from the gel with an HF solution. Samples made from both base catalyzed LUDOX/Kasil mixtures and acid catalyzed alkoxides were studied with tilt series of 10^4 to 10^5 magnification TEM micrographs of the surface replicas. The LUDOX/Kasil gels contained spherical particles with a raspberry-like morphology and the acid catalyzed alkoxides are composed of densely packed filaments and no apparent particles. The smallest structural entity in both gels are Pt/C coated filaments (12-14Å) estimated to be 1.5-2.7 Å in width. The pore sizes were characterized in both gel types, though TEM pore measurements were only made on the LUDOX/Kasil gels.

INTRODUCTION

The formation of glasses and ceramics by the "gel route" has advanced to a state where the preparations can be described in general terms of their processing, such as, by hydrolysis of alkoxides or by gelling LUDOX type materials[1,2,3]. It is currently accepted that the hydrolysis of alkoxides at lower pH (<8.5), low H_2O/Si (4-7.5) and low temperatures (~20°C) favor the formation of weakly crosslinked polymeric silica chains which, when dessicated and heated below their densification temperatures, yield "polymeric" type glasses. On the other hand, high pH (>9), high H_2O/Si (>10) and high temperatures (40-80°C) favor the formation of highly branched silica chains resulting in a more particle-like structure similar to those seen in LUDOX based glasses[1]. Despite the important relationships between structural properties, there has not been a great amount of effort to verify the proposed structural models, particularly in the 300 - 800° C temperature range. In this paper we describe our work of stereo-TEM visualization of surface replicated sol-gel silica glasses prepared by both acid and base hydrolized alkoxide solutions as well as from the gelling of LUDOX/Kasil mixtures.

EXPERIMENTAL

Alkoxide and LUDOX/Kasil sample preparation have already been described in this proceedings[4] (LUDOX is a trade name of DuPont colloidal silica in aqueous solution and Kasil is a potassium silicate solution which contains 38% K_2O and 62% SiO_2). The gels were characterized in terms of their pore sizes, surface areas, density and in some cases their particle size[4]. Mercury porosimetry was used extensively for the pore size and surface area measurements. Some samples were characterized by gas adsorption/desorption isotherm analyses using a Quantasorb Sorption unit. For high resolution work a specially constructed low temperature system was used[5]. The sol-gel samples were replicated at a 45° or a 90° angle with 9.7-11.3Å of Pt/C [6] at -178°C in a 5x10^{-8}torr. vacuum and backed with100-145Å carbon film. These experiments were performed in a Balzer's 301 freezing microtome modified with a cryopump and with a rebuilt specimen stage[7]. The replicas were separated from the sol-gel glasses with 49% aqueous HF, were rinsed on de-ionized water overnight and picked up on grids as previously described.[8] The microscope, tilt stage and operating conditions and the method for making reversal negatives for prints have been described.[8] Pictures of the sol-gel fine structure were taken at 96,850X and photographed at 10° tilt intervals (tilt axis orthogonal to the shadow direction) in a series of micrographs sometimes ranging from +50° to -50°. The filament measurements were made as previously described [9,13] except the measurement magnification was 6.9X10[6]. Similar replication procedures were used in fig. 1 to demonstrate the detection limits of the technique. Fig. 1a demonstrates that a 10Å high mica step[10] on a flat surface is visible around its full perimeter and in fig. 1b pectin polysaccharide chains about 7Å in width[11] are easily seen on the surface of a freeze-dried pectin gel[12].

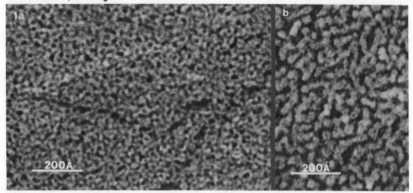

Figure 1a. A Pt/C replicated (10.6Å thick)10Å high arrow-head shaped mica step[10] is seen all the way around its perimeter. The background metal film filaments are 1-1.5Å thinner than the Pt/C film thickness measured on the quartz crystal monitor. Figure 1b. A freeze-dried purified pectin gel surface. This polymeric sugar chain contains long unbranched regions of helical galacturonosyl residues (13.4Å pitch[11]) which can make orthogonal bends at 1,2-linked rhamnose residues. The galacturonosyl residues are approximately 6.8X7.2Å in cross-section[11] and are 22-24Å wide with a 16.9Å thick coating of Pt/C[12].

RESULTS AND DISCUSSION

Table I summarizes the preparation and characterization data for several of the samples we examined. A range of average pore sizes from 40 to ~380Å was selected from Hg porosimetry measurements. For samples L15K - 10.5 and L5K - 11 BET analyses showed good agreement (within 8%) with the values given above. The lower values and wider distributions obtained from the stereo-micrographs are a result of estimating the pore size by measuring the longest and shortest chord of each pore and entering both values into a distribution (fig.2b). This kind of measurement estimates the anisotropy of pore shape and pore entrance limits. In order to correlate Hg porosimetry measurements with TEM pore measurements, pore perimeter measurements reduced to an average equivalent cylindrical diameter are

TABLE I Pore data for a number of sol-gel type glasses

Sample #@	Temp,C°	Average Pore Diameter Hg Porosimetry,Å *	TEM Pores,Å+	TEM Pore Axial Ratio++	Surface Area,M²/g
AM6-3	600	48±8			480
AM6-8.5	600	40±7			530
L5K-11	500	96±12	63±6	1.9±0.2	221
L15K-10.5	275	50±7			325
L15K-10.5	550	380±40	221±13	2.7±0.2	94
L25K-11	660	115±16			210

@ A refers to alkoxide, M to the methoxide, the 6 is the H_2O/Si ratio and the -3 is the pH. L5K is Ludox/Kasil = 95/5.
* 80-85% of the pores are within the measurement limits shown.
+ Pores were identified by stereo-viewing and were approximated by measuring each pore's longest and shortest chord which was recorded in a pore size distribution. Since the holes had irregular shaped openings, the average of the two chords will be an underestimate of the true pore equivalent diameter.
++The ratio of the two chords+ gives the axial ratio or pore opening asymmetry.

needed for a relevant comparison. It is clear from Fig. 2 that, although there are a number of pores in the 221±13Å range, most of the pore openings in preparations Pt/C shadowed at 90° to the top surface are noncylindical. An estimate of this asymmetry was made by calculating the average axial ratio (table I) of the longest to shortest chord measured for each pore. The top surface of the L15K-10.5 glass gave the same distribution mean (229±15Å) and pore axial ratio (3.0±0.3) as the cross-fractured glass surface measurements reported in table I.

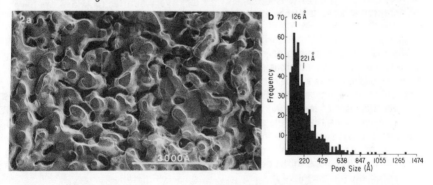

Figure 2a. Cross-fractured L15K-10.5 (550°C) Pt/C replicated (11.3Å thick) and backed with 136Å of carbon. The average pore size in this cross-fractured surface was 221±13Å and 229±15Å in the top surface. The gel interstices are cross-fractured in this surface whereas, they are smoothly rounded in the top surface
Figure 2b. This distribution was made from a cross-fractured surface of the gel shown in fig. 2a by measuring the longest and shortest chord of each pore and entering both values into the distribution. The mode of this distribution is 126Å, the mean and its 95% fractile error is 221±13Å and the standard deviation is 161Å (n=589).

In Fig. 3a we show micrographs of cross-fractured surface L5K - 11 (same morphology as top surface) and in fig. 3b the top surface of AM6 - 3. The structural differences are clearly seen. The L5K - 11 is LUDOX-based, prepared at pH 11, and is distinctly particulate or colloidal, while the AM6 - 3 glass, prepared from the methoxide at pH 3 shows no particle-like structure. The particle diameter is quite uniform and measures 178±4Å (S.D.=32Å, n=251). Since the starting LUDOX particle size measured 220Å, an obvious reduction occurred during gellation and/or firing.

Figure 3a. The cross-fractured surface of L5K-11 Pt/C coated with 11.3Å and backed with 136Å of carbon. This surface shows Pt/C coated spheres (189±4Å) with a raspberry-like morphology that are 178±4Å after correcting for the Pt/C coating[13]. Figure 3b. The top surface of AM6-8.5 coated with 9.7Å of Pt/C and backed with 127Å of carbon. This finely fiberous surface looks very similar to AM6-3. We have not detected spheres in either of these gels.

The high magnification stereo-micrographs shown in Fig. 4 (sample L5K - 11 on the left) reveals a filamentous substructure, which is found in both LUDOX and alkoxide-based glasses. The structural units are single straight and twisted filaments with diameters of 1.5-2.7Å. The single twisted filament diameter defines exterior diameter of the cylindrical path of the polysiloxane chain. However, careful examination shows that double and super twisted single twisted filaments are also present. Table II summarizes the size measurements on the various filaments seen in the preparations. This table only records filament types we have recognized from small sampling regions in the gel micrographs. The absence of a filament type in any gel is probably not meaningful.

Pt/C films deposited at 45° angle on ice and cleaved mica (fig.1a) generally formed filaments 1-1.5Å thinner than the Pt/C film thickness measured on the quartz crystal monitor. This apparently makes the polysiloxane chains 1.4 times thicker (9.7Å Pt/C) than the random metal filaments which were only seen (rarely) in the densified L15K-10.5 sample. Furthermore, the Pt-C coating on mica does not show filaments organized in the same pattern that we see in the glasses. In fact, all the replicated samples (i.e., those listed in Table 1) showed a structure suggesting that these filaments are the fundamental unit of sol-gel type glasses. Thus, the difference between colloidal and polymeric "glasses" is in the way the

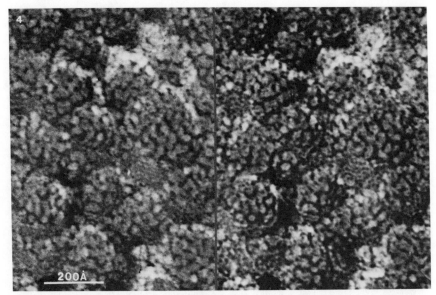

Figure 4. High magnification stereo-micrographs separated by a 10° tilt angle of L5K-11 described in fig. 3a . This stereo-pair clearly demonstrates that the raspberry-like morphology results from the Pt/C coated polysiloxane chains (12-13Å) projecting from its surface and separated by the "micropores". A spherical morphology is likely the result of polysiloxane chain branching.

TABLE II   Average Filament Diameter Corrected for Pt/C Coating[13]

Sample #	Single Straight (SS), Å	Single twisted (ST),Å	Super Twisted ST,Å	Double Filament (DF), Å	ST Twist Period,Å	Super Twist Period,Å
AM6-3	2.7±1.2	13±2.3	29.1±3.5		16±3	33±6
AM6-8.5	2.2±1.2	11±2.2			17±4	
L5K-11	1.5±1.4*	8.2±3.1*		9.7±3*	16±4	
L15K-10.5+	0.8±1.6*	6.6±3.1*		8.8±3*	16±5	

*Pt/C film thickness corrections are applied to filaments shadowed at a 45° angle to the average surface plane by subtracting the Pt/C film thickness measured on the quartz crystal monitor from an average filament width[13]. This correction may be too great by 1-1.5Å for filaments shadowed at a 90° angle to the surface.

+Densified sample; Unshadowed filaments measured 7.7±1Å which is in agreement with the objective aperture imposed diffraction limit of 6.6Å. Thus, the diffraction limit makes all filaments smaller than 6.6Å, this size.

filaments are or are not crosslinked, respectively, are interwoven, and their density of packing . From the filament diameters we measure, it is highly likely that a single filament is in fact a polysiloxane chain (i.e., $(SiO_4)_n$) because estimates of chain widths from bond distances can vary between 2.7-5.3Å[15]. The micropores or crevices between the filaments vary in width from 10-35Å and are in fact often referred to as "micropores" of the sol-gel glasses[14]. Although we have not yet followed the structure of the filaments all the way through the densification process, they appear to coalesce in the L15K-10.5 (550°C) and we would expect at sintering temperatures that individual filaments will no longer be detected above the Pt/C metal film background structure.

ACKNOWLEDGEMENT

We thank the Dartmouth EM facility for its support and G.C.R. thanks GEOM Co. for its financial support and W. Krakow for encouraging this work.

REFERENCES

1. C.J. Brinker and G.W. Scherer, J. Non. Cryst. Solids, 70, 301-332 (1985)
2. E.M. Rabinovich, J.B. Mac Chesney, D.W. Johnson, J.R. Simpson, B.W. Meagher, F.V. DiMarcello, D.L. Wood and E.A. Sigety, J. Non. Cryst. Solids, 63, 155-161 (1984)
3. R.D. Shoup, J. Colloid and Interface Science, 3, 63-69 (1976)
4. M.W. Shafer, V. Castano, W. Krakow, R. Figat and G.C. Ruben, in Mat. Res. Soc. Symp. Proc., Palo Alto, CA, Better Ceramics Through Chemistry, (in press,1986)
5. M.W. Shafer, D. Auschalom, and J.Warnock, in preparation
6. G.C. Ruben, 39th Ann. Meet. Elect. Microsc. Soc. Amer., G.W. Bailey, ed. (Claitors Publishing Division, Baton Rouge, LA) 566; 568; 570 (1981)
7. G.C. Ruben, J. Elect. Microsc. Tech. 2, 53 (1985)
8. G.C. Ruben and K.A. Marx, J. Elect. Microsc. Tech. 1, 373 (1984)
9. G.C. Ruben,K.A.Marx, and T.C. Reynolds, 39th Ann. Meet. Elect. Microsc. Soc. Amer., G.W. Bailey, ed. (Claitors Publishing Division, Baton Rouge, LA) 440 (1981)
10. W.A. Deer, R.A. Howie, and J. Zussman; in Rock-Forming Minerals, Sheet Silicates, (Wiley, NY, 1962-3) vol.3, pg. 4
11. M.D. Walkinshaw and S. Arnott, J. Mol. Biol., 153, 1055 (1981)
12. G.C. Ruben, G.H. Bokelman, and H.H. Sun, 44th Ann. Meet. Elect. Microsc. Soc. Amer., G.W. Bailey, ed. (San Francisco Press, S.F., CA) in press (1986)
13. G.C. Ruben and K.A. Marx , 42nd Ann. Meet. Elect. Microsc. Soc. Amer., G.W. Bailey, ed. (San Francisco Press, S.F., CA) 684 (1984)
14. R.K. Iler, The Chemistry of Silica, (Wiley, NY, 1979)
15. Friedrich Liebau, Structural Chemistry of Silicates, (Springer-Verlag, NY, 1985)

PART IV

Drying and Consolidation

MOVEMENT OF A DRYING FRONT IN A POROUS MATERIAL

T.M.Shaw
IBM
Thomas J. Watson Research Center
Yorktown Heights, NY, 10598

ABSTRACT

A technique is described for directly observing the movement of a drying front through a thin layer of porous material. The observations show that the front moves in abrupt steps and develops an irregular morphology as it advances. Comparison of the experimental images with computer simulations indicates that several features of the drying front can be accounted for by a model based on invasion percolation.

INTRODUCTION

The processing of ceramic powders and precursors by wet chemical and colloidal routes has been identified as an attractive route for obtaining chemically and physically homogeneous materials. An essential step in many of these processes is the removal of the liquid used as the working medium from a porous compact of the inorganic material being processed. Removal of the liquid is commonly achieved by evaporation. It is clear that the capillary forces that arise in such a drying process cause changes in the physical structure of the porous compact. On the one hand, the capillary forces can be desirable as they cause the compact to shrink and so lead to a more densely packed structure. Shrinkage, however, is frequently inhomogeneous and so leads to differential stresses in the compact which cause cracking of the compact on drying.

The origin of the capillary forces that arise during drying is the presence in the pores of curved liquid/vapor menisci that are left behind as a drying front moves through the porous material. Theoretical and experimental analysis of simple geometries have shown that the capillary force acting on the particles in a powder compact has two components [1 – 8] . One component arises directly from the the liquid/vapor interfacial tension and is proportional to area of liquid vapor interface per unit volume. A second component comes from the pressure difference across the curved menisci and leads to a term that is proportional to the product of the pressure difference and the volume fraction of liquid in the compact [8]. In the early stages of drying, when there is a substantial volume fraction of water in the pores, the pressure term dominates and the effective pressure acting on particles due to capillary forces depends mainly on the capillary pressure in the liquid and the local volume fraction of water in the compact. Analysis of the capillary forces acting on a drying body, therefore, requires a detailed knowledge of the spatial distribution in the compact of both the liquid phase and the capillary pressure. The approach that has been taken in the past is to assume a continuum transport coefficient for movement of the liquid in the porous body and to treat the problem as a diffusion type problem with a modified diffusion coefficient [9 – 11]. The diffusion approach predicts saturation profiles that vary smoothly throughout the compact.

In the present paper a matched refractive index technique is used to image the movement of a drying front through a porous material. The observations are used to show that, at the microscopic level at least, the movement of a drying front is not as a smooth diffusive type of front but rather as a highly irregular front whose morphology depends on the distribution of porosity in the material and on the rate at which the front moves. Ideas developed in the field of percolation theory are used to provide a basis for understanding several features of the movement of the drying front.

EXPERIMENTAL DETAILS

The porous medium used in the present experiments consisted of a thin layer of silica spheres packed between two glass slides. The refractive index of the spheres was sufficiently well matched to that of water that saturated regions of samples appeared transparent. On drainage of the pores, the sample became opaque and the drying front in the sample could be directly imaged by viewing it with transmitted light. 0.5 micron silica spheres for the experiments were prepared from tetraethyl orthosilicate by the Stober technique [12]. After the initial reaction to form the spheres, the spheres were repeatedly washed in distilled water to remove any remaining TEOS. Cells for containing the silica spheres were prepared by placing a 2.5 x 4.0 x 0.1 cm glass slide on a larger glass plate. The long edges of the slide were sealed to the plate using epoxy. This left an air gap between the slides that could be filled with distilled water. A dilute suspension of silica spheres in water was placed in a reservoir at one of the open ends of the cell and the water allowed to evaporate from the other open end. The flow of water through the cell caused by the evaporation drew the suspension of spheres into the gap between the glass slides. The spheres deposited as a dense but randomly packed layer between the slides, starting at the end of the cell from which evaporation occurred. Evaporation was continued until the entire gap between the slides was packed with spheres. To produce a uniform layer it was found that it was necessary to continually stir the reservoir of particles and to keep evaporation conditions constant during the formation of the particle layers. The measured thickness of the layers formed were in the range 10-15 microns or 20 - 30 particle diameters.

Once the layer of particles was complete the reservoir end was sealed with epoxy and evaporation allowed to continue from the open end of the cell. The propagation of the drying front from the open end was followed by viewing the cell with transmitted light in an optical microscope equipped with a video camera. Measurements of the movement of the drying front were made from video tapes of the front. In separate experiments the weight loss of cells was measured on a microbalance that remained zeroed to within a tenth of a milligram over the period of the experiment (1-2 hours). No attempt was made to control the external drying conditions used, but measurement of the evaporation rate from a free liquid surface showed that evaporation conditions were constant throughout the experiments.

RESULTS

The morphology of the drying front could readily be seen by viewing the thin layers of particles with transmitted light. A typical image of part of a drying front is shown in Fig.1. The light areas in the image correspond to regions of the sample that are fully saturated with water; the dark regions are where at least some of the pores in the section have drained. The drying front shown is moving down from the top of the micrograph. Fig.1b was taken about 5 seconds after that in Fig.1a. The most striking feature of the interface is its irregularity. The perimeter of the drying front at the magnification used in Fig.1 appears to have irregularities in it that range in scale from pore dimensions to the scale of the micrograph. When the front is viewed at a lower magnification, as in Fig.2, it is clear, however, that there is an upper limit to the scale of the irregularities in the perimeter of the drying front and that at some scale the drained regions form a stable front. A second feature of the drying front is that the drained region forms a continuous network that in places

217

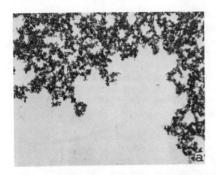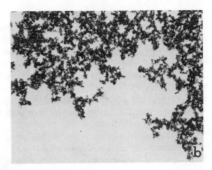

Figure (1) Images of a drying front moving through a thin layer
of silica spheres. The image on the right was taken approximately
five seconds after that on the left. The width of each image
is 0.5 mm.

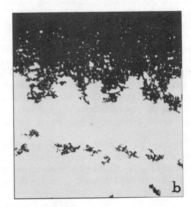

Figure (2) Comparison of the morphologies of a fast (on the left)
and a slow (on the right) moving drying front. The width of
each image corresponds to 5 mm.

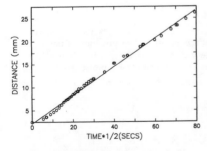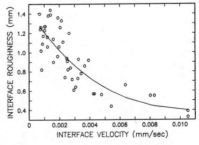

Figure (3) Plot of the average velocity
of the drying front against the square
root of time.

Figure (4) Variation of the peak to trough
distance on the drying front with the
average velocity of the front.

completely surround regions that are still saturated with water. In fig.2 it can also be seen that there are several groups of empty pores ahead of the front. These are believed to develop from defects in the original filling of the porosity with water as will be discussed later.

By comparing Fig.1a and Fig.1b, it can be seen that large areas of porosity have drained in the 5 second interval. The drainage of such regions frequently took place in the time of a single frame of the video recorder (about 1/20 of a second). Advance of the drying front in the region of the abrupt drainage then often halted for several seconds. The net effect was that the drying front moved by a sequence of abrupt events in which large numbers of pores emptied, rather than by a smooth process in which single pores drained one after another at a fixed rate.

Measurement of the average rate of advance of the drying front showed that it followed parabolic kinetics as can be seen from the linearity of plot of distance moved against the square root of time shown in Fig.3. The weight loss of the cell during the period in which the drying front was advancing through the cell was also parabolic. The parabolic kinetics suggest that rate of drying of the cell is controlled by a diffusion like process that is occurring through a diffusion barrier of increasing thickness. It is tempting to identify the diffusion barrier as the increasing volume of drained porosity that is formed as the drying front advances and the diffusion-like process as fluid flow through the residual liquid-filled channels in the partially dried material. If fluid flow through the partially drained porosity were rate-controlling then the rate of advance of the front should decrease with increasing distance from the open edge of the cell. In a number of cells where the homogeneity of packing of the silica spheres was poor it was observed that unstable propagation of the drying front resulted in portions of the drying front advancing much further down the cell than others. The weight loss in these cells was still parabolic, but the rate of advance of the drying front at any given instant was found to be independent of its distance from the open edge of the cell. The observations on inhomogeneously packed cells indicate that the transport of the water through the partially dried material behind the drying front is not controlling the rate of advance of the drying front. The overall effect of the parabolic kinetics on the drying front, however, is that the front is driven at an ever decreasing rate as it advances.

Slowing of the drying front has a systematic effect on its morphology. In the initial stages of drying, when the average velocity of the drying front is high, a compact front forms as shown in Fig.2a. As the front slows down the scale of irregularities on the front increases (see Fig.2b). In Fig.4 the peak to trough distance, a rough measure of the cutoff scale for irregularities on the front, is plotted against the average velocity of the front. From the plot it can be seen that in the course of the drying experiment there is a three to four fold increase in the scale of irregularities on the interface. At the slowest velocities the front is spread over several millimeters. On the scale of the particles the width of the interface corresponds to several thousand particle diameters.

MODELING AND DISCUSSION

The initial removal of liquid from a fully saturated porous material results in a curved meniscus forming at all the pores on the surface of the material. The curvature of the meniscus r, which is determined by the wetting angle that the liquid vapor interface makes with the surface of the solid and on the geometry of the pores, results in a pressure difference ΔP across the meniscus which is given by

$$\Delta P = 2\frac{\gamma_{lv}}{r}$$

For a wetting or partially wetting liquid the pressure difference is negative so that the liquid beneath the meniscus is in hydrostatic tension. Relief of the hydrostatic tension can occur in one of two ways, both of which lead to drainage of liquid from the pores. If the tension in the liquid builds to a large enough value, vapor-filled cavities could spontaneously nucleate in the bulk of the porous material and then expand into the surrounding pores. Although in the present experiments a few isolated groups of drained pores were observed in the bulk of the material, the number of groups was constant thoughout the experiment. This suggests that the pressure in the liquid-filled pores was insufficient to nucleate new clusters and that the existing clusters originated from pockets of air trapped in the pores during the initial filling of the cell. The dominant mechanism by which pores drain is believed to be a second mechanism for release of the hydrosatic tension, where the liquid/vapor meniscus moves in from the surface of the drying body by passing through the continuous network of pore channels in the porous material. It is this process that must be modeled to understand the movement of the drying front.

The problem of the movement of menisci through porous media is one that has been studied in considerable detail in relation to the extraction of oil from porous rocks by the immiscible displacement of oil by water [13 – 17]. The theoretical approach used is a modified form of percolation theory, known as invasion percolation, to predict the sequence in which menisci on a front advance through a pore structure. The model is based on the fact that an advancing meniscus must pass through restrictions in the pore channels in the porous material. A meniscus can advance though any given pore channel to which it has access provided the pressure difference across the meniscus exceeds a critical value. The critical pressure depends on the geometry of the restriction and the wetting characteristics of the fluids.

In the present experiments water is the the liquid that is being displaced and air or water vapor is the displacing or invading phase. At a given pressure the meniscus can pass through a certain fraction of the pore channels; the fraction depends on the distribution of pore channel diameters in the pore structure. Percolation theory then relates the extent to which the meniscus can advance into the pore structure at a given pressure to the fraction of channels that are passable [13]. The extent of penetration into the sample, as defined by the correlation length for percolation clusters, increases as the fraction of pore channels that are passable increases. When a critical fraction of passable pore channels is reached complete penetration of the drying front into the sample becomes possible. The disadvantage of the standard form of percolation theory is that it only describes the extent that the drying front can advance at a fixed capillary pressure, and, therefore, it says nothing about the sequence of meniscus movements that lead to formation of the drying front.

In invasion percolation the pressure difference across the liquid/vapor interface is allowed to adjust to a level such that at each stage of advance of the front a single meniscus can pass through a pore channel (i.e. the channel on the front that has the lowest breakthrough pressure). The sequence of the advances of menisci in idealized pore structures can then be simulated in a computer using a Monte Carlo algorithm based on the invasion percolation model [16].

Fig.5 shows the distribution of empty pores at one step in an invasion percolation simulation of the movement of menisci through a two dimensional square array of pores that are connected by pore channels with a distribution of sizes. The simulation qualitatively reproduces both the irregularity of the drying front and the isolated groups of saturated pores seen in the experimental images of the drying front. In the simulation of the displacement of oil from porous rocks, once a group of isolated group of oil filled pores forms it is said to be trapped. That is, no further invasion of the cluster can occur. In the case of the drying front the isolated clusters of filled pores are only isolated in the sense that

220

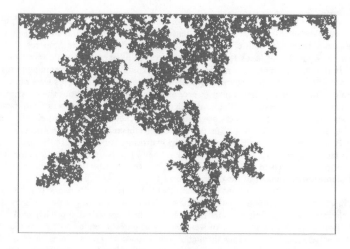

Figure (5). Invasion percolation simulation of the drainage of
pores in the absence of a pressure gradient. The simulation was
conducted on a 600x400 square lattice of pores.

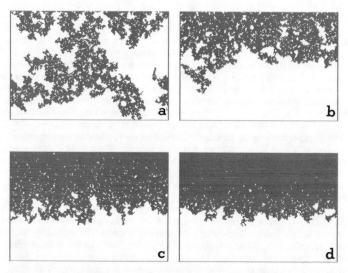

Figure (6). Invasion percolation simulations showing the effect
a linear pressure gradient has on the morphology of a percolation
front. In (a) there is no pressure gradient. In (b-d) The pressure
decreases from the top to the bottom of the figure by an amount
such that the probability of the meniscus being able to pass
through a pore channel decreases by .001, .0025 and .005 per pore
spacing. The simulations were conducted on 300x200 square
latices of pores.

they are they are completely surrounded by partially drained pores. Further drainage of the groups of pores, however,can occur and, therefore, the clusters of filled pores are not trapped as in the oil extraction case. Invasion percolation simulations conducted without trapping show that the invading phase, which in our case is the vapor phase, has a morphology similar to that of conventional percolation clusters. A simple measure that can be used to compare the the experimental images of the drying front to clusters produced by invasion percolation is the fractal dimension of the so called hull of the cluster. (The hull is the set of all the sites that lie on the external perimeter of the cluster which in the experimental images is the perimeter of the advancing front). For conventional percolation clusters the hull has a fractal dimension of about 1.74 for a two dimensional cluster [18]. Measurement of the fractal dimension of hull of the the drying front in Fig.1 by the yardstick method [19] indicates that it has a smaller fractal dimension of 1.38. There are several possible explanations for this discrepancy. The finite depth of the silica particle beds may result in behavior that is not truly two dimensional, or the behavior of the drying front may not be represented quantitatively by the invasion percolation model. More detailed quantitative analysis of the drying fronts is in progress to examine these possibilities.

The invasion percolation simulations suggest an explanation for the jerky movement of the drying front. Examination of the dynamics of invasion percolation fronts in the simulations show that frequently the passage of a meniscus through a single channel gives the front access to a long sequence of channels through which menisci can pass at relatively low capillary pressures. The result is that large local clusters of pores drain sequentially in the simulation before activity on the advancing front switches to other locations. The extreme rapidity with which the clusters of pores drain still needs to be explained, however,as it would be expected that the drainage of the pores would occur at a constant rate if driven solely by removal of water from the sample. A possible explanation is that drainage of the clusters is driven by local instability in the shape of the liquid-vapor interface. Such an instability could cause a local redistribution of the liquid from the pores to the liquid filled channels in the partially saturated regions of the sample. The redistribution of liquid would be analogous to so called Haines jumps that are associated with instability driven drainage of single pores [1]. The present case differs from Haines jumps, however, in that the instability leads to the drainage of large regions of the porous compact involving thousands of pores instead a single pore.

A difficulty with the invasion percolation model in its simplest form is that it cannot account for the upper limit observed for the scale of the roughness of the drying front. Invasion percolation simulations, in fact, indicate that irregularities at every scale larger than the pore size should occur on the drying front since invasion percolation clusters are fractal. The invasion percolation model in its present form, however, contains an unrealistic feature in that it that assumes that capillary pressure across the drying front is uniform at any given instant. In the drying experiment the liquid in the pores is being continually transported to the external surface of the sample as the pores drain. The transport of the liquid requires a pressure gradient. It is, therefore, likely that the drying front has a considerable pressure gradient across it as it moves.

To understand the origin of the pressure gradient in more detail it is necessary to consider how the liquid is removed from the drying front. On drainage from a pore on the drying front the liquid must be transported through the partially saturated portion of the sample to the external surface of the sample. Movement of the liquid can occur either by fluid flow or by vapor transport. As long as there is a continuous path for the liquid to flow through, fluid flow is likely to dominate the transport of liquid away from the front as it is the more rapid process. In particulate materials the overlap of the pendular rings of liquid retained at contacts between the particles maintains a continuous path for fluid flow at saturations that are as low as 10%. Transport of liquid in the partially saturated region of the sample near the drying front is, therefore, probably controlled by fluid flow through such channels. Fluid flow through a partially saturated porous medium is described by Darcy's law [20,10] which relates the velocity of flow (v) to the pressure gradient in the fluid (P) via the permeability (K) of the porous medium

$$v = - K\nabla P$$

As fluid flow occurs primarily perpendicular to the line of the drying front the form of Darcy's law indicates that magnitude of the capillary pressure increases (i.e. becomes more negative) in moving from the front to the edge of the sample from which drying is occurring. Menisci in the more advanced parts of the drying front, therefore, have a lower pressure difference across them than those closer to the edge of the sample. Experimental evidence for the presence of a pressure gradient across the front was found in the fact that the size of the clusters of drained pores on the saturated side of the front increased with distance from the front.

The effect of a pressure gradient on invasion percolation can simply be modeled by biasing the probability that the a meniscus can pass through a restriction in the porosity so that it increases as the magnitude of the capillary pressure increases [17] . The form of the biasing function depends on both the distribution of restrictions in the porosity and on the variation in pressure that occurs across the drying front. The spatial variation of the pressure must be obtained by solving the Darcy equation with the appropriate boundary conditions. At present a solution for the spatial distribution of the pressure can not be obtained due to the difficulty that the permeability of the medium is itself a function of saturation of the medium and hence of the capillary pressure.

The effect of a pressure gradient on the invasion percolation simulation can be qualitatively examined by assuming a functional form for the pressure gradient across the front. The simplest form of gradient is one in which the probability of a meniscus being able to pass through a restriction in the porosity decreases linearly across the front. This corresponds to assuming that the radius of the liquid vapor meniscus varies linearly across the front and that the pore channel distribution is uniform over a range of values. The model is identical to that proposed by Wilkinson to describe the effect of buoyancy forces on invasion percolation [17] .

Fig.6 shows four different simulations conducted using a two dimensional array of pores. In Fig.6a there is no pressure gradient and the invading cluster has the usual irregular morphology. In Figs.6b-d a pressure gradient has been applied so that the probability of the meniscus being able to pass through a pore channel is decremented by .0015, .0025 and .005 per pore spacing as you go down each figure. Each figure is 300 pore spacings wide and 200 pore pore spacings deep. It can be seen that the effect of the pressure gradient is to stabilize the front at some length scale. The cut off in length scale for irregularities on the front decreases as the pressure gradient increases. The behavior of the fronts in the simulations is consistent with that observed in experimental images of the drying front. The broadening of the drying front as it advances can be explained by the fact that as the drying front slows down the rate at which water is removed from the front must decrease. A slower rate of fluid flow corresponds to a decrease in the pressure gradient across the front and hence to a broadening of the drying front.

The effect of the pressure gradient in stabilizing the drying front has an important effect on the distribution of capillary forces during drying. With a fast drying rate rapid movement of the drying front will result in an abrupt transition from fully saturated porosity to drained porosity. As the effective pressure acting to compact the porous material is approximately equal to the product of the saturation and the capillary pressure, the abrupt transition corresponds to a steep stress gradient in the sample. Such a steep stress gradient is more likely to cause cracking than if the interface is diffuse and the stress gradient is more gradual. Cracking during drying therefore is likely to depend critically on the stability of the drying front as controlled by the pressure gradient across the drying front and the distribution of porosity in the compact.

SUMMARY

The movement of a drying front through a thin layer of silica spheres has been directly observed. The front is found to develop a highly irregular morphology as it as advances and to move by a series of abrupt jumps. Invasion percolation models can qualitatively account for the observations but do not adequately account for the fractal dimension of the perimeter or hull of the front. There is a cut off in the scale of irregularities on the front that changes with the average velocity of the front. The cut off is attributed to the presence of a pressure gradient across the front that is caused by the flow of water from drying front to the edge of the sample. A detailed quantitative picture of the drying front that combines both the invasion percolation description of the drainage of the pores and analysis of fluid flow across the drying front using Darcy's law is needed to understand fully the distribution of capillary forces that occur during drying.

ACKNOWLEDGEMENTS

I am grateful to Eric Liniger for his technical assistance in preparing the silica spheres used for the experiments, to Brian Stephenson for useful discussions and to George Onoda for a critical review of the manuscript.

REFERENCES

1. W.B. Haines, J. Agr. Sci 15, 529 (1925).
2. R.A. Fisher, J. Agr. Sci 16, 192 (1926).
3. G. Mason and W.C. Clark. Chem. Eng. Sci. 20 , 859 (1965).
4. T. Gillespie and W.J. Settineri, J. Colloid Interface Sci., 24, 199 (1967).
5. H.M. Princen, J. Colloid Interface Sci., 26 , 249 (1968)
6. R.B. Heady and J.W. Cahn, Metall. Trans., 1 , 185 (1970).
7. F.R.E. De Bisschop and W.J.L. Rigole J. Colloid and Interface Sci.,88, 117 (1982).
8. H. Schubert, Powder Tech. 37 , 105 (1984).
9. T.K. Sherwood, Ind. Engng. Chem., 24 , 307(1932).
10. E.E. Miller and R.D. Miller, J. App. Phys. 27 , 324 (1956).
11. A.R. Cooper, in Ceramic Processing Before Firing Edited by G.Y. Onoda and L.L. Hench,(John Wiley and Sons 1978) P.261.
12. W. Stober, A. Fink and E. Bohn, J. Colloid and interface Sci. 26 , 62 (1968).
13. P.G. de Gennes and E. Guyon, J. Mech. 17 , 403 (1978).
14. R. Chandler, J. Koplik, K. Lerman and J.F. Willemsen, J. Fluid. Mech., 119 , 249 (1982).
15. R. Lenormand and C. Zarcone, in Kinetics of Aggregation and Gelation, Edited by F. Family and D.P. Landau, (Elservier 1984) P.177.
16. D. Wilkinson and J.F. Willemsen J. Phys. A. 16 , 3365 (1983).
17. D. Wilkinson, Phys. Rev. A. 30 , 520 (1984).
18. R.F. Voss, J. Phys. A. 17 , L373 (1984).
19. B.B. Mandelbrot The Fractal Geometry of Nature (Freeman 1982).
20. H. Darcy, Les Fontaines Publique de la Ville de Dijon (Dalmont, 1856), P.570.

DRYING MECHANICS OF GELS

GEORGE W. SCHERER
E. I. du Pont de Nemours & Co., Central R & D Dept., Experimental Station
356/384, Wilmington, DE 19898 USA

ABSTRACT

A model is presented for the stresses and strains that develop in a
gel during drying. The driving force for shrinkage is assumed to be the
interfacial energy, and the gel is considered to be viscoelastic. The
liquid seeks to flow into the dry region of a gel in order to replace the
solid-vapor interface with a solid-liquid interface having lower specific
energy. This creates a "redistribution pressure" that causes the wet
region to contract. The free contraction rate can be calculated by
equating the decrease in surface energy with the energy dissipated in
viscous flow as the gel contracts. The permeability of the gel to the
liquid in the pores is especially important in the early stages of drying,
and may control the contraction rate. The model allows quantitative
predictions of contraction rate and stress during drying. In this paper,
the model is applied to a plate drying from both sides.

INTRODUCTION

The rate determining step in the preparation of monoliths from gels is
drying (i.e., solvent removal). This stage of the process is associated
with cracking and/or warping of the body. The difficulty of drying has
inspired the use of hypercritical drying [1,2] and chemical additives [3].
This paper presents a model that seeks to incorporate the essential physics
of the drying process in order to provide a quantitative explanation of
phenomena including shrinkage rate, stress, and warping. By establishing
the most important variables, the model may help to understand the function
of chemical additives, and to aid in their selection.

Existing models of drying assume that the rate of transport of liquid
in the pores controls the contraction rate of the gel. Cooper [4] analyzed
the drying of clay bodies assuming that the free strain is proportional to
the local liquid concentration, and that transport is governed by Fick's
law. In an open-pore material, such as a silicate gel, liquid transport
will obey Darcy's law, according to which the flux is proportional to the
pressure gradient. Models incorporating Darcy's law either neglect the
viscoelastic properties of the gel [5,6], or treat the gel as simply
elastic [7]. Organometallic gels exhibit finite viscosities, so the
present model includes both the elastic and viscous responses of the gel.
In contrast to earlier models, the driving force is explicitly related to
the microstructure of the gel. The permeability of the gel is shown to be
particularly important in the early stages of drying, when the viscosity of
the gel is relatively low.

DRIVING FORCES

A gel can reduce its energy by contracting so as to reduce its
interfacial area. For a particle compact, rearranging to a higher
coordination number causes a small change in area, but for a polymeric gel,
significant changes in area may be possible as a result of folding or
flexing of chains. A gel will contract within its solvent when evaporation
is prevented. This phenomenon, called syneresis, is driven by the solid-
liquid interfacial energy γ_{SL} of the gel. The solid-vapor interfacial

energy γ_{SV} is larger than γ_{SL} so the empty pores of the gel (the dry region) might be expected to shrink more rapidly than the full pores (the wet region). However, there is another factor driving the contraction of the wet region. If the wet region contracts, the liquid can move into the dry region and replace the solid-vapor interface with a solid-liquid interface. This permits a relatively large reduction in energy of the system with a small associated deformation, and is the major factor driving densification of the gel. The wet region contracts as if it were subjected to a compressive stress and the liquid experiences a tensile stress, which is the familiar capillary pressure that causes liquid to rise in a tube. This pressure, hereafter called redistribution pressure, acts even in particle compacts, since rearrangement of the particles allows movement of the liquid into the dry region. The rate of contraction of the gel is found by equating the energy gained by redistribution of the liquid into the dry region with the energy expended to deform (shrink) the wet region.

Other forces may contribute to shrinkage of gels. Organometallic gels contain reactive alkoxy and hydroxy groups that may undergo condensation reactions, and thereby reduce the chemical potential of the system. This may be treated as a contribution to interfacial energy, if the reactants are located on the internal surface of the gel, since the degree of reaction will be proportional to the decrease in surface area. Other forces, such as osmotic pressure and disjoining pressure do not seem to be important for silica gels, since the gels do not change dimensions when exposed to other solvents or to electrolytes.

VISCOELASTICITY

Gels made in this laboratory by hydrolysis of tetraethoxysilane (TEOS) have Young's moduli of ~1 MPa shortly after gelation, and the modulus increases by 1 to 2 orders of magnitude upon aging (without evaporation of solvent) for several weeks. Greater increases are observed if evaporation is permitted. The viscosity of the gel is ~10^{10} Pa.s shortly after gelation and rises considerably with time. These data will be presented in detail in a future report. If a gel is allowed to dry partially, it contracts; if it is then immersed in solvent, it does not expand. This indicates that the deformation is viscous, rather than elastic. Therefore, the following analysis considers only the viscous deformation to be expected during drying of the gel. The elastic and viscous deformations are analyzed in detail in ref. 8 by representing the microstructure of the gel by an array of cylinders intersecting at right angles. The array can be represented (9) by a cubic unit cell with edge length 1 ; the relative density, y, is related to the parameter x=a/1, where a is the radius of each cylinder. Such a model has been used successfully in the analysis of the sintering of porous glasses, including gels (10,11). The viscous strain rates during drying are found to be

$$\dot{\epsilon}_D = \frac{-\gamma_{SV}S\rho_S}{6\eta} \tag{1}$$

$$\dot{\epsilon}_W = \frac{-\gamma_{SL}S\rho_S}{6\eta} \tag{2}$$

$$\dot{\epsilon}_R = \frac{-(\gamma_{SV}-\gamma_{SL})S\rho_S}{6\eta}\left[\frac{3(2-3cx)^2}{2(1-y)}\right] \tag{3}$$

where $\dot{\epsilon}_D$ and $\dot{\epsilon}_W$ are the contraction rates of the dry and wet regions, respectively, caused by their own interfacial energies, and $\dot{\epsilon}_R$ is the contraction rate of the wet region caused by the redistribution pressure.

S is the specific surface area, ρ_S is the skeletal density, η is the viscosity of the solid phase of the gel, and $c = 8\sqrt{2}/3\pi$. The superscript dot indicates the derivative with respect to time.

PERMEABILITY

The strain rates presented above were derived without considering the rate at which the fluid can move through the pores of the gel. Since the pores are so small, the contraction of the gel could be limited by the difficulty of squeezing out the liquid. The transport of the liquid should obey Darcy's law,

$$J = \frac{D}{\eta_L} \frac{dP}{dx} \tag{4}$$

where J is the flux [volume/(area x time)], D is the permeability constant, η_L is the viscosity of the pore liquid, and dP/dx is the pressure gradient. During syneresis, contraction of the gel imposes compressive stress on the liquid, which causes it to flow toward the wet exterior surface, where the pressure is zero. The pressure, which results from the resistance of the liquid to flow, retards the contraction of the gel. For a plate, it can be shown [8] that if the gel would contract at the rate $\epsilon(0)$ when the liquid offers no resistance, then a gel with a finite permeability would contract at the rate

$$\epsilon(\alpha) = \epsilon(0)\cosh[\alpha(1-z/L)]/\cosh(\alpha) \equiv \epsilon(0)\ \Pi(\alpha) \tag{5}$$

where $\Pi(\alpha)$ is called the permeability factor, and

$$\alpha = [\eta_L L^2/DK_G]^{1/2} \tag{6}$$

where L is the thickness of the plate, z is the distance from the exterior surface, and K_G is a function proportional to the viscosity of the solid phase of the gel. This shows that the contraction rate will be smaller in the interior of the gel (large z), because of the difficulty of pushing the liquid to the (distant) exterior surface.

STRESS IN A PLATE

The free strain rates of the wet and dry regions of a gel are given by eqs. (1)-(3), modified by the permeability factor, as in eq. (5). The free strain rates can be used to calculate the stresses and strains in a drying gel by analogy to the analysis of thermal strains, where the free strain is the product of the thermal expansion coefficient and the temperature change. If a plate of gel dries by evaporation from both faces, so that there is no warping, the stresses and strains are easily expressed in terms of the free strain rates [12]. Stress development depends on the rates of three independent processes: evaporation of the pore liquid, viscous flow of the solid phase of the gel, and flow of the pore liquid. Consider first the case when the permeability is high (α=0). Figure 1 shows the calculated contraction of a plate of thickness L drying at a constant rate; w is the fraction of the pore liquid that has evaporated. If evaporation removes liquid slowly, compared to the rate at which the solid phase can contract, then no dry region forms and there is no stress. Eventually, the contraction will cause the viscosity of the gel to increase, so that its contraction cannot keep up with the rate of evaporation, and a dry region will form. The arrow in Fig. 1 indicates the point at which a dry region first appears in this simulation; Fig. 2 shows that there is no stress in the gel up to that

point. Since the rate of contraction of the wet region is faster than that
of the dry region, the exterior surface will be in compression, and the wet
interior will be in tension. The magnitude of the tension will increase as
the volume fraction occupied by the wet region decreases. As shown in Fig.
3, the final density of the gel will be greater in the interior, because
the wet region contracts faster; the longer a portion of the gel remains
wet, the more it contracts. When the permeability is low ($\alpha > 0$), the
exterior surface goes into tension as soon as contraction begins, because
the interior of the gel cannot contract as fast as the exterior. This is
illustrated in Fig. 4. The tension increases to a maximum at about the
time that the dry region develops, then the exterior goes into compression.
Since the liquid inhibits the contraction of the interior of the gel, the
final density may be smaller in the interior of the gel than at the
exterior surface, as shown in Fig. 5. [This depends on the rheology of the
dry layer and the rate of evaporation. If the dry layer is very stiff and
forms when the density of the gel is low, the continued contraction of the
wet interior may decrease or reverse this density gradient.]

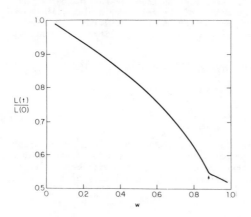

Fig. 1

Relative thickness of a
plate vs relative weight
lost (w) when $\alpha = 0$; arrow
indicates point at which
dry region forms.

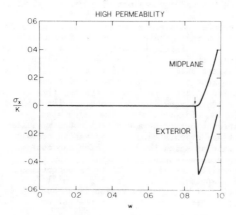

Fig. 2.

Stress at exterior surface
and at midplane of plate
versus relative weight
lost, when $\alpha = 0$; arrow
indicates point at which
dry region forms.

229

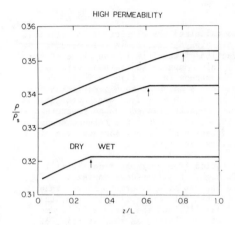

Fig. 3.

Relative density vs depth
in plate (z = 0 at
exterior, z = L at
midplane) at three stages
of drying when α = 0; arrow
indicates location of
liquid-vapor interface.

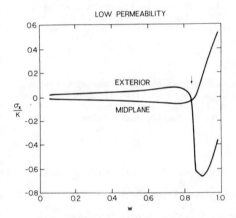

Fig. 4.

Same as Fig. 2, but α = 1.5

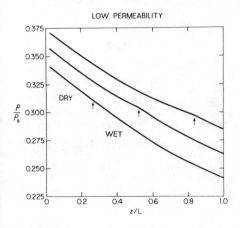

Fig. 5.

Same as Fig. 3, but α = 1.5

OTHER GEOMETRIES

The same approach can be used to calculate the stresses in a film on a rigid substrate [12]. The stresses are always tensile, in that case, because the substrate prevents contraction of the film in the plane of the substrate. For any layer that is very thin compared to the substrate, so that the substrate does not deform, the stress in the film is independent of its thickness. This seems inconsistent with the common observation that it is difficult to prepare films thicker than 0.5 μm from organometallic gels. The failure of thicker films is probably related to the fracture energy: if a crack forms in a very thin film, it does not release enough strain energy to compensate for the formation of the new surface area of the crack, so only relatively thick films can crack.

If a plate of gel dries by evaporation from only one side, it will tend to warp. If the permeability is high, it will become convex toward the direction of evaporation, because the wet side contracts faster than the dry side. In the more common case when the permeability is low, the drying surface will contract faster initially, and the plate will become concave toward the direction of evaporation. Later, as the viscosity of the gel rises and α decreases, the plate will reverse its curvature and become convex toward the drying side. This has been observed in our laboratory. If the drying rate is very slow, the pressure in the liquid equilibrates, and warping is prevented.

CONCLUSIONS

A model has been developed that allows predictions of the stresses and strains in a drying gel. The predictions are in qualitative agreement with observations, but quantitative tests of the model await independent measurements of the interfacial energies, viscosity, and modulus of gels. Such studies are currently underway.

REFERENCES

1. J. Zarzycki, M. Prassas, and J. Phalippou, J. Mater. Sci., 17, 3371 (1982)
2. P.H. Tewari, A.J. Hunt, and R.D. Loftus, Mater. Lett., 3 (9-10) 363 (1985)
3. S. Wallace and L.L. Hench in Better Ceramics Through Chemistry, eds. C.J. Brinker, D.E. Clark, and D.R. Ulrich, p.47, North-Holland, NY, 1984
4. A.R. Cooper in Ceramics Processing Before Firing, eds. G.Y. Onoda, Jr. and L.L. Hench, p. 261, John Wiley & Sons, NY, 1978
5. H.H. Macey, Trans. Brit. Cer. Soc., 41 (4) 73-121 (1942)
6. D.E. Smiles, Soil Science, 117 (3) 140-147 (1974)
7. T. Tanaka and D.J. Fillmore, J. Chem. Phys., 70 (3) 1214-1218 (1979)
8. G.W. Scherer, "Drying Gels: I. General Theory", submitted for publication in J. Non-Crystalline Solids
9. G.W. Scherer, J. Am. Ceram. Soc., 60 (5-6) 236 (1977)
10. G.W. Scherer, "Viscous Sintering of Inorganic Gels", to be published in Surface and Colloid Science, Vol.14, ed. E. Matijevic, John Wiley & Sons, NY
11. G.W. Scherer, C.J. Brinker, and E.P. Roth, J. Non-Crystalline Solids, 72, 369-389 (1985)
12. G.W. Scherer, "Drying Gels: II. Film and Flat Plate", submitted for publication in J. Non-Crystalline Solids

VISCOELASTIC PROBES OF SUSPENSION STRUCTURE

C.F. Zukoski[*], J.W. Goodwin[**], R.W. Hughes[**], and S.J. Partridge[**]
* Dept. of Chemical Engineering, University of Illinois, Urbana, Illinois 61801
** Dept. of Physical Chemistry, Bristol University, Bristol, U.K.

ABSTRACT

The use of the high frequency elastic modulus, G_∞, to probe particle interactions in weakly agglomerated, concentrated suspensions is discussed. We show that a model based on pair interaction potentials and a statistical description of pair spatial distribution yields accurate prediction of the volume fraction dependence of G_∞. The use of statistical treatments of particle distributions as predicted from particle interaction potentials is found to provide insight on how surface chemistry affects the uniformity of powder compacts.

Introduction

Numerous studies have shown that simply reproducing a set of processing conditions is not enough to insure component reliability (1-3). Instead, these studies indicate that powder processing must reproducibly generate uniform microstructures upon firing and that this uniformity is achieved by starting with monodisperse, unagglomerated powders and packing these powders in a homogeneous manner. Thus, the quest for reliable ceramics has begun to focus attention on the ways in which particle/particle interactions determine the structure (particle and pore distribution) within green powder compacts.

Methods of probing the arrangement of particles in a highly loaded suspension are limited. Here we describe a rheological measurement which can be used to follow the effects of processing conditions on structure development.

Several rheological techniques are available for these purposes. It should be noted, however, that different techniques do not provide equivalent information. For example, viscosities measured as a function of shear rate have been used to follow the breakdown of agglomerates (4). During these measurements a suspension is subjected to an infinite deformation which results in a continuous breakdown and reformation of aggregates. Consequently, from continuous shear measurements, it is difficult to infer properties of the suspension at rest. On the other hand, this technique provides information about the spatial distribution of particles under shear and thus, is valuable in assessing a suspension's moldability.

However, the structure maintained while the suspension is flowing can be quite different from that at rest. After flow ceases, the particle distribution relaxes into its equilibrium or static state. It is this static state which plays a dominant role in the microstructure developed during subsequent firing. A measure of particle interactions (and thus, implicitly, particle distribution) which does not disturb this equilibrium state is provided by the limiting high strain frequency storage modulus of the suspension, G_∞. This parameter is a measure of the energy stored in a suspension due to an infinitismal strain (5).
In this paper we describe experimental determinations of G_∞ on an idealized system. The goal of this work was to establish that particle distributions derived from molecular statistical mechanics could be used in conjunction with colloidal particle interaction potentials to evaluate the ordering within a suspension and its elastic response to a small deformation.

Experimental

Monodisperse polystyrene spheres were prepared following the methods of Goodwin et al (6). These particles were chosen because their size can be readily controlled and their surface chemistry has been well studied. They thus represent an ideal system with which to test modelling assumptions before proceeding to more complex situations. The latices were stabilized against aggregation by monolayer adsorption of a monodisperse surfactant, hexaethylene oxide dodecyl ether (C_6E_{12}) supplied by Nikko Chemicals Co., Ltd., Tokyo, Japan. Electrophoretic mobilities used to calculate surface potentials, were measured on a Penkem 3000 automated electrophresis apparatus. For our purposes, where the double layer thickness is much smaller than the particle radius ($a\kappa \gg 100$, where κ is the Debye-Huckel parameter), the Smoluchowski equation was used to convert mobilities to zeta potentials, ζ (7).

The suspension's wave rigidity modulus was determined as a function of volume fraction with a Rank Pulse Shearometer. For the systems studied here the loss modulus was small at the frequency of the Shearometer ($\sim$150 hz) and thus the wave rigidity modulus reduces to the high frequency limit of the storage modulus, G_∞ (5). More detailed descriptions of the experimental procedures are presented elsewhere (7).

Results

Due to space constraints, the results for only one of the particle sizes studied will be described. The interested reader is referred to Goodwin et al (7) for further discussion. The latex studied was composed of polystyrene spheres with an average radius of 487 nm and a standard deviation of 1.05. The surface potential of the latex covered with a monolayer of the surfactant and suspended in 0.5 M NaCl was 9.7 mV. Wave rigidity moduli were measured on suspensions maintained a this ionic strength.

Under these conditions, the particles are weakly flocculated. (The steric layer provided by the surfactant keeps the particles from falling into a primary Van der Waals minimum). Observation of dilute suspensions with an optical microscope showed that the particles clustered together into tight domains which were easily disrupted by tapping the coverglass.

In fig. 1, the volume fraction dependance of the wave rigidity modulus for this latex is presented. For all the latices studied, it was found that G_∞ increased monotonically with volume fraction.

Discussion

Due to the small strain applied by the shearometer, the suspension's equilibrium structure is not disturbed while measuring G_∞ and can be modeled in terms of pair interaction potentials and the equilibrium particle distribution. In this fashion Goodwin et al (7) develop an expression for G_∞ for concentrated suspensions based on a model originally derived by Zwanzig and Mountain (8). The final expression is written:

$$G_\infty = \frac{3C}{4\pi a^3} KT + \frac{3C^2}{8\pi a^6} \int_0^\infty g(r) \frac{d}{dr} \left[r^4 \frac{dU}{dr} \right] dr \tag{1}$$

This expression relates G_∞ to the suspension volume fraction, C, the particle radius, a, the product of Boltzmann's constant and the absolute temperature KT, the pair interaction potential, U, and the pair distribution function $g(r)$. The first term on the right hand side of eqn. (1) represents an entropic contribution where as the second term, which dominates for the systems at hand, represents the contribution due to particle/particle interactions.

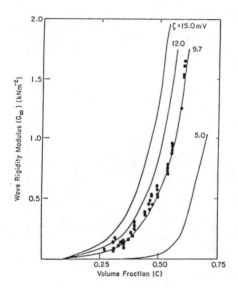

Fig. 1. Wave rigidity modulus, G_∞, as a function of volume fraction, C. Points are experimental values and the solid lines are model predictions discussed in the text.

        The interaction potential for the sterically stabilized system studied here can be described by an electrostatic repulsion and a truncated Van der Waals attraction. The surfactant layer extents 3.85 nm away from the particle surface and fixes the minimum approach two spheres can achieve at 7.7 nm. The Van der Waals force originates at the physical surface of the particle and, thus, the surfactant truncates the attractive potential. The electrostatic repulsion, on the other hand, originates at the outer edge of the surfactant layer, and, even though it is extremely short range in 0.5 M salt solutions (the double layer thickness is ~0.4 nm), this repulsive potential plays an important role in determining suspension elasticity (7). The interaction potential is written:

$$U = 2\pi \, \varepsilon \, a \, \zeta^2 \, \ln \left\{ 1 + \exp\left[-\kappa(h-2d) \right] \right\} + U_A + U_{HS}$$

where
$$U_A = - \frac{A}{12} \left\{ \frac{1}{x^2 + 2X} + \frac{1}{x^2 + 2X + 1} + 2 \ln \left[\frac{x^2 + 2X}{x^2 + 2X + 1} \right] \right\}$$

and $U_{HS} = \infty$ at $h \leqslant 2d$ and $U_{HS} = 0$ for $h > 2d$. (2)

for an adsorbed layer of thickness d, a surface to surface separation h, a Hamaker constant A, solvent dielectric constant ε and $x = h/2a$.

234

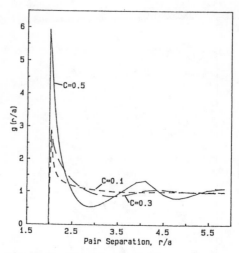

Fig. 2. Pair distribution functions, g(r/a), as a function of pair
separation, r/a, at three volume fractions. Note that for all C, g(r/a) = 0
for r/a < 2.0 and g(r/a) → 1.0 as r/a → ∞. As C increases the number of well
defined shells of particles (as indicated by maximima in g(r/a)) increases
showing increased ordering in the suspension.

The pair distribution function accounts for the ordering of the
particles in the suspension. The probability that there is a second particle
within the volume r + dr around a central particle is given by;

$$\frac{3r^2}{a^3} \, Cg(r)dr.$$

Studies of the pair distribution functions in colloidal systems with
light and neutron scattering techniques (9) indicate that like molecular
fluids, the pair distribution at high volume fractions is dominated by the
particle's rigid, noninterpenetrating properties. Thus it is the short range
repulsive portion of U which dominates the spatial distribution of particles
in a concentrated suspension. Goodwin et al (7) show how eqn. 1 can be
rewritten in terms of a pair distribution function derived for hard spheres
(where particles interact only through U_{HS} in eqn. (2)) and a perturbation to
this pair distribution function accounting for the electrostatic and Van der
Waals forces. Typical pair distribution functions are presented in fig. 2.
Numerically carrying out the integral in eqn. (1), Goodwin et al show that the
resulting predictions of G_∞ are relatively insensitive to the steric layer
thickness, d, or the Hamaker coefficient A. However, G_∞, is very sensitive to
variations in ζ. In fig. 1 the predictions of the model calculations for
various values of ζ using A = 9×10^{-21} J and d = 3.85 nm are shown. As seen
here, the model predicts the experimental values of G_∞ for a zeta potential
very close to that measured electrophorectically. For all the latices
studied, the zeta potential required to fit the G_∞ results was within 3–4 mV
of the measured value. This agreement suggests that the pair distribution
functions predicted from statistical mechanical models provide excellent
approximations for weakly flocculated systems and can be used to

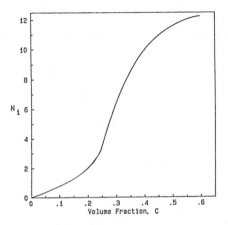

Fig. 3. Average number of nearest neighbors, N_1, as a function of suspension volume fraction, C. As N_1, approaches 12, the short range order in the suspension increases.

predict elastic properties of a suspension.

While G_∞ is important in evaluating the toughness of a powder compact, predictions of the pair distribution function can be used in developing methods of tailoring desired microstructures. As an example of this fig. 3 presents the number of particles in the first shell around a central particle, calculated from the pair distribution function used in the G_∞ predictions according to

$$N_1 = \frac{3C}{a^3} \int_{2a}^{r_1} r^2 g\,(r)dr$$

where r_1 is the position of the first minimum in $r^2 g(r)$ (9). Fig. 3 indicates that the number of nearest neighbors surrounding a central particle rapidly approaches 12 at volume fractions above 0.35. This is the value expected for a close packed array of spheres. That the number of nearest neighbors approaches 12 at volume fractions substantially below the close packed volume fraction of $\simeq 0.74$ reflects upon the short range order introduced into the suspension by the hard sphere repulsion and Van der Waals attraction. In order to maintain a space filling suspension, however, the order decays rapidly resulting in small domains of order interspersed with regions of disorder. Indeed, such clustering phenomena was observed. When dilute suspensions were studied under an optical microscope, particles formed small highly ordered regions interspersed with areas of much lower volume fraction reflecting the behavior predicted by the statistical model.

The consequences of suspensions which contain regions of short range order but long range disorder on microstructure development upon firing have been explored by Aksay (10). He found that while the close packed domains

236

sintered to completion at low temperatures, the interspersing voids
(representing regions of disorder) required much higher temperatures and
longer sintering times to result in a ceramic of near theoretical density.
Sintering out the disordered regions caused grain growth to occur and the fine
microstructure which arose from the close packed regions was lost.

In conclusion, we have shown that weakly agglomerated suspensions
acquire substantial elastic moduli and that these moduli can be predicted from
a model accounting for pair interactions and the static spatial distribution
of pairs. In addition, the perturbed hard sphere model for the pair
distribution suggests that the space filling, weakly agglomerated network
studied here consists of close packed regions interspersed with regions of
lower particle density. While the experimental system was chosen to test the
model developed for G_∞, the quality of the model has been verified suggesting
that experimental and modelling techniques developed here can now be applied
with greater confidence to ceramic precursor particles.

REFERENCES

1. Barringer, E., N. Jubb, B. Fegley, R. L. Pober and H. K. Bowen in
   Ultrastructure Processing of Ceramics, Glasses and Composites, L. L. Hench
   and D. R. Ulrich eds., John Wiley and Sons, N.Y. (1984)
2. Lange, F. F., J. A. Cer. Soc., 66, 396, (1983), F. F. Lange and M.
   Metcalf, ibid 66, 398 (1983), F. F. Lange, B. I. Davis and I. A. Aksay,
   ibid 66, 407 (1983)
3. Lange, F. F., J. Materials for Energy Systems, 6, 107 (1984)
4. Firth, B. A. and R. J. Hunter, J. Colloid and Interface Sci., 57, 266
   (1976); E. Carlstrom and F. F. Lange, Comm. of A. Cer. Soc. C-169, (1984)
5. Buscall, R. J. W. Goodwin, M. W. Hawkins and R. H. Ottewill, Chem. Soc.
   Faraday I, 78, 2889 (1982)
6. Goodwin, J.W., J. Hern, C. C. Ho and R. H. Ottewill, Colloid Polym. Sci.,
   252, 464 (1974)
7. Goodwin, J. W., R. S. Hughes, S. S. Patridge and C. F. Zukoski, accepted
   for publication in J. Chem. Phys. 1986.
8. Zwanzig, R. W. and R. O. Mountain, J. Chem. Phys., 43, 4464 (1965)
9. Van Megen, W. and I. Snook, Adv. Colloid and Interface Sci., 21, 119
   (1984)
10. Aksay, I., Adv. in Ceramics 9, Ceramic Forming, J. A. Mangles ed., Am.
    Ceramic Soc., Columbus, OH, (1984)

This work was supported in part by the Department of Energy (DOE DEAC02
76ER01198) and administered through the Micronanalysis Center in the Materials
Research Laboratory of the University of Illinois and in part by The Science
and Engineering Research Council, UK. The authors wish to thank Dr. R.
Buscall for valuable discussion and both ICI plc., Corporate and Bio-science
Laboratory and Schlumberger Cambridge Research.

SYNTHESIS OF HIGH PURITY SILICA GLASS FROM METAL ALKOXIDE

S.R. Su and P.I.K. Onorato
GTE Laboratories, Inc., 40 Sylvan Road, Waltham, MA 02254

ABSTRACT

High purity silica was synthesized by hydrolytic decomposition of tet-ramethylorthosilicate (TMOS), followed by polycondensation, supercritical drying and sintering. The key factors which govern the hydrolysis and poly-condensation were systematically studied. Gas chromatography/mass spectro-metry was used to monitor the hydrolysis rate. The hydrolysis of TMOS is the only reaction at ambient temperature in an acidic environment. The rate of polycondensation was also controlled by temperature as well as pH. The structural transformation from alkoxide to gel network and to glass was il-lustrated by FTIR. The structure of the TMOS derived supercritically dried gel sintered at 1000°C is identical to that of silica. The shrinkage kinet-ics and microstructural development of the gel heated at different tempera-tures was examined by dilatometry, B.E.T., TGA, and STEM. This combination of techniques showed that structural relaxation is not an important factor in the densification of these gels. Below 1000°C most shrinkage is due to condensation and water evolution. Above 1000°C, viscous sintering is the primary mechanism for shrinkage.

INTRODUCTION

The sol-gel process has been used commercially for about 30 years [1-5] to make thin film oxide coatings for automobile mirrors. Recently, however, additional applications of this process are beginning to be realized. For example, the sol-gel process has been used: 1) to form single layer antire-flection films on glass substrates for high power laser use, 2) to form oxi-dation resistant protective coatings on stainless steel, 3) to draw fibers for composite reinforcement, and 4) to prepare abrasive materials. The ad-vantage of the process is that the starting materials, which may consist of a wide variety of components, can be highly purified by distillation, and then, being in liquid form, they can be mixed on an atomic level. In gen-eral, this process consists of a series of steps:

1. complexation of alkoxides or inorganic salts in alcohol/water solution,
2. hydrolysis of the alkoxides to hydroxides,
3. condensation of the hydroxide to form a continuous network (gel),
4. drying of the gel to remove water and organic residue; and to form a mon-olith,
5. densification of the dried gel to form a dense compact body.

These steps are not entirely sequential, and the degree of completion of one step before the commencement of the next step and the relative rates of each reaction will determine the final structure and hence properties. The objective of this work was the systematic study of the chemistry of hy-drolysis of $Si(OCH_3)_4$ and the subsequent polymerization of $Si(OH)_4$. The drying and sintering methods were also investigated so as to develop a practical technique for the production of high purity amorphous silica at low processing temperatures.

EXPERIMENTAL PROCEDURES

Tetramethylorthosilicate (TMOS) was used as starting material without further purification. Doubly distilled deionized H_2O and reagent grade methanol were employed for the hydrolysis reactions. 1.0 N nitric acid and 1.0 N ammonium hydroxide were the catalysts. Gas chromatography HP5710A/mass spectrometer HP5980A was used to monitor the hydrolysis rate of TMOS. Samples of 0.1µl were taken at various intervals and injected into a SP1000 Carbopack B capillary column at 200°C isotherm. Helium was used as a carrier gas. The gel was dried supercritically in a Parr pressure reactor, and the dried gel was sintered in air as well as in oxygen.

RESULTS AND DISCUSSION

Hydrolysis and polymerization

Extensive studies of the hydrolysis rate and polymerization conditions were conducted on TMOS. This material was preferred because of its higher SiO_2 content (39.5 wt% vs. 28.8 wt% for tetraethyl orthosilicate) and chemical reactivity.

The overall hydrolysis and polymerization reactions of $Si(OCH_3)_4$ may be described schematically as:

$$n\ Si(OCH_3)_4 + 4\ n\ H_2O \rightarrow n\ Si(OH)_4 + 4\ n\ CH_3OH \tag{1}$$

$$n\ Si(OH)_4 \rightarrow n\ SiO_2 + 2n\ H_2O \tag{2}$$

The reactions which usually occur in the hydrolysis of a methanol solution of silicon methoxide, however, are considered to be more complicated.

Thus, the reactions occurring in the initial stage of sol-gel transition may be:

$$Si(OCH_3)_4 + n\ H_2O \rightarrow Si(OH)_n(OCH_3)_{4-n} + n\ CH_3OH \tag{3}$$

$$Si(OH)_n(OCH_3)_{4-n} + Si(OCH_3)_4 \rightarrow (OCH_3)_{4-n}\ (OH)_{n-1}\ Si\ O\ Si(OCH_3)_3 + \tag{4}$$
$$CH_3\ OH$$

$$2\ Si(OH)_n(OCH_3)_{4-n} + H_2O \rightarrow (OCH_3)_{3-n}\ (OH)_n\ Si\ O\ Si(OH)_n\ (OCH_3)_{3-n} + \tag{5}$$
$$2\ CH_3\ OH$$

$$2\ Si(OH)_n(OCH_3)_{4-n} \rightarrow (OCH_3)_{4-n}\ (OH)_{n-1}\ Si\ O\ Si\ (OH)_{n-1}\ (OCH_3)_{4-n} + \tag{6}$$
$$H_2O$$

These reactions are very sensitive to the experimental conditions such as the presence of acidic or basic catalysts, the gelling temperature, the molar ratio of alkoxides to H_2O. Consequently, the properties of the gels obtained by the sol-gel transition of $Si(OCH_3)_4$ depend on several experimental parameters. Two critical parameters, pH and temperature, which affect the rate of hydrolysis were systematically studied on the system TMOS: H_2O: methanol = 1:4:3. The results are tabulated in Tables I through VI.

This study has resulted in a clear mechanistic understanding of the hydrolysis and polycondensation steps. The data demonstrate that the hydrolysis of TMOS is the only reaction at ambient temperature and in an acidic environment (Table I). The hydrolysis reaction is complete in 4 hours at pH 4.4, whereas in alkaline conditions, polymerization starts to occur after an initial hydrolysis. The rate is also a function of H_2O concentration as

TABLE I.   HYDROLYSIS RATE OF $Si(OCH_3)_4$ vs. pH AT
AMBIENT TEMPERATURE

pH	Catalyst	Reaction Time (hrs)	wt % Unreacted TMOS
4.4	dil. HNO_3	4	none detectable
6.5	None	28	20
*8.0	dil. NH_4OH	1 1/2	65

* Sample gelled after 2 hrs reaction time.

shown in Table II. More than 6 hours are required for the starting TMOS to
be consumed at 1:2 molar ratio of TMOS to H_2O. Four hours are needed when
the theoretical amount of H_2O, i.e. four moles is used. However, less than
one hour is required when the molar ratio of H_2O: TMOS is 8:1. Furthermore,
the rate of hydrolysis is accelerated by raising the reaction temperature as
shown in Tables III and IV. When the acid is used as a catalyst, i.e.
pH=4.4, the reaction is complete in an hour at 75°C as compared to 4 hours
at ambient. In the neutral condition, i.e. pH=6.5, 10 wt% of TMOS remained
after 6 hrs reaction time at 50°C as compared to 20 wt% of TMOS after 28
hours at 25°C. A typical gas chromatogram is shown in Figure 1.

The amount of TMOS (retention time (t_R) = 1.4 min) decreases with reaction time, whereas the amount of methanol (t_R=0.2 min) increases with reaction time. This is consistant with reaction (1), in that four moles of
CH_3OH are produced for every hydrolyzed TMOS. The complete hyrolysis of
TMOS in an acidic environment was further evidenced by mass spectroscopic
analysis as shown in Figure 2. Species evolved from ambient temperature to
600°C were water and methanol. No organic fragment derived from unreacted
TMOS was identified. If the hydrolysis is carried out in a basic condition,
i.e. pH=8.0, polycondensation/polymerization starts immediately after the
initiation of hydrolysis of TMOS as shown in Table I, at this point, 65 wt%
of TMOS was unreacted. This would result in a partially hydrolyzed gel
which has detrimental effects on later drying and sintering steps because of
the trapped carbonaceous residue. More than 24 hrs are needed for gelation
to occur in an acidic environment as compared to less than 5 minutes in an
alkaline environment (Table V). The rate of gelation or polymerization is
controlled by the temperature as well as pH. The gelation rate increases
with temperatures as shown in Table VI. Only 45 minutes are required to gel
at 75°C, whereas 24 hours is needed at ambient after TMOS was hydrolyzed at
pH 4.4.

TABLE II.   HYDROLYSIS RATE OF TMOS vs. MOLAR RATIO OF TMOS
TO H_2O AT AMBIENT TEMPERATURE AND pH=4.4

Molar Ratio of TMOS:H_2O:MeOH	Time Required (hrs) for Complete Disappearance of TMOS
1:2:3	>6 hrs
1:4:3	4 hr
1:8:3	<1 hr

TABLE III. HYDROLYSIS RATE OF $Si(OCH_3)_4$ vs. TEMPERATURE AT pH=4.4

Temperature °C	wt % Unreacted TMOS After 1 hr	Time Required (hrs) for Complete Disapperaance of TMOS
25°	70	4
50°	10	1 1/2
75°	None detectable	1

TABLE IV. HYDROLYSIS RATE OF $Si(OCH_3)_4$ vs. TEMPERATURE AT pH=6.5

Temperature °C	Reaction Time (hrs)	wt % Unreacted TMOS
25°	28	20
50°	6	10

TABLE V. *EFFECT OF pH ON RATE OF GELATION AT AMBIENT TEMPERATURE

pH	Gelation Time	Quality of Gel
4.4	>24 hr	Clear
6.5	>20 minutes	Clear to Opaque
8.0	<5 minutes	Opaque

*TMOS was hydrolyzed at pH 4.4.

TABLE VI. EFFECT OF TEMPERATURE UPON RATE OF GELATION AT pH=4.4

Temperature °C	Gelation Time (hrs)
25°	>24
50°	4 1/2
75°	3/4

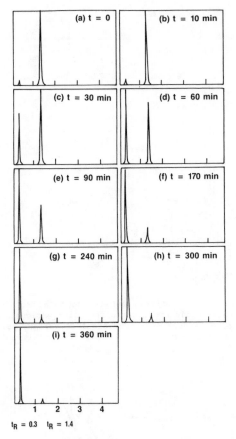

Figure 1. Gas chromatogram of $Si(OCH_3)_4:2H_2O:3CH_3OH$ a) t (reaction time) = 0; b) t = 10 min; c) t = 30 min; d) t = 60 min; e) t = 90 min; f) t = 170 min; g) t = 240 min; h) t = 300 min; i) t = 360 min.

Drying and Sintering

The wet gel consists of a network of SiO_2 with about 80 vol% liquid (methanol and water). If this liquid is allowed merely to evaporate into the atmosphere, the capillary forces on the gel would be substantial and the sample would crack. An alternative is to slow the rate of drying by slightly increasing the partial pressure of methanol over the sample and gradually allowing the methanol to escape. Drying by this technique is both slow and hard to control. For large bodies (2" in diameter, 1/2" thickness) one to two weeks is required. Therefore, the supercritical drying process described by Zarzycki [6] was adapted to expedite this drying process. Times of less than 24 hours were realized. The gel obtained from the supercritical drying process is porous with a surface area of 100–1100 m^2/gm depending on the conditions of hydrolysis and gelation.

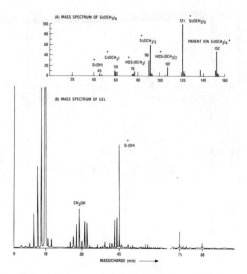

Figure 2. Mass spectrum of (A) $Si(OCH_3)_4$ and (B) air-dried gel.

The sintering of a dried gel was studied both isothermally and at a constant heating rate. Dilatometric studies using various heating rates and atmospheres were done to determine the effect of various parameters on sintering behavior. Some of these results are shown in Figure 3.

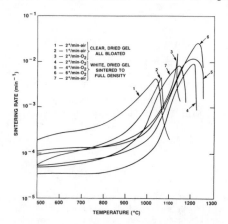

Figure 3. Sintering rate of TMOS-derived gels when sintered at constant heating rates of 1 to 6°C/min in O_2 and air.

The samples which were gelled in an acidic condition, clear as dried (curves 1-3) sintered at a lower temperature than those which were gelled in an alkaline condition, white or translucent as dried. However, the clear gels bloated immediately after sintering with continued heating. This is due to their much smaller pore size. The surface area of the clear, dried gels is 400-1000 m^2/gm. Samples that were white or translucent prior to sintering (curves 4-7) were stable against bloating. This is due to the larger pore size (surface area = 100-250 m^2/gm) which allows water to escape before the pores close. The effect of increasing the heating rate on the temperature of maximum sintering rate is small. In O_2, the sample is well dried before substantial shrinkage occurs, and the sintering rate increases as the heating rate increases. When the sample is heated in air, less drying has occurred before sintering and the higher water content lowers the viscosity and, therefore, the sintering temperature. During very slow heating in air (1°/min) the sample dries more than when it is heated rapidly, and the sintering temperature increases (curves 1 and 2). Some samples were also sintered in vacuum, these showed virtually no difference from the samples sintered in O_2. The results of dilatometric studies indicate the complexity of the sintering mechanism, major components, such as structural relaxation, viscous sintering and water evaluation, should be considered and separated. Therefore, isothermal sintering measurements were performed. The shrinkage of the samples was determined after isothermal sintering in O_2 for various times analyzed using the Scherer Model [7,8] for viscous sintering of a porous body. The shrinkage kinetics and microstructural development of the supercritically dried gel heated at different temperatures were also examined by dilatometry, B.E.T., TGA and STEM [9]. This combination of techniques showed that structural relaxation is not an important factor in the densification of these gels. Below 1000°C most shrinkage is due to condensation and water evolution. Above 1000°C, viscous sintering is the primary mechanism for shrinkage.

Properties of the Dense Glass

The viscosity calculated using the Scherer model is essentially the same as silica. The density of the sintered gel is 2.2 gm/cm^3 (measured by the Archimedes technique), and the average index of refraction is 1.457 (measured with the Abbe refractometer). The FTIR spectum is also the same as fused silica (Figure 4).

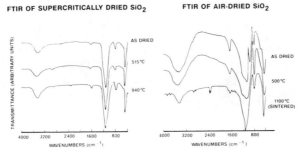

Figure 4. FTIR spectra of gel hydrolyzed from TMOS at various processing steps, indicating the progression to the structure of fused silica.

CONCLUSIONS

The parameters that affect the molecular network and structural morphology of the silica obtained by hydrolytic polymerization of tetramethylorthosilicate (TMOS) were systematically investigated. The effect of these structural and morphological modifications extends to the drying, sintering and subsequently, to the properties of the glass. The hydrolysis of TMOS is complete in 4 hours at pH 4.4 and at a molar ratio of TMOS/H$_2$O/methanol = 1:4:3. The reaction rate is accelerated by increasing the amount of H$_2$O as well as the temperature. The rate of gelation or polymerization is also controlled by the temperature and pH. In order to obtain a sinterable dried gel, the hydrolyzed clear sample should be polymerized in an alkaline condition. This not only expedites the reaction rate but also controls the pore size of the gel in the sinterable region. The gel prepared under these conditions can be sintered to dense amorphous silica at 1150°C which is identical in properties to silica glass prepared by conventional glass melting methods at 1900°C.

ACKNOWLEDGEMENTS

The authors are grateful to Ms. M. Coyne and Messrs. D. Carril, J. Daly for their able technical assistance. They also appreciate the support of Drs. L.J. Bowen and J.T. Smith, and Messrs. F. Avella and S. Natansohn's technical comments on the manuscript.

REFERENCES

1. H. Dischich, J. of Non-Crystalline Solids 57, 371-388 (1983).

2. J.D. MacKenzie, J. Non-Crystalline Solids 48, 1 (1982).

3. B.E. Yoldas, J. Non-Crystalline Solids 51, 105 (1982).

4. B.E. Yoldas, European Patent Application 000825 91978) Westinghouse Electric Corp., USA.

5. I. Matsuyama, K. Susa, S. Satah and T. Suganuma, Ceramic Bulletin, 63, 11, 1408-1411 (1984).

6. J. Zarzycki, J. Non-Crystalline Solids 48, 105 (1982).

7. G.W. Scherer, "Sintering of Low-Density Glasses: I, Theory", Journal of The American Ceramic Society, 60 [5-6], pp. 236-239 (1977).

8. G.W. Scherer and D.L. Bachman, "Sintering of Low-Density Glasses: II, Experimental Study", Journal of The American Ceramic Society, 60 [5-6], pp. 239-243 (1977).

9. P.I.K. Onorato to be published.

CALCULATED VS MEASURED VISCOSITIES OF SOL-GEL PROCESSED SILICA

T. A. GALLO AND L. C. KLEIN
Rutgers - The State University of New Jersey, Ceramics Department,
P.O. Box 909, Piscataway, N.J. 08854

ABSTRACT

The densification of sol-gel processed silica was studied by comparing calculated and measured viscosities between 700 and 850°C. Linear shrinkage was measured in a dilatometer. From shrinkage data, relative bulk densities were determined. The densities were used in a model for viscous sintering to calculate viscosities. Beam-bending viscosimetry was used to determine viscosities directly. The deflection rate of a centrally loaded beam was measured. Both the calculated and measured viscosities had the same time dependence. The temperature dependence was also determined.

INTRODUCTION

Up to this time, a sintering model has been used to calculate apparent viscosities [1-5]. These calculations have been used to model the densification behavior. The match of predicted behavior and observed behavior is quite good.

Calculated Viscosities

The sintering model used calculates isothermal viscosities when both structural changes due to loss of free volume and compositional changes due to dehydration are occurring. In an attempt to separate these effects, isothermal shrinkage data are analyzed in detail using this model [3,4]. This model is based on a geometry consisting of a cubic array of intersecting cylinders. According to the model [6], the quantity K is calculated using:

$$K = s/nl_o \ (p_s/p_o)^{1/3} \qquad (1)$$

where s is the surface tension, n is the viscosity, l_o is the initial cylinder length, p_s is the skeletal density, and p_o is the initial bulk density. The reciprocal of K is proportional to viscosity, and for this gel, n is between 5×10^{12} and 5×10^{14} poise.

Measured Viscosities

For comparison, a bending-beam viscosimeter, capable of measuring viscosities between 10^7 and 10^{15} poise, is used. The viscosity is determined from the rate of viscous deformation of a simple loaded beam. The load is applied to the sample with an alumina rod coupled to a linear variable differential transducer. The viscosimeter is calibrated using NBS reference materials.

The viscosity is calculated from the formula:

$$\dot{d} = W\, l^3/144\, n\, I \qquad (2)$$

where $\dot{d}$ is the deformation rate, W is the central load, l is the distance between supports and I is the moment of inertia.

The moment of inertia varies with the fourth power of the radius. During isothermal heat treatment, the moment of inertia may decrease by 25%. The reason that this geometry was selected was that the moment of inertia may change as the cylinder sinters but the distance between the supports remains the same. Also, it has been suggested that n be replaced by $n\, p_r/(3-2\, p_r)$ where p_r is the relative density. This factor corrects for the load bearing cross-section of the porous sample. This factor varies between 2 and 3.

EXPERIMENTAL TECHNIQUES

All samples used were gelled in a bell jar heated to 70°C, after which the samples were dried for forty days at 70°C. These samples are referred to as "D" gels and have 16 moles water per mole TEOS [7]. Their appearance is slightly cloudy monoliths. Bars were cut and dry polished from the dried "D" gel monoliths. Linear shrinkage was measured in an Orton Model 1500 Automatic Recording Dilatometer with an external programmable temperature controller. Samples were 1.27 cm long and 0.35 cm in diameter. About 22% linear shrinkage is needed to reach full density.

To measure viscosities directly, a beam bending viscometer was being used. The technique measures the deformation of a centrally loaded, end supported cylinder. The cylinders are 10 cm in length and 0.35 cm in diameter for viscosity measurements between 10^{12} and 10^{15} poise. The ends of the specimen are supported on a muffle in the central portion of a resistance furnace, equipped with a proportional temperature controller. The temperature variation in the specimen region over the time of experimental measurements is within $\pm$ 1 C.

Samples were held isothermally for 4 or 8 hours in both the dilatometer and viscosimeter. A flowing oxygen atmosphere was maintained. Also, samples were treated in steps such as an isothermal treatment at 700°C followed by an isothermal treatment at 800°C. Typically, a sample held at 700°C would have a measured surface area of 550 m^2/g and an average pore diameter of 27×10^{-10} m.

RESULTS

As the temperature is increased the viscosity decreases making viscous flow easier. At the same time, the viscosity is tending to increase due to decomposition of silanols. This decomposition also increases the surface tension which, in turn, increases the driving force for sintering. A plot of K^{-1}, the sintering parameter, versus time shows that the viscosity of the gel is increasing with time. Even after 6 hours, in most cases, a stable value has not been reached. Clearly water affects the viscosity of a gel or glass. Usually, over limited temperature ranges, the viscosity of a glass can be approximated with an equation of the type $n = n_0 \exp(E/RT)$. Differences in fictive temperature and water in the form of hydroxyls is expected to affect the activation energy E.

To assess the relative effects of dehydration and skeletal relaxation, a plot of log (K^{-1}) versus residual water gives a linear relationship for isothermal holds at lower temperatures [4]. This is taken to mean that the increase in activation energy is largely due to dehydration. However, at higher temperatures, water loss is no longer measurable, and the viscosity still increases due to slight skeletal changes in the gel.

Figure 1 is a plot of measured viscosity versus time for samples heated to 700, 750 and 850°C. At each temperature, the viscosity shows an initial rapid increase. Even after 4 hours it is still increasing.

Calculated and measured viscosities are plotted in Figure 2. The circles are measured viscosities for a sample heated to 850°C and held isothermally for 8 hours. The triangles and squares are calculated viscosities for a sample which underwent the same heat treatment schedule. The triangles were calculated using a surface tension of 250 erg/cm² and a pore diameter of

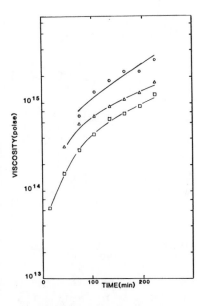

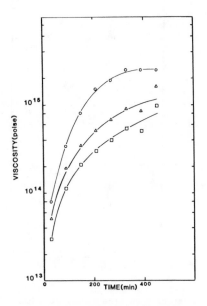

FIGURE 1
Measured viscosity vs time during isothermal heat treatments at 700 (circles), 750 (triangles), and 850°C (squares).

FIGURE 2
Viscosity vs time during isothermal heat treatments at 850°C: Measured (circles); calculated using 250 erg/cm² (triangles); and calculated using 150 erg/cm² (squares).

· 25 x 10^{-10} m. The squares were calculated using a surface tension of 150 erg/cm^2 and the same pore diameter. When the viscosity was recalculated using a surface tension of 250 erg/cm^2 [8] and a pore diameter of 35 x 10^{-10} m, the values were between those for the triangles and squares. The difference between measured and calculated viscosities is about a factor of 3 for 250 erg/cm^2 and a factor of 4 for 150 erg/cm^2. Previously, all calculated viscosities have been based on 250 erg/cm^2 and 25 x 10^{-10} m.

In Figure 3, the measured values of viscosity (closed symbols) and calculated values of viscosity (open symbols) are compared. In this case samples were soaked at 750°C for 4 hours then heated to 850°C and held for 4 hours. At 750°C the viscosity increases rapidly, followed by a more gradual increase at 850°C. At these temperatures, the difference between measured and calculated values is approximately a factor of 2.

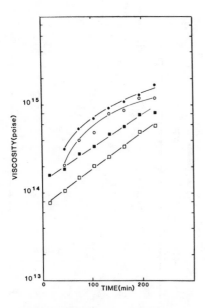

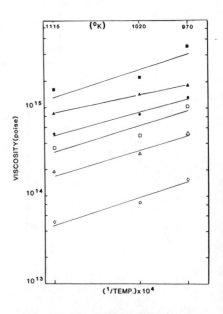

FIGURE 3

Viscosity vs time during step heat treatments at 750 followed by 850°C: measured at 750°C (filled circle); and calculated (open circle); measured at 850°C (filled square) and calculated (open square)

FIGURE 4

Log (viscosity) vs reciprocal temperature for isothermal heat treatment time (in minutes) of 30 (open circle), 90 (open triangle), 150 (open square), 210 (filled circle), 390 (filled triangle), 450 (filled square).

Finally, in Figure 4, for samples heated to 700, 750 and 850°C with 8 hour soaks, the log viscosity is plotted vs reciprocal temperature. Each line represents a constant time at these temperatures ranging from 30 min. to 450 min. The slope of these lines shows little time dependence. A value of 6 kcal/mole is calculated for these lines.

DISCUSSION

When a gel derived glass is held isothermally, the viscosity increases rapidly at first. When the sample is heated to a higher temperature, the initial increase in viscosity is not as great. This has been attributed to the fact that during an isothermal hold, a certain amount of excess free volume is annealed out. The continued increase in viscosity with time is largely due to dehydration.

If this is the case, it might be expected that the plots of log viscosity versus reciprocal temperature (Figure 4) would have slopes that increased with time. Yet the slopes in Figure 4 do not appear to be changing with time. The slope is approximately 6 kcal/mole. This value is much lower than the value for viscous flow in silica, so it represents a process other than viscous flow.

To explain the weak temperature dependence, consider instead the temperature dependence of the surface tension. The reduction of surface energy is the driving force for densification. The surface tension of silica varies from 130 erg/cm^2 for a fully hydrated surface to 260 erg/cm^2 for anhydrous silica [8].

When using the value for anhydrous silica, the calculated values of viscosity are consistently lower than the measured values by a factor of 2 to 3. However, the calculated and measured values show the same time dependence of viscosity and temperature dependence for short times.

For calculated viscosities, it is assumed that the energy released by reduction of surface area is dissipated in viscous flow and that this is the only source of energy. This does not take into account the energy from dehydration or structural relaxation processes. The energy from these processes obviously contribute to the low activation energy for viscous flow. Another way of putting this is that the calculated values would be equal to the measured values if an effective surface tension of about 500 erg/cm^2 was used in the model. Since this high value is not realistic, it is likely that the combination of silanols and the structural relaxation contribute to the temperature dependence for viscous flow in gels.

CONCLUSIONS

When calculated viscosities and measured viscosities are compared, the measured viscosities are higher by 2 or 3 times. The viscosities show the same time dependence and temperature dependence when calculated or measured. The difference in measured and calculated viscosities is that dehydration and structural relaxation provide energy for viscous flow in addition to the energy provided by the elimination of surface. This additional energy is not accounted for in the sintering model but is operating during direct measurements of viscosity. The calculated values can be made to equal the measured values by adjusting the surface tension term.

REFERENCES

1. G. W. Scherer, C. J. Brinker, E. P. Roth, J. Non-Cryst. Solids 72, 369 (1985).

2. G. W. Scherer, T. Garino, J. Am. Ceram. Soc., 68 [4], 216 (1985).

3. T. A. Gallo, L. C. Klein, "Apparent Viscosity of Sol-Gel Processed Silica," to appear in J. Non-Cryst. Solids (1986).

4. T. A. Gallo, C. J. Brinker, L. C. Klein, G. W. Scherer, in Better Ceramics through Chemistry, edited by C. J. Brinker, D. R. Ulrich. D. E. Clark (Elsevier, New York, 1984) p. 85.

5. G. W. Scherer, D. L. Bachman, J. Am. Ceram. Soc., 60 [5-6], 239 (1977).

6. G. W. Scherer, J. Am. Ceram. Soc., 60 [5-6], 236 (1977).

7. L. C. Klein, T. A. Gallo, G. J. Garvey, J. Non-Cryst. Solids 63 23 (1984).

8. R. K. Iler, The Chemistry of Silica, (John Wiley & Sons, Inc., New York, 1979), p. 544.

FLUORINE IN SILICA GELS

E. M. RABINOVICH AND D. L. WOOD
AT&T Bell Laboratories, Murray Hill, NJ 07974

ABSTRACT

Fluorine has a remarkable and diverse action in silica sols, gels and gel-derived glasses. Introduced in ionic form, it causes more rapid gelation, it reduces surface area and water adsorption of dry gels, and it helps to eliminate bubble formation during sintering and reheating of gel-derived glasses. The mechanisms for these actions of the fluoride ion are reviewed, with a more detailed discussion of the rate of gelation. Several previously held views of the way fluoride accelerates gel formation are compared with a new suggestion that it is caused by an electrostatic attraction between the proton of the OH of silicic acid and F^- present as Si-F groups.

INTRODUCTION

The fluoride ion has a dramatic effect on the formation and properties of high silica gels, and on the glasses formed from them. It accelerates the formation of gels from silicic acid, as well as those derived from colloidal silicas. It helps to dehydrate and dechlorinate gels after formation and to prevent foaming of glasses during sintering or reheating. It is the purpose of this paper to report some new results on gels and gel formation in the presence of fluoride, and to discuss what is known about the behavior of fluoride ion in gels and glasses in the light of various mechanisms which can be proposed to explain its behavior.

The polymerization of silicic acid has been reviewed by Iler [1] who found that traces of fluoride ion as low as 1 ppm can have a marked effect on the rate of polymerization in acid solutions (pH$<=$2). Tarutani, cited by Iler [1], stated that fluoride had no effect at pH=7. Lyakov and Samuneva [2] found that F^- has a catalytic effect on gelation mainly at low pH when polymerization is slow. They studied polymerization of hydrolyzed tetraethyl orthosilicate (TEOS) with F^- concentrations ranging from 0 to 0.5 mg/ℓ and found that the rate of gelation is proportional to the F^- concentration.

We studied the effect of higher fluoride concentrations on gelation and on the properties of silicic acid formed by hydrolysis of TEOS. We used two series of initial solutions. One series, called 1F, had molar ratios of TEOS:C_2H_5OH:(H_2O+HCl+HF) of 1:4:4, while the other, called 7F, was more dilute with molar ratios of 1:4:50. The number after F in each series indicates the amount of fluoride added as HF in grams of F per 100 g of SiO_2 equivalent. The infrared and Raman spectroscopy, thermomechanical and gas evolution studies of gels made from these starting solutions have already been reported elsewhere [3-5].

EXPERIMENTAL PROCEDURES

For preparation of sols and gels the required TEOS and ethanol were weighed, mixed, and then added to a solution of H_2O+HCl+HF at time zero. The original H_2O+HCl solution used for F-free compositions 1F-0 and 7F-0 had pH=1, but for other compositions F was added as a 49% solution of HF which replaced HCl on a weight basis until at higher F concentrations there was no longer any HCl in the solution. The mixtures were agitated with a magnetic stirring bar, and the gel time was taken to be that required for the funnel to disappear from the top of the liquid as the effectiveness of the rotation of the stirrer bar decreased with increasing viscosity of the solution. Heat evolution, viscosity, and light scattering measurements – all as a function of time – were made during hydrolysis and gel formation. The light scattering measurements were recorded photoelectrically at 90° to the incident radiation in a cylindrical cell. The heat evolution measurements were recorded from a thermocouple immersed in a 100 g batch of reactants in a vacuum insulated thermostat. The viscosities were measured in a Brookfield viscometer as the torque required to maintain the rotation of a cylinder immersed in an open vessel containing the mixtures.

Dry gels were prepared by stirring the mixed reactants for 30 min or until gelation, then aging in a closed container at 60°C for 24 h, and finally drying at 150°C for 48 h. The BET surface area measurements were performed on these samples by outside laboratories. The residual fluorine analyses were performed on dried gel samples which had been ground, fused in Na_2CO_3+ZnO or in K_2CO_3, and dissolved in water. The F concentration in the resulting solution was determined with an ion-specific electrode, or with a spectrophotometric method.

RESULTS

The dependence of gelation time on the amount of fluoride is shown in Fig. 1, where even 1% F^- causes a reduction in the gel time of 50x for the 1F series, and 660x for the more dilute 7F series. Larger additions of fluoride are increasingly less effective, and eventually further reduction is arrested. The poorer reproducibility in gel time for high fluoride sols in the 7F series is thought to be due to the effect of longer mixing times required to produce uniform sols. The mixing times were comparable to the gel times for these high fluoride mixes.

The curve of formation of a typical gel containing F^- is shown in Fig. 2 where the viscosity and the 90° light scattering are both plotted as a function of time. There is an induction period of about 10 minutes during which neither the light scattering nor the viscosity increase very much. Then there is a rapid increase for both properties, with the viscosity going to an off-scale value, and the light scattering continuing to increase long after the gel has solidified. The light scattering is more revealing of the molecular scale events in gel formation, since the measurements can conveniently bridge the transition from liquid to solid. The viscosity measurements for liquid samples are inappropriate for solids and cannot bridge the transition so well.

Fig. 3 shows the heat evolved, or temperature rise as a function of time during formation of the gel for isolated 100 g batches of 1F-0 and 1F-4. The most important feature of the curves is their steep and immediate rise from time zero. We associate this initial burst of heat corresponding to about 39 kJ/mole of TEOS with its hydrolysis. This means that the hydrolysis takes place during the induction period before a significant increase in light scattering or viscosity occurs. Thus, in these particular compositions, the hydrolysis step is separated in time from the polymerization step. In fact, 1F-0 containing only HCl+H_2O and no HF ordinarily takes more than a month to gel at room temperature (Fig. 1), but the heat is evolved in a minute or so. The slower evolution of heat in the case of 1F-4 is due to the reduced rate of

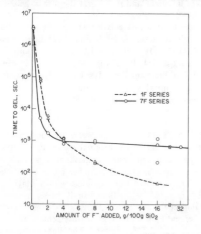

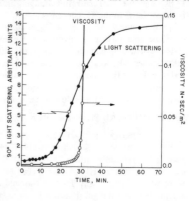

Fig. 1. Time of gelation for sols of series 1F (1:4:4) and 7F (1:4:50) as a function of the amount of F^- introduced into the initial solution.

Fig. 2. Light scattering and viscosity vs. time for sol 1F-4.

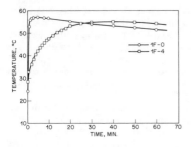

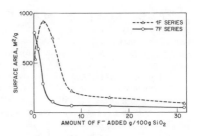

Fig. 3. Evolution of heat during hydrolysis of sols 1F-0 and 1F-4.

Fig. 4. BET surface areas of gels dried at 150°C for series 1F and 7F.

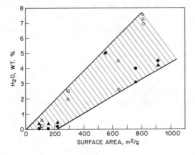

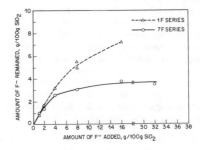

Fig. 5. Water content of gels dried at 150°C and then equilibrated with ambient air at room temperature as a function of the BET surface area. Filled symbols are for 1F series, empty symbols are for 7F series. Triangles: water content determined in June 1985 (as published in [3]); circles: data taken in March 1986. The samples were stored on the laboratory table and not dried additionally.

Fig. 6. Amount of F analyzed in gels after drying at 150°C vs. amount of F introduced into initial solutions as HF.

hydrolysis for HF which replaces about half of the HCl used in 1F-0. Note that the temperature rises to practically the same value for either the HCl or the HF+HCl catalyzed solutions. The temperature rise for a 100 g batch of the 7F series is proportionately smaller since the TEOS is more dilute.

With increase in F^- content the appearance of gels changed from clear and transparent or translucent (for 1-4% F) to a nearly opaque white for higher additions where gelling rates are more rapid. Upon drying, the latter formed white aggregates or powders, while the transparent gels formed relatively clear solid pieces 0.1 to 1 cm in size. Fig. 4 shows that the surface area was also sharply reduced with increased F content, although for the 1F series an initial increase was observed. A consequence of the reduced surface area is that the amount of H_2O adsorbed in the pores is reduced as Fig. 5 shows. This is a plot of previous infrared absorption data [3] against surface area for the same samples. Unfortunately, a significant dispersion of the data exists, but a general trend for both the 1F and 7F series is clear, and a linear relation is indicated.

Powders can be precipitated from TEOS without fluoride at high pH when NH_4OH is used as a catalyst. But a significant difference exists in that for the latter case the powder is formed as a precipitate without significant evolution of heat [6]. This led us to the conclusion that this process does not consist of hydrolysis followed by polymerization, but rather one in which the polymer forms directly without significant hydrolysis. In the case of low pH (<7) sols, addition of fluoride may change the external appearance of the process to one in which a precipitate forms and powder-like gels occur, but in such cases appreciable heat is evolved comparable to that evolved without fluoride.

It is known [1] that an increase in pH accelerates gel formation even without fluoride. However at pH=5 the time of gelation was shortened from 39 s for a sol containing 2% F to <1 s for 4% F. Introduction of F as NH_4F gave the same effect when the pH was maintained the same, and additions of NH_4F to solutions with pH=7 to pH=11 also sharply accelerated powder precipitations. Additions of NH_4Cl to solutions with low pH showed no effect. We conclude, therefore, that fluoride has an accelerating effect on reactions with TEOS at practically all pH values.

It has been shown that colloidal silica will also gel at a faster rate when suspended in water containing fluoride ion [1] even at low fluoride concentrations of 10-100 ppm. Rabinovich et al. [7] described gelation of colloidal silica having a surface area of 230 m^2/g in various acid and base aqueous solutions. Acid suspensions described in [7] showed increased fluidity over those with higher pH, but additions of HF or NH_4F to sols at pH<7 resulted in almost immediate gelation. It was also noted that addition of NH_4F to water used for dispersion resulted in the formation of ammonium fluosilicate even at room temperature.

The analyses for residual fluorine in dry gels are shown in graphical form in Fig. 6, and it is clear that drying at 150°C resulted in the loss of a significant fraction of the fluorine, especially for the higher additions. It is also evident from Fig. 6 that for the more concentrated formulations represented by the 1F series the fraction of fluorine retained in the dry gel is larger than for the more dilute 7F series.

DISCUSSION

1. Acceleration of Gelation

We shall discuss 5 possible mechanisms for the accelerated gelation in the presence of fluoride ion. They are: 1) Expansion of the coordination number of silicon from four to 5 or 6; 2) The formation of hydrophobic bonding; 3) Changes in hydrogen bonding; 4) Changes in electrostatic interactions; 5) Changes in the solubility of silica. The first mechanism applies to the gelation of silicic acid, while the last applies more directly to the gelation of colloidal silica in suspension. The other three apply to both silicic acid and colloidal silica. Iler has discussed the first two mechanisms [1], and the third has been considered by Nassau [8]. The other two are considered here for the first time.

The F^- ion, the OH^- ion, and the O^{2-} ion are all about the same size, and can substitute relatively freely for each other in the Si-O network of silica [9-11]. Krol and Rabinovich [4] studying dry gels of the 7F series by Raman spectroscopy found that fluorine in these gels is present as Si-F and various hydrogen-bonded H-F groups.

For the hydrolysis of TEOS in the absence of fluoride, one can write the formal reaction:

$$Si(OC_2H_5)_4 + 4H_2O = Si(OH)_4 + 4C_2H_5OH \tag{1}$$

When fluorine is present, for example as HF, it is possible to write an additional set of reactions taking place before polymerization occurs:

$$Si(OC_2H_5)_4 + xHF + (4-x)H_2O = Si(OH)_{4-x}F_x + 4C_2H_5OH \tag{2}$$

$$Si(OH)_4 + xHF = Si(OH)_{4-x}F_x + xH_2O \tag{3}$$

$$Si(OH)_4 + 6HF = 2H^+ + SiF_6^{2-} + 4H_2O \tag{4}$$

Then the polymerization reaction of silicic acid can be written, a) without fluoride:

$$Si(OH)_4 = SiO_{2-n}(OH)_{2n} + (2-n)H_2O \qquad (5)$$

or b) when fluoride ion is present:

$$Si(OH)_{4-x}F_x + Si(OH)_4 = 2SiO_{2-n}(OH)_{2n} + xHF + (4-2n-x)H_2O \qquad (6)$$

1.1. *Coordination number expansion.*

Equation (4) represents the formation of 6-coordinated fluosilicate groups from the 4-coordinated silicic acid, and the higher coordination is known to occur in the crystals of the alkali metal salts of fluosilicic acid which are isostructural with K_2PtCl_6 [12]. The group also occurs in solution, as pointed out by Iler [1]. It is reasonable to expect, therefore, that molecular structures like $[Si(OH)_kF_m]$ may exist in solution with k+m=5 or 6. From geometrical considerations alone this would lead to a higher probability that collisions would result in formation of -Si-O-Si- bonds and an acceleration of polymerization. However, this process would represent a rearrangement of the coordination around the silicon, and this may have an inhibiting effect on the polymerization, running counter to the proposed tendency to accelerate it.

1.2. *Formation of hydrophobic bonds*

Iler [1] explained the effect of fluoride on gelation of colloidal silica by the formation of Si-F groups partially replacing Si-OH groups on the silica surface, which then becomes hydrophobic. He referred to the more easily formed association of fluoride-containing silica micelles as hydrophobic bonding, because water molecules are more easily removed from the point of contact of the hydrophobic surfaces. The similar particles are then held together by strong surface tension forces formed at the negative radius of curvature of the zone of contact. Such an explanation can be applied also to the gelation of silicic acid with Si-F bonds formed according to reaction (3).

Hydrophobic bonding in proteins and in hydrocarbons was considered by Kauzmann [13]. These bonds are formed in aqueous environments between non-polar molecules which tend to adhere to each other. Their thermodynamic behavior is explained by the fact that entropy is reduced when, for example, hydrocarbon molecules are transferred from a non-polar solvent to water. Frank and Evans (cited by Kauzmann [13]) concluded that in the presence of a non-polar molecule the water molecules in their immediate vicinity must be arranged in a quasi-crystalline structure with better hydrogen bonding than in ordinary liquid water. The existence of this structure would result in reduction of the entropy. The clustering of such non-polar molecules in "hydrophobic bonding" would help to avoid this entropy reduction and accelerate gel formation.

Such an effect could occur for TEOS which is non-polar, and could perhaps contribute to polymerization at high pH where hydrolysis does not seem to occur first. However, it is less likely to apply to the acceleration of gelation by fluoride where hydrolysis occurs, as $Si(OH)_{4-x}F_x$ molecules cannot be considered as non-polar. Furthermore, the large effect of fluoride on the time to gel is observed at dilute F^- concentrations when the proximity of similar molecules is statistically unlikely. For example, at 1% F there will be one Si-F bond per 31.5 tetrahedra, and hydrophobic clustering is not expected. This should also be the case for dilute Si-F bonds on the surface of colloidal micelles.

1.3. *Hydrogen bonding and hydrophobicity*

The results of infrared studies reported earlier [3] showed that when dried gels are equilibrated with ambient air, the amount of adsorbed water decreased with increasing amount of added fluoride. This led us to consider whether the fluoride may create hydrophobic surfaces by a change in the character of hydrogen bonds. According to an earlier model [14] for gelation of colloidal silica, the separate micelles are initially held together by hydrogen bonds between silanol groups on the surface, or in the case of silicic acid between the $Si(OH)_4$ molecules, requiring that H_2O be expelled from the contact between them in order to form -Si-O-Si- bonds in the gel or the polymer. If the hydrogen bonds -Si-F $\cdots$ HOH and -Si-F $\cdots$ HF are weaker than the -Si-OH $\cdots$ HOH bonds, then H_2O could be expelled more easily when F^- is present, and gelation and polymerization would be accelerated.

But this mechanism suffers from the objection of the rarity of -Si-F bonds already discussed. In addition, Nassau [8], upon our request, conducted a detailed analysis of hydrogen bonds in silica gels. He concluded that -Si-F $\cdots$ HOH has slightly less hydrogen-bonding stabilization energy than either -Si-OH $\cdots$ OH_2 or -Si-HO $\cdots$ HOH, and that this indicates that the -Si-F group could be really less hydrophylic than the -Si-OH group. On the other hand, some contradiction arises from the fact that a high strength of the hydrogen bond between F^- and H^+ is usually considered the reason for the existence of the $(HF_2)^-$ ion [15].

Finally, the results shown in Fig. 5 do not support the existence of a significant effect of hydrophobicity of -Si-F groups. If fluoride caused a highly hydrophobic surface to form, there should be a departure from linearity of H_2O adsorbed from ambient air against surface area. Although our results are not very accurate or reproducible, there is good reason to assume that the effect of F^- on reduction of the water adsorption is mainly a secondary result of reduced surface area and not of superficial hydrophobicity.

1.4. Electrostatic interactions

We propose another explanation of the action of fluoride which does not involve expansion of the coordination number of silicon, but which does not exclude that mechanism. This is based on the asymmetry of the electrical fields of the participating ions. In the OH groups of the $Si(OH)_4$ molecule the electric field in the vicinity of the protons is positive, and this accounts for the formation of hydrogen bonds [15] with H_2O. When two identical molecules with this kind of electrical asymmetry meet, their first reaction should be repulsion because they have the same charge. Only in collisions with large energy will the barrier to reaction be overcome and the formation of the -Si-O-Si- bonds take place according to reaction (5), allowing polymerization and gelation to proceed.

It is different with fluoride present, where $Si(OH)_{4-x}F_x$ molecules appear according to reactions (2) and (3). When a molecule with at least one F^- ion meets locally the positive part of an $(OH)^-$ group from another molecule, attraction substitutes for repulsion, and the reaction of polymerization occurs even though the energy of collision may be small. This can be formally represented by the replacement of equation (5) by equation (6). The reaction (6) is shown schematically in Fig. 7 where the shape and charge distributions of the reacting $Si(OH)_4$ and $Si(OH)_3F$ molecules are represented in the top drawing as spheres of suitable sizes. The collision is represented in the center drawing, and the reaction products are represented in the lower drawing.

If one compares reaction (6) with reaction (3), it may be seen that all HF is returned to the solution. If this were indeed true, then HF could be considered as a true catalyst, appearing at the end of the reaction unchanged qualitatively and quantitatively. However, a part of the F^- remains in the structure as Si-F bonds, which were discovered by Raman spectra [4] after heating to 1000°C. They apparently can survive even heating to near 2000° [16], since the reduced refractive index due to fluoride doping [17] is preserved. Taking these groups into account, the product of reaction (6) should have a formula like $SiO_{2-n-y}(OH)_{2n}F_{2y}$. The amount of HF returned would be reduced to $(x-2y)$.

In addition to this, some HF may also be consumed in the formation of fluosilicic acid, as shown in reaction (4). This can decompose to SiF_4 and HF on heating, and both were evolved from gels as seen from the mass spectroscopy results shown in Fig. 8 taken from [5]. HF starts to evolve even below 200°C, while $(SiF_3)^+$ only appears above 450°C. This means that both free (formed according to reaction 6) and bonded HF (formed according to reaction 4) remain in gels after their formation. The significant amounts of loss of free HF during drying is reflected by the loss of overall F revealed by chemical analyses (Fig. 6).

This free HF regenerated during reaction (6) attacks again silicic acid still in solution as in reactions (3) and (4) to continue the accelerated polymerization, and even small amounts of

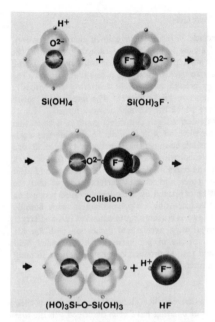

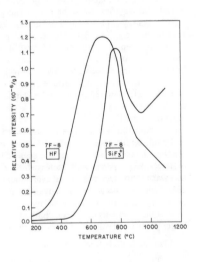

Fig. 7. Schematic diagrams of molecules of Si(OH)$_4$, Si(OH)$_3$F and the reaction between them. This figure was inspired by pictures of SiF$_4$ and HF molecules in a book by the Paulings [15].

Fig. 8. Emitted gas analysis for HF and (SiF$_3$)$^+$ in 7F-8 [5].

F$^-$ ions can significantly reduce gelation times with catalyst-like behavior. Of course, this process is limited by the incomplete return of all the F$^-$ as HF. At higher concentrations, collisions of molecules both containing fluoride become more probable, and the molecules may repel each other. The latter explains why in Fig. 1 the effectiveness of further F$^-$ additions is reduced at high concentrations.

Acceleration of gelation of colloidal silica with fluoride can be explained in a similar way. The mechanism of its gelation without F$^-$ was given by Wood et al. [14] and was mentioned above. The fact that (NH$_4$)$_2$SiF$_6$ forms when NH$_4$F added to water during colloidal silica dispersion proves that fluoride ion can be incorporated into the solid silica structure even at room temperature. We believe that this produces superficial substitution of Si-F for Si-OH. Then the attraction between Si-F groups on the surface of one micelle with Si-OH groups of another micelle can fix them together more rapidly than in case of micelles without fluoride substitution.

1.5. Solubility in the presence of fluoride

The formation of gels in the case of colloidal suspensions has been described [14] as taking place in two steps. The first step involves the initial fixation of the particles at a point of contact, as mentioned above. The second step, described by Iler [1], involves deposition of silica in the vicinity of the contact point due to decreased solubility with decrease of the radius of curvature of the surface presented to the aqueous solvent. If silica is more soluble in water when fluoride is present, especially as HF, this deposition step may be enhanced by a more rapid deposition of the strengthening neck at the point of contact, and a more rapid formation of the gel structure.

2. Surface Area and External Appearance of Dried Gels

It is clear that polymerization occurs because it results in a gain in energy. However, during the very first steps the rise in energy of a new formed surface exceeds the gain in the volume energy. This is a classical case of a two-step phase transition with nucleation and growth (see, e.g., in [18]). Nuclei can depolymerize back if their size is less than critical. The nucleation step is apparently reflected in a slow growth in the light scattering and viscosity during the first 15 min after the mixing of composition 1F-4 (Fig. 2). The later rapid growth in the scattering reflects the stage of growth during which the polymer network forms.

The change in BET surface area of alkoxide-derived gels is a direct result of acceleration of both nucleation and growth rates during polymerization with fluoride. A fluoride-free acid-catalyzed gel may be thought of as a continuous, three-dimensional network [19], or as one gigantic molecule $mSiO_{2-n}(OH)_{2n}$. Water and alcohol which are immiscible with this network are accommodated in pockets of fluid which become pores on drying. For a given ratio of solid to liquid, larger pockets of fluid lead to larger pores and smaller surface area, so that the relation of surface area to fluoride content must be explained by a tendency toward production of fewer large pores with increased fluoride. Acceleration of polymerization with fluoride necessarily produces more rapid evolution of H_2O and HF according to equation (6), and larger fluid inclusions occur in the network. In a similar way, accelerated nucleation with fluoride would promote the initiation of many separate regions of gel network with extended fluid regions included between them. Both effects would tend to produce larger fluid regions forming larger pores on drying and decreasing the surface area.

3. Fluorine in Gel–Derived Glasses

The most remarkable effect of fluorine in gel-derived silica glasses is that it prevents their swelling and bloating [5,16]. Two different modes of bloating have been described. In the first mode [5] gels like 1F-0 and 7F-0 increase in volume around the sintering temperature, typically below 1400°C. In the other mode [16] bubbles form in sintered glass which has been dehydrated in Cl_2 and then reheated at high temperatures around 1700-2000°C.

It was shown [5] that the first type of swelling is a result of water evolution due to broken silanol bonds. Since fluoride reduces the surface area to which water molecules are bound and replaces OH groups in the structure, it reduces the adsorption and then evolution of H_2O and the first type of swelling. However, rather a large amount of fluorine (more than 4% in the starting solution for the 7F series) is required to completely eliminate the effect.

The second type of bubbling can be effectively corrected even with minor amounts of fluoride retained in the glass structure [16]. Sintering of gels into OH-free glasses in accomplished by dehydrating the gels in Cl_2-containing atmospheres. Up to 0.4% Cl can be retained in the glass, apparently in Si-Cl bonds. These bonds cannot survive heating to high temperatures and they release Cl_2 gas which causes bubbling.

Fluoride introduced either as HF during gel dispersion or as NH_4F during firing, where it sublimes and reacts with silica at relatively low temperatures, may form Si-F bonds which occupy sites otherwise taken by Cl. If fluoride is introduced with Cl_2 during dehydration, the amount of retained chlorine can be reduced up to undetectable limit. The Si-F bonds are strong enough to survive reheating to more than 2000°C.

A further advantage of processing with fluoride during sintering is that it helps to dehydrate the glass [16], even though not as effectively as chlorine. The reduction of the index of refraction is another feature introduced in glass by fluorine [17] but this is not limited to gel-derived glasses.

CONCLUSIONS

1. Fluoride ion is a strong catalyst for polymerization and gelation of silicic acid and for colloidal silica. This effect is related mainly to the stronger attraction between Si-F and Si-OH groups than between two Si-OH groups. In the case of colloidal silica, fluoride ion in acid environments also changes the solubility of silica and thus has an additional accelerating effect on gelation through the mechanism of solution and redeposition.

2. Fluoride reduces surface area and water adsorption in dry alkoxide-derived gels by accelerating nucleation and polymerization during their formation.

3. Introduction of fluoride in different forms into different types of gels results in the reduction or elimination of swelling and bloating of glasses made from the gels during sintering or upon reheating to high temperatures near 2000°C.

ACKNOWLEDGEMENTS

The authors are grateful to S. C. Abrahams, K. A. Jackson, D. W. Johnson, Jr., C. R. Kurkjian, J. B. MacChesney, K. Nassau and F. H. Stillinger for helpful discussions. They thank L. D. Blitzer, T. Y. Kometani, C. McCrory-Joy and A. M. Williams for chemical analyses of gels. Several experiments with gels for this work were conducted by A. Rabinovich.

REFERENCES

1. R. K. Iler, The Chemistry of Silica (Wiley, New York, 1979).

2. D. Lyakov and B. Samuneva, presented at the 3d Internat. Workshop on Glasses and Glass-Ceramics from Gels, Montpellier, France, Sept. 12-14, 1985; to be published in J. Non-Cryst. Solids 82 (1986).

3. D. L. Wood and E. M. Rabinovich J. Non-Cryst. Solids 82 (1986).

4. D. M. Krol and E. M. Rabinovich, ibid.

5. K. Nassau, E. M. Rabinovich, A. E. Miller and P. K. Gallagher, ibid.

6. D. L. Wood and E. M. Rabinovich, Ref. 49-G-85F in Ceram. Bull. 64 (10), 1342 (1985).

7. E. M. Rabinovich, D. W. Johnson, Jr., J. B. MacChesney, and E. M. Vogel, J. Amer. Ceram. Soc. 66, 686-688 (1983).

8. K. Nassau, Ref. 50-G-85G in Ceram. Bull. 64 (10), 1342 (1985).

9. W. A. Weyl and E. C. Morboe, The Constitution of Glasses II, part 2 (Wiley, New York, 1967).

10. E. M. Rabinovich, Inorg. Mater. (English Transl.) 3 (5), 762-766 (1967).

11. E. M. Rabinovich, Phys. Chem. Glasses 24 (2), 54-56 (1983).

12. M. Remy, Lehrbuch der Anorganischen Chemie, B.1, (Akad. Verlagsgesellschaft Geest & Portig K.-G., Leipzig, 1960).

13. W. Kauzmann, in Advances in Protein Chemistry 14, edited by C. B. Anfinsen, Jr., et al., (Academic, New York, 1959) pp. 37-47.

14. D. L. Wood, E. M. Rabinovich, D. W. Johnson, Jr., J. B. MacChesney, and E. M. Vogel, J. Amer. Ceram. Soc., 66, 693-699 (1983).

15. L. Pauling and P. Pauling, Chemistry (Freeman, San Francisco, 1975) pp. 284, 287.

16. E. M. Rabinovich, D. L. Wood, D. W. Johnson, Jr., D. A. Fleming, S. M. Vincent and J. B. MacChesney, J. Non-Cryst. Solids 82 (1986).

17. J. W. Fleming and D. L. Wood, Appl. Opt. 22, 3102 (1983).

18. D. R. Uhlmann and H. Yinnon, in Glass: Science and Technology 1, edited by D. R. Uhlmann and N. J. Kreidl (Academic, New York, 1983) pp. 8-14.

19. E. M. Rabinovich, J. Non-Cryst. Solids 71, 187-193 (1985).

Raman Spectra of Rings in Silicate Materials*

D.R.Tallant, B.C.Bunker, C.J.Brinker, and C.A.Balfe
Sandia National Laboratories
Albuquerque, NM 87185

*This work performed at Sandia National Laboratories supported by the U.S. Department of Energy under Contract Number DE-AC04-DP00789.

ABSTRACT

Raman spectroscopic studies on gel-derived silicates have confirmed that narrow bands near 607 cm-1 and 492 cm-1, first observed in the Raman spectrum of fused silica, are associated with three- and four-fold siloxane rings. Using these results, we have identified three- and four-fold siloxane rings in other high-surface-area silica materials, including leached glasses and Cab-O-Sil. This Raman spectroscopic evidence not only shows that small siloxane rings are a common characteristic of a number of silica materials but also suggests that they form preferentially at silica surfaces. This paper reviews the Raman spectroscopic evidence that led to the identification of the vibrational frequencies of the small siloxane rings and presents the results of Raman experiments on high-surface-area silica materials in which the concentration of small siloxane rings is enhanced compared to fused silica.

INTRODUCTION

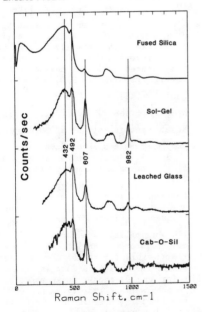

Figure 1. Raman spectra of silicate materials

Most silicate networks consist of interconnected rings composed of linked SiO_4 tetrahedra. The distribution of the sizes of these siloxane rings is of interest because it may influence diffusion through the network, chemical reactivity, and physical properties of the silica network. Random network models[20] predict that the majority of rings in fused silica contain five or more siloxane(Si-O-Si) linkages. The bands at 430, 800, 1065, and 1200 cm-1 in the Raman spectrum of bulk, fused silica(Figure 1) can be explained[1] on the basis of such a network. Only the narrow Raman bands near 492 cm-1 and 607 cm-1 are not explained by these calculations. From an initial classification of these bands as arising from "defect" structures, they have acquired the designations D1(492 cm-1) and D2(607 cm-1). As shown in Figure 1, the D1 and D2 bands are relatively more prominent in high-surface-area silica materials(e.g., gel-derived silicates, leached borosilicate glass, and dehydroxylated Cab-O-Sil) than in conventionally-synthesized fused silica. The silicate structures giving rise to the D1 and D2 bands appear, therefore, to form preferentially at silica surfaces and may be important in surface effects such as adsorption, catalysis, and stress-corrosion cracking. Models proposed to explain the origin of the D1 and D2 bands have included a) broken, elongated, or missing bonds[2], b) silanone($>$Si=O) species periodically arranged on surfaces of paracrystalline clusters as the source of D1 and rings containing four siloxane linkages(four-fold rings) formed by partial cluster fusion as the source for D2[3], and c) condensation of surface silanol species to form small siloxane rings[4,5,6]. Recent work[7] has confirmed Galeener's[8] assignment of the D1 and D2 Raman bands to vibrationally-decoupled[8] four-fold and three-fold siloxane rings, respectively, which form by condensation of surface silanol species as predicted by model c). In this paper we will review the Raman spectroscopic evidence for the presence of small siloxane rings in gel-derived silicates and present evidence for their presence in other high-surface-area silica materials.

CHARACTERIZATION OF SILOXANE RINGS IN GEL-DERIVED SILICATES

Silica materials formed by the sol-gel process[9] are especially useful for characterizing the "defect" bands D1 and D2 because the desiccated gels(xerogels) have large areas of internal surface[7] and because the development of intense D1 and D2 bands can be followed in detail through various stages of drying and heat treatment. Several authors[7,10,11] have noted that the development of the D1 and D2 bands coincides with the disappearance of Raman bands due to surface silanol($>$Si-OH) groups. In fact, the silanol bands at 980 and 3750 cm-1 decrease monotonically with increases in the D2 band(Figure 2) in the 350 C to 650 C temperature range. Experiments utilizing ^{18}O substitution into surface silanol groups have shown that the disappearance of the silanol band is more than coincidentally related to the development of the D2 band. Gel-derived silicates heated to 650 C were exposed to 100% relative humidity(RH) with H_2O or $H_2{}^{18}O$ to hydrolyze the D2 band[7,12]. The Raman spectrum of the $H_2{}^{18}O$-exposed sample shows a silanol Raman band at 3736 cm-1(in addition to one at 3750 cm-1), indicating incorporation of ^{18}O into silanol groups. On re-heating to 650 C, the D1 and D2 bands shift to lower frequencies(Figure 3), proving that ^{18}O originating as a surface silanol can be incorporated into the species responsible for the D1 and D2 Raman bands.

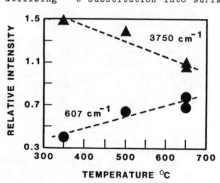

Figure 2. Variation of intensity of D2 (607 cm-1) and SiOH(3750 cm-1) bands in gel-derived silica with temperature

bands. This data proves that dehydroxylation of the silica surface is related to formation of the species responsible for the D1 and D2 Raman bands and suggests that they form by condensation of nearby surface hydroxyl groups.

Only two of the models previously proposed for the source of the D1 and D2 bands have silanol groups as precursors. The paracrystalline model[3] requires that the precursor species is a geminal silanol($>$Si$(OH)_2$ --> $>$Si=O + H_2O --> four-fold siloxane ring), while the silanol condensation model[4,5,6]

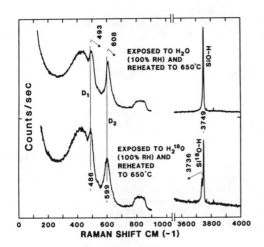

Figure 3. Raman spectra of hydrolyzed and re-heated gel-derived silicates

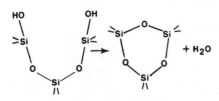

Figure 4. Condensation of three-fold siloxane ring from isolated silanol precursors

proposes that small siloxane rings are condensed directly from isolated[6,12] silanol groups. Two lines of evidence favor the silanol condensation model over the paracrystalline model. First, Krol and van Lierop[12] find evidence for isolated silanols but not geminal silanols in the Raman spectra of gel-derived silicates both after drying and during subsequent heat treatments. Second, the overall reaction to form four-fold siloxane rings(the source of the D2 band in the paracrystalline model) from silanone($>Si=O$) groups has been shown by molecular orbital calculations[13] to be highly exothermic. But the results of thermogravimetric analysis and differential scanning calorimetry[7] have shown that increases in the D2 band intensity are associated with a significantly endothermic reaction. Molecular orbital calculations[14] have shown that the direct condensation of isolated silanols to form either three-fold(Figure 4) or two-fold siloxane rings is a strongly endothermic reaction.

Therefore, the only model which is consistent with all the available evidence is that which proposes the direct condensation of small($<$five-fold) siloxane rings from isolated silanols. The hydrolysis of a heat-treated gel-derived silicate provides evidence that the D1 and D2 Raman bands represent two different species. At 25 C under conditions of 100% relative humidity, the species responsible for the D2 band reacts with a first-order rate constant of $5.2(\pm0.5) \times 10^{-3}$ min.$^{-1}$ ($t_{1/2}$=2.2 hours). The species responsible for the D1 band is stable with respect to hydrolysis under these conditions. Comparison of the frequencies of the D1(492 cm-1) and D2(607 cm-1) bands with the Si-O stretch modes of model compounds incorporating small siloxane rings(Table 1) clearly identifies the D1 band with the four-fold siloxane ring and the D2 band with the three-fold siloxane ring. Further confirmation of these assignments comes from kinetic data involving the hydrolysis of cyclo-siloxane model compounds in tetrahydrofuran[15], which shows that the four-fold siloxane ring is stable with respect to hydrolysis while the three-fold siloxane ring hydrolyzes with a first-order rate constant of $3.8(\pm0.4) \times 10^{-3}$ min.$^{-1}$ ($t_{1/2}$=3.0 hours).

TABLE 1
Si-0 Stretching Frequencies of Small Siloxane Rings
in Model Compounds

Four-Fold[a]	Three-Fold[b]	Two-Fold[c]
480 cm^{-1}	587 cm^{-1}	873 cm^{-1}

[a]Octamethyl cyclotetrasiloxane[16]   [b]Hexamethyl cyclotrisiloxane[15,16]
[c]Tetramesityl cyclodisiloxane[7]

With these identifications, characterization of the Raman bands observed in gel-derived silicates is complete. The observed Raman bands have been shown to be characteristic of the silica network structure($\geq$five-fold rings), of four-fold and three-fold vibrationally-decoupled siloxane rings, and of silanol species. Two-fold siloxane rings have not been observed in gel-derived silicates. We will use this information to characterize other high-surface-area silica materials.

SILOXANE RINGS IN LEACHED BOROSILICATE GLASS

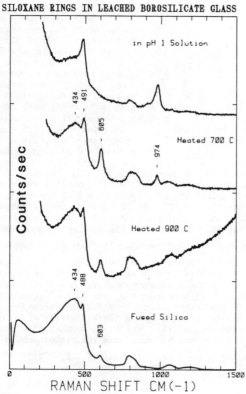

Figure 5. Raman spectra of pH1-leached borosilicate glass and fused silica

When sodium borosilicate glass is leached in aqueous solution, sodium and boron are preferentially removed, leaving a residual matrix of nearly pure silica. The structure of this residual matrix depends on the pH of the leachant solution and not on the $Na_2O:B_2O_3:SiO_2$ ratios in the glass prior to leaching[17]. Residual matrices from sodium borosilicate glasses leached in neutral or basic solution and dried at ambient temperature yield Raman spectra characteristic of network silica structures with some isolated silanols but with no evidence of four-fold(D1) or three-fold(D2) siloxane rings. On heat-treatment up to 700 C, weak D1 and D2 bands develop so that the Raman spectra resemble that of fused silica. Above 700 C the residual matrix crystallizes to crystabolite, a transformation probably catalyzed by residual surface alkali ions[18].

However, residual matrices resulting from leaching in pH1 solutions yield Raman spectra(obtained either in solution or after drying at ambient temperature) with prominent bands due to four-fold siloxane rings and to isolated silanols and relatively weak bands due to network silica structures(Figure 5). The Raman spectra of pH1-leached residual matrices resemble those of gel-derived silicates prior to heat treatment above 50 C[7]. On heat treatment, the changes occurring in the pH1-leached residual matrix(Figure 5) parallel the changes occurring in gel-derived silicates[7,10,11], with the band due to the three-fold siloxane ring(D2) increasing in intensity up to 700 C, and then decreasing at higher temperatures. By 900 C the spectrum resembles that of fused silica.

The differences in structure between the neutral/basic- and acid-leached residual matrices are believed to be due to differences in the kinetics of silicate ion reactivity in neutral/basic versus acid solution. Hydroxyl-catalyzed siloxane bond rupture and re-formation should be more rapid in neutral/basic versus acid solution. Thus, the surface siloxane and silanol species will have more opportunity in neutral/basic solution to close voids by re-forming as stable silica network structures. The relatively low surface area of the neutral/basic-leached residual matrices(~one-third that of their acid-leached counterparts[17]) is an experimentally observable result of hydroxyl-catalyzed re-polymerization leading to void closure. In acid solution the voids formed by leaching are meta-stable due to kinetic limitations on the siloxane reactivity. Apparently these void structures are conducive to the formation of four-fold siloxane rings and of silanol precursors to three-fold siloxane rings.

The four-fold siloxane ring apparently forms spontaneously in residual silicate matrices in pH1 leachant solution(Figure 5-top spectrum), an observation consistent with calculations[13,14] which show that it is an unstrained structure. It is also stable with respect to hydrolysis as part of an organo-siloxane molecule dissolved in tetrahydrofuran[15]. However, under conditions which enhance the kinetics of siloxane reactivity(high hydroxyl concentration or heat treatment above 900 C [7]) and in the presence of a silicate network which provides a framework for higher-order ring formation, the bonds comprising the four-fold siloxane ring tend to rupture and re-form into higher-order rings characteristic of the silicate network.

SILOXANE RINGS IN CAB-O-SIL

Cab-O-Sil is a high-surface-area commercial silica powder used as an adsorbent and filler. It is too fluorescent for Raman analysis in the as-received form. Since most of the fluorescence disappears with high-temperature(900 C) treatment in oxygen, its source is believed to be organic species adsorbed on the Cab-O-Sil surface. Unlike the gel-derived silicates and leached glasses, the heat-treated Cab-O-Sil was stored and analyzed in vacuum to prevent re-adsorption of contaminants and reaction with water molecules. The spectra of such "dehydroxylated" Cab-O-Sil include a relatively narrow fluorescence band that obscures the 800 to 1500 cm-1 Raman region(for 514.5 nm excitation). Overnight exposure to 0.25 Torr of water vapor causes this fluorescence band to disappear, yielding the Raman spectrum at the bottom of Figure 1. The presence of intense Raman bands due to the three-fold and four-fold siloxane rings in this Cab-O-Sil spectrum further reinforces the hypothesis that small siloxane rings are generally associated with "dehydroxylated" high-surface-area silica materials.

Fourier-transform infrared absorption(FTIR) studies[19] on Cab-O-Sil(heat-treated and in vacuum, but not exposed to water vapor) have identified absorption bands attributed to two-fold siloxane rings(edge-shared tetrahedra). These infrared bands disappear on overnight exposure to .25 Torr of water vapor. The fluorescence band observed in the Cab-O-Sil Raman spectrum does not appear to be related to these absorption bands, since its rate of disappearance with hydrolysis is different. However, it does obscure the Raman region(800 to 900 cm-1) in which two-fold siloxane rings are expected to yield Raman bands(Table 1), preventing Raman confirmation of the FTIR observations. Use of excitation lines other than 514.5 nm will shift this Raman region from the wavelengths of the fluorescence band, possibly revealing Raman bands due to the two-fold siloxane ring.

CONCLUSIONS

With the identification of the Raman bands associated with three-fold and four-fold siloxane rings, the characterization of the Raman spectra of silica materials is essentially complete. Other Raman features in the spectrum of fused silica(430, 800, 1065, and 1200 cm-1,Figure 1) are explainable as five-fold and larger rings in a continuous random network. The remaining Raman bands(980 and 3750 cm-1) observed in high-surface-area silica materials have been identified as surface silanol species. Further, the D1 and D2 Raman bands are representative of more than minor "defects" in bulk fused silica. The three-fold siloxane ring which gives rise to D2 Raman band is a major substituent of various heat-treated high-surface-area silica materials. The four-fold siloxane ring which gives rise to the D1 Raman band is present as a major substituent even in "wet" as well as dried and heat-treated high-surface-area silica materials. Both the three- and four-fold siloxane rings form by the condensation of isolated surface silanol groups. The two-fold siloxane ring has not yet been identified by Raman spectroscopy in silica materials. However, its Raman bands have been assigned in the spectra of model compounds, and infrared bands attributed to the two-fold ring have been observed in Cab-O-Sil. Silanol groups amd small siloxane rings are, therefore, important components of silica surfaces, and Raman spectroscopy is a technique well-suited to their study.

ACKNOWLEDGMENTS

The authors thank C. Ashley, K. Higgins, S. Martinez, T. Michalske, and W. Smith for their invaluable aid in obtaining the data reported in this paper.

REFERENCES

1. P.N.Sen and M.F.Thorpe, Phys. Rev. B15, 4030(1977);F.L.Galeener, Phys. Rev. B19(8), 4292(1979).
2. J.B.Bates,R.W.Hendricks, and L.B.Shaffer, J. Chem. Phys. 61, 4163(1974).
3. J.C.Phillips, J. Non-Cryst. Solids, 63, 347(1984).

4. F.L.Galeener, in The Structure of Non-Crystalline Materials 1982, eds. P.H.Gaskell, J.M.Parker, and E.A.Davis(Taylor and Francis Ltd., London,1982), p.337;J. Non-Cryst. Solids 49, 53(1982).
5. B.A.Morrow and L.A.Cody, J. Phys. Chem. 80, 2761(1976).
6. T.A.Michalske and B.C.Bunker, J. Appl. Phys. 56, 2686(1984).
7. C.J.Brinker, D.R.Tallant, E.P.Roth, and C.S.Ashley, Proceedings of the Fall 1985 MRS Symposium: Defects in Glass, eds. F.L.Galeener, D.L.Griscom, M.Weber, to be published, 1986; "Sol-Gel Transition in Simple Silicates III: Structural Studies During Densification", to be published, J. Non-Cryst. Solids.
8. F.L.Galeener, R.A.Barrio, E.Martinez, and R.J.Elliot, Phys. Rev. Lett. 53, 2429(1984).
9. C.J.Brinker and G.W.Scherer, J. Non-Cryst. Solids 70, 301(1985); C.J.Brinker, G.W.Scherer, and E.P.Roth, J. Non-Cryst. Solids 72, 345-368 and 369-389(1985).
10. D.M.Krol and J.G.van Lierop, J. Non-Cryst. Solids 63, 131(1984).
11. V.Gottardi,M.Guglielmi,A.Bertoluzza,C.Fagnano, and M.A.Morelli, J. Non-Cryst. Solids 63, 71(1984).
12. D.M.Krol and J.G.van Lierop, J. Non-Cryst. Solids 64, 185(1984).
13. T.Kudo and S.Nagase, J. Am. Chem. Soc. 107, 2589(1985).
14. M.O'Keefe and G.V.Gibbs, J. Chem. Phys. 81, 876(1984).
15. C.A.Balfe, K.J.Ward, D.R.Tallant, and S.L.Martinez, "Reactivity of Silicates 1. Kinetic Studies of the Hydrolysis of Linear and Cyclic Siloxanes as Models for Defect Structure in Silicates", to be published in THIS PROCEEDINGS.
16. A.F.Smith and D.R.Anderson, Appl. Spect. 38(6), 822(1984).
17. B.C.Bunker, and D.R.Tallant, to be published.
18. Z.C.Shan, J.Phalippou, and J.Zarzycki, "Influence of Trace Alkali Ions on the Crystallization Behavior of Silica Gel", presented at the Third International Workshop on Glasses and Glass Ceramics from Gels, Montpelier, France, September, 1985.
19. D.M.Haaland, T.A.Michalske, and B.C. Bunker, to be published in the J. Am. Cer. Soc.
20. R.J.Bell and P.Dean, Phil. Mag. Series 8, 25, 1381(1972).

CHEMICAL ROUTE TO ALUMINOSILICATE GELS, GLASSES AND CERAMICS

J.C. POUXVIEL,[*+] J.P. BOILOT,[*] A. DAUGER[**] and L. HUBER[***]
*Groupe de Chimie du Solide, Laboratoire de Physique de la Matière Condensée, Ecole Polytechnique, 91128 Palaiseau Cédex, France
+Also at Saint-Gobain Recherche, 38 Quai Lucien-Lefranc, 93304 Aubervilliers, France
**E.N.S.C.I., 47-73 Avenue A. Thomas, 87065 Limoges Cédex, France
***Laboratoire de Chimie de Coordination, Université de Nice, Parc Valrose, 06034 Nice Cédex, France.

ABSTRACT

Chemically ultra-homogeneous gels have been prepared, in the SiO_2-Al_2O_3 system, using new metal-organic precursors with Al-O-Si linkages. By thermal treatment at 900°C, transparent monolithic gels lead to aluminosilicate glasses and above 1000°C, to optically transparent glass-ceramics (for instance by homogeneous crystallization of mullite in an amorphous silica matrix).

I. INTRODUCTION

Aluminium and silicon oxides have widespread technical and industrial application because of their refractoriness and chemical stability. In the Al_2O_3-SiO_2 binary system, an important refractory material : Mullite ($3Al_2O_3$-$2SiO_2$) is also formed.

Most studies of the Al_2O_3-SiO_2 system involve preparation of material by high temperature reactions from oxide powders. By this conventional process, diffusion kinetics between oxides are often insufficiently fast to lead to complete reactions. Therefore, in the last years, sol/gel technique has been used to prepare reactive aluminosilicate gels and make glasses and ceramics at low temperature [1-6].

We report now the preparation of chemically ultra-homogeneous transparent gels, in the Al_2O_3-SiO_2 system by using new metal-organic precursors with Al-O-Si linkages : aluminosiloxanes and silicon-aluminium esters. The gelation mechanism has been studied by ^{29}Si and ^{27}Al NMR. Homogeneity and thermal behaviour (densification and crystallization) of gels were investigated by DTA, SAXS and X-ray diffraction techniques. Aluminosilicate glasses and optically transparent glass-ceramics can be prepared by thermal treatment of these gels.

II. ALUMINOSILOXANE $Al(O^iPr)_2$ $OSiMe_3$

1. Organometallic precursor

Aluminosiloxanes were prepared by transesterification method with trimethylacetoxysilane [7,8], following the reaction :

$$Al(O^iPr)_3 + x\ Me_3\ SiOAc \xrightarrow{C_6H_{12}} Al(O^iPr)_{3-x}\ (OSiMe_3)_x + AcO^iPr \quad (1)$$

with x = 1 - 3.

Pure aluminosiloxane with x = 2 was obtained by distillation. Mass spectra give evidence of the existence of a molecular complexity of 4, similar to the one observed in $Al(O^iPr)_3$. Figure 1a shows the ^{27}Al NMR spectrum for aluminosiloxane saturated solution in CH_2Cl_2. Line assignments were made in accordance with the literature [9]. The intense sharp line at 3.5 ppm (with respect to $Al(H_2O)_6^{3+}$) corresponds to 6-coordinated Al atoms. Broad line at 64 ppm is assigned to 4-coordinated Al atoms and weakly intense broad line at 38.9 ppm corresponds to 5-coordinated Al atoms. Taking into account previously proposed structures of aluminium alkoxides, the aluminosiloxane solution can be considered as a mixture of dimer (A), trimer (B) and tetramer (C) in the approximate molar ratio 2.-1.-2. In fact, this ratio is highly depending on the dilution and on the nature of the solvent.

A   B   C

$R = {}^iPr$, $R' = SiMe_3$.

2. Gelation

Aluminosiloxane (x = 2) was diluted in HO^iPr (0.1 Mole ℓ^{-1}) and ageing of the solution was carried out, at room temperature, in an atmosphere of 70% relative humidity, saturated with alcohol. After 4 hours of hydrolysis, ^{27}Al NMR spectrum (figure 1b) exhibits a broad band centered at 58 ppm corresponding to tetrahedral Al and the sharp peak at 3.5 ppm corresponding to octahedral Al. After 19 hours of hydrolysis, just prior to the gel point, NMR spectrum (figure 1c) only shows an intense broad line which indicates the conversion of octahedral Al to tetrahedral Al and the existence of an expected structural disorder in the gel.

Low angle region analysis of the SAXS curve obtained with this liquid leads to a radius of gyration of 26 Å. Analysis of the intermediate angle region gives a slope of -1.8, corresponding to the existence of randomly weakly branched chains [10].

All these results (^{27}Al NMR and SAXS) are in agreement with a chain polymerization mechanism where Al is 4-coordinated.

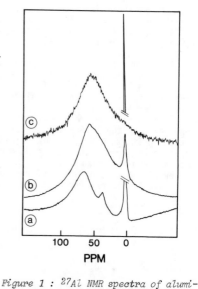

Figure 1 : ^{27}Al NMR spectra of aluminosiloxane solutions (ppm from $Al(H_2O)_6^{3+}$)
a) in CH_2Cl_2 (saturated solution)
b) in HO^iPr (0.1 Mole l^{-1}) - 4 hours of hydrolysis in an atmosphere of 70% R.H. (gelation time = 24 hours)
c) after 19 hours of hydrolysis.

After 24 hours of hydrolysis, the mixture becomes viscous and gelatinizes. Optically clear monolithic gels with a glass aspect are obtained. After drying (70°C) the organic-inorganic gels exhibit a density of 1.34 and IR spectrum gives evidence of the presence of siloxo groups (OSiMe3) in the solid.

Analysis of the intermediate angle region of the SAXS curve of the gel displays that the sol/gel transition is accompanied by the cross-over from a polymeric state with weakly branched chains to a state formed by particles with well defined boundaries [10] (The Porod slope is near -4).

3. Thermal behaviour

Above 300°C, the organic-inorganic polymer remains amorphous and transparent, but the DTA curve (figure 2) with a broad exothermic peak at 350°C, the IR spectra (with disappearance of the Si-CH3 stretching band at 1250cm^{-1}) and ionic microprobe analysis (Al/Si ratio of 22 at 700°C) clearly demonstrate the loss of the siloxo groups by removal of volatile siloxanes, probably (Me3Si)2O. This change of stoichiometry has been previously reported [11] for other aluminosiloxane gels.

At 500°C, analysis of the low angle region of the SAXS curve leads to a radius of gyration of 34 Å. The scattered intensity is probably essentially due to a microporous state. However a weakly crystalline spinel phase is detected on the X-ray diffraction pattern. At 957°C, DTA curve (Figure 2) exhibits an exothermic peak. Above this temperature and up to 1300°C a transparent monolithic glass-ceramic is obtained in which spinel and mullite phases are respectively the preponderant crystalline phases at 1100 and 1200°C.

Finally, another exothermic reaction is observed at 1416°C corresponding to the formation of the α alumina phase and the monolith becomes opaque and white.

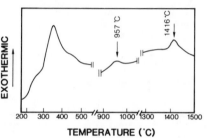

Figure 2 : DTA curve of the aluminosilicate gel prepared from aluminosiloxane precursor (heating rate 30°C/min.).

III. SILICON-ALUMINIUM ESTER : (EtO)3 Si-O-Al(OBu)2

1. Organometallic precursor

The silicon-aluminium ester has been supplied by Dynamit Nobel. ^{29}Si NMR spectrum shows the presence of two predominant Si resonances : Firstly, the one which can be assigned to the (RO)2(=AℓO)Si-O-Si(OAℓ=)(OR)2 group and secondly, the resonance corresponding to the (RO)2(=AℓO)Si-O-Si(OAℓ=)(OR)2 group.

This indicates a partial condensation of the starting silicic ester. In fact, a more complex ^{29}Si spectrum is generally observed resulting from an ester exchange reaction between ethoxy and butoxy groups.

2. Gelation

The starting mixture consists of silisic ester, H2O (pH = 2.5-HCℓ as catalyst) in the molar ratio 1-10, and iPrOH. The ageing of this solution was carried out, at room temperature, in a closed flask. Figure 3 shows the time evolution of the ^{27}Aℓ NMR spectrum. After 5 min. of hydrolysis, three ^{27}Aℓ

resonances are observed : an intense sharp line at 51 ppm corresponding to
4-coordinated Aℓ atoms of the starting organic precursor; two broadlines centered at 56 and 7 ppm in the intensity ratio 5-1, corresponding respectively
to tetrahedral Aℓ and octahedral Aℓ formed during the sol/gel polymerization.
When time increases, the sharp line progressively disappears and after 23
hours (t/tg ∿.96), only the two broad bands are observed. Using NMR peak intensities, one can deduce that 62% of Aℓ atoms are in tetrahedral and only
38% in octahedral sites. Therefore in this case, the gelation process is accompanied by a conversion of tetrahedral Aℓ to octahedral Aℓ and by an important change in the Aℓ environment, as indicated by the broad ^{27}Aℓ resonances.

In contrast, the ^{29}Si NMR spectrum is only slightly modified after 23
hours of hydrolysis. We have observed peaks corresponding to groups of the
starting organic precursor and new peaks assigned to $(RO)_2 (OH)Si^*-O-(Aℓ=)$
and $(RO)(HO)(=AℓO)Si^*-O-Si(OAℓ=)=$ groups resulting from the substitution of
one alkoxy group by OH. These NMR results clearly demonstrate that near the
gel point (24 hours) there is only a very weak concentration of silisic condensed species (Si atoms with bridging oxygens) and that the inorganic polymerization is essentially due to the formation of Aℓ-O-Aℓ linkages. In the
first stages of the polymerization process, it seems that a chain polymerization mechanism takes place where Aℓ atoms are mainly 4-coordinated (figure
3b), followed by a 3D mechanism associated with the conversion of tetrahedral
Aℓ to octahedral Aℓ (figure 3c).

After 24 hours of hydrolysis, optically clear monolithic gels, with a
glass aspect are obtained. After drying (70°C) the size of transparent pieces
is about of 1 cm^3).

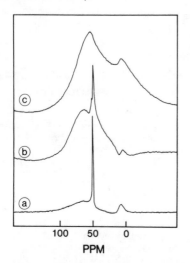

*Figure 3 : ^{27}Aℓ NMR spectra (ppm from
Aℓ $(H_2O)_6^{3+}$) of the silicon-aluminium
ester and H_2O (pH = 2.5) mixture, in
the molar ratio 1.10, in ^{i}PrOH (gelation time = 24 hours)*
  a) after 5 min.
  b) after 7 hours.
  c) after 23 hours.

Small angle X-ray scattering curves realized with these gels, exhibit
a well pronounced maximum at q = .08 Å^{-1} (figure 4). Taking into account that
electronic   densities of Aℓ and Si atoms are almost identical, the contribution of composition fluctuations to the X-ray scattering intensity can be
neglected. Therefore, the scattered intensity is due to the presence of
packing density fluctuations in xerogels. The gel structure can be seen as
dense amorphous particules with an average size of 20 Å, separated by low
density amorphous regions where organic residus (OR groups) remain mainly
bonded to Si atoms.

3. Thermal behaviour

Up to 980°C, the transparent monolithic gel remains amorphous and no variation of the chemical composition is observed. Figure 4 summarizes the evolution of the scattering curves during the thermal treatment of the gel. In the R.T.-900°C temperature range, there is no significative change of the dense particle average size. However, above 500°C, the scattered intensity progressively decreases, indicating a disappearance of density fluctuations by a densification process and the evolution of the gel towards an homogeneous dense amorphous state.

At 980°C a sharp intense exothermic peak is observed on the DTA curve, (figure 5) associated with a random formation of mullite crystallites detected by X-ray diffraction. This transformation leads at 1100°C to an interference peak on the SAXS curve which can be interpreted as the presence of islands of pure mullite in an amorphous matrix of silica. The gyration radius, deduced from the low angle analysis of the SAXS curve is about of 40 Å, in agreement with the domain size deduced from linewidth of the X-ray diffraction pattern. At 1200°C, the gyration radius becomes higher than 70 Å and a drastically increase of the scattered intensity is observed. In the optically transparent glass-ceramic, the growing of mullite particles is accompanied by the formation of a poorly crystallized cristobalite phase detected on the X-ray diffraction pattern.

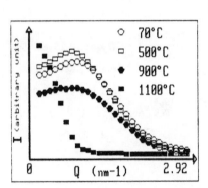

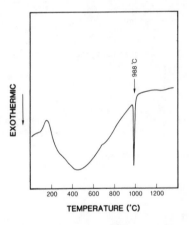

Figure 4 : Temperature evolution of the gel prepared from silicon-aluminium ester precursor : SAXS curves at different temperatures.

Figure 5 : DTA curve of the aluminosilicate gel prepared from silicon-aluminium ester (Heating rate : 10°C/min).

IV. CONCLUSION

We have prepared chemically ultra-homogeneous gels, using metal-organic precursors with Aℓ-O-Si linkages : aluminosiloxane and silicon-aluminium esters. With both precursors, the gelation mechanism leads to the formation of a polymer with Aℓ-O-Aℓ linkages. The organic-inorganic gel prepared from aluminosiloxane precursor gives, after thermal treatment and loss of siloxo groups, optically transparent glass and glass-ceramics with a chemical composition of 95 Aℓ-5 Si.

Gels prepared from silicon-aluminium ester exhibits density fluctuations corresponding to a packing of dense amorphous particles separated by low density regions with remaining OR groups. At high temperature (above 980°C) nucleation of mullite takes place in dense particles and optically transparent glass-ceramics are obtained with Aℓ/Si ratio of about 1, constituted of an homogeneous distribution of mullite crystallites in an amorphous silica matrix.

ACKNOWLEDGEMENTS

We wish to thank J.C. Beloeil (I.C.S.N., Gif-sur-Yvette) for NMR measurements, Saint-Gobain Recherche for DTA measurements and "Dyna France" for free samples of silicon-aluminium esters.

References

[1] A.K. Chakraborty, J. Am. Chem. Soc. 62, [3-4], 120 (1979).
[2] D.W. Hoffman, R. Roy and S. Komarneni, J. Am. Ceram. Soc. 67, [7], 468 (1984).
[3] Y.M.M. Al-Jarsha, K.D. Biddle, A.K. Das, T.J. Davies, H.G. Eimblem, K. Jones, J.M. McCullough, M.A. Mohd, A.B.D. Rahman, A.N.A. EL-M Sharf el deen and R.K. Wakefield, J. of Mat. Sc. 20, 1773 (1985).
[4] B.E. Yoldas, J. Mat. Sci. 12, 1203 (1977).
[5] B.E. Yoldas, Ceram. Bull. 59, 479 (1980).
[6] N. Blanchard, J.P. Boilot, Ph. Colomban, J.C. Pouxviel, J. Non-Cryst. of Solids (1986) in press.
[7] R.C. Mehrotra and B.C. Pont, Ind. J. Appl. Chem. 26, 109 (1963).
[8] C.G. Barraclough, D.C. Bradley, J. Lewis and I.M. Thomas, J. Chem. Soc. 2601 (1961).
[9] O. Kriz, B. Casensky, A. Lycka, J. Fusek and S. Hermanek, J. of Magn. Res. 60, 375 (1984).
[10] D.W. Schaefer and K.D. Keefer in Mat. Res. Soc. Symp. Proc. Vol. 32 (Elsevier Science Publishers, New-York 1984) pp. 1-14.
[11] A.G. Williams and L.V. Interrante, Mat. Res. Soc. Proc. Vol. 32 (Elsevier Science Publishers, New-York 1984) p. 151.

PART V

Structure of Random and Ordered Systems

FRACTAL ASPECTS OF CERAMIC SYNTHESIS*

DALE W. SCHAEFER[†] AND KEITH D. KEEFER[††]
Division 1152[†] and Division 1845[††], Sandia National Laboratories, P.O. Box 5800, Albuquerque, NM 87185

ABSTRACT

The concept of fractal geometry is used to describe the structure of silica polymers, colloidal aggregates, and critical systems. We illustrate the interpretation of scattering curves (X-ray, neutron and light) for fractal systems, and review simple growth models which generate fractal structures. We describe the polymerization of silica under various conditions and demonstrate that, depending on chemical conditions, polymerization maps onto simple fractal growth processes. The key factors which control growth are monomer-cluster vs. cluster-cluster growth, and reaction-limited vs. diffusion-limited growth.

INTRODUCTION

In spite of the importance of structural issues in ceramic science, little is known concerning the origin of geometric structures in glass or ceramic materials or their precursors. Although two classic growth models exist (nucleation-and-growth, and spinodal decomposition) these processes account for the structure of materials only under limited circumstances. The purpose of this paper is to introduce a third realm of growth, random or disorderly growth. We show that simple models of disorderly growth can successfully describe the structures observed in the solution precursors to some glass and ceramic materials. We concentrate on the silicate system and work within the general framework called sol-gel processing. The goal is to identify random structures in solution precursors and explain structure in terms of simplified growth models.

Advances in the area of random materials followed two critical developments. The first was the realization that random structures could be described by fractal geometry. Although the concept of fractal geometry [1,2] has been known to mathematicians for many years, it was a 1982 paper by Witten and Sander [3] that sparked applications to materials science. In this paper, Witten and Sander described a diffusion-limited aggregation model which they hoped would explain the geometry of smoke particles [4]. Their single-particle aggregation model does not correctly describe gas-phase aggregates, but the model does reproduce many structures observed in nature [5].

The second development was the application of scattering techniques to random structures [6-12]. Schaefer et al. demonstrated that the scattering curves from silicate polymers [10], aggregates of colloidal silica [11], and silica aerogels [6,7] could be interpreted in terms of fractal geometry. Bale and Schmidt [12] showed that the surface scattering from pores in coal could be described by fractal geometry. We call these two classes of structures mass fractals and surface fractals. As described later, the distinction between these two classes is easily made on the basis of scattering curves [6,7].

*This work performed at Sandia National Laboratories, Albuquerque, NM and supported by the U.S. Department of Energy under Contract NO. DE-AC-04-76DP00789.

278

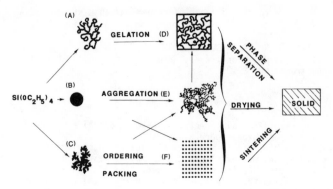

Fig. 1. Schematic diagram [9] of the growth processes and structures
expected in the precursors of materials. Although the diagram is specific
to silicate systems, the general ideas discussed in this paper should have
far wider applications.

FRACTAL STRUCTURES IN GLASS AND CERAMIC PRECURSORS

Fig. 1 is a schematic representation [9] of structures and processes in
precursors to glass or ceramic materials. Several growth processes and
structures can be identified. Consider, for example, the sol-gel process
where one starts with a soluble silicate $Si(OC_2H_5)_4$, TEOS. This material is
polymerized in solution to form either compact colloidal particles or
ramified polymeric structures. More complex structures can develop by
gelation, aggregation, or by an ordering process. These large scale
structures are then converted to a solid material typically by drying,
sintering, and/or phase separation. The resulting material may be anything
from a foam to a dense structural ceramic. Proper understanding and control
of the precursor growth processes is essential to the generation of the
desired material.

Nucleation-and-growth [13] is one
of the simplest and most successful
models which relates growth and
structure. In this model, compact
structures are formed by growth on the
surface of a sufficiently large
nucleus. This model can describe
phase separation in thermodynamically
metastable systems and can also
describe the growth of compact
colloidal particles. Droplets formed
by nucleation and growth have a very
distinct scattering curve as
illustrated in Fig. 2, which shows the
scattering curve for colloidal silica
(Ludox™ SM). The essential feature of
this curve is the limiting slope of -4
when plotted on a log-log plot. The
slope of -4 is called Porod's law [14]
and is the signature of a particle
with a distinct surface.

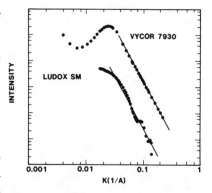

Fig. 2. Scattering curves for
materials prepared under conditions
which lead to nucleation-and-growth
(Ludox) and spinodal decomposition
(Vycor 7930). The oscillations in
the Ludox curve are characteristic
of monodisperse spheres.

Spinodal decomposition occurs in the thermodynamically unstable regime [15]. In this process nucleation is unnecessary and phase separation occurs by growth of an unstable Fourier component of the density fluctuation spectrum. Because certain spatial Fourier components grow at the expense of others, these systems have a characteristic peak in their scattering curves. Fig. 2, for example, shows the scattering curve for a borosilicate glass (Vycor 7930) after dissolution of the soluble phase [16]. The characteristic feature is the peak which moves to smaller angles, corresponding to larger distances, as growth proceeds. The limiting slope of -4 indicates sharp boundaries.

Random structures grown by fractal processes also have characteristic scattering curves [6]. Fig. 3, for example, shows the scattering curves for a series of structures found by the polymerization of TEOS. In general, these structures do not show power-law exponents of -4. The purpose of the rest of this paper is to understand and interpret these slopes in terms of simplified growth processes.

Fig. 3. Scattering curves for silicates. 3a and 3b are silicates produced by a two-stage polymerization process [10,29]. The first stage is acid catayzed, the second stage is acid catlyzed (3a) or base catalyzed (3b). Curve (3c) is for a rough colloidal particle prepared by single-stage growth under base-catalyzed conditions [27]. Curve (3d) is for by single-stage acid-catalyzed growth [25]. Curves 3c and 3d are "analytically" desmeared Kratky data (observed intensity divided by K).

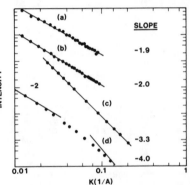

SCATTERING FROM FRACTALS

Fractal objects differ from ordinary crystalline systems in that they have dilation symmetry as compared to translation or rotation symmetry. Dilation symmetry means that if a portion of the object is magnified, the resulting magnified structure looks identical to the original structure. The fractal dimension, D, is consistant with the common notion of dimension expressed in the relation $M \sim R^D$, where M is the molecular weight and R is the radius. Note that D is a dimension in the sense that if the object were a rod, disc or a sphere, then the exponent in this expression would be 1, 2, or 3. Fractal objects, however, may have non-integer dimensionality. Many objects grown by random processes have dilation symmetry and are therefore fractal objects.

Scattering experiments measure structure in reciprocal space. At a given scattering angle θ one measures the amplitude of a certain spatial Fourier component of the density, the spatial frequency being $K = 4\pi\lambda^{-1}(\sin \theta/2)$ where λ is the incident wavelength. At large K one probes short wavelength fluctuations and at small K one probes long wavelength fluctuations. Typically one plots the intensity, I, as a function of K.

In the weak scattering limit, mass-fractal objects are characterized by an intensity distribution which is power-law

$$I(K) \sim K^{-D} \qquad (1)$$

where D is the fractal dimension described above. Thus the fractal dimension of these mass-fractal objects can be extracted from the slope of the scattering curve when plotted on a log-log plot. Exceptions to this rule can occur for power-law polydisperse systems [17,18].

Objects with rough surfaces can also be described by fractal geometry. If a small section of a fractal surface is magnified then the resulting structure looks like the original. Dilation symmetric surfaces can be represented mathematically by the equation $S \sim R^{D_s}$, where R is measure of length and S is the surface area and D_s is the surface fractal dimension. D_s would equal 2 for a euclidian three-dimensional (3-d) object since the surface of a uniform 3-d object is two-dimensional. In general, however, if the object is fractally rough, D_s can be between 2 and 3. These structures are called surface fractals [19].

Surface fractals have a distinctly different scattering curve compared to mass fractals. Bale and Schmidt [12] show that

$$I(K) \sim K^{D_s-6} \qquad (2)$$

This equation indicates that for three-dimensional objects, scattering laws with powers between -3 and -4 are expected. Objects with smooth surfaces yield slopes of -4 (Porod's law) [14], while rough objects with fractal surfaces have slopes between -3 and -4.

Under certain circumstances scattering curves with slopes less than -4 are expected. For example, objects with broad interfaces [20] give rise to profiles which, although they are non power-law, often appear power-law when plotted on log-log axes and give slopes less than -4. In addition, sub fractal [21] surfaces are possible. These objects are rough, but the surface area does not diverge with distance. In other words, D_s is less than 2, so "roughness" decreases with length [22]. We expect such objects to yield power-law scattering curves with slopes less than -4.

The above analysis indicates that power-law scattering curves are the signature of fractal objects. Curves with slopes greater than -3 indicate mass-fractal or polymer-like objects (objects where the surface is proportional to the mass), whereas slopes greater than -4 indicate objects with uniform cores and rough surfaces. If the slopes lie between -3 and -4, the objects have fractally rough surfaces, whereas a slope of -4 indicates a smooth surface. Finally, slopes less than -4 indicate either a sub-fractal surface or a broad interface.

EDEN GROWTH (GROWTH FROM MONOMERS)

A simple growth model which has relevance to ceramics science was that proposed in 1961 by Eden to describe the growth of cancers [23,24]. One starts with a seed (one site on a lattice) and randomly picks and occupies one of the neighboring sites. This process is repeated over and over, with every empty site which abuts a filled site being a potential point of growth. The result in structure is shown in Fig. 4.

Fig. 4   Eden cluster produced by monomer-cluster growth.

The geometric characteristics of the clusters grown with Eden rules are: 1) They have smooth surfaces relative to their radius, and 2) the interiors are uniform. It is interesting to note that this non-equilibrium model generates smooth-surface particles without reference to surface tension.

The Eden model is realized when, 1) growth occurs from monomers, and 2) the functionality of the monomers is large. We expect this type of growth, for example, for base-catalyzed polymerization of silicic acid. Keefer has argued that the most facile silicate condensation reaction is nucleophilic substitution in which one of the polymerizing partners undergoes an inversion [25]. This mechanism alone is sufficient to evoke the rules of Eden growth, because only a monomer can invert easily. Cluster-cluster reactions are inhibited because, being parts of clusters, neither of the two participating species can invert. In addition, monomer-cluster growth is expected when condensation is limited by a slow, preceding hydrolysis reaction.

POISONED EDEN GROWTH

We recently proposed a modification of the Eden model to explain the structure of silica particles produced under mild base-catalyzed conditions [27]. Curve c in Fig. 3, for example, shows a scattering curve for silica produced under base-catalyzed conditions from TEOS with the stochiometric water-silica ratio, $[H_2O]/[Si] = 2$. The slope of -3.3 is indicative of an object with a compact core and a rough surface ($D_s = 2.7$). Poisoned Eden growth can account for this structure.

Poisoned Eden growth is a modification of the Eden model in the sense that growth also occurs from monomers. It differs from the original Eden model, however, in that there is a distribution of monomer functionality. That is, growth occurs say from a 50/50 mixture of tetrafunctional and difunctional monomers. A non-uniform distribution is reasonable on the basis of the chemistry of the hydrolysis reaction and has in fact been confirmed by NMR measurements by Kelts et al. [28].

Fig. 5 shows a 2-d cluster grown by the poisoned Eden cluster process using a 50/50 mixture of di- and tetrafunctional monomers. Although this cluster is porous, the pores do not occur on every length scale so the interior of the object is uniform and non-fractal. The external surface of the cluster, however, is exceedingly rough. The surface fractal dimension of this object can be obtained from the radius dependence of the perimeter and it is found to be 1.75 (intermediate between 1 and 2) indicating a fractally rough object.

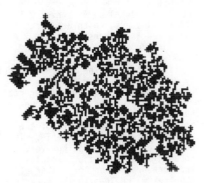

Fig. 5   Poisoned Eden cluster

The full ramifications of the poisoned Eden model are quite complex and are beyond the scope of this review. It is not clear, for example, whether the clusters like Fig. 5, remain fractally rough as the cluster becomes infinitely large. In addition, slight changes in the rules of growth can lead to mass-fractal objects.

PERCOLATION AND CLUSTER-CLUSTER GROWTH

The models discussed so far, the Eden and Poisoned Eden, both yield structures which have uniform interiors. Now we consider polymeric objects which have holes on every length scale (mass fractals). Depending on the application, any of these structures may be desirable. If one is interested in making powders as precursors for structural ceramics, then one would grow uniform particles by an Eden process. If, on the other hand, one wanted a rough surface for a catalytic application, one would seek to modify the chemical processes to produce poisoned Eden growth. Under other circumstances (for example, the generation of a film by sol-gel process [29]), polymeric structures will probably be the most desirable. Here we describe a model called percolation which generates ramified polymeric structures. We suggest that this type of process is realized in the silicate system when growth from monomers is precluded.

In its simplest form, percolation processes occur by randomly occupying either sites or bonds on a lattice [30]. This realization does not represent a growth process in the sense that one does not start from a seed. If one randomly occupies 50 percent of the sites of a square 2-d lattice, then clusters of all sizes will appear which are distinguished by their connectivity. The structure and connectivity of percolation clusters is an extremely well studied problem [30,31].

Percolation can also be realized by a growth process [31]. One starts with a seed, located at one site on a lattice, and then using a random number generator one picks growth sites on the perimeter of the existing seed with the certain probability p. At the first stage of growth, all the growth sites are occupied. All non-growth sites on the perimeter are dead forever. In the next step, one takes the new perimeter (defined as the cluster plus dead sites) and again randomly chooses growth sites with probability p and occupies them. This process is repeated until no growth sites are available. If this process is executed with the probability $p < p_c$, p_c being the percolation threshold, one finds that the clusters will always die. That is, at some stage in the growth no growth sites are available. For $p \geq p_c$ one finds that growth may continue forever. The enormous cluster which occurs when $p = p_c$ is called the percolation cluster and has a fractal dimension of 2.5. Below the percolation threshold the largest clusters, which are rare, are called lattice animals. These objects have a fractal dimension of 2.0. A 2-d cluster generated by percolation is shown in Fig. 6.

In principle, the same structures may be grown by either the Eden or percolation process. The difference between these methods can be found in the weighting function [32]. For lattice animals all isomers of the same mass are equally probable. In the Eden growth process, however, the probability of realizing a certain structure depends on the number of sequences which lead to the structure. This biased weighting leads to a strong preference for compact clusters. Percolation clusters are intermediate between Eden clusters and

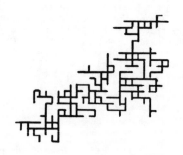

Fig. 6 Small cluster produced by a percolation process (from ref. 2).

lattice animals. Renormalization-group analysis shows that percolation clusters are weighted with a prejudice against those with large surface areas [30,33].

In order to generate ramified clusters, it is necessary to inhibit the reactions of monomers. Whenever monomer-cluster reactions are possible, then growth sites at the interior of a particle may be occupied at the later stages of growth, and the particle will tend toward a uniform core. Thus, ramified growth depends on the inhibition of reactions between monomers and clusters.

We have suggested that lattice-animal structures are realized in the two-step polymerization process originally developed by Brinker [29] as a precursor for sol]-gel films [10]. In the first stage small clusters, approximately 5 Å in radius are generated by acid-catalyzed polycondensation of TEOS with substochiometric amounts of water. Under acid-catalyzed condition, hydrolysis occurs rapidly [34] with monomers being more likely to be hydrolyzed than sites on clusters [25]. As the reaction goes to completion, most monomers are incorporated into clusters with few stragglers left behind.

In the second stage of the reaction, excess water is added and available ethoxy groups on the small clusters are hydrolyzed. Since hydrolysis takes place on monomers which are already attached to clusters, the normal inversion reaction does not take place, and all sites on the cluster are approximately equally probable hydrolysis points. In addition, there is no prejudice toward monomers in the polycondensation reaction so that all potential crosslinking points are equally probable, apart from steric restrictions. Thus, we realize the rules necessary for lattice animals (equal bonding probability).

The steric restrictions which occur when two clusters are joined together provide the second rule necessary for lattice-animal structures. Namely the interior holes in the structures will not fill in. We expect to realize lattice-animal structures whenever reactive sites join with equal probability, being limited only by steric restrictions.

The growth process described above has sometimes been called reaction-limited cluster-cluster aggregation [35,36], and has been simulated on the computer by Kolb, Jullien [35], and Brown and Ball [36]. F. Leyvraz [37] has given a detailed argument supporting the idea that reaction-limited cluster-cluster aggregation, RCLA, will lead to lattice-animal structures.

It should be noted that the above argument is highly simplified and ignors several factors which are known to effect real systems. Solvent quality, for example, can effect swelling and modify observed D's. In addition, we have already demonstrated that bonding is not completely random and that catalytic conditions do effect structure on the 100 Å scale [10]. Finally, several recent experiments show that more compact percolation-like clusters (D=2.5) are realized in the late stages of growth in gelling systems. At best, our argument applies to the early stages of cluster-cluster growth. The point we are trying to make is that non-compact structures are expected when monomer-cluster growth is precluded. Finally, we do not expect the classic lattice-animal cluster mass distribution($M^{-1.5}$) since we have described a growth process and not the equilibrium ensemble.

Fig. 3a,b shows the scattering curves for silicates grown by the two-stage process [10]. Although the second stage is carried out under different catalysis conditions, the slope of the scattering curves is close to -2, the result expected for lattice animals.

The data like Fig. 3a,b do show systematic deviations from a slope of -2 depending on catalysis conditions. Under acid-catalyzed conditions, the slopes tend to be slightly greater than -2, and under base-catalyzed conditions they tend to be slightly less. We associate these small differences with a higher degree of branching being realized in the base-catalyzed conditions [10,25]. Renormalization arguments of Family [38], however, suggest that even under widely different branching ratios, lattice-animal geometry obtains at sufficiently long length scales.

The deviations described above demonstrate that the mapping to the lattice-animal problem is not exact. Although the acid and base-catalyzed structures have nearly the same fractal dimension, they differ substantially in other properties, such as their behavior on dilution and sintering. Unfortunately, the fractal dimension carries no specific information on topology. Although mathematical descriptions of topology (based on fractal like exponents) exist [39], the experimental techniques to extract these topologic exponents do not exist. Experimental characterization of the topology of branched molecules is one of the primary unsolved problems of polymer physics. In spite of the above caveats, it seems clear that two-stage growth leads to ramified structures.

NON-FRACTAL GROWTH PROCESSES

The two growth processes already described, Eden (reaction-limited monomer-cluster) and percolation (reaction-limited cluster-cluster), lead to structures with well behaved power-law scattering curves. In many cases, however, power-law scattering is not observed and it is the purpose of this section to show that even under these circumstances, the concept of fractal geometry is helpful in qualitatively characterizing these structures. Fig. 3d, for example, shows Keefer's data [25,27] for a single stage acid-catalyzed polymerization of TEOS. The data indicate a broad crossover from ramified structures at large length scales (corresponding to the region $K=.01A^{-1}$ in the scattering curve) to compact structures at small length scales (corresponding to $K=.1A^{-1}$ in the scattering curve). This behavior is expected based on the concepts outlined above.

Under acid-catalyzed conditions, one expects rapid hydrolysis and growth from monomers during the early stages. Compact structures are possible because 1.) growth can occur from monomers and 2.) most monomers are likely to be at least partially hydrolyzed. As polymerization proceeds, however, a crossover from monomer-cluster growth to cluster-cluster growth must occur. We have already argued that cluster-cluster growth leads to ramified structures. Thus, at the larger length scales we expect the scattering curves to display a slope more positive than -3.

The above argument leads us to believe that ramified structures will always exist at sufficiently large length scales under kinetic growth conditions. That is, if the system is sufficiently dilute, eventually all monomers will be consumed and if reactive sites are still available, the polymerization process must proceed by cluster-cluster growth. If these later stage reactions are reaction-limited, we predict lattice-animal like structures. On the other hand, if the reactions are exceedingly rapid, the reactions may be diffusion-limited in the later stages. Under diffusion-limited conditions, structures with the fractal dimension of 1.7 are anticipated [40,41].

The above results suggest that the three models of growth discussed, Eden, poisoned Eden, and percolation can be realized by proper choice of

285

polymerization conditions. In addition, crossovers between the models are expected if the polymerization conditions change from monomer-cluster to cluster-cluster growth. It should be emphasized that the models discussed are highly simplified and the mapping of real chemical conditions on to these models is at best approximate. We hope, however, that the models discussed and the general perspective on these problems will help others in designing chemical processes to achieve structural goals.

COLLOID AGGREGATION

Aggregation phenomena represent a second area where fractal concepts apply in ceramic precursors. Aggregation is a desirable phenomenon if high surface area powders are the goal. Since aggregation precludes effective packing, however, aggregates are generally undesirable in the precursors for structural materials. Finally, aggregation may play a key role in determining the structure of porous materials such as aerogels [6,7].

Fig. 7 shows the development of the scattering curve for aggregates of colloidal silica particles [11]. These data represent a combination of light and x-ray scattering data. The composite curve shows two power-law regimes. At large K the slope of -4 is indicative of a smooth particle. The slope of -2.1 at low K, however, is indicative of a ramified structure with a fractal dimension of 2.1. The crossover between the two regimes occurs at the condition K x a = 1, where a is the radius of gyration of the primary colloidal particle.

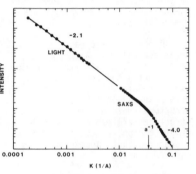

Fig. 7. Combined light and x-ray scattering curves from colloidal silica aggregates (from ref. 11).

To understand the fractal structure of the colloidal aggregate in Fig. 7 it is necessary to introduce a simple model for colloid aggregation. This model is called cluster-cluster diffusion-limited aggregation (the same model referred to above) and has been studied extensively by computer simulation [40,41]. In this model, particles are placed on a lattice and allowed to move by random motion. Each time clusters approach within one lattice constant, they stick permanently and then move as a unit. The larger units then also stick on contact with other clusters. The result of this process is an open highly ramified object with a fractal dimension of 1.7, substantially smaller than that observed in Fig. 7. Schaefer et al. suggested that the more compact structure results from residual repulsive forces between particles [42] so the clusters do not stick on every contact. Simulations of diffusion-limited aggregation with very low sticking probability yield a fractal dimension of 2.05 ± .05 close to the value of 2.1 observed in Fig. 7 [35,36].

Diffusion-limited aggregation with low sticking probability, of course, is not a diffusion-limited process at all, but a reaction-limited process. This type of aggregation is now called reaction-limited cluster aggregation (RLCA). RLCA is nothing other than the process described above for the generation of lattice animals by a polymerization process. Because of the low sticking probability, all geometrically allowed unions between clusters are weighted with equal probability and thus the aggregation problem also maps onto the lattice animal problem. It is interesting to note that the seemingly different processes of colloid aggregation and polymerization have been reduced to essentially the same simplified growth model.

More ramified colloidal aggregates (and polymers) are expected under conditions of aggressive aggregation. The original studies by Forrest and Witten [4], for example, of iron oxide soot, are consistent with the fractal dimension of 1.7 expected for cluster-cluster diffusion-limited aggregation [4]. Martin, Schaefer, and Hurd [6,43] also studied commercial fumed silica aggregates and found a fractal dimension of 1.8-1.9 reasonably close to the diffusion-limited aggregation result. Cannell and Aubert [44] found certain conditions under which silica aggregates in solution follow the diffusion-limited process. Finally, Weitz et al. [45], reported diffusion-limited aggregation in metallic gold particles. Wilcoxon, Martin, and Schaefer [46], however, have also studied this system and found evidence for strong inter- and intracluster multiple scattering, which they believe compromises the analysis of the light scattering curves.

PHASE SEPARATION

The final process we wish to discuss where fractal concepts apply is phase separation. Although several theoretical papers have appeared which predict ramified structures in the early stages of phase separation, experimental evidence is very scanty. Herrmann and Klein [47] predicted the existence of ramified structures in the early stages of nucleation-and-growth and Klein [48] and Desai and Denton [49] have predicted ramified structures during spinodal decomposition. In both cases, these ramified structures are expected only in the early stages before surface tension drives the phase-separated regions to compact geometry. We expect, therefore, that these structures would only be observed in systems with low surface tension.

Fig. 8 shows the data of Kotlarchyk, Chen, and Huang [50], for the neutron scattered intensity from a microemulsion system near the critical point. Although they interpreted these data in terms of a distribution of droplets, the data are also interpretable within the fractal scheme. When plotted on a log-log plot it is clear that these data have two power-law regimes. One with the slope of -1.4 and one with the slope of -5.0. The steeper slope at large K is indicative of a structure with a broad interface. Such a structure is reasonable in a microemulsion system where the detergent lies at the surface and would indeed give rise to a smeared interface. The slope of -1.4 at lower K suggests that one has a structure much like a colloidal aggregate in which droplets are organized in a fractal way. The fractal dimension, however, is surprisingly small. Abnormally low D's were also observed by Hurd and Schaefer [51] for interfacial aggregates.

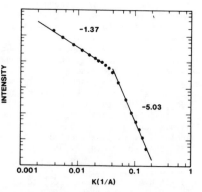

Fig. 8. Scattering curve for a microemulsion near the critical point (from ref. 50).

CONCLUSION

The value of fractal concepts in ceramic science lies in their simplicity. Without the simplifying concept of fractal geometry all the structures discussed above would have been hopelessly complex. Other models not based on fractal geometry obviously exist for polymerization aggregation and phase separation. It is our belief, however, that the general rules for understanding the relationship between structure and growth processes would not have been obvious within these more complex models. Although our analysis is over simplified, and further modification is necessary, it seems clear that the general issues of monomer-cluster versus cluster-cluster growth and diffusion-limited versus reaction-limited aggregation represent the broad categories necessary to understand random growth processes. These classifications, in fact, cover most of the structures which have been measured. In cases where they do not adequately represent structures, the deviations from these simple models can be identified and, in some cases, the underlying cause of the deviations can be identified.

REFERENCES

1. B. B. Mandelbrot, The Fractal Geometry of Nature, (Freeman, San Francisco, 1982).
2. H. E. Stanley and N. Ostrowsky, On Growth and Form (Martinus-Nijhoff, Boston, 1986).
3. T. A. Witten and L. M. Sander, Phys. Rev. Lett. $\underline{47}$, 1400 (1981).
4. S. R. Forrest and T. A. Witten, J. Phys. A $\underline{12}$, L109 (1979).
5. R. C. Ball, Ref. 2, p. 69.
6. D. W. Schaefer, J. E. Martin, A. J. Hurd, and K. D. Keefer in Physics of Finely Divided Matter, M. Boccara and M. Daoud, (Springer Verlag, New York, 1985) p. 31.
7. D. W. Schaefer and K. D. Keefer, Phys. Rev. Lett. $\underline{56}$, 2199 (1986).
8. D. W. Schaefer and K. D. Keefer, Mat. Res. Soc. Symp. Proc. $\underline{32}$, 1 (1984).
9. D. W. Schaefer and K. D. Keefer in Fractals in Physics, edited by L. Pietronero and E. Tosatti, (Elsevier, Amsterdam, 1986) p. 39.
10. D. W. Schaefer and K. D. Keefer, Phys. Rev. Lett. $\underline{33}$, 1383 (1984).
11. D. W. Schaefer, J. E. Martin, P. Wiltzius, and D. S. Cannell, Phys. Rev. Lett. $\underline{52}$, 2371 (1984).
12. H. D. Bale and P. W. Schmidt, Phys. Rev. Lett. $\underline{53}$, 596 (1984).
13. W. D. Kingery, H. K. Bowen, and D. R. Uhlmann, Introduction to Ceramics (Wiley, New York, 1976).
14. G. Porod, Kolloid Z. $\underline{124}$, 83 (1951).
15. J. W. Cahn, Trans. Met. Soc. AIME $\underline{242}$, 166 (1968).
16. D. W. Schaefer, B. C. Bunker, and J. P. Wilcoxon (to be published).
17. J. E. Martin, J. Appl. Cryst. $\underline{19}$, 25 (1986).
18. J. E. Martin and A. J. Hurd, "Scattering from Fractals," J. Appl. Cryst. XX, XX (1986).
19. P. Pfeiffer and D. Avnir, J. Chem. Phys. $\underline{79}$, 3558 (1983).
20. W. Ruland, J. Appl. Cryst. $\underline{4}$, 70 (1971).
21. D. W. Schaefer, K. D. Keefer, J. H. Aubert, and P. B. Rand, in Science of Chemical Processing, edited by L. L. Hench and D. R. Ulrich (J. Wiley, New York 1986).
22. P. Z. Wong, Phys. Rev. B $\underline{32}$, 7417 (1985).
23. M. Eden, Proc. 4th Berkeley Symposium on Math., Stat. and Prob., F. Neyman, Editor (University of California Press, Berkeley, 1961), Vol. 4, p. 223.
24. P. Meakin, Ref. 2, p. 111.
25. K. D. Keefer, Mat. Res. Soc. Symp. Proc. $\underline{32}$, 15 (1984).
26. K. D. Keefer, Viz. Vol.

27. K. D. Keefer and D. W. Schaefer, Phys. Rev. Lett. 56, 2376 (1986).
28. L. W. Kelts, N. J. Efinger, and S. M. Melpolder, "Sol-gel Chemistry studied by ^{1}H and ^{29}Si NMR," J. Coll. Int. Sci. XX, XXX (1986).
29. C. J. Brinker, K. D. Keefer, D. W. Schaefer, C. S. Ashley, J. Noncryst. Solids 48, 47, (1982).
30. D. Stauffer, Introduction to Percolation Theory, (Taylor and Francis, London, 1985).
31. D. Stauffer, Ref. 2, p. 79.
32. H. J. Herrmann, Ref. 2, p. 3.
33. H. E. Stanley in Structural Elements in Particle Physics and Stat. Mech., edited by J. Honerkamp, K. Pohlmeyer, and H. Romer (Plenum, New York, 1982) p. 1.
34. R. A. Assink and B. D. Kay, Mat. Res. Soc. Symp. Proc. 32, 301, (1984).
35. M. Kolb and R. Jullien, J. Phys. Lett. (Orsay) 45, L977, (1984).
36. W. B. Brown and R. C. Ball, J. Phys. A 18, L517 (1985).
37. F. Leyvraz, "Chemically Limited Cluster-Cluster-Aggregation and Lattice Animals", Preprint.
38. F. Family, J. Phys. A 15, L583 (1983).
39. J. E. Martin, J. Phys. A 18, L207 (1985).
40. P. Meakin, Phys. Rev. Lett. 51, 119, (1983).
41. M. Kolb, R. Botet and R. Jullien, Phys. Rev. Lett. 51, 1123, (1983).
42. D. W. Schaefer, J. E. Martin, P. Wiltzius, and D. S. Cannell in Kinetics of Aggregation, edited by F. Family and D. P. Landau, (North-Holland, New York, 1984).
43. J. E. Martin, D. W. Schaefer, and A. J. Hurd, Phys. Rev. A XX, XXX (1986).
44. C. Aubert and D. S. Cannell, Phys. Rev. Lett. 56, 738 (1986).
45. D. A. Weitz, J. S. Huang, M. Y. Lin, and J. Sung, Phys. Rev. Lett. 53, 1416 (1985).
46. J. P. Wilcoxon, J. E. Martin, and D. W. Schaefer, in Mat. Res. Soc. Ext. Abs. Fractal Aspects of Materials, edited by R. B. Laibowitz, B. B. Mandelbrot, and D. E. Passoja (Mat. Res. Soc., Pittsburgh, 1985) p. 33.
47. D. W. Herrmann and W. Klein, Phys. Rev. Lett. 50, 1062 (1983).
48. W. Klein, Phys. Rev. Lett. 47, 1569 (1981).
49. R. C. Desai and A. R. Denton, ref. 2, p. 237.
50. M. Kotlarchyk, S. -H. Chen, and S. Huang, Phys. Rev. A 28, 508 (1983).
51. A. J. Hurd and D. W. Schaefer, Phys. Rev. Lett. 54, 1043 (1985).

SAXS STUDY OF SILICA SOLS AND GELS

GERARD ORCEL, ROBERT W. GOULD, LARRY L. HENCH
Department of Materials Science and Engineering, University of Florida,
Gainesville, Florida 32611

ABSTRACT

Small angle X-ray scattering was used to characterize the structure of sols and gels in the $TMOS-MeOH-H_2O-CHONH_2$ system. The scattering curves were analyzed in both the Porod and Guinier regions. A fractal analysis shows that the structure evolves with time and temperature from a linear type polymer to a more highly branched polymer and then towards a "particle". However, these structural units disappear when the temperature increases above a critical value.

INTRODUCTION

Previous work has shown how drying control chemical additives (DCCA's) can affect the hydrolysis and polycondensation reactions of metal alkoxides, resulting in different gel structures [1-4]. The ability of a gel to withstand processing treatments without cracking all the way through densification depends on a careful control of the experimental conditions as well as on the gel structure. Small angle X-ray scattering (SAXS) is a convenient means to follow the evolution of the structure of gels from the liquid sol to the densified material. SAXS has been used to assess structural changes during the gelation process of silica and silicate gels [5-10]. The purpose of this work is to characterize the evolution of the structure of a silica gel derived glass, from the formation of a sol to densification, and to quantify how the process is affected by addition of different amounts of formamide DCCA.

EXPERIMENTAL PROCEDURE

The theory of small angle X-ray scattering has been worked out in detail by Guinier [11]. It was shown that for dilute systems the scattered intensity I(h) is proportional to the square of the structure factor F(h). As a first approximation:

$$F^2(h) = n^2 \exp\left(\frac{-h^2 Ro^2}{3}\right) \qquad (1)$$

where:

$$h = \frac{4\pi}{\lambda} \sin\theta \qquad (2)$$

with 2θ being the scattering angle and λ the X-ray wavelength, Ro the electronic radius of gyration, and n the total number of electrons in the particle. This assumption is well verified at low angles. At higher angles, the scattering curves obey a power law:

$$I(h) = A h^{-D} \qquad (3)$$

where D is the fractal dimension. Both quantities, Ro and D will be used to characterize our samples.

We used the facilities of the National Center for Small Angle Scattering Studies associated with the Oak Ridge National Laboratories. Since there was no need for high resolution at low angle, the 10M SAXS system was set up for a sample detector distance of 2.13 m. The X-ray source was 2 KW and the incident beam wavelength 1.54Å, characteristic of Cu Kα radiation. The run time was 600 sec., resulting in integrated intensities around 100,000. The liquid samples were inserted in Al cells with Kapton 300 V windows (Mylar 0.075 mm thick). The path length was about 1 mm. Correction routines for background, beam and detector sensitivity were used. The transmission coefficient of the samples was around 0.3.

The sample preparation has been described elsewhere [1]. SW55 corresponds to a sample prepared by hydrolysis and polycondensation of TMOS in MeOH. The films were treated one hour at the specified temperature.

RESULTS

Isointensity Curves

Figure 1 represents the planar distribution of the scattered intensity on the detector at different stages of the gelation process of a SW55 sample. The circular symmetry of the equi-intensity contours indicates that there is no anisotropy in the sample. The development of these curves gives relative information on the hydrolysis and polycondensation rates of the solutions. Significant scattering develops more rapidly for samples without formamide and for those catalyzed by HCl. This trend is in very good agreement with the values of the rate constants calculated from NMR data [1]. Formamide slows the hydrolysis but speeds up the polycondensation reaction.

This behavior is confirmed by the shape of the curves: integrated intensity Q versus reaction time t. The acid catalyzed sample, which hydrolyzes more rapidly, has the steepest curve at short reaction times. In contrast, the sample with the highest concentration of formamide is the first to reach a plateau, illustrating a higher condensation rate.

Additional information can be obtained from scattering curves, Fig. 2. By plotting the evolution of this curve with time, one can assess the variation in size and number of the scattering units as the chemical reactions proceed [8]. For the samples studied, there is first an increase in size and number of scatterers, but after a critical time, there is no production of new scattering centers. When formamide is present, the critical reduced time is higher.

Structure

The two main entities accessible by SAXS to characterize sols are the electronic radius of gyration Ro and the fractal dimension D. The fractal dimension is one way to assess the structure of the "particles" in suspension in the sol. The structure of these particles is very much dependent on the hydrolysis and polycondensation reactions. It can range from a linear polymer to a dense particle, and D then varies from 1.6 to 3 respectively. Dense colloids have a Porod Slope of -4 [8], but are non fractal objects. For the samples analyzed in this study, D increases with time and reaches a value around 2.2-2.3 at the gelation point (Figure 3). These values are intermediate between one characteristic of a branched polymer (2.16) and that of a diffusion limited aggregate (2.4). Higher D values are attained for samples containing formamide.

291

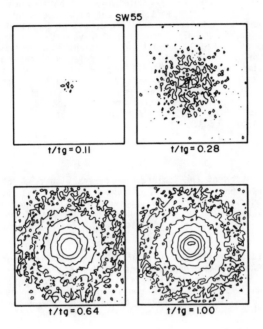

Fig. 1   Isointensity contours of a SW55 sol as a function of reduced
time.

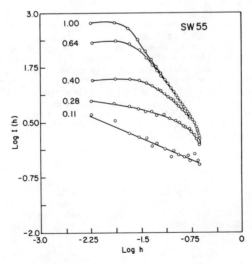

Fig. 2   Variation with time of Log I(h) vs. Log h curves for a SW55 sample
(no formamide).

The radius of gyration was determined for both regular sols and sols diluted in MeOH (volume ratio 1:20). For both experimental conditions Ro does not change except for the sample prepared by substituting 25 vol.% of MeOH with formamide. All the samples have then a radius of gyration which varies as in Fig.3. So all the solutions but SF25 are non-interacting up to the gelation point and the correlation length measured in the Guinier regime (60Å at t/tg=1) approaches the radius of gyration. Previous experiments [3] have shown "particle" radius in the range of 25Å for the same solutions. Similar results are observed for the gel films. This suggests that the gel structure is formed of primarily branched "particles" of about 25Å, which agglomerate at the gelation point in structural units of 60Å.

Gel Films

The evolution of the isointensity curves with temperature of a SW 55 sample is represented in Fig. 4. As the temperature increases, less scattering is observed. There is a progressive disappearance of the scattering centers, which can be explained by the collapse of the gel structure and by the ongoing polycondensation which leads to a more "meltlike" structure.

There is a greater uncertainty in Ro and D in the higher temperature samples. However, the order of magnitude of the electronic radius of gyration is 25Å and it remains fairly constant for most samples up to 1000°C. Though the gels which contain the higher amount of structural water foam below 1000°C and have a Ro near 90Å, the value of D increases with temperature from about 2.5 to a value close to 4.

SUMMARY

Silica sols and gels prepared under different electrolytic conditions were analyzed by small angle X-ray scattering. An analysis of the Porod and Guinier domains provides data (electronic radius of gyration and fractal dimension) which are comparable with results published by other groups [6,7]. Furthermore, these values are in very good agreement with previous characterization by other techniques [1,2]. Formamide decreases the hydrolysis rate and increases the polycondensation rate of $Si(OCH_3)_4$ in MeOH. Higher scattering exponents and radii of gyration are obtained when formamide is added to the solution. Comparison of the values of D and Ro of the sols and the gels suggest that the morphology consists of branched polymer units of silica aggregated in larger structures. The growth mechanism is qualitatively represented by the cluster aggregate or diffusion limited aggregate models.

ACKNOWLEDGEMENTS

The authors gratefully acknowledge the partial financial support of the Air Force Office of Scientific Research, Contract #F49620-83-C-0072. They also thank Dr. J.S. Lin from the National Center for Small Angle Scattering Studies at Oak Ridge National Laboratory for his valuable help.

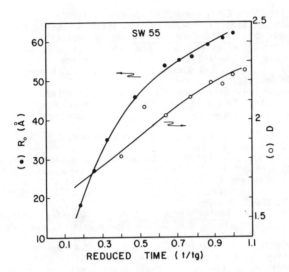

Fig. 3 Time evolution of electronic radius of gyration (Ro) and fractal dimension (D) of a SW55 solution.

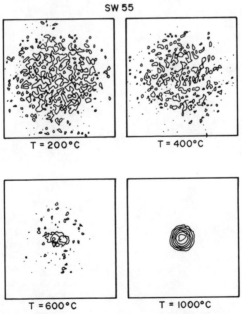

Fig. 4 Variation with temperature of isointensity contours of SW55 gel film.

REFERENCES

1) Gérard Orcel and Larry Hench, J. Non-Cryst. Solids 79, 177 (1986).
2) I. Artaki, M. Bradley, T.W. Zerda, J. Jonas, G. Orcel, and L.L. Hench, in Science of Ceramic Chemical Processing, edited by L.L. Hench and D.R. Ulrich (J. Wiley and Sons, New York, 1986) pp 73-80.
3) Larry Hench and Gérard Orcel, presented at the Third International Workshop on glasses and Glass-Ceramics from Gels, Montpellier, France, 1985 (in press).
4) Larry Hench, G. Orcel and J.L. Nogues, presented at the 1986 MRS Spring Meeting, Palo Alto, CA, 1986 (Proceedings).
5) M. Yamane, S. Inoue and A. Yasumori, J. Non-Cryst. Solids 63, 13 (1984).
6) C.J. Brinker, K.D. Keefer, D.W. Schaefer and C.S. Ashley, J. Non-Cryst. Solids 48, 47 (1982).
7) J.C. Pouxviel, J.P. Boilot, A Dauger and L. Huber, presented at the 1986 MRS Spring Meeting, Palo Alta, CA, 1986 (Proceedings).
8) D.W. Schaefer and K.D. Keefer, in Better Ceramics through Chemistry, edited by C.J. Brinker, D.E. Clark and D.R. Ulrich, (North Holland, New York, 1984) pp 1-14.
9) K.D. Keefer, in Better Ceramics through Chemistry, edited by C.J. Brinker, D.E. Clark and D.R. Ulrich (North Holland, New York, 1984) pp 15-24.
10) C.J. Brinker, K.D. Keefer, D.W. Schaefer, R.A. Assink, B.D. Kay and C.S. Ashley, J. Non-Cryst. Solids 63, 45 (1984).
11) A. Guinier and G. Fournet, in Small Angle Scattering of X-Rays, (John Wiley and Sons, New York, 1955).

GROWTH AND STRUCTURE OF FRACTALLY ROUGH SILICA COLLOIDS

KEEFER, K.D., Sandia National Laboratories, Albuquerque, NM 87185

ABSTRACT

Small-angle x-ray scattering measurements on partially hydrolyzed silicon tetraethoxide solutions indicate the formation of colloidal particles which have fractal structures or fractally rough surfaces. The structures and growth kinetics are consistent with chemically limited nucleation and growth of the particles from slowly generated reactive silanol species. Two dimensional computer simulations of nucleation and random growth of clusters from partially hydrolyzed monomers generate the same range of non-fractal, fractally rough and fractal clusters observed in the experiment.

INTRODUCTION

Ceramic precursors often form by random growth and aggregation processes. One of the major advances in the study of the structure of these precursors has been the application of fractal geometry to describe them [1,2]. Through the use of fractal geometry, random structures can be classified and quantified permitting the comparison of structures produced by different processes. In many cases growth processes generate structures with characteristic fractal dimension. By measuring the fractal dimension of ceramic precursors, it is sometimes possible to determine the process by which they were formed.

In real systems these processes are often the result of complex competing forces which are difficult to separate and study. For this reason, computer simulation of idealized growth processes is employed in order to study their essential features and to observe what types of structures may be generated. Some processes which have been studied in this manner are diffusion limited aggregation [3], in which clusters grow from a seed by the attachment of randomly diffusing monomers, and cluster-cluster aggregation, which is a similar model, but in which the clusters themselves are free to move and aggregate. These diffusion-limited processes generate structures of relatively low fractal dimension because reaction occurs on the first encounter of two particles.

Another class of particle-cluster growth processes is not diffusionally limited, but chemically limited. In these processes the probability of a monomer attaching to a cluster is low with respect to the frequency with which it encounters the cluster. All potential growth sites are therefore sampled by the monomers, and the probability for attachment is determined by local structure, which determines the probability of attaching to a particular site per encounter, rather than large scale structure, which governs the probability that a monomer will encounter a given site.

Chemically limited growth can occur when silicon tetraethoxide is hydrolyzed slowly and polymerized in neutral to alkaline solution. Under these circumstances, silica particles are nucleated and grow from orthosilicate monomers. A variety of structures may be produced in such a process depending on the extent of hydrolysis of the silicate monomers [4,5,6,7]. These structures range from highly ramified fractal objects, to homogeneous, colloidal particles with fractally rough surfaces, to smooth, colloidal particles. This paper will describe computer simulations of this chemically limited nucleation and growth process which generates structures which are qualitatively similar to those observed in the small-angle x-ray scattering experiment made on these systems.

*This work performed at Sandia National Laboratories supported by the U.S. Department of Energy under contract number DE-AC04-76DP00789.

FRACTAL GEOMETRY

Fractal structures are generated by computer simulations of many growth processes which occur in polymer and colloidal systems. A fractal dimension quantitatively describes the average structure of these objects by the relation

$$M \alpha R^{D_f} \tag{1}$$

where M is the mass of the object, R is some measure of the radius and D_f is the fractal dimension. For Euclidian (nonfractal) objects, D_f equals the dimension of space (i.e. in three dimensions the mass of a sphere scale is the radius cubed). For fractal objects D_f is less than the dimension of space and is generally not an integer.

There is another class of fractal structures, surface fractals, which have homogeneous, if porous, cores and whose surface areas are defined by the relation

$$S \alpha R^{D_s} \tag{2}$$

where S is the surface area and D_s is the surface fractal dimension.

Although fractals mathematically exist only in the limit of infinite size, real objects can only exhibit fractal behavior over a finite length scale. This length scale is small with respect to the overall dimensions of the object (the radius), but large with respect to the size of the monomer units of which the object is composed. In this range, fractal structures exhibit "universal" behavior because their properties are independent of both the overall shape and the composition of the object. Any physical measurement on a real system (such as the scattering measurements discussed below) will therefore exhibit a crossover to Euclidian behavior at both large and small length scales.

Qualitatively, mass and surface fractals can be distinguished by visual observation. Since mass fractals have a density less than space, they will have holes on all length scales, up to the size of the object itself. For the same reason, they are also "all surface" and the surface fractal dimension is equal to the mass fractal dimension. Surface fractal behavior is exhibited only by objects that are not mass fractals. These objects may be porous, but the holes do not get as large as the object itself so the object is uniform (if less than theoretically dense) on length scales larger than the largest hole. It is the periphery (external surface or "hull") that is fractal and will exhibit roughness on the scale of the object itself. This periphery is not the same as open porosity, as pores of small size are not counted.

SCATTERING MEASUREMENTS

Electron density fluctuations on the order of 10-1000 Å are probed in a small-angle x-ray scattering experiment. For both types of fractals, these fluctuations have a power-law correlation function which causes the scattering curve to also be a power law. For mass fractals, the exponent of the power law is simply the fractal dimension of the object

$$I(h) \alpha h^{-D_f}, 1 \leq D_f \leq 3 \tag{3}$$

where $h = 4\pi \sin \theta / \eta$, 2θ is the scattering angle and λ is the wavelength of the radiation. Such a curve will be a straight line with a slope of $-D_f$ when plotted log-log.

If an object is uniform (i.e. $D_f = 3$), "scattering" from the bulk is parallel to the incident beam, and scattering at finite angles arises only from the surface. This scattering also follows a power law [8]:

$$I(h) \, \alpha h^{D_s - 6} \; ; \; d-1 < D_s < d. \tag{4}$$

For uniform objects with smooth, non-fractal surfaces, $D_s = 2$ and Equation 4 gives the familiar Porod's Law, $I(h) \, \alpha h^{-4}$. Equation 4 is therefore a generalization of Porod's law to fractal surfaces. Scattering measurements can distinguish between mass fractals, whose scattering curves will be straight line with slopes of -1 to -3 when plotted log-log, and surface fractals, whose curves will have slopes of -3 to -4.

A complication occurs when scattering curves are recorded with a slit, rather than a point, geometry. The integration of the scattered intensity from the measured angle, θ, out to the maximum angle passed by the length of the slit causes a "smearing" of the scattering curves. For the particular case of "infinitely" long slits, this integration causes power-law forms of $I(h)$ to be increased by a factor of h, hence exponents measured with an infinite slit geometry are one greater than would be measured in point collimation (i.e. mass and surface fractals will generate slopes of 0 to -2 and -2 to -3, respectively).

EXPERIMENT

The solutions studied were 1 M TEOS in ethanol with water concentration of 1,2,3 and 4 M and .01M NH_4OH as a catalyst. The details of the preparation are reported elsewhere [4]. For the kinetic experiments, the x-ray scattering curves were measured periodically over a span of 15 days.

The small-angle scattering measurements were made with an Anton-Parr compact Kratky camera adapted to a 12kW Rigaku rotating anode x-ray generator. The beam path was evacuated and graphite monochromatized CuK_g radiation was used. Scattering curves were recorded with a TEC model 205 position-sensitive proportional counter having 47µm resolution with a sample-to-detector distance of 211mm and an entrance slit of 100µm. The samples were contained in 1mm capillaries and pure solvents were used as backgrounds.

At low angles, the scattering was analyzed with Guinier's law:

$$I(h) = I_e(h)N(\Delta\rho v)^2 \exp(-h^2 R_g^2/3) \tag{5}$$

where I_e is the scattering from a single electron, N is the number of scattering particles, $\Delta\rho$ is the electron density between the particle and matrix, v is the volume of the particle and R_g is the electronic radius of gyration. Samples were diluted to reduce the effects of interparticle interference. It should be noted that the prefactor in Guinier's Law weight the derived R_g toward the larger particles.

RESULTS

The small-angle x-ray scattering curves measured from the solutions are shown in Figure 1 and have been reported elsewhere. All four curves give reasonable power-law behavior over nearly a decade of reciprocal space. Generally speaking, real objects will produce power-law scattering curves only in the range $3/R_g < h < 3/a$, where a is the size of a monomer, so much more than one decade of real-space fractal behavior is required to generate a power-law scattering curve. These curves are interpreted as arising from mass fractals of $D_f = 1.84$ for R=1 mole H_2O/mole TEOS and surface fractals of $D_s = 2.45$ to 2.71 for R=2 to 4 moles H_2O/mole TEOS [7,9]. Since there is no known chemical hiatus between R=1 and R=2, it is believed that the system is

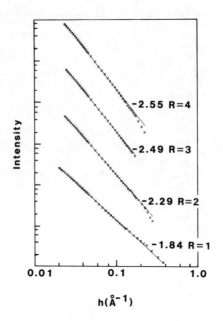

Figure 1. Small-angle scattering curves (slit smeared) from partially hydrolyzed TEOS in alkaline solution. R is the molar ratio of water to TEOS. The slope for R=1 is consistent with mass fractals with D_f=2.84, and the slopes for R=2, 3 and 4 are consistent with surface fractals with D_s=2.81, 2.51 and 2.45 respectively.

crossing over continuously from mass fractal to surface fractal behavior. That is, as the water concentration is increased, the objects become less and less ramified and eventually become uniform but with highly dissected surfaces as R approaches 2.

The kinetics of the particle growth was studied by measuring the scattering as a function of time and analyzing the scattering curves with Guinier's Law. From inspection of Guinier's Law, it is apparent that if the total mass in solution remains constant and the particle grow, that I(0) will scale with the molecular weight (NxM remain constant, so I(0) increases with M). A log-log plot of I(0) vs R_g measured for each solution at different times is shown in Figure 2. Although the errors are rather large, the data are consistent with a mass fractal dimension of 3 (the difference between D_f=2.85 and D_f=3 is less than the scatter in the data, so the results for R=1 appear to follow the same trend). This observation is consistent with the interpretation of surface fractals.

In Figure 3, the log of the radius of gyration of the particles is plotted as a function of the log of reaction time for the different solutions. At all water concentrations, R_g has a cube root dependence on the reaction time (a line of slope 1/3 is shown of the graph). The rate of particle volume increase (hence mass increase, since D_f=3) is therefore linear in time, implying the reaction rate is not increasing with the surface area of the particles. An independence of the reaction rate and the surface area would imply that the rate-limiting step is the generation of reactive monomers by the hydrolysis of TEOS and not the reaction with the surface.

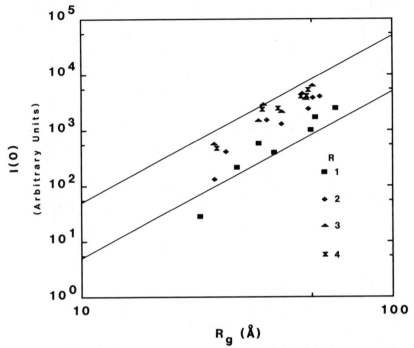

Figure 2.   The intensity at zero angle as a function of R_g for the different
ratios of water to TEOS. The lines have a slope of 3, implying non-mass
fractal particles.

SIMULATIONS: COMPLETE HYDROLYSIS

In the pH range of about 5-10, silica undergoes condensation
polymerization by a nucleophilic substitution of an HO- by a deprotonated
silanol group (SiO-). This mechanism favors the reaction of weakly acidic
monomers with more highly polymerized units, particularly the surface of
colloidal particles [10]. The essential process is therefore one of particle
nucleation and growth, since a monomer is more likely to attach to an
existing particle than it is to react with another monomer to form a new
nucleus. This favors a single nucleation event, particularly if monomers are
generated slowly, with subsequent growth of the nuclei and produces a
relatively monodisperse system of particles. The growth of the particles is
reaction, rather than diffusion, limited, as evidenced by the fact that
particle growth occurs over several hours or days even in highly hydrolyzed
systems, whereas diffusion-limited reactions are typically over in a few
seconds. In chemically limited growth, the large scale structure of the
particle does not affect the growth probability (in the absence of long range
forces) because the monomer can sample many potential growth sites and find
those most favorable for growth, rather than simply reacting with the first
one that it encounters. The simplest rule for chemically limited growth is
that all potential growth sites have an equal probability of reaction,
regardless of long or short range structure.
A growth model which incorporates these two essential features,
nucleation and chemically limited growth, was first proposed by Eden [11] to
describe the growth of cell colonies. This mode of growth can be simulated

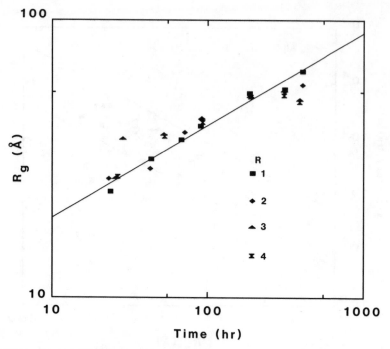

Figure 3. Particle R_g as a function of time. The data are consistent with nucleation and growth of particles with $D_f=3$ with the generation of monomer as the rate limiting step.

in a computer model by designating a site on a lattice as the nucleus and all lattice sites adjacent to the nucleus as potential growth sites. The particle is then grown by selecting one of the potential growth sites at random and occupying it. All sites adjacent to this newly occupied site then become potential growth sites. The process is then repeated until the particle reaches any desired size. The result of such a growth process in two dimensions is shown in Figure 4. The particle is nucleated in the sense that it grows from the initial "seed" placed on the lattice.

The particles that result in the Eden model of growth are neither mass nor surface fractals, but are homogeneous, and their surface roughness is small with respect to the size of the particle. These simple rules of growth, therefore, generate the same sort of structures that arise when silica colloids are grown from fully hydrolyzed monomers such as occurs in solutions of TEOS at high concentrations of water and base [12].

The reaction limited nature of the Eden model means that diffusion both to and within the cluster is ignored. "Lakes" can fill up even when they are no longer connected to the outside, and relaxation processes, such as reduction of surface area owing to surface tension do not occur. Lack of diffusion from outside the particle is primarily an artifact of two dimensional simulations, since in three dimensions closed pores are small and rare. It is an interesting result that, despite the lack of "ripening" or annealing, Eden clusters still become smooth as they get large, demonstrating that surface energy relaxation is not an essential process in the generation

Figure 4. A cluster of mass 25600 generated by the pure Eden model.

of smooth colloidal particles. The two simple rules for Eden growth are sufficient to account for the structure of silica colloids grown under conditions of slow generation of completely hydrolyzed monomer.

SIMULATIONS: INCOMPLETE HYDROLYSIS

The correlation of fractal dimension with water concentration led Keefer and Schaefer [7] to propose that incomplete hydrolysis was responsible for the development of fractal surfaces and the crossover to mass fractal character observed in this experiment. Since hydrolysis is the rate limiting step in particle growth in these experiments, it is likely that the silicate monomers are not fully hydrolyzed when they attach to the particle. Further, because of steric hindrance, it is reasonable to assume that alkoxide groups bound to the particle are more resistant to hydrolysis than those on the monomer, hence they would continue to block condensation on the sites they occupied. Computer simulations of these conditions indicate that certain distributions of partially hydrolyzed monomers do generate fractal structures in two dimensions.

Chemically limited nucleation and growth from incompletely hydrolyzed monomers can be simulated by a modification of the rules to the Eden model. When a growth site is occupied, instead of designating all of the adjacent uncommitted sites new growth sites, some of them are "poisoned" and are prohibited from ever being occupied. The growth sites then represent silanol groups and the "poisoned" sites alkoxide groups. Growth from a specified distribution of partially hydrolyzed species is simulated by picking in advance the number of sites on the monomer which will be poisoned, the exact number being chosen at random, but with a probability determined by the distribution. The available growth sites are then examined at random, and when one is found which has a number of uncommitted adjacent sites equal to or greater than the number of poisoned sites on the monomer, that site is occupied. The uncommitted adjacent sites are then poisoned at random, with any excess becoming growth sites. If the number of growth sites on the cluster drops to zero, the cluster is declared "dead". If the number of growth sites is not zero, but there is no site which can accomodate the particular functionality of monomer chosen, there are two recourses: one is to declare the cluster "dead;" the other is to "cheat" and try a monomer of different functionality for which there are growth sites available. The latter is physically more reasonable, and in practice the number of monomers "thrown back" is very small with respect to the total mass.

One of the important features of computer simulations is that quantities such as the mass, radius and perimeter of a cluster are readily measured. The mass is, of course, simply the number of occupied sites. The radius of gyration (the root mean square distance of the monomers from the center of mass) and the length of the perimeter (of two dimensional clusters) can be readily measured at different values of the mass as the cluster is grown. From the scaling behavior of these quantities, measured for a large number of very large clusters, fractal dimensions can be determined directly. The asymptotic slope approached as the mass goes to infinity on a log-log plot of the mass versus R_g will be the mass fractal dimension, and the slope of the corresponding plot of the perimeter ("hull") will be the surface fractal dimension.

The structure of the clusters formed in this growth process depends not only on the average number of sites poisoned per monomer, but also on the distribution of these sites among the monomers, in particular the variance of the distribution. Figure 5 shows a cluster of 25600 monomers grown from an equal number of tetra-, tri- and di-functional monomers (denoted 1:1:1:0), a distribution which has an average of one unhydrolyzed site per monomer. As can be seen, the cluster is porous and its surface is rougher than that of the pure Eden cluster composed of all tetrafunctional monomers. However, it is obviously neither a mass nor a surface fractal, an observation confirmed by the scaling behavior of the mass and perimeter in Figure 6.* If the average number of unhydrolyzed sites is maintained the same, but the distribution is changed to have equal numbers of di- and tetrafunctional monomers (1:0:1:0), the surface starts to exhibit fractal behavior, with $D_s = 1.1$. Unfortunately, this value is within experimental error of being Euclidian and may merely be due to finite-size effects (i.e. the largest clusters grown, mass of 204800, are not sufficiently close to infinite). If a small amount of monofunctional monomers are added to these systems, the fractal character of the clusters becomes quite pronounced. Figure 7 shows a cluster grown from a distribution of 3:3:3:1 tetra- to tri- to di- to monofunctional units, in essence the same distribution which generated Figure 5 with the addition of 10% chain terminating groups. As can be seen in Figure 7, the surface is extremely rough, although there are no large holes in the structure. The graph of the mass and perimeter vs. the radius (Figure 6) confirms that this is a surface fractal with $D_s = 1.3$. The three-dimensional analogue of the structure would generate a scattering exponent between -3 and -4 (-2 and -3, slit smeared) as observed in the experiment. Yet other distributions generate mass fractals in two dimensions. Figure 8 was generated from a mixture of 1:3:3:1 tetra- to tri- to di- to

 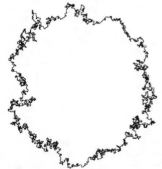

Figure 5. A cluster of mass 25600 generated by Eden growth from a 1:1:1 mixture of 4, 3 and 2 functional monomers and its perimeter. The cluster is porous but is neither a mass nor a surface fractal.

---

* The fractal dimensions for the 1:1:1:0 and 1:0:1:0 monomer distributions differ from those reported in Ref. 9 due to a change in the algorithm. In this work, an existing growth site cannot be poisoned, whereas in Ref. 9, it could.

303

monofunctional units and is clearly a mass fractal (D_f=1.82). The perimeter encounters almost all of the monomers in the cluster. It appears that other distributions generate clusters with a range of fractal dimensions. Thus it is quite possible that partial hydrolysis of the silicate monomers in the experiment is responsible for the implied crossover from mass fractals to surface fractals and for the apparent continuous variation in the scattering exponent.

CONCLUSION

The small-angle scattering measurements are consistent with the interpretation that colloidal particles which are mass fractals or have fractally rough surfaces are formed in these solutions. The particles appear to form by a nucleation and growth process, with the generation of reactive monomer as the rate limiting step. Computer simulations demonstrate that the postulated growth mechanism can in fact generate the range of structures observed, and that the fractal dimension of the particles increases as the number of active silanols decreases due to incomplete hydrolysis.

ACKNOWLEDGEMENTS

The author is grateful for countless hours of discussions with Drs. Dale W. Schaefer, James E. Martin and Alan J. Hurd, the indefatigable efforts of Mr. R. Zane Lawson for the sample preparation and scattering measurements and the preparation of this monograph by Mrs. Julie Walker, all of Sandia National Laboratories.

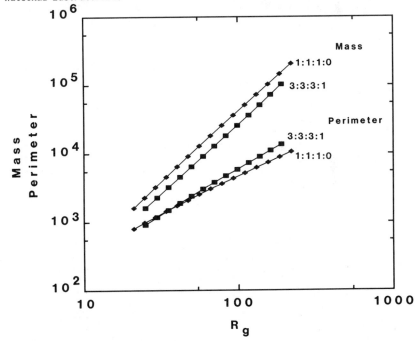

Figure 6. The average mass and perimeter as a function of R_g for simulation of 200 clusters of the types shown in Figures 5 and 7.

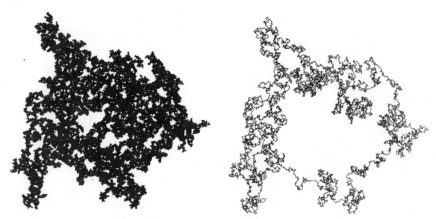

Figure 7. A cluster of mass 25600 generated from a 3:3:3:1 mixture of 4, 3, 2 and 1 functional monomers. This is a surface fractal with D_s=1.3 and D_f=2.

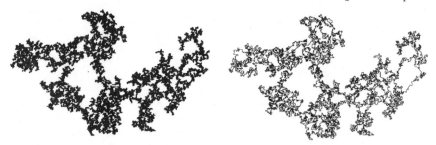

Figure 8. A cluster of mass 10000 generated for a 1:3:3:1 mixture of 4, 3, 2, and 1 functional monomers. This is a mass fractal with D_f=1.82. Note that almost every site in Figure 8 is on the perimeter.

REFERENCES

1. B. B. Mandelbrot, The Fractal Geometry of Nature, (Freeman, San Francisco, 1982).
2. H. E. Stanley and N. Ostrowsky, On Growth and Form, (Martinus-Nijhoff, Boston, 1986).
3. T. A. Witten and L. M. Sander, Phys. Rev. Lett. 47, 1400 (1981).
4. K. D. Keefer, Mat. Res. Soc. Symp. Proc. 32, 15 (1984).
5. D. W. Schaefer and K. D. Keefer, Mat. Res. Soc. Symp. Proc., 32, 1 (1984).
6. D. W. Schaefer and K. D. Keefer, Phys. Rev. Lett., 33, 1383, 1984.
7. K. D. Keefer and D. W. Schaefer, Phys. Rev. Lett., 56, 2376, 1986.
8. H. D. Bale and P. W. Schmidt, Phys. Rev. Lett. 53, 596 (1984).
9. K. D. Keefer, Science of Ceramic Chemical Processing, edited by L. L. Hench and D. R. Ulrich (J. Wiley, New York, 1986).
10. R. K. Iler, The Chemistry of Silica, John Wiley & Sons, New York (1979).
11. M. Eden, Proc. 4th Berkeley Symp. Math. Stat. and Prob., 4, 223 (1961).
12. W. Stöber, A. Fink and Ernst Bohn, J. Colloid Interface Sc., 26, 62 (1968).

GENERAL SCALING PHENOMENA IN POROUS SILICA GELS
AS PROBED BY SAXS AND MOLECULAR ADSORPTION (TILING)

J. M. Drake*, P. Levitz*,†, and S. Sinha*

*Exxon Research and Engineering Co., Annandale, NJ 08801
†on leave from C.N.R.S.-C.R.S.O.C.I., 45071-Orleans Cedex 2, France

Abstract

Using N_2 adsorption (B.E.T.), small angle X-ray scattering and molecular adsorption (tiling) we attempt to describe the internal pore morphology and surface structure of three porous silica gels. The results are analyzed from both a classical euclidian and a fractal scaling viewpoint. The applicability of both viewpoints are shown to be limited but important in describing the structure of porous silicas.

Introduction

We have investigated the pore morphology of a series of porous silica gels (prepared by E. Merck, Darmstadt, Germany). The purpose of this work has been to characterize the internal structure of the pore network placing specific emphasis on the geometrical disorder of the pore network and the surface. We have attempted to establish the possibility of describing the pore morphology using the fractal scaling model proposed for silica gels in the recent work of Avnir, Pfiefer and Farin [1,2].

Two approaches have been taken to study the silica; small angle X-ray scattering (SAXS) and physical adsorption (tiling). The SAXS technique provides structural details at length scales from 2Å (atomic scale) to 1000Å (macroscopic scale). The physical adsorption process, using a series of alkane (C_1-C_4) and alcohol (C_1-C_4) molecules, was used to probe the surface structure at length scales up to 10Å. Both techniques proved useful in probing the pore morphology although the results obtained by each technique are shown to be strongly dependent on the nature of the models used to interpret the data.

Characterization of Silica Gel

The three silica gels used in this work are referred to a silica Si-40, Si-60 and Si-100. Electron microscopy shows that each of these silicas exist as particles of irregular shape with a diameter of near 120 microns (μ) (see Fig. 1). These particles are composed of beads of radii between 1000-2000Å as shown in Figure 2. We refer to these particles as the secondary building blocks, while the primary particles are within the secondary building blocks and are not seen directly by microscopy. We have determined the apparent size of the primary building blocks by assuming that the mean of the pore size distribution (R_p) is a good estimation of the size of the primary particle (R_B) (see Fig. 3). This assumption can be supported by considering the relationship between silica gel porosity (Φ_v) and average coordination number (n) of these aggregated spheres. The mass pore volume (V_p) measured using nitrogen adsorption is 0.66 cc/g (Si-40), 0.924 cc/g (Si-60) and 1.25 cc/g (Si-100). We may calculate the porosity using Eqn. (1) [3]

$$\Phi_v = \frac{V_p}{V_p + 0.455} \tag{1}$$

1. Electron Micrograph of Si-60 at 600X.

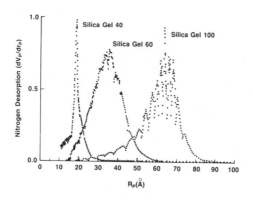

2. Electron micrograph of Si-60 at 20,000X.

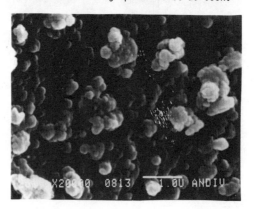

3. Pore size distribution for nitrogen desorption.

which gives 0.59 (Si-40), 0.67 (Si-60) and 0.73 (Si-100). Based on these
porosity values, we estimate the average coordination number (n) to be near
4 using the general equation proposed by Meissner [4],

$$n = 2 \exp (2.4 (1-\phi_v)). \qquad (2)$$

We then use the average coordination number and the porosity to relate the
radius of the primary particle (R_b) and the radius of the pore (R_p) (see
Iller [3] page 483). Our general conclusion is that for these silica gels
the radius of the primary particle would appear to be nearly equivalent to
the radius of the pore. Therefore, the primary particles have an average
radius given by the mean of the pore size distribution which for these
silicas is 18A(Si-40), 35A(Si-60) and 60A(Si-100). These primary particles
aggregate to give the secondary structure shown in Fig. 2.

The surface within the secondary particles make up the major surface
area of the silica. Using this simple euclidian model for uniforming
aggregating spheres, we can compare the surface area measured by N_2 adsorption (B.E.T.) and that calculated by Eqn. (3) [3] using the particle radius
estimated by N_2 desorption:

$$Sq = \frac{2750}{2 R_p} \qquad (3)$$

	N_2 (B.E.T.)*	Sq*
Si-40	768	764
Si-60	391	393
Si-100	281	229

* area in M^2/g, R_p in nm.

The agreement is surprisingly good between the B.E.T. results and the
simple geometrical picture of aggregated spheres. The only significant
deviation is seen for Si-100 where we under-estimate the area by 20%.
Looking at the pore size distribution for Si-100 (Fig. 3) we see a near
periodic oscillation in the desorption. The origin of this apparent
oscillation is not clear to us, although we would suggest that it is due in
part to two factors which have been before now ignored by our simple
geometrical approach: the explicit role of surface roughness on a length
scale much less than R_p(2-10A) and the extent to which the sphere interfaces have smeared together due to sintering at the contact points. The
question of sphere surface roughness at small length scales will be addressed in the next two sections which deal with SAXS and adsorption measurements. However, if roughness is present it would result in an inferior
estimation of the area using R_p. Sphere contacts would lead to a loss of
area, the order of magnitude of which can be estimated using Eqn. (4) [3].

$$S = Sq (1 - n \frac{d}{4R_p}) \qquad (4)$$

Where S is the accessible specific surface area (M^2/g), Sq is the specific
surface area of the original separate spheres (M^2/g), n is the coordination
number, d is the diameter of the nitrogen molecule (nm) and R_p is the
radius of the primary sphere (nm). This correction is less than 5% for Si-
100, but would be more important for Si-40 and Si-60 because the area correction is sensitive to 1/Rp. It is only the limit of sphere contact which

is estimated in by Eqn. (4), not the area loss due to sintering. We have no direct evidence of the extent of sintering at the contact points. Although none of the arguments are a clear proof of the hierarchical morphology of these silicas, it is at least suggestive that there is utility in the more classical geometrical model. The limitations of such an approach are apparent when we consider the assumptions implicit in these arguments (Eqns. 1-4) and their dependence on the validity of the nitrogen B.E.T. model and the results it yields.

The nitrogen B.E.T. analysis produces two pieces of data important to our conclusions: surface area (adsorption) and pore size distribution (desorption). The surface area depends on the fit of the B.E.T. equation to the pressure data to obtain a measure of a monolayer coverage. As pointed out by Brunauer [5], the B.E.T. method still gives a true value for the specific surface area provided there are no pores with radii smaller than 12Å. This would include particles with a surface roughness on a length scale between 2-12Å. A roughness would result in monolayer values superior to the real monolayer value.

The pore size distribution data which is obtained from N_2 desorption data is dependent on both the surface roughness of the particles and their morphological organization. However, to obtain the pore size distribution from the N_2 desorption data, a specific pore model is needed (cylindrical pore geometry is used) together with the Kelvin equation. In this case, we are using data obtained assuming one pore morphology to establish the existence of second morphology without admitting the possibility of significant error due to the inconsistency presented by these different morphologies at short length scales.

Based on the microscopy, N_2 B.E.T. surface areas and pore size distribution data, we suggest that all three silica gels have approximately the same hierarchical pore morphology. A hierarchical morphology which is made up of two size particles each of which retain some of the original features of spheres.

Small Angle X-ray Scattering of Silicas

The small angle X-ray scattering (SAXS) experiments were done at the Exxon line x 10A at the national synchrotron light source. The in-plane resolution was achieved using a Bonse-Hart set up [6] consisting of a double bounce monochromator and a triple bounce analyzer. In all cases, we utilized perfect crystals of silicon cut for reflection from the (111) planes. The full width at half maximum of the instrumental resolution function was 0.0002Å^{-1}. The wave vector resolution normal to the scattering plane was achieved by utilizing a series of slits before and after the sample and was measured to be 0.0016Å^{-1}. Thus, the experiments were done under "point geometry" rather than "slit geometry" conditions.

SAXS experiments were carried out on samples of silica gel 40, 60, 100 and silica gel (licrosphere) 4000. In all cases these samples were first baked out and sealed in a vacuum in 1.5mm diameter quartz capillary tubes, which were mounted on the spectrometer for the scattering experiments. The measurements were all performed at room temperature. After each sample was measured, the quartz capillary was opened at the end, emptied of the silica gel sample, and the empty capillary measured again for the purpose of obtaining the background under the true scattering curve. Appropriate corrections were made for the transmissions of the full and empty capillaries in each case.

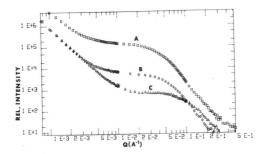

4. SAXS for Si-100 (A), Si-60 (B) and Si-40 (C) from 1×10^{-3}–5×10^{-1} $Q(\text{Å}^{-1})$.

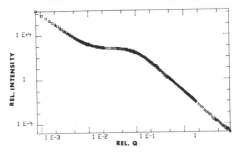

5. Scaled SAXS data for Si-40, Si-60 and Si-4000 (S (q)/So(ro) versus F (qα) where α = ro/(ro)Si-40. α = 2.112 for Si-60 and α = 44.47 for Si-4000.

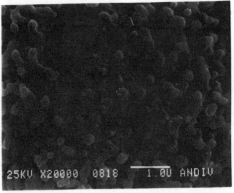

6. Electron micrograph of Si-4000 at 20,000X.

The results for the scattering function S(q) (apart from the arbitrary normalization factor) for several different samples of silica gel are shown in Fig. 4, plotted on a log-log scale. One may see that they all basically have the same form, namely an initially steeply decreasing region at small q where the intensity drops as a power of q which is close to -3, then a fairly flat region, and at still larger q value another rapidly decreasing region. Interestingly, by scaling these curves both in q and intensity, one finds that all of these curves can be represented by a single general curve of the form

$$\frac{S(q)}{So(ro)} = F \ (qro) \tag{5}$$

where F is a general scaling function and So(ro) and ro are appropriate scaling constants for each individual sample. This is illustrated in Fig. 5. The parameter ro turned out to scale reasonably well with the "pore size" obtained from N_2 desorption measurements as will be shown below. The only case where this scaling did not work very accurately was in the small q region and in the very large q region. Thus, we may say that the morphology of these different silica gels on the length scale of the nominal pore sizes are very similar and scale with the nominal pore size. In fact, in this regime, the scaling function may be approximated approximately by the form obtained by Debye et al. [7] for scattering from a random interface between two otherwise uniform media, i.e.,

$$F \ (q \ ro) \sim (\frac{A}{1 + q^2 ro^2})^2 \tag{6}$$

Here ro is a length characteristic of a mean spacing between interfaces. Representative values for ro deduced from Eqn. (6) are given in Table 1. In the approximation of a random interface, ro can be written as [7]

$$ro = \frac{4V}{S} \ \Phi_v(1-\Phi_v) \tag{7}$$

where V is the volume of the scatterer; S the interface between the solid and the void volume; and Φ_v the porosity. Knowing the B.E.T. surface (S_{BET}), the pore volume (V_p), and using Eqn. (1) for the porosity Φ_v, ro becomes

$$ro = \frac{4\Phi_v(1-\Phi_v) \ (V_p + \frac{1}{\rho})}{S_{BET}} \tag{8}$$

where ρ is the density of bulk silica ($2.16g/cm^3$). As shown in Table 1, a good accordance is generally obtained between the value of ro calculated from Eqn. (8) and using data obtained from N_2 adsorption desorption measurements and the experimental one deduced from the SAXS experiments. However, a divergence appears for Si-100 where the calculated characteristic length ro overestimates the experimental one. As we will see below, this disagreement can be explained in a roughness of the interface on a length scale of 10Å or less.

Solid	S_{BET} M^2/g	V_p cm^3/g	Φ_v cm^3/cm^3	Experimental ro(Å)	Calculated ro(Å)
Si-40	768	0.66	0.59	13.7	14
Si-60	391	0.92	0.67	29	31
Si-100	281	1.25	0.73	39.3	47

Table 1

We turn now to the small q regime where an upturn is seen in S(q). This is obviously due to the existence of much larger objects in the medium whose minimum threshold size appears to scale with the "nominal pore size" as witnessed by the fact that the upturn occurs at a value of q which scales with the rest of the curve. These objects are most probably the voids (such as seen in Fig. 6), and while it is tempting to think of them as occuring on all length scales greater than the threshold size as witnessed by the power law behavior which seems to be observed, we note that the range of values over which this behavior is observed is too small to make any definitive statement about this distribution other than the threshold size.

Large q regime shows Porod-like or q^{-4} behavior consistent with Eqn. (6), indicative of the fact that the internal surfaces are basically smooth. However, for Si-100, a deviation away from q^{-4} behavior to an asymptotic behavior more like $q^{-3.5}$ is observed at values of $q > 0.1\text{Å}^{-1}$ (Fig. 7). We note that this is suggestive of scattering from a fractal surface, which must behave asymptotically like $q^{-(6-\bar{d}s)}$ [8], $\bar{d}s$ being the surface fractal dimension. This would be consistent with a silica gel having an interface which was fractally rough on length scales of 12Å or less, with $\bar{d}s \approx 2.5$. We also point out that the characterization of a surface as "fractal" only on length scales below 10Å, but smooth above these length scales, is not a very satisfactory way of describing such surfaces, but simply a way of implying that the large q scattering function from this surface cannot be described by Porod's law.

To conclude this section, at the scale of the pore size, the surface of silica gels Si-40 and Si-60 can be considered as a random interface with no clear evidence of self-similarity. The sintering effect between the primary building block appears to be relatively efficient to smear the original spherical shape and to avoid strong interference pattern between neighbor primary building blocks. In the case of Si-100 the surface appears rough on length scales below 12Å.

Physical Adsorption as a Probe of
Pore Morphology and Surface Structure

Recent work by Avnir, et al. [1] suggests that for porous silicas there is a unique scale relationship between the number of molecules needed to form a surface monolayer (N_m) and the cross-sectional area of that molecule (σ, "yardstick scaling"),

$$N_m \propto \sigma^{-\bar{d}s/2} \tag{9}$$

These tiling experiments are proposed to be a direct measure of the fractal dimension ($\bar{d}s$) of the pore surface, which is suggested to be a space filling rough surface. This relationship (Eqn. 9) is said to be equivalent to the statement,

$$N_m \propto R_b^{-\bar{d}_s} \tag{10}$$

where R_b is the radius of the silica particle. Using these two apparently connected relationships, a series of tiling experiments have been done using an anologous group of alcohol molecules [2]. The conclusion put forth from these experiments appear consistent with a fractal interpretation as generalized by the previous two equations. The statement is made that the $\bar{d}s$ for silica is ~ 3.0 with a self-similar range extending from 37Å^2 to greater than 250Å^2 in sigma.

312

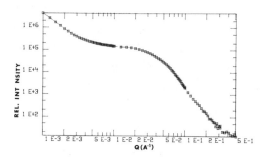

7. SAXS data for Si-100, showing departure from q^{-4} at large q.

8. Cartoon for tiling on a porous surface.

PORE STRUCTURE OF SILICA

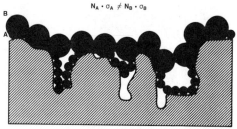

$N_A \cdot \sigma_A \neq N_B \cdot \sigma_B$

Cross-Section of Pore

    The tiling method is a technique which depends in part on a detailed understanding of the adsorption process for a binary mixture. It also is a technique which requires that it is possible to measure accurately the exact number of molecules needed to cover the surface.

    The fractal model of the tiling experiment on the internal surface of a porous silica is appealing because it is consistent with our intuition. The cartoon, shown in Fig. 8, represents the intuition inherent to the tiling experiment. It suggests that many properties associated with the physical characterization of porous silicas will be "yardstick" dependent (i.e., surface area, pore volume, porosity, and pore size distribution). The question which seems important to answer relates to the relationship between the physical disorder and the chemical heterogeniety of the surface. It is important to recognize that the tiling experiment, while it may be sensitive to the local disorder or roughness of the pore surface (as suggested by Fig. 8), may also be inherently sensitive to the chemical nature of the surface. Certainly, in the case of the adsorption of alcohol molecules (ROH) on silica gels (Si-OH) there is likely to be a stochiometric relationship between the number of adsorbed ROH molecules and the number of surface hydroxyl groups accessible. In fact, the critical aspect of the tiling experiment is in part related to establishing if there is a correlation between the value of σ and the actual size of an adsorption site. This leads us to ask if the distribution of adsorption sites is sufficient to probe directly the topological disorder of the pore surface over the self-similar range.

Tiling Experiments on Silica Gel

We performed two types of tiling experiments on all three silica gels (Si-40, Si-60 and Si-100). The first experiments were an attempt to repeat the measurements reported by Hoffman, et al. [9] and Avnir, et al. [2] using a series of alcohol molecules. These experiments were done from solutions of alcohol and toluene. The second experiments were done using a series of alkane molecules from the gas phase. In each case, the silicas were dried at 150°C at 10 millitorr for 24 hours and transferred to a clove box where samples were manipulated in a dry nitrogen atmosphere. The alcohols and the toluene were dried and stored over 4A molecular sieves to eliminate interference by adsorbed water. The alcohol adsorption was measured using two techniques: Refractive index (RI) changes at the NaD wavelength and gas chromatographic separation with F.I.D. detection. We report here only the RI analysis as this is the technique used by previous investigators and therefore establishes a limited concentration range for which the adsorption may be followed.

9. Adsorption isotherm for methanol from toluene on Si-40, $(n_2^S$-vs-$C_2)$, solid lines only to guide the eye.

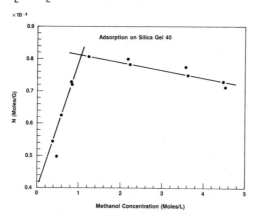

10. Same data as in Fig. 9 with change in variables, C_2/n_2^S-vs-C_2, solid line fit to data at low conc.

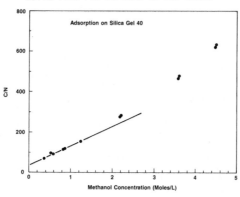

The alkane adsorption was done using four different methods: continuous volumetric, static volumetric, flow gravimetric and static gravimetric [3]. What we will attempt to show are the limitations of the adsorption process as a probe of the surface features of silicas. This is best done by looking at the constraints implied by using the Langmuir model to measure the monolayer coverage (N_m) for different yardsticks (σ_i). The Langmuir treatment is a model based on the assumption that the surface consists of adsorption sites with an area per site given by σ_i. This site-based model has four limiting assumptions about the adsorption process: (1) that all surface sites have the same activity for adsorption, with surfaces that are energetically uniform; (2) there is no interaction between adsorbed molecules; (3) all the adsorption takes place by the same mechanism, with each adsorbed complex having the same structure; and (4) the extent of adsorption is less than one monolayer.

Alcohol Tiling Measurements

The apparent adsorption isotherm was obtained for each of the alcohols on each silica. The data was analyzed assuming a Langmuir adsorption model using Eqn. 11,

$$\frac{C_2}{n_2^S} = \frac{1}{n_m b} + \frac{C_2}{n_m} \tag{11}$$

The terms are: C_2 is the equilibrium solution concentration of alcohol, n_2^S is the apparent number of moles of alcohol adsorbed per gram of silica, n_m is the number of moles of adsorption sites per gram of silica and is equivalent to the monolayer adsorption capability under the appropriate limiting conditions, and b is related to the interaction strength between the alcohol and surface site. The use of this model for the adsorption process assumes that the solution is sufficiently dilute, the solute and solvent activity coefficients are constant. If these conditions are not met at all alcohol concentrations, then what is observed is the composite isotherm where the apparent adsorption n_2^S is related to the surface excess (Γ_2^S). In this case, Γ_2^S is a measure of the difference between the individual adsorption isotherms of the alcohol and the toluene. Under these conditions it would not be possible to measure the surface monolayer capacity (n_m) from a plateau in the isotherm because the isotherm shows a peak followed by a negative slope [10]. This type of behavior has been observed in all our isotherms (see Fig. 9). To obtain a value of N_m for alcohol we change variables according to Eqn. 11 fitting it to the points before the peak in the isotherm (see Fig. 10). From the slope we calculate N_m and plot this value versus the value of σ estimated by Hoffman [9]. A log-log plot (as shown in Fig. 11, line B) suggests a power law with a slope of 2.96 ± 0.2, 2.95 ± 0.2, and 2.60 ± 0.2 for Si-40, Si-60 and Si-100 respectively. If we were to propose that the power is describing the fractal scaling behavior of rough surface, as given by Eqn. 9, then the apparent $\overline{d}_s$ is approximately 6, 6 and 5.2 for Si-40, Si-60 and Si-100. This is a physically unreal result as the limiting ranges of $\overline{d}_s$ are between 2 (flat) and 3 (space filling). It is also clear that the power law scaling with σ does not include the nitrogen value for all three silicas. This is inconsistent with the assumption that such a tiling experiment is only sensitive to the topology of the surface and therefore the local roughness of the surface while being independent of the nature of the adsorbate/surface site interaction. This is clear evidence that the tiling experiment is sensitive to both the chemical (site) and physical disorder of the pore surface. The tiling experiment is a counting problem in that we must

obtain an accurate count of the number of molecules needed to achieve a
monolayer. The counting is done indirectly by assuming that the adsorption
isotherm evolves from the low surface coverage regime without change in the
adsorption mechanism and that this process is identical for each alcohol.
This is likely not to be the case.

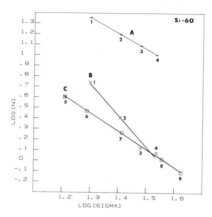

11. Tiling data for alkanes (C), and alcohols (B) on Si-60. (1) methanol,
(2) ethanol, (3) 1-propanol, (4) 1-butanol, (5) nitrogen, (6) methane,
(7) ethane, (8) propane, (9) butane. Curve (A) Gurvitch data for
alcohols filling the pore volume, calculated from liquid density at
25°C. X = Hoffman Sigmas (σ) [9] and Nay [12] Y = N_m (mmoles/g).

Alkane Tiling Measurements

The alcohol adsorption process is in the strong thermodynamic limit
for adsorbate/surface interactions. This adsorption process involved a
competition between alcohol and toluene molecules which can lead to
complications due to alcohol/alcohol complexation, etc.. So, we have
performed tiling experiments with alkanes in the gas phase. This
represents the weak thermodynamic limit for adsorbate/surface interaction
while elminates the complexity of dealing with a binary solution.

The alkane tiling using the C_1-C_4 linear series also gave a power
relationship between N_m and σ (see Fig. 12). The slope for each slilica
(Si-40, Si-60 and Si-100) was 1.51 ± 0.2, 1.71 ± 0.2 and 1.45 ± 0.2,
respectively. If we interpret this using the fractal model, the values of
$\bar{d}_s$ are 3.02 ± 0.4, 3.42 ± 0.4, and 2.90 ± 0.4. These seem to be consistent
with the picture of a space filling fractally rough surface. Yet these
results are for the same silicas that gave $\bar{d}_s \sim 6$ for alcohol tiling.
Certainly the alkanes are less sensitive to the nature of the chemical
(site) disorder of the surface but this by itself doesn't explain what
surface feature the tiling process probes. Based on the physical
characterization data (N_2 B.E.T. area, pore volume, porosity, etc.) and the
SAXS data it is impossible to corrobate $\bar{d}_s$ = 3.0 as suggested by the alkane
tiling results (at the length scale of this adsorption process).

316

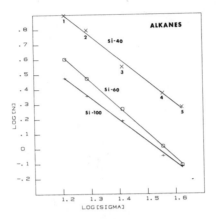

12. Tiling data for Si-40, Si-60 and Si-100 (alkanes 2-5, nitrogen1).
N_m-vs-σ.

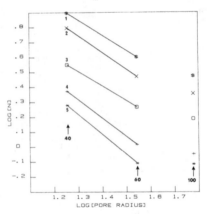

13. Alkane tiling data as in Fig. 12 with change of variable N_m-vs-R_p.
(1) nitrogen, (2) methane, (3) ethane, (4) propane, (5) butane.

Surface Volume Model of Tiling

Based on the results of the tiling experiments on these silicas, we have developed a heuristic model (picture) which is consistent with the data. We begin by suggesting that the tiling probes a surface layer with volume (V_a)

$$V_a \propto R_b^{\overline{d}_s} \xi \, (\ell) \tag{12}$$

where V_a is adsorbed surface volume per primary building block after sintering, R_b is radius of the "primary particle", $\overline{d}_s$ is the fractal dimension of the surface at the scale of the pore radius, ξ (ℓ) which is the thickness of the adsorbed layer when $\overline{d}s = 2$, and ℓ is the characteristic size of the tile. The number of adsorbed molecules per unit mass N_m is given by

$$N_m \; \alpha \; \frac{R_b^{(\overline{d}_s-3)} \cdot \xi \; (\ell)}{\ell^3} \tag{13}$$

assuming the mass of the primary building block goes as R_b^3 . A generally accepted expression for ξ (ℓ) is

$$\xi \; (\ell) \; = \; \ell^{3-\overline{d}_s} \tag{14}$$

Therefore, the expression for N_m becomes,

$$N_m \; \alpha \; R_b^{(\overline{d}_s-3)} \cdot \ell^{-\overline{d}_s} \tag{15}$$

If we now analyze the tiling data by plotting N_m versus R_p for each tile (ℓ) (see Fig. 13) there is a contradiction with the previous analysis using N_m versus σ (ℓ^2). (Both analyses assume that ξ depends on ℓ.) For N_m -vs-R_p (assumed to be equivalent with R_b) for Si-40 and Si-60 we find $\overline{d}_s \sim 2.0$ while for N_m -vs-σ we find $\overline{d}_s \sim 3$. If we assume that the contradiction arises because of the assumption that ξ depends on ℓ is not correct for this limited series of small molecules (tiles) but is independent of ℓ, then

$$N_m \; \alpha \; \frac{R_b^{(\overline{d}_s-3)} \; \xi}{\ell^3} \tag{16}$$

is able to account for tiling results. There is a strong deviation from $\overline{d}_s \sim 2$ for Si-100 (see Fig. 13). This suggests that the surface of the primary building block of Si-100 may be rough on a length scale important to the adsorption process. This result is supported by the SAXS results which show a departure from Porod (q^{-4} at large q (very small length scales < 12Å)). This may be due to surface roughness ($\overline{d}_s \sim 2.5$, over a very limited range of q).

318

The question that still remains is how do we explain this apparent independence of ξ on the size of the adsorbed molecule. At present we don't know nor do we have any independent evidence to support this assumption. We do believe that the tiling experiments are suggestive of a real surface adsorbed volume and therefore the adsorption process might be better analyzed using a potential theory which deal with surface volumes for the adsorbed layer [11].

Using nitrogen B.E.T., SAXS and surface tiling measurements, we have attempted to characterize the internal pore morphology and surface of three porous silica gels. The results are in part consistent with a picture of silica gels composed of a hierarchical structure of randomly associated particles on two length scales. A primary building block, with a radius similar to the mean pore size are associated to form a secondary particle with a radius near 1000Å. These secondary particles associate to form an irregularly shaped macroscopic particle of $\sim 100\mu$ diameter. We suggested that both the primary and secondary particles retain some of the spherical features while showing sintering at the particle-particle contacts. The majority of the data is consistent with a pore surface which is smooth down to $\leqslant 12$Å and only the Si-100 results (SAXS) are suggestive of roughness below 12Å. Whether this roughness is self-similar or self-affine has not been determined.

The tiling experiments with alcohol molecules appear to contradict the data presented by Avnir, et al. [1]. The origin of the contradiction is not clear but may be due in part to the method of data analysis. It may be that estimating the monolayer capacity (N_m) from the plateau in the isotherm leads to a value of N_m which is related to an adsorbed surface volume, containing more molecules that would exist in a monolayer. In the case of the alkanes, the systematic over-estimation of N_m seems to lead to a value of $\bar{d}_s$ (3.0), which is inconsistent with the apparent surface structure of these silica gels.

The heuristic model (picture) proposed to explain the tiling results for the alkanes corroborates the SAXS data.

Acknowledgments

We would like to thank Dr. A. F. Venro, of Omicro Technology Corp, for work on the alkane adsorption and Mr. John Swirczewski for his work on the alcohol adsorption. A special thanks to Dr. J. Klafter for many useful discussions.

References

1. David Avnir, Dina Farin and Peter Pfeifer, Nature 308, 261 (1984).
2. Peter Pfeifer, David Avnir and Dina Farin, J. Statistical Physics, 36 (5-6), 699 (1984).
3. Ralph K. Iler, "The Chemistry of Silica", John Wiley & Sons, Inc. (1979).
4. P. Meissner, A. A. Michaels, and R. Kaiser, Ind. Eng. Chem. Process Des. Div., 3 202 (1964).
5. S. Brunauer, J. Colloid Interface Sci., 41, 613 (1972).
6. U. Bonse and M. Hart, Appl. Phys. Lett., 6, 155 (1965).
7. P. Debye and A. M. Bueche, J. Appl. Phys., 20, 518 (1949).
8. H. D. Bale and P. N. Schmidt, Phys. Rev. Lett., 53, 596 (1984).
9. R. L. Hoffmann, D. G. McConnell, G. R. List and C.D. Evans, Science 157, 550 (1967).
10. J. J. Kipling and D. A. Tester, J. Chem. Soc., 4123 (1952).
11. Robert S. Hansen and Walter V. Fackler, Jr., J. Phys. Chem. 57, 634 (1953).
12. M. A. Nay and J. L. Morrison, Can. J. Res. B27, 205 (1949).

FRACTAL ASPECTS OF SURFACE SCIENCE - AN INTERIM REPORT

DAVID AVNIR
Department of Organic Chemistry, The Hebrew University of Jerusalem,
Jerusalem 91904, Israel

ABSTRACT

The activity in our laboratory and in collaboration with other
laboratories in applying fractal geometry for the characterization of
irregular surfaces and materials and for the analysis of molecular
interactions with such objects are summarized in this report. Most of the
studies were performed on the xerogel of silica. Various theories and
experimental techniques have been employed for this purpose. They include
adsorption studies, computerized image analyses in one (boundary lines),
two (textures) and three (proteins) dimensions; non-radiative, Förster-
type, one step electronic energy transfer (EET); small angle X-ray
scattering; and analysis of chemical reactivity of fractal objects. The
relation between porosity and fractal dimension has been studied by EET, by
the photophysics of adsorbed pyrene and by monolayer adsorption studies. A
direct method for the determination of adsorption conformations has been
developed. Limitations of the fractal approach have been identified and
outlined.

1. INTRODUCTION

Two years ago, we summarized our findings on the wide occurrence of
fractal surfaces in a large table (Table 2 in Ref. 10). Our findings were
based on the analysis of adsorption data. This state of art has changed
significantly in the past two years. Many more materials and objects have
been analyzed and the methods we use for fractal analysis have diversified.
In addition to adsorption studies, they now include computerized image
analysis techniques (of boundary lines, textures and of three dimensional
protein X-ray data), analyses of photophysical processes, scattering
experiments, analyses of surface chemical reactivities, and flow analyses
[1-19]. The time has come therefore to compile a new comprehensive table,
which is the heart of this report. Following the table, we briefly
summarize some topics in which we make use of the measured fractal
dimensions, D. These include the question of chemical reacivity of fractal
objects, the question of porosity vs. surface fractal dimension, and the
question of cross-sectional areas of adsorbates.
Space limitations force me to be very brief on all of these issues. I
also wish to emphasize that this report is not a review of the field, but
only an updated summary of the title activity in our research group and in
collaboration with other groups [1-19]. The interested reader can find
full details of our work as well as description of contributions by other
researchers in our cited papers.

2. METHODS FOR DETERMINING FRACTAL DIMENSIONS and other Explanations for
    the Summarizing Table

2.1 General Information

The table is divided into five columns:

First Column - All entries are numbered for ease of future reference.

Second Column - The materials are grouped as follows: a) Oxides. Most of our activity and most of the various techniques we used were concerned with the family of silicon oxides. b) Organic Materials. These include carbon adsorbants, polymeric materials and organic crystals. c) Proteins. These are three dimensional analyses of X-ray atomic coordinates. d) Soil and Rocks. e) Simulated objects.

Third Column - Fractal dimensions. Error bars usually indicate the standard deviation from a least-square fit to a log/log straight line. In a few cases it indicates the experimental error. D values in the range 1-2 refer to boundary line image analyses. Ranges of fractality for each of the methods are described in Section 2.2

Fourth Column - The Methods. See Section 2.2

Fifth Column - References. The starred references indicate experiments and computations performed in our laboratories. Unstarred references indicate our analyses of data published by other laboratories.

2.2 The Methods (Fourth Column)

Method 1 - Surface Area/Molecular Probe Size Relations [1,3,8,10,19]
    In this method, the surface is tiled with n moles/g of adsorbent molecules (indicated in the Table) with cross-sectional area σ (or radius r). Provided simple physisorption occurs, the material has a fractal surface if

$$n \propto \sigma^{-D/2} \tag{1a}$$

$$n \propto r^{-D} \tag{1b}$$

within a range which is at least $\sigma_{min} - \sigma_{max}$. Typically, the range obtained by this technique covers one decade on molecular-size scales (10-100 Å^2). If polymers are used, the range increases to four decades. See Ref. 10 for other forms of Eq. 1.

Method 2 - Surface Area/Particle Size Relations [8-10]
    The resolution analysis of Method 1 can be carried out alternatively by varying the size of the object (particle radius, R), and measuring the surface areas, n, with a fixed probe-molecule (indicated in the Table; usually N_2 by the BET method):

$$n \propto R^{D-3} \tag{2}$$

    An assumption is made that for most mechanically stable powders, the number of particles/g scales like R^{-3} (i.e., that the scale of surface fractality is much smaller than mass-fractality, if such exists). If the features of surface irregularities are at least of the size of the probe-molecule, σ_{min}, then the range of fractality is $\sigma_{min} < \sigma < \sigma_{min}(R_{max}/R_{min})^2$ [19]. Two to three decades are covered by this method.

Method 3 - Computerized Image Analysis [6, 14]
    The ultimate goal in such analyses is to have three-dimensional coordinates of the object to be analyzed (Method 4). This is not always possible, and therefore image analyses in material sciences are also carried out on lower grade information, i.e., analyses of boundary lines and of two-dimensional textures.
    There are several ways to determine the fractal dimension of a boundary line. We found that the following one is convenient: Count the number of yardsticks, y, of size p (in pixels) necessary for measuring the perimeter length:

$$y = kp^{-D} \qquad (3)$$

where k is a unit constant. Unlike Eq. 1b in which the yardsticks are placed at a distance r from the surface, the yardsticks, p, are placed on the line itself. One of the limitations of boundary line analysis is that it is carried out on a single object. Unlike the previous two methods in which a thorough statistical averaging occurs by virtue of the use of ~10^{20} yardsticks on 10^6 particles, such averaging does not exist in the image analysis techniques. We have shown that under such circumstances the log/log analysis of Eq. 3 leads to artifacts such as indicating fractality where such definitely does not exist. We have suggested two methods of analysis which are much more sensitive and overcome the log/log difficulty. The first is to plot dD/dp vs. p, in which the "local fractal dimension" is calculated for every three consecutive points. A fractal object should reveal a horizontal straight line. The second sensitive method is to analyze the increase in length as a function of yardstick size 1. If p_{i+1}/p_i is kept constant and if the object is fractal, then:

$$p_i - p_{i+1} = k'p_i^{-D} \qquad (4)$$

The applications and limitations of boundary line analyses are discussed in detail in Ref. 6. The range of fractality by this technique is about two orders of magnitude. The analysis is limited by the number of details and the resolution of a single picture. Larger ranges are obtainable in principle by taking pictures at various magnifications.

Texture analysis [14] is performed by converting the grey-levels of the texture into a three-dimensional rough surface, in which the roughness is created by the grey-level heights. If fractal, then the area, A, of that imaginary surface is

$$A(\epsilon) = F\epsilon^{2-D} \qquad (5)$$

where F is a constant and 2ϵ is the thickness of a "blanket" which covers the surface from the top and from the bottom. $A(\epsilon)$ is computed from $(V_\epsilon - V_{\epsilon-1})/2$ where V is the volume of the blanket.

Method 4 - Analysis of Atomic Coordinates of Proteins [7,12]
We have investigated the question whether the irregular surfaces of proteins (and other polymeric bio-molecules) can be characterized by applying fractal geometry. We found (Part C in the table) that, indeed, over a limited range (0.3-20 $Å^2$), fractality can be attributed to protein surfaces. The interpretation of this result is quite unclear: one expects that protein bio-activity, especially of enzymes, is dictated by highly specific, geometrical arrangements of the atoms. In spirit, the concept of fractal dimension is as far as possible from the idea of specific key/lock geometries. For instance, the active site of trypsine has a D value which is significantly higher than the D of the surface of the whole protein (2.80 ± 0.04 and 2.62 ± 0.2, respectively). Does this difference carry any useful information? We discuss it in detail in Ref. 12.

Protein analysis is the only case where full three-dimensional image analysis could be carried out. X-ray coordinates are available from databases. The apparent surface area was determined by counting how many spheres of radius r are necessary to cover it, similarly to Eq. 1b. Connoly's commercial computer program [21] was used for that purpose.

Method 5 - Electronic Energy Transfer Between Adsorbates [2,4,5]
A number of silicas with varying pore-size distributions were investigated by Förster-type, single step electronic energy transfer (EET) between an excited adsorbed donor (rhodamine B) to adsorbed acceptors (malachite green). It has long been known that EET processes behave

TABLE: SUMMARY OF FRACTAL ANALYSES

No.	Material	D	Method	Reference
a. Oxides				
1	Silica gel-60	2.97±0.02	(1) Spherical alcohols (C_{1-7})	[1]*
2	- " -	3.00±0.01	(2) N_2	[2]*
3	- " -	3.03±0.09	(2) Naphthalene	[1]*
4	- " -	2.96±0.04	(2) tert-amyl-alcohol	[1]*
5	- " -	3.04±0.05	(2) EtOH	[1]*
6	- " -	3.0 ±0.1	(5)	[2]*
7	- " -	3.0 ±0.1	(6)	[2]*
8	Silicic acid	2.94±0.04	(1) N_2, 5 alcohols	[3]
9	Silica gel-200	3.0 ±0.1	(2) N_2	[4]*
10	Silica gel-1000	3.02±0.14	(2) N_2	[4]*
11	- " -	3.0 ±0.2	(5)	[5]*
12	Silica gel-2500	2.0 ±0.2	(5)	[5]*
13	Silica gel-4000 (LiChrospher)	1.36±0.03	(3)	[6]*
14	Silica gel-5000	2.0 ±0.2	(5)	[5]*
15	Precipitated silica (Hi-Sil-243LD)	3.0 ±0.1	(2) N_2	[7]*
16	Aerosil-non porous fumed silica-Degussa	2.02±0.06	(2) N_2	[8]
17	"	2.0 ±0.2	(5)	[2]*
18	Snowit-ground fine Belgian quartz glass of high purity	2.15±0.06	(2) N_2	[9]
19	Snowit, after etching with HF	2.02±0.07	(2) N_2	[9]
20	Madagascar quartz from Thermal Syndicate	2.14±0.06	(2) N_2	[9]
21	Madasgascar quartz after etching with HF	1.84±0.27	(2) N_2	[9]
22	Quartz Min-U-Sil	2.13±0.13	(2) N_2	[7]
22a	Quartz	2.21±0.01	(7) HF	[9]
22b	Ottawa sand	2.02±0.01	(7) HF	[9]
23	Glass spheres	2.13±0.08	(8)	[7]
24	Crushed Corning 0100 lead glass	2.35±0.11	(2) Hexadecyl-1-^{14}C-trimethyl ammonium bromide	[8]
25	Activated alumina Grade F-20	2.79±0.03	(1) Polystyrene fractions	[10]
26	Alumina (Merck)	2.98±0.03	(2) Phenol	[7]*
27	- " -	3.0 ±0.1	(2) N_2	[7]*
28	- " -	3.0 ±0.1	(6)	[11]*
29	Periclase-electrically fused and crushed magnesite	1.95±0.04	(2) N_2	[9]
30	Synthetic Faujasite $(Na_2O \cdot Al_2O_3 \cdot 2,67 \cdot SiO_2 \cdot xH_2O)$	2.02±0.05	(1) N_2, n-alkanes $(C_{1,3-7})$	[8]
31	Porous α-FeOOH pigment for magnetic tapes	2.57±0.04	(1) N_2	[9]
b. Organic Materials				
32	Granular activated carbon Fujisawa B-CG, from coconut shell	2.80±0.16	(1) Various organic molecules	[7]
33	Granular activated Tsurumi HC-18, from coconut shell	2.71±0.14	(1) Various organic molecules	[7]

No.	Material	D	Method	Reference
34	Charcoal (BDH) from animal origin	2.78±0.21	(1) Styrene-methyl-methacrylate random copolymers	[8]
35	Porous coconut charcoal	2.67±0.16	(1) Nitrogen, ethane, ethylene acetylene, methane, butane, 1-butane,propane	[8]
36	Coal mine dust from Pennsylvania	2.52±0.07	(2) N_2	[9]
37	- " -	2.33±0.08	(2) N_2	[9]
38	Carbon black	2.25±0.09	(1) N_2, benzene, napthalene, anthracene, phenanthrene	[9]
39	Slightly porous coconut charcoal	2.54±0.12	(1) Nitrogen, ethane, ethylene, acetylene, methane, butane, i-butane, propane	[8]
40	- " -	2.30±0.07	(1) - " -	[8]
41	Coconut shell charcoal	2.60±0.20	(1) 13 various organic molecules	[7]
42	Coal fly ash from Corrette power plant	2.53±0.02	(2) N_2	[7]
43	Graphite-Vulcan 3G (2700)	2.07±0.01	(1) N_2, C_{22}, C_{28}, C_{32}	[8]
44	Graphon-partially graphitized carbon black formed by heating MPC black to 3200C	2.04±0.16	(1) N_2, MeOH, EtCl, Benzene cyclohexane	[8]
45	Graphon-graphitized carbon black (Cabot Corp.)	2.13±0.16	(1) Styrene-methyl-methylacrylate random copolymers	[8]
46	Active (non-porous) coconut charcoal	2.04±0.04	(1) N_2, ethane, ethylene, acetylene, methane, butane, i-butane, propane	[7]
47	- " -	1.97±0.02	(1) - " -	[8]
48	Western fly ash	2.12±0.06	(2) N_2	[7]
49	Carbon black aggregate	1.10±0.01	(3)	[6]*
50	Cellophane	2.27±0.04	(2) 4 alkanes (C_{7-10})	[7]
51	Cellulose	2.06±0.04	(2) Butanol, dioxane 3-alkanes (C_8, C_{10}, C_{12})	[7]
52	Sulfonamide drug-sulfisomezole	1.89±0.04	(7) Gastrointestinal fluid	[7]
53	Sulfonamide drug-sulfamethizole	2.07±0.04	(7) - " -	[7]

c. Surfaces of Proteins and Biomolecules

No.	Material	D	Method	Reference
54	Trypsine	2.62±0.02	(4)	[7,12]*
55	Trypsine-active site	2.80±0.04	(4)	[7,12]*
56	Trypsine inhibitor	2.34±0.03	(4)	[7,12]*
57	Lysozyme	2.51±0.02	(4)	[7,12]*
58	DNA (B)-cisplatin complex	2.05±0.04	(4)	[7,12]*

d. Soils and Rocks

No.	Material	D	Method	Reference
59	Goodland high calcium rock from Idabel, OK	2.97±0.01	(2) Kr	[9]
60	Upper Columbus dolomitic rock from Bellevue, Ohio	2.91±0.02	(2) Kr	[9]
61	- " -	2.15±0.10	(7) NH_4Cl	[7]
62	Carbonate rock from ground water test well, Nevada	2.90±0.01	(2) N_2	[9]

No.	Material	D	Method	Reference
63	Granitic rock from SHOAL nuclear test, Nevada	2.88±0.02	(2) N_2	[9]
64	Halimeda skeletal carbonate	3.02±0.07	(2) Kr	[7,13]
65	- " -	2.05±0.08	(7) Sea water	[7,13]
66	Igneous rock sample from SHOAL site	2.73±0.05	(2) Ar	[9]
67	Coral skeletal carbonate	3.02±0.07	(2) Kr	[7,13]
68	- " -	1.98±0.07	(7) Sea water	[7,13]
69	Echinoid skeletal carbonate	2.69±0.04	(2) Kr	[7,13]
70	- " -	2.15±0.07	(7) Sea water	[7,13]
71	Mosheim high calcium from Stephens City, VA	2.63±0.03	(2) Kr	[9]
72	Niagara (Guelph) Dolomite Woodville, OH	2.58±0.01	(2) Kr	[9]
73	- " -	2.07±0.06	(7) NH_4Cl	[7]
74	Glassy melted rock from Rainier nuclear zone, Nevada	2.46±0.11	(2) Ar	[9]
75	Hybla alkali feldspar	2.36±0.02	(2) Ar	[7]
76	Hybla alkali feldspar after etching with HCl	2.09±0.03	(2) Ar	[7]
77	Iceland spar, Massive calcite from Chihuahua, Mexico	2.16±0.04	(2) Kr	[9]
78	Iceland spar	2.04	(7) $Fe(ClO_4)_2$	[7]
79	Georgia Marble ($CaCO_3$)	2.09±0.01	(2) N_2	[7]
80	- " -	2.16±0.05	(7) Calcination	[7]
81	Fredonia Valley ($CaCO_3$)	2.13±0.06	(2) N_2	[7]
82	- " -	2.18±0.06	(7) Calcination	[7]
83	Soil (Kaolinite, trace Halloysite)	2.92±0.02	(2) Malachite green	[9]
84	Soil (mainly feldspar quartz and limonite)	2.29±0.06	(2) - " -	[9]
e. Simulated Objects				
85	Simulated floc (601 particles)	1.29±0.01	(3)	[6]*
86	- " -	1.28±0.01	(3)	[6]*

differently in three and two dimensions. Recently, this was generalized to any fractal dimension [22]. The survival probability, P(t), of the donor is given by

$$P(t) = \exp[-P_1(t/\tau)^{D/6} - t/\tau] \qquad (6)$$

where τ is the radiative life time of the donor and P_1 is a function of several system characteristics (critical Förster radius, acceptor surface coverage, etc.). The range of this analysis is determined by the donor-acceptor distances in which EET is detectable, i.e., from molecular scales up to ~100Å.

Method 6 - Small Angle X-Ray Scattering [2]
We use the method developed by Schmidt, et al. [23]. The results for two materials are given in the table (entries 7, 28). Other material analyses are in progress. It should be noted that silica-60 was comparatively examined by Methods 1, 2, 5, 6. The range of analysis here is similar to that of Method 5.

Method 7 - Analysis of Chemical Reactivity of Surfaces [7,13,17]
One of our recent important conclusions has been that the surface sub-fractal dimension, $\overline{D}$, as experienced by species reacting with or at a surface, is usually (but not always) lower than the geometric fractal dimension D: A screening effect operates. For many surface reactions, we observe:

$$\text{rate} \propto \text{exposed or reactive surface area} \propto R^{\overline{D}-3} \qquad (7)$$

It has been shown [13] that the empirical dissolution reaction order, m, is

$$m = (\overline{D} - 3)/(D - 3) . \qquad (8)$$

In the table, where Method 7 is used, the sub-fractal dimension is reported and the reactive chemical indicated. Range considerations are similar to those of Method 2. Simulations of these processes [13] are in progress.

Method 8 - Air Permeability [7]
The interstitial surface fractal dimension of particulate materials can be determined from air permeability measurements. As in the previous method, the D value refers to surface irregularity as experienced by flowing air.

$$M^3 t/(1-M)^2 \rho^2 d = R^{\overline{D}-3} \qquad (9)$$

where M is the interstitial porosity, t is flow time, ρ is powder density and d is flow path length. This approach is only at the exploratory stage and only one entry is given [23].

3. STUDIES ON FRACTAL SURFACES

3.1 Photophysical Studies
We made several interesting observations:

Intramolecular excimerization of 1,3-di(1-pyrenyl) propane indicated that the second layer of adsorbed 1-octanol on silica-60 already has the properties of bulk 1-octanol. The high D value (3.0) causes fast structural memory loss in multilayers [20].
The rate of dispersion of adsorbed, aggregated pyrene molecules on silicas with varying pore size distributions depends on this parameter and not on D. We concluded that, at very low coverages, diffusional pathways do not follow the geometric irregularity details of the surface [4].
Efficient dual fluorescence was obtained from adsorbed N,N-dimethylaminobenzonitrile on silica-60 indicating efficient solvation by the surface. This has been attributed again to the "volume-like" behavior of this D=3 surface[24].
EET and adsorption experiments give preliminary indications for a step-function relation between D and average pore size in silicas: a drop from D=3 to D=2 occurs between 1000-2000Å [4,5].

328

3.2 Conformational Aspects of Adsorbed Molecules

Fractal analysis enables the direct gain of information on the conformation of adsorbed molecules, a notorious poblem in surface science. This has been discussed in detail in previous papers, both for small molecules and for polymers [1,8,10]. In a recent study [15], which we regard as a major support for our fractal approach, the cross-sectional areas and adsorption conformations of adsorbed n-alkanoic acids on silica were determined by two completely independent approaches: fractal theory and a novel theory of molecular shape. The results are identical, both in interpretation (i.e., that the acids, unlike alcohols [1], lie flat on the surface) and in numerical values.

3.3 Difficulties with the Fractal Approach

None of the methods described above are free of difficulties: some are general, others refer to specific cases. These difficulties, described in our papers and those of other groups, need a detailed discussion which is beyond the scope of this brief report. I do, however, wish to draw attention to an alarming tendency which I find in recent literature: to regard irregular objects or phenomena describable by fractal geometry as "superior" to those cases where the fractal description does not fit. Because it is so easy to get a "good fit" to power laws, exponential laws, and straight log/log lines (especially for weak irregularities), one tends to find fractals everywhere. I am not certain that this is the case, but what remains solid is the notion that much useful information can be gained from an investigated object by performing a resolution analysis in general. The outcome may or may not be fractal and there is obviously no superiority to any of these outcomes.

ACKNOWLEDGEMENTS

The table was compiled by D. Farin, who is also responsible for the majority of the entries. The persons who made it all possible are listed below and I thank them all. This work was sponsored by the Israel Academy of Sciences and by the US-Israel Binational Fund, and was supported by the F. Haber Research Center for Molecular Dynamics.

REFERENCES

1. D. Farin, A. Volpert, and D. Avnir, J. Am. Chem. Soc. 107, 3368, 5319 (1986).
2. D. Rojanski, D. Huppert, H. B. Bale, X. Dacai, P. W. Schmidt, D. Farin, A. Seri-Levy and D. Avnir, Proc. 2nd Int. Conf. Photoactive Solids, H. Scher, ed., Plenum Press, 1986; idem, Phys. Rev. Lett., submitted, 1986.
3. D. Avnir and P. Pfeifer, Nouv. J. Chim., 7, 71 (1983).
4. E. Wellner, M. Ottolenghi, D. Avnir and D. Huppert, Langmuir, submitted, 1986.
5. D. Huppert, D. Rojanski and D. Avnir, manuscript in preparation.
6. D. Farin, S. Peleg, D. Yavin and D. Avnir, Langmuir, 1, 399 (1985).
7. D. Farin, Ph.d. Thesis, THe Hebrew University of Jerusalem, 1986.
8. D. Avnir, D. Farin and P. Pfeifer, J. Chem. Phys. 79, 3566 (1983)
9. idem, J. coll. Interface Sci., 103, 112 (1985).
10. idem, nature, 308, 261 (1984).
11. P. W. Schmidt, D. Avnir et al., unpublished.
12. A. Goldblum, D. Farin and D. Avnir, manuscript in preparation.
13. M. Silverberg, D. Farin, A. Ben-Shaul and D. Avnir, Ann. Israel Phys. Soc., 1986, in press (Proc. F³ Conference, Israel, January 1986, eds. R. Engleman and Z. Yaeger.)
14. S. Peleg, J. Naor, R. Hartley an D. Avnir, IEEE Trans. Pattern Anal. Machine Intel, PAMI-6, 518 (1984).
15. A. Meyer, D. Farin and D. Avnir, J. Am. Chem. Soc., submitted 1986.

16. D. Farin, D. Avnir and P. Pfeifer, Particulate Sci. Tech. 2, 27 (1984).
17. P. Pfeifer, D. Avnir and D. Farin, J. Stat. Phys., 36, 699 (1984); 39, 263 (1985).
18. P. Pfeifer and D. Avnir, Surface Sci., 569 (1983).
19. P. Pfeifer and D. Avnir, J. Chem. Phys., 79, 3558 (1983); 80, 4573 (1984).
20. D. Avnir, R. Busse, M. Ottolenghi, E. Wellner and K. Zachariasse, J. Phys. Chem., 89, 3521 (1985).
21. M. L. Connolly, J. Am. Chem. Soc. 107, 1118 91985).
22. J. Klafter and A. Blumen, J. Chem. Phys. 80, 875 (1984).
23. H. D. Bale and P. W. Schmidt, Phys. Rev. Lett., 53, 596 (1984).
24. A. Levy, D. Avnir and M. Ottolenghi, Chem. Phys. Lett. 121, 233 (1985).

STRUCTURAL OBSERVATIONS OF POROUS SILICA GELS

M. W. SHAFER, V. CASTANO, W. KRAKOW, R. A. FIGAT and G. C. RUBEN*
IBM Thomas J. Watson Research Center, P. O. Box 218, Yorktown Heights, New York 10598
*Dept. Biology, Dartmouth College, Hanover, New Hampshire 03755

ABSTRACT

Silica gels were prepared by two methods; from the alkoxides under cc ditions where moderate degrees of crosslinking are expected and from LUDOX - alkaline silicate sols from which the alkali was subsequently removed. By selective thermal treatment from 300 to 850°C, pore diameters (excluding micropores) in the range 35Å to 400Å are seen. The pore size increased with temperature while the surface area and total porosity remained essentially constant. Extensive high resolution TEM analyses showed both colloid and polymeric structures with particle sizes between 60Å and 250Å which decreased on heating. An intra particle structure, presumably micropores of 10Å to 30Å is also seen. By using a shadow cast Pt-C replica technique, stereoimages show the particles to have a filamentry type structure.

INTRODUCTION

Single oxide glasses such as silica can be prepared rather routinely from gels formed by a variety of techniques, i.e. hydrolysis of the alkoxides [1], double dispersion of silica fume [2], LUDOX-alkali silicate [3], etc. However, many of the properties of such glasses depend on the conditions of gelation and also how the resulting alcogels and hydrogels are converted to xerogels and subsequently to fully densified glass. As a result, much of the work aimed at developing the sol-gel process to one where reproducible shapes and sizes can be formed has focused on the drying and densification steps. From these studies, for "pure silica" for example, it is seen that in the temperature range ~300° to ~950° highly porous, reactive "silica gels" are formed. Pore sizes ranging from less than 2nm to several hundred nm with rather broad distributions as well as large surface areas have been measured in these gels [4]. Little of this previous work was primarily concerned with forming "glasses" with controlled well defined pore sizes. Thus, the intent of this paper is to show some results of an investigation aimed at producing porous silica glasses with well defined pore sizes which can be used as hosts to study physical phenomena and materials behavior in well characterized confined geometries. For example, a series of these controlled pore size glasses was recently used [5] to verify a recent dynamic theory [6] of capillary condensation and liquid behavior in porous media.

EXPERIMENTAL

Sample Preparation

Because the ultimate goal of this study was to prepare porous glasse· with a broad range of pore sizes, e.g. <20Å to >400Å diameter, but having a narrow distribution within each range, it was necessary to prepare and desiccate the gels under a variety of conditions.

332

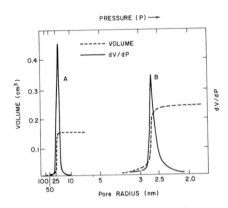

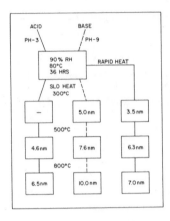

Fig. 1. Typical Hg porosimetry tracings for two samples; A, pore diameter of ~40nm, B~5.2nm.

Fig. 2. Flow diagram showing pore size dependence on processing conditions. Ethoxide, $H_2O/Si = 10$.

Two general methods were used: the hydrolysis of the alkoxides and the gelling of LUDOX with Kasil and formamide and the subsequent leaching of the alkali with acid. Basically, we used conditions, which according to the present lore, would produce a range of gels from essentially "polymeric" to strictly colloidal. Thus, our range of conditions were from acid catalyzed low H_2O/SiO_2 ratios to high PH high H_2O/SiO_2 ratios for the alkoxide derived gels. LUDOX gels were formed at different PH's and "Kasil" concentrations.*

In general, rather conventional techniques were used for the preparation and drying of the alkoxide gels while the LUDOX-Kasil gels were prepared by a modification of the technique described by Shoup [3]. The flow-diagram in Fig. 2 summarizes the procedure for one set of samples, i.e. an alkoxide $H_2O/SiO_2 = 10$. The gels were characterized in terms of their pore sizes, surface areas, density and in certain cases, their particle size. Mercury porosimetry was used extensively for pore diameter and surface area measurements. Some samples were characterized by gas adsorption/desorption isotherm analysis using a Quantasorb absorption unit. For high resolution work, a specially constructed low temperature system was used [7]. Extensive TEM analyses were also made. Two techniques were used; first, the conventional one of depositing powdered samples, suspended in water, on microscope grids and carbon coating. The other technique involved the direct TEM imaging of top and cross fractured surfaces which were replicated (45° angle) with 10-11Å of Pt/C at -178°C in 10^{-8} torr vacuum and backed with 100-145Å carbon film [8]. The replicas were separated from the sol-gel glasses with 49% aqueous HF, rinsed and picked up on grids. The microscopy equipment and conditions and the method for making reversal negatives for prints are described elsewhere [9].

---

*Space does not permit us to tabulate the details of preparation for all the samples studied. These data are available and will be given in a more detailed publication [7].

RESULTS

Figure 1 shows the mercury porosimetry data for two typical samples with different pore sizes. Sample B was prepared from the alkoxides and A from LUDOX-Kasil. In both cases a narrow pore distribution is seen. Because of the logarithmic type pore radius scale, the dv/dp peak of sample A appears sharper indicating a narrower pore distribution. This is not the case - B has a slightly narrower distribution. Although the mercury penetration method has frequently been questioned as to its reliability for pore diameters less than ~100Å, we found very good agreement between results obtained by it and the nitrogen adsorption isotherm method - for both pore size and surface areas. Thus, most of the results which follow are based on mercury porosimetry measurements.

In the process diagram, Fig. 2, we show the pore size dependence on PH and temperature for an alkoxide glass with a H_2O/SiO_2 ratio of 10:1. Note that larger pore sizes occur in samples prepared at higher PH and temperatures. Surprisingly, rapidly heated samples tend to have smaller pore sizes, and in most cases have narrower distributions than those heated slowly. The reason for this is not clear at this time. The increased pore diameter with temperature and with the H_2O/SiO_2 ratio is more evident in Fig. 3 which shows a series of glasses prepared at a PH of 8.5. The increase is rather linear until ~850°, at which point there is a rather sharp decrease due to densification. Likewise the temperature dependence of pore size for LUDOX-Kasil compositions is shown in Fig. 4. In this particular case LUDOX HS40 was used and the PH was adjusted to 10.4. Linear type increases are also seen here for LUDOX/Kasil ratios less than 70/30. At higher Kasil concentrations a sharp increase in pore diameter is seen above 500°C. The PH of gelation also has a significant effect on pore size, e.g. for a LUDOX/Kasil = 80/20 fired at 500°C pore size decreased from 90 to 55Å as the PH was lowered from 12 to 6.5. This is a result of a corresponding decrease in particle size, as confirmed by TEM measurements.

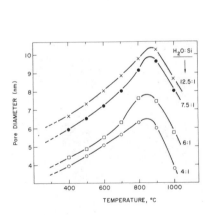
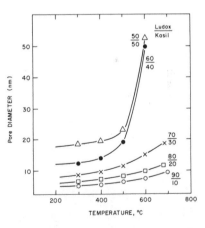

Fig. 3. Pore size diameter vs. firing temperature for alkoxide glasses formed at PH=8.5.

Fig. 4. Pore size vs. firing temperature for LUDOX/Kasil glasses.

Figure 5 shows how the total porosity (P) and surface area (S) vary with temperature - both for alkoxide and LUDOX based glasses. It should be noted, particularly for the alkoxide gels that P and S vary little with temperature until densification begins around 800-850° This is in contrast to the pore size which increases linearly in the same temperature range. This is not easily understood and similar previous studies [10,11,12] are contradictory in that they showed either no or slight increases in pore diameter with temperature. For the "colloidal type" gels, this behavior can possibly be explained by a combination of two effects. One is that at favorable inter-particle contact points that "necks" grow during initial stages of sintering and the overall structure changes from strictly a random sphere type to one which has some characteristics of an array of intersecting cylinders. Such a rearrangement could conceivably enlarge the inter-particle channels. An indication that this occurs is seen in Fig. 6, which shows the necking at 850° which is clearly absent at 600°C. The other effect is an overall reduction in particle size with temperature. This is seen by TEM in many different samples and is about 15% between 500° and 850°C. Both Figs. 7 and 8 clearly show particles of 180-200Å diameter while the starting LUDOX was measured to be ~220Å. A different explanation is necessary to explain the pore size dependence on temperature for the polymeric type gels since we see no evidence of individual particles. Presumably it involves a rearrangement and/or size reduction of the interpenetrating polymeric chains. It should be noted that there is a somewhat smaller pore size increase with temperature for the polymeric type than those showing particle morphology. Porosities between 70 and 80%, as measured by Hg intrusion, are quite common for gels heated in the temperature range 400-700°C. Slightly higher values for the total porosity are obtained by weighing well defined geometries (the triangles in Fig. 5), which is further evidence of micropores in the 10-30Å range which cannot be seen by the Hg intrusion. It should also be noted that for the LUDOX-Kasil gels, S and P have a greater temperature dependence and sintering occurs at a lower temperature.

Rather extensive characterization was done by TEM and some of the results are shown in Figs. 6-10. Although the particle diameter of LUDOX HS40 is listed as ~120Å, we found it, as received, to be consistently in the 210-220Å range. Figure 7 shows replicated

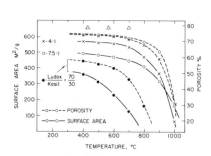

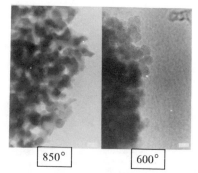

Fig. 5. Surface area (S) and porosity (P) vs. firing temperature for both alkoxide and LUDOX type glasses.

Fig. 6. TEM of LUDOX/Kasil=90/10 glasses fired at 600 and 850°. Mark=20nm.

surfaces of both colloidal and polymeric type glasses. The LUDOX/Kasil is obviously particle-like, having ~180-200Å diameter particles. The sample on the right prepared at PH=8.5 is also polymeric but appears to be more highly crosslinked than the sample prepared at PH=3 which is definitely polymeric. This is consistent with the generally accepted ideas that low PH tends to produce samples which are weakly crosslinked and more polymeric [1].

Figure 8 shows a high magnification (2 million times) replicated surface of a colloid type glass (LUDOX/Kasil = 95/5) heated to 700°C. Even at this magnification the spherical nature of the particulates is seen; their diameters being roughly 200Å. The substructure of the individual particles is definitely filamentry where the individual filaments are more or less randomly interwoven and even twisted in some cases. The filament widths are in the 2-10Å range although they appear larger in the photo because they were coated with a 10-11Å thick metallic film. This particular sample had a very sharp pore distribution at 96Å, as determined by BET and Hg porosimetry measurements, which is not seen here due to the high magnification but what is clearly seen are the micropores within the individual particles. These appear to be rather uniform and interconnected with most having diameters between 10-30Å. If one compares the substructure shown in Fig. 8 with that seen by regular TEM, e.g. the 600° sample in Fig. 6, the pore morphology is very similar but the diameters seen in the replicated samples are normally of the order of 30-40% larger. This is because in the regular TEM we are observing an image formed by transmission through an inhomogeneous array of particles and voids while in the replicated samples we are looking directly into the individual crevices. Thus, we believe that the microporosity seen in most gel derived glasses is in fact more in the 10-25Å range than the 5-10Å which is frequently reported by TEM measurements.

In Fig. 9, we show a TEM replica of a LUDOX/Kasil 85/15 sample heated to 600°C and cross fractured. This particular sample had a pore size distribution where 85% were 380±40Å as measured by Hg porosimetry. It is seen in both micrographs that the pore structure can be described as one containing a number of intersecting cylindrical type channels of rather uniform diameter.

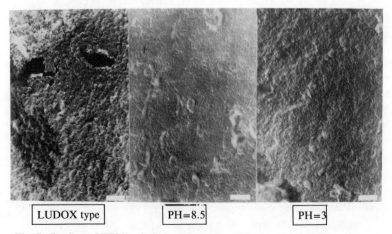

| LUDOX type | PH=8.5 | PH=3 |

Fig. 7. Replicated TEM's of glasses prepared at PH=3 and 8.5. Mark=100nm.

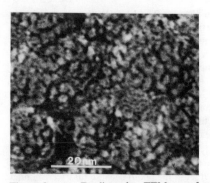

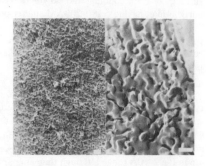

Fig. 8. Replicated TEM of LUDOX/Kasil=95/5; 700°C.

Fig. 9. Replicated TEM's of LUDOX/Kasil=85/15; 600°C. Right bar = 100nm; Left = 200nm.

In conclusion, we have shown that porous silica glasses with pore diameters ranging from less than 20Å to greater than 400Å can be prepared reproducibly by the proper selection of the chemical system, e.g. LUDOX, alkoxides, etc. by careful control of the preparative chemistry and by selective thermal cycling. The latter process has a larger effect than previously reported on the pore size and on the sharpness of the distribution. Our TEM measurements clearly show the detailed differences between the colloidal and polymeric type glasses. Well developed interwoven filaments with width of ~2-10Å are clearly seen and are thought to be the "structural entity" of the individual particles as well as the polymeric type material. Two types of pores are seen; the so-called micropores with diameters of between 10-30Å which occur between the interwoven filaments and the larger pores, which are the interstices between packed spheres and/or filaments. The latter are highly process dependent and can range from 20Å to many hundred angstroms. Their morphology is clearly seen in replicated TEM stereoimages.

REFERENCES

1. C. J. Brinker and G. W. Scherer, J. Non-Cryst. Solids **70** 301-332 (1985).

2. E. M. Rabinovich, J. B. MacChesney, D. W. Johnson, J. R. Simpson, B. W. Meagher, F.V. DiMarcello, D. L. Wood and E. A. Sigty, J. Non-Cryst. Solids **63** 155-161 (1984).

3. R. D. Shoup, J. Colloid and Interface Science **3** 63-69 (1976).

4. R. K. Iler, The Chemistry of Silica, John Wiley and Sons, New York (1979).

5. D. Awschalom, J. Warnock and M. W. Shafer and J. Warnock, D. Awschalom and M. W. Shafer (in press).

6. M. W. Cole and W. F. Saam, Phys. Rev. Lett. **32** 985 (1974).

7. M. W. Shafer, D. Awschalom and J. Warnock (to be published).

8. G. C. Ruben and K. A. Marx, J. Elect. Microsc. Tech. **1** 373-385 (1984).

9. G. C. Ruben and M. W. Shafer, Mat. Res. Soc. Symp. Proc. Palo Alto, CA, in press (1986).

10. Gallard-Hasid, H. Jones and B. Imelik, Bull. Soc. Fr. 1959, 608 (1959).

11. W. O. Milligan and H. H. Hachford, J. Phys. Colloid Chem. **51** 333 (1947).

12. R. A. Van Nordstrand, W. E. Kreger and H. E. Ries, Jr., J. Phys. Colloid Chem. **55** 641 (1951).

PACKINGS OF UNIFORM DISKS HAVING BOND
ORIENTATIONAL ORDER

George Y. Onoda
Thomas J. Watson Research Center, IBM, P.O.Box 218, Route 134,
Yorktown Heights, NY 10598

ABSTRACT

Packings of uniform disks with bond orientational order (BOO) are described. These are distinct from both ordered structures, that have translational correlations as well as BOO, and fully disordered structures that lack both features.

INTRODUCTION

Packings of uniform spheres (in 3-D) and disks (in 2-D) are generally assumed to have structures that are either ordered or random. An ordered structure can be described completely by a unit cell, many of which can be stacked together in an array to fill space. Given a set of particles at some location, the position of a particle far from this location can be predicted with certainty, assuming no defects exist. Ordered structures possess both translational and bond orientational long range correlations. In contrast, a random structure has no unit cell; the position of a particle far from a given group of particles can at best be described by a probability function (e.g., a radial distribution curve). Random structures lack both types of long range correlations (positional and orientational).

There exists another class of structural arrangements that is neither ordered nor random. These are structures that may lack long-range translational order but possess long-range bond orientational order (BOO). Structures of this class will be referred to as BOO structures. In this context, a bond is a line drawn between the centers of neighboring particles that are in contact with each other. In a BOO structure, only a given set of bond directions exists. Thus, given a small group of particles, the location of a neighboring particle to one in the group must lie along one of the possible set of directions. While ordered structures also have BOO, the BOO structures are distinguished by the absence of long-range translational order. BOO structures lack unit cells; consequently, the relative positions of individual particles cannot be predicted with certainty. There are, however, a finite set of discrete sites that have a finite probability of occupancy, while the probability is zero at all other sites.

The purpose of this paper is to present some very simple examples of BOO structures in two dimensions, using the packings of uniform disks to generate the structures. For this purpose, the structures are created by a simple computer algorithm, as described shortly. This algorithm, depending on the starting variables, creates a wide variety of structures that are useful for assessing the variable properties of BOO structures.

COMPUTER ALGORITHM

The algorithm [1,2] starts with a seed, a small cluster of three or more disks in a simple arrangement (Fig. 1). Disks are then added sequentially to the cluster at sites where the new disk touches at least two others without overlapping any existing disks. As there will be many sites of this kind along the perimeter of the growing cluster, another rule must be added which determines which of these sites is selected. For present purposes, this rule consists of selecting a site that is the closest possible to some specified origin, usually located within the original seed. If two or more possible sites exist that are equidistant to the origin, one of them is selected randomly.

The arrangement of disks in a packing can be visualized readily by drawing lines between the centers of touching disks. This procedure creates a two-dimensional graph consisting of a tiling of polygons that fill space. This graph representation will be used in some of the illustrations that follow. It should be remembered that the disks are centered at the vertices in this graph.

A trivial example of the results of this algorithm is when the seed consists of three disks in a triangular arrangement (Fig. 1a) and we select the closest sites to the center of the triangle when we add disks. The subsequent addition of disks results in the formation of an ordered, triangular packing (the densest possible structure) having long range translational and BOO correlations.

The various structures created by this general method are classified as being deterministic or indeterministic. If exactly the same structure is formed each time the computer program is run , the structure is defined as being deterministic. This occurs because there are no random choices in the sequence of adding disks that affect the structure. If, on the other hand, a different structure is formed each time the program is run, such structures are defined as being indeterministic. Various random choices during growth create numerous possible structures.

INDETERMINISTIC STRUCTURES

In Fig. 2 is seen a typical arrangement of disks initiated by a square arrangement of four disks (Fig. 1b) in which new disks are selected based on their being closest to the center of the square. We say "typical" because a different structure is formed each time the computer program is run. In all instances, a graph filled with equilateral triangles and squares is formed by drawing bond line between the centers of touching disks. While the arrangements of triangles and squares appear random, bond orientational order is always preserved. This case clearly illustrates a structure

339

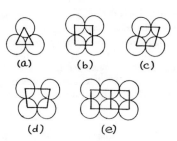

Figure 1. Examples of Seeds for Generating Structures: (a) triangle seed; (b) square seed; (c) rhombus seed; (d) distorted square seed, in the form of a trapezoid with one side longer than the other three; and (e) rectangle seed.

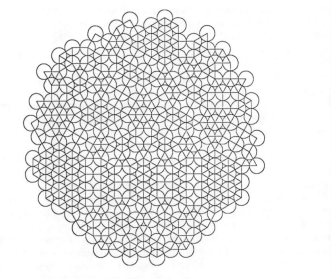

Figure 2. Structure Initiated with a Square Seed.

having BOO. That is, bonds are always in directions that are in this case multiples of 30°; however, the exact position of a disk at some distance away from the seed cannot be predicted with certainty.

A rhombus arrangement of disks (Fig. 1c) is actually a more general configuration that includes the square and the triangular arrangements as special cases. If ϕ is the acute angle of the rhombus, ϕ can vary from 90° (giving a square) to 60° (giving two triangles). A structure generated when $\phi = 75°$ is illustrated in Fig. 3. It is seen that the graph is filled with equilateral triangles and rhombi of the same type is the seed. Again, BOO is preserved but at angles that depend only on ϕ. As with the case with the square seed, a different arrangement is formed each time that the computer program is run.

As mentioned already, the structures shown in Figs. 2 and 3 are indeterministic; they vary each time that the computer program is run. The reason for this can be seen by considering the cluster in Fig. 4, which represents growth from a square seed after several disks have been added. The dots around the seeds represent some of the locations where the centers of new disks might be added (where they touch two existing disks). There are eight sites that are equidistant from the center of the square, grouped in pairs. For any pair, the distance between the sites is shorter than the diameter of a disk, so if one is occupied, the other cannot be occupied. Thus, either site of a pair is selected randomly. With four pairs and two possible sites, the next four disks can be added in 16 ways. At subsequent stages of growth, there is usually a number of random choices at any given time. This results in patterns that are different each time. The same argument can be made for rhombus-shaped seeds.

When one of the two equivalent sites of a pair is occupied, a new site is created which eventually fills, creating a square (or rhombus) instead of a triangle (Fig. 4b). This effect causes squares as well as triangles to form continuously as more disks are added to the cluster.

These structures, in addition to having BOO, also exhibit other notable features. Firstly, the rhombi(or squares) , when clustered, exist in larger blocks having overall shapes that are parallelograms, rather than in other more complex shapes that have more than four sides (e.g., L-shaped configurations). Examples of such parallelograms can be seen in Figs. 2 and 3). Secondly, the individual rhombi or blocks of rhombi are all joined at their corners to other rhombi or blocks of rhombi. The reason for these features can be understood by considering the polygons surrounding any vertex. There are only three general ways that rhombi and equilateral triangles can meet at a vertex. They can meet as six triangles ($6 \times 60° = 360°$), as four rhombi (two acute and two obtuse angles, which add up to 360°) or as two rhombi (an acute and an obtuse angle) with three triangles ($180° + 3 \times 60° = 360°$). No other combinations result in a sum of 360°. Returning now to the issue of the L-shaped cluster of blocks, there would exist a concave angle within the L, which can only be filled with one of the angles of the rhombus. Thus, all such concave angles are filled with rhombi until the cluster becomes a larger parallelogram.

341

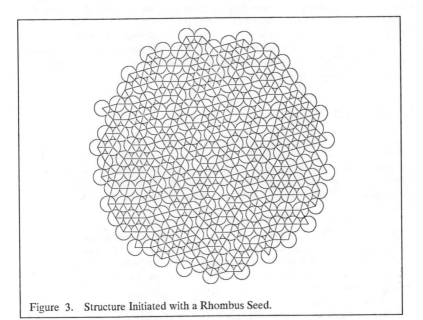

Figure 3. Structure Initiated with a Rhombus Seed.

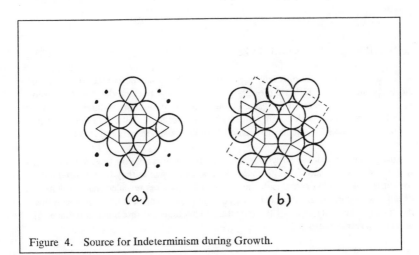

(a)   (b)

Figure 4. Source for Indeterminism during Growth.

Concerning the issue of clusters being joined at corners, this occurs because no cluster can be completely surrounded by triangles. If so, any vertex at the corner of the cluster would be surrounded by only one rhombus, which is impossible. It must have be surrounded by two or four rhombi. If the attached rhombi shared a face with the cluster, it forms an L shape cluster; the cluster would have to grow to a larger parallelogram. Thus, clusters must be joined at corners so that two rhombi meet at their common vertex without sharing a face.

DETERMINISTIC STRUCTURES

Deterministic structures form when situations are created such that equivalent, overlapping choices are not allowed to arise. This is accomplished by breaking the symmetry of the seed relative to the origin. One way to do this is to move the center, displacing it by the smallest amount recognizable by the computer. When this is done, similar structures to Figs. 2 and 3 are formed, but now the same structures are formed each time.

Another way to break the symmetry is to alter the symmetry of the seed. For example, a seed can be distorted very slightly such that two of the disks are not quite touching (Fig. 1d). With this change for a square seed, the structure shown in Fig. 5 is formed. This structure is deterministic and also exhibits mirror symmetry around the x and y axis.

A third way to break the symmetry is to start with seeds with more than four disks, arranged in an asymmetric configuration. For example, we could start with two squares sharing an edge (a 2 by 1 array of squares), as illustrated in Fig. 1e, and locate the center at the center of the rectangle. This creates the packing in Fig. 6, which is similar to Fig. 5 in that the structure has mirror symmetries around the x and y axes.

DENSITIES VERSUS AGGREGATE SIZE

A examination of the different aggregates suggests that the defects (rhombi) in chaotic structures are rather uniformly distributed. The packing density, expressed as an area fraction covered by disks, of the structure containing squares (Fig. 2) is found to remain near 0.85 during growth [2]. This compares with the value of 0.907 for the close-packed triangular lattice and 0.785 for a square array of disks.

On the other hand, the structures with deterministic symmetry (e.g., Fig.6) seem to have islands of triangular regions that increase with size as they form further from the center. A detailed analyses of these types of structures revealed that the density of defects (squares or rhombi) decreases during growth with a fractal dimension that depends on the acute angle of the rhombus. This aspect is discussed in detail elsewhere in a recent paper [3].

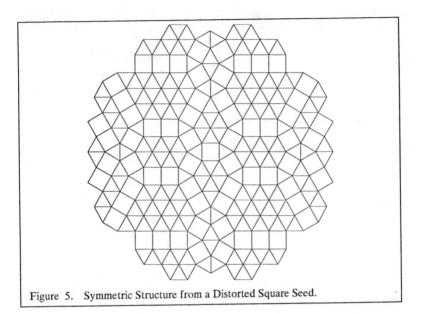

Figure 5. Symmetric Structure from a Distorted Square Seed.

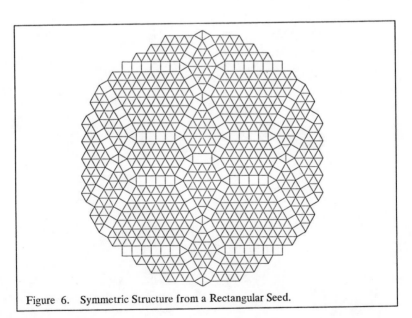

Figure 6. Symmetric Structure from a Rectangular Seed.

344

ACKNOWLEDGEMENTS

The author wishes to thank J. Toner, R. Cook, D. Clarke, T. Shaw, B. Stephenson, C. Thompson, and P. Horn for many stimulating discussions and suggestions.

REFERENCES

1. M. Rubinstein and D. R. Nelson, Phys. Rev. B 26, 6254 (1982).

2. L. D. Lindsay, Bureau of Mines Bulletin 658 (1971).

3. G. Y. Onoda and J. Toner, submitted to Phys. Rev. Lett.

PARTICLE COMPACTION WITH ALTERNATING ELECTRIC FIELDS:
THE EFFECTS OF ELECTRODE GEOMETRY

ALAN J. HURD
Sandia National Laboratories, Albuquerque, NM 87185

ABSTRACT

A technique for inducing ordered, close-packed arrangements of various symmetries among colloidal particles is discussed. An external alternating electric field applied to the colloid induces dipole interactions of variable strength by polarizing either the dielectric material of the particles or their electrostatic double layers. Ordering in various symmetries can be obtained by switching the field rapidly between pairs of electrodes, thereby changing the orientation of the induced dipoles. A small dc bias serves to deposit and compact the aligned particles.

INTRODUCTION

There is a need to control particle packing in ceramic "green bodies" in order to enhance their sinterability and the ultimate strength of the resulting ceramic.[1] Defects in this packing, such as aggregates, tend to shrink during sintering at different rates from ordered regions. Large defects and packing inhomogeneities have been implicated with reduced strength and failure of the final part. Also, since the sintering rate increases with the average number of nearest neighbor contacts, lower temperatures are required to sinter ordered arrays than disordered arrays. These considerations have lead to the proposal[1] that a perfect, closest packed arrangement of monodisperse particles is the ideal starting point for obtaining the highest strength at the lowest cost.

High purity submicron ceramic precursors are generally formed in solution. Typically, high packing density is achieved by drawing out the solvent and applying high pressure on the molded part. There is little free energy to gain to transform from random close-packing to closest packing since the free energy is proportional to the packing fraction, which merely changes from 0.64 to 0.74. It would be desirable to turn on a controlled, anisotropic interaction between particles that enhances the ordering by long range forces rather than relying on short range isotropic steric repulsion.

In an aqueous environment, many colloidal particles are charged and the stability of a suspension against spontaneous aggregation, which is quite important to the rheology of a slurry and therefore its handling requirements, is often dominated by electrical repulsion. Manipulation of this dominant interaction can be achieved by adjusting the chemical environment of the particles, eg. with pH or ionic strength, however no chemical change can be made to vary the attractive (Van der Waals) interaction or the directionality of the electrical interactions.

The application of external ac electric fields allows such manipulations through the voltage, direction and frequency of the field. The basic effect is to induce dipoles on the particles along the field direction, providing an attraction longitudinally and a repulsion transversely.[2] That is to say, the particles become like little bar magnets. At low particle concentrations and sufficiently high field voltages, the result of this interaction is to form chains of particles that are touching; this demonstrates that the interaction is larger than both the thermal energy and the repulsion energy of the charged particles with no applied field (the Debye-Huckel term). At these "interesting" field strengths the voltage drop across a particle is comparable to the surface potential of the charged particles. When the field is turned off, the chains break up and the particles diffuse away. At high particle concentrations, the chains interact side-to-side (as well as continuing to interact end-to-end to form longer chains) in such a way as to form closest packed, two-dimensional arrays. The transverse repulsion between individual particles is partly negated when the particles reside in chains because the dipoles are saturated by interactions with their longitudinal nearest neighbors. The remaining higher order terms in the interaction between chains give rise to an attraction that has the nice feature that it aligns the chains to come into contact with the correct registry to form closest packed arrays.

Removal of the field, without first forcing the particles to freeze into position some way, allows the array to come apart. Arrays have been observed literally to fly apart from their stored energy. Further discussion of the interactions and appropriate references to the physics of the interactions are found in Ref. 2.

Several practical problems arise in applying this technique to compacting ceramic green structures. Perhaps it is desired to mold oddly shaped objects or alter the packing density in some controlled way. Finally, it is necessary to "solidify" the array to prevent it from coming apart. These two issues are addressed here.

IMPORTANT TIME SCALES

Ordering in more than one direction at once can be accomplished by taking advantage of the widely separated time scales of the interaction and of the particle diffusion. Fortunately the colloidal dipoles can be induced in any direction within a reponse time that is very short compared to the time it takes the particles to diffuse far enough to destroy any order that has developed. For close-packed order, this distance is roughly the particle radius. By rapidly switching the direction of the ac field, then, some degree of aligned-dipole order can be induced along different directions. If the preferred structures for each direction "fit" together, they will be enhanced on each switch of the field's direction; if not, they will be "frustrated" in some way. If T is the switching period, t is the period of the ac field applied between switches of the direction, τ is the response time for the dipole moments to build up after the application of a field, and a is the radius of the particles with diffusion constant D, then the criterion for ordering in multiple directions is

$$\tau \ll t \ll T \ll a^2/D \qquad (1)$$

If the mechanism of dipole induction is the polarization of the electrostatic double layer, then $\tau \sim 10^{-6}$ s.[2] In this case, τ is limited only by the mobility of the counterions in the double layer. Another mechanism of dipole induction is through the polarization of the dielectric material of the particle against a background of solvent dielectric with a different polarizability; this is an even more rapid process.

For 1 μm spheres in water, $a^2/D \sim 10^{-1}$ s characterizes the time to diffuse one radius. Thus the field frequency t^{-1} and switching frequency T^{-1} can be set in a large window of $10 - 10^6$ Hz.

EXPERIMENTAL

The details of the apparatus have been described previously.[2] The colloid is confined in a thin gap of 10 μm between glass plates, one of which has transparent, conductive electrodes on it in contact with the fluid. Voltages are applied between electrodes while the particles are observed by an optical microscope (Fig. 1). In the present work, several electrode tips converge onto an open area of 200 - 500 μm extent. Electrodes with four-fold symmetry are shown in Fig. 2a. Concentrated latex particles (2a = 1.053 μm and 0.412 μm @ 7% weight fraction) were placed in the thin gap and computer-controlled voltages (±15 v max.) were applied with a typical ac frequency of 10 kHz and a switching frequency of 1 kHz. A small (0.1 v) dc voltage was set up in some cases to deposit the particles on one or two electrodes.

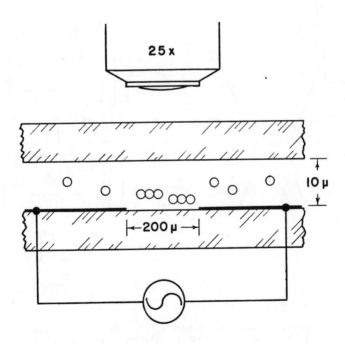

Figure 1. Schematic of thin-film sample and electrodes.

348

OBSERVATIONS AND DISCUSSION

A search was made for an efficient sequence of voltages in order to bring about the best order in the array. An acceptable sequence is illustrated in Fig. 2b for the four-fold case. Opposing electrodes were activated together in order to produce dipole chains running diagonally.

Ideally, two orthogonal sets of straight chains would "fit" to make a square lattice in two dimensions. The pointed electrodes did not seem to be ideal for this, however, because the field lines, made visible by the dipole chains, tended to be bowed from tip to tip (Fig. 2a). Except for the very center of the pattern, the field did not appear to have the desired symmetry; however, weakly formed chains did form in orthogonal directions at the center.

Similar experiments were performed with three-fold symmetry using a pair of electrodes operating at equal voltages as the return path for the third. Again, the expected order occurred only at the center.

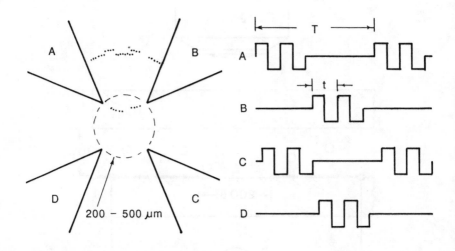

Figure 2. Scheme for four-fold symmetry. (a) Electrode geometry showing field lines followed by dipole chains. (b) Applied voltages.

With a dc bias, the particles were deposited in closest packed arrays at one or two electrodes, as shown in Fig. 3. Individual chains are not visible in the photograph owing to blurring by Brownian motion, but through the microscope they could be easily seen radiating out from the curved crystal as they deposited. When the bias voltage was too large, these chains collapsed into haphazard amorphous packings, but, with a small voltage, their alignment was preserved. Low-angle grain boundaries formed to accomodate the different chain-alignment directions going around an electrode; these boundaries can be seen in Fig. 3 as lines of dark dots between uniform crystalline areas. Bragg reflection of the incident light demonstrated that the grains had adopted similar orientations. Striated grains, also seen in Fig. 3, occasionally formed. These grains have highly visible surface dislocations (at the glass-colloid interface) that follow a given crystal plane. The crystal planes can be seen to curve, following the local field direction.

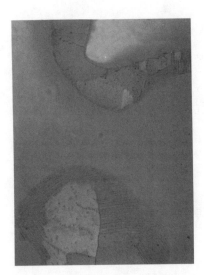 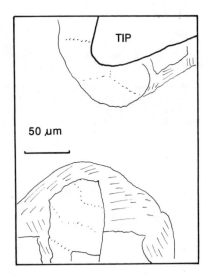

Figure 3. Structures formed in ac field with small dc bias. The tip of one of the four electrodes is visible at the top and a large mass of compacted, crystalline arrays is growing from the bottom. Striated grains indicate defects in packing near the cell walls and dotted lines indicate low-angle grain boundaries. Although the particles are too small to resolve in the arrays, Bragg scattering colors showed qualitatively that the neighboring grains share a similar alignment.

FUTURE DIRECTIONS

As mentioned in Ref. 2, an electrohydrodynamic instability creates problems with maintaining chain alignment. This instability tends to form eddies in the plane of the sample, breaking up chains as they try to lengthen. When the sample is very thin (5 μm or less) or the field strength is high, the instability is suppressed. Further understanding of this problem is necessary before it will be possible to scale up the experiments to make macroscopic, three-dimensional objects. Improved electrode design and higher voltages will help to make larger areas with the desired field symmetry. Interesting "frustrated" packings that don't fill space are expected for five-fold symmetries. Future studies will address these issues.

ACKNOWLEDGMENTS

This work evolved from studies performed at Brandeis University, under the direction of Robert B. Meyer whose helpful comments contributed greatly. I would like to thank Seth Fraden at Brandeis, who built the first version of a multiple-electrode switching device, for numerous fruitful discussions. Experiments were performed at Sandia National Laboratories, Albuquerque, NM and supported by the U.S. Department of Energy under Contract NO. DE-AC-04-76DP00789.

REFERENCES

1. H.K. Bowen, Mat. Res. Soc. Symp. Proc. Vol. 24, 1 (1984); Materials Science and Engineering 44, 1 (1980).

2. A.J. Hurd, S. Fraden and R.B. Meyer, in Proceeding of the Second International Conference on Ultrastructure Processing of Ceramics, Glasses, and Composites, edited by L.L. Hench and D.R. Uhlrich (Palm Coast, Florida, 1985).

FRACTAL PORES OR A DISTRIBUTION OF PORE SIZES:
ALTERNATIVE INTERPRETATIONS OF POWER-LAW SMALL-ANGLE SCATTERING*

PAUL W. SCHMIDT
Physics Department, University of Missouri, Columbia, MO 65203 USA

ABSTRACT

The intensity $I(q)$ of the small-angle x-ray or neutron scattering has been calculated for a system of randomly oriented, independently scattering pores with a number distribution of pore diameters which has the form of a power law. As has already been shown, [P. W. Schmidt, J. Appl. Cryst. 15, 567-569 (1982)], when the number distribution of the maximum diameters a of the pores is proportional to $a^{-\gamma}$, $I(q)$ is proportional to $q^{-(7-\gamma)}$, where $q = 4\pi\lambda^{-1}\sin(\theta/2)$, θ is the scattering angle, and λ is the wavelength. The coefficient of the power-law intensity has been expressed in terms of some of the constants which determine the diameter distribution. Equations have been obtained for the scattered intensity $I(q)$ at q values larger and smaller than those at which power-law scattering occurs. The intensity scattered by this system is compared with the intensity from a system of pores with fractal pore-boundary surfaces which have a fractal dimension D.

INTRODUCTION

The small-angle x-ray and neutron scattering from many porous materials is often [1--5] proportional to a negative power of the quantity $q = 4\pi\lambda^{-1}\sin(\theta/2)$, where λ is the scattered wavelength, and θ is the scattering angle. There are two ways to explain this power-law scattering. Recently Harold Bale and I found [5] that the intensity of the scattering from a system of pores with an average maximum diameter $\bar{a}$ was proportional to $q^{-(6-D)}$ when $q\bar{a} \gg 1$ and the pore boundaries are fractal surfaces which have a fractal dimension D. (The maximum diameter a, which is defined to be the length of the longest line segment which will fit inside the pore, is the quantity which I will use as a measure of the size of a pore.) The small-angle scattering also obeys a power law for a polydisperse system (i.e., a system with a distribution of maximum diameters) of smooth-walled, randomly-oriented, independent scatterers [6] when the number distribution of the maximum diameters of the pores is proportional to $a^{-\gamma}$. The scattered intensity for this system is proportional to $q^{-(7-\gamma)}$. According to Pfeifer and Avnir [7], a system of pores with fractal boundary surfaces which have a fractal dimension D can be considered to be equivalent to a polydisperse system composed of pores with smooth walls when the pore-diameter <u>number</u> distribution function is proportional to $a^{-(1+D)}$. This power-law diameter distribution gives [6] a scattered intensity proportional to $q^{-(6-D)}$, just as in Bale's and my calculation of the scattering from pores with fractal boundary surfaces. The same exponent for the power-law small-angle scattering thus is obtained both for porous systems with fractal pore boundaries and for the equivalent polydisperse system of smooth-walled pores.

I have recently extended my calculations of the small-angle scattering from a polydisperse system of smooth-walled pores, in order to compare the scattering from these systems with that from the equivalent system of pores with fractal boundary surfaces. In these studies, I felt that two subjects were of special interest. First, I wanted to see how the quantities which determine the power-law maximum-diameter distribution are related to the properties of the corresponding system of pores with fractal boundary surfaces. I also wanted to know how the scattered intensity is affected when the number distribution of the maximum pore diameters is a power law only in the interval $a_1 < a < a_2$, rather than for all a.

The results of the calculation are summarized below. A complete description will be published later.

II. Some Properties of the Scattered Intensity

The scattering sample is assumed to be a material which has a uniform density δ and which contains smooth-walled pores. (The atomic-scale structure can be neglected, since it affects the scattering only at angles too large to be studied by small-angle scattering [8].) The distribution function $\rho(a)$ of the maximum diameters is defined so that $\rho(a)\,da$ is the probability that a pore has a maximum diameter between a and $a + da$. For $a_1 < a < a_2$ the distribution function will be written

$$\rho(a) = \rho_0(a),$$

where

$$\rho_0(a) = N_a D \, (a_1)^D \, a^{-(1+D)} \, , \tag{1}$$

$$N_a = \frac{N_0}{1 - (a_1/a_2)^D} \, ,$$

and N_0 is the number of pores in the sample which have maximum diameters in the interval $a_1 < a < a_2$. Since $2 < D < 3$ for fractal pore boundaries, and because $a_1 \ll a_2$ whenever the power-law diameter distribution is useful for analyzing the scattering data, N_a will rarely differ appreciably from N_0.

The distribution is normalized so that

$$\int_0^\infty \rho(a)\ da = N_0 + N_1 + N_2 . \tag{2}$$

In (2), N_2 is the number of pores with maximum diameters not smaller than a_2, and N_1 is the number of pores which have maximum diameters not greater than a_1. The pores are assumed to be randomly oriented and to scatter independently, so that the total scattered intensity $I(q)$ is the sum of the scattering from the individual pores, averaged over all orientations of the pores. All pores are considered to have the same shape. That is, the maximum diameter is assumed to be sufficient to specify the boundary surface of the pore and thus to give all information necessary for calculating the scattering from this pore.

For a polydisperse system [6],

$$I(q) = \int_0^\infty \rho(a)\ i(q,a)\ da, \tag{3}$$

where $i(q,a)$, the intensity scattered by a pore with maximum diameter a, is given by the equation [9]

$$i(q,a) = 4\pi I_e \delta^2 V_0 a^3 \int_0^a r^2 g(r/a) \frac{\sin qr}{qr}\ dr .$$

In this expression, I_e is the intensity scattered by one electron, δ is the electron density of the material in which the pores are located, the volume of the pore is $V_0 a^3$, and $g(r/a)$ is the pair-correlation function [10] of the pore. [Instead of using the conventional notation $g(r)$, I have written the pair correlation function as $g(r/a)$, in order to emphasize the dependence of this function on the maximum diameter a.] By a change of the variable of integration, $i(q,a)$ can be expressed

$$i(q,a) = 4\pi I_e \delta^2 a^6 V_0 \int_0^1 t^2 g(t) \frac{\sin qat}{qat}\ dt . \tag{4}$$

For $D > 2$, by substitution of (11) in (5), $I(q)$ can be written

$$I(q) = I_0(q) + I_1(q) + I_2(q), \tag{5}$$

where

$$I_0(q) = 2\pi^2 I_e \delta^2 F_0(D) \frac{N_a a_1^D}{q^{6-D}} , \tag{6}$$

$$F_0(D) = - \frac{D(4-D)\ V_0 \int_0^1 t^{D-3}\ g^{(1)}(t)\ dt}{\Gamma(D-2)\ \{\sin[\frac{\pi}{2}(D-2)]\}} , \tag{7}$$

$$I_1(q) = \int_0^{a_1} [\rho(a) - \rho_0(a)]\ i(q,a)\ da , \tag{8}$$

and

$$I_2(q) = \int_{a_2}^\infty [\rho(a) - \rho_0(a)]\ i(q,a)\ da . \tag{9}$$

Although Equations (1)--(9) are valid for $2 < D < 6$, the only meaningful values of D for fractal pores lie in the interval $2 \leqslant D < 3$.

DISCUSSION

At q values for which $I(q)$ can be approximated by a power law, $I_1(q)$ and $I_2(q)$ are negligible compared to $I_0(q)$. As q becomes larger, $I_0(q)$ decreases, while $I_1(q)$ varies slowly, if at all, provided that, as I will assume to be the case, a_1 is small enough that $I_1(q)$ does not differ greatly in magnitude from $I_1(0)$. Then $I_1(0)$ can serve as a convenient estimate of the magnitude of $I_1(q)$ for all q. Power-law scattering breaks down when q has become so large that $I_0(q)$ and $I_1(0)$ are of the same magnitude. After q has increased until $I_0(q)$ has become negligible relative to $I_1(q)$, $I(q) \simeq I_1(q)$ if $I_1(q)$ is positive. When $I_1(q) < 0$, $I(q)$ decreases more rapidly than $I_0(q)$ for large q. Although I know of no reason why $I_1(q)$ or $I_1(0)$ cannot be negative, $I_1(q)$ is positive in all scattering curves where it can be identified in References 1--5.

I would like to mention two ways in which the scattering from polydisperse systems of pores differs from that for fractal pores. First, Equations (6)--(10) cannot be used when $D = 2$ (that is, when the corresponding fractal pore boundaries are smooth). Second, the coefficient of the term proportional to $q^{-(6-D)}$ in Equation (8) of Ref. 5 becomes zero when $D = 3$. This equation therefore cannot be used to analyze the scattering data from samples for which the fractal pore surfaces are so rough and convoluted that the fractal dimension approaches 3. However, there is no rapid change in $I_0(q)$ as D approaches 3 in (6). As yet I am unable to explain this difference. Some of my recent unpublished calculations suggest that the power-law scattering is proportional to q^{-4} when $D = 3$. Perhaps when $D = 3$, the intensity from the polydisperse system is very nearly equal to $I_2(q)$ and thus is positive and proportional to q^{-4} when $qa_2 \gg 1$.

The constant of proportionality in equation (6), which gives the power-law scattering, is

$$2\pi^2 I_e \delta^2 F_0(D) \; N_a a_1^{\;D}.$$

The value of I_e can be determined in a scattering experiment [11,12], and δ can be evaluated from the chemical composition and mass density of the solid material in the sample. The fractal dimension D can be obtained from the slope of the straight line in a plot of the logarithm of $I(q)$ as a function of the logarithm of q. If enough is known about the material to suggest a reasonable pore shape, $g(t)$ can be calculated [13,14], and

$F_0(D)$ can be evaluated from (7). Then $F_0(D)$, I_e, δ, and the scattering data can be used to find the product $N_a a_1^D$. Multiplication of this product by D gives the constant of proportionality in the power-law maximum-diameter distribution (1). Additional information, however, is required to obtain N_a and a_1 from $N_a a_1^D$.

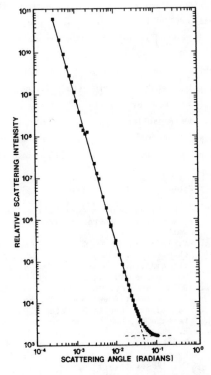

Fig. 1. The scattered intensity from Beulah lignite coal. [Reproduced from Reference 4 by permission of the American Chemical Society]

Some of the ideas which I have discussed can be applied to the small-angle scattering curve used in Ref. 3 to illustrate the scattering from the fractal pores in Beulah lignite coal, for which [5] D = 2.56 ± 0.03. Because the intensity in the outer part of the curve seems to approach a constant value, I assume that $I_1(q) = I_1(0)$. The dotted lines, which show the extrapolations of $I_0(q)$ and $I_1(0)$, meet at a scattering angle of about .045 radians, corresponding to the q value q_1 at which $I_1(0) = I_0(q)$. It can be shown that

$$q_1 a_1 = \left\{ F_1(D) \left[\frac{(6-D)\phi_1 N_1}{D N_a} - 1 \right] \right\}^{-1/(6-D)},$$

where

and

$$F_1(D) = \frac{2\Gamma(D-2)\,\sin[\frac{\pi}{2}(D-2)]\,\int_0^\infty t^3 g^{(1)}(t)\,dt}{3\pi(4-D)(6-D)\int_0^\infty t^{D-3} g^{(1)}(t)\,dt},$$

356

$$\phi_1 N_1 = \phi_1 \int_0^{a_1} \rho(a)\, da = a_1^{-6} \int_0^{a_1} a^6 \rho(a)\, da \ .$$

From the definitions of ϕ_1 and N_1, $0 \leqslant \phi_1 \leqslant 1$. Since $6 - D$ is between 3 and 4, a rough but useful approximation is simply to assume that $q_1 a_1 = 1$. Then $a_1 = 5$ Å. According to this estimate, the power-law maximum-diameter distribution extends almost to atomic dimensions. As I_e was not evaluated for this sample, $N_a a_1^D$, which is proportional to the constant N_0 in Ref. 5, cannot be calculated.

When $I(q)$ is described by a power law at the smallest angles for which intensity data are available, as was true for most of the porous samples which I have studied, no information can be obtained about $I_2(q)$. However, in the relatively few cases where $I_2(q)$ was appreciable, it was negative. Just as is true for $I_1(q)$, however, there is no reason that $I_2(q)$ must always have the same sign. When $I_2(q)$ is positive, the power-law exponent in $I(q)$ will change from $-(6-D)$ to -4 as q decreases.

REFERENCES

* Acknowledgement is made to the Donors of the Petroleum Research Fund, administered by the American Chemical Society, for support of this work.
1. M. Kalliat, C. Y. Kwak, and P. W. Schmidt, in New Approaches in Coal Chemistry, edited by B. D. Blaustein, B. C. Bockrath, and S. Friedman, ACS Symposium Series No. 169 (American Chemical Society, Washington, D. C., 1981), pp. 3-22.
2. P. W. Schmidt, M. Kalliat, and B. E. Cutter, Preprints, Div. Fuel Chem., Amer. Chem. Soc., 29, No. 2, pp. 154-159 (1984).
3. H. Kaiser, and J. S. Gethner, in Proceedings of the 1983 Intl. Conference on Coal Science, Pittsburgh, Pa. August, 1983, pp. 301-303.
4. H. D. Bale., M. D. Carlson, M. Kalliat, C. Y. Kwak, and P. W. Schmidt, in The Chemistry of Low-Rank Coals, H. H. Schobert, ed., ACS Symposium Series No. 264 (American Chemical Society, Washington, D. C., 1984), pp. 79-94.
5. H. D. Bale and P. W. Schmidt, Phys. Rev. Lett. 53, 596-9 (1984).
6. P. W. Schmidt, J. Appl. Cryst. 15, 567-569 (1982).
7. P. Pfeifer and D. Avnir, J. Chem. Phys. 79, 3558-3565 (1983).
8. A. Guinier, G. Fournet, C. B. Walker, and K. L. Yudowitch, Small-Angle Scattering of X-Rays, (Wiley, New York, 1955), pp 3--4.
9. Ref. 8, p. 7 and Eq. 21, p. 12. (As is explained on pages 38-40 of Ref. 8, according to the small-angle scattering analogue of the Babinet principle, properties of the correlation function g(r/a) for particles of uniform density will also apply to pores of the same shape.)
10. Ref. 8, pp. 12-16.
11. R. W. Hendricks, J. Appl. Cryst. 5, 315-24 (1972).
12. Ref. 1, p. 13.
13. O. Glatter and O. Kratky, Small-Angle X-Ray Scattering, (Academic Press, New York, 1982), Chapter 5.
14. A. Miller and P. W. Schmidt, J. Math. Phys. 3, 92-96 (1962).

PART VI

Non-Oxides

STUDIES OF ORGANOMETALLIC PRECURSORS TO ALUMINUM NITRIDE

LEONARD V. INTERRANTE, LESLIE E. CARPENTER II, CHRISTOPHER WHITMARSH, and
WEI LEE, Dept. of Chemistry, Rensselaer Polytechnic Institute, Troy, NY
12180-3590; MARY GARBAUSKAS and GLEN A. SLACK, General Electric Corporate
Research and Development, P.O. Box 8, Schenectady, NY 12301.

ABSTRACT

The reaction of trialkylaluminum compounds with ammonia has been
examined as a potential route to high purity AlN powder and to AlN thin
films. This reaction proceeds in stages in which the initially formed Lewis
acid/base adduct undergoes thermal decomposition to a series of intermediate
alkylaluminum-amide and -imide species with increasing Al-N bonding, i.e.,

$$R_3Al + NH_3 \rightarrow R_3Al:NH_3 \rightarrow \rightarrow \rightarrow AlN + 3RH$$
$$(\text{where } R = CH_3, C_2H_5, C_4H_9, \text{ etc.})$$

The structure and properties of several of these species have been
studied using various physical and chemical methods, leading to a better
understanding of the chemistry of this novel AlN precursor system. The
structure of the intermediate organoaluminum amide, $(CH_3)_2AlNH_2$, has been
determined by single crystal X-ray diffraction methods and found to contain
molecular trimer units with a six-membered Al-N ring structure similar to
those which make up the wurzite structure of AlN. This compound is readily
volatile and has been used to deposit AlN thin films on Si surfaces by a
low-pressure CVD process. This approach has also been used to prepare AlN
as a high surface area, high purity powder.

INTRODUCTION

The high refractory and chemically resistant nature of aluminum nitride
coupled with its large energy gap (ca. 6 eV), its high intrinsic thermal
conductivity (3.2 W/cmK), its closely matched thermal expansion to silicon,
and the fact that it is piezoelectric with a high acoustic wave velocity,
make it an attractive prospective material for a wide range of applications
in electronics, including substrates for Si-based integrated circuits as
well as both passive and active components in various electronic devices
(1-8). Moreover, its optical transparency throughout the visible and the
near infrared (2,3) provide additional materials application opportunities.
However, it is well known that small concentrations of impurities such as
oxygen as well as vacancies and other defects resulting from off
stoichiometry can appreciably degrade the thermal conductivity to a small
fraction of the "intrinsic" value (3,7,8). These impurities can also have a
significant effect on other physical properties, including the conductivity
and light transparency, rendering AlN of little real value for many
prospective applications unless it can be economically prepared in an
exceptionally high state of purity.

The most frequently used procedures for the preparation of AlN powder
involve the direct reaction of aluminum with either nitrogen or ammonia, or
the carbothermal reduction of alumina in the presence of nitrogen (2,9).
These methods generally yield a gray AlN powder which often contains some
unreacted Al as well as high concentrations of oxygen (>1%) and other
impurities. Moreover, it is difficult to control the particle size of the
AlN powder, rendering consolidation to a dense ceramic by sintering or hot
pressing more difficult. AlN powder of high purity has been prepared by
reaction of AlF_3 (or its precursor $(NH_4)_3AlF_6$) with NH_3 gas (3,10);

however, this method requires high temperatures, is inherently costly and yields the highly corrosive HF as a byproduct.

Aluminum nitride has also been prepared in the form of thin films by various chemical and physical vapor deposition procedures including the reaction of trimethyl aluminum (TMAL) and NH_3 (11-13). This latter approach employs the reactants in the vapor state at near atmospheric pressure with hydrogen as a carrier gas. Under proper conditions, this method and various other CVD and PVD methods have been used successfully to generate both poly-crystalline and epitaxial AlN thin films. However, these methods generally employ substrate temperatures well in excess of 1000°C and, are therefore, inappropriate for many of the prospective applications in electronics.

The reaction of trimethyl aluminum with ammonia was apparently first studied by Wiberg in Germany during WWII and is reported in summary form in a review of WWII German Science published in 1948 (14). The basic reactions involved include the formation of a Lewis acid/base adduct between $Al(CH_3)_3$ and NH_3 and its stepwise conversion to AlN and CH_4 on heating.

$$Al(CH_3)_3 + NH_3 \rightarrow (CH_3)_3Al:NH_3$$
$$(CH_3)_3Al:NH_3 \rightarrow (CH_3)_2AlNH_2 + CH_4$$
$$(CH_3)_2AlNH_2 \rightarrow CH_3AlNH + CH_4$$
$$CH_3AlNH \rightarrow AlN + CH_4$$

The first reaction was carried out under unspecified conditions to yield a crystalline solid which was identified as the $(CH_3)_3Al:NH_3$ adduct. This adduct was found to melt at 56.7 C with the concurrent evolution of methane leading, after 1 hour heating at 70°C, to the intermediate $(CH_3)_2AlNH_2$ species, which was reported to be a dimer in liquid ammonia solution. This crystalline intermediate had an appreciable volatility (vapor pressure at 70°C = 1 mm Hg) and melted at 134.2°C. Further heating at 160 - 200°C resulted in the loss of more methane and, via a "CH_3AlNH" intermediate, eventually gave what was reported to be "pure" AlN. A detailed experimental procedure was not provided in this brief account nor was a full description of the intermediates and form of the AlN final product given. Subsequent work on organoaluminum-nitrogen compounds has focused largely on the products of the reaction of the aluminum trialkyls with organic amines and, for the most part, has not dealt with the conversion of these compounds to AlN. The apparent sole exception is a 1978 Japanese patent in which the preparation of AlN powder by pyrolysis of various organoaluminum amides and imides was claimed (15). The preferred embodiment of this method employed a reaction involving triethylaluminum, aniline and ammonia in hexane solution, followed by heating of the precipitate formed at up to 500°C to yield a gray-white powder which was identified as AlN by elemental analysis, X-ray diffraction and its infrared spectrum.

A detailed study of the TMAL + NH_3 reaction sequence was carried out in order to evaluate the possible utilization of this approach for the preparation of high purity AlN powder. This study and the subsequent investigation of the analogous reactions of the higher alkyl homologues of TMAL has resulted in a better understanding of the chemistry of the R_3Al + NH_3 system as well as the development of an improved procedure for the preparation of high purity AlN powder and a new method for the CVD of AlN films.

RESULTS AND DISCUSSION

Studies of the TMAL + NH_3 Reaction Sequence

Due to the extreme oxygen and moisture sensitivity of the alkylaluminum compounds and their ammonia reaction products, all reactions and

361

manipulations were carried out in an inert atmosphere (N_2 or Ar) with rigorous exclusion of oxygen and water vapor. Initial studies of the reaction between TMAL and NH_3 were carried out at low temperatures (ca. -78°C) in hydrocarbon solvents by bubbling electronic grade gaseous ammonia through the cold TMAL solution. The $(CH_3)_3Al:NH_3$ adduct was obtained on evaporation of the solvent in vacuo and either used as is for subsequent reactions or recrystallized from pentane when a higher purity product was required. This white, crystalline adduct was found to melt, as previously reported, at ca. 55°C with the evolution of a molar equivalent of methane on further heating up to 70°C, as determined by weight loss measurements and analysis of the gaseous product by gas chromatography (g.c.).

This reaction was also followed by both visual observation and measurements of the pressure of a small sample of $(CH_3)_3Al:NH_3$ in a sealed Pyrex tube connected to a pressure gauge which was heated in an oil bath. The resultant pressure/temperature curve obtained (Figure 1) shows a substantial increase in pressure between ca. 60 and 75°C accompanying the observed melting and eventual re-solidification of the sample in this temperature range. The further increase in pressure observed between 75 and ca. 135°C corresponds to that expected from the inert gas equation, indicating a well-defined decomposition reaction leading to an intermediate which is stable up to 135°C. At this point the intermediate melted with a further evolution of gas up to the temperature limit of the experimental apparatus employed (ca. 170°C).

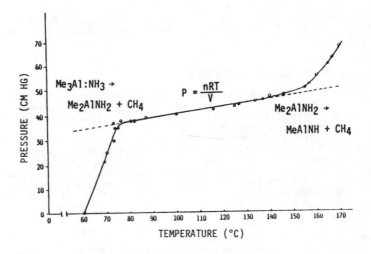

Figure 1. Changes in pressure accompanying the thermolysis of $(CH_3)_3Al:NH_3$

Separate analysis of the gas evolved on pyrolysis of the $(CH_3)_3Al:NH_3$ compound up to 300°C by g.c./mass spec showed a small amount of ethane as the only gaseous product other than methane detected. This ethane may have been formed from the dimerization of methyl radicals; however, it is more likely that it derives from the small proportion of triethylaluminum (ca. 0.4%) present in the TMAL samples used.

The conversion of the $(CH_3)_3Al:NH_3$ compound to $(CH_3)_2AlNH_2$ can also be carried out in solution by maintaining the temperature of the solution above ca. 70°C. This provides a more convenient route to the $(CH_3)_2AlNH_2$ intermediate, which can be obtained simply by bubbling excess ammonia through a refluxing solution of TMAL in a solvent that boils above 70°C, such as benzene (b.p. 78°C). The product is then collected after removal of the solvent by evaporation and purified by sublimation in high vacuum.

The white, crystalline product so obtained, as reported by Wiberg, is quite volatile and readily sublimes on heating above 70°C, even at atmospheric pressure. Small amounts were heated up to 200°C, in a covered Petrie dish in a glovebox, resulting in significant loss of material by sublimation but with eventual conversion of most of the sample to an amorphous white solid of variable composition (presumably the "CH_3AlNH" intermediate reported by Wiberg). Further heating was conducted in a furnace up to 1000°C in a nitrogen atmosphere, yielding a high surface area gray-black powder which microscopic and X-ray diffraction measurements indicate is mainly AlN but with numerous black inclusions. These black inclusions are probably carbon resulting from the direct thermolysis of the last remaining fraction of $-CH_3$ groups to C and H_2 as the concentration of the N-H groups decrease and the AlN lattice becomes more rigid.

To test the possibility that this undesireable side reaction could be avoided by providing an alternative hydrogen source during this final step of the conversion, a small quantity of the powder obtained by heating the adduct up to 200°C was slowly heated (over 24 hr) to 1000°C in a furnace while passing gaseous ammonia over it. The product of this reaction was a light gray powder which was essentially free of the carbon inclusions observed in the previous sample. This product was identified as AlN by its X-ray diffraction pattern and its surface area was measured by the BET method using N_2. A value of 250 m^2/g was obtained, indicating an average particle size of around 75 A.

Preparation of High Purity AlN Powder

Due to the problems with loss of material during unconfined pyrolysis, as well as the higher cost of TMAL compared to other trialkylaluminum compounds, subsequent efforts to develop a route to high purity AlN powder have focused on the higher alkyl aluminum derivatives and, in particular, the use of triethylaluminum (TEAL) in place of TMAL in the reaction with ammonia. TEAL is currently used in large quantities in the chemical industry and is available at relatively low cost (ca. $2.50/lb) making its use for the preparation of high purity AlN powder not only economically feasible but, in fact, competitive with current AlN prices.

The preliminary results of the investigation of the TEAL + NH_3 reaction have indicated that high surface area (ca. 40-80 m^2/g), white AlN powder of exceptional purity (0 <0.3%; C = 0.06%; Si = 20 ppm; Fe and other metallic elements below detectable limits) can be obtained in essentially quantitative yield by this method. This powder picks up oxygen readily on exposure to air at ambient humidity, resulting in incorporation of over 5% O after 24 hr exposure. A small sample of this powder was hot pressed to yield a translucent specimen with a thermal conductivity of 0.8W/cmK (the oxygen content of this particular sample was ca. 2% due to exposure of the AlN powder to air on handling).

Similar results were obtained using tri-iso-butyl aluminum [$(C_4H_9)_3Al$] in place of TEAL, which also gave a white AlN powder of high surface area and purity. In contrast to the R = CH_3 derivative, both the R = C_2H_5 and C_4H_9 systems yielded a liquid R_2AlNH_2 intermediate which was obtained on removal of the solvent under reduced pressure after addition of ammonia to the trialkylaluminum compound at room temperature.

The initial loss of alkane from the $(CH_3)_3AlNH_3$ adduct is accompanied by a sizeable exotherm (ca. 82 kJ/m by DSC); however, the apparent lack of a stable adduct with R = C_2H_5 and C_4H_9 has prevented a detailed comparison of the three systems. Differential scanning calorimetry studies of the corresponding R_2AlNH_2 compounds were carried out indicating an exothermic process for the thermolysis reaction, $R_2AlNH_2 \rightarrow RAlNH + RH$, with similar heat effects (71-83 kJ/mole) and with peak decomposition temperatures which ranged from ca. 165°C for R = CH_3 to 193°C for R = C_4H_9. The final thermolysis step, leading to AlN, is much less well-defined, occurring over a broad temperature range with no obvious heat effect evident in DSC measurements of the CH_3AlNH derivative up to 400°C. TGA studies (in flowing nitrogen at 10°C/min) indicate that most of the weight loss in this case occurs before 400°C with the last few percent occurring between 400° and 750°C; however, in the case of the R = C_2H_5 derivative, the weight loss was apparently complete by 400°C. For both systems the weight loss for the last thermolysis reaction, RAlNH $\rightarrow$ AlN + RH, was close to that expected for the elimination of one mole of RH. Due presumably to the loss of starting material by volatilization or physical transfer from the TGA pan, the overall weight loss on heating to 1000°C from the R_3AlNH_3 or R_2AlNH_2 stages was somewhat greater than that expected for the elimination of the alkane.

Characterization of the $(CH_3)_2AlNH_2$ Intermediate

The high volatility of the intermediate compound, $(CH_3)_2AlNH_2$, in the TMAL + NH_3 reaction sequence suggested another possible area of application for these AlN precursors. In particular, the continuing interest in AlN thin films for a variety of prospective applications in electronics, including SAW devices operating in the high frequency region as well as electrically insulating and passivating layers for semiconductors (5,6,11-13), led to our investigation of the $(CH_3)_2AlNH_2$ compound as a potential vapor phase precursor for the chemical vapor deposition of AlN.

In order to determine the actual structure of the $(CH_3)_2AlNH_2$ species, which was reported to be a dimer by Wiberg, single crystals were grown from heptane by slow cooling of a hot, saturated solution. One of these crystals was sealed in a glass capillary and a structure determination was carried out using X-ray diffraction methods, revealing a trimeric $[(CH_3)_2AlNH_2]_3$ composition for the constituent molecules. The structure of the $[(CH_3)_2AlNH_2]_3$ trimers is illustrated in Figure 2. There is a general resemblance to the structure of the organic molecule cyclohexane, with a six-membered $(Al-N)_3$ ring in a skewed-boat arrangement and a quasi-tetrahedral arrangement of C and N about the Al atoms and H and Al around the nitrogens. These $(Al-N)_3$ rings also resemble closely the fused six-membered rings which make up the wurzite structure of AlN itself. High resolution mass spectroscopic studies of this compound indicate that these trimer molecules survive on volatilization and constitute the major species present in the gas phase.

The Possible Structure of "CH_3AlNH"

The structure of the final precursor to AlN in this series, nominally "CH_3AlNH", but of variable composition depending on the thermal history of the sample, is not yet known; however, it is likely that this insoluble species is a polymeric substance with a broad molecular weight distribution. A possible structure for this polymer can be derived from that of the $[(CH_3)_2AlNH_2]_3$ molecules by extending these rings into two-dimensional sheets of fused six-membered rings. These sheets would contain a CH_3 and a H at each internal Al and N, respectively, with additional CH_3 and H groups around the periphery of each "macromolecule". Further pyrolysis of this polymeric mixture could then lead to the further extension of the macromolecular sheets in the third dimension, resulting finally in the formation of the AlN structure.

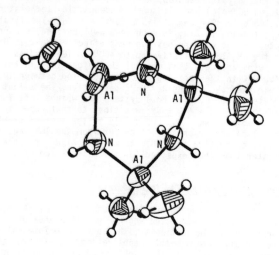

Figure 2.   Ortep view of the molecular structure of $[(CH_3)_2AlNH_2]_3$.

The Chemical Vapor Deposition of AlN Using $[(CH_3)_2AlNH_2]_3$

The possibility of using the intermediate $[(CH_3)_2AlNH_2]_3$ compound as a single component precursor for the CVD of AlN films was tested by subliming a sample of the compound into an evacuated furnace tube containing Si wafer sections which was heated to 700°C. A low pressure (ca. 50 μm) was maintained in the furnace tube by continuous pumping. In this initial experiment AlN was deposited as a rough, grainy deposit on the first wafer piece, close to the entrance of the furnace. Further into the furnace the less rapid build up of AlN on the Si wafer sections resulted in progressively more fine-grained and dense deposits, which ranged from 10 to 1 μm in thickness. A micrograph of a 1 μm layer of AlN on a Si surface produced in this manner is shown in Figure 3. This apparently dense, uniform deposit was examined by TEM and found to consist of polycrystalline AlN with a typical grain size of 0.05-0.15 μm. The electron diffraction pattern gave d-spacings which could be indexed to the AlN structure. This initial experiment has been repeated several times with similar results and a systematic study of the effect of variations in the precursor sublimation rate and furnace temperature, as well as the effect of added NH_3 as a buffer gas, is in progress. Preliminary results from this study suggest that the use of ammonia has a beneficial effect on the purity of the AlN deposit and that AlN can be deposited at temperatures at least as low as 600°C.

365

In the context of the promising preliminary results from the CVD studies and successful development of what appears to be an economically viable approach to high purity powder, the $R_3Al + NH_3$ system holds considerable promise as a future source of AlN for applications in electronics and other areas where high purity material of controlled morphology is needed.

<u>Figure 3.</u>　SEM (x 20K) of AlN deposit on Si wafer, viewed in cross-section.

REFERENCES

(1) D. Gerlich, S. L. Dole, and G. A. Slack, GE CRD Report No. 84CRD244, Oct. 1984.

(2) N. Kuramoto and H. Taniguchi, J. Mats. Sci. Ltrs., 3, 471 (1984)

(3) G. A. Slack and T. F. McNelly, J. Cryst. Growth, 34, 263 (2976).

(4) G. A. Slack and S. F. Bartram, J. Appl. Phys., 46, 89 (1975).

(5) A. Fathimulla and A. A. Lakhani, J. Appl. Phys., 54, 4586 (1983).

(6) Y. Pauleau, A. Bouteville, J. J. Hantzpergue, J. C. Remy, and A. Cachard, J. Electrochem. Soc., 129, 1045 (1982).

(7) G. A. Slack and T. F. McNelly, J. Cryst. Growth, 42, 560 (1977).

(8) G. A. Slack, J. Phys. Chem. Solids, 34, 321 (1973).

(9) A. Rabenau, Chapt. 19, in "Compound Semiconductors", Vol. 1, edited by R. K. Willardson and H. L. Goering (Reinhold Publ. Corp., New York, 1962, pp. 174-176.

(10) I. Huesby, J. Amer. Ceram. Soc. 66, 217 (1983).

(11) H. M. Manasevit, F. M. Erdman, and W. I. Simpson, J. Electrochem. Soc., 118, 1864 (1971).

(12) M. Morita, N. Uesugi, S. Isogai, K. Tsubouchi, and N. Mikoshiba, Jap. J. Appl. Phys., 20, 17 (1981).

(13) U. Rensch and G. Eichhorn, Phys. Stat. Sol. (a), 77, 195 (1983).

(14) E. Wiberg, in G. Bahr, FIAT Review of German Science, Vol. 24, Inorganic Chemistry, Part 2, W. Klemm, ed. (1948), p. 155.

(15) T. Maeda and K. Harada, Sumimoto Chem. Co. Ltd., Japan Kokai 78 68,700, 19 June 1978, Appl. 76/145,137, 01 Dec 1976; Chem. Abs. 89, 165623f (1978).

ACKNOWLEDGEMENTS

The authors are grateful to the General Electric Company for their support of this work, in part through a contract to the RPI Chemistry Department. We also acknowledge support received for the CVD studies from the Office of Naval Research under Contract No. N0001485K0632. We thank N. Lewis and C. Robertson of the GE Materials Characterization Operation for TEM and SEM studies, respectively, of the AlN films and S. Dorn and W. Ligon for their g.c./mass spec investigation of the $(CH_3)_3AlNH_3$ system.

A LOW TEMPERATURE CHEMICAL ROUTE TO PRECURSORS OF BORIDE AND CARBIDE CERAMIC POWDERS

Joseph J. Ritter
Ceramics Division
National Bureau of Standards
Gaithersburg, MD 20899

ABSTRACT

The controlled reductive dehalogenation of elemental halides has been studied as a low temperature approach to boride and carbide precursors. We have shown that $SiCl_4$ and CCl_4 can be reacted with metallic sodium at 130° in a non-polar solvent to give the precursor to SiC. Similar type reactions with $TiCl_4$ and BCl_3 or with BCl_3 and CCl_4 have produced the precursors to TiB_2 and B_4C respectively. This procedure has also been used to generate the precursors to the two-phase composites SiC/TiC and SiC/TiN. The method is generally applicable to any combination of elemental halides which can be reduced by alkali metals or alkali metal alloys.

INTRODUCTION

Over the past thirty years considerable research has been focused on various low-temperature chemical processes which lead to oxide ceramics [1]. The search for appropriate low-temperature chemical procedures leading to non-oxide ceramics is just now beginning to gain momentum. In this publication I shall describe the application of an inorganic redox chemistry, known to chemists for well over a hundred years, to the synthesis of carbide, boride and carbide particulate composite powders. Within the past decade, Rieke and co-workers [2] have successfully exploited the reduction of metal halides with alkali metals as a route to very finely divided, reactive metal powders. Our work demonstrates that this approach can be extended even further to include the interactive reduction of two or more metalloid or non-metal halides to generate precursors to boride and carbide ceramic powders. Recent interest in the particulate composite SiC/TiC [3,4], has led us to explore the low-temperature synthesis of this system as well. A U.S. patent application for our process has been filed.

EXPERIMENTAL

As noted above, the chemical process used to generate the boride and carbide precursors is the reduction of elemental halides by an alkali metal. For example, the reaction to produce the SiC precursor may be written:

$$SiCl_4 \; + \; CCl_4 \; + \; 8Na \; \xrightarrow[\; 130° \;]{\text{n-heptane}} \; \text{"SiC"} \; + \; 8NaCl \qquad (1)$$
$$\text{(amorphous precursor)}$$

In our initial experiments, we found that simply refluxing the reaction systems while agitating with a magnetic stir bar gave poor yields of precursor. For our multicomponent systems the best results were obtained by combining amounts of elemental halides calculated to provide 20-40% excess of halogen with respect to Na. The reactants were heated to 130° in sodium-dried heptane with concomitant high shear stirring in a sealed reactor, shown in Figure 1. This reactor was designed and constructed at NBS.

The stirring module consists of a central stainless steel housing which contains a stainless steel shaft supported on carbon-filled polytetrafluoroethylene bushings. The shaft is hermetically sealed within the reactor, magnetically coupled to an external variable speed electric motor and is fitted with a 3 cm dia. homogenizer blade.

Standard dry-box and vacuum line techniques are used to load the alkali metal and to degas n-heptane in the main reactor chamber. Elemental halides were either diluted with heptane and loaded into the smaller side bulb or distilled directly into the reactor. The side bulb is particularly useful for the preparation of two-phase composites where the halides for one phase are placed directly in the reactor and either partially or totally consumed. The second phase halides are introduced subsequently from the side bulb. Because alkali metal reduction reactions of this type do present a possible EXPLOSION or FIRE hazard, the loaded reactor was operated in a hood, behind a safety shield and over a bed of vermiculite.

In a typical reaction to synthesize the SiC precursor, 40 millimoles each of $SiCl_4$ and CCl_4 are combined with 290 millimoles of Na and 270ml of Na-dried n-heptane in the reactor. The stirrer is operated at ~3000 rpm while the contents of the reactor are brought to ~130°C with an oil bath. As soon as the Na melts (~100°) the reaction begins with the gradual development of a fine, black precipitate. After 72-96 h of reaction time, volatile materials are vacuum distilled from the reactor and the remaining solids dried at ~200° with cold trapping under dynamic vacuum. A clear, relatively involatile oil (c) is collected in the cold trap. A weighed portion of the solids is transferred to a baffled graphite tube (Figure 2) which is inserted into a larger fused silica tube with an appropriate closure and valve. The assembly is evacuated and heated under dynamic vacuum to ~900°C over a 15-25 h period. Some caution is required as the temperature of the material approaches 300-400°, since a quantity of, as

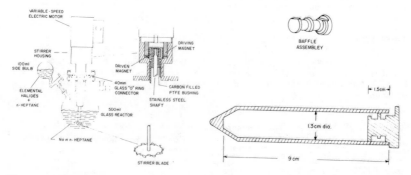

Figure 1. Stirred Reactor   Figure 2. Graphite tube with baffle

yet unidentified, non-condensible gas is released precipitously. The baffle assembly helps to minimize powder expulsion from the graphite tube when this gas-release stage is reached. Sodium chloride (b) sublimes to the cool portions of the fused silica tube while the precursor (a) remains in the graphite tube. The black precursor (a) is transferred under dry N_2 to a SiC crucible and heated to ~1750° for 15 h under Ar/5% H_2. The sublimate (b) is treated with distilled water and the clear extract analyzed for Cl^- by precipitation as AgCl [5]. Data generated from these procedures were used to estimate precursor yields. The TiC, TiB_2 and boron carbide precursors are generated in a similar fashion from the appropriate elemental halides and sodium.

The preparation of the particulate composite SiC/TiC was conducted as follows. The 500 ml reactor was loaded with 39.8 millimoles CCl_4, 39.9 millimoles $SiCl_4$, 290 millimoles Na and 230 ml dry n-heptane. The 100 ml side bulb was loaded with 8.38 millimoles $TiCl_4$, 8.77 millimoles CCl_4 and 40 ml dry heptane. The $SiCl_4$, CCl_4 and Na in the 500 cc bulb were reacted with stirring at ~130°C for 24 h. The reactor was cooled, and the $TiCl_4/CCl_4$ solution slowly added from the side bulb. The reactor was again brought to 130°, and stirred continuously until no sodium could be detected visually in the slurry (usually 4-6 da). Solvent and NaCl separation were conducted with the distillation and sublimation techniques mentioned earlier.

Surface areas of the SiC powders were determined with a commercial B.E.T. apparatus using N_2 as the adsorbed gas. Microchemical analyses for trace Cl were performed by a commercial laboratory.

RESULTS AND DISCUSSION

The reaction product arrays consist of three separable components, the ceramic precursor (a), NaCl(b), and a small quantity (0.2-3 ml) of relatively involatile, clear oil (c).

Typical yields of precursor are of the order of 85-90% of theoretical based upon the amount of NaCl recovered in the reactions. X-ray diffraction (XRD) patterns taken of the crude reaction mixtures after vacuum drying at 200°C show NaCl as the only crystalline phase. Initially, all of the precursors are amorphous to x-rays. After separation of the NaCl by vacuum sublimation at 900°C, the TiC and TiB_2 materials are crystalline while the silicon carbide and boron carbide materials remain amorphous. Silicon carbide crystallizes from its precursor between 1450 and 1750°, whereas B_4C has been identified as the crystalline phase detectable after heating the boron carbide precursor in vacuum for ~5h at 1980°. Scanning electron microscope examination of the crystallized powders suggests typical particle sizes in the 1-5μm range.

In the SiC/TiC composite system, cubic TiC is the only crystalline phase identifiable by XRD after separation of the precursor from NaCl at 900°C. The SiC remains amorphous. When this composite precursor is heated to 1460°C for 4.5h under vacuum (p ~ 2 x 10^{-6} torr) cubic (8F) SiC and cubic TiC are the only crystalline phases detected by XRD. Heating this precursor to 1500° under N_2/5% H_2 for ~65h gives a material whose XRD shows sharp intense peaks for TiN and broad weak peaks for cubic SiC. There is no evidence for α-Si_3N_4. Moreover, when the precursor is heated at 1500° for ~70h under Ar/5% H_2 (which can contain up to 3 ppm N_2) the XRD patterns show both SiC and TiN to be well crystallized. An abundance of nitrogen during the heat treatments seems to attenuate the crystallization of SiC by a process not well understood.

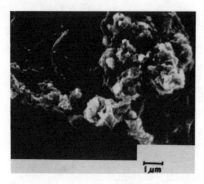

Figure 3. S.E.M. micrograph of black SiC precursor after 20h, 900°

Figure 4. S.E.M. micrograph of grey SiC powder after 20h, 1750°

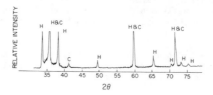

Figure 5. XRD of grey SiC powder after 20h, 1750° H = 4H hexagonal, C = 8F cubic

Of the materials prepared, only the SiC has been characterized in any detail up to the present time. The amorphous, black SiC precursor exhibits a sponge-like appearance in the S.E.M. (Figure 3). When this material is heated for 5 to 20 h under Ar/5% H_2 at ~1750° in a SiC crucible it undergoes a considerable volume reduction and becomes grey in color. S.E.M. examination of the resultant powder (Figure 4) shows both crystallite and whisker-like material. XRD indicates the grey material to be a mixture of 4H (hexagonal) and 8F (cubic) SiC (Figure 5). Measured B.E.T. surface areas for precursor and crystallized powders are 69.5 and 26.4 m ^{2}/g, respectively. Chemical analysis on precursor and crystallized SiC for residual Cl gave the respective values of 0.090 and 0.086%.

A cursory examination of the involatile oils (c) derived from these reactions leads to some interesting observations. With the exception of the boron carbide system, only small amounts (~0.2 - 0.5 ml) of these oils, just sufficient for liquid film infrared studies, were recovered. The predominant features of the spectra were infrared bands in the saturated C-H region, 2800-3000 cm^{-1}. Considerably more oil (~3 ml) was obtained from the boron carbide synthesis. The infrared spectrum of this material showed a number of bands which are broadly classified as follows: 2870,2950 cm^{-1}, C-H; 1110-1145 cm^{-1}, B-C; 920-970 cm^{-1}, B-Cl; 600,665 cm^{-1}, C-Cl. Moreover, this oil reacts violently with water at 25° giving a clear solution and a white solid. Quantitative Cl analysis [3] on the clear solution indicates that 25% of the sample weight is hydrolyzable Cl. In this system, hydrolyzable Cl is that Cl which is bonded to boron. The combined infrared, relative involatility and chemical analysis data leads us to speculate that this oil may be a mixture of >C$_7$-hydrocarbons, probably resulting from some solvent degradation and chlorinated boron-carbon intermediates of a type similar to (I) mentioned below.

The mechanism by which the synthesis reaction occurs is not understood at this time. Our observation that high-speed, high-shear stirring is essential to produce good yields of precursor suggests that the reaction probably occurs at or near the surfaces of the fine sodium particles thus generated. The continuous removal of adhered reaction products with the resultant exposure of fresh metal surface is an important factor in these reactions. While in the simplest terms one may envision the simultaneous reduction of two elemental halides, in fact, the reaction mechanism is almost certainly much more complex. The related Wurtz-Fittig reaction, familiar to organic chemists, is thought to proceed by nucleophilic displacement [6].

Applying a similar rationale to our systems, one may postulate that the initial interaction of an elemental halide with metallic sodium generates the nucleophile $Cl_3M^{\ominus}$ and NaCl.

$$
\begin{array}{c}
Cl \\
| \\
Cl-M-Cl \\
| \\
Cl
\end{array}
+ \; 2Na \; \longrightarrow \; Na^{\oplus} \;
\begin{array}{c}
\ominus \\
\end{array}
\begin{array}{c}
Cl \\
| \\
M-Cl \\
| \\
Cl
\end{array}
+ \; NaCl \qquad (2)
$$

The nucleophile seeks out and forms a complex with another elemental halide molecule and in so doing displaces a halogen from the receptor molecule, giving a coupled M-M' product (I) and another molecule of NaCl.

$$
\begin{array}{c}
Cl \\
| \quad \ominus \\
Cl-M \\
| \quad \oplus \\
Cl \quad Na
\end{array}
\cdots\cdots
\begin{array}{c}
Cl \quad Cl \\
\backslash \; / \\
M' \\
/ \; \backslash \\
Cl \quad Cl
\end{array}
\; \longrightarrow \;
\begin{array}{c}
Cl \quad Cl \\
| \quad | \\
Cl-M-M'-Cl \\
| \quad | \\
Cl \quad Cl \\
(I)
\end{array}
\; + \; NaCl \; (3)
$$

For the Si-C system, this initial coupled product (I) is a very reasonable intermediate, since $Cl_3Si - CCl_3$ is a known, relatively stable compound. However, each element-halogen site on the coupled product (I) is subject to attack by either Na or another nucleophile with the final result that all of the halogens are eliminated to form NaCl and a three dimensional M-M' matrix is developed. This elemental matrix is the ceramic precursor.

Summary and Conclusions

The feasibility of generating precursors to boride and carbide ceramic powders using low-temperature reductive dehalogenation has been demonstrated. The amorphous precursors of SiC, TiC and TiB_2 can be crystallized by heating between 900 and 1750°C to give particulates in the 1-5μm size range. Crystallization of the boron carbide precursor to B_4C has been achieved at temperatures near 2000°.

The chemical reaction to form these precursors is thought to proceed by a series of nucleophilic displacement reations which lead to coupled intermediate species. Infrared and chemical data on the oils derived from the boron-carbon system are consistent with the presence of chlorinated B-C species which comprise reasonable intermediates in this system.

The extension of this chemistry to the synthesis of the two-phase composites SiC/TiC and SiC/TiN suggests some interesting possibilities for further development. The approach can be modified to include any combination of elemental halides which are reduced by alkali metals so that a variety of ceramic-ceramic metal-metal and metal-ceramic compositions are accessible. The possibilities for synthesizing semiconductor materials by this method using the halides of Group IV_A, V_A and VI_A should not be overlooked.

Acknowledgements. The author would like to thank Dr. R. S. Roth for his guidance in the interpretation of XRD patterns, Mr. Carl Robbins for the surface area measurements, and Mr. N. K. Adams for the S.E.M. work. Special thanks are due to Mr. Sam Schneider who encouraged a search for low temperature chemical procedures which would be useful in the synthesis of non-oxide ceramic powders.

References

1. David W. Johnson Jr., Am. Ceram. Soc. Bull., 64, 1597, (1985), and references therein:

2. Ruben D. Rieke, Accts. Chem. Res., 10, 301, (1977) and references therein:

3. G. C. Wei and P. F. Becker, J. Amer. Ceram. Soc., 67, 571, (1984).

4. Mark A. Janney, Am. Ceram. Soc. Bull., 65, 357, (1986).

5. I. M. Kolfhoff and E. B. Sandell, Textbook of Quantitative Inorganic Analysis, (MacMillan Co, N.Y. 1956), p. 307

6. E. Earl Royals, Advanced Organic Chemistry, (Prentice-Hall Inc, N.J., 1959) p. 117.

THE PYROLYSIS OF A TUNGSTEN ALKYNE COMPLEX AS
A LOW TEMPERATURE ROUTE TO TUNGSTEN CARBIDE

Richard M. Laine[*] and Albert S. Hirschon
Contribution from the Organometallic Chemistry Program
SRI International, Menlo Park, CA 94025

ABSTRACT

The synthesis of designed organometallic compounds and their selective activation and transformation into materials of high purity (for electronic applications), high strength and/or high temperature stability (for refractory or structural applications), represents a potential area of extreme growth in organometallic chemistry. Research in this area could provide entirely new, inexpensive, fabrication methods for common and exotic materials.

In this paper, we develop design principles for the preparation of organometallic precursors, "premetallics" that can be selectively converted, in high yields, to a desired refractory metal. We also describe our preliminary efforts to prepare tungsten carbides (WC_x) from a premetallic. Pyrolysis studies using $Cp_2W_2(CO)_4(DMAD)$ [DMAD = Dimethylacetylene dicarboxylate], 5, as the premetallic demonstrate that 5 will decompose at temperatures of ~ 700°C to give good yields of W_2C. 5 is soluble and decomposes fully in 10-20 minutes, without the need of another reactant.

INTRODUCTION

Although the industrial production of tungsten carbide (WC) is founded on an extremely mature, well developed technology, tungsten carbide production is quite energy and equipment intensive because of the lengthy process times and high temperatures necessary to first form tungsten carbide powder and then transform it into a finished product. The current industrial process[1] (see Figure 1) begins with careful preparation of the tungsten oxide ore, followed by hydrogen reduction, under exacting conditions, to metal powder. The metal is then blended with carbon powder and carburized at high temperatures (1700-2200°C) to generate the carbide. This process always leads to WC powder which must be further transformed through alloying, sintering and pressing to obtain a finished piece.

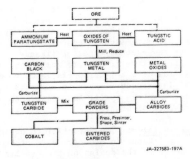

FIGURE 1 SCHEMATIC FLOW CHART FOR THE PRODUCTION OF CARBIDE POWDERS

Mat. Res. Soc. Symp. Proc. Vol. 73. ©1986 Materials Research Society

We believe that there is now an alternative method of producing refractory metals that could, upon further refinement, eliminate the long
process times, the costly equipment and high energy consumption required in
the current process. This method relies on the use of designed organometallic compounds as precursors to refractory metals. In the process we
envision, a tractable (soluble, meltable or malleable) organometallic compound is designed such that it can be shaped in the same fashion as one
uses to shape organic polymers (e.g., spinning, coating, injection molding
etc.) and then briefly pyrolyzed at moderate temperatures to generate the
desired refractory metal. We present here, both a conceptual approach to
the development of designed organometallic precursors for the production of
refractory metals, and our preliminary efforts to establish the validity of
these concepts through the production of tungsten carbide.

Background

Our interest in the development of designed organometallic precursors,
premetallics, to refractory metals derives from our interest in the preparation of preceramic precursors to silicon nitride[2] and from our work on
the laser pyrolysis of organometallic compounds. These latter studies were
originally undertaken to develop a means of establishing the bond dissociation energies (BDE's) for various ligand–metal bonds in the gas phase.[3-5]
Table 1 provides some examples of this work

Table I

BOND DISSOCIATION ENERGIES FOR ORGANOMETALLIC
COMPOUNDS, AS DETERMINED BY LASER PYROLYSIS

Metal Carbonyl	BDE (kcal/mol)[a]	Compound	BDE (kcal/mol)[b]
$Fe(CO)_6$	41.5	$(CH_3)_4Ge$	83
$Cr(CO)_6$	36.8	$(CH_3CH_2)_3P$	63
$Mo(CO)_6$	40.5	$(\pi-C_5H_5)_2Fe$	90
$W(CO)_6$	46.0	$(CH_3CH_2)_4Pb$	54

[a]For loss of the first CO ligand.
[b]For loss of the first alkyl group.

The results presented in Table 1 are of interest for two reasons. The
first concerns the metal carbonyl BDE's, where pyrolysis always leads to
full decarbonylation, with the highest metal carbonyl BDE always being for
loss of the first carbonyl. The second reason concerns the considerable
differences in BDE's for the various types of ligands. The first result is
of importance because it confirms literature results wherein both solvent
and gas phase decomposition of metal carbonyls leads solely to metal.[6,7]
The second result is of importance because it suggests that there are some
pyrolysis conditions under which it will be possible to selectively remove
some ligands in the presence of others.

Molecular Design Principles: The potential to design molecules that
can selectively lose some ligands in preference to others, under pyrolysis
conditions, is a key factor in the design of premetallic precursors to
refractory metals or other materials. We have developed a set of synthetic

guidelines or design principles based on both organometallic literature and common sense that we believe can serve as the basis for designing precursors to a given material. We present these principles without attempting to fully detail the logic that went into their formulation.

- The overall goal of the design is to create the most favorable situation in the premetallic molecule such that, during pyrolysis, only one nonmetallic element will remain bound to the metal, to the exclusion of all others, to give the desired product.

- Thus, multiple bonding between the metal and the preferred non-metallic element, in the premetallic, is desirable.

- Furthermore, the stoichiometry of the metal and the element(s) desired in the pyrolysis product should be closely approximated in the premetallic.

- A secondary goal is to synthesize the simplest premetallic which is still tractable (soluble, meltable or malleable).

- Polynuclear or oligomeric premetallics are preferable to mono-nuclear species because of the need to avoid loss of yield through volatilization of the premetallic during pyrolysis.

- Oxygen- or halogen-metal bonds should be avoided to prevent the facile incorporation of these elements in the products.

- High product yields are predicated by both the mass and type of ligands lost. A gross difference between a premetallic's molecular weight and the formula weight of the desired pyrolysis product will result in low absolute product yields. Thus, ligand size and chemical makeup must be carefully considered in synthesizing a premetallic.

These preliminary design principles represent working guidelines for the selection of premetallics. They have not been tested in a formal sense; although they are based, in part, on our experience with the synthesis and pyrolysis of silicon nitride precursor polymers.[2] These principles may also be extended to the design of precursors to alloys or to ternary materials.

Tungsten Carbide Premetallics: Compounds 1-5 are potential sources of WC and serve to illustrate the application of the above design principles in the search for suitable precursors to WC. With the exception of 5 each complex

$\underline{1}^8$ $(tBuO)_3W\equiv CH_3$ $\underline{2}^9$ $CH_3C\equiv W(CO)_4X$

$\underline{3}^{10}$ $CH_3C\equiv W((CO)_2Cp$ $\underline{4}^{11}$ $(PhC\equiv CPh)_3WCO$

$$C_p = \pi - C_5H_5$$

$\underline{5}^{12}$ $Cp(CO)_2W\overset{\displaystyle R}{\underset{\displaystyle R}{\overset{|}{\underset{|}{\overset{\textstyle C}{\underset{\textstyle C}{\diagup\!\!\diagdown}}}}}}WCp(CO)_2$ $R = -CO_2CH_3$

contains one carbon atom triply bonded to tungsten. Although, there are no useful BDE's reported for metal-carbon triple bonds, we assume that the carbyne BDE's in 1-4 are 40-60 kcal greater than what one would expect for a tungsten-carbon single bond. Thus, it is reasonable to speculate that during pyrolysis of 1-4, the last species left bound to the metal could easily be the carbyne carbon which satisfies the stoichiometry requirement for WC.

While this is a good beginning, there are several potential problems associated with 1-4. Complex 1 is a discrete, tractable compound readily prepared by the method of Schrock et al.[8] Unfortunately, this complex has three tungsten-oxygen bonds which are likely to lead to tungsten oxide formation on pyrolysis. Moreover, this monometallic complex is likely to volatilize during pyrolysis. The E.O. Fischer carbyne complex, 2, has similar drawbacks in that it contains a halogen and is volatile.[9] Complex 3, which is readily prepared from type 2 complexes,[10] does not contain direct oxygen- or halogen-tungsten bonds and appears to be quite promising as a precursor to WC; however, it is volatile. Complex 4[11] contains little oxygen, but represents a low yield compound given that it contains 42 carbons-far more than required to form WC. It's pyrolysis is likely to lead to excessive amounts of carbon.

Of the complexes shown above, 5 appears to be the best choice despite the fact that it does not contain a tungsten carbyne feature. Of prime importance is that it has no noncarbon-tungsten bonds and is not volatile. The tetrahedral unit formed by the two tungsten bonds and the two carbon atoms of the alkyne ligand could actually provide some unusual integrity/-stability relative to the rest of the molecule. The unit also has the 1:1 W to C stoichiometry necessary(?) for a WC precursor. Finally, the R = $-CO_2CH_3$ complex is easy to make, tractable, and unlike 1-4, it is stable in air for months.[12]

One potential problem we have purposely ignored, but which may be the most difficult barrier to the successful design of premetallics, is the possibility that during pyrolysis, the various ligands can either react with each other prior to bond homolysis to give new ligands or eliminate in some unplanned for mode. For example, 5 is formed by addition of an alkyne to a metal-metal triple bond, reaction (4). It is conceivable that under pyrolysis

$$Cp(CO)_2W \equiv W(CO)_2Cp + RC \equiv CR \longrightarrow 5 \qquad (1)$$

conditions the reaction is reversible[13] and destroys the integrity of the tetrahedron. Alternately, complex 2 could rearrange as shown in reactions (2)-(4), eliminating ketene[14] and leaving a ready source of tungsten metal, $W(CO)_3$:

$$RC \equiv W(CO)_4X \longrightarrow R(X)C = W(CO)_4 \qquad (2)$$

$$R(X)C = W(CO)_4 \longrightarrow [\eta^2 - R(X)C = C = O]W(CO)_3 \qquad (3)$$

$$[\eta^2 - R(X)C = C = O]W(CO)_3 \longrightarrow \eta^2 - R(X)C = C = O + W(CO)_3 \qquad (4)$$

Unfortunately, the pyrolysis chemistry of organometallics is only sparsely studied[15-17] and in general, the studies that have been reported

5

77

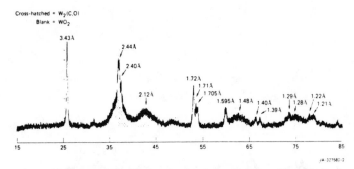

FIGURE 2   X-RAY POWDER DIFFRACTION PATTERN FOR Cp$_2$W$_2$(CO)$_4$(DMAD) PYROLYZED AT 750°C FOR 10 MIN

are concerned with the organometallic products that survive pyrolysis rather than the decomposition products—the sole interest of the work here. Therefore, we have chosen, for the moment, to ignore potential design complications, as postulated above, until a reasonable base of knowledge concerning the pyrolytic decomposition chemistry of organo-metallics is available. Thus, predicated on the above design arguments, we proceeded to examine the pyrolysis chemistry of 5 where R = -CO$_2$CH$_3$.

Results and Discussion

Our initial objective, for the pyrolysis studies, was to perform a series of simple pyrolyses, changing very few of the reaction parameters, in an effort to determine what kinds of products formed during pyrolysis and to establish the reproducibility of our techniques. Our goal was to work at temperatures well below those normally needed to form bulk tungsten carbides (1700°C) in order to reduce the number of specialized pieces of apparatus required for our experiments. For example, standard tube furnaces can reach ~ 1000°C and quartz pyrolysis tubes will not deform at these temperatures.

In our initial series of experiments, samples of 5 were sealed, under vacuum, in quartz tubes and pyrolyzed at 750°C for 10, 30 and 90 minutes. The x-ray powder diffraction patterns of these samples, shown in Figures 2-4, indicate the formation of two major products, WO$_2$ and W$_2$(C,O). The longer

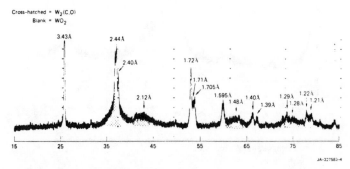

FIGURE 3   X-RAY POWDER DIFFRACTION PATTERN FOR Cp$_2$W$_2$(CO)$_4$(DMAD) PYROLYZED AT 750°C FOR 30 MIN

378

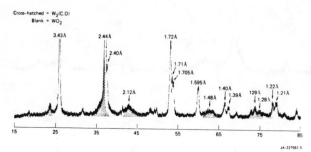

Cross-hatched = $W_2(C,O)$
Blank = WO_2

FIGURE 4   X-RAY POWDER DIFFRACTION PATTERN FOR $Cp_2W_2(CO)_4$ (DMAD) PYROLYZED
AT 750°C FOR 90 MIN

pyrolysis times clearly reveal the increased amounts of WO_2 that form
relative to $W_2(C,O)$ with increasing pyrolysis time. These increases can be
ascribed simply to sintering of WO_2 microcrystals or to a back reaction of
the pyrolysis gases formed in the sealed tubes with the pyrolysis products.
A simple test of the latter possibility was to run the pyrolyses in open
quartz tubes under a nitrogen atmosphere. The results of these studies
were significantly different from the closed tube reactions in that only
one major product was formed, $W_2(C,O)$, according to x-ray powder diffractometry, as seen in Figure 5. Figure 6 shows the corresponding Auger
spectrum.

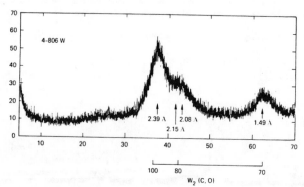

FIGURE 5   X-RAY POWDER DIFFRACTION PATTERN OF $Cp_2W_2(CO)_4$ DMAD PYROLYZED
UNDER N_2 AT 750°C FOR 20 MIN

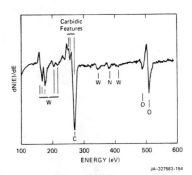

FIGURE 6 AUGER ANALYSIS OF $Cp_2W_2(CO)_4$ DMAD PYROLYZED
UNDER N_2 AT 750°C FOR 20 MIN, 95% $W_2(C, O)$

The extremely broad x-ray powder diffraction pattern seen in Figure 5 illustrates a recurring problem in these types of studies, the production of amorphous or microcrystalline products. While microcrystallinity or small grain size may actually be quite desirable in the production of refractory metals, the need to analyse bulk product properties by x-ray powder diffractometry is of utmost importance to the success or failure of synthetic studies such as those presented here. Consequently, our studies in this area often have a lower temperature limit, necessitated solely by the need to have sufficient crystallinity to run x-ray diffractometry studies.

For comparative purposes, we also pyrolyzed $W(CO)_6$ under the same conditions. Figure 7 illustrates the difference between pyrolysis of volatile premetallics and nonvolatile premetallics. The tube initially containing $W(CO)_6$, now has a mirror-like skin/coating along a considerable portion of the tube's length. The tube initially containing 5 now contains a cylindrical, porous, shiny-grey metallic cake of approximately the same dimensions as the original premetallic charge. Figure 8 is the x-ray powder diffraction pattern

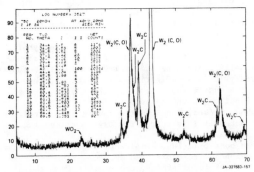

FIGURE 7 X-RAY POWDER DIFFRACTION PATTERN OF $W(CO)_6$ PYROLYZED AT 750°C
FOR 20 MINUTES

FIGURE 8

of the $W(CO)_6$ pyrosylate. The sample, which is taken from the top of the tube in Figure 7, reveals a mixture of products including WO_2, $W_2(C,O)$ and W_2C. We propose that these products are generated when CO, produced by decomposition of $W(CO)_6$ in the bottom of the tube, reacts with a tungsten metal coating produced during the earlier part of the pyrolysis to carburize the metal. The fact that pyrolysis of 5 leads exclusively to $W_2(C,O)$, rather than the mixture obtained with $W(CO)_6$, strongly suggests that the $W_2(C,O)$ was formed from the component elements in 5 rather than through back reaction with the pyrolysis gases. If this result is valid and generally applicable to materials synthesis, then it could eliminate the need to mix and react gases as done in CVD.

The fact that pyrolysis of 5 leads exclusively to formation of $W_2(C,O)$ despite our rationalization that the molecular design should lead to WC provided us with the first test of the validity of our design principles. Our efforts to understand why this disparity should arise led us to examine the pyrolysis tube more carefully. Examination of the pyrolysis tubes shown in Figure 7 reveal that there is a white frosting around the bottom of the tube originally containing 5. If the intermediate tungsten species formed during pyrolysis is an effective oxygen getter, then it is quite conceivable that the quartz tube itself is the source of the oxygen in the $W_2(C,O)$ and the frosting is the byproduct of this reaction. In addition, the nearly amorphous product found in the quartz tube pyrolyses of 5 (Figure 5) might actually result from the oxygen scavenging.

To test these possibilities, samples of 5 were pyrolyzed (at 750°C) in aluminum and nickel boats commonly used in combustion analyses. While aluminum melts at these temperatures, the product from the nickel boat is, as shown in Figures 9 and 10, reasonably pure W_2C.

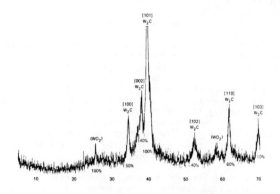

FIGURE 9   X-RAY POWDER DIFFRACTION PATTERN OF Cp₂W₂(CO)₄ DMAD
PYROLYZED AT 750°C FOR 20 MINUTES IN A NICKEL TUBE

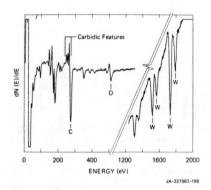

FIGURE 10   AUGER OF Cp₂W₂(CO)₄ DMAD PYROLYZED AT 750°C
FOR 20 MINUTES IN A NICKEL TUBE

We have tried to lower pyrolysis temperatures further, but the products are
amorphous or microcrysalline as shown by x-ray powder analyses. Thus,
establishing a lower temperature limit for conversion of 5 to carbide must
await some refinements in our analytical techniques. However, the fact
that we can generate W_2C at temperatures of 750°C, especially from a
selected precursor that contains all of the elements required in the final
product, is extremely encouraging.

Finally, while we have not developed a route to WC, the common bulk
carbide currently in use, W_2C has the same relative hardness as WC and is
just as effective when used as a coating material.[18] Therefore, in the
future, it may be possible to form W_2C coatings on wide varieties of

382

substrates simply by painting on a solution of 5 or some simple congener, and briefly heating the surface to a temperature of 500°C or less for periods of 10-20 minutes.

Conclusions

The work reported here demonstrates that organometallic compounds can serve as precursors to useful refractory metal materials. We have preliminary evidence that shows that both the pyrolysis conditions and the choice of precursor can effect the product selectivities. Furthermore, our results show that it is possible to generate refractory metals from organometallics, at moderate temperatures, without the need for other reactants. This result may portend significant changes in both CVD and PVD methodology. Additionally, our reaction temperatures are at least 1000°C below those used industrially. Finally, we strongly believe that the principles developed above are generally applicable to the synthesis/preparation of all types of common and exotic materials.

ACKNOWLEDGMENTS: We would like to thank Mr. Eldon Farley for performing the x-ray powder diffraction studies and Mr. Bernard J. Wood for performing the Auger work. We would also like to thank Dr. David J. Rowecliffe of SRI International's Materials Laboratory for several useful discussions concerning tungsten carbides and Drs. Robert Caliguiri, Lawrence Eiselstein and Donald Shockey for their enthusiastic support for this work. This work was supported by SRI International's Advanced Materials Program.

REFERENCES

[1] K.J.A. Brookes, "World Directory and Handbook of Hardmetals," 2nd Ed., Pub. by Engineers Digest, 1979, Hertfordshire, U.K.
[2] Y.D. Blum and R.M. Laine, Organomet. Chem. in press (1986).
[3] G.P. Smith and R.M. Laine, J. Phys. Chem. 85, 1620 (1981).
[4] G.P. Smith, D.M. Golden and K.E. Lewis, J. Am. Chem. Soc. 106, 3905-3911 (1984).
[5] K.E. Lewis and G.P. Smith, J. Am. Chem. Soc. 106, 4650 (1984).
[6] P. Hess and P.H. Parker, J. Appl. Polym. Sic. 10, 4650 (1966).
[7] D.S. Lashmore, W. A. Jesser, D. M. Schladitz, H.J. Schladitz, and H.G.F. Wilsdorf, J. Appl. Phys. 48, 478 (1977).
[8] R.R. Schrock, M.L. Listemann, L.G. Sturgeoff, J. Am. Chem. Soc. 104, 4389 (1982).
[9] E.O. Fischer, U. Schubert, J. Organomet. Chem. 100, 59-81 (1975).
[10] E.O. Fischer, T. L. Linder and R. R. Kreissl, J. Organomet. Chem. 112, C27-C30 (1976).
[11] R.M. Laine, R.E. Moriarty, and R. Bau, J. Am. Chem. Soc. 95, 1402 (1972).
[12] R.M. Laine and P.C. Ford, J. Organomet. Chem. 124, 29 (1977).
[13] S. Slater and E.L. Muetterties, Inorg. Chem. 19, 3337-3342 (1980).
[14] (a) H. Fischer, A. Motsch, R. Markl, and K. Ackermann, Organomet. 4, 726-735 (1985).
     (b) M. Wolfgruber, W. Sierber, F. R. Kreissl, Chem. Ber. 117, 427-433 (1984) and references therein.
[15] H-Y. Parker, C.E. Klopfenstein, R.A. Wielesek and T. Koenig, J. Am. Chem. Soc. 107, 5276-5277 (1985). This and the following two references are examples of the type of transition metal organometallic pyrolysis work currently being reported in the literature.
[16] N.T. Allison, J.R. Fritch, P.C. Vollhardt, and E.C. Walbrosky, J. Am. Chem. Soc. 105, 1384-1386 (1986).
[17] R.D. Adams, P. Mathur, and B.E. Segmmuller, Organomet. 3, 1258-1259 (1983).
[18] J.J. Yee, Internat. Metals, Rev. 1, 19-42 (1978).

PRECURSORS TO BORON-NITROGEN MACROMOLECULES AND CERAMICS

C.K. NARULA, R.T. PAINE AND R. SCHAEFFER
Department of Chemistry, University of New Mexico, Albuquerque, NM 87131

ABSTRACT

Boron nitride has been prepared in the past from classical high temperature reactions and more recently by CVD methods. Few attempts have been made to prepare this important material from pyrolyses of preceramic oligomers or polymers. In the present study oligomerization reactions of substituted borazenes with silyamine crosslinking groups have been found to provide useful gel materials which upon pyrolysis form boron nitrogen materials. Selected aspects of this chemistry and some characterization of the materials is presented.

INTRODUCTION

Hexagonal boron nitride is an increasingly important ceramic material with a wide variety of traditional, as well as new applications. In the past, h-BN was commonly prepared in powder form by high temperature pyrolysis of mixtures of simple boron and nitrogen bearing starting materials, e.g., $B(OH)_3$ and $(NH_2)_2CO$, $B(OH)_3$, C and N_2 and KBH_4 and NH_4Cl [1,2]. Boron nitride bodies were subsequently obtained from the powders by high temperature sintering. More recently, chemical vapor deposition (CVD) methods have begun to displace the simple pyrolysis reactions and of course the synthetic chemistry has also been modified. For example, CVD syntheses have typically employed combinations of simple gaseous reagents such as BCl_3 and NH_3, BF_3 and NH_3 or B_2H_6 and NH_3 at a hot substrate [3]. The CVD processing technique is quite flexible, and it has been found that both high purity, high density BN bodies and films can be prepared. It is interesting to note here that detailed physical characterizations reveal a strong dependence of h-BN microstructure on preparative methodology [3,4]. Continuation studies of processing procedures will likely reveal systematics in microstructures not yet realized.

A third general synthetic procedure, pyrolysis of preceramic polymers, has been attracting increasing attention. Wynne and Rice [5] have pointed out that the polymer pyrolysis technique offers several advantages over other preparative techniques, and the advantages have been well studied for the formation of carbon materials from pyrolyses of carbon-based polymers. More recently, the technique has been found especially suitable to syntheses of silicon carbide and silicon nitride [5]. In the same review Wynne and Rice have suggested that preceramic pyrolyses chemistry should find applications in the formation of boron nitride materials; however, until recently, few descriptions of the technique applied to BN materials have appeared.

A partial explanation for the slow growth of preceramic pyrolytic syntheses of BN materials may be found in the history of boron-nitrogen molecular chemistry. This sub-area received a good deal of attention during the 1950's and 1960's, and many simple amine boranes and amino boranes were reported [6]. In addition, numerous attempts were made at that time to prepare polymeric boron-nitrogen compounds [7]. However, very few successes were published in the open literature. Since those years this preparative chemical sub-area has remained nearly dormant in the USA.

More recently, there has been a rekindling of interest in boron-nitrogen compound monomers as ceramic precursors and several intriguing results have been published utilizing borazene compounds. For example, Taniguchi and co-workers [8] noted that B-triamino, N-triphenyl borazene $(H_2NBNPh)_3$ formed a polymeric melt which was drawn into fibers or formed as a film and pyrolyzed to make boron nitride materials. Constant and Feurer [9] have pyrolyzed hexachloroborazene $(ClBNCl)_3$ at 900-950°C on silicon substrates and observed the formation of an amorphous boron-nitride material which could be converted, in part, to h-BN by electron bombardment. Rice and co-workers [10] have also pyrolyzed a series of borazene materials and a variety of products apparently rich in boron and poor in nitrogen have been obtained.

Our own prior experience in boron-nitrogen chemistry which included studies of the chemistry of aminoboranes, borazenes and borazanes suggested potential useful routes to the formation of oligomeric or polymeric boron nitrogen compounds which might serve as preceramic materials for BN materials. We outline here several aspects of our synthesis program devoted to the synthesis and utilization of BN ceramic precursors.

EXPERIMENTAL

Preparation of Preceramic Materials

One synthetic strategy which we have exploited involves the crosslinking of cycloborazene rings with silyl amine groups. This reaction scheme depends upon the formation of B-N-B bonds with concomitant elimination of trialkylsilyl halide, a reaction which is well known to be facile in the formation of monomeric aminoboranes (Equation 1) [11]. In order to test this concept utilizing borazene monomers a series of model compound reactions were explored. For example, combination of B,B'-dimethyl

$$R_2BCl + R_2'NSiMe_3 \rightarrow R_2BNR_2' + Me_3SiCl \tag{1}$$

$$\text{R = H, Me} \qquad \underset{\sim}{1} \tag{2}$$

B"-chloro N-trimethyl borazene with hexamethyldisilazane or heptamethyl disilazane (Equation 2) resulted in the formation, in high yield, of the crosslinked diborazene compounds, 1, [12], which have been fully characterized by mass, infrared and NMR spectroscopic techniques. More importantly, the reaction course confirmed the synthetic concept.

Extending the scheme outlined above, combination of B-methyl B',B"-dichloro N-trimethyl borazene with disilazane reagents (Equation 3) resulted in the formation of heavy oils which deposit film materials. These compounds have indefinte stability in methylene chloride solutions, and the solutions species have been partially characterized by ^{11}B and ^1H NMR spectroscopy. The molecular weight distribution for the oligomer has not yet been accurately determined. The film materials can be pyrolyzed and TGA analyses (Figure 1) show loss in weight from 25°C to

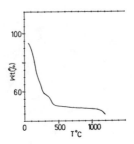

Figure 1. TGA in air for borazene film material 2.

250°C with minor weight loss between 300-400°C. The total weight loss is about 50% out to 1000°C, and a significant amount of initial weight loss is due to expulsion of trapped solvent. Further characterization of the film before and after pyrolysis is in progress. Although the characterization of these film materials is incomplete, the results are consistent with formation of a two-dimensional oligomer containing monomer units 2' with the chains terminated by B-Cl and N-SiMe3 groups.

Based upon these results we next examined the condensation reactions of several tri-chloroborazenes with disilazanes anticipating that the borazene fragments might oligomerize via crosslinking at all three boron atoms in each ring. Indeed, combination of B-trichloroborazene 3 with

hexamethyldisilazane (Equation 4) in a series of solvents (C_6H_5Cl, Et_2O or CH_2Cl_2) resulted in the formation of colorless gels. Extensive pumping and agitation of the gel resulted in the formation of a hard, granular solid. TGA of the solid material obtained from chlorobenzene is shown in Figure 2. The white solid was also dissolved in liquid ammonia and the material obtained from the evaporated filtrate showed the TGA trace shown in Figure 3. Clearly solvent treatment affects pyrolysis profiles and this behavior is being examined in detail.

386

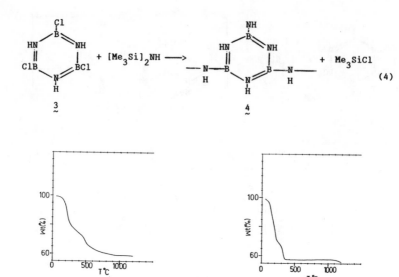

Figure 2. TGA in argon for borazene
oligomer 4 prepared in
CH_2Cl_2.

Figure 3. TGA in air for borazene
oligomer 4 prepared in
CH_2Cl_2 and redissolved
in NH_3.

Bulk samples of the gel materials with Me3SiCl and free solvent
removed were then heated in vacuo to 275°C. The volatiles were
continuously removed into a cold trap and analyzed by mass spectrometry and
infrared spectroscopy. The volatiles were found to be NH_3 and NH_4Cl.
The remaining solid material was examined by FT-IR, powder X-ray
diffraction.and TEM techniques. The infrared spectrum shows two bands,
1425 and 1065 cm^{-1}, in the B-N stretch region and a single band,
710 cm^{-1}, in the B-N-B bending region, and the spectrum is similar to
spectra which have been published for borazenes and boron nitride [13]. In
addition, there is a strong, broad absorption centered at ~3300 cm^{-1}
which can be assigned to a terminal N-H stretch. This observation on its
own suggests that there are numerous N-H bonds in the idealized $BN_{1.5}H_{1.5}$
material 4 which have not undergone pyrolysis and condensation with
neighboring boron atoms.

X-ray powder diffraction patterns for the thermolyzed 4 indicated that
this material is a weak scatterer; however, a broad relatively strong
reflection, $2\theta(Cu, K_\alpha)=24°$, a weak, sharp reflection, $2\theta=33°$, and
a weak, broad reflection, $2\theta=42°$, were observed. Furthermore, these
reflections can be compared with the 2θ values for the first three lines
of authentic h-BN: $2\theta=26.8, 41.7$ and $43.9°$. The appearance of the X-ray
powder patterns is reminiscent of patterns published for thick CVD films on
iron [14]. Lastly, TEM and electron diffraction photographs of finely
powdered samples of thermolyzed 4 confirmed that the majority of the
material is amorphous.

Samples of the thermolyzed borazene oligomer 1 were subsequently pyrolyzed for 1–3 days at 825°C under dynamic evacuation or under nitrogen. During this period additional NH_3 and NH_4Cl were evolved and identified by IR and mass spectrometric methods. Infrared spectra (Figure 4) of the resulting cream colored solid were found to be very similar to spectra recorded for CVD films of h–BN although there remained a band, at much lower intensity compared to the band in the spectrum of the thermolyzed material, centered at 3300 cm^{-1}. The X-ray powder pattern for the pyrolyzed material still displayed rather broad reflections at $2\theta=22°$ and 42°. TEM and electron diffraction photographs again indicated a highly amorphous material.

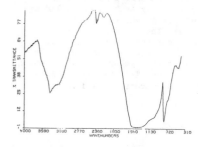
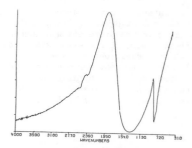

Figure 4. a) Infrared spectrum for pyrolyzed (825°C) borazene oligomer 4.

Figure 4. b) Infrared spectrum for commercial h–BN.

Elemental analyses for the thermolyzed and pyrolyzed BN materials were difficult to perform. Conventional microanalyses performed by commercial laboratories typically produced incomplete accounting of mass and low values for nitrogen composition. In our laboratory we have employed Kjeldhal techniques as well as a methodology developed at Sandia National Laboratory [15] which provided improved results. Even still small amounts of material remained unconverted in the dissolution step. Typical results to date indicate very low silicon contents, and, unlike the borazene pyrolysis products reported by Rice, reasonably large nitrogen contents, e.g., 43% vs 56.4% (theoretical for BN) are obtained.

Some additional characteristics of the gels, 4, have been examined. For example, 0.5 g samples of 3 have been combined with hexamethyldisilazane in solvent and the mixture placed in a 2.5 cm diameter teflon container capped with a septum. The septum was then punctured with a hypodermic needle and the container placed in a dessicator flushed with dry nitrogen. Solvent and Me_3SiCl slowly escaped through the needle and the gel shrunk and released from the sides of the container. This behavior resembled silica gel consolidations. In the final stage (3–7 days) a clear, transparent slightly yellow colored button was obtained with a volume approximately 20% of the original gel. The origin of the yellow color and the thermal chemistry of the buttons are under study. Preliminary critical point solvent exchange with CO_2 indicates that the original included solvent can be removed and the resulting buttons are also under study.

388

Several other preceramic oligomeric gels have been prepared from
N-substituted borazenes. For example, B-trichloro N-trimethyl borazene
combines with hexamethyldisilazane and heptamethyldisilazane to form
colorless gels. These materials have been treated and characterized in a
fashion similar to that outlined above. In these cases, the pyrolysis
products are grey or black and inclusion of carbon is indicated.

CONCLUSIONS

    Based upon these early results from our program it appears that
ceramic materials rich in boron and nitrogen can be prepared from thin film
and gel precursors. Further studies are required and are in progress to
complete the chemical and physical characterizations of the pyrolyzed
products. The nature of the thin film and gel precursors suggests that
these materials may be amendable to other processing techniques including
microwave heating and critical point solvent drying which are being
utilized in the preparation of other ceramics. The amorphous, porous
microstructure of the pyrolysis products also suggests several applications
of the materials where foams are required. The synthetic methodology is
relatively straight forward and lends itself to the systematic introduction
of dopant groups. Chemistry which introduces Group IVA, VA and VIA main
group elements and transition metals in the ring structures and in
crosslinking groups is in progress at this time.

ACKNOWLEDGEMENT

    The authors wish to acknowledge the generous support of this research
by Sandia National Laboratory. We also wish to thank W. Hammeter of SNL
for assistance in obtaining TGA data, and Professor A. Datye of the UNM
Chemical Engineering Department for obtaining initial TEM analyses.

REFERENCES

1. A. Meller in Gmelin Handbuch der Anorganische Chemie, Boron Compounds
    2nd Suppl. Vol.1 (Springer-Verlog, Berlin, 1983) p.304.
2. H. Sumiya, T. Iseki and A. Onodera, Mat. Res. Bull. 18, 1203 (1983).
3. N.J. Archer, Chem. Soc. (London) Special Publ. No. 30 1977, 167 and
    references therein.
4. T. Matsuda, N. Uno, H. Nakae and T. Hirai, J. Mat. Sci. 21, 649 (1986).
5. K.J. Wynne and R.W. Rice, Ann. Rev. Mater. Sci. 14, 297 (1984).
6. K. Niedenzu and J.W. Dawson in The Chemistry of Boron and its
    Compounds edited by E.L. Muetterties (J. Wiley Publishing Co.,
    New York, 1967) p.377.
7. M.F. Lappert, in Developments in Inorganic Polymer Chemistry, edited
    by M.F. Lappert and G.J. Leigh (Elsevier Publishing Co., New York,
    1962) p.20.
8. I. Taniguchi, K. Harada, T. Maeda, Jpn. Kokai, 7653, 000 (1976); Chem.
    Abstr. 85:96582v.
9. G. Constant and R. Feurer, J. Less-Common Met. 82, 113 (1981).
10. B.A. Bender, R.W. Rice and J.R. Spain, Cer. Eng. Sci. Proc. 6, 1171
    (1985).
11. H. Nöth, Z. Naturforschg. 16b, 618 (1961).
12. C.K. Narula, R.T. Paine and R. Schaeffer, to be published.
13. M.J. Rand and J.F. Roberts, J. Electrochem. Soc. 115, 423 (1968).
14. T. Takahashi, H. Itoh and A. Takeuchi, J. Cryst. Growth 47, 245 (1979).
15. S.L. Erickson and F.J. Conrad SNL Rept. SC-RR-710073, Feb. 1971.

A NEW CATALYTIC METHOD FOR PRODUCING PRECERAMIC POLYSILAZANES

YIGAL D. BLUM, RICHARD M. LAINE, KENNETH B. SCHWARTZ, DAVID J. ROWCLIFFE,
ROBERT C. BENING, AND DAVID B. COTTS
Contribution from the Organometallic Chemistry and the Ceramics Programs and
the Polymer Sciences Department, SRI International, Menlo Park, CA 94025

Abstract

A transition metal (e.g., $Ru_3(CO)_{12}$, Pt/C) catalyzed process for Si-N
bond formation is discussed that provides a new route to mono-, oligo-, and
polysilazanes. The catalysts function by activating Si-H bonds in the presence of ammonia. Polymeric silazanes can also be produced from oligomers in
the presence of ammonia at low temperatures. This method allows us to control
or modify the composition of the polysilazane during or after the polymerization. A variety of polysilazanes were prepared and converted to Si_3N_4 with
ceramic yields ranging from 55%-85%. By varying the monomers and reaction
conditions, we can control the nitrogen and carbon content in the preceramic
polymers, which enables us to obtain ceramic products that are primarily Si_3N_4
and simultaneously minimizes the coproduction of SiC and C.

Introduction

The feasibility and potential advantages of using inorganic or organometallic polymers as ceramic precursors were first suggested by Chantrell and
Popper in the early sixties [1]. The concept represents a potential solution
to the difficulties encountered in fabricating advanced ceramic shapes normally prepared by high temperature and pressure forming of bulk powders.
Tractable (soluble, meltable, or malleable) inorganic or organometallic polymer ceramic precursors (preceramics) can be formed (shaped, coated, or spun)
at low temperatures and pressures and then pyrolytically converted to ceramic
materials [2]. They are also potentially advantageous when used as binders
[3] for ceramic powders in comparison with the organic binders currently in
use because pyrolysis converts the preceramic binder to ceramic, reducing void
space in the green body. Additionally, better ceramic structures are feasible
because one can control the molecular structure and the chemical makeup of the
precursor [4], which could readily carry over into the ceramic product.

In a recent review, Wynne and Rice [5] report the existence of preceramic
precursors to Si_3N_4, SiC, B_4C, and BN. However, as these authors note, very
little experimental work has been done on the subject, and almost no scientific basis has been established for the design and development of new preceramics. Consequently, considerable emphasis must be placed on developing all
aspects of preceramic chemistry: from the design and synthesis of preceramics, to studies of their reaction chemistry during pyrolysis, and to the
detailed characterization of the ceramic products.

This article describes the use of transition metal catalysts as a mean of
synthesizing polysilazane ceramic precursors to silicon nitride. The method
described can be used to synthesize, modify, and crosslink polysilazane precursors to silicon nitride and mixtures of silicon nitride and silicon carbide. A companion paper by Schwartz et al. describes the pyrolysis chemistry
of selected polysilazanes [6].

Background

Synthesis routes to polysilazanes (mostly low molecular weight oligomers)
have been available for sixty years and are based on the ammonolysis of chlorosilanes [7]. The advent of high utility polysiloxanes or silicone rubbers
some thirty years ago renewed interest in the synthesis of polysilazane
analogs which were potentially more useful because nitrogen can form one more

bond than oxygen. This impetus resulted in the development of acidic and basic methods [8,9] for the oligomerization/polymerization of cyclotrimeric, and tetrameric silazanes (ring–opening polymerization). However, the products of these reactions are almost always somewhat higher molecular weight oligomers that are often intractable [9] and/or unsuitable as ceramic precursors because of their high volatility or ease of decomposition to volatile products.

The rare exception is found in the work of Verbeek and Winter [10,11], in which $MeSi(NHMe)_3$ (produced by aminolysis of $MeSiCl_3$) gives a tractable resin on heating to 520°C. Pyrolysis gives moderate (60%) ceramic yields of $Si_3N_4/$ SiC mixtures. Recently, Penn et al. [10c] showed that the material can be spun and transformed into good quality ceramic fibers. Seyferth and Wiseman then built on the original work of Fink [12] to catalytically couple cyclo-silazanes of the form $-[RSiHNH]_x-$ to form tractable sheet preceramics as illustrated by reaction (1) [13].

$$\left[RSiHNH \right]_x \xrightarrow{\quad catalyst \quad} \left[\begin{array}{c} \end{array} \right]_x \qquad (1)$$

The catalysts, which act as strong bases, dehydrogenate the reactant causing the selected crosslinking necessary to provide high ceramic yields when these oligomers are pyrolyzed.

In all the published and patented studies, polysilazane pyrolysis always results in the formation of mixtures of Si_3N_4, SiC and often produces carbon. The apparent exception can be found in the work of Seyferth and Wiseman [14] in which the pyrolysis of $-[SiH_2NMe]_x-$ gave no SiC; nor was there evidence for carbon formation. Taken overall, these results lead us to the empirical conclusion that the presence of Si–C bonds in preceramic polysilazanes results in the formation of SiC and carbon in addition to the desired Si_3N_4 during pyrolysis.

Results and Discussion

Several years ago, Zoeckler and Laine [15] reported that transition metal clusters [e.g., $Ru_3(CO)_{12}$ and $Rh_6(CO)_{16}$] activated Si–N bond cleavage and catalyzed ring–opening oligomerization of octamethyltetrasilazane, $[SiMe_2NH]_4$, 1, reaction (2).

$$\underset{1}{[SiMe_2NH]_4} + (Me_3Si)_2NH \xrightarrow{catalyst/135°C} \underset{x = 1-12}{-[SiMe_2NH]_x-} \qquad (2)^*$$

catalyst = $Ru_3(CO)_{12}/H_2$, $H_4Ru_4(CO)_{12}$

*$(Me_3Si)_2NH$ is used as capping agent, but is not necessary for Si–N bond activation.

Recently, we discovered that the catalytic efficiency of reaction (2) increases by two orders of magnitude when the reactions were conducted under as little as 1 atm of H_2 or in the presence of a metal hydride [16]. The major nonvolatile products obtained from vacuum distillation (180°C/300 μm) were linear oligodimethylsilazanes with number average molecular weights ($\bar{M}n$) from 500 to 2600 daltons as determined by gas chromatography–mass spectrometry (GC-MS) and by vapor pressure osmometry (VPO). In addition, much smaller amounts of bicyclic and tricyclic compounds were also obtained. The selectivities,

$\overline{M}n$, and yields of products could be controlled to some extent through choice of catalyst and reaction conditions [16].

The important observation, the enhanced activity exhibited in the presence of hydrogen or metal hydrides, led us to propose a preliminary reaction mechanism for reaction (2) involving metal hydride catalyzed hydrogenation of the Si-N bond as shown below:

$$\lfloor[SiMe_2NH]_4\rfloor + MH_2 \rightleftharpoons HM-[SiMe_2NH]_4-H \qquad (3)$$
$$\underset{2}{}$$

$$HM-[SiMe_2NH]_4-H \rightleftharpoons H-[SiMe_2NH]_4-H + M \qquad (4)$$
$$\underset{3}{}$$

$$H-[SiMe_2NH]_4-H + HM-[SiMe_2NH]_4-H \rightleftharpoons MH_2 + H-[SiMe_2NH]_8-H \qquad (5)$$

$$H-[SiMe_2NH]_8-H \xrightarrow{M/termination} cyclomers \qquad (6)$$

$$H-[SiMe_2NH]_8-H \xrightarrow{M/propagation} higher\ oligomers \qquad (7)$$

Because all the reactions are reversible, this type of reaction cannot be used for synthesizing high molecular weight preceramic precursors. However, the proposed intermediacy of the H-M-Si group, compound 2, in the catalytic cycle is exactly analogous to the intermediates formed metal catalyzed hydrosilation [17]. Compound 3 is actually a special form of silane, $HSiR_2---NH_2$, that contains a terminal NH_2 group. It is apparently fairly reactive because it is not observed in the mass spectral analyses of the reaction products (even using low temperature field ionization MS). These results suggested that the reaction might proceed using simple silanes and amines, as illustrated by reaction (8):

$$R_3SiH + H_2NR \xrightarrow{catalyst} R_3SiNHR + H_2 \qquad (8)$$

Sommer and Citron [18] were able to catalytically synthesize compounds containing single Si-N bonds from the reactions of silanes with amines as suggested by reaction (8), although no evidence was found for formation of oligomers.

We attempted to synthesize oligomers based on the implications of reactions (4)-(7) and the original work of Sommer and Citron, as shown in reaction (9):

$$HSiMe_2NHSiMe_2H + NH_3 \xrightarrow{Ru_3(CO)_{12}/60°C} H-[Me_2SiNH]_x-Me_2SiH + cyclomers + H_2 \qquad (9)$$

Indeed, a very fast reaction ensues even at 60°C and an initial turnover frequency (TF = moles of H_2 or Si-N bonds formed/moles catalyst/h) of 1000-1300 as determined by gas chromatography and NMR. Cyclomers (D.P. = 3-7), linear oligomers (D.P. = 2-11), and small amounts of branched oligomers form as evidenced by GC-MS results [16]. Distillation (180°C/300 μm) of volatile products from the product mixture provides a 20% yield of nonvolatile products with $\overline{M}n$ = 1000-2000 daltons, depending on the ammonia pressure and reaction times. NMR and chemical analyses suggest that the nonvolatiles are mostly linear with between one and two HMe_2Si- branches per thirty $-Me_2SiNH-$ units [16].

Pyrolysis of these polymers provides only traces of ceramic materials including ruthenium silicide. This result supports the general concept presented by Wynne and Rice [5] that linear oligomers/polymers that are incapable of crosslinking will decompose to volatile products during pyrolysis, giving negligible yields of ceramic materials. Linear polymers that do not crosslink have sufficient mobility to loop back on themselves during pyrolysis and split out volatile cyclomers.

Substitution of Et_2SiH_2 in reaction (9) also results in a fast uptake of ammonia (initial TF = 1570 at 60°C) to produce considerable quantities of both linear and cyclic products as

$$H_2SiEt_2 + NH_3 \xrightarrow{Ru_3(CO)_{12}/60°C} H-[Et_2SiNH]_x-Et_2Si-H + \overline{[Et_2SiNH]_x} +$$

$$x = 1-3 \text{ (major)} \qquad x = 3-5 \text{ (major)}$$

$$+ \quad H-[Et_2SiNH]_x-H + H_2 \qquad (10)$$

$$x = 2-5 \text{ (minor)}$$

determined by GC-MS [19]. Unfortunately, the $\bar{M}n$ for this reaction is only 340.

Reactions (9) and (10) suggest that catalytic formation of Si-N bonds via the dehydrogenation of Si-H and N-H bonds is feasible under conditions where Si-N bond cleavage is not prevalent. Moreover, the fact that a variety of both homogeneous and heterogeneous catalysts (e.g., Pt/C [6,19,20]) promote both reactions suggests that the catalytic formation of Si-N bonds represents a general route for synthesizing of polysilzanes. The following example demonstrates the application of this process for synthesizing of a useful preceramic polysilazane.

Preceramic Polysilazanes

As mentioned above, Seyferth and Wiseman [13] found that oligomers of the type $-[SiH_2NMe]_x-$, prepared via aminolysis of H_2SiCl_2 [reaction (11)],

$$H_2SiCl_2 + MeNH_2 \longrightarrow MeNH-[SiH_2NMe]_x-H + \overline{[SiH_2NMe]_4} \qquad (11)$$

$$66\% \qquad\qquad 34\%$$

$$\text{Crude } \bar{M}n = \sim 320 \text{ daltons; } x = \sim 10$$
$$\text{Distilled } \bar{M}n = \sim 570 \text{ daltons}$$

gave only α Si_3N_4 and no SiC or carbon when pyrolyzed at 1000°C and then heated to 1200°C, as determined by x-ray powder diffraction; however, the ceramic yields were low ($\sim 38\%$). The presence of Si-H bonds in these oligomers, when coupled with our observations of catalytic formation of Si-N bonds and Wynne and Rice discussions of the importance of latent reactivity/crosslinking in the conversion of preceramic polymers, suggested that we might be able to improve the ceramic yields obtainable from these polymers through catalytic crosslinking reactions.

The existence of both Si-H bonds and terminal N-H bonds provides the opportunity to perform both crosslinking and chain extension reactions. The latter operation offers a method of increasing the molecular weight of these compounds, which is very important if one wants to spin preceramic fibers from them. We have now examined the reactions of the products from reaction (11) with $Ru_3(CO)_{12}$ as catalyst under a variety of conditions. We find that we can catalyze chain extension, and that this results in greatly improved ceramic yields.

The results of our studies are recorded in Table I. The pertinent observations that can be made from these results are as follows:

- Ratios of catalyst (as ruthenium) to silazane unit (Si-N) can be as low as 1:5400 and still give good crosslinking and high ceramic yields.
- Oligomer $\bar{M}n$ can be increased from 560 daltons up to ~ 1600 with a concomitant increase in viscosity.

- The 1600-dalton product is still soluble in toluene and THF.
- On extended low temperature curing, the oligomers thermoset to give soft rubbers.

Table I. Polymerization Degree and Ceramic Yields of
$-[H_2SiNMe]_x-$ in the Presence of $Ru_3(CO)_{12}$

Si-N/Cat.[a]	Temp. (°C)	Time (h)	Product		Ceramic Yield (% at 900°C)
1360	60	4	Sol. visc. oil $\bar{M}n$ 1180		68 (Si_3N_4)
1360	135	2	Soft rubber		75 (Si_3N_4)
			$\bar{M}n$	Viscosity[b] (poise)	
5420	90	8	1560	143	64
5420	90	20	1620	6450	68

[a]Moles of silazane units per mol of catalyst.
[b]The viscosity of the reactant is 12 poise.

Some of the current problems that remain to be solved include the need to further increase polymer molecular weights and/or viscosities while maintaining polymer tractability. The reason for our limited success in obtaining further increases in both viscosity and molecular weight is likely a direct consequence of the early increases in viscosity we do obtain. We believe that transport of the active catalyst to reactive sites in the polymer becomes increasingly more difficult as polymer viscosity increases. In future studies we plan to examine the use of solvents to dilute the precursor, providing lower initial viscosities, so that we can obtain additional chain extension and higher viscosities.

The pyrolysis studies described by Schwartz et al. [6] demonstrate that this approach is successful in providing preceramics that give significantly higher ceramic yields (~ 64-75 wt%) than the 38% reported for the starting mixture of oligomers; (the volatile oligomers are not distilled prior to the catalysis and the pyrolysis processes). We find that our ceramic products are mostly α silicon nitride (> 80 wt%) as determined by x-ray powder diffraction, with the remainder being carbon (2.5-18 wt%) and traces of oxygen (< 0.5 wt%) and ruthenium as determined by chemical analyses. We do not observe, in any instance, the formation of SiC, which seems to confirm our empirical observation that the absence of Si-C bond in the polymer precludes the formation of SiC on pyrolysis. However, further studies must be undertaken to confirm these observations. We must comment that the elemental analyses for carbon appear somewhat inconsistent with our TGA results, especially in view of the mass spectral analyses that correlate product gas formation with TGA weight losses.

Conclusions

Our Si-H bond activation studies provide potential solutions to several general problems in the synthesis of tractable silicon nitride preceramic polymers:

- Ammonolysis of halosilanes can be replaced by transition metal catalyzed treatment of hydridosilanes, thereby avoiding problems due to chlorine impurities and ammonium chloride separation and disposal.
- Low molecular weight oligosilazanes containing Si-H and/or N-H bonds can be catalytically chain extended and crosslinked to increase viscosity and molecular weight.
- Very low concentrations of transition metal catalysts are also valuable because they introduce thermosetting features into polysilazanes at low temperatures.
- The absence of Si-C bonds eliminates the formation of large quantities of SiC where silicon nitride is the desired product.
- The silicon-to-nitrogen ratio of the preceramic can be controlled through choice of amine and silane or combinations of amines and silanes (copolymers).

Experimental details of the work reported here are available elsewhere [6,19,20].

References

[1] P. G. Chantrell and P. Popper, Special Ceramics, E. P. Popper, Ed.; New York: Academic (1964), pp. 87-103.
[2] B. E. Walker, Jr., R. W. Rice, P. F. Becker, B. A. Bender, and W. S. Coblenz, Am. Ceram. Soc. Bull. (1983) 62, 916.
[3] R. W. Rice, ibid., 89.
[4] a. S. P. Mukherjee, J. Cryst. Solids (1980) 42, 477.
     b. K. S. Mazdigazni, Ceram. Int. (1982) 8, 42.
[5] K. J. Wynne and R. W. Rice, Ann. Rev. Mater. Sci. (1984) 14, 297.
[6] K. B. Schwartz, D. J. Rowcliffe, Y. D. Blum, and R. M. Laine, in this publication.
[7] a. A. Stock and K. Somieski, Ber. Dtsch. Chem. Ges. (1921) 54, 740-758.
     b. S. D. Brewer and C. P. Haber, J. Am. Chem. Soc. (1948) 70, 361.
     c. R. C. Osthoff and S. W. Kantor, Inorg. Syn. (1957) 5, 61.
[8] a. R. Kruger and E. G. Rochow, J. Poly. Sci. A (1964) 2, 179.
     b. G. Redl and E. G. Rochow, Angew. Chem. (1964) 76, 650 and references therein.
[9] K. A. Andrianov, B. A. Israilov, A. M. Kononov, and G. V. Kotrelev, J. Organomet. Chem., (1965) 3, 129.
[10] a. W.Verbeek, U.S. Patent 3,853,567 (December 1974).
     b. G. Winter, W. Verbeek, and M. Mansmann, U.S. Patent 3,892,583 (July 1975).
     c. B. G. Penn, F. E. Ledbetter III, J. M. Clemons, and J. G. Daniels, J. App. Poly. Sci. (1982) 27, 3751.
         See also related work in reference 11.
[11] R. R. Wells, R. A. Markle, and S. P. Mukherjee, Am. Ceram. Soc. Bull., (1983) 62, 904.
[12] W. Fink, Neth. Pat. Application 6,507,996.
[13] D. Seyferth and G. H. Wiseman, U. S. Patent 4,482,669 (1984).
[14] D. Seyferth and G. H. Wiseman, in Ultrastructure Processin og Ceramics, Glasses and Composites, L. L. Hench and D. R. Ulrich, Eds. (1984), pp. 265-271).
[15] M. T. Zoeckler and R. M. Laine, J. Org. Chem., (1983) 48, 2539.
[16] Y. D. Blum and R. M. Laine, paper submitted to Organometallics.
[17] J. P. Colleman and L. S. Hegedus, Principles and Applications of Organo-transition Metal Chemistry, University Science Books, (1980), pp. 384-401.
[18] L. H. Sommer and J. D. Citron, J. Org. Chem., (1967) 32, 2470.
[19] R. M. Laine and Y. Blum, U.S. Patent Application No. 06/727,414 (patent has been allowed).
[20] R. M. Laine and Y. Blum, U.S. Patent Application submitted March 1, 1986.

PYROLYSIS OF ORGANOSILICON GELS TO SILICON CARBIDE

JOSEPH R. FOX, DOUGLAS A. WHITE, SUSAN M. OLEFF, ROBERT D. BOYER
AND PHYLLIS A. BUDINGER, Standard Oil Research and Development Center,
Warrensville Heights, Ohio 44128

ABSTRACT

Sol-gel precursors to silicon carbide have been prepared using trifunctional
chloro and alkoxysilanes which contain both the silicon and carbon necessary for
SiC formation. Crosslinked gels having the ideal formula $[RSiO_{1.5}]_n$ have been
synthesized by a hydrolysis/condensation scheme for a series of saturated and
unsaturated R groups. The starting gels have been characterized by a variety of
elemental analysis, spectroscopic and physical measurements including IR, XRD,
TGA, surface area and pore volume. A particularly powerful method for
characterizing these gels is the combination of ^{13}C and ^{29}Si solid state NMR which
can provide information about the degree of crosslinking as well as residual
hydroxy/alkoxy content.

The controlled pyrolysis of these gels has been used to prepare silicon carbide-
containing ceramic products with surface areas in excess of $600 m^2/gm$. The
pyrolysis products are best described as a partially crystalline, partially amorphous
mixture of β-SiC, silica and carbon. The effect of carbon chain length and the
degree of unsaturation in the R group on the composition and surface area of the
product has been determined. The origin of the high surface area of the pyrolysis
products has been identified and its implications on potential uses of these materials
is discussed.

INTRODUCTION

The use of sol-gel and inorganic polymer precursors to prepare both oxide and
nonoxide ceramics has received considerable attention in recent times [1]. In this
work, the controlled pyrolysis of polymeric organosilicon gels has been investigated
as a means of preparing high purity, high surface area silicon carbide powders. A
series of organosilicon polymers of the general formula $[RSiO_{1.5}]_n$ have been
synthesized and pyrolyzed to form SiC. The gels were prepared from trifunctional
alkoxy and chlorosilanes which contained both the silicon and carbon necessary for
silicon carbide formation. The R groups (R=methyl, ethyl, propyl, hexyl, vinyl, allyl,
phenyl) were selected to examine the effect of carbon chain length as well as
saturated vs. unsaturated character on the yield and the properties of the SiC
produced. One of the goals of this research was to identify the R group which
would provide the optimum yield of silicon carbide with little or no residual carbon
or silica.

RESULTS

Synthesis of Gels

The gels were synthesized via the hydrolysis and condensation of trifunctional
silanes with the general formula $RSiX_3$, where R was an alkyl or aryl substituent
with less than 7 carbons and X was a hydrolyzable group such as alkoxy or chloro.
Three different synthetic routes to the $[RSiO_{1.5}]_n$ gels were investigated: 1) a base-
catalyzed route employing the alkoxysilanes; 2) an acid-base route which also
employed the alkoxysilanes and 3) hydrolysis of the chlorosilanes.

Mat. Res. Soc. Symp. Proc. Vol. 73. ©1986 Materials Research Society

396

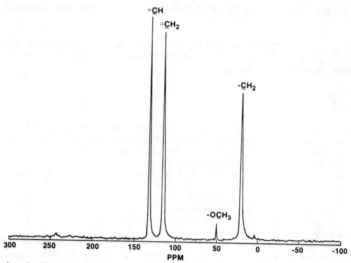

Fig. 1. Solid state ^{13}C NMR spectrum of $[CH_2=CHCH_2SiO_{1.5}]_n$ gel prepared from $CH_2=CHCH_2Si(OCH_3)_3$.

The effectiveness of each procedure was dictated by the nature of the R group. Simple base-catalyzed gellation of trialkoxysilanes worked well for R=methyl and phenyl. In general however, the preferred approach for the alkoxysilanes in terms of gel yield and homogeneity used acid-catalyzed hydrolysis of the -OR groups prior to base-catalyzed gellation. The acid-base route also proved to be particularly amenable to the production of mixed gels, not only for combinations of trifunctional silanes but also for trifunctional and difunctional silane mixtures. The chlorosilane route typically provided high gel yields and employed less expensive starting materials, although the reactions were more exothermic and less controllable than syntheses with alkoxysilanes.

Characterization of Gels

The most useful characterization tool for the precursor gels was the combination of both ^{13}C and ^{29}Si solid state NMR. ^{13}C NMR was useful for determining the residual alkoxy content, while ^{29}Si NMR provided structural information about the silicate framework within the gel, especially the extent to which crosslinking had occurred. For the alkoxysilane-derived gels, solid state ^{13}C NMR was particularly useful in determining the extent to which hydrolysis had occurred. Whereas infrared spectroscopy was incapable of providing information about the residual alkoxy content due to the overlap in the Si-O vibration frequencies, solid state ^{13}C NMR clearly distinguished the carbons present in residual alkoxy groups from those present in the R group directly attached to silicon. In the solid state ^{13}C NMR spectrum in Figure 1 of an allyl gel prepared by the acid-base route, the resonance at 50 ppm is attributable to residual-OCH_3 groups, whereas the stronger resonances at 130, 115 and 20 ppm are due to the carbons in the allyl group itself.

The ^{13}C NMR results showed that the use of an acid catalyst and higher water to silane ratios during hydrolysis led to lower residual alkoxy content in the precursor gels. Solid state ^{13}C NMR spectra of mixed organosilicate gels were also obtained and allowed a quantitative estimate of the ratio of R groups. ^{13}C NMR

397

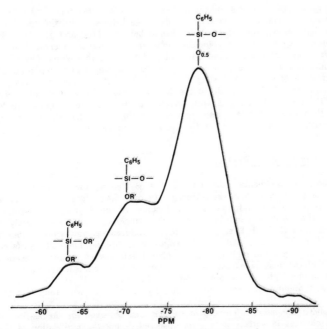

Fig. 2. Solid state ^{29}Si NMR spectrum of $[C_6H_5SiO_{1.5}]_n$ gel prepared from $C_6H_5Si(OCH_3)_3$.

also proved extremely helpful in identifying when undesirable side reactions had occurred during gel synthesis.

Whereas ^{13}C NMR was particularly useful for determining the extent of hydrolysis during synthesis, ^{29}Si NMR provided information about the degree of crosslinking in the gels. Figure 2 illustrates the solid state ^{29}Si NMR spectrum of a phenyl gel synthesized by the base-catalyzed route. The resonances at -80, -70 and -63 ppm can be attributed to crosslinked, monosubstituted, and disubstituted units respectively, where R'=H or CH_3 [2]. The ^{29}Si NMR data indicated that all of the gels deviated from the ideal $[RSiO_{1.5}]_n$ stoichiometry and contained numerous uncrosslinked units with residual hydroxy and/or alkoxy groups.

Gel Pyrolysis

For the pyrolysis experiments, the polymeric gel precursors to SiC were prepared by the optimum route identified for each R group. Typically, the gels were fired at 2°C/min under a flowing argon atmosphere to 1500°C with a soak time of two hours. The most important variable during pyrolysis was the heating rate. With a heating rate of 10°C/min, dramatic weight losses were observed upon pyrolysis and the char yield was quite low.

Pyrolysis products heated to 1500°C typically consisted of large, shiny black particles which were quite brittle. X-ray diffraction indicated that in essentially every case, the low temperature β form of silicon carbide had formed as a result of gel pyrolysis. The relative amount of silicon carbide and its crystallinity varied dramatically from sample to sample. The x-ray data for gels heated to 1000 and 1200°C did not show any evidence for the formation of crystalline silicon carbide.

Thermal gravimetric analysis indicated that gel decomposition occurred in essentially two major stages--a low temperature stage between approximately 500 and 600°C and a high temperature stage which began between 1400 and 1500°C. The high temperature weight loss was coincident with the appearance of crystalline SiC in the x-ray diffraction patterns and can be attributed to the carbothermic reduction of silica to produce silicon carbide. The low temperature decomposition step produced an intimate mixture of carbon and silica, and the size of the weight loss was dramatically influenced by the nature of the R group. Elemental analysis of residues heated to 900°C indicated that gels with unsaturated R groups decomposed to leave more carbon than their saturated counterparts, with the result that the carbothermic reduction reaction proceeded further at the elevated temperatures.

Characterization of Pyrolysis Products

Surface area measurements indicated that the as-fired residues could have a range of surface areas, from less than $1m^2/gm$ to over $600m^2/gm$. The surface area of the pyrolysis products was found to be a function of the R group, the synthesis conditions and the pyrolysis conditions, i.e., firing temperature and time. Pyrolysis at 1500°C of gels with R=methyl, ethyl and propyl produced residues with surface areas less than $5m^2/gm$. Vinyl and allyl gels decomposed to give residues with surface areas an order of magnitude higher, whereas phenyl gels always produced residues at 1500°C with surface areas in excess of $100m^2/gm$. Mercury porosimetry data indicated that in the high surface area products obtained from $[C_6H_5SiO_{1.5}]_n$ gels, most of the pore volume was associated with very small pores with radii less than 100Å.

In addition to the large differences in surface area observed for the as-fired gels, very dramatic changes in surface area were observed during the purification of the residues to remove residual carbon and silica. These changes were accompanied by changes in the pore volume and pore distribution as indicated by the data in Table I for a pair of $[C_6H_5O_{1.5}]_n$ gels fired at 1500°C. Treatment of the fired residues with HF usually resulted in an increase in surface area, especially for samples with lower surface areas. On the contrary, the oxidation step to remove excess carbon usually resulted in a large decrease in surface area, almost an order of magnitude for the samples with surface areas in excess of $500m^2/gm$. The samples with lower surface areas exhibited relatively smaller losses in surface area upon oxidation.

TABLE I. Changes in surface area, pore volume and pore distribution during purification of $[C_6H_5SiO_{1.5}]_n$ gels.

	Surface Area (m²/gm)	Pore Volume (cc/gm)	Pore Distribution (Å)
As-fired	226	0.110	r < 100
After HF	877	0.173	r < 100
As-fired	612	0.392	r < 100
After Oxidation	62	0.939	150 < r < 250

Figure 3a shows an SEM photograph of a pyrolyzed $[C_6H_5O_{1.5}]_n$ gel residue with a surface area in excess of $600m^2/gm$. The morphology consisted of large (>100μ) irregularly-shaped particles with a smooth, glass-like surface devoid of visible pores. The large particles did not appear to be physical agglomerates of much smaller particles. Upon treatment with HF, the sample experienced a 3.8% weight loss, but no dramatic changes in morphology or surface area were observed. When the as-fired residue was oxidized at 1000°C, the sample experienced a 20.8% weight loss and an order of magnitude drop in surface area to $62m^2/gm$. Once again however, SEM revealed no dramatic changes in morphology.

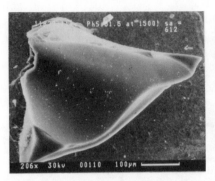 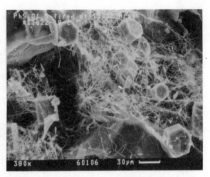

Fig. 3. Scanning electron micrographs of a) as-fired $[C_6H_5SiO_{1.5}]_n$ gel and b) pyrolysis product after oxidation and HF treatment.

Figure 3b shows an SEM photograph of the purified residue after both oxidation and subsequent HF treatment. There were distinct changes in morphology observable after both purification steps. However, it is important to note that a significant fraction of the sample still consisted of the large, irregularly-shaped particles, and that Figure 3b indicates portions of the sample where morphology changes were apparent. The SEM results indicated that HF treatment after carbon burnoff was more successful at etching away glassy portions of the large particles. In those regions where significant etching had occurred, two distinct morphologies of β-SiC were observed -- 1) whiskers with a diameter less than 1μ and lengths of up to 100μ; and 2) extremely well-defined hexagonally-shaped crystallites which reached 30μ in size.

Although controlled pyrolysis of certain gels yielded residues with a high silicon carbide content, it was found that the glass-like nature of the pyrolysis products made complete carbon and silica removal difficult. The most difficult problems encountered were associated with carbon removal. It was found that the ease of carbon removal from the residues was closely linked to the porosity of the solid. If a sample was highly porous, oxidation of the carbon was quite facile, but carbon removal from low surface area, low porosity materials proved to be extremely difficult. In such cases, an HF wash prior to oxidation facilitated carbon removal by opening up porosity in the residue.

DISCUSSION

One of the most touted advantages of inorganic polymer and sol-gel routes to ceramics and glasses is the homogeneity afforded on a microscopic level in the precursor and ultimately in the final product. In this research, polymeric organosilicon gels which contained both the carbon and silicon necessary for silicon carbide formation were synthesized and pyrolyzed to produce SiC. Strictly speaking, these precursors are best described as inorganic polymers which proceed through a gel stage during synthesis. In this respect, they can be distinguished from sol-gel routes to nonoxidic ceramics which utilize an external carbon source such as sucrose [3,4]. The pyrolysis products from the organosilicon polymers were not similar in their chemical or physical makeup and appearance to such sol-gel derived SiC powders. The latter consist of discrete submicron-sized particles and the high surface area is associated with small particle size.

On the basis of microscopy and x-ray diffraction analysis and the purification results, the gel pyrolysis products are best described as a partially crystalline, partially amorphous mixture of β-SiC, silica and carbon, whose surface area is associated with the porosity/pore distribution in the glassy matrix. The porosity in the residues is generated by carbon monoxide evolution during the carbothermic reduction to produce SiC and during the purification steps to remove residual carbon and silica.

The present work has shown how the composition and the porosity of the pyrolysis product is dramatically affected by the structure of the precursor gel, the nature of the substituents attached to the silicon atoms and the pyrolysis conditions. Gels fired at 1000 and 1200°C had not undergone the carbothermic reduction reaction and as a result produced very low surface area, low porosity residues, whose x-ray diffraction patterns showed no evidence for crystalline SiC. Gels with unsaturated R groups decomposed to leave more carbon available for the carbothermic reduction reaction and produced residues with higher SiC content and higher surface areas than their saturated counterparts. Of the R groups investigated, it was determined that only the gels with R=phenyl decomposed to provide enough carbon to convert all of the silica into silicon carbide. The pyrolysis of highly crosslinked $[C_6H_5SiO_{1.5}]_n$ gels provided extremely high surface area residues with SiC content as high as 75%. These residues contained micropores with radii less than 100Å, whose presence translated into the high surface areas observed for these materials.

During purification, the removal of carbon and/or silica often had dramatic effects on the surface area of the residues. These changes could be linked to changes in the porosity/pore distribution of the material. Whereas silica removal opened up smaller pores resulting in an increase in surface area, carbon burnoff opened up considerably larger pores translating into a decrease in surface area. Thus, it appears that it was the distribution of pores which ultimately determined the surface area of these products.

In view of the difficulties associated with the removal of the carbon and silica as well as the irregular morphology of the products, further refinement of the pyrolysis schedule and the purification procedure is necessary before the sinterability of these powders is investigated at higher temperatures. The present work has shown that the pyrolysis of the $[RSiO_{1.5}]_n$ gels can produce SiC-containing residues with surface areas ranging from less than $1m^2/gm$ to almost $900m^2/gm$. This wide range may provide potential opportunities for these materials as binders, infiltrants or high temperature catalyst supports.

REFERENCES

1. K.J. Wynne and R.W. Rice, "Ceramics via Polymer Pyrolysis," Ann. Rev. Mater. Sci. 14, 297-334 (1984).

2. G. Engelhardt, H. Janke, E. Lippmaa and A. Samoson, "Structure Investigations of Solid Organosilicon Polymers by High Resolution Solid State ^{29}Si NMR," J. Organomet. Chem. 210, 295-301 (1981).

3. G.C. Wei, C.S. Morgan, C.R. Kennedy and D.R. Johnson, "Synthesis, Characterization and Fabrication of Silicon Carbide Structural Ceramics," Annu. Conf. Mater. Coal Convers. Util. (Proc.) 7, 187-219 (1982).

4. T. Kuramoto, H. Ono and T. Miyoshi, Japanese Patent No. J5,8104-010 (11 December 1981).

PRECURSORS OF TITANIUM CARBONITRIDE

L. MAYA
Chemistry Division, Oak Ridge National Laboratory, Oak Ridge, TN 37831

ABSTRACT

A series of novel compounds was isolated in the course of exploratory work on the chemistry of titanium halides in liquid ammonia. This work was undertaken to study synthetic approaches to titanium-containing precursors of ceramic materials. Representative of these compounds is a mixed valence Ti(III)-Ti(IV) tetramer, $[NH_4^+ \cdot NH_3]_2[Ti_4Br_4(NH_2)_{12}]^{-2}$, which was produced by the reaction of potassium borohydride and titanium IV bromide in liquid ammonia at room temperature. Similar ammonium salts of either Ti(IV) or Ti(III) were also prepared. The reaction of the ammonium salts with sodium acetylide in liquid ammonia evolves acetylene in an amount equivalent to the ammonium ion present. This provided the charge of the complex and yielded novel titanium acetylide derivatives. The acetylides convert into titanium carbonitrides upon thermal treatment to 800°C.

The reaction of titanium halides of their ammonolytic products with sodium acetylide in liquid ammonia to yield halogen-free acetylide precursors having a relatively high titanium content appears to be a convenient synthetic approach. This is made possible by the fact that the alkali metal halide by-products are soluble and easily separated in that reaction medium. This approach appears to be a generalized route applicable to a number of transition metal elements of interest.

INTRODUCTION

Current interest on advanced ceramic materials centers around materials such as carbides, nitrides, borides and silicides which have properties such as hardness, corrosion and thermal stability that cannot be matched by metallic alloys or other structural materials. The preparation of appropriate precursors for conversion by thermal or photolytic means into ceramic materials is an area that can benefit from the contribution of chemists. A recent review [1] described the evolution of this approach and points out the opportunities for developments in this area of research. Alternative preparative approaches to advanced ceramic materials include the use of plasmas, chemical vapor deposition and high temperature syntheses.

The preparation of advanced ceramic materials requires, in general, the exclusion of oxygen since the presence of this element can have detrimental effects in the properties of some of the above mentioned materials. In view of this requirement, use of oxygen-free reactants and solvents is almost a necessity in the preparation of precursors. For this reason liquid ammonia is a useful reaction medium for conducting this type of synthetic chemistry.

Titanium halides are convenient starting materials because of their availability and degree of purity. Titanium IV halides undergo ammonolytic reactions in liquid ammonia to yield initially $TiX(NH_2)_3 \cdot 2NH_3$, however the ammonia of solvation is lost upon continued pumping at room temperature [2-4]. The ammonium halide by-product of the ammonolysis is readily separated from the insoluble titanium products thus providing intermediates with a relatively higher titanium content for subsequent chemical transformation. The titanium III halides do not ammonolyze [5] but form a number of solvates of which, $TiX_3 \cdot 4NH_3$ is the room temperature phase. Most of the past efforts in this field were concerned with the stoichiometry of the ammonolytic reactions and the composition of the solvates as a function of temperature. In the present study the ammonolytic products were subjected to a series of reactions with alkali borohydride, amide and acetylide in order to obtain halogen free derivatives. The nature of the products of these reactions are revealing of the structure of the ammonolytic products. The conversion of the derivatives by thermal means into ceramic materials was also studied.

RESULTS AND DISCUSSION

Mixed Valence Compounds

A mixed valence Ti(III)-Ti(IV) compound $[NH_4^+ \cdot NH_3]_2[Ti_4Br_4(NH_2)_{12}]^{2-}$ was obtained from the reaction of Ti(IV) bromide and potassium borohydride in liquid ammonia at room temperature. The reaction was conducted in a sealed ampoule provided with a side arm into which the soluble fraction was separated leaving the product in the straight portion of the ampoule. The analysis of this material, compound A, showed titanium, bromide and ammonia in the ratios of 1:1:4 respectively. The oxidation state of the titanium was established by back titration of excess Ce(IV) that showed an average oxidation number of 3.5. This value suggested the presence of at least a dimer but the tetrameric character of compound A was suggested by the following evidence.

Treatment of compound A with potassium amide at -50°C in liquid ammonia to replace the bromide by amide produced only partial substitution obtaining a product, compound A1, containing K, Ti, Br and NH_3 in a 0.25:1.0:0.75:3.0 relation. Total substitution of a bromide by amide was observed in a separate experiment conducted at room temperature that pro-duced a condensed material, compound A2, with an empirical formula corresponding to $KTi(NH)_2$. Additional evidence regarding the tetrameric character of compound A was derived from the composition of purple solids, A3, isolated from the initially soluble fraction of the $TiBr_4$-KBH_4 reaction mixture. These materials contained Ti, Br, B and NH_3 in the relation 1:0.75:0.25:2.5. An additional product, A4, was isolated from the $TiBr_4$-KBH_4 reaction conducted at -50°C. In this case the analyses showed Ti, Br and NH_3 in the ratios of 1:0.75:5.25. Compound A4 showed an average oxidation number for the titanium of 3.75 corresponding to one Ti(III) out of four titanium atoms.

The above evidence not only suggests that compound A is tetrameric but also that this structure exists already in a common Ti(IV) precursor of all the described derivatives. This material is most likely the initial ammonolytic product $[TiBr(NH_2)_3 \cdot 2NH_3]_4$.

The ionic character of compound A is derived from the fact that the IR spectrum shows the presence of ammonium ions. Reaction of compound A with monosodium acetylide in liquid ammonia provided a way to establish the charge of the tetramer since ammonium ions react with acetylide to release acetylene in a direct stoichiometric relation described by

$$NH_4X + NaC \equiv CH \xrightarrow{Liq.\ NH_3} NaX + HC \equiv CH + NH_3 \tag{1}$$

Treatment of compound A with excess sodium acetylide produced acety-lene corresponding to a charge of -2.34 for the tetramer. The slight excess produced was within the experimental variability for such deter-mination for this and similar compounds. The procedure was validated using ammonium bromide and finding 96% of the theoretically anticipated amount of acetylene.

The product from the acetylide treatment, compound A5, showed an idealized composition corresponding to $Ti_4(C \equiv C)(NH)_6$. The formulation of this compound as an acetylide derivative is based on the chemical analysis and its IR spectrum which shows weak bands at 2030 and 2180 cm^{-1}. These values are in agreement with observations of acetylide derivatives of other acetylides of transition elements [6]. The relative intensity of the $C \equiv C$ stretching bands suggests that the acetylide ligands are bridging and perhaps form the backbone of a polymeric structure [7]. This view is supported by the fact that dissolution of the acetylide requires extended digestion with acid.

Additional physicochemical characterization of compound A included IR and optical spectroscopy as well as ESCA. The latter confirmed the presence of Ti(III) and Ti(IV) in equal proportions as found by chemical analysis. The binding energies for $2p_{3/2}$ peaks of Ti(IV) and Ti(III) were 458.2 and 456.3 eV respectively. These compare with 458.7 eV for Ti(IV) in TiO_2 and 454.8 eV for Ti(II) in TiO reported by Ramqvist [8]. It is seen that the Ti(III) value is intermediate between TiO and TiO_2 but somewhat higher than 455.3 eV reported for TiN [8] which could formally be considered as a Ti(III) compound. However, it seems that the electron transfer from titanium to the nitrogen is less in this compound than that predicted by the stoichiometry [8].

The course of the reaction between compound A and potassium amide at -50°C suggests that the ammonia molecules of solvation are associated with the ammonium ions. This leaves a core consisting of $Ti_4Br_4(NH_2)_{12}^{2-}$ which is similar to other tetrameric units of transition metal elements such as the $Zr_4(OH)_8^{8+}$ ion for which the structure is known [9] and consists of double hydroxo bridges joining the metal centers. The eight coordinate environment around the metal is completed by four moles of H_2O and the charge of the whole unit is balanced by eight halide atoms. Quite similar to these systems is the tetrameric unit $Ru_4(OH)_{12}^{4+}$ known as a solution species [10] which shows a redox chemistry, similar to compounds A and A4, whereby either one or two of the ruthenium atoms is reduced to Ru(III) while preserving the integrity of the unit. The question arises then as to what would be a plausible structure for the titanium tetramer. A double amido bridge structure is attractive but it cannot be assembled with octahedral centers, such as those most likely occurring in titanium. This leaves a tetramer made of two double bridges and two single bridges or as in a similar chromium mixed-valence system, Cr_2F_5, which contains a double bridge and three single bridges [11]. A choice favoring either possibility and the nature of the bridges whether amido or bromo or both cannot be made at this point.

Titanium III Compounds

Titanium III chloride was treated with sodium borohydride in liquid ammonia. In this case, and in spite of the fact that Ti(III) does not ammonolyze, the product of the reaction, compound B, contained a Cl/Ti ratio of 2.0. Evidently, the presence of borohydride promoted substitution of an equivalent of chloride by amide, quite possibly, through the formation of a borohydride substituted intermediate. A purple coloration was observed in the supernatant of the reaction mixture.

Compound B is formulated as $[NH_4^+NH_3]_4[Ti_4Cl_8(NH_2)_8]^{4-}$ on the basis of the chemical analysis and the result of the acetylide treatment which liberated acetylene corresponding to a charge of -3.88. The titanium acetylide derivative, compound B1, showed a composition corresponding to $Ti_4(C \equiv C)_2(NH)_4$. It is interesting to note that the sodium acetylide treatment applied to pre-formed $TiCl_3 \cdot 4NH_3$ produced $Na_4Ti_4(C \equiv C)_4N_4$. This is a compound similar to a series complex carbide and nitrides having empirical

404

formulas corresponding to $Ti_4M_2^{II}[C,N]$ where M^{II} = Zn, Cd or Sn [12-14]. On the other hand the acetylide treatment of $TiCl_3$ produced $NaTi_4(C\equiv C)_2(NH)_4NH_2$. The derivatives of the Ti(III) compounds showed the characteristic band of the $C\equiv C$ moiety as in the case of the acetylide derivative of compound A.

Treatment of compound B with sodium amide at -50°C which was expected to produce partial substitution of the halide by amide led instead to total substitution to yield a condensed product $NaTi(NH)_2$.

Titanium IV Compounds

The Ti(IV) compound with a nominal formula of $TiBr(NH_2)_3$, compound C, proved to also contain ammonium ions as found by IR spectroscopy. The sodium acetylide treatment of this compound evolved acetylene corresponding to a charge of -2.02 for the titanium tetramer hence leading to a formulation of $2[NH_4^+][Ti_4Br_4(NH)_4(NH_2)_6]^{2-}$ for this material. The acetylide derivative thus produced, compound C1, showed a composition corresponding to $Ti_4(C\equiv C)(NH)_7$.

Pyrolysis of Titanium Acetylide Derivatives

Representative acetylide derivatives were pyrolyzed by heating to 800°C under dynamic vacuum. The condensible volatile products of the reaction were trapped at -196°C. The volatile products were mostly acetylene and ammonia and a small fraction consisting of hydrogen cyanide. The solid residues of the pyrolysis were titanium carbonitrides with the exception of that derived from $Na_4Ti_4(C\equiv C)_2N_4$. This material converted into $Na_4Ti_4C_2$ which is similar to the compounds prepared by Jeitschko et al. [12-14] The compound $NaTi_4(C\equiv C)_2(NH)_4NH_2$ produced a carbonitride with an empirical formula corresponding to $Ti_4N_2C_2$. The sodium initially present volatilized and deposited as a mirror in the colder parts of the tube in which the pyrolysis was conducted. The compounds $Ti_4(C\equiv C)(NH)_7$ and $NaTi_4(C\equiv C)(NH)_7NH_2$ produced Ti_4N_3C upon pyrolysis. The latter one also produced a sodium mirror. Weight losses during pyrolysis of all the above mentioned acetylides were in the range of 20-30%. Micrographs of the $Ti_4N_2C_2$ and Ti_4N_3C carbonitrides are given in Fig. 1 and 2, respectively.

CONCLUSIONS

Ammonolysis of titanium halides either directly or promoted by borohydride provides intermediates that can be converted into precursors for titanium nitride or carbonitride. This synthetic approach yields precursors of relatively high metal content which convert at <800°C into the desired ceramic materials.

ACKNOWLEDGMENT

This research was sponsored by the Division of Materials Sciences, Office of Basic Energy Sciences, U.S. Department of Energy under contract DE-AC05-84OR21400 with Martin Marietta Energy Systems, Inc.

405

Fig. 1. Titanium carbonitride containing 9.2% C
derived from Ti(III) amido-acetylide.

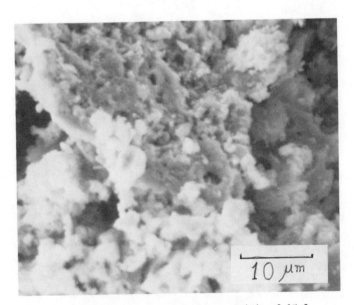

Fig. 2. Titanium carbonitride containing 3.1% C
derived from Ti(IV) imido-acetylide.

REFERENCES

1. R. W. Rice, Bull. Ceram. Soc. $\underline{62}$, 889 (1983).
2. G. W. A. Fowles and D. Nicholls, J. Chem. Soc., 990 (1959).
3. G. W. A. Fowles and D. Nicholls, J. Chem. Soc., 95 (1961).
4. T. Kottarathil and G. Lepoutre, J. Chim. Phys. (France) $\underline{73}$, 849 (1976).
5. D. Nicholls and T. A. Ryan, J. Inorg. Nucl. Chem. $\underline{39}$, 961 (1977).
6. N. Hagihara, Mem. Inst. Sci. Ind. Res. Osaka Univ $\underline{35}$, 61 (1978).
7. L. J. Bellamy, The Infrared Spectra of Complex Molecules, Metheun & Co. Ltd., London (1960).
8. L. Ramqvist, K. Hamrin, G. Johanson, A. Fahlman and C. Nordling, J. Phys. Chem. Solids $\underline{30}$, 1835 (1969).
9. G. M. Muha and P. A. Vaughan, J. Chem. Phys. 33, 194 (1960).
10. L. Maya, J. Inorg. & Nucl. Chem. $\underline{41}$, 67 (1979).
11. H. Steinfink and J. H. Burns, Acta Cryst 17, 823 (1964).
12. W. Jeitschko, H. Holleck, H. Nowotny and F. Benesovsky, Monatch Chem $\underline{95}$, 1004 (1964).
13. W. Jeitschko, H. Nowotny and F. Benesovsky, Monatch Chem $\underline{95}$, 178 (1964).
14. W. Jeitschko, H. Nowotny and F. Benesovsky, Monatch Chem $\underline{94}$, 672 (1963).

THERMAL CONVERSION OF PRECERAMIC POLYSILAZANES TO Si_3N_4: CHARACTERIZATION OF PYROLYSIS PRODUCTS

KENNETH B. SCHWARTZ, DAVID J. ROWCLIFFE, YIGAL D. BLUM AND RICHARD M. LAINE
SRI International, 333 Ravenswood Avenue, Menlo Park, CA 94025

ABSTRACT

Characterization of the pyrolysis products of preceramic polysilazanes synthesized by reactions of the oligomer $-[H_2SiNMe]_{\overline{x}}$ in the presence of $Ru_3(CO)_{12}$ catalyst has demonstrated that these polymers have great potential as precursors of Si_3N_4 for several applications. The polysilazanes studied are viscous liquids that can be converted to ceramic material with a yield of >65 wt %. The vitreous product contains regions of fully dense material and large cavities that indicate of considerable gas evolution. A preliminary pyrolysis sequence has been constructed based on a combination of TGA of the polymer, SEM investigation of the product, and mass spectroscopic analysis of the gases and heavy fragments released during conversion of the polysilazanes to ceramic material. This understanding of pyrolysis mechanisms will aid in developing even more effective polymeric precursors to Si_3N_4 and in optimizing pyrolysis procedures for a variety of useful applications.

INTRODUCTION

Organometallic precursors that can be converted to ceramic material upon pyrolysis continue to show great promise for synthesizing high technology ceramics. One particularly useful feature of many such preceramic polymers is that their mechanical and rheological properties can be controlled during synthesis. This feature can be advantageous for a variety of processing applications. For example, polymeric precursors to Si_3N_4 or SiC that can be fabricated as viscous liquids are excellent candidates for use in preparing ceramic fibers. A preceramic polymer in such a form could also be used as a binder in the processing of ceramic powders.

The conversion of polymeric precursors of Si_3N_4 or SiC to ceramic material is a complicated and poorly understood process [1,2]. It is invariably associated with the rapid evolution of various gases and results in a ceramic product containing unwanted impurities, most notably carbon and oxygen. Both these factors are serious impediments to the successful application of preceramic polymers in ceramic technology. Gas evolution during pyrolysis will tend to damage any fabricated fiber or shape formed using such polymers. Any presence of carbon and/or oxygen can affect the high temperature stability of Si_3N_4 and SiC, thus degrading their high temperature performance. Therefore, it is essential to understand the pyrolysis mechanisms of such polymers to determine if these and other problems can be ameliorated during pyrolysis.

In a companion paper by Blum et al. [3], we describe the production of high-yield preceramic polysilazanes. These polymers have mechanical and rheological properties that make them candidates both as forming aids for use with ceramic powders and as precursors for Si_3N_4 fibers. In this paper, we discuss the pyrolysis of some promising polymers that have been studied with a variety of analytical tools in order to discern the details of the pyrolysis reactions. The increased understanding of pyrolysis mechanisms is highly desirable both for developing an optimal pyrolysis procedure and for designing improved preceramic polymers.

EXPERIMENTAL

Several polysilazanes synthesized by the methods of Blum and Laine [3,4] were examined in this study. Oligomers having [H_2SiNCH_3] units, first synthesized by Seyferth and Wiseman [5], were polymerized by transition metal catalytic activation of $Ru_3(CO)_{12}$ under a variety of conditions. To efficiently evaluate the new preceramic polysilazanes, we formulated a heirarchy of experimental procedures to identify polymers that are candidates for further study.

Initially, the preceramic polymer is judged on yield of ceramic products after pyrolysis. This screening is performed in a system constructed for bulk pyrolysis of preceramic polymers under conditions of controlled temperature and atmosphere. Atmospheric control is important, because many of the preceramic polysilazanes are air- and moisture-sensitive and improper handling is a major cause of oxygen contamination. The pyrolysis schedule used for this study was a continuous temperature ramp at 0.5°C/min from room temperature to ~850°C and a 3-hour hold at this elevated temperature. All bulk pyrolysis reactions for this study were performed under atmospheric pressure of flowing nitrogen. Some products of pyrolysis were heat treated at 1600°C under 600 kPa of N_2.

In addition to determining ceramic yield, the products of polysilazane pyrolysis were analyzed for chemical composition and the presence of crystalline phases. Silicon content was determined by the inductively coupled plasma technique, and direct measurement for N, O, and C-H was performed at 1600°C in a LECO induction furnace. Crystalline phases were identified by x-ray diffraction. Thermogravimetric analysis (TGA) and scanning electron microscopy (SEM) were used to track the pyrolysis reactions and to obtain detailed information on the texture and morphology of the pyrolysis products. Previous work on polysilazanes [6,7] and results from this study had demonstrated large-scale volatile release during pyrolysis. Therefore, we used mass spectroscopy to directly identify the mass of components released during pyrolysis. TGA experiments were performed under flowing nitrogen with a heating rate of 0.5°C/min. For pyrolysis experiments using mass spectroscopic detection of volatile components, samples were heated under flowing He or a high vacuum with more rapid heating rates of up to 5°C/min.

RESULTS

The results of bulk pyrolysis experiments showed that preceramic polymers synthesized by reactions of [H_2SiNMe]$_x$ in the presence of $Ru_3(CO)_{12}$ repeatedly produced ceramic yields of 65 wt % and higher when pyrolyzed at 0.5°C/min in flowing nitrogen. The polymers had varying physical properties, ranging from liquids to waxes to soft rubbers. The ceramic products were typically black solids, ranging from coarse powders to irregular-shaped fragments (up to 5 mm) having a vitreous appearance. The larger pieces contained large vesicles, indicating of the release of trapped gas during pyrolysis.

The ceramic products of pyrolysis were noncrystalline, yielding no discernible x-ray diffraction pattern. Upon further processing above 1550°C in nitrogen, the XRD examination showed that α- and β-Si_3N_4 were the only crystalline phases present. Chemical analyses determined that carbon was the major elemental impurity, with carbon contents between 2.5-18 wt %. Oxygen was present only in trace amounts (< 0.5 wt %) and ruthenium was detected at the 2700 ppm level.

In accordance with our intended applications for polysilazane precursors as binders or preceramic fibers, our interest in the polymers focused on those that could be synthesized as viscous liquids. The candidate chosen for more intensive examination was a golden, high-viscosity liquid synthesized by the reaction

$$-[H_2SiNMe]_x^- \xrightarrow[\text{8 hr/90}^\circ\text{C}]{Ru_3(CO)_{12}} \text{viscous liquid polymer} + H_2$$

This polymer (BL530) had a ceramic yield of 65.6 wt % after bulk pyrolysis and produced a material with a glassy appearance. Chemical analysis of the product material showed the amorphous product to be ~82 wt % Si_3N_4, with 17 wt % carbon, < 1 wt % hydrogen, and < 0.1 wt % oxygen as the dominant impurities. XRD studies on material subsequently treated above 1600°C showed α- and β-Si_3N_4, with α-Si_3N_4 as the dominant phase.

SEM studies on the pyrolysis product of BL530 (Figs. 1 and 2) show that, in accordance with XRD examinations, the material is amorphous after treatment to 860°C. The bulk material has a glassy, vitreous appearance, demonstrates concoidal fracture, and shows no microstructural evidence of crystallization down to the resolution limit of 500 nm (Fig. 1). Much of the material appears fully dense with only isolated micrometer-sized pores. In Fig. 2, at lower magnification, evidence of very large gas vesicles is also apparent.

Fig. 1

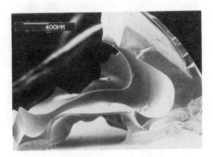

Fig. 2

A representative TGA profile of BL530 heated in flowing nitrogen at 0.5°C/min is shown in Fig. 3. This profile shows that the pyrolysis process

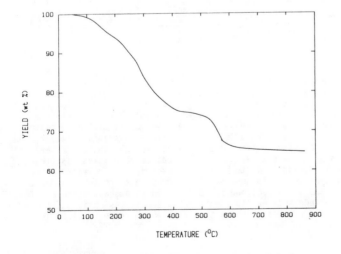

Fig. 3

occurs in three stages: the first occurring below 250°C, the second between 250°-450°C, and the third between 500°-600°C. These stages suggest that different chemical reactions are occurring at different stages of polymer pyrolysis. Comparison of these results with TGA data on other polymers synthesized from $[H_2SiNMe]_x$ shows that the behavior is similar, although some other polymers have a more distinct boundary between the first and second stages. Any differential in final ceramic yield among the various polymers studied by TGA can be correlated with differences in the pyrolysis process between 300°-400°C. The behavior of different polymers below 300°C and above 500°C is virtually identical.

Several mass spectroscopic methods were used to identify the volatile gases and heavier compounds that were released during pyrolysis of BL530. For detection of low mass components, the polymer was heated at 5°C/min in flowing He in a quartz microreactor. This reactor was attached to a quadrupole mass spectrometer by a gas line heated above 100°C. This system allowed for identification of light volatile components and the approximate temperature range during which they evolved. However, due to the presence of a long gas line in the system, heavier fragments tended to condense before reaching the mass spectrometer and were therefore not detected. The results of this experiment are shown in Fig. 4. This temperature profile correlates well with TGA results, showing the release of material with mass 31 amu (possibly $MeNH_2$ or SiH_3) and lighter fragments of this compound during stage 2. It also demonstrates that the weight loss in stage 3 is characterized by the release of CH_4 (16 amu) and CH_4 fragments.

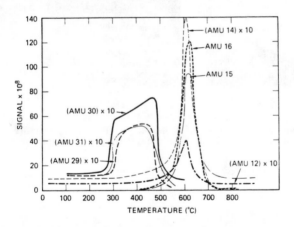

Fig. 4 Mass spectroscopic detection of light volatile release for pyrolysis under He.

To identify heavier components that are released during pyrolysis, we are also obtaining mass spectroscopic data using direct sampling methods. For these experiments, the polymer is vaporized directly into the ion source from an effusion cell so the material evolved is sampled in the absence of collisions, increasing the likelihood that heavy fragments will be detected. This work in progress will be reported in a future publication. However, preliminary results confirm that: (1) species of mass 31, 30, and 29 amu dominate the mass spectrum between 300°-450°C, (2) methane makes its first appearance above 400°C and grows to be the dominant species released above 500°C, and (3) many heavy, silazane-containing fragments dominate the mass spectrum during stage 1 and continue to be released throughout stage 2. The heavy fragments include some easily identifiable species, such as $[H_2SiNCH_3]_3$ (176 amu), and many others that still need to be identified.

DISCUSSION

The ultimate goal of any investigation of preceramic polymers is to obtain information relevant to their applications in the fabrication of high-technology ceramics. Our characterization of polysilazanes, synthesized from $[H_2SiNMe]_x$ by a catalytic route, and their pyrolysis products gives promising results that these organometallic compounds can be developed into a variety of useful Si_3N_4 materials. These polymers can be synthesized as viscous liquids that are soluble in various organic solvents. These two characteristics are important in binder applications. Liquid polymers can be directly combined with ceramic powder to produce a polymer/powder mixture that can be compression molded. Alternately, the polymer can be put into a solution to which the powder is added. In this case, the solvent improves the homogeneous dispersion of the mixture before its evaporation. We have had success in producing simple shapes using such polymer/powder systems with both methods.

The polysilazane preceramics described here appear to meet the initial criteria required in terms of the quality of pyrolysis products. For example, ceramic yields are 65 wt % and above, consisting of > 80 wt % Si_3N_4 with carbon as the principal impurity. Control of impurities in the pyrolyzed polymer is important because impurities may affect crystallization and thermal stability. The virtual absence of oxygen in our materials might mean that they are not susceptible to breakdown at temperatures above 1000°C, in contrast to current commercial SiC fibers, which are intentionally exposed to oxygen during processing [8] and degrade readily at high temperatures. It has not yet been established whether, in the absence of oxygen, free carbon is necessarily deleterious to the stability of Si_3N_4. Free carbon in the Si_3N_4 product could potentially reduce it to SiC with consequent changes in microstructure or properties. However, no evidence for crystalline SiC has been found in the limited annealing experiments made so far. In general, though, carbon is held to <0.1% in Si_3N_4 powders because of the possible formation of unwanted grain boundary phases through reaction with oxide sintering aids [9]. We believe that residual carbon may be controlled to some degree by modifying the polymer and perhaps by adjusting the pyrolysis scheme.

The morphology of the pyrolyzed products, as seen in the SEM studies (Figs. 1 and 2), indicates that gas evolution is a major factor during pyrolysis. Control of this gas evolution will be essential for the production of coherent fibers and undamaged bodies using these polymers. The large gas pockets suggest that significant amounts of gas are being generated while the polymer is in a plastic form, possibly during stage 2 when the components of mass 30 and 31 amu are dominating the mass spectrum. These components have not been positively identified, but either of the likely candidates, monomethylamine or silane (and lighter fragments with H^+ removed), would be a gas capable of causing such cavities. The SEM studies also show that the ceramic product contains extensive regions of the order of 100 μm thick that contain only isolated pores of ~1 μm. The mechanism of formation of these pores is unknown. However, the observation of extensive pore-free material is encouraging because spun fibers of these polymers in the 10-20 μm range, or polymers dispersed in ceramic powder on a smaller scale, might also pyrolyze to fully dense material.

Though still rudimentary, data obtained from SEM, TGA, and mass spectroscopy allow us to begin the construction of a sequence of events during pyrolysis. During stage 1 (below 250°C), weight loss is slow and continuous and is dominated by the loss of heavy, silazane-containing fragments. During the transition to stage 2 (250°-450°C), weight loss accelerates and the evolution of the light volatile components with masses 29, 30, and 31 amu increases, becoming dominant above 300°C, although heavier fragments are still being lost. After a period of relative dormancy between 450°-500°C, the final surge of weight loss occurs in stage 3, due

412

almost entirely to the release of methane. No heavy fragments can be
detected in this stage, though a small signal (<5%) due to masses 27-30 amu
can be seen in the mass spectrum.

The loss of heavy, silazane-containing fragments during stage 1 and
stage 2 of pyrolysis is a problematic aspect of polymer conversion that can
seriously reduce the ceramic yield. Therefore, a method must be found to
prevent early loss of heavy silazane units while enhancing carbon release.
This can be accomplished either by altering the polymer structure during
polymer synthesis or by changing the pyrolysis reaction by adjusting the
conditions during pyrolysis. Such adjustments could entail changes in the
atmosphere [5] or the pyrolysis schedule. Note, though, that the loss of
heavy silazane units in mass spectroscopic experiments could be exacerbated
by the fact that these experiments take place under a high vacuum.

This current work has established the principal features of
polysilazane pyrolysis. Future work will focus on refining our
understanding of the pyrolysis mechanisms and the role of processing
parameters, such as pyrolysis heating schedule and atmosphere, in producing
a purer product and limiting the damage due to gas evolution. Achieving
these goals is essential to the future production of ceramic fibers from
spun or extruded preceramic polysilazanes. Additional knowledge must be
obtained on optimizing polymer/powder systems for binder applications
[10]. In conclusion, this preliminary work has shown promising results and
indicates the potential importance of polysilazanes as a precusor to Si_3N_4
fibers and monolithic shapes.

ACKNOWLEDGMENTS

This work was supported by the Office of Naval Research (Contract No.
N00014-85-C-0668) and the Advanced Materials Program at SRI International.
Mass spectroscopy experiments were performed by Gilbert Tong, Gilbert St.
John, and Robert Brittain at SRI International. Chemical analyses were
performed by Galbraith Laboratories.

REFERENCES

1. R.R. Wills, R.A. Markle, and S.P. Mukherjee, Am. Ceram. Soc. Bull. 62, 904-911, 915 (1983).
2. K.J. Wynne and R.W. Rice, Ann. Rev. Mat. Sci. 14, 297-334 (1984).
3. Y.D. Blum, R.M. Laine, K.B. Schwartz, D.J. Rowcliffe, R.C. Bening and D.B. Cotts, in this volume.
4. Y.D. Blum and R.M. Laine, submitted to Organometallics.
5. D. Seyforth and G.H. Wiseman, in Ultrastructure Processing of Ceramics, Glasses, and Composites, L.L. Hench and D.R. Ulrich, eds., John Wiley, New York, pp. 265-271 (1984).
6. D. Seyferth, G.H. Wiseman, and C. Prud'homme, J. Am. Ceram. Soc. 66, C-13-C-14 (1983).
7. D. Seyferth and G.H. Wiseman, J. Am. Ceram. Soc. 67, C-132-C-133 (1984).
8. S. Yajima, Phil. Trans. R. Soc. Lond. A294, 419-426 (1980).
9. H. Knoch and G.E. Gazza, J. Am. Ceram. Soc. 62, 634-635 (1979).
10 K.B. Schwartz and D.J. Rowcliffe, J. Am. Ceram. Soc., in press.

CONVERSION OF CHEMICALLY-DERIVED POLYMERIC PRECURSORS
TO HIGH PERFORMANCE CERAMIC FIBERS

M. ISHAQ HAIDER AND TERENCE J. CLARK
Celanese Research Company, 86 Morris Avenue, Summit, NJ 07901

ABSTRACT

In recent years, there has been steadily increasing research activity directed towards preparation of ceramic materials via polymer pyrolysis. Among the most thermomechanically stable structural ceramics are SiC, Si_3N_4, and their solid solutions. These materials are widely used as ceramic fiber reinforcements in composites owing to their high mechanical strength, stiffness, and oxidative stability. Processes have now been developed to make these fibers by utilizing melt spinning. Fine diameter continuous filaments are formed, and further processing involves the steps of mild reactive cure and high-temperature pyrolysis. The approach in this review is to summarize the fiber making process with emphasis on improving fired properties by utilizing known mechanisms of fiber failure and process modification procedures. Fiber characteristics, applications of high performance fiber, and perceived trends will also be discussed.

INTRODUCTION

As a result of the increasing demands for high-temperature structural components, new fibrous materials, exhibiting a high level of mechanical properties, have been developed. Emphasis has been directed towards preparing fibrous materials characterized by a high strength-to-weight ratio, thermal stability, and oxidation resistance up to 1200°C. Silicon carbide, silicon nitride, and their solid solutions are among the most thermomechanically stable structural ceramics. Silicon carbide, in particular, has long been of interest as a high-temperature engineering material; its most attractive properties include chemical inertness and extreme hardness.

As a fiber, silicon carbide is a prime candidate for high-temperature applications [1]. In the early seventies, the preparation of these refractory fibers by pyrolyzing melt-spinnable organosilicon polymers effected a dramatic change in production technology. Prior art had been to either deposit a vaporized reactant on a heated core (chemical vapor deposition, C.V.D.) [2] or heat carbon fibers in the presence of a carbide forming metal or vaporized compound of the metal (e.g., a halide of silicon). These processes have disadvantages in that relatively thick silicon carbide fibers are made which are difficult to handle. Moreover, high production rates cannot be reached.

The exciting prospect of polymer pyrolysis in ceramic processing was first proposed by Chantrell and Popper [3]. However, the successful development of continous silicon carbide fiber by pyrolysis of a polymer precursor fiber was demonstrated by Yajima [4] and co-workers at Tohoku University. The pyrolysis concept itself is not new, and it should be stated that previous experience in carbon-graphite technology by the authors' company has resulted in a build-up of considerable technical capability for the preparation of polymeric fibers and this has aided process development of ceramic fiber.

Mat. Res. Soc. Symp. Proc. Vol. 73. ©1986 Materials Research Society

The importance of making ceramic fibers by the pyrolysis of organo-metallic preceramic fibers and the growing technology to make suitable reinforcing ceramic fibers are well documented (5-9). One major problem which persists is the scarcity of such materials. To date, the only ceramic fiber commercially available is the polycarbosilane derived silicon carbide (Yajima et al. [7,10]) produced by Nippon Carbon Company under the trade name NICALON. NICALON is commercially marketed; yet, because of the complex nature of the fiber processing, its availability seems to be limited. This paper will provide an in-depth review on the fabrication of silicon carbide fibers from organosilicon polymers. Melt spinning and some of the problems related to the spinning characteristics of polycarbosilane will be highlighted. Principally, this review is concerned with only the fiber making process for generation of precursor polymeric fibers. Down-stream processing steps such as curing and pyrolysis will be discussed; however, they are major topics in themselves and, therefore, will receive limited treatment. The chemistry of preceramic polymers is not within the scope of this paper and, for a comprehensive coverage, the reader is referred elsewhere [11,12].

FIBER PROCESS

Production of ceramic fibers consists of spinning of a preceramic polymer followed by curing and a high-temperature heat treatment. A precursor, such as polycarbosilane (PC), is first spun into a continuous fine-diameter filament, and then cured by pre-heating in an oxidizing atmosphere at temperatures ranging from 110°C to 190°C [7,8]. The cured fibers are then heat treated in an inert atmosphere at a temperature of 1200-1400°C to make high-performance SiC fibers [7-9].

In general, the methods of handling and fabrication of ceramic fibers are similar to those used in graphite, glass, mineral wool, asbestos, and other fibers although, in most cases, ceramic fibers may require modification or further development of a present conventional technique. While there has been a plethora of spinning techniques available, only three major methods are the most common processes with great potential for ceramic applications. Melt spinning, a process in which a molten polymer is extruded through holes in a spinneret, is the predominant and the most inexpensive technique to make fibers [9]. Solution or dry spinning, where the polymer is first dissolved in a suitable solvent and then extruded through a spinneret in a dry spinning apparatus, has also been used successfully [7]. Wet spinning, in which a solution of the fiber-forming substance is extruded into a liquid coagulating medium where the polymer is regenerated, as in the manufacture of rayon, has not been reported in ceramic fiber processing. The technique, however, has promise as a viable method by combining spinning and a chemical cure step for the improvement of fiber-handling characteristics. Key variables affecting fiber properties in any wet spinning process include: solvent composition, solid level, extrusion temperature and kinetics, spin-pack design, spin-bath composition, (non-solvent to solvent ratio), spin-bath temperature, and fiber draw ratio. These parameters affect the rate of coagulation of the fiber and, therefore, determine fiber diameter, density (porosity), tensile strength, elongation, and modulus.

In dry spinning, when the solid-to-solvent ratio exceeds 90% by weight, it is also referred to as slurry spinning. In slurry spinning, i.e., spinning of polyacrylonitrile, the polymer/solvent slurry is extruded through the spinneret followed by solvent extraction to coalesce the filaments prior to sintering and densification. Although dry and slurry spinnings are versatile, established ways of producing fibers, (i.e., DuPont currently manufactures alumina fiber by slurry spinning) there remains associated with the process the removal of residual solvents plus the added problem of working with a two-phase system. Whatever the process used, it is appropriate to point out that the morphology of the fiber produced is largely dependent on the processing it receives.

In the fiber-forming operation of melt spinning, polymer chips or granules are fed from a hopper to a melter grid where they are fused at a temperature approximately 20°C to 50°C above the polymer melting point. A metered amount of the molten polymer is delivered to the spinneret where it is extruded through spinneret holes into filaments and cooled (quenched). After quenching, finish is applied to lubricate the filaments to prevent breakage and deformation by the various guide surfaces directing the yarn to the windup. An extruder combined with a positive displacement metering pump is normally used to provide a molten polymer supply to the spinneret assembly, referred to as spinpack. In the spinpack, the polymer is passed through a heated filtration media, where it is sheared and filtered to remove microgels and insoluble contaminants which are harmful to spinning [13]. The filtration media in spinpacks are usually metal powder fractions of varying densities designed for maximum filtration without inoperably high pressure. It would be desirable to filter out even the smallest inhomogeneity (< 10 microns in size) because such particles can only weaken fibers and promote extrusion discontinuities. The fiber must be free of bubbles (dissolved gases) or other deleterious contaminants at the point of emergence from the spinneret, otherwise threadline instability (breakage) will plague fiber production.

In the melt spinning of conventional polymers such as PET or Nylon, the fiber emerging from the spinneret is accelerated through the quenching zone by a godet roll. The surface speed of the godet roll, combined with the polymer throughput, determines the as-spun fiber diameter (or spun denier). The fibers are cooled between the spinneret and the windup; beyond a certain point on the line, the fiber behaves like a solid. In the case of brittle polymers such as ceramic precursors, where the as-spun filaments are weak and fragile and cannot tolerate the surface friction presented by the godet roll, the fibers are drawn and wound directly on a rotating windup assembly in a single step. The speed of rotation of the windup assembly (takeup speed) and the throughput (volume of polymer passing through spinneret assembly per unit time) of material are the two variables which can govern the diameter of the fiber.

MECHANICAL AND MORPHOLOGICAL PROPERTIES

Organosilicon processing approaches to ceramic fibers has been plagued with the problem of low as-spun fiber tensile strength. Yajima and co-workers have reported tensile strength figures of 9.8 MPa [4], 4.8 MPa [9], and 5.0 MPa [10]. Thus the as-spun fiber is very fragile, which poses a great problem in handling. Polymer morphology and molecular structure seem to be the primary causes for the fiber weakness. In contrast to the conventional organic polymers, the molecular chains on preceramic polymers are highly branched and non-linear. Only short chain lengths with low molecular weights (MW) are spinnable. Increasing the MW increases the chain branching and crosslinking and, as a result, spinning performance will deteriorate [18]. Although ceramic precursors are referred to as polymers, the available materials are not of sufficient chain length and should, instead, be classified as oligimers. Furthermore, it should also be mentioned that, in linear polymeric structures, the molecular chains can be oriented along the fiber axis to develop crystallinity during the fiber-making operation. With preceramic polymers, on the other hand, to induce any degree of parallelism of the chain molecules to develop orientation is almost impossible. Henceforth, a major global goal for ongoing ceramic fiber research has been to establish a way to consistently obtain maximum tensile properties for a low molecular weight polymeric precursor. For example, polycarbosilane is an oligomer with a molecular weight of 1000-2000 compared to 40,000 to 50,000 for polymers of synthetic fibers such as polyesters [9]. Therefore, the spinning stability of such low molecular weight and heavily branched polymers, as reported [8], are very poor when compared to the more linear and high MW polyesters.

Spinning problems related to the weak and fragile nature of the ceramic precursors can best be envisioned by the use of Trouton's classic experiment on the flow of a free descending polymer stream, performed in 1906 [14]. A conventional polymer, such as PET with an average molecular weight of 50,000, could carry its suspended weight several hundred feet; whereas a ceramic precursor can be shown to carry <15 feet in a descending freefall stream. This illustration indicates that fibers spun from materials such as PC are extremely weak.

The major strength-limiting defects that have been found in as-spun preceramic polymeric fibers are (a) surface defects, (b) internal voids, (c) foreign particle inclusions, and (d) interfilament fusion. All these defects have been found in fibers from other precursors, such as in the case of PAN-based carbon fiber [15]. In the following sections, an attempt will be made to address the strength-limiting factors of the as-spun preceramic fibers and establish the influence of flaws caused during processing upon the strength of the pyrolyzed (silicon carbide) fiber.

FACTORS AFFECTING FIBER STRENGTH

Fiber processing and as-spun fiber handling in silicon-based ceramic fibers seem to be an important factor in determining the final pyrolyzed properties. Based on fracture surface analysis and fractography studies, strength-limiting failures initiate at either internal or surface-related defects [16,17]. Therefore, to make high property fibers, it is essential to minimize surface flaws and granular defects. Surface flaws can be controlled by improving fiber processing, whereas internal flaws can be chemistry related.

In an attempt to reduce process related surface defects, it is important to know that (a), the apparent strength of most fibers falls considerably below the theoretical values and (b), processing factors can play important role in determining the internal atomic and microscopic structure of the fiber, the surface perfection, and whether or not the fiber will be a crystalline or amorphous material. Each step of the fiber processing -- spinning, curing, and pyrolysis -- can have great impact on fiber properties and, therefore, cannot be ignored. Spinning variables that can affect ceramic properties are spinning temperature, quench conditions, spinpack and spinneret design, polymer throughput, draw ratio and takeup speed. Furthermore, polymer variables, such as impurity level and melt viscosity, will also influence both the downstream mechanical properties and the spinnability of the materials.

Curing is a crosslinking step. For PC molecules, when heated in air (oxygen) at mild temperatures (<300°C), a crosslinked/cured material forms. This step is a necessity if polycarbosilane decomposition at high temperatures, into inorganic materials, is to occur without softening. Significant influences to fiber properties appear to include the following: fiber handling techniques; cure time, temperature, and environment; and removal of by-products from cure reactions.

The reduction of fiber strength by the presence of flaws, formed either during spinning or as a result of subsequent handling, has been well documented [16,17,18]. A way to increase fiber strength is by reducing the fiber diameter. The classic work of Griffith [19] has led to a plausible explanation of effect of increasing strength with decreasing fiber diameter. Since the tensile strength is influenced by the presence of flaws, it could be immediately inferred that any variation in stength with fiber diameter could be explained as a volume effect, since more flaws are expected to be present in a unit length of thick fiber than in a thin fiber. Removal of all polymer contaminants and impurities should therefore lead to improved properties. Flaws caused by the reaction of impurities with the fibers during processing provide another satisfactory explanation of any observable strength-gauge length effect.

The internal flaws become important after heat treatment to high temperatures. When cured polycarbosilane polymer is heat treated at 1200-1250°C in a nitrogen gas atmosphere, the resulting strength of the ceramic material is greatest [4]. Beyond 1300°C, the tensile strength then decreases. The deterioration may be related to SiC crystallization or grain growth. The high strength at 1200°C may be because the resultant structure is principally amorphous (it also includes ultra-fine β-SiC crystals and C particles). Significant quantities of oxygen and excess carbon are also found in the fiber, leading to the popular belief that silica and free carbon are mixed with the SiC. Indeed, the thermomechanical stability has been found to diminish in a manner analogous to vitreous silica, at 1200°C [20].

A major limiting step in the processing of Si-based preceramic fibers is the spinline instability at takeup conditions required to make low diameter fibers. As indicated [9], this is possibly due to the brittle and weak nature of the polymer. The filaments emerging from the spinneret capillary cannot tolerate the steady application of high takeup force, and this results in spinline interruption. This is a critical spinning problem since low diameter filaments are necessary for high tensile properties and this demands high takeup speeds.

CONCLUSIONS

From these considerations, it becomes quite evident that the apparently simple process of melt spinning preceramic polymers involves complexities. Not completely understood is the problem of threadline instability, although the present work sheds light on some important influences. In the quest for fine diameters, poor spinnability has been observed to become a major problem. This is because of the necessary use of relatively high takeup speeds and low throughput. Also, polymer structure and spinning variables such as takeup speed and spinpack design affect the quality of as-spun fibers and these, in turn, dictate in part the ultimate mechanical properties of the end product.

In considering approaches that can be pursued in achieving the desired morphological features and properties of fiber, several gross variables that can be adjusted were discussed, while other significant variables were not described. Unfortunately, space does not permit any discussion of the following key fabrication parameters, yet the reader should be made aware of their existence: polymer melt viscosity at spinning conditions, shear in the spinpack, polymer orientation and crystallization, spinneret capillary design, fiber drawdown ratio as well as rheological surface tension, areodynamic and gravitational forces acting on a threadline.

An encouraging note is that from the work that has addressed this area, many influences from the gross process variables were identified. The ability to achieve the high tensile properties predicted by many is still, in all likelihood, a possibility. The goal of future studies will be to establish how to consistently obtain these properties from yarn to yarn, lot to lot, etc.

REFERENCES

1. C.H. Andersson and R. Warren, Composites, Vol. 15, No. 1 pp. 16-24 (Jan. 1984).
2. S.R. Nutt and F.E. Wawner, J. Mat. Sci., 20, pp. 1953-1960 (1985).
3. P.G. Chantrell and P. Popper, in Special Ceramics, ed. Popper, Academic Press (1965).
4. S. Yajima, Handbook of Composites, Vol. 1, ed. Watt and Perov, pp. 202-237.
5. S. Yajima, J. Hayashi, and M. Omori, Chem. Lett., pp. 931-934 (1975).
6. S. Yajima, J. Hayashi, M. Omori, and K. Okamura, Nature, 261 pp. 683-685 (1976).
7. S. Yajima, K. Okamura, J. Hayashi, and M. Omori, J. Am. Ceram. Soc., 59, pp. 324-327 (1976).
8. S. Yajima, Am. Ceram. Soc. Bull., 62, 893 (1983).
9. T. Ishikawa and H. Teranishi, New Materials and New Process, Vol. 1, pp. 36-42 (1981); Vol. 2, pp. 379-385 (1983).
10. Y. Hasegawa, M. Iimura, and S. Yajima, J. Mat. Sci., 15, pp. 720-28 (1980).
11. R.R. Wills, R.A. Markle, and S.P. Mukherjee, Am. Ceram. Soc. Bull., 62, 904 (1983).
12. R.H. Baney, J.H. Gaul and T.K. Hilty in Emergent Process Methods for High-Technology Ceramics, ed. Davis, Palmour, and Porter, Materials Science Research, Vol. 17, Plenum Press, pp. 253-262 (1984).
13. R.M. Lodge, Fibers from Synthetic Polymers, ed. R. Hill, Elsevier, p. 363 (1953).
14. F. Trouton, Proc. Roy. Soc., A77, pp. 426-440 (1906).
15. J. H. Ross, Applied Polymer Symposium No. 29, pp. 151-159, Wiley and Sons (1976).
16. H. Funkunnaga and K. Goda, SAMPE J., pp. 27-31 (Nov/Dec 1985).
17. L.C. Sawyer, R.M. Arons, F. Haimbach IV, M. Jaffe, and K. D. Rappaport, Cer. Eng. Sci. Proc., Vol. 6, No. 7-8, pp. 567-575 (1985).
18. R.M. Arons, Symp. on Mat. Emerging Tech., Am. Chem. Soc. Annual Mtg., 1985.
19. A.A. Griffith, Trans. Royal Soc., London, A221, pp. 163-169 (1920).
20. T.J. Clark, R.M. Arons, J. Rabe, and J. Stamatoff, Cer. Eng. Sci. Proc., Vol. 6, No. 7-8, pp. 576-588 (1985).

PART VII

Comparisons of Chemically and Conventionally Derived Ceramics

DIFFERENCES BETWEEN GEL-DERIVED MELTS AND THOSE PRODUCED BY BATCH MELTING

ALFRED R. COOPER*

* Case Western Reserve University, Metallurgy and Materials Science Department, Cleveland, Ohio 44106

ABSTRACT

Properties of a system at equilibrium depend on pressure, temperature and composition. Thus for a melt produced by the sol-gel (SG) process to be different from an identical composition melted from batch (MB), both cannot be at equilibrium. Non equilibrium melts can be associated with structural differences or homogeneity differences. The former have been suggested for SG melts while the latter is always possible in MB melts. Appropriate relaxation times are presented for structural and heterogeneity relaxation. From this it is concluded that structural differences will not persist unless the SG melt is metastable with respect to the equilibrium melt. A method for testing this unlikely premise is proposed.
Melts which are imperceptibly different from equilibrium may have non-equilibrium crystal embryo distributions that relax toward equilibrium with a time constant longer than that for structural relaxation. A difference in embryo and stable nuclei distribution will result in different crystallization kinetics.

INTRODUCTION

Research and development on the sol gel (SG) method for making glass has been one of intense activity in the past decade. The ability to produce homogeneous glasses at temperatures far lower than required by conventional melting of batch (MB), the ease of production - little more than a beaker and some chemicals being necessary - the opportunity to produce glasses with very low impurity content, have been among the factors causing the excitement.

SG methods are part of the new wave toward "high tech ceramics" where chemicals are used as "raw" material. Their high* costs are justified by the special characteristics achieved. Glass optical wave guides are perhaps a prime example of the necessity to use chemical materials to achieve the required result. Even though, as yet, SG methods have found relatively little application in optical wave guides, the SG process provides an opportunity to use the same philosophy to produce devices which can not be made by conventional means. Confidence in this approach sustains the boom of activity in SG.

There is also a scientific basis for the boom. While the processing techniques for making glasses by the SG route are straightforward, the gelation of silica and silicate sols is intricate and a suitable object for description by percolation theory. [1] [a] There is a challenge to describe the evolution of the sol into a gel and the gel into a glass by use of spectroscopic, diffraction, and thermal measurements. [2] [b] Comparison of the structure, chemical composition, properties and crystallization kinetics of SG glasses and melts with those of equivalent MB melts and glasses is an active field of study. [3] [c] Since gels can typically shrink by several hundred volume percent upon elimination of water and alcohol during drying, large strains develop

---

*Per SiO_2, tetra ethyl orthosilicate costs several hundred times more than quartz sand.

which, if non-uniform, can cause residual stress, warping, and fracture. Accurate description of the drying behavior presents another challenge. [4] There is motivation to seek in the SG processing route molecular configurations not attainable by conventional glass melting. The purpose of this paper is to investigate the conditions under which this is possible. The approach will be to examine the thermodynamic and kinetic conditions that apply. Consistent with this approach, attention will be directed at melts rather than glasses since glasses are in the frozen state and neither equilibrium nor kinetic conditions are of use. Further, it is well known that processing beneath Tg can cause structural changes. For example, application of high pressure [5] or neutron radiation [6] to SiO_2 glass in the frozen state causes marked irreversible structure alteration. It is to be expected, then, that if the gel-glass transition can be carried out sufficiently below the glass transition of the equivalent MB glass, different structures will be produced and maintained in the frozen state.

OLD WISDOM

Prior to the advent of sol gel technology it was generally accepted that the properties of a melt depended only on its average composition, and its homogeneity. The property differences between silica glass produced by melting quartz and that produced by flame hydrolysis of $SiCl_4$ were thought to be attributed to compositional differences. Density and index of refraction changes from high pressure compaction [5] and neutron radiation [6] of glass are erasible by annealing at a melt temperature resulting in a melt indistinguishable from one that has neither been compacted nor irradiated. On a personal note my undergraduate thesis with M.R. Nadler at Alfred University in 1948, was directed at studying the effect of substitution of Wollastonite for equivalent amounts of sand and limestone in a soda lime silica glass. When we reported to Professor S.R. Scholes that we found no significant difference in properties, he scoffed at us for even suspecting a difference.

Despite this old wisdom there have been persistent recent reports of differences between SG and MB melts. After a brief presentation of the appropriate fundamentals, we then examine a group of pertinent experimental findings from this viewpoint.

THERMODYNAMICS

When a system of specified composition is at equilibrium with appropriate external variables, say pressure, P and temperature, T, all of its properties are determined. Since properties depend only on the current external variables, it follows that all effects from previous history are erased in an equilibrium system. Glass forming liquids are no exception. They have the additional feature that the molten state can readily be extended to temperatures beneath the melting temperature. Fig. I displays schematically the free energy, G as a function of temperature, T for various phases in a unary isobaric glass forming system. Curve C refers to the crystalline state. It is the stable equilibrium state at temperatures less than the equilibrium melting temperature, $T_m(MB)$. At temperatures such that $T > T_m(MB)$, the crystalline state is unstable because there is no nucleation barrier to surface melting. Hence in this region curve C represents an instantaneous value of the free energy. The stable phase in this region is that produced by an equilibrium transformation from the crystal. The usual assumption is that a melt produced by a non equilibrium transformation almost instantly relaxes to the equilibrium state. Consistent with this the MB curve above the melting point represents the stable equilibrium state and the equilibrium melting temperature has been labeled $T_m(MB)$.

Below T_m(MB) the MB melt is supercoded. In this region it is stable with regard to all small fluctuations because of the existence of a nucleation barrier between the super cooled melt and the crystal C, but unstable relative to a fluctuation large enough to overcome the barrier. It is therefore said to be in a metastable equilibrium state. Seeding with crystals converts the MB metastable melt into an unstable phase which transforms to the crystal. Experimental evidence [7] in super cooled MB melts that finite deviations in different directions all decay to the same state confirms that super cooled MB melts are indeed in metastable equilibrium.

If the SG melt has a different structure than the equilibrium melt, MB, it must have a higher free energy as indicated by the curve labeled SG on Fig. 1. The SG melt can either be unstable or metastable relative to the MB melt. If it is unstable, the curve SG represents an instantaneous value of the free energy. While the author is not familiar with any example of a melt which is metastable with respect to another of the same composition, and while it seems unlikely that a nucleation barrier can be present between two melts of the same composition, the possibility is included to achieve maximum generality.

Fig. 2(a) is a schematic equivalent to Fig. 1 for a simple binary system, AB, at a specified mol fraction of, B, X_B^0. Figure 2(a) shows the G(T) curves for an MB melt, an SG melt and crystals (MC) mixed in the proper ratio to give mol fraction X_B^0. In the temperature interval between the liquidus, T_ℓ(MB), and the solidus, T_s(MB) equilibrium occurs between crystal A and an MB liquid whose mol fraction, X_B^0, varies with temperature. This equilibrium is represented by the fine dashed line drawn between the MB and MC curves. Likewise, the fine dashed line drawn between the SG and MC curves represents equilibrium between an SG melt and crystal A. Fig. 2(B) shows schematically a phase diagram of the same binary system with only an MB melt in equilibrium with crystal A. Figure 2(a) and 2(b) both show that T_ℓ and T_s are higher for the SG melt. Thus, crystals precipitated from an SG melt between T_ℓ(MB) and T_ℓ(SG) will melt to form MB.

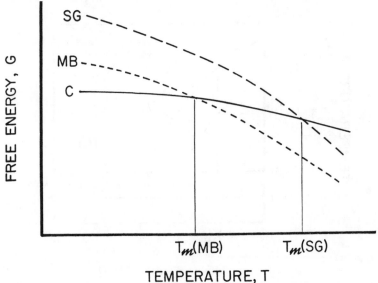

Figure 1
Schematic of G(T) for Possible Phases in an Unary System

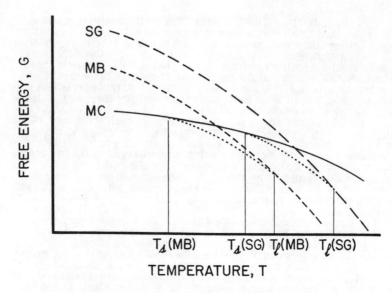

Figure 2a
Schematic Showing G(T) of Several Phases at a Specified
Composition in a Binary System....Represents a 2-Phase Region

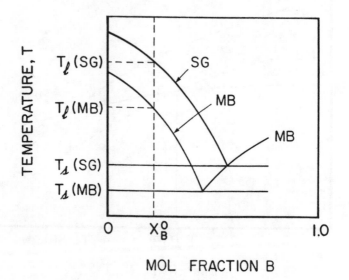

Figure 2B
Schematic Phase Diagram for a Binary System Whose
Free Energy at $\bar{X} = X_B^\circ$ is Given in Figure 2.

KINETICS OF STRUCTURAL RELAXATION

Unstable structures in glass melts relax toward equilibrium. Using a Williams Watts function [8] to describe the time dependence of the free energy of an unstable SG melt yields:

$$(G(t) - G(MB))/(G(SG) - G(MB)) = \exp(-t/\tau_s)^\beta \qquad (1)$$

where $0.4 \leq \beta \leq 1$ and τ_s, the structural relaxation time, is not very different from the shear relaxation time. Hence:

$$\tau_s \sim \eta/G \sim \eta/7 \times 10^{10} \qquad (2)$$

where η is the shear viscosity in Pas and G, the shear modulus, in Pa. A typical value of G for silicate melts is substituted to give second equality of equation 2.

Regardless of the value of β relaxation to the equilibrium, structure is approximately 2/3 complete when $t = \tau_s$. Thus well into the glass transition range (i.e. $\eta = 10^{14}$Pas) relaxation of an unstable melt is substantial after some 1500 s. At higher temperatures relaxation becomes so-fast as to be instantaneous. The obvious conclusion is that only in or below the glass transition range can unstable structures persist long enough to be detectable.

On the other hand if an SG melt is metastable it can exist almost permanently well above the glass transition. It is important, therefore, to be able to test for metastability of SG melts.

A METHOD FOR TESTING FOR THE EXISTENCE OF A METASTABLE MELT

Suppose it is suspected that an SG melt is metastable relative to an MB melt because property differences between the melts persist for times very long compared to τ_s.

To test the metastability premise:

1. Measure a sensitive property e.g., density or viscosity of the SG melt.
2. Thoroughly stir into this melt a small volume fraction of MB melt of the same nominal composition, producing an abundance of nuclei of the MB phase, without perceptibly changing the SG melt composition.
3. Remeasure the properties at the same temperature.
4. If the properties are perceptibly changed, this is evidence that the suspected SG melt was indeed metastable and had a structure different from the equilibrium liquid. If properties are unchanged, it is likely that the initial property differences between the SG and the MB melt were due to composition or homogeneity differences.

HOMOGENEITY

Two multicomponent melts with identical average composition, but with different homogeneity will have different properties due in part to the contribution of the gradient energy [9] to the free energy. While heterogeneity's contribution to property differences are usually small and may be imperceptible by ordinary measurement techniques, they are considered here for completeness.

The homogeneity of an equilibrium melt is determined by thermodynamics, particularly by the heat of mixing. Equilibrium concentration fluctuations in the gel precursor will in general be different from those in the SG melt, but in both cases the scale of segregation will typically be of the order of the molecular size.

MB melts have much greater heterogeneity than SB melts because their scale of segregation is of the order of the batch particle size and the composition differences are large.

The decay of composition non-uniformity is described by a time constant, τ_H, defined as follows:

$$\tau_H = \delta^2/D \sim (6\Pi a/kT)\,\delta^2\eta \sim 10^{11}(N^{-1})\delta^2\eta \qquad (3)$$

where D is the smallest diffusion matrix eigen value and δ is the scale of segregation, roughly the average distance between adjacent concentration maxima. The second equality results from substitution of the Stokes Einstein relation where a is the ionic radius of the least mobile cation. The third equality is obtained by setting $a = 10^{-10}$m and T = 1500K. The decisive importance of scale is evident. Since $\delta(SG)/\delta(MB) \sim (10^{-3} - 10^{-7})$ equilibrium homogeneity is typically achieved much more rapidly in SG melts. This is consistent with Mukherjees' [10] comparison of the homogeneity of SG and MB glasses by scattering measurements.

However, the value of δ for an MB melt is under the control of the investigator, who by starting with fine particle size, a well mixed batch, and stirring intermittently with a scheme that causes convection in the entire volume, can reduce δ to $\leq 100\mu$m and hence by melting at a high enough temperature ($\eta \sim 1$ Pas) nearly achieve equilibrium homogeneity in an MB melt in less than an hour. This is consistent with Mackenzie's [11] comment that the homogeneity of optical glass is excellent.

CRYSTAL NUCLEATION

While bulk properties of a melt, e.g. enthalpy, volume, viscosity, etc. depend more or less equally on the environment of each of the ions, crystallization depends on the existence of relatively few stable nuclei of the crystal phase. Such nuclei typically occupy an extremely small ($\approx 10^{-10}$) volume fraction of the melt. They produce a change in melt structure, far too small to be perceived by measurement of bulk properties, but sufficient to cause a super cooled melt to be unstable rather than metastable with respect to crystallization.

The approach to equilibrium of the embryo and stable nuclei distribution in the bulk of the melt is determined [12] by a time constant, τ_e, which is given by

$$\tau_e = \eta(\Omega_c)^{1/3}/\sigma_{mc} \sim \eta/10^8 \qquad (4)$$

where σ_{mc} is the surface energy of the melt crystal interface and Ω_c is the crystal molecular volume. Substituting reasonable values $\Omega_c \sim 5 \times 10^{-30}$m^3 and $\sigma_{mc} \sim 0.02$J/m^2 gives the second equality. Thus:

$$\tau_e/\tau_s \sim 10^3 \qquad (5)$$

At high temperatures ($\eta \leq 10^8$Pas) τ_e and τ_s are both so short as to be indistinguishable. However, at temperatures in the glass transition range the distinction between τ_e and τ_s is easily recognized. Here a melt may be at apparent "structural" equilibrium yet possess a decidedly non equilibrium distribution of embryos.

Nucleation may be easier in a gel that is being heated to convert it to a melt than in an equivalent MB glass at the same temperature because of (i) the existence of interior surfaces in the gel, (ii) the structural differences between the melt and the partially converted gel resulting in a larger driving force for nucleation and (iii) the fact that the viscosity of the incompletely transformed gel can be much less than that of the glass. [13]

If an SG melt is heated to high temperature ($\eta \leq 10^6$Pas) without crystallization, or heated above T_m if crystals were formed at lower temperatures, then the equilibrium distribution of embryos is quickly

obtained and SG and MB melts should show the same crystallization behavior if
their chemical compositions are identical.

RECENT EXPERIMENTAL RESULTS ON BULK PROPERTIES

As reviewed by Mackenzie [11], Dislich [14], Yamane et al. [15] and Sakka and
Kamiya [16] have shown that property values are similar (though not
necessarily identical) when SG borosilicate glass, silica glass, and titania
-silica glass are compared to MB glasses of similar composition.

Gottardi et al. [17] performed calorimetric measurements on MB and SG
sodium borophosphate melts of nominally the same chemical composition.
While they found approximately the same values for Tg, the DSC curves were
sharper for the SG melt. They attributed this to the SG melt being more
homogeneous.

Brinker et al. [18] found from DSC measurements on cooling that the
specific heat of alkali-alumino borosilicate melts as a function of
temperature was within experimental error identical for MB and SG melts
when both had been heated to 1600K. Repeating such experiments [19] with
care taken to measure the OH content of both SG and MB glass they concluded
"no significant difference could be detected from examination of the
relaxation kinetics of a MB glass and a SG glass of the same composition and
OH content."

These authors [18] also found that Raman spectra of SG silica glass are
virtually indistinguishable from silica glass made in a more conventional
manner if the SG melt is preheated to high enough (~1300K) temperature.
Brinker [20] found the FTIR spectra of SG and MB glasses are likewise
indistinguishable if the OH content and the rest of the chemical composition
are the same. These results do not suggest the existence of a metastable SG
melt.

Reports [21] on liquidus temperature increase of a SG sodium silicate
melt compared to an MB melt of nominally the same chemical composition can
not refer to melts of identical composition, because as can be observed on Fig.
2 - if a crystal is precipitated from a higher free energy phase at a higher
temperature than the melting point of the stable equilibrium phase, then the
precipitated crystals will transform to the stable equilibrium melt. The
observed increase in liquidus temperature must therefore be due to a
composition difference between SG and MB melts.

There are also reports by the same authors [22] that the miscibility
temperature of an SG sodium silicate melt is ~ 100K higher than that of an MB
melt of nominally the same chemical composition. However, as shown by
Gupta [23], a melt of higher free energy will have a lower miscibility
temperature than a melt of the same composition that is at local equilibrium.
It appears therefore that the observed decrease in miscibility temperature
must have been caused by a composition difference between the SG and MB
melts. This possibility was admitted by the authors.

Yoldas [24] has advanced the concept that gel structure is primarily
influenced by the ratio R, mols H_2O/mols TEOS. He showed that silica gel at
low temperatures and silica gel sintering at temperatures within the glass
transition were different for different values of R. While different properties
for glasses with different values of R may be of practical importance, these
results also confirm that if two different SG glasses are never heated above
the glass transition, they may have different structures which will persist so
long as the glasses remain in the frozen state. [25]

Yoldas also prepared three soda lime silica melts by both SG and MB
processing. He states that in all cases the viscosity of the SG melt was less
than that of the corresponding MB melt. As an example, Figure 3 adopted from
his Fig. 8 shows a significantly lower temperature at log (η/Pas) = 6.65 for
the SG melt. As is clear the difference persists after soaking the melt well
above the liquidus temperature for 50 hours. These results appear to be

evidence of the existence of a metastable SG melt structure, especially since the reported water contents are only marginally higher in the SG glass. An easy way to confirm that the SG melt is metastable has been described above. Such a verification is crucial because this observation of different viscosities leads to a conclusion contrary to that deduced from the rest of the experimental evidence.

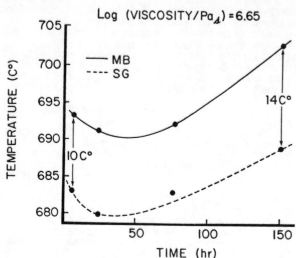

Figure 3
Temperature at Fixed Viscosity MB and SG Melts
as a Function of Soaking Time at 1838K

CRYSTALLIZATION OF SG GLASSES

Avoidance of crystallization on hydrolysis, drying and compaction is vital to permit formation of glass at temperatures below the liquidus.

Neilson and Weinberg [26] found that when a sodium silicate gel of soda, N silica S, ratio NS_4 is heated to 993K NS_2 crystals are abundant, while no such crystals are found if the SG precurser or powdered batch is heated to 1830K under conditions sufficient to assure homogeneity and equilibrium prior to the soak at 993K.

Grassi et al. [27] compared the crystallization kinetics of powdered SG glass, of NS_2 composition, dried at 700K for 24 hours and powdered MB glass of same composition melted at 1373K for 6 hours. From DTA curves they calculated that the activation energy for crystal growth was nearly the same for both powders, the crystal phase was the same, and that SG melt crystallizes at a lower temperature. The latter observation was attributed to the finer particle size of the SG glass, although easier nucleation would give the same result.

Mukherjee [28] has found that an SG sodium borosilicate glass monolith of molar composition 0.04 N - .10 B - .87 S has a strong crystallization tendency during the gel to glass conversion and hence greater crystallization tendency than the equivalent MB glasses. He also found that this crystallization tendency can be affected by low temperature heat treatment and by leaching. This is probably due to the influence of these processes on crystal nucleation during the gel to glass conversion.

Zarzycki [13], who analyzed the crystallization of SG melts and glasses, has pointed out that ease of nucleation in the SG route permits homogenous nucleation of crystal phases without addition of nucleants.

Mukherjee and Zarzycki [29] studied crystallization behavior of various melts in the system La_2O_3-SiO_2 from 0 to 5.79 mol % La_2O_3 and found that after dehydrating, drying and heating at 1073K for 2h, none of the specimens showed crystallinity, but that on subsequent heating to temperatures near the lower part of the glass transition, there was always some crystallization. The lowest temperature where crystallization was observed and the extent of crystallization depended on the type of gel.

Consistent with the discussion of crystal nucleation, the experimental results indicate that precursor gels are more susceptible to crystal nuclei formation during their conversion to glass. Once they have been heated to the equilibrium state to remove the effects of their history, SG melt crystallization behavior is virtually indistinguishable from that of the same MB melt.

CONCLUSIONS

Theory and most experiments suggest that if SG melts are of different structure from MB melts, they are unstable, not metastable. Yoldas' viscosity report is a solitary exception that needs confirmation. If SG melts are initially unstable, then except at and below the glass transition they will rapidly relax to the equilibrium state and be indistinguishable from the corresponding MB melt. SG glasses produced below the glass transition are in a frozen state. Hence there is no requirement that their structure be identical to that of a MB glass of identical composition and homogeneity.

Excessive (compared to MB glass) crystal nucleation at low temperatures is an understandable consequence of the SG process. Its avoidance is vital to the SG method of glass formation.

ACKNOWLEDGEMENT

Acknowledgement appreciation is expressed to P.K. Gupta for a critical reading and constructive suggestions and to NSF DMR 83-12301 for financial support.

REFERENCES

1. Richard Zallen, The Physics of Amorphous Solids, (John Wiley & Sons, New York, 1983). p. 174-183.

2. G. Kordas and R.A. Weeks, in Effects of Modes of Formation on the Structure of Glass, edited by R.A. Weeks, D.L. Kinser, and G. Kordas (North-Holland, Amsterdam, 1985), p. 327; A.A. Wolf, E.J. Friebele, and D.C. Tran, ibid., pp. 345.

3a. Ibid, p. 331.

3b. Ibid, p. 317.

3c. Ibid, p. 187.

4a. J. Zarzycki, J. Mat. Sci. 19, 1656 (1984).

4b. See also papers by G. Scherer and T. Shaw this meeting.

5. J. D. Mackenzie, J. Amer. Cer. Soc. 46, 461-70 (1963). Ibid 47, 76-80, 1964.

430

6. W. Primak, L.H. Fuchs, and P. Day, J. Amer. Cer. Soc. 38, 135-139 (1955).

7. M. Hara and S. Suetoshi, Report Research Lab, Asahi Glass Co. 5, 126-135, 1955. Also see George W. Scherer, Relaxation in Glass and Composites, (John Wiley & Sons, New York, 1986), p. 118.

8. G. Williams and D.C. Watts, Trans. Far. Soc. 66, 80-85 (1970).

9. J.W. Cahn and J.E. Hilliard, J. Chem. Phys. 28, 228 (1958).

10. Shyama P. Mukherjee, J. Non-Cryst. Sol. 63, 35-43 (1984).

11. John D. Mackenzie, J. Non-Cryst. Sol. 48, 1-10 (1982).

12. D. Kashchiev, Surface Science 14, 209 (1969).

13. J. Zarzycki, J. Non Cryst. Sol. 48, 105-162 (1982).

14. H. Dislich, Angew Chem. Int. Ed. 10, 363-70 (1971).

15. M. Yamane, S. Aso, S. Okano and T. Sakaino, J. Mat. Sci. 14, 607 (1979).

16. S. Sakka and K. Kamiya, J. Non-Cryst. Sol. 42, 403 (1980).

17. V. Gottardi et al., in Thermal Analysis, (Edition Wiedeman Birkhauser, Basel, 1980), p. 493.

18. C.J. Brinker et al., J. Non-Cryst. Sol. 71, 171 (1985).

19. G.W. Scherer, C.J. Brinker and E. Peter Roth, presented at the III International Workshop on Glasses and Glass Ceramics from Gels, to be published.

20. C. Jeffrey Brinker (private communication).

21. M.C. Weinberg and G.F. Neilson, J. Amer. Cer. Soc. 66 (2), 132-134 (1983).

22. G.F. Neilson and M.C. Weinberg, in Materials Processing in the Reduced Gravity Environment of Space, edited by Guy E. Rindone (Elsevier Science Publishing Co., New York, 1982), p. 333.

22a. M.C. Weinberg and G.F. Neilson, J. Mat. Sci. 13, 1206 (1978).

23. P.K. Gupta, in Effects of Modes of Formation on the Structure of Glass, edited by R.A. Weeks, D.L. Kinser, and G. Kordas (North-Holland, Amsterdam, 1985), p. 29 (see eqn. 8, p. 32).

24. B.E. Yoldas, J. of Non-Cryst. Sol. 51, 105-121 (1982).

25. A.R. Cooper, J. Non-Cryst. Sol. 71, 5 (1985).

26. G.F. Neilson and M.C. Weinberg, J. of Non-Cryst. Sol. 63, 365-374 (1984).

27. A. Grassi et al., Thermochemica Acta 76, 133-138 (1984).

28. S.P. Mukherjee, J. De Physique 43, C9-265 (1982).

29. S.P. Mukherjee and J. Zarzycki, J. of Amer. Cer. Soc. 62, 1-4 (1979).

ARE GEL-DERIVED GLASSES DIFFERENT FROM ORDINARY GLASSES?

M. C. WEINBERG, Jet Propulsion Laboratory, California Institute of Technology, Pasadena, CA 91109

ABSTRACT

A review is presented of some of the previously reported differences and similarities between comparable gel glasses (and gels) and ordinary glasses. In this regard, considerations are made with respect to such factors as structure, physical and thermal properties, and phase transformation behavior. A variety of silicate glass compositions are used for illustrative purposes. The discussion is roughly divided into two sections; low and high temperature behavior. At low temperatures one anticipates that differences between gel and conventional glasses will exist, but such dissimilarities are not expected to persist to high temperatures. However, experimental evidence is presented which indicates the perpetuation of such differences to very high temperatures. A partial resolution for this anomalous behavior is offered.

INTRODUCTION

Gels may be produced by the hydrolysis and poly-condensation of metal alkoxides. Often these gels may be converted to gel glasses via relatively low temperature heat treatments. The conventional (or ordinary) method of glass preparation involves the high temperature (above $T\ell$, the liquidus temperature) processing of crystalline powders followed by undercooling the liquid to a point below T_g, the glass transition temperature. Hence, the thermal histories experienced by gel and ordinary glasses differ in important respects. Conventionally prepared glasses undergo high temperature treatment and are formed "from above" (i.e. by cooling), while gel glasses do not experience high temperatures and are formed "from below" (i.e. by heating).

Since thermal history may influence the structure and properties of a glass, one may be led to the following inquiry. Are gel prepared and conventionally produced glasses the same? This question has been addressed frequently in the literature [1-20], often with contradictory conclusions. As will be illustrated, part of the confusion and discrepant findings may be attributed to the lack of preciseness used in phrasing the question posed above. For example, it is well documented that the structure and properties of the dried, but undensified gel, can be quite different from that of the glass. On the other hand, gels which have been converted to glasses employing long duration heating at high temperatures are anticipated to exhibit the behavior and properties characteristic of ordinary glasses. Hence, in investigating the similarities and differences between gel derived and ordinary glasses, one is actually inquiring as to the thermal histories (the times and temperatures) which are and are not adequate to insure the equivalence of these glasses. A second important issue is the criteria employed in testing the equivalence (or non-equivalence) of the glasses. Here one must distinguish between the detection and recognition of gross and subtle differences between gel and ordinary glasses. If gross differences exist between these glasses, in a given compositional system, then they could probably be detected by a variety of property and/or structural measurements. Furthermore, more delicate questions, such as the variation of properties and structure of ordinary glasses with fictive temperature and the effects of minor concentrations of impurities, most likely may be ignored. The detection of subtle dissimilarities between gel and ordinary glasses, however, requires a more painstaking analysis and a meticulous experimental approach. In reviewing previous studies, considerations will given to the points discussed above.

EVIDENCE FOR DISPARATE GEL GLASS BEHAVIOR

The primary source of evidence for dissimilarities between gel and ordinary glasses derives from the study of glass phase transformation behavior. Several experimental studies have revealed qualitative differences between the phase separation and crystallization characteristics of gel and ordinary glasses.

Mukherjee, Zarzycki, and Traverse [1] observed that the phase separation behavior produced in ordinary and gel La_2O_3-SiO_2 glasses were notably different. They reported that in the gel glasses phase separation occurred with a uniform microstructure and with the production of small droplets, while the oxide glass exhibited a phase separation characterized by a non-uniform distribution of large and more irregularly shaped particles. The crystallization behavior of gel and ordinary glasses in the La_2O_3-SiO_2, La_2O_3-Al_2O_3-SiO_2, and La_2O_3-ZrO_2-SiO_2 systems was investigated, too, in the same study. It was noted that, in general, there was a greater degree of crystallization in the gel glasses than in the ordinary glasses for any given heat treatment. In addition, it was observed that the crystallization morphology in the gel La_2O_3-SiO_2 glass was more fine-grained than that appearing in the corresponding ordinary glass. Two features concerning these results warrant comment. First, there were notable differences between the phase transformation behavior of the gel and ordinary glasses. Also, the gels were heated well in excess of Tg (and most likely above Tℓ) during the gel glass preparation.

Weinberg and Neilson investigated the phase separation behavior of ordinary and gel glasses in the Na_2O-SiO_2 glass system [2,4]. Two aspects of the phase separation behavior of these glasses were studied; the morphologies of the developing second phase upon heat treatment and the location of the immiscibility temperature. The developing minor phases in both glasses were investigated using replication electron microscopy (REM) and small angle X-ray scattering (SAXS). Immiscibility temperature measurements were made by noting the clearing temperature of a phase separated sample and by the use of REM. The REM of ordinary and gel glasses, which were given identical heating schedules, exhibited very different morphologies. SAXS studies corroborated the REM results. Also, it was found that the immiscibility temperature of the gel glass was at least 100°C higher than that of the corresponding ordinary glass. The gel glasses used in this study were prepared by high temperature melting of the gel precursor.

Using property measurements, Yoldas [3,5] has presented evidence that gels heated to high temperatures may exhibit anomalous behavior in comparison to ordinary glasses. For example, he measured the viscosity of three soda-lime-silicate compositions, and found that there were significant differences between the values of the viscosities for the ordinary and gel glassses. The viscosities of gel and ordinary glasses were measured for a series of glasses for which the melting times were varied. It was observed that the difference between the gel and conventional glass viscosities increased with longer melting times.

In the same study Yoldas reported additional evidence implying that gel glass peculiarities can persist to the molten state. He found that SiO_2 prepared by the gel method in which large amounts of water were used for hydrolysis resulted in a more fluid melt and a lower melting point than those obtained when small amounts of water were used for hydrolysis. Also, IR studies indicated that the connectivity of the SiO_2 network differed for the polymers produced by these two procedures. In addition, it was noted in this same investigation that the heating schedule chosen for the gels can affect the resulting glass properties. In particular, Yoldas reported that the glass transition temperature of a pyroceram glass composition could be altered by 50°C by varying the heat treatment of the gel.

Finally, Yoldas indicated that the electrical properties of glass may be affected by the preparation procedure. Significant differences in both dielectric loss and resistivity were noted for a particular glass prepared by gel and conventional melting techniques.

Weinberg and Neilson [6] determined the liquidus temperatures of ordinary and gel 19 wt% soda-silica glass via x-ray diffraction measurements. It was found that the liquidus temperature of the gel glass was 33°C higher than that of the conventional glass.

The examples cited above relate to situations where rather dramatic differences between gel and ordinary glasses have been detected and where the gels were heated well in excess of Tg. This section will conclude with a discussion of several comparisons in which homogeneous gel glasses were produced by heating in the vicinity of Tg, but which could not be prepared as uniform glasses via conventional methods.

Yamane and Kojima [21] prepared homogeneous glasses in the SrO-SiO$_2$ system of composition 1,5 and 10 wt.% SrO. An immiscibility gap is known to exist in this system extending from 3 to 30 wt.%, and thus the 5 and 10 wt.% SrO compositions prepared by conventional techniques are phase separated. The work of Yamane and Kojima illustrates that unique glasses may be prepared in systems with rapidly occurring phase separation if at low temperature the rate of gel densification exceeds the rate of phase separation. Similar results have been obtained by Hayaski and Saito [22], who prepared CaO-SiO$_2$ glasses by the gel method. They were able to produce monolithic CaO-SiO$_2$ glasses in the compositional region where homogeneous glasses cannot be produced by conventional melting methods due to rapid liquid-liquid immiscibility.

EVIDENCE FOR EQUIVALENCE OF GEL AND ORDINARY GLASSES

A good deal of experimental data has been gathered which seems to indicate that gel and conventional glasses are virtually identical. This data has been obtained via comparative measurements of properties, structure, and kinetic behavior for a wide variety of gel and ordinary glasses. In some of these studies the gels were heated in excess of Tg. However, in some cases it was found that gels heat treated below Tg closely resembled their ordinary glass counterparts.

Discussions of glass structure will be given subsequently . Here the reported similarities between gel and conventional glasses will be reviewed with regard to property measurements and kinetic behavior.

The gel glass which has received the most intensive study is SiO$_2$. Yamane [9] and Mackenzie [8] compared several properties of commercial silica and gel prepared SiO$_2$. These included density, refractive index, Vickers hardness, thermal expansion, Young's modulus, and shear modulus. In all cases the measured properties were identical or nearly identical for the gel and ordinary glass. However, the gel was heated only to 1070°C, which is below Tg for SiO$_2$.

Similarities between properties of gel and conventional glasses have been noted in more complex glasses, too. Sakka and Kamiya [10] and Mackenzie [8] compared the densities and thermal expansions of several gel and ordinary TiO$_2$-SiO$_2$ glasses, and found them to be quite similar. The density, thermal expansion, and glass transition temperature of a complex borosilicate glass were measured by Dislich [11] and reported by Mackenzie [8] for ordinary and gel glass. They were found to be nearly identical. In this case the borosilicate gel was converted to a glass by pressing at temperatures 30-60°C in excess of Tg.

Nogami and Moriya [12] compared the properties of several gel and ordinary glass compositions in the B$_2$O$_3$-SiO$_2$ system. They made measurements of glass transition temperature, density and Vickers hardness. They found minor differences in these properties for the glasses prepared by the two techniques.

In two separate studies Brinker, Scherer, and coworkers [13,14] investigated the structural relaxation kinetics of a complex borosilicate glass prepared by gel and ordinary methods. Differential scanning calorimetry (DSC) measurements were made to obtain constant pressure specific heats (Cp) and to obtain parameters required for the fitting of the specific heats to the relaxation model of Debolt et.al [23]. In the first study the fitting parameters were determined from the DSC data obtained from a densified gel which had received a prior heat treatment at 1100°C for 30 minutes. These parameters, also, provided a very good fit to the Cp data of a conventionally prepared glass and a gel which had been heated to 1600°C. Thus, it was found that gel and melted glass exhibited identical behavior within experimental error. In the more recent study [14], the fitting parameters were determined for an undried gel, a desiccated gel, a melted gel, and a melted ordinary glass. The Cp curves constructed for the melted glasses (gel and ordinary) were well fit by the plots constructed using the relaxation parameters obtained for the desiccated gel. Hence, it was concluded that no difference exists between the structural relaxation kinetics of the ordinary and gel glasses.

Neilson, Weinberg, and Smith [15] studied the kinetics of phase separation of a gel and ordinary soda-silica glass employing SAXS measurements. It was observed that the phase separation rates of comparable gel and ordinary glasses are quite similar.

Neilson and Weinberg [16] investigated the crystallization behavior of melted ordinary and gel glass of a particular Na_2O-SiO_2 composition. These glasses were found to show qualitatively similar crystallization behavior. In particular, the crystalline phases produced in both glasses for the same heat treatment conditions were identical.

Ravaine, et. al. [17] investigated the electric properties of dried gels in the $Na_2O - SiO_2$ system. They concluded that thermal treatments as low as 200°C were adequate to reproduce the electric properties of conventionally prepared glasses.

A number of other investigators have reported strong similarities between gel and ordinary glasses based upon observations of glass properties and structure. A complete description of all such studies would be too lengthy to report here. However, the results of a few investigations of glass structure will be given in the final section.

In summary, the results obtained for the properties and kinetic behavior of a wide variety of gel glasses seem to indicate that in all respects these glasses are equivalent to their conventional glass counterparts. In some instances these similarities are quite remarkable since they are evidenced even in situations where the gel was densified at temperatures below Tg.

POTENTIAL ORIGINS OF GEL GLASS ANOMALIES

Several examples of dramatic differences between gel and ordinary glasses have been discussed. These differences are considered dramatic since 1) the discrepancy between gel and ordinary glass behavior or property is substantial and 2) they are observed even after the gels have been heated to temperatures well in excess of Tg. On the other hand, in the previous section it was illustrated that many gel glasses are essentially equivalent to ordinary glasses. Hence, cases of unusual gel behavior (or characteristics) may be considered atypical, and will be given the designation of gel glass anomalies.

Here several possible explanations for the appearance of gel glass anomalies will be considered. It will be argued that the anomalies most likely have more than one origin, and that in some instances the source of the disparate gel behavior can be identified.

It is well known that water in a glass may affect its kinetics of crystallization [24] and phase separation [25]. Since gel glasses may contain large amounts of bound hydroxyl (unless particular efforts are made for removal of OH), elevated OH concentrations could be responsible for gel glass anomalies.

The elevation of the immiscibility temperature reported for a Na_2O-SiO_2 gel glass was explained in these terms [4]. IR spectral measurements of the ordinary and gel glass revealed that the gel glass had a water concentration about 3.5 times that of the ordinary glass. Water is to a large extent incorporated in this glass as SiOH. Thus, glasses containing enhanced water content would be expected to have a lower degree of structural coherency since they would possess fewer bonding oxygens. It was argued that this reduction in connectivity could account for the elevation of the immiscibility temperature. However, further studies [19] demonstrated that enhanced water content was not responsible for the unusually high immiscibility temperatures reported for the gel glasses.

Similarly, Yoldas [5] has argued that the anomalous values of viscosity found in the gel soda-lime-silicate compositions could not be accounted for by excess hydroxyl content. The data which he presented on water content of ordinary and gel glasses appears to support this claim.

Mukherjee et. al. [1] attributed the enhanced gel glass nucleation and crystallization rates to elevated OH concentrations. However, hydroxyl determinations were not reported for either the gel or the conventional glasses.

Thus, although gel glasses may often contain elevated OH concentrations, it does not appear that the anomalous behavior of gel glasses discussed herein is attributable to incorporated hydroxyl groups.

One might argue that trace impurities might play some role in the dramatic differences in behavior between some ordinary and gel glasses. However, this conjecture appears implausible. Weinberg and Neilson performed analyses of gel and ordinary Na_2-SiO_2 glasses for certain impurities which are known to have a large influence on the immiscibility temperature. They found that these impurities were present in too low a concentration (in all glasses) to account for the observed large difference in immiscibility temperature. Also, one would anticipate that impurities would tend to hasten crystallization kinetics. Thus, the accelerated crystallization kinetics of the gel glasses found by Mukherjee et. al. can't be attributed to impurities since one anticipates the impurity levels to be higher in the ordinary glass than in the gel glass. Also, it is difficult to imagine how one could explain the temperature dependence of the anomalous behavior of soda-lime-silicate gel glasses reported by Yoldas in terms of impurity effects. In summary, it seems that one may dismiss impurity ion concentrations as being responsible for the anomalous behavior of gel glasses.

Yoldas has argued [3,5] that structural differences between ordinary and gel glasses persist to high temperatures and probably are the cause of the disparent gel glass properties. Although a more thorough discussion of gel and conventional glass structure will be presented in the final section, it is important to present here a cursory evaluation of the likelihood that structural deviations present in the gel can survive at high temperatures. One may employ the value of the structural relaxation time as a guide to the time scale required for the structure to come to equilibrium. At temperatures well in excess of Tg, the relaxation time is proportional to the shear viscosity. One may easily demonstrate that for melt viscosities of about 10^3 poise, the relaxation time is a fraction of a microsecond [26]. Hence, gels which have been melted (and whose melt viscosities are not excessively large) are anticipated to have reached their equilibrium structures at high temperatures. Thus, it is implausible to expect that in such cases the gel glasses can retain "memory" of structural deviations which were present in the gel.

Up to this point, several potential explanations for some of the more dramatic differences between gel and ordinary glasses have been considered and rejected. Now, a factor will be considered which in all likelihood is the cause of the anomalous gel glass behavior observed in many instances. It is recognized that gross compositional differences between glasses could easily account for differences in properties and behavior. In addition, it is realized that some glass properties and behavioral characteristics are more sensitive to compositional shifts than others. For example, some aspects of the transformation properties of glasses are very composition dependent. It was indicated by Weinberg and Neilson [19] that the differences in phase separation and crystallization behavior between gel and ordinary glasses reported in [1] might be attributable to compositional variations. The glasses in this study were prepared using a focussed heating source which produced vaporization of the samples. In addition, chemical analyses of conventional and gel glasses were not reported. In the immiscibility temperature studies performed by Weinberg and Neilson the compositions of the gel and ordinary glass were determined employing atomic absorption methods. The ordinary and gel glasses were deemed to be quite close in composition, and thus compositional variations were not seriously considered as a probable source of the anomalous gel behavior. However, this was an important oversight since recent investigations indicate that the "elevation" in the immiscibility temperature of the Na_2O-SiO_2 gel glasses were due to differences in gel and conventional glass composition [19]. Two factors played a role in obscuring the origin of the discrepancies. First, it must be recognized that in the Na_2O-SiO_2 system the immiscibility temperature is quite dependent upon glass composition (especially at compositions near the edges of the immiscibility dome). This feature implies that very accurate compositional analyses are required in comparative studies of gel and ordinary glass liquid-liquid phase separation studies. Secondly, the reliable and consistently accurate chemical analysis of glass is difficult.. Atomic absorption, which is one of the favored techniques for such purposes, can give fairly inaccurate results. This was demonstrated in the case of the Na_2O-SiO_2 system with the aid of refractive index measurements. It is interesting to note that similar conclusions concerning the need for very accurate compositional measurements were reached by Brinker and Scherer [14] in their studies of the kinetics of structural relaxation of gel and ordinary glasses. For the borosilicate glass which they studied they found that the relaxation kinetics were extremely sensitive to composition. As a result it was concluded that differences in kinetic behavior between the gel and conventional glasses could be inferred erroneously if glass compositions are not known precisely.

Yoldas [5], however, has claimed that the anomalous viscosity behavior,which he observed in soda-lime-silicate compositions could not be attributed to differences in alkali or hydroxyl content. However, he did not report the precise compositions of the gels and conventional glasses, nor did he provide data on the composition dependence of the viscosity. Thus, it appears that a careful study of the composition dependence of the viscosity and precise compositional analyses of gel and conventional glasses would be warranted. The results of such investigations might indicate that compositional variations are the source of the anomalous gel properties noted in the work of Yoldas, too.

Thus, in summary, there are strong indications that the reported dramatically deviant behavior and properties of gel glasses are spurious and due to compositional differences between gel and conventional glasses.

SUBTLE DIFFERENCES

Although it has been concluded that the dramatic differences between gel and ordinary glasses found by several investigators were probably "artifacts", it is quite plausible that subtle distinguishing features may exist for gel and ordinary glasses. In particular, gels which have been converted to glasses by heating at temperatures in the vicinity of Tg may retain "memory" and exhibit structure and property characteristics associated with that of the gel. In this section several examples will be cited in which small property differences have been noted.

Nogami and Moriya [12] have studied the glass formation and properties of SiO_2-B_2O_3 glasses prepared by the gel method. They measured the Vickers hardness, density, and glass transition temperature of several compositions using gel and conventionally prepared glasses. Although they concluded that, "the difference in the properties of the glasses prepared by the different processes is hardly noticeable", differences can be observed (particularly for the Tg data). Of course, as indicated previously, precise compositional analyses of gel and ordinary glasses are required to establish that such differences are "real".

Mackenzie [8] compared the densities and thermal expansions of gel prepared and conventionally melted TiO_2-SiO_2 glasses. Although the data reported for the gel and ordinary glasses were similar, differences were definitely observable.

Another case of an unexplained difference between conventional and gel glasses occurs in the work of Brinker et.al. [13]. It will be recalled that they found that the same set of fitting parameters would provide an excellent representation of Cp for the gel or conventional glass. They took this as evidence for the equivalence of the gel and ordinary glass. However, Brinker and coworkers noted that the Cp vs. temperature curve for the gel glass exhibited a distinct inflection point at about 800°K. This inflection was absent from the corresponding curve for the ordinary glass. It would have been informative if the Cp data for gel and ordinary glass had been plotted on the same set of axes so that one could ascertain directly the correspondence of these two curves.

Thus, the question of whether minor differences can exist between ordinary glasses and gel glasses which have been heated to temperatures in the vicinity of Tg has not yet been resolved. If these departures in behavior and properties are "real", one may inquire as to whether they have their origins in structural variations of the gel glass. This topic will be addressed next.

GEL AND ORDINARY GLASS STRUCTURE

The discussion of glass structure given here must be limited in scope since several volumes could be filled just reviewing the topic of ordinary glass structure. Hence, descriptions of structure determinations per se will be omitted, and attention will be centered solely on the observed similarities and differences between gel and ordinary glasses. Furthermore, to restrict the length of the presentation, only several prototype compositions will be discussed, and attention will be limited primarily to IR and Raman spectroscopic data. Also, no attempt will be made to review the information on the development of the gel structure upon heating. Comparisons will be made solely between ordinary glasses and gels which have been heated as close as possible to Tg without the intervention of crystallization.

The opportunities for structural diversity and the types of structural modifications anticipated will depend upon the number of components in the glass and the nature of these components (network former or modifier, coordination number possibilities, etc.). For example, one would expect a greater chance for structural variations in a system containing two network formers (M_1,M_2) than in a glass such as SiO_2 or GeO_2. This is because of the possible formation of M_1-O-M_1, M_1-O-M_2, and M_2-O-M_2 groupings. Also, the possibilities for producing structurally unique borate gel glasses might be thought greater than that for silicate compositions since boron can assume 3 or 4 coordination in borate glasses. These considerations will motivate the choices of compositions selected for illustration.

Although vitreous silica gel may not offer the multitude of possibilities for structural diversity alluded to above, it is prudent to consider its structure first for the following reasons: 1) It has been the most intensely studied composition, 2) It has been prepared by a large variety of gel methods, 3) there is no possibility for compositional effects (i.e. structure differences due to small discrepancies between conventional and gel composition).

Comparisons were made between the IR spectra of vitreous silica and vitreous silica gels heated at 800°C [27,28]. Decottignies et.al. [27] prepared a silica gel glass by "flash" hot pressing. The density of this gel glass was found to be that of vitreous silica. They concluded that apart from the appearance of water and Si-OH stretch bands in the gel (absent in the ordinary glass) the IR spectra of gel and ordinary glass were virtually identical. However, there were small shifts in the peak positions of the major IR absorption in silica. The band centered at 1000 cm^{-1}, which is due to Si-O stretch, was shifted about 20 cm^{-1}, and the O-Si-O bending vibration centered at 468 cm^{-1} was shifted about 12cm^{-1}. No information was reported regarding the relative intensities of the bands obtained from the spectra of the two glasses.

Bertoluzza et.al [28] have compared the structure of silica gel and ordinary glass employing IR and Raman spectroscopy. Although spectra were presented of gels heated at several temperatures, here only the gel spectra obtained after heating to 800°C (maximum temperature) will be reviewed. Bertoluzza and coworkers concluded that the latter spectra are very similar to those obtained from ordinary silica glass. However, minor differences may be detected in both IR and Raman spectra. For example, the IR band in the region of 800cm^{-1} is broader in the gel spectrum and does not exhibit the shoulder evident in the ordinary silica spectrum. According to Bertoluzza et.al. [28] this band has been assigned to Si-0-Si symmetric band stretching vibration and to vibrational modes of ring structure in silica. The Raman bands in the vicinity of 600cm^{-1} and 487cm^{-1}, attributed to network defects [29], appear in both gel and ordinary glass spectra. However, the structure of these bands appears to differ somewhat in the latter spectra.

Krol and van Lierop have investigated the densification of silica gel using Raman spectroscopy [30]. They, too, observed a strong similarity between the spectra of the SiO_2 gel heated at 1000°C and ordinary silica glass. However, they noticed an interesting behavior of the defect bands in the gel spectra. Krol and van Lierop measured the intensity of the ~600cm^{-1} band (relative to the 800cm^{-1}band) as a function of gel heat treatment temperature. For gel heat treatment temperatures at or above 800°C the intensity of this defect band fell far below that found for ordinary silica glass. They explained this result in the following manner. The defect bands correspond to high energetic structures (postulated to be 3 and 4 membered rings). A silica glass melt is anticipated to possess a significant number of these structures as a result of its high temperature. During quenching a non-negligible number of these structures are anticipated to survive and hence remain in the glass. The number of these defects appearing in the ordinary glass will depend upon the fictive temperature. Lower fictive temperature glasses should be more defect free. Thus, in essence the gel SiO_2 glass may correspond to a uniquely low fictive temperature silica glass. Stolen and Walrafen [29], indeed, have studied the effects of fictive temperature (and hydroxyl concentration) upon the defect concentration in ordinary silica glass. They observed that the intensities of the defect bands in the vicinity of 600cm^{-1} and 490cm^{-1} increased with larger fictive temperatures and decreased with higher hydroxyl content. From their results, however, it appears that the variation in intensity of defect bands with fictive temperature is rather small (compared to the dramatic difference found in [30]). In addition, it should be noted that the hydroxyl content of the gel glass heated to 900°C by Krol and van Lierop was significantly higher than that of the ordinary glass. Hence, the low intensity of the 600cm^{-1} band in the gel reported in [30] might be attributable to OH. However, it appears as though further studies are required.

In summary, it may be concluded that IR and Raman spectroscopic measurements of ordinary and gel SiO$_2$ glasses indicate strong structural similarity. These strong resemblances are seen even though in most cases the gel has been heated at temperatures well below Tg for ordinary silica. Minor differences in the gel and glass spectra have been noted, however. Whether or not these differences are "real" or caused by differing OH content, experimental procedures, or some other uncontrolled factor is not yet clear.

The structure of two component gels and gel glasses consisting of network formers is of interest for reasons mentioned previously. Several such gels have been prepared as monoliths [12,31].

Nagami and Moriya [12] probed the structural development of SiO$_2$-B$_2$O$_3$ gels employing IR spectroscopy. They investigated compositions containing from 15 to 50 mol % B$_2$O$_3$. The spectra of the gels which were heated in excess of Tg appeared very similar to the spectra of the corresponding glasses. Also, absorption bands were not in evidence at 720cm^{-1}. Absorption bands in this region are indicative of B-O-B bonds. However, a band in this region was detected in the 40 mol % gel which was heated at lower temperatures. Although minor discrepancies were apparent between the gel and conventional glass spectra, there were no indications of major differences such as regions of B-O-B, Si-O-B, and Si-O-Si linkages. Nevertheless, more detailed structural studies appear warranted.

Woignier et. al. [31] studied the production and characterization of monolithic SiO$_2$-B$_2$O$_3$ and SiO$_2$ - P$_2$O$_5$ gels. They found that subsequent to heat treatment at 1100°C, the gels exhibit IR spectra very similar to those of the corresponding ordinary glasses. However, comparative spectra were not presented.

Mukherjee and Sharma [32] have investigated the structure of gel derived glasses in the SiO$_2$-GeO$_2$ system employing Raman spectroscopy. They studied several compositions in this system using a variety of gel preparation procedures. They observed that the Raman spectra of low GeO$_2$ concentration gels prepared by different procedures exhibited different structural details even when the gels were heated as high as 1300°C. In particular, the intensities of the Si-OH and silica network defect bands could be altered via gel synthesis procedure. As mentioned previously, this could be a reflection solely of the relative hydroxyl content of these gels. Comparisons between the spectra of these gel glasses and conventional glasses of corresponding composition were not presented, but the authors claimed that they appear similar. The gels containing substantial concentrations of GeO$_2$ crystallized when heated to high temperatures, and hence will not be discussed.

Therefore, it appears that the structural data gathered concerning binary gels involving network formers indicates the following. The gel structure initially produced may be a strong function of synthesis procedure. However, subsequent to heat treatment near Tg the gross structural features of the gel are quite similar to those of the conventional glass. Minor difference between gel glass and conventional glass, though, seem to persist. However, it is yet unclear whether or not these minor differences result from an enhanced OH content in the gel glass and/or small compositional variations between the ordinary and gel glass.

CONCLUSION

The question of the equivalence of gel and conventional glasses has not been resolved, nor have all of the aspects of this query been addressed herein. However, it is possible to summarize the current state of affairs regarding the points which we have considered.

In all probability gel glasses which have been prepared by melting or heating to temperatures well in excess of Tg are equivalent to the corresponding conventional glasses if the OH concentration in each are the same. The recent evidence of Brinker and Scherer and Weinberg and Neilson strongly indicate this to be the case. However, Yoldas' findings remain unexplained, especially his results on SiO_2 where compositional variations can't be the source of difficulty. Nonetheless, it seems as though we are on the verge of dispelling the notion of gel glass anomalies (in the context used herein).

The situation is entirely different with respect to gel glasses produced by hot pressing or thermal heat treatments at low temperatures. Here one may speculate that glasses with novel structures and properties could be formed. Rabinovich [20] has noted that gel glasses prepared in this manner may correspond to 'low fictive temperature glasses'. He has argued that such glasses could not be produced by rapid quenching of the melt since this procedure could not come close to obtaining glasses with a fictive temperature as low as those which can be produced by gel methods. He has stated that one would expect such differences to be most noticable in multicomponent glasses. Rabinovich's arguments are quite reasonable. Yet, there seem to be few instances in which "low temperature gel glasses" have been found to be dramatically different from the corresponding conventional glasses. Notable exceptions are the gel glasses produced by Yamane [21] and Hayashi [22] which could not be produced free from phase separation by conventional methods.

Undoubtedly, the final word has not been said concerning this topic. Much more work is needed to carefully probe the gel glass structure, especially for binary and multicomponent compositions. The effects of gel synthesis procedure upon the resulting structures and the role of hydroxyl also need further elucidation.

ACKNOWLEDGEMENTS

This article was written at the Jet Propulsion Laboratory, California Institute of Technology, under contract with the National Aeronautics and Space Administration (NASA). Gratitude is extended to the Microgravity Sciences and Applications Division of NASA for the support of this effort. Also, appreciation is given to Dr. George Neilson, of the Jet Propulsion Laboratory, for his valuable comments.

REFERENCES

1. S. P. Mukherjee, J. Zarzycki, and J. P. Traverse, J. Mater. Sci. 11, 341(1976).

2. G. F. Neilson and M. C. Weinberg, in Materials Processing in the Reduced Gravity Environment in Space, edited by Guy E. Rindone (Elsevier Science Publishers, New York, 1984),p.333.

3. B. E. Yoldas, J. Non-Crystalline Solids 63, 145 (1984).

4. M. C. Weinberg and G. F. Neilson, J. Mater. Sci. 13, 1206 (1978).

5. B. E. Yoldas, J. Non-Crystalline Solids 51, 105 (1982).

6. M. C. Weinberg and G. F. Neilson, J. Amer. Ceram. Soc. 66, 132 (1983).

7. C. J. Brinker, E. P. Roth, and G. W. Scherer, presented at the 1985 Annual Ceramic Society Meeting, Cincinnati, OH., May 1985 (unpublished).

8. J. D. Mackenzie, J. Non-Crystalline Solids 48, 1 (1982).

9. M. Yamane, S. Aso, S. Okano, and T. Sakaino, J. Mater. Sci. 14, 607 (1979).

10. S. Sakka and K. Kamiya, J. Non-Crystalline Solids 42, 477 (1980).

11. H. Dislich, Agnew. Chem. Int. Ed. 10, 363 (1971).

12. M. Nogami and Y. Moriya, J. Non-Crystalline Solids 48, 359 (1982).

13. C. J. Brinker, E. P. Roth, G. W. Scherer, and Dr. R. Tallant, J. Non-Crystalline Solids, 71, 171 (1985).

14. G. W. Scherer, C. J. Brinker, and E. Peter Roth, J. Non-Crystalline Solids (in press).

15. G. F. Neilson, M. C. Weinberg, and G. L. Smith, J. Non-Crystalline Solids (in press).

16. G. F. Neilson and M. C. Weinberg, J. Non-Crystalline Solids 63, 365 (1984).

17. D. Ravaine, J. Traore, L. C. Klein, and I. Schwartz, in Better Ceramics Through Chemistry, edited by C. J. Brinker, D. E. Clark, and D. R. Uhlrich (Elsevier Science Publishers,New York, 1984),p.139.

18. C. J. Brinker, W. D. Drotning, and G. W. Scherer, in Better Ceramics Through Chemistry, edited by C. J. Brinker, D. E. Clark, and D. R. Uhlrich (Elsevier Science Publishers, New York, 1984), p.25.

19. M. C. Weinberg and G. F. Neilson, in Sol Gel Technology, edited by L. Klein (Noyes Publications, New Jersey) (in press).

20. E. M. Rabinovich, J. Non-Crystalline Solids 71, 187 (1985).

21. M. Yamane and T. Kojima, J. Non-Crystalline Solids 44, 181(1981).

22. T. Hayashi and H. Saito, J. Mater. Sci. 15, 1971 (1980).

23. M. A. Debolt, A. J. Eastel, P. B. Macedo, and C. T. Moynihan, J. Amer. Ceram. Soc. 59, 16 (1976).

24. P. F. James, in Advances in Ceramics, Volume 4 , edited by J. H. Simmons, D. R. Uhlmann, and G. H. Beall (American Ceramic Society, Columbus, OH., 1982), p.1.

25. N. J. Kreidl and M. S. Maklad, J. Amerc. Ceram. Soc. 52, 508 (1969).

26. G. W. Scherer, in Relaxation in Glass and Composites, (John Wiley and Sons)(in press).

27. M. Decottignies, J. Phalippou, and J. Zarzycki, J. Mater. Sci 13, 2605 (1978).

28. A. Bertoluzza, C. Fagnano, M. A. Morelli, V. Gottardi, and M. Guglielmi, J. Non-Crystalline Solids 48, 117 (1982).

29. R. H. Stolen and G. E. Walrafen, J. Chem. Phys. 64, 2623 (1976).

30. D. M. Krol and J. G. van Lierop, J. Non-Crystalline Solids 63, 131 (1984).

31. T. Woignier, J. Phalippou, and J. Zarzycki, J. Non-Crystalline Solids 63, 117 (1984).

32. S. P. Mukherjee and S. K. Sharma, J. Non-Crystalline Solids 71, 317 (1985).

443

A COMPARISON OF STRUCTURES AND CRYSTALLIZATION
BEHAVIOR OF GELS, GEL-DERIVED GLASSES AND CONVENTIONAL GLASSES

SHYAMA P. MUKHERJEE
IBM Corporation, Endicott, New York 13760

ABSTRACT

Studies on the structures and crystallization behavior of gels in TiO_2, GeO_2, SiO_2-GeO_2 and PbO-GeO_2 systems are presented to analyze the differences in structures and crystallization behavior of gel-derived glasses and conventional glasses. The kinetic parameters for crystallization such as activation energies and frequency factors of gels, gel-derived glasses and conventional glasses in the GeO_2-PbO system are presented to elucidate their difference in nucleation and crystallization behaviors. The influence of gel processing parameters such as pH and water concentration on the crystallization behavior of gels are discussed.

INTRODUCTION

Studies aiming at elucidating the structural and microstructural differences of gels, gel-derived glasses and conventional glasses are very limited [1-5]. However, it should be noted that the multicomponent gels which tend to crystallize during gel-to-glass transformation would produce glasses of different structures where larger and more ordered building blocks might exist [5-7]. The nature and distribution of the blocks will depend on the chemical polyermization conditions and thermal history of gel-derived glass. Moreover, the structures which could be achieved by chemical polymerization cannot be obtained by melting oxides where primarily the thermal disorder, i.e., the melt structure and, the cooling conditions dictate the glass structures [6]. Experimental works indicate that [1,5-10] the kinetics of the ordering process for the gels and gel-derived glasses appears to be different from those of the vitreous solids produced from the melts.

Earlier work of Mukherjee, et al. [1] on the nucleation and crystallization of gel-derived glasses and conventional glasses in the $SiO_2-La_2O_3$ system indicate that the nucleation rate of gel-derived glass, i.e., glass obtained by "flash melting" of gels is much higher than that of the glass melted from the oxide mixtures. The higher nucleation rate was explained on the basis of formation (in gels) of extremely small La^{+3}-rich clusters [7] in higher concentration which were inherited in the gel-glass structures and acted as the nucleation centers. In present work, structural and crystallization studies on the following systems are discussed: (a) TiO_2, (b) GeO_2, (c) SiO_2-GeO_2 and (d) GeO_2-PbO.

POLYMERIC TITANIUM OXIDE GEL

Hydrolytic polycondensation of titanium isopropoxide using 2 mols H_2O gives polymeric TiO_2 gel which can be dried at 70°C to a transparent amorphous solid. During initial stages of thermal treatment below 250°C (depending on the gel preparation and thermal history) no crystallinities were observed when examined by X-ray diffraction or electron diffraction technique [8]. When heated to 370°C the presence of anatase crystal and amorphous phase are observed [8,10]. Results of transmission electron microscopic study of these gels by Mukherjee [8] indicate that it is possible to detect the presence of some extremely fine particle (<100A) in the as-prepared dired gel, but it is extremely difficult to get quantitative data about the size distribution of inhomogenities and practically impossible to monitor the microstructural evolution during thermal aging. Recently, Wright, Mukherjee and Epperson [11] generated small angle neutron scattering data with these titania gels as a function of thermal aging in the absence and presence of H_2O which act as a reactant in microstructural/structural evolution. Results of their work reveals that a texture on a length scale of 30A exists in the as-prepared transparent non-crystalline gel. These islands of pure TiO_2 develop within a matrix of non-crystalline polymeric TiO_2, containing chemically bound organics. When heated to about 400°C in vacuum, these islands of TiO_2 grow up to 80A and pure TiO_2 islands can be reorganized as anatase. A continuity in evolution of the SANS data from 70°C to 400°C (Figure 1) suggest that the 30A texture of the dry gel and well-defined anatase crystal, developed at 370°C and above are closely related. Moreover, variation of $I(Q=0)$ with the particle volume shows a linear relationship which passes through the origin. The results correspond to an Ostwald ripening process [12] where the total volume fraction of

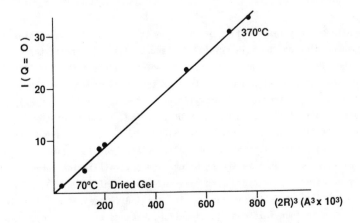

Figure 1. Correlation between I(Q=0) and the particle volume
corresponding to an Ostwald ripening process.

dispersed phase remain constant. Data for the dry gel at 70°C also fit this
equation. Thus, TiO_2 islands formed at 70°C have the same volume fraction
and density as anatase but do not give identifiable diffraction peaks. This
might be due to their poor crystalline order. The amorphous matrix seems to
be impure TiO_2 having chemically bounded residual organics which hinders the
crystallization and ripening process. Scattering studies in presence of
excess water at above room temperature show a sharp increase in scattering
intensity compared to the dry thermal aging in vacuum. This phenomenon
indicates that water plays a strong role in the structural rearrangement of
gel-network and thus enhances the ripening process. In this context, it
should be pointed out that the molecular complexity of the polymeric species
produced by the hydrolytic polycondensation of titanium isopropoxide changes
with the degree of hydrolysis. The origin of the formation of TiO_2 islands
in the metal-oxide alkoxide polymeric matrix might be due to the role of
concentration and reactivity of water which influence the molecular
complexity of the metal-oxide alkoxide polymers that were initially produced
[13]. It may be anticipated that the polymeric structures produced
initially rearrange on thermal aging to two different structures: one is
amorphous pure TiO_2 another is the polymeric network containing residual
organics.

GERMANIUM DIOXIDE GELS

Inspite of many similarities between the structures of silicon dioxide and germanium dioxide, both in crystalline and in vitreous state, the work done in the field of GeO_2 gels is very limited. Recently Mukherjee, et al. [14] reported the synthesis and structure of two types of GeO_2 gels synthesized by two methods (i) hydrolytic polycondensation of $Ge(OC_2H_5)_4$ and (ii) reactive dissolution of GeO_2 in NH_4OH solution. Results of these preliminary studies indicate that unlike SiO_2 gels, GeO_2 gels have a strong tendency toward crystallization during gel-to-glass transformation. In the present work, we will analyze the structure and crystallization behavior of GeO_2 gels prepared by the hydrolytic polycondensation of $Ge(OC_2H_5)_4$. This type of gel is structurally similar to vitreous GeO_2, because Ge^{+4} ions in these gels are in four-fold coordination (i.e., hexagonal structure). GeO_2 gels were synthesized by hydrolytic polycondensation of $Ge(OC_2H_5)_4$ in absolute ethanol at a temperature $0°C$. The amount of water added during gel synthesis was 1 mol per mol of $Ge(OC_2H_5)_4$. Since the gel preparation was not done in a dry glove box, the actual concentration of water when gel is formed was more than 1 mol per mol of $Ge(OC_2H_5)_4$ [14]. A translucent gel formed at room temperature. The X-ray powder diffraction pattern of a wet gel-sample after aging at room temperature for two weeks in a closed container is shown in Figure 2. The d-spacings corresponding to the broad peaks are 3.53A and 2.34A which corresponds to the d-values (3.425A and 2.364A) of hexagonal GeO_2 crystals. The intensity and the sharpness of the peaks increase with increasing thermal aging temperature and time. The presence of moisture in an open system during thermal aging enhances the crystallinity. However, if the thermal treatment is done under vacuum with a sample which was evacuated at room temperature before thermal treatment, the crystallization is less pronounced. DSC data of the gel (whose X-ray powder diffraction pattern is shown in Figure 2) are given in Table I. The DSC curve with heating rate 15°C/min is shown in Figure 3. It is evident from the DSC data that the gel losses the solvent, C_2H_5OH at around 74-90°C and the structural OH hydroxyl groups are removed in the temperature range of 387 to 410°C. The crystallization rates and the extent of crystallization increases with the decrease of heating rate; the occurrence of endothermic peak (at 600°C) due to T_g, when heating rate is high, indicates that a glass-like non-crystalline phase exists in the gel. GeO_2 gel produced by the hydrolysis of $GeCl_4$ looses its water simply on aging in air at room temperature and the water is given off without any intermediate steps; and the gel crystallizes readily during aging [15]. It is

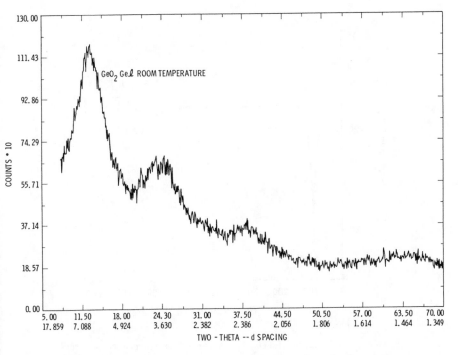

Figure 2. X-ray powder diffraction pattern for GeO$_2$ gel after aging
at room temperature for two weeks in a closed container.

Table I. DSC Data of GeO$_2$-Gel as a Function of Heating Rate
Sample Size: 30 to 32 mg. Thermal History: Wet gel
after aging at 25°C for 2 weeks.

Heating Rate (°C/min)	Endothermic Peak Temperatures (°C)			Exothermic Peak Temperatures (°C)
	T_s	T_d	T_g	T_c
5	74	387	-	633.6, 675 (small)
10	85	375	600 (v small)	650, 687
15	90	410	600	695

T_s = Temp for the removal of solvents.
T_d = Dehydroxylation temperature.
T_c = Crystallization temperature.
T_g = Glass transition temperature.

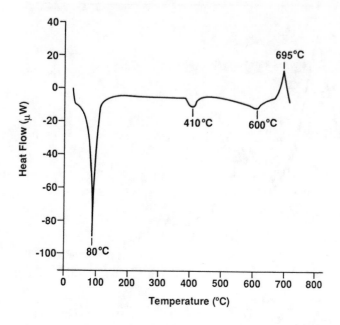

Figure 3.   DSC curve of GeO$_2$ gel after aging at room temperature
for 2 weeks: heating rate 15°C/min.

anticipated that the elimination of structural hydroxyl groups is an
important parameter in crystallizing the GeO$_2$ gels in solid state, in
otherwords, the removal of structural hydroxyl groups and chemically bonded
organic groups below T$_g$ causes a structural rearrangement which leads to
crystallization.   Infrared absorption bonds and Raman spectra of GeO$_2$ gel,
vitreous GeO$_2$ and hexagonal GeO$_2$ crystals are shown in Table II and
Table III respectively.   IR absorption bands and Raman spectra of GeO$_2$ gels
have similarities with those of hexagonal crystalline GeO$_2$ and of vitreous
GeO$_2$.   However, the triplets at 520, 547 and 586 cm^{-1} which indicates
crystallinity of the gel is replaced by a broad peak at 570 cm^{-1} in vitreous
GeO$_2$.   It should be noted that under certain processing conditions (lower
temperature and less aging), a broad peak at around 570 cm^{-1} is obtained
instead of the triplets at 520, 547 and 586 cm^{-1}.   The sharpness of the
Raman bands also indicates the formation of extremely fine crystallites in
the gel.

Table II. IR Absorption Bands in Wave Numbers (cm^{-1}) of GeO_2
Gels and of Crystalline and Vitreous GeO_2

GeO_2-Gel 1 After Drying at 25°C	Hexagonal Polycrystalline GeO_2	Vitreous GeO_2
520 s	333 vs	295 s
547 s	355 sh	315 s
586 s	515 s	334 s
752 w	555 s	569 s
886 vs	587 s	878 vs
961 sh	885 vs	966 sh
1331 w	957 sh	
1444 w	1105 w sh	
1627 w		
3480 vs br		

Abbreviations: w = weak, s = strong, v = very, m = medium, sh = shoulder,
br = broad

Table III.   Raman Frequencies (cm^{-1}) of GeO_2 Gels and
Crystalline and Vitreous GeO_2

GeO_2-Gel After Drying at 25°C	Hexagonal Polycrystalline GeO_2	Vitreous GeO_2
332 vw	112 s	327 m
444 s	116 s	418 vs
510 w	212 w	497 s
576 w	246 w	577 s
590 w	262 w, sh	856 w
738 sh	286 vw	880-884 vw, sh
772 w	326 w	969 w
860 sh	360 vw	
880 w	442 vs	
962 vw	516 w	
1044 vw	590 w	
	698 vw	
	585 vw	
	880 w	
	960 vw, sh	
	970 w	

Abbreviations: w = weak, s = strong, v = very, m = medium, sh = shoulder,
br = broad

450

SiO_2-GeO_2 SYSTEM

Mukherjee [15] has reported the synthesis and stability of the gels in this system. Subsequently, Mukherjee and Sharma [5,16] investigated the effect of preparation procedures on the structure of glasses prepared by both sol-gel and conventional melt techniques using Raman and infrared spectroscopic techniques. Results indicate that the structure, hydroxyl content and the crystallization behavior of gels produced depend on the gel processing conditions such as the type of alkoxides, pH of the sol, etc. Figure 4 shows the room temperature spectra of 94 SiO_2 6 GeO_2 gel-derived glasses obtained by sintering of gels prepared at pH $\simeq 2$ and at pH $\simeq 8$.

It is evident that when the gel is produced at low pH, the gel monoliths are transformed into glass without any crystallization by sintering, at around T_g. Raman spectra of the gels and gel-derived glasses closely resemble to the corresponding spectra of glasses prepared by the melting of conventional batches. When the gel is prepared at higher pH = 8;

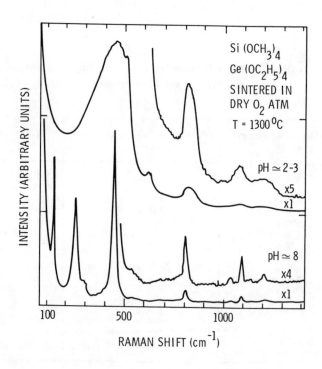

Figure 4. Room temperature Raman spectra of 94 SiO_2 · 6 GeO_2 gel prepared at different pH and sintered in dry O_2 atmosphere at ~1300°C.

the gel crystallizes during gel-to-glass transformation. We anticipate that at low pH, a random molecular scale distribution of GeO_4 networks in silica network takes place, whereas at high pH, the clustering of GeO_4 network occurs due to high polymerization rate of $Ge(OC_2H_5)_4$ during gelling. The GeO_2 rich phases subsequently dehydroxlate and crystallize and produce sharp bands at 112 (s), 226 (s), 288 (wsk), 422 (vs), 512 (w), 788 (m), 1012 (w), 1074 and 1184 (w) cm^{-1}. The positions and spectral characteristics of these bands are close to those bands observed in the Raman spectrum of cristobalite polmorph of SiO_2. The X-ray diffraction powder pattern also indicates the same. Similar phenomenon is observed with the gel compositions having high GeO_2 content, prepared even at low pH. Figure 5 depicts the effect of heat treatment on the Raman spectra of 58 SiO_2 42 GeO_2 gel. The presence of the sharp low-frequency band at 116, 160 and 182 cm^{-1}, indicates that extremely fine crystallites of hexagonal GeO_2 are present in the germania-rich phase. On heating the gel to 365°C for 4 hours the intensity of the low-frequency bands of hexagonal GeO_2 crystallites increases. This is due to the removal of structural OH groups leading to the crystallization of GeO_2. On heating the gel to 500°C above the intensities of the low-frequency bands of hexagonal GeO_2 crystallites decrease indicating that small hexagonal GeO_2 crystallites react with silica gel matrix and condensation reaction leading to the formation of Ge-O-Si bonds. The Raman spectrum of partially crystallized gel heated to 1500°C for 1/4 hr and quenched shows that the presence of broad bands at 326 (sh), 428 (vs bd), 542 (sh), 666 (w bd), 864 (vw, bd) and 1122 (vw) and sharp bands at 988 (m) and 1286 (w) cm^{-1}. The broad bands are characteristics of binary SiO_2-GeO_2 glass and closely resemble bands observed in the spectrum of conventional 50 SiO_2 50 GeO_2 [17,18]. It should be noted, however, though vibrational spectra are similar, the structure of the gel-glass obtained at this stage is anticipated to be significantly different from the conventional glass in terms of the size of -GeO_4- building blocks which should be larger in the gel-glass at this stage.

The Raman spectroscopic studies [15,16] also shows that the removal of structural hydroxyl groups form Ge^{+4} ions takes place at lower temperatures when compared with removal of OH groups from SiOH. Hence, the distribution of structural OH groups between two network formers is not uniform at different stages of thermal treatment. The structure, depicting orderness and size of building blocks of GeO_2 of the gel-derived glasses of certain compositions and/or prepared under certain conditions are significantly different from the glasses made by the conventional technique.

452

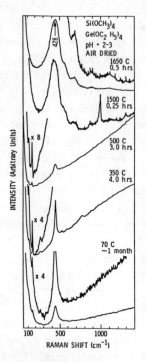

Figure 5. Raman spectra of 58 SiO$_2$ · 42 GeO$_2$ gels recorded
at room temperature after thermal treatment at
various temperatures for a given period.

GeO$_2$-PbO SYSTEM

Recently, Mukherjee [19] reported the synthesis of 90 GeO$_2$ 10 PbO
(mol %) gel by the hydrolytic polycondensation of Ge (OC$_2$H$_5$)$_4$ with basic
lead acetate in ethanolic solution (see Table IV). He also investigated
[19] the kinetics of crystallization of gels, gel-derived glasses and
conventional glasses of 90 GeO$_2$ 10 PbO composition by the non-isothermal
method using differential scanning calorimetry. The kinetic parameters were
determined using the following equation derived from Johnson-Mehl-Arvami
(JMA) equation for solid state phase transformation [20,21]: $\ln(T_p^2/\phi) = \ln$
(E/R) - $\ln \nu$ + E/RT$_p$, where T$_p$ is the temperature corresponding to the DSC
peak, ϕ is the heating rate, E is the effective activation energy for
overall crystallization process, ν is the frequency factor. A plot of
$\ln(T_p^2/\phi)$ vs. 1/T$_p$ gives a straight line; E and ν are obtained from the slope

Table IV. Gel Preparation Parameters for 90 GeO_2 · 10 PbO Composition

| Gel No. | Molar Ratios | | | Reaction Conditions | | | | | |
| | $\frac{HNO_3}{Ge(OC_2H_5)_4}$ | $\frac{H_2O}{Ge(OC_2H_5)_4}$ | $\frac{C_2H_5OH}{Ge(OC_2H_5)_4}$ | Temp (°C) | Mixing Time (Min) | pH | | Gelling Time (Hrs) |
						Initial	Final	
A	0.014	0	34	-10	90	2	4	120
B	0.014	1	34	-10	90	2	4	96
C	0.014	2	34	-10	90	2	4	0.08

and intercept of the line. The results of DSC experiments are given in Table V. The activation energy for crystallization E and the frequency factor were calculated from the slope and intercepts of the lines obtained from plots of $\ln(T_p/\phi)$ vs. $1/T_p$. Results are given in Table VI.

Results show that the activation energy of gel is considerably higher than that of gel-glass and conventional glass. The frequency factor for gels is exceedingly higher ($\sim 10^{14}$) than that of gel-derived glass and conventional glass, whereas the activation energy and frequency factor of gel-glass is slightly higher than that of conventional glass.

Infrared spectra of the gel, the gel-derived glass and the conventional glass are shown in Figure 6. Results of infrared spectroscopie study are summarized in Table VII. An analysis of the spectroscopic data indicates that the spectra of conventional glass and gel-derived glass are almost identical, whereas the spectrum of the gels is different in terms of the presence of sharp peaks (triplet) at 530 cm^{-1}, 560 cm^{-1} and 590 cm^{-1} which indicates crystallization due to hexagonal GeO_2. In vitreous GeO_2, a broad at 570 cm^{-1} peak instead of triplet is observed. The development and sharpness of this triplet depends on the processing conditions particularly the concentration of water.

Considering the spectroscopic and X-ray diffraction results (given in Table VIII) along with the results of crystallization kinetic studies, it may be stated that the nucleation rate in gels is much higher when compared with the gel-derived glasses and the conventional glasses. The similarities in the values of kinetic parameters of gel-derived glass and conventional glass might be attributed to the gel melting condition (3 hrs at 1200°C), which was severe enough in disrupting the structures evolved during the thermal treatment of gel. Presumably gel-derived glasses having higher nucleation rates could be produced under melting/sintering conditions which will not disrupt the evolved gel-structures.

The nature of ordered structures/crystallites that develops at gel state depends on the chemistry of the polymeric solution and also on the metal-oxide composition. The X-ray diffraction powder pattern of a 50 GeO_2 50 PbO (mol %) gel after drying at 70°C is shown in Figure 7. The broad peaks (d-spacings: 3.486A and 2.864A) correspond closely to the d-spacings of $PbGeO_3$. The crystallization behavior of this gel during gel-to-glass transformation, as evident from the DSC studies at different heating rates (not presented here) is very complex and is significantly different from the crystallization behavior of the conventional glass of same composition.

455

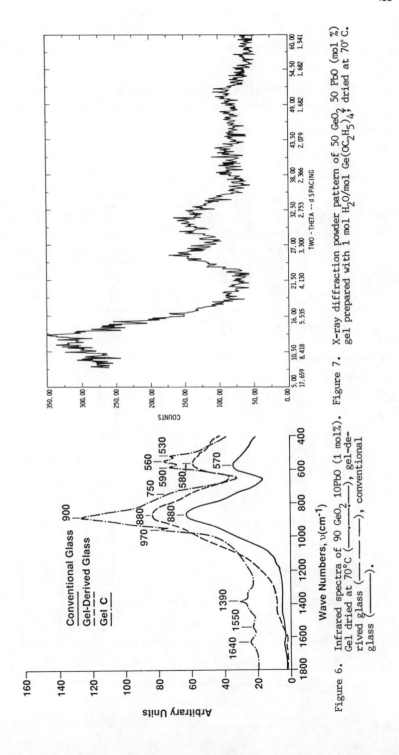

Figure 6. Infrared spectra of 90 GeO₂ 10PbO (1 mol%). Gel dried at 70°C (—·—·—), gel-derived glass (— — —), conventional glass (————).

Figure 7. X-ray diffraction powder pattern of 50 GeO₂ 50 PbO (mol %) gel prepared with 1 mol H₂O/mol Ge(OC₂H₅)₄; dried at 70°C.

Table V. DSC Data of Gel A, Gel-Derived Glass and Conventional Glass
Composition: GeO_2 90, PbO 10 (mol %), Sample Size = 16
to 19 mg

Heating Rate (°C)/Min	Conventional Glass		Gel-Glass		Gel A (After 70°C Drying)	
	T_g(°C)	T_p (°C)	T_g (°C)	T_p (°C)	T_g (°C)	T_p (°C)
10	460	654.6	465	646.9	462	538.7
15	-			654.7	-	-
20		676.66		665.9		549.4
25		684.35		673.9	-	-
30						552.76

T_g = Glass Transition Temperature

T_p = Exothermic Peak Due to Crystallization

Table VI. Kinetic Parameters for Crystallization

Composition	Activation Energy (E) (kJ/Mol)	Frequency Factor, ν
Conventional Glass	209.7	3.14×10^9
Gel-Glass	220.9	1.95×10^{10}
Gel A	403.0	1.04×10^{24}

Table VII. IR Absorption Bands in Wave Numbers (cm^{-1}) of Conventional
Class, Gel-Glass and Gels. Composition: 90 GeO_2 · 10 PbO.

Conventional Glass	Gel-Glass	Gels		
		Gel A After 70°C Drying	Gel C After 70°C Drying	Gel C After Super-Critical Drying
580 (s, br)	570 (s, br)	550 (s, br)	530 (s)	520 (s)
		750 (sh)	560 (s)	540 (s)
880 (vs, br)	880 (vs, br)	830 (vs, br)	590 (s)	580 (s)
			750 (s)	750 (s, sharp)
1630 (v, w)	1630 (v, w)	1040 (v, w)	890 (vs, nr)	860 (sh)
		1390 (w)	970 (sh)	880 (vs, nr)
			1390 (w)	960 (sh)
		1630 (w)	1550 (w)	-
			1630 (w)	-

vs=very strong   s=strong   w=weak   vw=very weak   br=broad   nr=narrow   sh=shoulder

Table VIII. Crystallinity and Major Crystalline Phases Detected with
Gels Having Different Thermal Treatments.
Composition: 90 GeO_2 · 10 PbO.

Gel	As-Prepared Wet Gel at 25°C	After Drying at 70°C	After Drying at 500°C/½ Hr	After Drying at 600°C/½ Hr	After 1000°C No Holding	After Super-Critical Drying at 250°C
A	Non-Crystalline	Non-Crystalline	Partially Crystallized GeO_2*, $PbGeO_3$	-	-	-
B	Non-Crystalline	Non-Crystalline	Crystallized GeO_2, $PbGeO_3$	Crystallized GeO_2, $PbGe_4O_9$	Crystallized GeO_2, $PbGe_4O_9$	
C	Broad Peak Due to GeO_2	Partially Crystallized GeO_2	-	-	-	Completely Crystallized GeO_2, $PbGe_2O_5$ $PbGe_4O_9$

*GeO_2 crystallized in hexagonal form.

458

SUMMARY

In multicomponent oxide systems having a metal-oxide gel which tends to crystallize during gel-to-glass transformation, the structure of gel-derived glasses can be significantly/uniquely different from the structure of glasses made by the melting of oxides. The structural differences originate from the following phenomena:

(a) The orderness and the distribution of network former or modifier in larger blocks (GeO_2, TiO_2 or La_2O_3) which tend to crystallize.

(b) Ostwald's ripening process during thermal treatment can change the microstructure of the gel, and subsequently, the gel-derived glass can inherit the evolved microstructures (e.g., TiO_2, TiO_2-SiO_2, La_2O_3-SiO_2 systems) of gels.

(c) During hydrolythic polycondensation process a preferential reaction between two reactants can produce an ordered structures/crystallites of a compound which can persist after gel-to-glass transformation (e.g., $PbGeO_3$ gel).

(d) The presence of polymeric metal-oxide alkoxide/chemically bonded organics, and structural hydroxyl groups can play a strong role in stabilizing the disordered structures in as-prepared gels.

ACKNOWLEDGEMENTS

Publication support for the paper was provided by IBM Corporation, Endicott, NY.

REFERENCES

1. S. P. Mukherjee, J. Zarzycki, J. P. Traverse, J. Mater. Sc 11 341, (1976).

2. M. C. Weinberg, and G. F. Neilson, J Amer. Ceram Soc 66 132, (1983).

3. S. P. Mukherjee and R. K. Mohr, J. Non-Crystalline Solids, 66 523, (1984).

4. C. J. Brinker, E. P. Roth, G. W. Scherer, D. R. Tallant, J. Non-Crystalline Solids 71 171, (1985).

5. S. P. Mukherjee and S. K. Sharma, J. Non-Crystalline Solids 71 317, (1985).

6. S. P. Mukherjee, J. Non-Crystalline Solids, 63, 35, (1984); 73 639, (1985).

7. S. P. Mukherjee and J. Zarzycki, J. Am. Ceram Soc, 62 1, (1979).

8. S. P. Mukherjee, in "Emergent Process Methods for High Technology Ceramics," edited by R. F. Davis, H. Palmour, R. L. Porter, Plenum (1984) p. 95.

9. M. Decottingnies, S. P. Mukherjee, J. Phallippou, J. Zarzycki, C. R. Acad Paris, 289, C285 (1977).

10. A. F. Wright, S. P. Mukherjee and J. E. Epperson, to be published in the Conf on "The Structure of Non-Crystalline Materials," held in Grenoble, France, July 8-12 (1985).

11. A. F. Wright, A. N. Fitch, F. B. Hayter, B. E. F. Fender, Phys Chem Glasses (in press).

12. D. C. Bradley, R. C. Mehrotra, D. P. Gaur, Metal Alkoxides, Academic Press (1978) p. 160.

13. S. P. Mukherjee, A. Glass and M. J. D. Low, Presented at the Amer. Ceram Soc Glass Div Meeting, held at Corning, USA Nov 6, (1985) (to be published).

14. S. P. Mukherjee in "Better Ceramics through Chemistry," eds C. J. Brinker, D. E. Clark, D. R. Ulrich (Elsevier-North-Holland), New York (1984).

15. S. P. Mukherjee and S. K. Sharma, J. Amer. Ceramic Soc (to be published).

16. J. B. Bates, J. Chem Phys, 57 4042 (1972).

17. S. K. Sharma, D. W. Matson, and J. A. Philpotts, J. Non-Crystalline Solids, 68, 99, (1986).

18. S. P. Mukherjee, in the Proc of the 3rd Int Natl Workshop on "Glasses and Glass-Ceramics from Gels" Montpellier, France, Sept 12-14, (1985), (to be published).

19. N. P. Bansal, R. H. Doremus, A. J. Bruce and C. T. Moynihan, J. Amer. Ceram Soc, 66 (4) 233, (1983).

20. C. S. Ray and D. E. Day, J. Amer. Ceram. Soc, 67 806, (1984).

ELECTRON-SPIN RESONANCE AND OTHER SPECTROSCOPIES
USED IN CHARACTERIZING SOL-GEL PROCESSING

L. C. KLEIN* AND G. KORDAS**
*Rutgers-The State University of New Jersey, Dept. of Ceramics, P.O. Box 909,
Piscataway, NJ 08854. **Vanderbilt University, Dept. of Mech. and Mat.
Engineering, Nashville, TN 37235.

ABSTRACT

The molecular structure of sol-gel processed silica can be studied
using electron-spin resonance (ESR) spectroscopy. Spectra reveal changes in
structure from sol to gel and from gel to glass. Gels prepared under various
conditions contain intrinsic and extrinsic defects. Other spectroscopies
that have been used to investigate gels are nuclear magnetic resonance (NMR),
Raman, and visible.

INTRODUCTION

Now that a good number of formulations exist for sol-gel silica, it is
time to study this class of materials with all of the available techniques
commonly used to study fused silica. Since silica finds application largely
in optics, it is its optical properties that draw concern. Whether the
silica is used in the geometry of a fiber, thin film or lens, what is needed
is a full characterization of its interactions with electromagnetic radiation. These interactions depend on electronic, molecular and defect structures. Characterization of these structures is achieved with four techniques: ESR, NMR, Raman and visible spectroscopy.

The purpose of this paper is to review what is available in the literature on these four techniques. The number of papers which have appeared
using these spectroscopies on gels has increased rapidly in the last five
years. Some of the results are preliminary and any conclusions are still
speculation. An attempt will be made to sort out what is known and what
remains unknown.

A question often asked is whether or not sol-gel silica and fused silica are equivalent. This question can be answered in terms of chemical properties or mechanical properties, as well as other bulk phenomena. While the
density, microhardness, thermal expansion and elastic moduli of fused silica
may be duplicated in sol-gel silica, this may not answer the question entirely. Alternatively, this question can be posed in terms of structure.

When it is posed in terms of structure, the techniques listed in Table
I provide clues to the answer. These techniques may be used during the sol
to gel transition or the gel to glass transition. The sol-gel transition is
determined by inspection. It is the time when a viscous liquid changes
irreversibly to a viscoelastic solid. Both NMR and Raman can be used to
monitor the progress of hydrolysis and polymerization. In this way, reaction
kinetics can be obtained leading to the sol-gel transition. Once the viscoelastic gel is dried, ESR will discriminate extrinsic defects (e.g. organic
residue from incomplete hydrolysis) from intrinsic defects. In this way,
structural differences remaining from solution chemistry can be observed.
Dried gels can be heat treated to bring about a gel to glass conversion. For
these materials, visible spectroscopy will expose effects of heat treatment
conditions on the oxidation state and coordination of transition metals in
gels and gel-derived glasses. In short, the combination of spectroscopies

reveal reaction mechanisms, reaction kinetics and structural evolution. Now each technique will be treated in turn.

Table I

Spectroscopies for Sol-Gel Structure Analysis

Technique	ESR	NMR	Raman	Visible
Sample Preparation	Irradiation Spin labels	Isotopic Mixture	Transparency	Transition Metal Dopant
Excitation	Microwaves	rf-Field	Laser Light	Visible Light
Mechanism	Resonance	Resonance	Inelastic Phonon Scattering	Electronic Transition
Value Determined	g-value	Chemical Shift ppm	Raman Shift cm^{-1}	Transmitted Intensity %
Specificity	Molecular Environment	Coordination	Molecular Structure	Probe Environment

ELECTRON SPIN RESONANCE

ESR (or EPR) spectra have been obtained from silica gels prepared with a wide range of water contents (1,2,3). Spectra have been obtained after heat treatments from 100 to 1000°C. Primarily, these spectra are searched for evidence of peroxy linkages (4). These defects relate to the incorporation or expulsion of water from gels.

Bridging oxygens in gels come about from condensation reactions. At low temperatures, it might be expected that the silica network in a gel would not be the same as the silica network in a melted glass. The mechanism and rate of condensation is known to depend on type of catalyst and amount of water. That is acid catalyzed and base catalyzed gels are structurally distinct. Similarly, low water and high water gels are distinct. However, all gels evolve towards melted silica as they are heated. It is interesting to see how gel structures evolve and under what conditions structural differences are erased.

Using ESR-spectroscopy on gels is an extension of ESR-spectroscopy on fused silica. For several years now, ESR-spectroscopy has been successfully employed for the characterization of the defect centers in fused silica. A major fraction of the defect centers in glasses are not paramagnetic. For example, an oxygen vacancy occupied by two electrons is a diamagnetic center. This defect center can be converted into a paramagnetic state by exposing the glass to ionizing irradiation (e.g. gamma-rays). This way, E'_1-centers can be formed and detected with the ESR-method. In conventional fused SiO_2, the E'_1-center concentration is a function of the temperature at which the melt was equilibrated, the cooling rate and the oxygen partial pressure (5). Their formation is not linear with respect to equilibration temperature or cooling rate. Besides the E'_1-centers, a variety of other irradiation induced paramagnetic centers such as non-bridging oxygens and peroxy radicals are detected in SiO_2 (6). Formation of these various defects cannot be described by a single thermally activated process.

In preliminary studies, all gels were prepared from tetraethyl-ortho-silicate and acid catalyst. Paramagnetic states were investigated after irradiation of vacuum dried SiO_2 based gels produced with various values of the ratio of water to TEOS (1), and during the gel-to-glass transformation of the gels with 16 moles water (R=16) or 4 moles water (R=4) per mole TEOS. These samples were heat treated to temperatures between $400^{\circ}C$ and $1000^{\circ}C$ at a constant heating rate (2).

In vacuum dried gamma ray irradiated gels, the ESR-spectra of O^-_2-ions were observed probably in interstitial positions. The same molecular ions were observed in gels with R=16 densified up to $500^{\circ}C$. At $900^{\circ}C$ the irradiated R=16 gels exhibited the signal of the E'_1-centers and non-bridging oxygens. After heat treatments above $900^{\circ}C$ three distinguishable types of O^-_2-centers were obtained probably in three different interstitial sites.

These O^-_2-ions-signals were not detected in the densified gels produced with R=4. In these densified gels, the signals of several organic inclusions (e.g. methyl) were recorded at temperatures below $1000^{\circ}C$. At $1000^{\circ}C$ CO^-_2 and E'-centers were observed (2).

Consistently, ESR studies on gels point to the prevalence of peroxy linkages or peroxy radicals (1,2,3). The peroxy linkages are likely the remnants of the chemical approach to putting together a silica network, as opposed to conventional melting. More experiments are required before the consequences of the sol-gel process are fully realized.

From spectra already collected, there is information in line shape, line width and intensity as well as g-value. Existing models for defects in fused silica are a starting point for identifying defects in gel glasses, as long as subtle differences are not overlooked. Also, there are many extrinsic defects from residual organics or the acid catalyst.

Experiments need to be performed during gellation. Since gel times in acid-catalyzed TEOS-water solutions may be hours to days, it is experimentally practical to detect radicals involved in polymerization and estimate average molecular weight. If the experiment, a sweep of 200 G, is carried out in 10 sec, most available spectrometers will have the sensitivity needed, assuming one in a thousand atoms is paramagnetic. The time of the experiment is much shorter than the gel time. Successive experiments can be performed.

Experiments also need to be performed using spin-labeling. The technique is based on the reaction between silanol groups and paramagnetic organic radicals. Isolated silanols can be distinguished from vicinal silanols. This should lead to a discrimination of linear from planar polymers. This should also lead to discrimination of open from closed porosity.

Another way to trace the origin of the peroxy linkages is to compare spectra from gels dried by evaporation (xerogels) and gels dried under hypercritical conditions (aerogels). When aerogels are prepared in an autoclave, the water-rich solvent phase is removed without affecting the interior surface of the gel. When xerogels are dried by natural evaporation, this mechanism may be responsible for peroxy linkages (4):

$$Si-OH + HO-Si - - - Si-O-O-Si + H_2 \qquad (1)$$

Quantitative analysis with spin labeling should help to resolve whether or not such linkages result from condensation and dehydration. Also, the quantity of these species should indicate how gentle or how stressful the structural evolution is in xerogels.

Going back to the question of when is sol-gel silica structurally equivalent to fused silica, at this point the answer is not known. However, the ESR technique has the capability of answering this question. With ESR, the answer to when the two structures are equivalent can be found. In addition, the stages and mechanisms in the structural evolution can be traced.

NUCLEAR MAGNETIC RESONANCE

NMR spectra have been obtained for ^{1}H and [29]Si on solutions, gels and dried gels (7,8,9,10,11). Primarily, these spectra are useful for confirming reaction mechanisms (12,13) and obtaining reaction kinetics (14).

First of all, there is proton NMR. When using the multi pulse cycle to suppress proton dipolar interactions, the chemical shift for CH_3, CH_2, H_2O and OH can be measured. These species reflect degree of hydrolysis of the TEOS monomer. As monomers convert to dimers and trimers and various oligomers, the NMR spectra can be interpreted completely. An exact structure for the oligomer can be determined that identifies the position and quantity of groups bonded to a Si (6).

This experiment provides confirmation for the mechanism proposed some time ago (12) where the rate constant for the hydrolysis of TEOS is proportional to the proton concentration. This mechanism is used to predict linear polymers under acid and low water conditions (13). At some time during hydrolysis a species with 5 or 6 coordinated silicon may occur, but so far this intermediate has not been detected. If the over-coordinated silicon intermediate has a life time amenable to detection, the NMR technique may yet confirm its existence:

$$H_2O + Si(OH)_4 + OH^- - - - [(H_2O)Si(OH)_5]^- \qquad (2)$$

Next there is ^{29}Si NMR. This spectroscopy can be used to follow the time evolution of polymer structures during the sol-to-gel transition. An experiment was performed to find out if pressure could be used to shorten the gelation without altering the texture of dried gels (9). The finding was that pressure enhanced the condensation rate without changing the condensation mechanism. The enhancement was explained in terms of the effect of pressure on the over-coordinated silicon intermediate. This explanation may be simplistic at this time, but it is worth pursuing. Multipulse spectroscopy loses its sensitivity as the sol-gel transition is reached.

For solid samples such as dried silica gels, there is magic angle spinning NMR (7). Using this technique, the nature of protons largely in silanols can be probed. The result is a discrimination between surface silanols and bulk silanols. Magic angle spinning has also been used to determine the coordination of [27]Al in single phase and diphasic aluminosilicate gels (11).

In combination, ^{1}H, [29]Si and magic angle spinning, the NMR technique gives information about composition, location of species and their mobility. The technique gives time evolution from sol to gel and gel to glass. An important controversy would be resolved if the existence of the over-coordinated silicon intermediate could be confirmed.

RAMAN SPECTROSCOPY

Raman spectra have been obtained for solutions, wet gels, dry gels and heat treated gels (10,15,16,17,18,19,20,21,22). Primarily, these spectra are compared to those from melted silicates (23) to detect defect structures associated with rings (24,25,26).

Micro-Raman spectroscopy is used to examine the kinetics in the ethanol-TEOS-water system during hydrolysis. The reactants are water and TEOS in the solvent ethanol. The time it takes for these reactions to produce a gel depends on the ratio of water to TEOS, the concentration of acid catalyst and temperature.

Rather than study the high frequency part of the spectra between 3000 and 4000 cm^{-1}, one can look primarily at low frequencies, between 0 and 1600 cm^{-1}. In the low frequency range, it is possible to observe the disappearance of bands due to TEOS and observe the appearance of bands related to the structure of the developing gel.

The optical geometry for micro-Raman makes sample handling easier (16). The 514.5 nm line from an Ar laser is passed through a line filter and into the incident illumination port of a microscope which has been modified for use in micro-Raman spectroscopy. Scattered light collected by the microscope is dispersed by a double monochromator. Dispersed photons are detected by a photomultiplier tube, counted by a photometer, and stored in a digital computer.

Assignments for Raman bands observed in the ethanol-TEOS-water system are listed in Table II.

Table II.

Observed peak positions in the Raman Spectra for Ethanol-TEOS-Water System

cm^{-1}	Assignment
440	Silica glass
495-500	Hydrated silicate (probably chain structure)
650-660	TEOS
810-825	Hydrated silicate monomer
885-895	Ethanol
950-985	Hydrated silicate chain
1055-1060	Ethanol
1090-1110	Ethanol, TEOS
1280-1300	Ethanol, TEOS
1455-1480	Ethanol, TEOS

The band at 823 cm^{-1} is assigned to the breathing mode of silicate monomer while the 977 cm^{-1} band is assigned to a similar mode of silicate structures having an average of two non-bridging oxygens per tetrahedra (i.e. chains). The band at 497 cm^{-1} is now associated with cyclictetramers (26).

Micro-Raman spectroscopy can be used in a semiquantitative way to watch the disappearance of reactants and the appearance of products. Just as well, this technique can be used after the sol-gel transition has been reached. The solvent filled gel, the dried gel and the heat treated gel can be studied. In fact, most Raman spectroscopy studies (17,18,19) report spectra for silica gels heated to temperatures as high as 800°C and fairly dense states (22). The effect of water to TEOS ratio on the sol-gel transition was evaluated, by comparing gels prepared with a ratio greater than six to gels prepared with a ratio less than six. By observing the Raman band at 980 cm^{-1}, it was found that the condensation of hydroxyl groups to give crosslinking occurred above 600°C in the high ratio gels, while pyrolysis below 400°C in the low ratio gels produced more Si-O-Si bonds. Regardless of the preparation method, by 800°C all of the silica gels have been largely converted to a silica network. The spectra for these materials show Raman bands at 490 and 600 cm^{-1}.

Raman-spectra for fused silica exhibit two scattering bands at 490 cm^{-1} (D_1-line) and 600 cm^{-1} (D_2-line) which have been attributed to defects (24). The intensity of the D_1- and D_2- lines can be used to determine the fictive temperature. In attempting to explain these defects, peroxy radicals, threefold and non-bridging oxygens, three and four-fold planar rings (25), have been suggested. If they are peroxy radicals and non-bridging oxygens, they can be detected with the ESR-method after exposing the glasses to ionizing irradiation. If any of the paramagnetic defect centers are also Raman active centers, one can expect that the variation of the processing variables of the sol-gel glasses would result in the variation of the intensity of their ESR- and Raman-signals to the same extent. The characterization of ESR- and Raman- active centers in the same samples as a function of the processing variables should yield a correlation with these defects. Such experiments have yet to be performed.

VISIBLE SPECTROSCOPY

Visible spectra have been obtained for silica gels doped with transition metals Cr, Mn, Fe, Co, Ni and Cu (27,28,29,30,31). Most often, the gels are prepared using iron nitrates with up to 30 mole %. In one case, a rare earth (Eu) was incorporated (32). Primarily, these spectra are compared to those from melted silicates (33). Crystal field effects are interpreted in terms of oxidation state and coordination.

In dopant amounts (less than 1 wt %) transition metals ions can be used as probe ions in silica glasses to show differences in glass structure. In this study, small amounts of Fe^{3+} (acetyl acetonate) were dissolved into TEOS-methanol-water solutions (31). The structure and properties of this compound are shown in Figure 1. Solutions were prepared by mixing-methanol, TEOS and water for 2 hours at 65°C. Nitric acid was added. The amount of water represented 4,6,8 or 16 moles water per mole TEOS. The Fe^{3+} (acetyl acetonate) was then added to the previously hydrolyzed solution. For comparison the compound was added to solutions before mixing for 2 hours.

Thin films were spun onto glass cover slides. Visible spectra were collected in transmission. Dried gel films prepared from various solutions were used to probe structural differences due to hydrolysis conditions, alcohol dilution and aging. Absorption in the visible range was used to estimate film thickness.

Adding the compound before mixing gave some reaction of the iron with the TEOS. This reduced transmission at 700 nm. Adding the compound after mixing gave films that had the same transmission at 700 nm for all water levels. The spectra for films prepared with 4,6,8 and 16 moles water per mole TEOS are shown in Figure 2.

For the xerogel films, the highest transmission was for the film with 16 moles water and lowest for 6 moles water. With 16 moles water, the solution is diluted and takes a long time to gel. With 6 moles water, the solution gels more quickly than either 16 moles or 4 moles. The film with 6 moles water is the thickest, and the absorption at 440 nm scales with the thickness.

467

Bright, orange-red rhombic pyramid or rhombic
bi pyramid crystalline powder

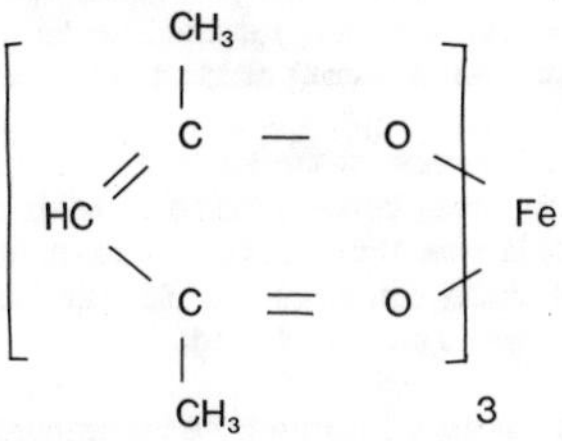

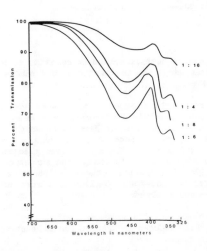

Figure 1

Iron III Acetylacetonate

$Fe(C_5H_7O_2)_3$

Source: Amspec, Gloucester City, NJ

Figure 2

Water Content Effect

Pre-hydrolyzed Series

A simple Lambert-Beer relation between film thickness and absorption would be expected if the iron compound dissolves in the solution and does not react with the network. This appears to be the case. In the study comparing Co^{2+} and Fe^{3+} (30), it was found that cobalt reacted with the gel network and became complexed with the hydroxyl. However, the Fe^{3+} did not. Instead, the iron compound segregated to colloidal particles in the pores of the gel. The color arising from the iron was a combination of scattering and absorption.

There may be some difference between gels prepared with iron nitrate and gels prepared with iron acetyl acetonate. With iron acetyl acetonate, there is a solubility limit in methanol that means the level of this compound trapped in the pores of the gels is small. If the limit is exceeded, the iron compound escapes when the gel is dried.

In addition to differences observed between gels doped before hydrolysis and doped after hydrolysis, there are differences between low water and high water films. Low water films are largely transparent and high water films show some scattering especially when the solutions are allowed to age before coating.

Comparison of transmission spectra in Figure 2 with spectra for melted silicates show that the iron is mostly Fe^{3+} in tetrahedral coordination (33). This is deduced from the absorptions at 440 and 350 nm.

It appears that the iron compound does not react with the prehydrolyzed TEOS solution, nor is the structure of the xerogel films an equilibrated silica network after drying. The silica structure traps the Fe^{3+} in tetrahedral coordination rather than octahedral. If the films are heated, they lose their orange color but do not reach an Fe^{2+}/Fe^{3+} ratio in equilibrium with air until heated above $250^{\circ}C$.

What the transition metal dopants indicate about xerogel structure is hard to say at this time. The study with iron acetyl acetonate is a way of checking thin film thickness and uniformity. More studies need to be performed and spectra extended in the ultraviolet.

SUMMARY

Four techniques have been described in relation to sol-gel silica. Studies which have been performed are mentioned and studies which should be performed are suggested.

Starting with the early studies with infrared spectroscopy (e.g. 34), there has been talk about possible differences between sol-gel derived silica and conventional fused silica. Yet, in dense glasses produced using sol-gel processing many bulk properties, such as microhardness, index of refraction and bulk density are equivalent to those in melted glasses. Polymerization and structural relaxation continue to change the structure at the higher temperature ranges (35). With further studies of these gel materials, it may be possible to say when the gel-derived glass is equivalent to melted glass and how its structure has evolved.

Acknowledgement - A. Thayer-Cohen is thanked for collecting the visible spectra on doped thin films.

REFERENCES

1. G. Kordas, R. A. Weeks, L. C. Klein. Electron-spin-resonance (ESR) study of sol-gel glasses. J. Non-Crystal. Solids 71 (1985) 327-333.
2. G. Kordas and L. C. Klein. The effects of the water content on the structure of gels and sol-to-gel transformations. Ultrastructure Processing of Ceramics, Glasses and Composites II, John Wiley (1985).
3. A. A. Wolf, E. J. Friebele, D. C. Tran. EPR spectra of gel-to-glass reactions. J. Non-Crystal. Solids 71 (1985) 345-350.
4. F. Freund. Conversion of dissolved "water" into molecular hydrogen and peroxy linkages. J. Non-Crystal. Solids 71 (1985) 195-202.
5. G. Kordas, R. A. Weeks, D. L. Kinser. The influence of fusion temperature on the defect center concentration of GeO_2 glass. J. Appl. Phys. 54 (1983) 53-54.
6. D. L. Griscom. Electron spin resonance in glasses. J. Non-Crystal. Solids 40 (1980) 211-272.
7. H. Rosenberger et al. Characterization of the SiO_2-gel glass forming process by high resolution ^1H NMR in solids. Colloids and Surfaces. 12 (1984) 53-58.
8. L. W. Kelts, N. J. Effinger, S. M. Melpolder. Sol-gel chemistry studied by ^1H and ^{29}Si nuclear magnetic resonance. To appear in J. Non-Crystal. Solids (1986).
9. I. Artaki, S. Sinha, A. D. Irwin, J. Jonas. ^{29}Si NMR study of the initial stage of the sol-gel process under high pressure. J. Non-Crystal. Solids 72 (1985) 391-402.
10. I. Artaki, M. Bradley, T. W. Zerda, J. Jonas. NMR and Raman study of the hydrolysis reaction in sol-gel processes. J. Phys. Chem. 89 (1985) 4399-4404.
11. S. Komarneni et al. Solid state ^{27}Al and ^{29}Si magic-angle spinning NMR of aluminosilicate gels. J. Am. Ceram. Soc. 69 (1986) C42-C44.
12. R. Aelion, A. Loebel, F. Eirich. Hydrolysis of ethyl silicate. Am. Chem. Soc. 72 (1950) 5705-5712.
13. K. D. Keefer. The effect of hydrolysis conditions on the structure and growth of silicate polymers. In Better Ceramics through Chemistry, ed. by C. J. Brinker, D. R. Ulrich and D. E. Clark (Elsevier, New York, 1984) pp. 15-24.
14. C. J. Brinker et al. Sol-gel transition in simple silicates II. J. Non-Crystal Solids 63 (1984) 45-59.
15. T. W. Zerda, M. Bradley, J. Jonas. Raman study of the sol to gel transformation under normal and high pressure. Mater. Letters 3 (1985) 124-126.
16. L. C. Klein, C. Nelson, K. L. Higgins. Micro-Raman spectroscopy of fresh and aged silica gels. In Ref. 11, pp. 293-299.
17. A. Bertoluzza et al. Raman and infrared spectra on silica gel evolving toward glass. J. Non-Crystal. Solids 48 (1982) 117-128.
18. V. Gottardi et al. Further investigations on Raman spectra of silica gel evolving toward glass. J. Non-Crystal. Solids 63 (1984) 71-80.
19. D. M. Krol and J. G. van Lierop. The densification of monolithic gels. J. Non-Crystal Solids 63 (1984) 131-144.
20. D. M. Krol and J. G. van Lierop. Raman study of the water adsorption on monolithic silica gels. J. Non-Crystal. Solids 68 (1984) 163-166.
21. S. P. Mukherjee and S. K. Sharma. A comparative Raman study of the structures of conventional and gel derived glasses in the SiO_2-GeO_2 system. J. Non-Crystal. Solids 71 (1985) 317-325.
22. C. J. Brinker, E. P. Roth, G. W. Scherer, D. R. Tallant. Structural evolution during the gel to glass convesion. J. Non-Crystal. Solids 71 (1985) 171-185.

470

23. P. McMillan. Structural studies of silicate glasses and melts - applications and limitations of Raman spectroscopy. Am. Mineralogist 69 (1984) 622-644.

24. S. H. Garofalini. Defect species in viteous silica - a molecular dynamics simulation. J. Non-Crystal. Solids 63 (1984) 337-345.

25. F. L. Galeener. Planar rings in vitreous silica. J. Non-Crystal. Solids 49 (1982) 53-62.

26. C. J. Brinker, D. R. Tallant, E. P. Roth, C. S. Ashley. Defects in gel-derived glasses. In Defects in Glass, MRS Vol. 61, ed. by F. L. Galeener (Elsevier, New York, 1986).

27. M. Guglielmi, A. Maddalena, G. Principi. Moessbauer investigation on $SiO_2 - Fe_2O_3$ glasses obtained from gel. J. Mater. Sci. Lett. 2 (1983) 467.

28. S. Sakka, K. Kamiya, K. Makita, Y. Yamamoto. Optical absorption of transition metal ions in silica coating films prepared by the sol-gel technique. J. Mater. Sci. Lett. 2 (1983) 395-396.

29. S. Sakka, S. Ito, K. Kamiya. Electronic spectra of transition metal ions in gel derived and melt-derived glasses. J. Non-Crystal. Solids 71 (1985) 311-315.

30. A. Duran, J. M. Fernandez Navarro, P. Casariego, A. Joglar. Optical properties of glass coatings containing Fe and Co. To appear in J. Non-Crystal. Solids (1986).

31. A. Thayer-Cohen, L. C. Klein, M. Guglielmi. Transition metal ions in sol-gel films. Am. Ceram. Soc. Bull. 64 (1985) 492, Abstract 6-GBP-85.

32. D. Levy, R. Reisfeld, D. Avnir. Fluorescence of europium (III) trapped in silica gel-glass as a probe for cation binding and for changes in cage symmetry during gel dehydration. Chem. Phys. Lett. 109 (1984) 593-597.

33. P. C. Schultz. Optical absorption of the transition elements in vitreous silica. J. Am. Ceram. Soc. 57 (1974) 309-313.

34. M. Nogami and Y. Moriya. Glass formation through hydrolysis of $Si(OC_2H_5)_4$ with NH_4OH and HCl solution. J. Non-Crystal. Solids 37 (1980) 191-201.

35. G. W. Scherer, C. J. Brinker, E. P. Roth. Structural relaxation in gel-derived glasses. To appear in J. Non-Crystal Solids.

THE NEED FOR CONTROLLED HETEROGENEOUS NUCLEATION IN CERAMIC PROCESSING

GARY L. MESSING, JAMES L. MCARDLE AND RICHARD A. SHELLEMAN
The Pennsylvania State University, University Park, PA   16802

ABSTRACT

The principles underlying controlled heterogeneous nucleation are reviewed. The application of these principles for the preparation of ceramic powders with equiaxed morphology and narrow size distribution is illustrated with examples from the literature and our recent work on alumina. It is concluded that controlled heterogeneous nucleation will be an important factor in fully realizing the benefits of chemically derived polycrystalline ceramics.

INTRODUCTION

With increasing emphasis on material performance in industry, it is clear that efforts to improve the precursor to the material are of considerable importance. This is particularly true in the ceramics industry where microstructure development and densification, and therefore the final properties, are inextricably related to the initial powder characteristics. Today it is well accepted that desirable ceramic powder characteristics for high performance ceramics include chemical homogeneity and purity, submicron particle size, equiaxed particle shape and freedom from agglomeration.

Traditionally, ceramic powders have been prepared through a series of process steps to ready them for forming and sintering operations. These process steps include (1) heating (calcination) to obtain the thermodynamically stable form of the material before sintering, (2) grinding and milling to reduce the particle size and/or modify the aggregate character resulting from calcination, and (3) dispersion to control the agglomeration processes associated with submicrometer size particles. Reasons for the thermal treatment process and examples of ceramic materials that have been traditionally synthesized by these processes include

$$\text{Decomposition} \quad\quad Al(OH)_3 \rightarrow \gamma\text{-}Al_2O_3$$
$$\text{Solid State Reaction} \quad MgO + Al_2O_3 \rightarrow MgAl_2O_4$$
$$\text{Phase Transformation} \quad \gamma\text{-}Al_2O_3 \rightarrow \alpha\text{-}Al_2O_3$$

It is important to note that there are also many examples of chemically synthesized ceramic materials that must undergo these processes including

$$BaTiO(C_2O_4)_2 \cdot 4H_2O \rightarrow BaTiO_3 \quad\quad\quad\quad [1]$$
$$\text{Alkoxide derived } BaTiO_3 \quad\quad\quad\quad\quad\quad\quad [2]$$
$$[(Me_2Si)_n(PhMeSi)]_x \rightarrow SiC \quad\quad\quad\quad [3]$$

While many of the so-called chemically derived ceramics are pure, chemically homogeneous and submicrometer in particle size, they also are aggregated from the high temperature calcination step still required to obtain the powder in the desired state prior to forming and sintering. Alternatively, these chemically-derived powders or sols can be shaped into a monolith although substantial shrinkage occurs on heating to the finished stable form. This approach to ceramics preparation is reminiscent of the so-called calsintering or reactive phase sintering

process wherein densification is enhanced by having the calcination and
densification processes occur simultaneously.

When one considers ceramics derived by either a traditional or a
chemical approach, it is clear that the high temperature calcination
process represents a fundamental process step in the development of
powders with controlled physical characteristics, for it is during this
process that the stable particle is developed. Indeed, it is often a
nucleation and growth processes during calcination that determine the
process temperature and time, particle size and chemical phase
development. In this paper the rationale for controlling the nucleation
process, and thereby growth, will be presented as a means for more fully
realizing the benefits of chemical processing of ceramic powders. The
means for developing this control will be established through a review of
the relevant literature and examples from the authors' recent research on
seeded nucleation of alpha alumina.

HOMOGENEOUS NUCLEATION

The desire to produce mono-sized particles is not unique to the
ceramics field but has formed the basis of a rather large literature. The
work of LaMer and Dinegar [4] is fundamental to the development of
techniques for preparing monosized particles by homogeneous nucleation.
Figure 1 shows the increase in sulfur concentration as a function of time
for the decomposition of sodium thiosulfate in a dilute HCl solution.
Upon reaching the critical supersaturation of sulfur, spontaneous
nucleation occurs and the solution concentration is rapidly reduced below
the concentration condition for nucleation and into the growth controlled
regime. By controlling the balance between the nucleation and growth
processes, the nuclei generated in the initial step can grow without the
development of any additional nuclei. In this manner it is possible to
obtain monosized particles from liquid or vapor synthesis processes. The
utilization of homogeneous nucleation for the direct synthesis of
monosized particles has been particularly well applied by Matijevic [5]
and more recently Barringer and Bowen [6]. However, homogeneous
nucleation is limited to rather simple chemistries and is not readily
amenable to complex, multicomponent ceramics. Furthermore, because of the
requirement to limit concentration to avoid secondary nucleation,
homogeneous nucleation processes are limited in output.

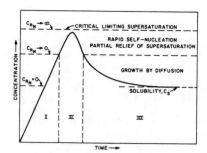

Figure 1. Schematic representation of the concentration of molecularly
dissolved sulfur before and after nucleation as a function of
time (From LaMer and Dinegar [4]).

The study and control of purely homogeneous nucleation becomes extremely problematical, since true homogeneous nucleation requires a truly homogeneous system. By definition, this implies that no impurities or heterogeneities may be present as they would interfere with the delicate balance of surface and volume free energies in the homogeneously nucleated system. In reality, of course, some form of intrinsic or extrinsic impurities are always present and therefore most systems are heterogeneously nucleated.

HETEROGENEOUS NUCLEATION

It is generally known that certain solid bodies, distinct from and extraneous to a system, promote phase transformations and reactions within the system, especially condensation and crystallization reactions. Fahrenheit, Gay Lussac, Gernez, and others [7] experimented with precipitation and crystallization from solution, noting the effectiveness of different substances in "catalyzing" the reactions. In the early 1950's, Turnbull, Vonnegut, and others developed theories quantifying the relationships between crystal structure and "catalytic potency" of nucleation enhancing heterogeneities [8]. The term "nucleation catalyst" was used to distinguish deliberately chosen, carefully controlled heterogeneous additions to a system from uncontrolled, random impurities and contaminants.

Heterogeneous nucleation requires the presence of foreign bodies or substrates upon which stable crystalline nuclei may form. In general, heterogeneities act to replace high energy substrate-matrix interfaces with lower energy substrate-nucleus interfaces, thus lowering the overall surface energy contribution to the nucleation barrier, as depicted in Figure 2. In liquid systems, the magnitude of the surface energy

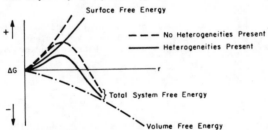

Figure 2. Effect of heterogeneities on nucleation free energy.

contribution depends on the relative interfacial energies of the three phases (substrate-solution, substrate-nucleus, and solution-nucleus) and the resulting contact angle. In addition, the heterogeneous surface affects the volume free energy change for the growing nucleus by imposing lattice distortions at the substrate-nucleus interface. A certain degree of lattice mismatch may be accommodated at the interface, but increasing strains associated with a mismatch must be accounted for in the volume free energy term:

$$\Delta G = 4/3\pi r^3 [\Delta G_v + \chi] + 4\pi r^2 [\gamma_{cl} + \gamma_{cs} - \gamma_{sl}] \tag{1}$$

γ_{cl} = nucleus-liquid interfacial energy

γ_{cs} = nucleus-substrate interfacial energy

γ_{sl} = substrate-liquid interfacial energy

χ = lattice distortion factor

CHARACTERIZATION OF INTERFACES

The crystallographic nature of a heterogeneous surface is a critical factor in determining its effectiveness for controlling heterogeneous nucleation. The degree of lattice matching between two crystals on either side of an interface, as between a substrate and a nucleating phase, determines the structure of the interface. In a coherent interface, planes of atoms are continuous across the interface. One-to-one registry occurs between the two structures at the interface, so that only the second coordination of individual atoms is affected, and a relatively low energy interface results. If lattice mismatch is such that perfect registry cannot be maintained, a certain amount of disregistry can be accommodated by strain of the lattices at the interface. A semi-coherent interface, then, consists of regions of strained registry connected by line dislocations. The elastic strain energy and dislocation energy contribute to interfacial energies on the order of several hundred ergs/cm^2. If crystallographic differences or planar misorientation cause a high degree of lattice mismatch, lattice strains cannot accommodate the disregistry. The highly distorted, incoherent interface results in interfacial energies up to thousands of ergs/cm^2.

From Eq. 1 and Fig. 2, it is clear that the energy barrier to heterogeneous nucleation in solutions and in solid state reactions should scale with the crystallographic "fit" between substrate and nucleating crystal, i.e. with increasing coherency of the interface. The lattice distortion factor of Eq. 1 now takes on real meaning. The coherency of an interface can now be quantified as [9]:

$$\chi = E\delta^2 \tag{2}$$

where E represents the elastic modulus of the forming nucleus at the interface, and δ the disregistry between substrate and nucleus. δ is expressed in terms of lattice spacings of the substrate and forming nucleus [9]:

$$\delta = \frac{a_o^{sub} - a_o^{nuc}}{a_o^{sub}} \tag{3}$$

Nucleation catalysis, according to Turnbull and Vonnegut [7], predicts the effectiveness of a heterogeneous catalyzing substance, based on crystallographic similarity between the catalyst and nucleating crystal. In general, catalysis and therefore the effectiveness of controlled nucleation scales inversely with lattice disregistry, Eq. 3, for low index planes of catalyst and nucleating phase. Consequently, coherent and semi-coherent interfaces should form in systems with small δ, and evidence indicates that highly coherent nuclei form only for $\delta < 0.15$ [7]. When δ is zero, as when systems are seeded with discrete particles of the desired phase, a coherent interface is formed and development of the new crystalline phase commences with a greatly reduced nucleation barrier.

APPLICATION

Since the early work outlined in the previous sections there has been considerable progress in the application and understanding of controlled nucleation for materials synthesis. Table I summarizes some of the varied and complex processes where principles of controlled heterogeneous nucleation naturally occur or are routinely practiced to obtain materials of specific phase composition, crystallography and microstructural

Table I. Applications of controlled heterogeneous nucleation.

Crystallization	$Al(OH)_3$, $MgSO_4 \cdot 7H_2O$
Synthetic Zeolites	Faujasite, Analcime
Microelectronics	Si/SiO_2, $Ge/MgAl_2O_4$
Crystal Growth (melt)	Sapphire, Silicon, Ge
Hydrothermal Crystal Growth	Sapphire, Quartz
High Pressure Synthesis	Diamonds
Biological Mineralization	Kidney Stones, Tooth and Bone Mineralization
Corrosion Films	$Fe/FeO/Fe_2O_3$
Glass Ceramics	$Li_2O \cdot Al_2O_3 \cdot SiO_2$, $MgO \cdot Al_2O_3 \cdot SiO_2$

character. For most of these applications a seed crystal was carefully
selected for which there was no lattice mismatch between the developing
nucleus and the heterogeneous surface. This particular type of controlled
nucleation is more commonly known as epitaxy and has played an important
role in the development of these fields.

    Although single crystal growth may not appear to be important to
ceramic powder applications, one only has to envision the addition of
numerous crystals to a reacting matrix to appreciate the potential utility
of applying the principles of epitaxy. This is well illustrated in Figure
3 which shows the progressive change in faujasite particle size that
Kacirek and Lechert [10] obtained when they hydrothermally reacted 0.23 um
diameter faujasite particle seeded sodium alumnosilicate gels with
increasing seed particle concentration.

          (a)   (b)   (c)

Figure 3. Photomicrographs of NaY zeolite particles hydrothermally grown
           from aluminosilicate gel with different faujasite seed crystal
           concentration (a) 0.044%, (b) 0.44%, (c) 4.4% (From Kacirek and
           Lechert [10]).

    The utilization of the above principles has only recently been
reported for the control of polycrystalline ceramic reactions (Table II).
It is important to note that the applications are not restricted to oxide
systems but also include non-oxides. The benefits of controlled

Table II. Applications of controlled heterogeneous nucleation in
ceramics.

Si_3N_4	H. Inoue, K. Komeya and A. Tsuge (1982)
	F. Lange (1979, 1983)
	P. Morgan (1980)
$\beta\text{-}Al_2O_3$	A. Jatkar et al. (1978)
	S. Heavens (1984)
ZnO	K. Eda, M. Inada and M. Matsuoka (1983)
$BaO\cdot6Fe_2O_3$	C. Lacour and M. Paulus (1975)
$BaTiO_3$	D. Hennings, R. Janssen and P. Reynen (1985)
$\alpha\text{-}Al_2O_3$	M. Kumagai and G. Messing (1984, 1985)
	Y. Suwa et al. (1985)
	R. Shelleman, G. Messing and M. Kumagai (1986)
	J. McArdle and G. Messing (1986)
$9Al_2O_3\cdot2B_2O_3$	J. McArdle and G. Messing (1986)
$3Y_2O_3\cdot5Al_2O_3$	L. David and T. Takemori (1986)

heterogeneous nucleation for silicon nitride powder synthesis is
particularly well illustrated in Figure 4. Here the authors [11] have
added alpha silicon nitride seed particles to a SiO_2-C mixture that was
reacted in nitrogen at 1400°C to obtain alpha Si_3N_4 powders with
controlled size and uniform shape.

Figure 4. Photomicrographs of Si_3N_4 powders produced by seeding with
increasing concentrations of alpha Si_3N_4 seed particles (From
Inoue, Komeya and Tsuge [11]).

Our recent work [20-22] on the seeded transformation of boehmite
(γ-AlOOH) to alpha alumina is particularly illustrative of the potential
benefits of controlled heterogeneous nucleation. Boehmite sols were
seeded with polycrystalline alpha alumina particles of known size before
gelation. The gels were heated at 500°C for 60 min and then heated to
various temperatures. The reaction kinetics and reaction products were
examined by DTA, X-ray diffraction and scanning electron microscopy.
The incubation period for most reactions is attributed to nucleation
with a long incubation being a result of poor nucleation kinetics. Figure
5 shows the effect of alpha alumina seeding on the incubation period for
the gamma to alpha alumina transformation. The unseeded system has an
incubation time of ~270 sec. whereas a nucleation concentration or
frequency of 2.5×10^{13} seeds per volume of anhydrous alumina reduces the
incubation period to ~65 sec. A further increase of nucleation frequency
results in instantaneous reaction at 1050°C and an incubation period of
~4 sec. at 1025°C. Thus, the effect of controlled heterogeneous
nucleation in this case is to significantly reduce the time required for

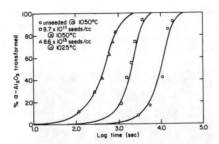

Figure 5. Isothermal Kinetics for the gamma to alpha Al_2O_3 transformation as a function of alpha Al_2O_3 seed particle concentration.

nucleation and to reduce the activation energy for nucleation resulting in a lowering of the transformation temperature. Of course, the amount of temperature reduction is a function of thermodynamics; thus, in some systems there would be no change in the transformation temperature. Lowering the transformation temperature for the growth of the alpha alumina into the theta alumina matrix results in the controlled development of the microstructure and important consequences on subsequent densification [20].

The reduction in the transformation temperature can be more readily detected by thermal analysis than by isothermal reaction kinetics studies. In Figure 6 the peak transformation temperature is clearly lowered with increasing seed concentration for reasons discussed above. It is interesting to note that the transformation begins as low as 900°C at the higher seed concentration. Thus, when this sample is heated at 1°C/min the peak temperature is lowered to ~1000°C. In general, if the reaction temperature were lowered by as much as 100 to 200°C, it would significantly reduce the degree of aggregation and particle growth during calcination reactions and would have important consequences on particle development.

The particle size obtained after high temperature reaction is controlled by thermally activated growth processes and the number of nucleation sites. Generally, the growth processes dominate the development of particle characteristics because of the high temperatures

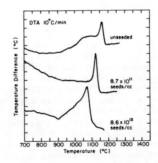

Figure 6. Shift in the theta to alpha alumina transformation temperature as a function of alpha Al_2O_3 seed concentration.

required to sufficiently reduce the incubation period. If a reaction is nucleated then it may be possible to observe the effect of nucleation frequency. Figure 7 shows an alpha alumina microstructure formed after seeding boehmite with 2.6×10^{14} seeds/cc and heating at $1180^{\circ}C$ for 4 min.

Figure 7. Alpha Al_2O_3 particles formed by seeding a boehmite gel with 2.4×10^{14}/cc alpha Al_2O_3 particles and heating at $1180^{\circ}C$ for 4 min.

Assuming that one alpha alumina particle is obtained after transformation for each seed particle, the grain or particle size (d) can be calculated from the nucleation frequency (f);

$$d = k(f)^{-1/3} \qquad (4)$$

where k is a particle shape and transformation shrinkage dependent constant. For the sample shown in Figure 7 the calculated particle size is ~0.18 um after accounting for the volume change associated with the transformation to alpha alumina. The agreement between the calculated value and the observed particle size demonstrates the ability to directly control the particle size by seeding. Furthermore, because all particles are nucleated within a short period of time, impingement of growth fronts limits grain growth to yield a powder with a narrow particle size distribution. The potential of this method of particle size control is clearly illustrated in Figures 3, 4 and 7 for three different routes of powder synthesis.

SUMMARY

The principles of controlled heterogeneous nucleation are routinely practiced in many other fields to obtain materials of specific crystallography, phase composition and microstructure. It has been demonstrated through examples in the literature and our recent work on alumina that these same principles can have dramatic effects on reaction conditions and kinetics with a consequent effect on microstructure development. For synthesis of powders with equiaxed particle morphology and narrow size distribution, controlled heterogeneous nucleation has obvious practical implications. Although seed particles with epitaxial relations in the reacting system were illustrated in this paper, similar findings would be observed for controlled heterogeneties having the requisite crystallographic characteristics (i.e. $\delta < 0.15$). Finally, although significant progress is resulting from a chemical approach to inorganic materials synthesis, it will only be after one also controls crystalline particle and microstructure development that the benefits of chemical processing will be fully realized. It is believed that the application of the principles underlying controlled heterogeneous nucleation will play an important role in realizing the objectives of chemically synthesized ceramics.

ACKNOWLEDGEMENT

The financial support of this research by the Ceramics and Electronic Materials Division of the National Science Foundation under Grant No. DMR-8509422 is gratefully acknowledged.

REFERENCES

1. W. S. Clabaugh, E. M. Swiggard and R. Gilchrist, J. Res. Natl. Bur. Std. $\underline{56}$ (5), 289 (1956).

2. K. S. Mazniyasni, R. T. Dolloff and J. S. Smith II, J. Am. Ceram. Soc. $\underline{52}$ (10) 523 (1969).

3. R. West, in Ultrastructure Processing of Ceramics, Glasses and Composites edited by L. L. Hench and D. R. Ulrich (J. Wiley and Sons, NY, 1984), p. 235.

4. V. K. LaMer and R. H. Dinegar, J. Am. Chem. Soc. $\underline{72}$ (11), 4847 (1950).

5. E. Matijevic, Ultrastructure Processing of Ceramics, Glasses and Composites edited by L. L. Hench and D. R. Ulrich (J. Wiley and Sons, NY, 1984), p. 334.

6. E. A. Barringer and H. K. Bowen, J. Am Ceram. Soc. $\underline{65}$ (12), C199 (1982).

7. V. K. LaMer, Ind. Eng. Chem. $\underline{44}$ (6), 1270 (1952).

8. D. Turnbull and B. Vonnegut, Ind. Eng. Chem. $\underline{44}$ (6) 1292 (1952).

9. A. G. Walton, The Formation and Properties of Precipitates (Robert Krieger Publishing Co., Huntington, NY, 1979), p .8.

10. H. Kacirek and H. Lechert, J. Phys. Chem. $\underline{79}$ (15), 1589 (1975).

11. H. Inoue, K. Komeya and A. Tsuge, J. Am Ceram. Soc. $\underline{65}$ (12), C205 (1982).

12. F. F. Lange, J. Am. Ceram. Soc. $\underline{62}$ (7-8), 428 (1979).

13. F. F. Lange, Am. Ceram. Soc. Bull. $\underline{62}$ (12) 1369 (1983).

14. P. E. D. Morgan, J. Mat. Sci. Lett. $\underline{15}$ 791 (1980).

15. A. D. Jatkar, I. B. Cutler, A. V. Virkar and R. S. Gordon, Mat. Sci. Res. $\underline{11}$ 421 (1978).

16. S. W. Heavens, J. Mat. Sci. $\underline{19}$ 2223 (1984).

17. K. Eda, M. Inada and M. Matsuoka, J. Appl. Phys. $\underline{54}$ (2) 1095 (1983).

18. C. Lacour and M. Paulus, Phys. Stat. Solid (a) $\underline{27}$ (2) 441 (1975).

19. D. Hennings, R. Janssen and P. Reynen in Sintering - Theory and Practice - II, eds., H. Palmour and G. Kuczynski, Plenum Press, New York (1986).

20. M. Kumagai and G. L. Messing, J. Am. Ceram. Soc. 67 (11) C230 (1984); 68 (9) 500 (1985).

21. R. A. Shelleman, G. L. Messing and M. Kumagai, J. Non-Cryst. Solids 48 (1986).

22. J. L. McArdle and G. L. Messing, J. Am. Ceram. Soc. 69 (5) (1986).

23. Y. Suwa, R. Roy and S. Komarneni, J. Am. Ceram. Soc. 68 (9) C238 (1985).

24. J. L. McArdle and G. L. Messing (unpublished work).

25. L. David and T. Takemori, J. Am. Ceram. Soc. accepted for publication (1986).

COMPARISON OF CLAY-WATER SYSTEMS WITH ALUMINUM HYDROXIDE
GELS PREPARED IN ACIDIC MEDIA

A. C. Pierre, Aérospatiale, 33165 St-Médard-en-Jalles, France
D. R. Uhlmann, Department of Materials Science and Engineering, M.I.T.,
Cambridge, Mass. 02139, USA.

ABSTRACT

Alumina sols made by the hydrolysis of aluminum sec-butoxide in an
acidic medium have been found to show behavior during drying and aging which
appears similar to that reported for gels in a number of clay-water systems.
Structural and phenomenological characteristics of these two classes of
materials are compared. The extension to aluminum hydroxide systems of the
models proposed to explain the gelation of clay sols are discussed, and a
model is presented to describe the gelation behavior of aluminum hydroxide
sols.

INTRODUCTION

In previous studies [1, e.g.] of stable sols obtained from aluminum
sec-butoxide, it has been found that two types of gels could be obtained,
both related to the boehmite variety of aluminum monohydroxide. One type,
termed RT gels, is obtained when the processing is carried out entirely at
room temperature. The model proposed for its structure is that of boehmite
in which extensive folding of the (020) layers has occurred. The other type,
termed HT gels, is obtained over a wider range of processing temperature
(above 50°C). The proposed structural model is that of boehmite with preferred orientation of the (020) layers, which are unfolded. It appears that
the HT alumina gels belong to the ordered lamellar gel category by Flory [2].
Also in this category are clay gels, composed of well-defined lamellae
arranged in parallel stacking over comparatively long range.
Obtaining large monoliths from clay is a common process, for which much
information has been gathered. A comparison between the boehmite HT gels
and clays therefore appears promising for improving the sol-gel processing
of alumina ceramics. In previous publications we have reported three different types of experimental results which show a close relationship between
boehmite gels and monotmorillonite gels. The evidence is: (1) the existence
of a long time (6 months) aging evolution of the sols, in which rigid
thixotropy continues to develop [3]; (2) the existence of a minimum in the
gel point (or sedimentation) volume for a particular electrolyte proportion
in the sol [4]; and (3) the existence of a drying rate which is unchanged on
passing through the gel point, down to a point called the leatherhard point
[5].

EXPERIMENTAL PROCEDURE AND RESULTS

As in clays, the layer structure in HT boehmite gels arises from their
atomic organization, which consists of oxygen octahedral layers linked by
weak hydrogen bonds. Cracking and peeling of successive layers can be
observed when a flat, dry piece of gel is fracture by bending.
The behavior during drying is important with respect to many ceramic
applications. Hence, we have extended our previous results by comparing of
drying over a layer of mercury [6, e.g.] with drying in a Pyrex dish. Such
a comparison provides insight into the effects of sticking to the container
on the rearrangements which take place in the gel between the gel point and

the leatherhard point, and on the various types of cracking which develop.

It has been found that adherence to the container favors unidirectional shrinkage in the thickness direction, with the boehmite platelets sliding across each other to rearrange in a thinner solid with enhanced preferred orientation, not only on the bottom of the container but also on the side walls. The samples dried over Pyrex are more transparent than those dried over mercury; and the final pores are smaller and more easily retain water.

Regarding the formation of monoliths, we observed that three different types of cracking occur at different stages of drying.

1. From the gel point down to the leatherhard point, adhesion to the walls and bottom of the container is the main source of cracking. Cracking can be avoided by separating the gel from the side walls to greater depth as gelation proceeds from the top surface, and by using a liquid such as mercury.

2. Near the leatherhard point, internal restraint of the gel sample develops. Stresses result from the lower moisture content of the free surface of the gel, which reaches first the leatherhard point, compared to that of the bulk. Two opposite cases have been observed. When the bulk gel sticks to the bottom of the container, the upper part of the sample, which has reached its leatherhard point, is maintained in a state of tensile stress. In contrast, when the bulk gel does not stick to the bottom of the container, shrinkage occurs in all directions, and the upper hard layer of the sample is in compression. All these stresses tend to relax with time. When drying is completed, the samples dried over Pyrex assume a concave-upward curvature; while samples dried over mercury display convex-upward curvature (Fig. 1). In the former case, the curvature is very pronounced. The stresses in some cases lead to radial cracking, which on occasion is spiral in form.

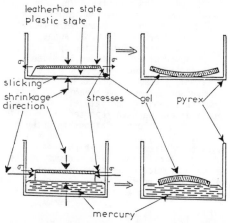

Fig. 1 Influence of restraint by the bottom of a container on the shape of the dry sample

In this stage of drying, reducing the gradient of water content in the gel, so that the sample reaches its leatherhard state as uniformly as possible throughout the thickness, favors a monolith. Before the leatherhard point, water transport in the gel is suggested to occur not by capillarity, but by the "osmosis-suction" mechanism, which is well known in clays [7]. No capillarity stresses arise in the sample, and water evaporation is the rate limiting step in drying.

3. After the leatherhard point, the solid network of the gel stops deforming plastically. Through capillarity, the large pores lose their free water, which is replaced by air. The stresses responsible for cracking are of capillary origin. Under these conditions, the surface tension of the

solvent in the pores becomes important, and techniques such as hypercritical drying can be helpful.

Modification of the gelation-drying behavior with various additives has also been studied. The modifiers were added either in a high volume proportion (of the same order as water) or in a very small proportion (surfactants). The additives shown in Table 1 were tested with aqueous sols hydrolyzed in excess water and peptized at 90°C with 0.07 mole HNO_3 corresponding to the minimum gelation volume.

Table 1 Effects of various additives on the relative mass at the gel point, of sols peptized at 90°C with 0.07 HNO_3

ADDITIVES for 150 cm³ of sol	moles	dipolar moment μ (Debyes)	surface tension γ(dyne/cm)	$\frac{m(gel\ point)}{m(alkoxide)}$
Water alone		1.85	73	2.30
-Ethanol	1.64	1.65	23	1.90
-methanol	2.25	1.70	23	2.47
-Ethylene glycol	1.69	2.28	47.7	5.52
-N,N dimethyl formamid	1.17	3.82		7.37
-Formamid	1.28	3.73	58.2 ·	10.72
CATIONIC SURFACTANTS 1 drop for 100 cm³ of sol				
-N,N',N'-tris(2-hydroxyethyl)-N-Tallow-1,3 propane diamine (Ethoduomene T/13® AKZO)			34.3	4.23
-bis-(2-hydroxyethyl)-coccamine (Ethomeen C12® AKZO)			34	4.86
-cocoamine (Mazeen C2® MAZER)			28	5.65
-Alkylamine-guanidine-popyoxyethanol (Aerosol C61® CYANAMID)			34	4.89
ANIONIC SURFACTANTS -disodium N-octadecyl sulfosuccinate (Aerosol 18® CYANAMID)			41	4.24
-sodium diamyl sulfosuccinate (Aerosol AY 100® CYANAMID)			29	4.83
NON-IONIC SURFACTANT -POE(40) Stearyl Ether (Macol SA40® MAZER)			17.4	2.2

Most of the additives with a different surface tension or a different dipole moment did not change the drying and cracking behavior between the gel point and the leatherhard point. In contrast, some additives such as formamide or ethylene glycol have a strong effect on the gelation volume. This likely reflects hydrogen bonding between the sol particles. Such bonding can be strong enough to increase the gelling volume in direct proportion to the amount of additive employed (see the data with formamide shown in Fig. 2). Additives with a relatively high dipole moment (e.g., DMF) or with an ionic structure (cationic and anionic surfactants) also lead to increased gelling volume; while non-ionic surfactants, or those with a low dipole moment (e.g., C_2H_5OH) result in low volumes at the gel point.

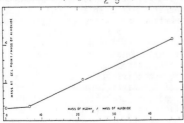

Fig. 2 Variation of the mass at the gel point for increasing proportions of formamid

DISCUSSION

Since the comparison between boehmite clay and gels can apparently be extended to drying and gelation, it seems attractive to extend various models used to explain the gelation of clays to describe the gelation of boehmite. The most striking differences are associated with the electric charges on the flat faces of the platelet particles. These charges are negative in clays, and positive in boehmite.

Four types of models have been suggested for clay-water systems:

1. The model of hydration linking between the particles [8, 9, e.g.]. In this theory, the repulsion between particles is associated with the orientation of water dipoles in the electric field around the charged particles. The degree of orientation of the dipoles should greatly decrease with increasing distance from the charged surface. While the usual hydration shell in clays corresponds to a particle separation of the order of 10 Å or less, the interparticle distances in clay gels are often very large, beyond any sensible range of hydration forces; and this model seems insufficient to explain observed gelation behavior.

2. The model of heterogeneous chain formation of the particles with solvent bridges. This model can explain the volume at the gel point increasing in proportionally to the solvent volume when the solvent can provide hydrogen bonding (e.g., formamide). With water as the solvent, Ford [7] has proposed that the counterions (e.g., Na^+) could provide gel bonding between short water dipole chains adsorbed onto the clay particles. The equivalent for alumina would involve the formation of bridges by the anions NO_3^-. Such a large scale organized structure of water is difficult to imagine.

3. Electrical double layer models. Macey [10] and Norrish [11] suggested that the repulsive forces caused by interacting electrical double layers are responsible for the gel structure. Good agreement is obtained between D.L.V.O. theory and the variation of the applied pressure with compression, and suggests that parallel alignment of the platelets is produced by application of pressure.

The suggestion that the approach of least repulsion between two particles would be edge-to-face is, however, not obvious; and suggesting that electrical double layer repulsion is responsible for gelation stands in contrast to regarding such repulsion as responsible for stabilization in a fluid sol. Further, it is not clear how this model can explain elasticity in tension after the gel point.

4. Electrostatic Linking Model. In the case of montmorillonite or kaolinite clays, Van Olphen [9, 12] considered the sedimentaion volume after centrifugation as well as the rheological behavior of stable sols. Given the platelet shape of the particles, with their edges charged positively and their flat faces charged negatively, as long as the two types of double layers are present, the particles should associate by a face-to-edge linking, producing a gel with a "cubic cardhouse" texture and a high sedimentation volume.

When an electrolyte such as NaCl is added to the sol, both types of double layers are progressively reduced; and for some proportion of added NaCl, they disappear. The first to disappear is that at the edge. Electrostatic edge-to-face linking becomes inoperative, producing a fluid and stable sol with repulsion between the negatively charged faces. At this stage, a minimum sedimentation volume is observed. The negative double layers associated with the faces of the platelets also disappear upon further addition of electrolyte. Attractive Van der Waals forces can then produce flocculation; and the sedimentation volume again increases, due to random edge-to-face or face-to-face aggregation of the particles.

To adapt this electrostatic binding model to boehmite, in which the flat faces of the particles are charged positively, a negative charge on the edges of the boehmite plates, due, e.g., to adsorption of NO_3^- ions or to an excess negative charge on the edge oxygens, can be postulated. Starting with an HNO_3/alkoxide ratio above 1, boehmite platelets with electric layers of opposite signs on the edges and faces favor gelation by electrostatic linking face-to-edge in a cardhouse structure and a high volume at the gel point.

With decreasing HNO_3 content, the negatively charged layers on the edges should disappear first. Stabilization of sols would be easier, since only repulsion between the positive faces would remain. The expected lower volume at the gel point is observed.

Provided the boehmite particles have a narrow distribution of platelet sizes, a phase transition in the sol of the smectic-nematic type (Fig. 3) can be imagined. The smectic phase can be regarded as the equivalent of an opal-type material with platelet-shaped particles.

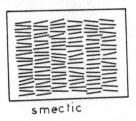
smectic

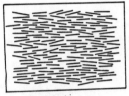
nematic

Fig. 3 Suggested texture of a boehmite sol with preferred orientation.

With further decreases in HNO_3 content, it should become increasingly difficult to stabilize the sol; and the particles should be arranged in a less dense packing at the gel point due to random aggregation, so that the sedimentation volume again increases.

Concerning the RT gels, which are suggested to consist of folded boehmite plates [1], the linking should also result from the opposite charge on the flat portions (positive) and the sites of oxygen octahedron stacking defects (negative). The stacking defects are suggested to be lost, however, by the unfolding of the boehmite planes during the RT-to-HT transition. The suggested impossibility of the reverse transition, from the HT to RT sol, agrees with observations. The folding represents a defect which is introduced during the initial formation of the platelet structure. Removed by treatment at elevated temperatures, such defect structures should not reappear on re-exposure to room temperature.

Another phenomenon common to clays and HT boehmite sols is their irreversible behavior on long time again. In clays, Worrall [13] found that aging could be greatly accelerated by ultrasonic vibration of the sol. This was found to reflect a breaking down of the individual particles. The question of present interest is how can the irreversible breaking down of particles produce gelation at rest in excess water, and then be distinguished from reversible thixotropy, which corresponds to the reversible building of links between particles at rest and destruction of these links under motion.

We propose here a mechanism for both clays and boehmite gels which is based on the particles having a layer structure and exhibiting two different types of break-up (Fig. 4): (1) breakage of a boehmite platelet into 2 platelets of the same thickness but smaller lateral size, and (2) cleavage between layers of the platelets.

The second type (cleavage) could explain the observed irreversible aging behavior, even if it were accompanied by breakage into shorter platelets. Cleavage would produce a larger number of platelets, which by edge-to-face linkage would produce longer "card house" ribbons, and would result in gelation at rest in excess water. The formation of large flocs would also explain the white appearance of the sol; and the aggregation and disaggregation of the particles (thixotropy) would involve edge-to-face association and disassociation of the platelets.

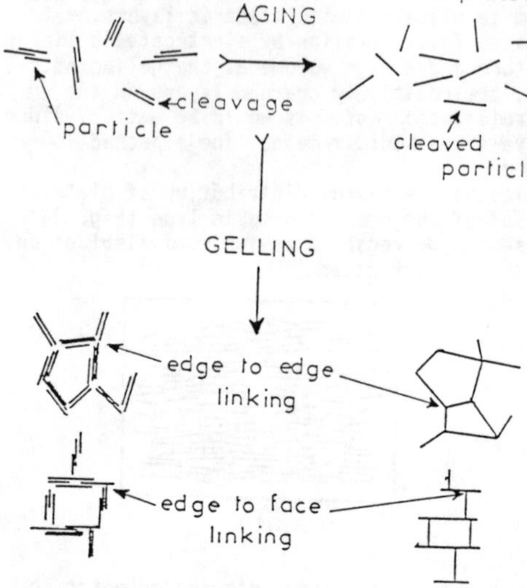

Fig. 4 Aging and gelation of clay and boehmite suspensions

The boehmite HT gels can also be compared to many clays (e.g., mont-morillonite) with respect to their swelling behavior in water. Such swelling can increase the volume by a large factor, to a point where the interlayer distance is of the order of 100 Å [11]. The existence of a limit is diffi-cult to distinguish fron a gradual transition to a sol. The swelling behavior depends on the type of cations in the clay; and the specific atomic structure, including the nature of the cations, is suggested as central to the gelation behavior of boehmite as well as clays.

CONCLUSIONS

It is proposed that gelation of boehmite occurs by electrostatic linking between negatively charged edges and positively charged faces of platelet particles. The negative charges on the edges of the plates can be associated with factors such as the attraction of anions to the Al cations at the edges. The proposed mechanism can explain the increased gelation volume under highly acidic conditions, the existence of a minimum in the gel point volume at low acid concentrations, as well as the observed aging behavior.

ACKNOWLEDGEMENTS

Financial support for the MIT portion of the present work was provided by the Air Force Office of Scientific Research. This support is gratefully acknowledged.

REFERENCES

1. A.C. Pierre, D.R. Uhlmann, Int. Conf. on Glass and Glass-Ceramics from Gels, Montpellier, France, Sept. 12-14 (1985).
2. P.J. Flory, Disc. Far. Soc. 57, 7 (1974).
3. A.C. Pierre, D.R. Uhlmann, 87th Annual Conference of the American Ceramic Society, Cincinnati, May 5-9, 264 (1985).

4. B.E. Yoldas, J. Mat. Sci. 10, 7 (1975).
5. A.C. Pierre and D.R. Uhlmann, Matls. Res. Soc. Symp. 32. 119 (1984).
6. W.O. Milligan and H.B. Weiser, J. Phys. Chem. Colloid 55. 490 (1984).
7. R.W. Ford, Drying (MacLaren, London, 1964).
8. W.C. Laurence, in Ceramics Processing Before Firing, ed. G.Y. Onoda, Jr. and L.L. Hench (N.Y., 1978), pp. 193-210.
9. H. Van Olphen, Clay Colloid Chemistry (Interscience, N.Y., 1963)
10. H.H. Macey, Trans. Brit. Ceram. Soc. 41, 73 (1942).
11. K. Norrish, Disc. Far. Soc., 18, 120 (1954).
12. H. Van Olphen, Disc. Far. Soc. 11, 82 (1951).
13. W.E. Worrall, Ceramics, 20, 10 (1969).

STRUCTURE AND PROPERTIES OF CERAMIC FIBERS
PREPARED FROM POLYMERIC PRECURSORS [1]

J. LIPOWITZ AND H. A. FREEMAN, Dow Corning Corporation, Midland, MI and
H. A. GOLDBERG, R. T. CHEN AND E. R. PRACK, Celanese Research Company,
Summit, NJ.

Ceramics can be prepared by pyrolysis of organosilicon polymers.
Advantages of this method of ceramics preparation are; the ability to
prepare shapes difficult to achieve by other methods such as fibers and
films; the ability to achieve high purity because reagents used to prepare
the polymer can be purified by well established chemical methods;
processing at lower temperature than conventional methods [2].

This approach is being used to prepare ceramic fibers, which may be used
for reinforcement of ceramic, metal and plastic matrices. The primary goal
is development of a family of Si-C, Si-C-N and Si-N fibers which can be
used to fabricate ceramic composites with high fracture toughness, high
temperature performance and ease of formation of complex shapes.
Attainment of this goal would overcome the poor design reliability of
existing ceramics due largely to catastrophic brittle failure.

An integrated, multistep process scheme is used to prepare ceramic fibers.
An organosilicon, thermoplastic polymer is synthesized, converted to a
fiber by melt-spinning, crosslinked in the solid state to cured fiber, and
pyrolyzed to 1100-1400 C in an inert atmosphere.

This paper characterizes the chemical structure, microstructure, morphology
and some properties of such ceramic fibers.

Ceramic fibers which were characterized include a standard grade of
NICALON™ Fiber (SGN) and a ceramic grade of NICALON™ Fiber (CGN),
commercial products of Nippon Carbon Company [3]. These are Si-C-O fibers.
An Si-C-N ceramic fiber developed in this program is prepared by air cure
and pyrolysis of MPDZ-PhVi melt-spun fiber, prepared by reaction of
chloromethyldisilanes and hexamethyldisilazane, which also contains
phenylvinylsilazane functionality [4]. HPZ ceramic fibers are prepared by
melt-spinning, cure and pyrolysis of a polymer prepared from
trichlorosilane and hexamethyldisilazane [5]. HPZ ceramic fiber is
predominantly a Si-N composition. Polymer compositions are shown in
Table I.

Characterization includes total elemental analysis of ceramic fibers
(100 ± 5 w/o). Table I shows compositions for the ceramic fibers based on
thermodynamic rule-of-mixtures criteria, and ceramic densities. Oxygen is
first assigned to silicon as SiO_2; then nitrogen is assigned to Si as
Si_3N_4; finally carbon is assigned as SiC. Carbon is found in excess in all
these ceramics. It cannot be readily oxidized as can free carbon in
ceramics prepared conventionally.

Ceramic fibers do not actually consist of this composition of crystalline
materials as is clearly shown by x-ray diffractograms (Figure 1). The
NICALON™ fibers contain microcrystalline β-SiC as shown by the broad
$\langle 1,1,1 \rangle$ line at $2\theta \sim 36°$. Crystallite sizes calculated from line
broadening are 0.8 nm for SGN and 1.7 nm for CGN. Crystallites are less
than 50 w/o of SGN and CGN ($\sim 1/3$ by volume) by a quantitative x-ray method
[6]. CGN contains less O than SGN and is closer to the SiC composition.

Table I.   Composition of Pre-Ceramic Polymers

NICALON™ Fiber (Polycarbosilane)
$(MeHSiCH_2)_x$

MPDZ-PhVi (Methylpolydisilylazane)
$(Me_{2.6}Si_2NH_{1.7})_x(PhViSiNH)_y(Me_3SiNH_{1/2})_z$

HPZ (Hydridopolysilazane)
$(HSiNH_{3/2})_x(Me_3SiNH_{1/2})_y$

Fig. 1.   X-ray diffractograms for powdered ceramic fibers

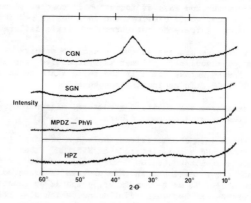

Fig. 2.   Infrared spectra (KBr pellet) of powdered ceramic fibers

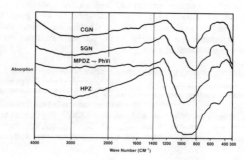

CGN thus gives a larger crystallite size and somewhat greater v/o
crystallinity. Selected area diffraction measurements by TEM confirm β-SiC
microcrystallinity. Darkfield TEM gives crystallite sizes that are
somewhat larger than x-ray measurements, as expected (2.6 nm for SGN and 4
nm for CGN). Crystallites appear to be surrounded by a continuous Si-C-O
amorphous phase.

Infrared spectra, Figure 2, obtained as KBr pellets on finely ground
ceramics, show only the presence of broad adsorption bands covering the
primary stretching region for SiOSi, SiCSi, and SiNSi bonds (700 to
1100 cm^{-1}).

Solid state, magic angle sample spinning 29-Si NMR spectra were obtained on
powdered ceramic fibers (Figure 3). These spectra provide strong evidence
for a predominant amorphous phase with nearest neighbor bonding about
silicon (to C, N and O) approaching random.

There are five possible tetrahedra for carbon and oxygen groups bonded to
silicon which give rise to the five signals observed in the SGN and CGN
(not shown) fiber spectra. Model compound chemical shifts are superimposed
on the spectra to demonstrate a good fit in chemical shifts. There are
fifteen possible tetrahedra for C, O, and N groups bonded to silicon. A
broad envelope covering this chemical shift region is expected, and found,
since individual signals are too broad to be resolved. Note that the
signal peak for SGN at -11 ppm and MPDZ-PhVi fiber at -26 ppm corresponds
closely to the signals for α-SiC (-14, -20 and -25 ppm) and for β-SiC (-18
ppm) [7].

Rule-of-mixtures compositions in Table I predict a maximum in composition
for SGN, CGN and MPDZ-PhVi fibers at the SiC composition (SiC$_4$ tetrahedral
structure), which is found. Note that the chemical shifts of these maxima
do not correspond exactly to chemical shifts for α-SiC or β-SiC. This is
likely a result of second order effects on chemical shifts. That is, Si
atoms bonded to SiC$_4$ tetrahedra are not necessarily bonded to three
additional C atoms as in silicon carbide, but rather bond to all possible
combinations of C and O atoms (in NICALON fibers) and C, N and O atoms (in
MPDZ-PhVi fiber). Such second order bonding effects are expected to
contribute to shifting and broadening of the first order signals in all of
these ceramic compositions.

HPZ fiber rule-of-mixtures composition is a maximum at the Si$_3$N$_4$
composition (Table I). The spectrum shows a chemical shift maximum at
-46 ppm, corresponding closely to the reported signals of -48 and -50 ppm
for α- and -48.5 ppm for β-Si$_3$N$_4$ [8]. A broad peak is observed for HPZ and
MPDZ-PhVi fibers resulting from the 15 possible Si (C, N, O) tetrahedra.
Chemical shifts for model organosilicon compounds superimposed on the
spectra are taken from [9].

Such random, glassy silicon oxycarbide or silicon oxycarbonitride
structures have not been previously described. They can be referred to as
a "ceramic alloy" structure. This "ceramic alloy" structure can be
represented as SiC$_a$O$_b$N$_c$. Depending on the relative values of subscripts a,
b or c, statistical bonding to C, N and O approaching random predicts
undifferentiated amorphous structure (for a ∿ b ∿ c; or a ∿ b, c —> 0; or
a ∿ c, b —> 0, or b ∿ c, a —> 0); regions rich in crystalline or amorphous
Si (low a, b and c), C (high a), SiC (high a), SiO$_2$ (high b), Si$_3$N$_4$ (high
c) or Si$_2$N$_2$O (high b and c). Only SiC (in NICALON fiber) appears to
develop crystalline regions of homogeneity large enough (∿1 to 4 nm) for
identification as microcrystallites by x-ray diffraction (XRD) and high

492

Fig. 3. ^{29}Si MASS-NMR spectra of ceramic fibers

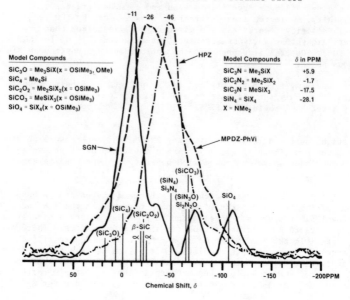

Table II. Rule-of-Mixtures Composition and Density of Ceramic Fibers[a]

	MPDZ-PhVi Fiber	HPZ Fiber	CGN Fiber	SGN Fiber
SiO_2	0.35	0.21	0.18	0.29
Si_3N_4	0.39	1.10	<0.01	<0.01
SiC	1.0	1.0	1.0	1.0
C	2.63	0.79	0.47	0.63
ρ (Mg/m^3)[b]	2.18	2.32	2.55	2.52

[a] moles or equivalents relative to SiC
[b] measured by density gradient method

Fig. 4. Raman spectra of ceramic fibers

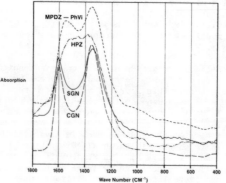

resolution TEM under typical pyrolysis conditions (short times at 1000 to 1400 C). These structures are thermodynamically metastable and revert to more stable crystalline structures on thermal aging at high temperatures (>1300 C) with evolution of CO, SiO and N_2 (if present).

Raman spectroscopy was used to obtain information on the structure of the excess carbon. Raman spectroscopy shows two distinct bands (at 1570 to 1650 cm^{-1} and a band at 1350 cm^{-1}) for all fibers except HPZ fiber (Figure 4). HPZ fibers show a Raman spectrum in this region, but it appears to be a single broad band. The band at 1575 to 1610 cm^{-1} has been assigned to graphite-like carbon and the band at 1350 cm^{-1} to disordered graphitic structure or to very small graphitic crystals [10]. From the ratio of the intensities of the two bands, a domain size of ~5 nm in the graphitic plane can be calculated for the microcrystallites. The crystallites may be only a few graphitic layers in thickness (few tenths nm in the c axis), thus accounting for their non-detectability by high resolution TEM. Unfortunately, due to self adsorption of the already weak Raman scattered radiation by the fibers, it is not possible to make quantitative intensity measurements of excess C by this method.

Figure 5 is a schematic diagram comparing characteristic distances of homogeneity of ceramics prepared from pre-ceramic polymers with those prepared by conventional ceramics processing. The amorphous cured polymer is highly homogeneous on a scale of <1 nm, in contrast to conventionally processed ceramics in which grain sizes are generally >1 μm. Homogeneity is used here to define a minimum characteristic radial distance over which elemental composition approaches that of the bulk composition. Pyrolysis of cured polymer produces a glassy, amorphous ceramic until temperatures high enough to lead to crystallization are reached. This temperature is generally greater than 1200 C. Crystallization does not appear to occur at lower temperatures unless the ceramic composition is close to stoichiometric for a crystalline species such as SiC, Si_3N_4 or SiO_2. Crystallization of these ceramics to coarse SiC crystals (β + α) occurs at temperatures of 1300 C and above in inert atmospheres (vacuum, Ar, or He) with loss of gases such as CO, SiO and N_2. The composition upon the loss of these gases then approaches stoichiometric SiC. Cristobalite forms as a surface layer on exposure to air above 1300 C [11].

Considerable surface-connected microporosity appears to be present in these ceramic fibers. During pyrolysis, hundreds of volumes of gas are evolved per volume of cured fiber undergoing pyrolysis, producing surface-connected porosity after pyrolysis is completed. Typical pore volume fractions vary up to 0.20 (CGN fiber) to 0.33 (MPDZ-PhVi fiber). Porosity is measured by N_2 desorption, mercury porosimetry and comparison of bulk, gradient column, and theoretical rule-of-mixture densities. Porosity of this magnitude will reduce Young's modulus and tensile strength considerably below the fully dense values [12, 13].

The fibers exhibit the classical tensile fracture behavior of brittle materials and follow the Griffith equation for tensile strength. Predominant strength-limiting critical flaws are interior granular defect regions and spinning-induced or mechanical damage-induced surface defects [14, 15]. Critical flaw sizes range from less than 0.1 μm to several μm.

Table III shows representative tensile strengths and Young's modulus of these ceramic fibers, which are generally 10 to 20 μm in diameter. Tensile strength and Young's modulus values continue to increase as better control of each of the process steps and their overall integration is achieved.

494

Fig. 5. Homogeneity of ceramics

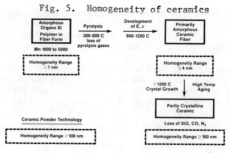

Table III. Typical Properties of Ceramic Fibers[a]

	Tensile Strength, GPa/ksi	Young's Modulus, GPa/Msi
SGN	2.57/372	200/29.0
CGN	2.26/327	207/30.0
MPDZ-PhVi	2.28/330	172/25.0
HPZ	2.41/350	186/27.0

[a] Typical diameters 10 to 20 μm.

REFERENCES

1. Work performed under DARPA Contract F33615-83-C-5006, Administered by the Air Force, Wright Aeronautical Laboratories.

2. R. W. Rice, "Ceramics from Polymer Pyrolysis, Opportunities and Needs - A Materials Perspective", Cer. Bull., 62, 889 (1983).

3. S. Yajima, J. Hayashi and M. Omori, "Silicon Carbide Fibers Having a High Strength and a Method for Producing Said Fibers", U.S. Pat. 4,100,233, July 11, 1978.

4. J. H. Gaul, Jr., "Process for the Preparation of Poly(Disilyl)Silazane Polymers and the Polymers Therefrom", U.S. Pat. 4,340,619, July 20, 1982.

5. J. H. Gaul, Jr., "Silazane Polymers from $(R'_3Si)_2NH$ and Organochlorosilanes", U.S. Pat. 4,312,970, January 26, 1982.

6. L. K. Frevel and W. C. Roth, "Semimicro Assay of Crystalline Phases by X-Ray Powder Diffractometry", Anal. Chem., 54, 677 (1982).

7. G. R. Finlay, J. S. Hartman, M. F. Richardson and B. L. Williams, "^{29}Si and ^{13}C Magic Angle Spinning N.M.R. Spectra of Silicon Carbide Polymorphs", J. Chem. Soc., Chem. Commun., 159 (1985).

8. R. Dupree, M. H. Lewis, G. Leng-Ward, D. S. Williams, "Coordination of Si Atoms in Silicon-Oxynitrides Determined by Magic-Angle-Spinning NMR", J. Mater. Sci. Lett., 4, 393 (1985).

9. P. Diehl, E. Fluck and R. Kosfield, eds, "NMR-Basic Principles and Progress", vol 17, J. P. Kintzinger and H. Marsmann, "Oxygen-17 and Silicon-29", Springer-Verlag, NY, 1981, p. 65.

10. F. Tuinistra and J. L. Koenig, "Raman Spectrum of Graphite", J. Chem. Phys., 53, 1126 (1970).

11. T. J. Clark, R. M. Arons, J. B. Stamatoff and J. Rabe, "Thermal Degradation of NICALON™ SiC Fibers", Ceramic Engineering and Science Proceeds, 9th Annual Conference on Composites and Advanced Ceramics Materials, p. 576 (1985).

12. D. C. Larsen and J. W. Adams, "Property Screening and Evaluation of Ceramic Turbine Materials", AF Contract F33615-79-C-5100 (1983).

13. W. D. Kingery, H. K. Bowen and D. R. Uhlmann, "Introduction to Ceramics", 2nd ed., J. Wiley and Sons, NY, 1976, pp 808-811.

14. L. C. Sawyer, R. Arons, F. Haimbach, M. Jaffe, and K. D. Rappaport, "Characterization of NICALON™: Strength, Structure, and Fractography", Proceedings of the 9th Annual Conference on Composites and Advanced Ceramic Materials, Cocoa Beach, FL, January 1985, p. 567.

15. Unpublished data, L. C. Sawyer, Celanese Research Company, and C. T. Li, Dow Corning Corporation.

PART VIII

Applications of
MO/MD Calculations

APPLICATIONS OF MOLECULAR DYNAMICS SIMULATIONS TO SOL-GEL PROCESSING,*

S.H. Garofalini and H. Melman, Department of Ceramics,
Rutgers University, Piscataway, NJ 08854

ABSTRACT

The molecular dynamics computer simulation technique has been
used to study silicic acid and pyrosilicic acid molecules (H_4SiO_4 and
$H_6Si_2O_7$, respectively). The structure of the simulated molecules are
compared to those found by molecular orbital calculations as well as
structures inferred from silicate hydrates. The potentials used to
simulate the molecules were also used in simulations of bulk silicates
and compared with experimental data. Results indicate good
correlation.

INTRODUCTION

In order to understand sol-gel processing and the dynamics of
the early stages of polymerization at an atomic level, the molecular dynamics (MD) computer simulation technique is being used to
study the interactions between silicic acid molecules, $Si(OH)_4$, and
the growth of these molecules into larger clusters. Although sol-
gel processing of silica involves polymerization of molecules which
may be only partially hydrated, such as $Si(OH)_2(OR)_2$, where R
indicates alkyl groups, the silicic acid monomer is nonetheless a
useful starting point for two reasons. First, additional complexity
would be introduced into an already complex system of Si, O, and H
ions if the alkyl groups were also included. Reasonable potential
functions already exist for Si-Si, Si-O, O-O, O-H, and H-H interactions, as will be shown below. The additional potentials required
to describe all such pair interactions with the ions in the alkyl
group was not deemed necessary at this stage. Second, the silicic
acid monomer and the pyrosilicic acid 'dimer', $H_6Si_2O_7$, and other
Si-O-H containing molecules have been studied using molecular
orbital, MO, calculations(1-3). The molecular dynamics simulations
of the silicic acid and pyrosilicic acid molecules could then be
compared to the MO calculations. In addition, the potentials used
to simulate these molecules in MD can also be used to simulate bulk
glasses using several hundred to several thousand atoms, a feature
which cannot be done using MO calculations. The results of the
simulations of bulk silicate systems can then be compared to the
experimental data available for silicate glasses and minerals as a
means of verifying the applicability of using such potentials in
simulations of these systems.

Molecular dynamics simulations have been used for the past
twenty years in simulations of liquids (4). MD simulations of
silica glasses began about nine years ago (5), and only in the last
three years have MD simulations of silica surfaces been reported
(6). Although the simulations of silicate glasses use potentials
which are only approximate, as will be commented on below, the
simulations reproduce many of the structural and dynamic features
of these glasses. In particular, the MD simulations reproduce the

---

*The authors acknowledge funding from the Center for Ceramic Research
and computation time from CCIS, Rutgers University

498

tetrahedral coordination of O around the Si, an appropriate radial distribution function (5), Si-O-Si bond angle distribution (7), and frequency spectrum (8,9). Even the Si-O bond length increase with decreasing Si-O-Si bond angle observed in a large number of crystalline silicates as well as molecular orbital calculations (10) is also reproduced in the MD simulations (11). In the simulations of silica glass surfaces, the MD results reproduce the known (12-14) predominance of oxygen as the outermost species , with the presence of non-bridging oxygens (NBO), and strained siloxane bonds (6). More recent simulations indicate the nature of these 'strained' siloxane bonds at the surface (15,16). Additional simulations of multicomponent glasses (17-19) also correlate well with experiment.

Given the ability of the MD technique to reproduce a number of structural and dynamic properties of silica and silicate glasses, the technique was applied to studies of silicic acid molecules and polymerization, in which the first step was to evaluate the ability of the technique and the selected potentials to reproduce the appropriate molecular structures.

COMPUTATIONAL PROCEDURE

The molecular dynamics technique involves solving Newton's equation of motion for a system of interacting particles (atoms) via an assumed interatomic potential function. From the coordinates, forces, and time derivatives of the positions at time t_o, a subsequent configurational and dynamic state of the system at a time $t_o + \Delta t$ can be determined. Continuation of these calculations creates a series of configurations which represent a time evolution of the system.

In the work presented here, the modified Born-Mayer-Huggins (BMH) potential (5-7) and the revised Rahman-Stillinger-Lemberg (RSL2) potential are being used to simulate silicic acid, pyrosilicic acid, and water molecules. The modified BMH equation gives the potential between two atoms, i and j, separated by a distance r_{ij}, as

$$\phi_{ij} = A_{ij} \exp\left(-\frac{r_{ij}}{\rho_{ij}}\right) + \frac{z_i z_j e^2}{r_{ij}} \quad \xi \left(\frac{r_{ij}}{\beta_{ij}}\right),$$

where A_{ij} is a short range repulsive parameter based on ion sizes, Z is the ion valence, and ρ_{ij} and β_{ij} are constants. The ξ(x) function in the coulomb term is equal to erfc(x), where erfc is the complimentary error function. The BMH equation is used for all Si-Si, Si-O, and O-O pairs in the non-water molecules and has been used in simulations of silica. The BMH equation is also used for Si-H interactions in these simulations. The Si-H potential is chosen as purely repulsive in the current simulations in order to prevent the Si-O-H bond angle from becoming unreasonably small.

This angle is believed to be 113° at the hydroxylated silica surface (21), and 129° in the H_4SiO_4 molecule (3). The full BMH potential is useful in describing ionic systems, in which a short range
repulsive term, a coulomb term and dispersion terms are used. In
the application of the modified version of this equation to simulations of vitreous silica, the dispersion terms are normally ignored
and only the first term or real space summation of the Ewald sum of
the coulomb term is employed. Since silica is partially covalent,
the BMH equation used here should clearly be viewed as an effective
potential in describing the pair interactions in silica and the
molecules studied here. Interestingly, although the full ionic
charges of +4 and -2 are used for Si and O, respectively, in the
BMH equation used here, the $\mathcal{E}$ function can be seen as altering
the formal charges in the coulomb term. As a result, the effective
charge on the O ions at an O-O distance of 2.6 A is about -0.7e,
which is similar to that found in MO calculations (22). This
effective charge varies slightly depending upon the value of βij
used in the $\mathcal{E}$ function. More importantly, this β ij parameter
can be useful for incorporating the importance of non-bonded
forces into the calculations. Ab initio calculations indicate that
the overlap populations between nonbonded Si-Si increased much more
rapidly with decreasing Si-O-Si bond angle than in nonbonded O-O
interactions (23). Such differences in nonbonded repulsions can be
qualitatively taken into account using this $\mathcal{E}$ function.

The RSL2 equations were developed as effective central force
equations for simulations of water and are used for all O-H and H-H
interactions, as well as for O-O interactions between O ions in
water molecules. The RSL2 equations are given (in kcal/mol and Å
units) as:

$$\phi_{HH}(r) = \frac{36.1345}{r} + \frac{18}{1 + \exp.[40r - 2.05]} - 17\exp[-7.62177(r-45251)^2].$$

$$\phi_{OH}(r) = \frac{-72.269}{r} + \frac{6.23403}{r^{9.19912}} - \frac{10}{1+\exp[40(r-1.05)]} - \frac{4}{1+\exp[5.49305(r-2.2)]}$$

$$\phi_{OO}(r) = \frac{144.538}{r} + \frac{26758.2}{r^{8.8591}} - 0.25 \exp[-4(r - 3.4)^2] - 0.25 \exp[-1.5(r-4.5)^2].$$

Although the O-H interaction may be different in an Si-O-H combination (as in H_4SiO_4) than in an H-O-H combination, for which the
potential was originally derived, the RSL2 potential should nonetheless be a close starting approximation. Changes were only made
to the BMH potential, not the RSL2 potential in simulations of the
H_4SiO_4 and $H_6Si_2O_7$ molecules. Any such changes in the potential
were subsequently evaluated in simulations of bulk v-SiO_2 and

compared with previous simulations of v-SiO_2 and experimental data.
Thus, the potential which reproduces the silicic acid molecules
would still accurately reproduce bulk silica and enable polymerization and growth of the molecules to bulk glass without redesign of
the potential. Runs from 20,000 to 100,000 moves, or time steps,
were used in the simulations, with each time step being $1.0x10^{-4}$
psec. Structural data was averaged over every tenth configuration
over at least 20,000 moves.

RESULTS AND DISCUSSION

Figure 1 shows a schematic of the H_4SiO_4 and $H_6Si_2O_7$ molecules, with the smallest circles representing H, the next largest
representing Si, and the largest circles representing O. Table I
shows the bond length changes in the monomer with changes in some
parameters in the BMH equation. The β_{ij} parameter in the BMH
equation was $2.50x10^{-8}$ cm for all pair interactions for the data
reported in Table I. The A_{Si-O} value of $2.96x10^{-9}$ ergs is used in
simulations of bulk silicates and gives Si-O distances in bulk
silica equal to 1.62 Å, similar to the value observed experimentally. In the monomer, this A_{Si-O} value results in shorter Si-O distances, as seen in Table I. Here, O is a non-bridging oxygen (NBO)
attached to one Si and one H. The shorter Si-NBO bond length (in
comparison to Si-BO (bridging oxygen) bond length) is consistent with
data reported in the literature (10). Also, molecular orbital
calculations of an $H_4Si_2O_6$ molecule with edge sharing tetrahedra give
an Si-NBO bond length 0.065 Å shorter than the Si-BO (bridging oxygen)
bond length (24). However, other MO calculations indicate that the Si-
NBO bond length should be 1.622 Å (3). The A_{Si-O} value of $3.25x10^{-9}$
ergs reproduces this 1.62 Å value in the monomer, as seen in Table I.
The Si-O-H bond angle was between 125° to 130° in all of the runs
shown in Table I, which is similar to the 129° value found in MO
calculations (3).

The energy difference between the H_4SiO_4 molecule and the
$H_3SiO_4^-$ ion was calculated to be around $1.5x10^{-11}$ to $1.6x10^{-11}$
ergs/molecule in the simulations using the various sets of β_{ij}
values. This result compares fairly well to the value of $2.62x10^{-11}$
ergs/molecule obtained from MO calculations of the energy
difference between these molecules (3). This is considered good
because the total energy of an individual molecule calculated by MO
differs by two orders of magnitude from that calculated by MD,
whereas the energy differences are within a factor of two.

Simulations of the $H_6Si_2O_7$ dimer were done using the A_{Si-O}
value of $2.96x10^{-9}$ ergs and a β_{ij} value of $2.50x10^{-8}$ cm for all
Si-Si, Si-O, Si-H, and O-O interactions. The resultant dimer structure
indicated collapse of the dimer to Si-O-Si bond angles less than 120°.
This collapse is very dependent upon the presence of the hydrogen ions
attached to the oxygen; without them, no such collapse occurs.
Nonetheless, in order to prevent such a collapse of the dimer, the
β_{ij} parameters were altered so as to take into account the nonbonded
repulsive forces, in which Si-Si repulsions were considered to be
stronger than O-O nonbonded repulsions, as reported in the MO

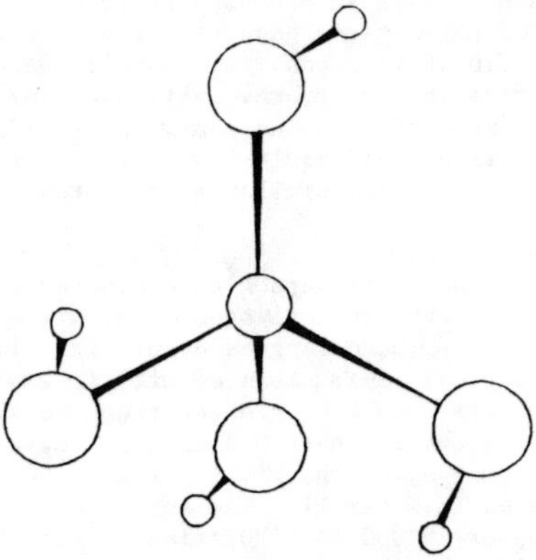

SILICIC ACID (MONOMER)
H_4SiO_4

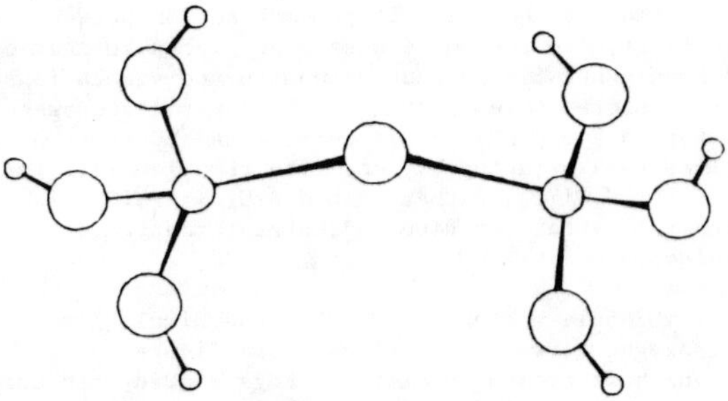

PYROSILICIC ACID (DIMER)
$H_6Si_2O_7$

FIGURE 1

calculations (23). Table II shows the results of different β_{ij} combinations on bond lengths and bond angles in the dimer. Results indicate that with appropriate combinations of β ij parameters, the dimer structure does not collapse and appears to be fairly insensitive to the various sets of parameters. Each set of parameters were also used in simulations of bulk vitreous silica and compared to previous simulations of silica as well as to the available experimental data. Results showed that combinations using the $\beta_{Si-Si} = 2.6x$ ($x10^{-8}$) cm series gave the best structural results, with few overcoordinated defects and a frequency spectrum similar to that reported previously (9).

The tetrahedral bond angles were also evaluated in the dimer. Previous experimental and theoretical studies of silicate structures indicated that shorter Si-O bond lengths occur with wider O-Si-O bond angles (tetrahedral angles) (10). As a result, it was reported that the NBO-Si-NBO bond angle should be greater than the NBO-Si-BO bond angle (10). Figure 2 shows the tetrahedral bond angles resulting from a simulation of the dimer using the β_{ij} series 2.60, 2.55, 2.53 ($x10^{-8}$cm), as presented in Table II. The angles were averaged every tenth configuration over 20,000 configurations. The NBO-Si-BO bond angle is ϕ_1; the NBO-Si-NBO bond angle is ϕ_2. Note that ϕ_2 is larger than ϕ_1, in accord with the simulation results which show that the Si-NBO bond length is shorter than the Si-BO bond length (see Table II).

In order to simulate the onset of polymerization between two silicic acid monomers, two such molecules were placed in near contact to each other in an otherwise perfect vacuum in the simulations. The molecules were attracted to each other, with a resultant associative reaction which formed two pentacoordinate Si attached to each other by one (and sometimes two) bridging oxygen. The reaction was, therefore, $H_4SiO_4 + H_4SiO_4 \rightarrow H_8Si_2O_8$, rather than $H_4SiO_4 + H_4SiO_4 \rightarrow H_6Si_2O_7 + H_2O$, in which the latter reaction indicates formation of the dimer plus a water molecule.

The $H_8Si_2O_8$ molecule was often found in the simulations to involve edge sharing of oxygen between the silicon (see figure 3), rather than corner sharing which is common in silica. Edge shared structures exist in silica W and are believed to exist at dehydroxylated glass surfaces (16,25). The potentials used to simulate this reaction were the same as the $\beta_{ij} = 2.6x$ ($x10^{-8}$cm) series used to simulate the dimer, which did not collapse. The presence of the hydrogen enhanced the interactions between the monomers. The dissociation of the H_2O molecule from the $H_8Si_2O_8$ molecule was probably inhibited by the activation barrier to the dissociation and the perfect vacuum conditions of the simulations.

CONCLUSION

The molecular dynamics computer simulation technique has been used to simulate silicic acid, H_4SiO_4, and pyrosilicic acid, $H_6Si_2O_7$, molecules using a combination of classical, effective, central force potentials. The simulation results were compared to molecular orbital

Table I

Effect of Changes in Repulsive Parameter on the Bond Lengths in the $Si(OH)_4$ Molecule.

$A_{Si-O}(x10^{-9}ergs)$	$A_{Si-H}(x10^{-9}ergs)$	$T^{O}K$	$r_{Si-O}(Å)$
4.00	9.95	100	>1.70
3.50	9.95	100	1.65
3.25	9.95	50	1.62
		100	1.62
3.00	9.95	100	1.60
		300	1.55
2.96	9.95	100	1.57

$r_{O-H} = 0.96(Å)$ throughout

Table II

Effect of β_{ij} Parameters in BMH Potential on Dimer Structure.

$\beta_{Si-Si}(x10^{-8}cm)$	$\beta_{Si-O}(x10^{-8}cm)$	$\beta_{O-O}(x10^{-8}cm)$	Average Bond Lengths (Å)		Si-O-Si Bond Angles
			Si-O	Si-NBO	
3.0	2.7	2.7	1.67	1.55	149°
3.0	2.7	2.6	1.65	1.54	139°
2.9	2.7	2.7	1.66	1.55	143°
2.9	2.7	2.6	1.64	1.54	137°
2.7	2.7	2.7	1.63	1.55	153°
2.7	2.6	2.5	1.64	1.54	151°
2.65	2.55	2.50	1.63	1.55	138°
2.65	2.56	2.53	1.63	1.55	135°
2.63	2.55	2.53	1.63	1.55	147°
2.61	2.55	2.55	1.63	1.55	139°
2.61	2.55	2.50	1.63	1.55	137°
2.61	2.55	2.49	1.63	1.55	136°
2.60	2.55	2.60	1.63	1.55	136°
2.60	2.55	2.53	1.63	1.55	138°
2.60	2.55	2.50	1.63	1.55	140°
2.50	2.50	2.50	1.63	1.55	<120°

504

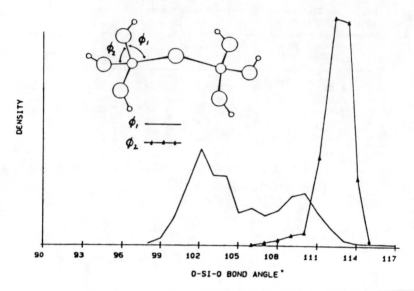

TETRAHEDRAL BOND ANGLE

FIGURE 2

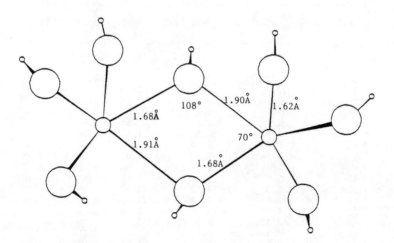

FIGURE 3: Bond lengths and angles for $H_8Si_2O_8$ product molecule after a 100,000 move run from two monomers. Note that structure differs considerably from dimer structure shown in Table II.

calculations of these molecules. In addition, the potentials, and
parameters therein, were also used to simulate bulk glasses in order
to compare with available experimental data from bulk systems. The
simulations reproduce the Si-BO bond length observed in MO
calculations and bulk silicates and gives appropriate Si-O-Si and Si-
O-H bond angles. The simulations give an Si-NBO bond length
approximately 0.06A-0.08A less than the Si-BO bond length. The O-Si-O
tetrahedral angles correlate with this difference in Si-BO versus Si-
NBO bond lengths observed in the pyrosilicic acid molecule.

Reaction between two silicic acid monomers in an otherwise
perfect vacuum led to pentacoordinate silicon formation, which may act
as an intermediate in the formation of the dimer. The perfect vacuum
condition and short real time of the simulations may have inhibited
the completion of the reaction.

<u>REFERENCES</u>

1. G.V. Gibbs, Amer. Mineral <u>67</u>, 421 (1982).
2. G.V. Gibbs, E.P. Meagher, E.P. Newton, M.D., and E.K. Swanson,
    in <u>Structure and Bonding in Crystals</u>, edited by M. O'Keeffe
    and A. Navrotsky (Academic Press, New York, 1981), <u>1</u>, p. 195.
3. E.J. O'Keeffe, B. Domenges, and G.V. Gibbs, J. Phys. Chem. <u>89</u>,
    2304 (1985).
4. A. Rahman, Phys. Rev. <u>136</u>, A405 (1964).
5. L.V. Woodcock, C.A. Angell, and P. Cheeseman, J. Chem. Phys. <u>65</u>,
    1565 (1976).
6. S.H. Garofalini, J. Chem. Phys. <u>78</u>, 2069 (1983).
7. T.F. Soules, J. Chem. Phys. <u>71</u>, 4570 (1982).
8. S.H. Garofalini, J. Chem. Phys. <u>76</u>, 3189 (1982).
9. S.H. Garofalini, J. Non-Chryst. Solids <u>63</u>, 337 (1984).
10. G.V. Gibbs, M.M. Hamil, S.J. Louisnathan, L.S. Bartell, and H.
    Yow, Amer. Mineral <u>57</u>, 1578 (1972).
11. S.H. Garofalini, presented at the Pacific Coast Meeting of the
    American Ceramic Society, San Francisco, CA, 1984.
12. R.C. McCune, Anal. Chem. <u>51</u>, 1249 (1980).
13. R.K. Iler, <u>The Chemistry of Silica</u> (John Wiley and Sons,
    New York, 1979).
14. C.G. Pantano, J.F. Kelso, and M.J. Suscavage, in <u>Advances</u>
    <u>Materials Characterization</u>, edited by D.R. Rossington, R.A.
    Condrate, and R.L. Snyder (Plenum, 1983).
15. S.M. Levine and S.H. Garofalini, <u>Defects in Glass</u>, MRS
    1985.
16. S.M. Levine and S.H. Garofalini, submitted to J. Chem. Phys.
17. A. Rosenthal and S.H. Garofalini, submitted to J. Non-Cryst. Sol.
18. T.F. Soules and R.F. Busbey, J. Chem. Phys. <u>75</u>, 969 (1981).
19. S.H. Garofalini and S.M. Levine, J. Am. Ceram. Soc. <u>68</u>,
    376 (1985).
20. a. F.H. Stillinger and A. Rahman, J. Chem. Phys. <u>68</u>, 666 (1978).
    b. A. Rahman, F.H. Stillinger, and H.L. Lemberg, J. Chem. Phys.
       <u>63</u>, 5223 (1975).
21. J.B. Peri, J. Phys. Chem. <u>70</u>, 2937 (1966).
22. M.D. Newton and G.V. Gibbs, Phys. Chem. Minerals <u>6</u>, 221
    (1980).
23. M.D. Newton, M. O'Keeffe, and G.V. Gibbs, Phys. Chem. Minerals
    <u>6</u>, 305 (1980).
24. M. O'Keeffe and G.V. Gibbs, J. Chem. Phys. <u>81</u>, 876 (1984).
25. T.A. Michalske and B.C. Bunker, J. Appl. Phys. <u>56</u>, 2686 (1984).

MOLECULAR DYNAMICS MODELING
OF THE STRUCTURAL FAILURE OF GLASS

JOSEPH H. SIMMONS* AND CHARLES J. MONTROSE**
*University of Florida, Gainesville, FL 32611
**Catholic University of America, Washington, DC 20064

ABSTRACT

Molecular Dynamics (MD) methods for modeling the structure and kinetic behavior of ceramics and glasses are reviewed and discussed while emphasizing the effect of various potential functions on the structure developed and the relevance of model calculations. Two dynamic experiments from the authors' research are used as examples of modeling technique and interpretation of results. In the first, a Lennard-Jones force function is used to model the behavior of viscous liquids under large deformations. In the second, the brittle response of amorphous SiO_2 modeled with an ionic Born-Mayer-Huggins interparticle potential is investigated at very high deformation rates. A comparison of results with experiment show the limitations and the great promise of the MD approach.

GENERAL DISCUSSION OF MOLECULAR DYNAMICS MODELING

Molecular dynamics (MD) computations provide a simulation of materials behavior by giving the "exact" dynamics of relatively small number of mutually interacting particles (atoms, molecules, ions) which strictly obey Newton's laws of motion. These particles are taken to represent a portion of a material (generally a cubic section) and their motions are taken to represent the thermodynamic and kinetic behavior of the material. The forces that operate between and among the particles are chosen to mimic, as accurately as is reasonably possible, those that are present in the real material. Both equilibrium and non-equilibrium situations can be considered, the latter resulting from the introduction of sources or sinks of whatever controls the desired independent variable (particles, velocity, kinetic energy, momentum, stress, etc.) [1].

Molecular dynamics simulations offer the opportunity for bridging the gap between experimentally observed macroscopic behavior and analytically modeled microscopic processes. By combining the MD results with a few relevant experiments, one can hope to unify a wide variety of experiments in a field by determining the controlling microscopic processes. Such a determination enables one to develop realistic models that accurately portray the actual material behavior. For these reasons, molecular dynamics (MD) calculations are most fruitful when conducted in areas where a broad data base is available to guide the selection of parameters and to provide a solid experimental test of the computer model.

Molecular Dynamics Technique

In general terms, molecular dynamics is a computer simulation technique for probing the microscopic behavior of systems (liquids, solids and gases) under a rather wide range of conditions. The technique allows one to account for the many-body nature of the atomic and molecular processes that are operative in equilibrium, steady-state, and non-equilibrium situations. The basic idea involves following some "representative" number of molecules N ($\leqslant 1000$) by solving their classical equations of motion on a high speed computer.

508

The basic character of the MD system is determined by specifying some model potential function, or functions, that describes the interations among the atoms and/or molecules that comprise the system. Commonly employed forms used to simulate liquids and amorphous solids include the Lennard-Jones (LJ) 6-12 potential for monatomic systems [2], and the Born-Mayer interaction for polyatomic and charged ion systems [3]. Initial positon coordinates and velocities are chosen -- thereby fixing the density and internal energy (or temperature) of the system -- and a suitable algorithm is used to integrate numerically the equations of motion. In order to obtain trajectories, calculations of position and velocity are made in time steps t ($\leq 10^{-14}$s). These "data" -- essentially a record of the system's path through phase space -- enable one to work out any measurable property of the system, as well as many that are inaccessible in conventional experiments. The details of these procedures, as well as the computer programs for executing them, are well known to those practiced in molecular dynamics calculations. We send the interested reader to a number of excellent reviews [4].

Relevance of MD Calculation to Materials Behavior

Because molecular dynamics is a simulation technique one must continually assess the relevance of MD results to "real-material" problems. Certain limitations are obvious, but can generally be circumvented or, at least, their effect can be minimized. For instance, to avoid the situation where boundary effects, rather than material behavior, dominate the results (as would certainly be the case for a 1000-particle system confined to a rigid box), periodic boundary conditions are imposed. The MD cell (of volume V = L^3) is considered to lie at the center of a large (thermodynamically infinite) system composed of exact replicas of itself. An atom located at coordinates (x,y,z) in the MD cell implies the presence of atoms at the coordinates ($x\pm n_x L$, $y\pm n_y L$, $z\pm n_z L$) where n_x, n_y, n_z = 1, 2, 3 .... in the replica cells. The situation is shown schematically in the two-dimensional sketch below.

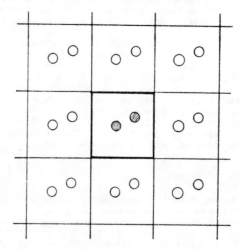

Figure 1

The use of periodic boundary conditions obviously prohibits studying properties with characteristic wavelengths or correlation lengths greater than or comparable to the box length L; it also proscribes against examining propagation effects on time scales longer than that required for the traverse of a box lenth.

In addition to these systemic limitations, there is the more serious question as to whether the potential energy interaction function(s) chosen for the simulation gives a meaningful representation of real material. In certain respects this can only be judged a posteriori, by comparing the behavior of the MD system with that of some real system.

However, as a practical matter, there are certain guidelines that can be used to enable one to make a reasonable first estimate of a suitable interaciton. One knows that, discounting, for the moment, any angularly varying terms, a potential form composed of a relatively short-range attractive term and a sharp repulsive core will give a fairly good representation of general material behavior. The Born-Mayer-Huggins form, for example, is known to account for many of the equilibrium properties of silicate and fluoride glasses [3-5].

In some cases, the property being investigated is relatively insensitive to the exact form of the potential function; any interaction with a "hard" repulsive core and a short-range attraction will give reasonably good results. This appeared to be the case in viscous flow calculations as shown below, where a Lennard-Jones potential reproduces the basic features of the behavior of inorganic glasses [6]. The pair correlation function in dense systems depends only on the existence of repulsive core of approximately the right range. On the contrary, however, for some properties, there is a strong correlation with the specific form of the potential function as in calculations of the distribution of bond angles. For example, calculations of defect distributions (as discussed above) rely strongly on lattice vacancies, and coordination numbers. Therefore, clear differences will arise from the angular form of the potential [covalent or central force (ionic)] or from the screening distance of the attractive force or from the size (repulsive term) of the particles. In all cases, however, only approximate forms of the potential function are used and one needs therefore to assess the effect of a particular potential function on the obtained results.

Time scaling is another important assumption which is inherent in the application of MD modeling to the behavior of viscous or glassy materials. Since molecular dynamics calculations model elemental atomic motions, the basic time-step unit used in the computations can be no greater than about 10^{-14} sec -- the limitation is set by the requirement that this step be less than the vibration period of the Debye cutoff frequency. For fluid systems, where characteristic relaxation times are no more than a few thousand time steps (i.e. on the order of picoseconds), no scaling need be done. However, in viscous liquids or glassy solids, characteristic relaxation times may be of the order of seconds. Direct computation in these cases would require 10^{14} integration steps -- far outside the practical range of today's computers. However, it is often possible to extend the results of calculations conducted on fast processes to more realistic experimental times by suitable scaling procedures. Such procedures are acceptable when considering properties and mechanisms which can be reduced functionally by the relaxation time characteristic of that process. Flow, diffusion and electrical conduction processes appear to scale adequately [4,7].

Assessing the relevance of different potential functions or scaling extrapolations must come from the proper experimental support. It is necessary to determine critical experiments which can differentiate between alternate approaches, and which can uniquely support the results of the MD calculations. Often, as in our work on high shear rate flow processes, critical experiments are devised as a result of questions discovered in the MD calculations. The synergistic relationship between MD calculations and selective critical experiments cannot be overemphasized in the attempt to assure relevance of model developments to actual material behavior.

Because of the strong support that experimental results provide to MD model development, it is most fruitful to study fields where sufficient critical experiments have been reported to provide well-defined guidelines. Insight into microscopic behavior is essential at the start since MD calculations can only generally differentiate between alternate models. Once predictions obtained from the developing model show agreement with the data base, then further insight may be gained from MD calculations to continue model development.

This approach is demonstrated in the two examples, presented below, of the use of molecular dynamics to probe certain fundamental atomistic mechanisms which control the behavior of amorphous materials under stress. These examples show the results of investigations into the two extreme failure modes possible in these materials: ductile in the first and brittle in the second.

The first example consists of a study of the fluid response of a molten glass to applied rapid strains which drive the melt into a non-Newtonian viscosity region and then to ultimate structural failure [6-8]. This work is supported by very specific experimental investigations which provide an unexpected demonstration of the effects predicted by the simulation and allow the combination of both analysis methods to understand the microscopic processes underlying the ductile failure process in glasses [7,8].

The second example consists of the study of the brittle failure of a modeled "silica glass" when strained at extremely rapid rates. This investigation probes the relationship between structure and strength, and examines the origin of material failure under stress [9]. The conditions of the test are experimentally inaccessible, since, in order to test a kinetically frozen structure, the strains must be applied at a rate faster than the material's atomic thermal vibrations.

HIGH SHEAR RATE STUDIES OF VISCOUS FLOW OF A LENNARD-JONES AND A SODA-LIME-SILICA GLASS

The flow of amorphous materials under an applied stress has generally been observed to be linear (Newtonian) for low stresses or strain rates. Organic materials have exhibited large deviations from Newtonian behavior both in the pseudoplastic direction (below Newtonian viscosity) and the dilatant direction (above Newtonian viscosity) [10]. Since a correspondence between structural changes and the non-Newtonian response has not been determined for these materials, it has been generally surmised that the non-Newtonian behavior is a result of complex molecular chain kinetics such as unfolding, stretching, cross-linking, etc.

In the studies presented here, computer "experiments" were conducted on a simulated molten glass using Molecular Dynamics calculations [6], to investigate the flow behavior of the molten glass under high shear deformation rates. The model system under investigation consisted of an assembly of 108 identical particles, interacting through an interatomic force derived from the Lennard-Jones 6-12 potential function:

$$V(r) = -4u \left[(a/r)^6 - (a/r)^{12} \right] \qquad (1)$$

where r is the interparticle separation and u and a are constants with dimensions of energy and distance, respectively. The particles were placed in a cubic box with periodic boundary conditions and dimensions selected to produce a density well above the triple point thus simulating a subliquidus viscous glass. Details of the calculation technique are found in Ref. 6. The results show that when the material is subjected to pure shear motion at various strain rates, and when the resulting stresses are measured, the onset of non-Newtonian viscous flow behavior is observed under increasing strain rates. For all but the very low normalized strain rates (normalized by multiplying the strain rate with the average shear isothermal relaxation time of the glass in the Newtonian region), the viscosity exhibits a strong dependence on strain rate, decreasing drastically with increasing strain rate. The resulting behavior is shown in Fig. 2, where the ordinate consists of the measured viscosity normalized by the measured Newtonian value, and the abscissa is the normalized strain rate. Since in Lennard-Jones glasses the atoms interact only via central forces, and no chain-like structures can be formed, it seems clear that the observed non-linear behavior in the MD calculations is a fundamental property of the liquid or glassy state.

Laboratory experiments were conducted on a soda-lime-silica glass using a fiber elongation rheometer. The details of the experiment are given in Ref. 7. Similar conditions of constant strain rates were reproduced in the laboratory experiments. The results showed complete agreement in almost every aspect with the MD calculations. The glass showed an increasing deviation from Newtonian behavior with increasing strain rate (see Fig. 2). Additional support was found in measurements on a Rb_2O-SiO_2 glass [11], which also exhibited the same shear thinning and non-Newtonian viscosity behavior. Figure 2 compares the three results and shows the dramatic agreement obtained between the MD simulation and the laboratory experiments.

Measurements were conducted on the soda-lime-silica glass as a function of time for constant applied shear strain rates. This allowed a study of the resulting stress development as a function of time, whereby the stress first increased to a maximum and then decreased to a lower steady-state value for times greater than the isothermal shear relaxation time. This stress overshoot behavior was indicative of a structural breakdown in the glass as the strain rate is applied. Greater strain rates would produce greater structural breakdown and thus a lower non-Newtonian viscosity.

Having established the excellent correspondence between the MD modeling experiments and the laboratory experiments, it was possible to study the behavior of the amorphous structure of the glass by looking at the atomic rearrangements in the MD system. This observation, only possible in the MD experiment, showed a distinct structural rearrangement of the atoms leading to the measured shear thinning behavior. As the system was strained, the atoms appeared to form into distinct layers parallel to the flow direction, thus clearly easing flow and reducing the viscosity of the liquid. This structural rearrangement of the glass at high shear rates was subsequently demonstrated [8]. These results show the synergistic effect of MD experiments and a well founded laboratory data base.

512

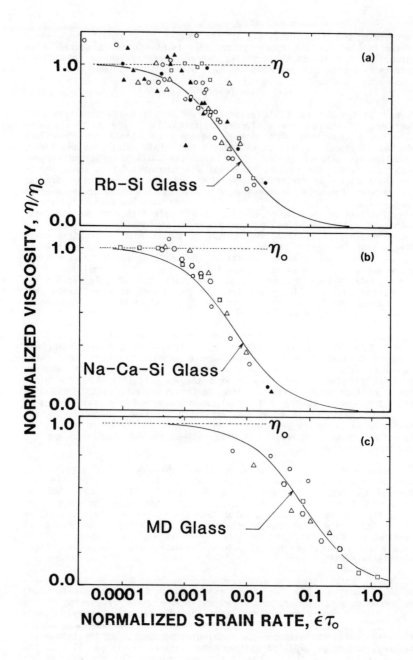

Figure 2

The non-Newtonian behavior of the liquid at high strain rates is reminescent of the behavior observed in sol-gels near the gelation point. Sacks and Sheu [12] have shown that as the sol-gel suspension begins to form a 3-dimensional structure, the viscosity exhibits a marked shear rate dependence. The exhibited shear thinning behavior is most likely to be a result of a breakdown of the 3-dimensional structure by the applied shear strain rate. These effects are presently under investigation to determine whether the shear thinning behavior of the sol-gel solution follows the same analytical form which accurately described the MD glass and the soda-lime-silica glass. (This analytical expression is plotted in Fig. 2 as the solid line.) Should good correspondence exist, it will be possible to estimate the strength of the growing 3-dimensional structure in the sol-gel solution.

HIGH STRAIN RATE STUDIES OF BRITTLE FAILURE IN A SILICA GLASS

A molecular Dynamics study was conducted to simulate the structure of silica glass. In this case, the sample was composed of two kinds of atoms representing Si^{4+} and O^{2-} which interact via a Born-Mayer-Huggins potential leading to a truncated interatomic pair wise force, $F(r)$, as follows:

$$F_{ij}(r) = [pA \exp(-pr) + Z_i Z_j / r^2] [1 - (r/R_{max})^6] \qquad (2)$$

where the truncation distance $R_{max} = 5.5A$. A sample was generated at 10,000K with 1908 particles (1272 O^{2-} and 636 Si^{4+}), equilibrated then cooled to form a solid glass at 500K.

The resulting sample exhibited many of the characteristics of fused silica such as tetrahedral symmetry, and a radial distribution function, a compressibility and a phonon spectrum similar to silica glass. The density of 2.35 g/cm³ is higher than silica by almost 10% but this was found to have little effect on the results obtained. The sample, after equilibration as a solid, (i.e. no measurable atomic diffusion) was elongated in a uniaxial fashion at varied rates. Two classes of behavior were observed, associated with elongation rates faster than the speed of sound, and elongation rates slower than the speed of sound. In all cases, however the strain rates were much faster than any atomic diffusive motion. The speed of sound was determined by the atomic thermal vibrations. At the high rates, the sample was elongated to failure in a period shorter than an atomic vibration, while the period of the low rates was much longer than atomic vibration periods.

These high rates were chosen to determine the strength of the glass structure without the weakening effect of atomic diffusion (i.e. pure brittle behavior). While these experiments can have no counterpart in the laboratory, the results yielded a significant insight into the initial process of brittle failure in materials. For example, it was unexpectedly found that atomic vibrational motions can significantly rearrange the structure of the strained glass and will alter the failure mode. The results reported in Ref. 9 show a decrease in brittle strength of the material by 60% when going from fast pulling rates to slower pulling rates. This strength reduction results from structural rearrangements induced by atomic thermal vibrations. The failure mechanisms also differed. At high shear rates, it was observed that the SiO_4 tetrahedra separated under the applied stress and failure occurred by an increase in the Si-O bond length. At shear rates below the sound velocity, the SiO_4 tetrahedra remained essentially unchanged; the average Si-O distance did not increase and a distinct rotation of the tetrahedra about the bridging

514

oxygens is observed. Failure occurs by the separation of undeformed tetrahedra through the breakup of some oxygen bridges.

The reduction in strength produced by the atomic thermal vibrations was clearly unexpected, as was the difference in failure mode. Experiments are presently underway where the samples are heated to develop atomic diffusive motions. A further decrease in strength associated with this localized flow is expected. Finally at sufficiently high temperatures viscous flow will occur and the two studies will merge as the failure mode will go from brittle to ductile.

REFERENCES

1. D. Levesque, L. Verlet and J. Kurkijarvi, "Computer Experiments on Classical Fluids," Phys. Rev A 7, 1690-1700 (1973); also W. G. Hoover, A. J. C. Ladd, R. B. Hickman, "Bulk Viscosity Via Non-Equilibrium and Equilibrium Molecular Dynamics," Phys. Rev. A 21, 1756-60 (1980).
2. B. L. Holian, "Cell Model Prediction of Melting of a Lennard-Jones Solid," Phys. Rev. B 22, 1394-1404 (1980).
3. T. F. Soules and Busbey, "Sodium Diffusion in Alkali Silicate Glass by Molecular Dynamics," J. Chem. Phys. 75, 969-975 (1981); also T. F. Soules and A. K. Varshneya, "Molecular Dynamics Calculations of a Sodium Borosilicate Glass Structure," J. Am. Ceram. Soc. 64, 145-150 (1981).
4. C. A. Angell and P. Cheeseman, "Molecular Dynamics Studies of the Vitreous State: Simple Ionic Systems and Silica," J. Chem Phys. 65, 1565-77, (1976); also T. F. Soules, "MD Calculations of Glass Structure and Diffusion in Glass." J. Non Crystalline Solids 49, 29-52 (1982).
5. C. A. Angell, "Structure and Dynamics of Inorganic Fluoride Glasses by Ion Dynamics Calculations," presented at the Amer. Ceram. Soc. Annual, Cincinatti, 1982, Bull. Amer. Ceram. Soc. 61, 376 (1982). S. A. Brawer and M. J. Weber, "MD Simulations of the Structure of Rare-Earth Doped Beryllium Fluoride Glass," J. Chem. Phys. 75, 3522-41 (1981).
6. D. M. Heyes, J. J. Kim, C. J. Montrose and T. A. Litovitz, Non-Linear Shear Stress Effects in Simple Liquids, An MD Study," J. Chem. Phys. 73, 3987-3996 (1980); also D. M. Heyes, C. J. Montrose and T. A. Litovitz, "Viscoelastic Shear Thinning of Liquids, An MD Study," to be published in J. Chem. Soc. Faraday Trans. II.
7. J. H. Simmons, R. K. Mohr and C. J. Montrose, "Non-Newtonian Viscous Flow in Glass," J. Appl. Phys. 53, 4075-4080 (1982).
8. J. H. Simmons, R. K. Mohr and C. J. Montrose "Viscous Failure of Glass at High Shear Rates," J. de Physique C9, 439-442 (1982).
9. R. Ochoa and J. H. Simmons, "High Strain Rate Effects on the Structure of a Simulated Silica Glass," J. Non-Crystalline Solids 75, 413-418 (1985).
10. D. W. Hadley and I. M. Ward, Rep. Prog. Phys. 38, 1143 (1975).
11. J. H. Li and D. R. Uhlmann, J. Non-Crystalline Solids 3, 127 (1970).
12. M. D. Sacks and R. S. Sheu, "Rheological Characterization During Sol-Gel Transition," 2nd Int. Conf. on Ultrastructure Processing of Ceramics, Glasses and Composites, John Wiley (to be published).

MOLECULAR MIMICRY OF STRUCTURE AND ELECTRON DENSITY DISTRIBUTIONS IN MINERALS

G.V. GIBBS* AND M. B. BOISEN, JR.**
* Dept. of Geological Sciences and ** Dept. of Mathematics, Virginia Polytechnic Institute and State University, Blacksburg, VA 24060

ABSTRACT

Molecular orbital calculations on hydroxyacid molecules with first- and second-row X-cations ($X = Li$ through N and Na through S) yield bond lengths and angles that mimic those of chemically similar minerals. These bond lengths are used to find a formula giving bond length as a function of a bond-strength parameter that reproduces XO bond lengths in crystals with main-group X-cations from all six rows of the periodic table within 0.05Å on average. The molecular orbital calculations also provide insights into reaction energies, physical properties of crystals such as electron density distributions, and data not amenable to direct measurement. They also provide a basis from which computational models for mineral structures may be constructed.

INTRODUCTION

The chemistry of the mineral world is dominated by silicon and oxygen with silicates comprising more than 95% by weight of the earth's crust and mantle. If we are to improve our understanding of their crystal chemistry, their physical properties, and their manifold uses in the manufacture of glasses, molecular sieves, silicone polymers and electronic devices, it is essential that we have a good understanding of the bonding in these minerals. Of all the properties of a silicate that depend in some way on the nature of its binding forces, only a handful can be uniquely identified with an individual bond. Of these, bond length is special in the sense that it provides a reliable measure of the strength of a bond: the shorter a particular bond, the greater its strength. Indeed, studies of bond length variations have played a key role in the development of a bonding theory for molecules and crystals [1]. Although such studies have done much to advance our knowledge of the SiO bond, our understanding of this bond is still far from complete [2,3,4,5,6,7,8,9,10,11 and references therein].

During the past 25 years, with the development of sophisticated tools for collecting and processing diffraction data, the structures of many silicate minerals and siloxane molecules and crystals have been determined. One of the most important experimental facts provided by these studies is that the bond lengths and angles of a disiloxy ($SiOSi$) bond in a crystal like α-quartz are virtually the same as those of a gas phase molecule like disiloxane, $H_3SiOSiH_3$ [12,13,14,15,11]. Thus, an α-quartz crystal, regarded as a polymer of silicate SiO_4 tetrahedra appears to be held together by the same forces that comprise the disiloxy bond of the molecule. Therefore, it is not surprising that molecular orbital (MO) calculations on disiloxane [16,17,11] and other molecules with SiO and disiloxy bonds have accurately reproduced the bond lengths and angles of comparable units in silicates. The calculations have also reproduced the bulk modulus of α-quartz as well as providing support for Galeener's [18] assignment of the defect bands in the Raman spectra of vitreous silica. In addition, calculations on hydroxyacid molecules containing first- and second-row cations have reproduced Shannon's [19] bond lengths for oxide crystals as well as the Baur [7] and the Brown and Shannon [8] bond strength-bond lengths curves. Also, electron density maps calculated for these molecules have yielded bonded radii that mimic those provided by electron density maps measured for crystals. The aim of this report is to summarize the (MO) calculations that have provided these results and to examine some of their implications.

BOND LENGTH AND ANGLE VARIATION IN SILICATES

Of the more than 3000 known minerals, more than 900 are silicates. Almost all contain tetrahedral SiO_4 groups while about 20 are known to contain octahedral SiO_6 groups. Those with tetrahedral groups either contain monomeric and/or polymeric silicate groups joined by disiloxy bonds [20]. Our insight into the nature of the bonding in these tetrahedral and octahedral groups has been provided by MO calculations on H_4SiO_4 and H_8SiO_6 molecules whereas that of the disiloxy bond has been provided by calculations on $(OH)_3SiOSi(OH)_3$ and $H_6Si_3O_3$. In addition, calculations on hydroxyacid molecules with first- and second-row X-cations ($X = Li$ through N and Na through S) have provided insights into the other kinds of bonds that may occur in silicates. Unless otherwise stated, the molecular orbital calculations described here were completed with near-Hartree-Fock 6-31G* basis sets with polarization functions on the X-cations using Gaussian 82 software [21,22].

Monosilicic Acid and Silicates with Monomeric Tetrahedral SiO_4 Groups :

Monosilicic acid, H_4SiO_4, is the simplest molecule with a monomeric SiO_4 group. A calculation of the minimum energy geometry of this molecule within the constraints of S_4-point symmetry with d-type polarization functions on the oxygen atoms, yields four SiO bond lengths of 1.63Å, two $OSiO$ angles of 105°, four $OSiO$ angles of 112°, four OH bond lengths of 0.94Å and four $SiOH$ angles of 118°. The minimum energy SiO bond length is statistically identical with the average SiO bond length (1.635Å; Figure 1) observed for a variety of monosilicates. The narrow range of bond lengths observed for monosilicates like olivine, humite, zircon and garnet conforms with the relatively large force constant (680 N/m) calculated for the SiO bond in H_4SiO_4. An examination of the relationship between SiO bond length, $R(SiO)$, and $OSiO$ angle for a variety of silicates has shown that $R(SiO)$ is inversely related with the fraction of s-charcter of the Si atom, $f_s(Si)$ [23]. Figure 2 shows that this correlation is similar to that provided by the minimum energy geometries of H_4SiO_4 and $H_3SiO_4^-$ [24].

Orthosilicic Acid Dihydrate and Silicates with Octahedral SiO_6 Groups :

While the vast majority of silicates contain SiO_4 groups, some like $K_2Si_4O_9$ contain both SiO_4 and SiO_6 groups [25] whereas others like the high pressure silica polymorph stishovite contain only SiO_6 groups. In an MO study of the geometry of the SiO_6 group in H_8SiO_6, Gibbs *et al.* [26] calculated a minimum energy $R(SiO)$ value of 1.76Å in

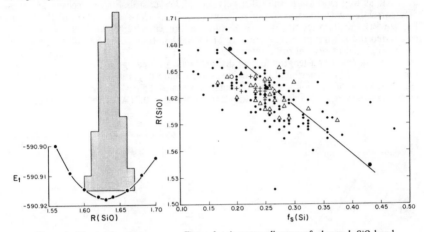

Figure 1: The total energy, E_T, of the monosilicic acid molecule vs. SiO bond length, $R(SiO)$. The histogram was prepared with SiO bond length data recorded for monosilicates.

Figure 2: A scatter diagram of observed SiO bond lengths, $R(SiO)$, in silicate crystals vs. the fraction of s-character, $f_s(Si)$, of the bond. The smaller solid bullets represent one point, the open triangles two, the plus signs three, the open bullets four, and the crosses five and the close triangles six data points. The larger closed bullets are data points calculated for H_4SiO_4 and $H_3SiO_4^-$ and the line is drawn to fit these three points.

agreement with the Shannon [19] SiO bond length of 1.76Å. Figure 3 compares a total energy curve calculated for the molecule as a function of $R(SiO)$ with a frequency distribution of experimental bond lengths for silicates with SiO_6 groups. As the histogram shows, the experimental bond lengths are clustered about the minimum with a mean value of 1.768Å. Also, a deformation map calculated by Hill *et al.*, [27] for a molecule with a SiO_6 group shows a modest but significant accumulation of electron density in its SiO bonds as observed for the group in stishovite. They also observed that the net atomic charge calculated for Si is $+1.78e$ compared with that ($+1.71e$) determined experimentally from a population and κ-refinement of the stishovite diffraction data. It was also found that the net charge calculated for the Si atom of H_4SiO_4 ($+1.10e$) agrees with that ($+1.00e$) obtained in a similar analysis of the diffraction data of α-quartz [28].

Disilicic Acid and Silicates with Disiloxy Bonds:

As observed above, the polymeric group of tetrahedra in a large variety of silicates consist of SiO_4 groups joined by disiloxy bonds. O'Keeffe *et al.*, [24] have optimized the geometry of the disilicic acid molecule, $(OH)_3SiOSi(OH)_3$, with d-type polarization functions on the bridging oxygen in an exploration of the bond. The calculation yielded a minimum energy SiO bridging bond length of 1.62Å and an $SiOSi$ angle of 141°. These values agree with values (1.612Å; 142.4°) obtained by Lager *et al.*, [29] in

518

a refinement of the structure of α-quartz with diffraction data recorded at 13K. Newton and Gibbs [30] have found that the experimental SiO bond lengths in α-quartz as well as those in the silica polymorphs low-cristobalite and coesite are linearly correlated with the fraction of s-character, $f_s(O) = 1/(1\text{-sec} < SiOSi)$, of the bridging O of the disiloxy bond [31]. In a study of this correlation, O'Keeffe et al., [24] calculated the minimum energy bridging SiO bond length of $(OH)_3SiOSi(OH)_3$ with its $SiOSi$ angle set at several different angles ranging from 115° to 180°. Figure 4 shows the resulting SiO bond lengths for the disiloxy bond plotted against $f_s(O)$ together with the corresponding experimental data for several carefully refined silica polymorphs [32]. The slope of the line for the experimental data is slightly steeper than that of the calculated data with the $R(SiO)$ values agreeing within 0.01Å at 180° and exactly at 135°.

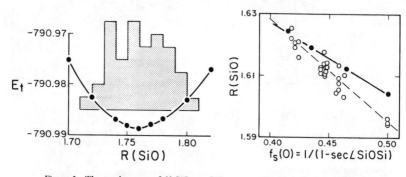

Figure 3: The total energy of H_8SiO_6 vs. SiO bond length, $R(SiO)$, for 6-coordinate Si. The histogram was made with the SiO bond lengths observed for silicates with SiO_6 groups (After [11]).

Figure 4: A scatter diagram (open circles) of SiO bond lengths, $R(SiO)$, observed for the silica polymorphs α-quartz, coesite and low crystobalite vs. the fraction of s-character of the bond $f_s(O)$. The data plotted as solid circles were calculated for $(OH)_3SiOSi(OH)_3$ [24,32].

A total energy curve for $(OH)_3SiOSi(OH)_3$ calculated as a function of its $SiOSi$ angle is displayed in Figure 5 with a histogram of observed angles in silicates. The curve is broad and asymmetric in shape rising relatively slowly in energy (10 KJ/mole) as the angle is opened to 180° and more rapidly as the angle is closed to 120° where the energy is seen to increase rapidly. The histogram of $SiOSi$ angles conforms in general with the broad and asymmetric shape of the curve by showing a wide range of values, by rising steeply to a peak at about 140° and then falling off more gradually to 180°. The subsidary peak at 180° is at variance with the curve. However, 180° angles have been reported for disiloxane molecules where intermolecular interactions apparently overcome the small energy expended by the molecule in adopting a wide angle [33,34 and references therein].

In a study of the energetics of the disiloxy bond, O'Keeffe et al., [35] found that the symmetric bending force constant of the $SiOSi$ angle calculated using a STO-3G basis is in rough agreement with that reported for α-quartz. When the SiO stretching (850 N/m) and the $OSiO$ bending force constants (60 N/m) calculated for $(OH)_3SiOSi(OH)_3$ were found to be much larger than the $SiOSi$ bending force constant (10 N/m), they concluded that the bulk modulus of α-quartz might be controlled in large part by the compliance of the the disiloxy bond. A modeling of the bulk modulus of the mineral using the *SiOSi* bending force constant calculated for $(OH)_3SiOSi(OH)_3$ yielded a value of 0.397 megabars compared with that observed (0.393 megabars). Ross and Meagher [36] have completed similar calculations on an $(OH)_3SiOSi(OH)_3$ molecule surrounded by a "pressure medium" of He atoms in a simulation of the bond length and angle variation of coesite at high pressures. The trends given by their calculations mimic those provided by a structure analysis of the mineral undertaken at high pressure [37] and suggest, as observed, that pressure changes in the mineral should have a relatively large effect on the "soft" *SiOSi* angle and little effect on the stiff *SiO* bond.

As noted by Newton and Gibbs [30], the broad nature of the curve in Figure 5 indicates that the *SiOSi* angle in a silicate or silica glass can be easily deformed from its equilibrium value without excessively destabilizing the resulting structure [38]. Thus, the compliant nature of the angle may account at least in part for the the wide range of angles exhibited by silicates and silica glass; it may also account for the wide diversity of polymerized tetrahedral groups and structure types exhibited by silicates [39,40,20].

In a study of the Raman spectra of vitreous silica, Galeener [18] has ascribed sharp bands in the spectra to symmetric vibrations of the oxygen atoms in tricyclosilicate rings. O'Keeffe and Gibbs [41] obtained a minimum energy geometry of a tricyclosiloxane $H_6Si_3O_3$ molecule in an *MO* study of defects in vitreous silica and obtained a *SiOSi* angle (137°) and a *SiO* bond length (1.65Å) that agrees with experimental values (132.9°; 1.630, 1.648Å) obtained for the ring of the tricyclosilicate benitoite [42]. A calculation of the frequency of the symmetric vibrations of the oxygen atoms in the ring yielded a value of $650 cm^{-1}$ compared with a measured value of $600 cm^{-1}$. As 6-31G* basis sets generally give calculated frequencies some 10% higher than those observed, the calculated frequency is considered to be in agreement with the observed value and to support Galeener's assignment of the bands to such rings. O'Keeffe and Gibbs [43] also examined the energetics of an *Si*-hydroxyacid molecule specially chosen to simulate an $Si=O$ double bond in silica [44] and concluded that such siliconyl bonds are not plausible defects in vitreous silica.

As energies provided by MO calculations close to the Hartree-Fock limit provide good estimates of differences in heats of reaction, O'Keeffe and Gibbs [41,43] also calculated the hydration energies of corner, edge and face sharing silicate tetrahedra in an exploration of the relative stabilities of these structural features in silicate crystals and glasses. In accordance with Pauling's [2] third rule, they found, relative to a dimer of corner sharing tetrahedra, that the sharing of an edge destabilizes a structure by 210 KJ/mole whereas face sharing destabilizes a structure by 710 KJ/mole. The large energy difference associated with each of these features (compared with -460 KJ/mole, the enthalpy of the *SiO* bond) explains their absence in silicate crystals. It also casts doubt on their being high concentration defects in silica glass.

In a study of bond length and angle variations in silicate and aluminosilicates framework structures, Geisinger et al., [45] have found that minimal basis *MO* calculations on representative molecules yield bond length and angle variations that mimic those exhibited by the *AlO* and *SiO* bonds in such frameworks. In addition to reproducing *SiO* and *AlO* bond length-bond strength-sum curves established by Baur [7,46], the calculations provide insight into the geometry of a framework structure given the composition and the ordering of the tetrahedral atoms into particular sites. As observed by Navrotsky et al., [47], these calculations also shed light on the ease of glass formation in silicate and aluminosilicate systems, the pattern of immiscibility of

520

borosilicate glasses and the thermochemical mixing properties of a variety of aluminosilicate glasses.

BOND LENGTH-BOND STRENGTH VARIATIONS IN OXIDE CRYSTALS

In the previous section, we saw that the *SiO* bond length and angle variations in the silica polymorphs are essentially the same as those calculated for chemically similar molecules. Comparable calculations on phosphate, borate and sulphate molecules have been equally successful in generating bond length and angle variations that match those in chemically similar crystals [24,48,49,50]. More recently, Gibbs *et al.*, [26] calculated theoretical *XO* bond lengths, R_t, for hydroxyacid molecules with first- and second-row *X*-ions. These bond lengths together with *BO*, *SiO*, *PO* and *SO* bond lengths calculated in earlier works are plotted in Figure 6 against the Shannon [19] *XO* bond lengths, R_o. The plot illustrates that the calculated bond lengths reproduce the experimental values to within 0.02Å on average. A statistical analysis shows that 99% of the variation in the experimental bond length can be explained in terms of a linear dependence on those calculated.

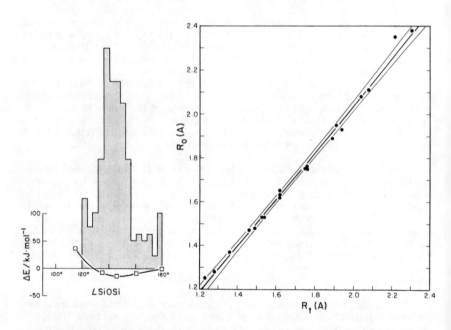

Figure 5: The total energy of $(OH)_3SiOSi(OH)_3$ plotted as a function of the *SiOSi* angle, < *SiOSi*. The histogram was constructed with *SiOSi* angles taken from a wide variety of silicates with disiloxy bonds.

Figure 6: A scatter diagram of R_o vs. R_t where R_o denotes mean experimental bond lengths for oxide crystals [19] and R_t denotes the theoretical *XO* bond lengths calculated for hydroxyacid molecules. The heavy line is the least-squares line and the light curves define its 99% confidence bands.

Calculations similar to those described above for molecules involving X-cations from rows in the periodic table beyond the second row have yet to be reported. Consequently, an attempt was made to extend the results from the first two rows to the remaining rows of the table using a formula. This formula is based on the Pauling [2] bond strength of a bond involving a cation defined to be $s = z/v$ where z is the valence of the cation and v is its coordination number. To adjust s for the various rows, a bond strength parameter $p = s/r$ was defined where $r = 1,2,3,...$ for first-, second-, third-,..., row main group X-cations, respectively. When the minimum energy bond lengths R_t calculated for the molecules are plotted against this bond strength parameter, we see that p ranks R_t with shorter bonds involving larger bond strengths (Figure 7). A plot of the bond strength parameter against the Shannon [19] XO bond lengths for the first through sixth rows shows that the parameter ranks the experimental bond lengths in the same way (Figure 8). In a regression analysis of the R_t data calculated for the hydroxyacid molecules, Gibbs $et\ al.$, [26] found that this curve is best fit (among the exponential functions) by

$$R = 1.39p^{-0.22} \tag{1}$$

Using R as the predicted bond length corresponding to an XO bond having a bond strength parameter p, more than 99% of the variation of the Shannon [19] XO bond lengths reported for rows 1 through 6 can be explained in terms of a linear dependence upon the corresponding R-values given by Eq. 1. Furthermore, 56% of the R-values agree within 0.05Å of those observed, 83% within 0.10Å and 94% within 0.15Å.

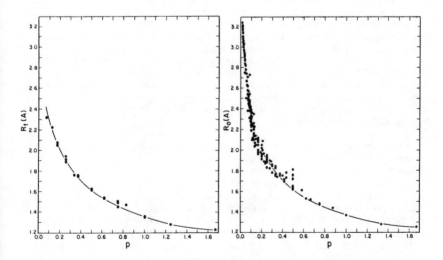

$Figure\ 7$: A scatter diagram of minimum energy XO bond lengths, R_t, plotted against the bond strength parameter p.

$Figure\ 8$: A scatter diagram of Shannon [19] XO bond lengths for main-group X-cations vs. the bond strength parameter p.

CRYSTAL AND BONDED RADII

The structures of many oxide minerals have been described as closest-packed arrays of large oxide anions with smaller cations in tetrahedral and octahedral voids [51,52]. Given the positional coordinates of the ions, it is usually a simple job to calculate the distances between the ions in such an array. On the other hand, a calculation of the radii of the ions in this array requires a number of simplifying assumptions. In some definitions, for example, the ions are assumed to be hard spheres and the number of anions surrounding each cation is assumed to be maximal subject to the constraint that the cation makes "contact" with each anion. With these assumptions, several sets of ionic radii have been derived assuming that the radius of the oxide ion is constant and that the XO bond length is equal to the sum of the radii of the X-cation and the oxide ion [53,54,55]. More recently, Shannon and Prewitt [56] and Shannon [19] derived a set of crystal radii from a study of more than 1000 accurately determined bond lengths, assuming a 6-coordinate oxide ion radius of 1.26Å. Their crystal radius of a given ion takes into account coordination number, oxidation state and spin state but is independent of the ionicity of the bond. For example, an oxide ion with a given coordination number is assumed to have a fixed radius regardless of whether it is bonded to an electronegative atom like S or a less electronegative one like Na. With considerable success, crystal radii have been summed to generate XO bond lengths for oxide crystals, matching those obtained in recent structure analyses within 0.01Å on average. While these radii reproduce bond lengths, they are not directly related to some important physical properties. In particular, they do not relate to the electron density distribution of an ion.

One difficulty in defining a radius that is related to the electron density distribution of an ion is to decide upon a feature of that distribution that, in some sense, determines the outer "surface" of the ion. Since, theoretically, the distribution of the electron density of an ion extends to infinity, the outermost excursions of its electrons is not a defined feature. One reasonable feature of an isolated ion is the maximum radial charge density of its outermost electrons. The atomic radii as proposed by Bragg [57] and modified by Slater [58] relates to this feature. One advantage of these radii is that they generate bond lengths equally well for covalent, metallic and ionic bonds in both crystals and molecules. However, since coordination number and other factors affecting bond lengths are not considered, these radii do not reproduce bond lengths as accurately as do crystal radii [58].

Within the context of an XO bond, a feature that reasonably defines the point of "contact" between bonded ions is the point along the bond at which the total electron density is a minimum. Consequently, the "bonded radius" of an ion is defined to be the distance from the center of the ion to this minimum point [59,60]. Unlike the definition of crystal or atomic radii, bonded radii make sense only in the context of a bond and so the bonded radius of the oxide ion in an XO bond may vary when bonded to different X-cations [61,62]. Such a variable radius is consistent with arguments that the oxide ion radius should vary with the field strength of the X-cation [63]. Note that because the bonded radius only describes the "size" of an ion in the direction of the bond, the term bonded radius should not be taken to imply that the ion is spherical with the same radius in all directions.

As deformation density maps calculated by quantum mechanical methods for molecules representing borate, silicate, and sulphate crystals reproduce experimental maps within a reasonable tolerance [49,27,64,50], Finger and Gibbs [65] have derived bonded radii from total electron density maps calculated for optimized $H_{8-m}X^{m+}O_4$ hydroxyacid molecules with 4-coordinate first- and second-row X-cations (Figure 9).

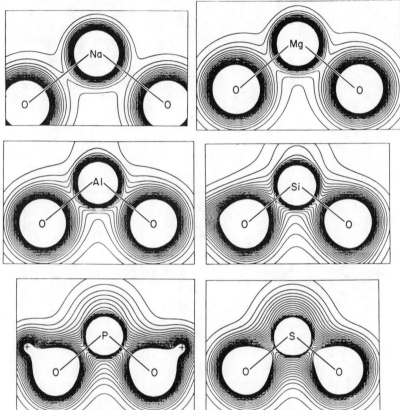

Figure 9: Total electron density maps calculated through the OXO bonds of geometry optimized $H_{8-m}X^{m+}O_4$ hydroxyacid molecules containing second-row 4-coordinate X-cations (See [65] for maps of first-row X-cation hydroxyacid molecules). The contour interval is $0.1e/Å^3$

Figure 10a shows that the bonded radii of the cation, r_x, and oxide ion, r_o, comprising the XO bonds increase linearly with the optimized bond length, R_t. In particular, the bonded radius of the oxide ion from row r can be expressed by the equation

$$r_o = (0.35 - 0.1r) + R_t/2 \tag{2}$$

The presence of the term $R_t/2$ in this equation implies, for a given row, that the change in R_t is shared equally between the bonded radius of the cation and the oxide ion as one X-cation is replaced by another. The bonded radii of these cations are plotted against their crystal radii in Fig. (10b) where it is seen that these two sets of radii are highly correlated.

While there is a paucity of data, a few comparisons can be made between the bonded radii calculated for the hydroxyacid molecules studied by Finger and Gibbs [65] and those measured from experimental electron density maps of chemically similar crystals. In particular, the value (1.08Å) calculated for the oxide ion in the orthosilicic acid dihydrate molecule is only 0.04Å larger than that provided by the total electron density maps of stishovite [66]. Similar agreement exists between the radius of the oxide ion calculated for the monosilicic acid molecule (0.95Å) and that (0.94Å) measured for coesite [32]. On the basis of these results, it seems that the radii obtained for hydroxyacid molecules mimic those measured for chemically similar crystals to within about 5%.

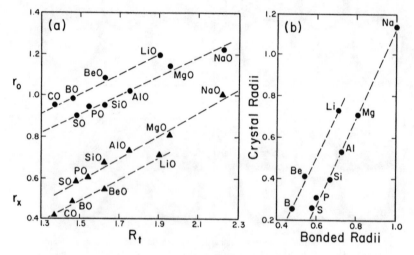

Figure 10: A plot of bonded radii for the oxide ion and for first- and second-row X-cations obtained from theoretical electron density maps: (a) Shows the bond radii (plotted as triangles) of 4-coordinate X-cations, r_x, and those of the oxide ion, r_o, both plotted against the minimum energy XO bond length, R_t; (b) Shows the bonded radii of the cations plotted against Shannon's [19] crystal radii.

The bonded radii of the oxide ion derived by Finger and Gibbs [65] are intermediate in value between Slater's [53] atomic radius (0.60Å) for the oxygen atom and Pauling's [1] ionic radius (1.40Å) for the oxide ion. As one might expect from simple bonding arguments, the bonded radius of the oxide ion increases in a regular way with increasing ionicity of the XO bond from 0.90Å for the more covalent SO bond to 1.22 Å for the more ionic NaO bond (i.e., with increasing charge transfer from the X-cation to the oxide ion). Hence, radii obtained from electron density maps indicate that a definition of radii that assumes a constant value for the oxide ion is open to question.

MOLECULAR MIMICKING OF CRYSTAL STRUCTURE

As we have seen in this review individual bond lengths and angles within a crystal structure, like α-quartz , can be mimicked by *MO* calculations on a molecule like $(OH)_3SiOSi(OH)_3$. Furthermore, the total energy surface calculated for the molecule gives insight into crystal properties such as the bulk modulus. The next natural step is to try to reproduce the crystal structure itself using the molecular model. Recently, Lasaga and Gibbs [67] developed a computational model for the structural energy of α-quartz. from a total energy surface calculated for the disilicic acid molecule with an STO-3G basis. By minimizing this energy with respect to the structural variables, the atomic coordinates of the mineral were reproduced within 0.02Å on average of the observed values. Not only did the calculations yield a bulk modulus of 0.394 megabars which is in agreement with the observed value, but they also yield an an equation of state that matches that measured for α-quartz up to pressures of 5 kilobars. The success of these calculations bodes well for the development of new computational models for mimicking the structures and properties of minerals using energy surfaces derived for carefully selected molecules.

ACKNOWLEDGEMENTS

The National Science Foundations supported this effort with Grant EAR 8218743 awarded to GVG for studying bonding and charge density distributions in minerals. We are pleased to thank our University for supporting this study with generous grants of monies to defray computing costs. We also wish to thank Sharon Chiang for drafting the figures, Virginia Chapman for preparing a camera copy of the manuscript, C. G. Lindsay and J. A. Stuart for reading an early version of the paper, and Dr. C. Jeffery Brinker for inviting one of us to present this paper at The Materials Research Society Symposium: Better Ceramics Through Chemistry.

REFERENCES

[1] Pauling, L., *The Nature of the Chemical Bond*, 3rd ed. (Cornell University Press, Ithaca, New York, 1960).

[2] Pauling, L., J. Am. Chem. Soc. **51**, 1010 (1929).

[3] Cruickshank, D. W. J., J. Chem. Soc. **1077**, 5486 (1961).

[4] Cruickshank, D. W. J., J. Mol. Structure **130**, 177 (1985).

[5] Smith, J. V. and S. W. Bailey, Acta Crystallogra. **16**, 801 (1963).

[6] Noll, W., Angewandte Chemie, 2 **2**, 73 (1963).

[7] Baur, W. H., Trans. Am. Cry. Assoc. **6**, 125 (1970).

[8] Brown, I. D. and R. D. Shannon, Acta Crystallogr. A **29**, 266 (1973).

[9] O'Keeffe, M. and B. G. Hyde, Acta Crystallogr., B **32**, 2923 (1976).

[10] Tossell, J. A., Trans. Am. Cry. Assoc. **14**, 47 (1979).

[11] Gibbs, G. V., Am. Mineralogist **67**, 421 (1982).

[12] Almenningen, A., O. Bastiansen, V. Ewing, K. Hedberg and M. Traetteberg, Acta Chemica Scandinavica **17**, 2455 (1963).

[13] Noll, W., in *Silicons, Chemistry and Technology of Silicones* (Academic Press, New York, 1968), 702 pp.

[14] Tossell, J. A., and G. V. Gibbs, Acta Crystallogr. A **34**, 463 (1978).

[15] Barrow, M. J., E. A. V. Ebsworth and M. M. Harding, Acta Crystallogr. B **35**, 2093 (1979).

[16] Oberhammer, H. and J. E. Boggs, J. Am. Chem. Soc. **102**, 7241 (1980).

[17] Ernst, C. A., A. L. Allred, M. A. Ratner, M. D. Newton, G. V. Gibbs, J. W. Moskowitz and S. Topiol, Chem. Phys. Lett. 81 3, 424 (1981).

[18] Galeener, F. L., J. Non-Cryst. Solids **49**, 53 (1982).

[19] Shannon, R. D., Acta Crystallogr. A **32**, 751 (1976).

[20] Liebau, F., *Structural Chemistry of Silicates* (Springer-Verlag, New York, 1985).

[21] Hehre, W. J., L. Radom, P. V. R. Schleyer and J. A. Pople, in *Ab Initio Molecular Orbital Theory* (John Wiley & Sons, 1986).

[22] Binkley, J. S., M. J. Frisch, D. J. DeFrees, K. Raghavachari, R. A. Whiteside, H. D. Schelgel, E. M. Fluder and J. A. Pople, *Gaussian 82, for the IBM/MVS system* Copywrite c 1984 (Carnegie-Mellon University, 1984).

[23] Boisen, Jr. M. B. and G. V. Gibbs, (In preparation, 1986).

[24] O'Keeffe, M., B. Domenges and G. V. Gibbs, J. Phys. Chem. **89**, 2304 (1985).

[25] Swanson, D. K. and C. T. Prewitt, Am. Mineral. **68**, 581 (1983).

[26] Gibbs, G. V., L. W. Finger and M. B. Boisen, Jr., Am. Mineral. (Submitted).

[27] Hill, R. J., M. D. Newton and G. V. Gibbs, J. Solid State Chem. **47**, 185 (1983).

[28] Stewart, R. F., M. A. Whitehead and G. Donnay, Am. Mineral. **65**, 324 (1980).

[29] Lager, G. A., J. D. Jorgensen and F. J. Rotella, J. Appl. Phys. **53**, 6751 (1982).

[30] Newton, M. D. and G. V. Gibbs, Phys. Chem. Minerals **6**, 221 (1980).

[31] Newton, M. D., in *Structure and Bonding in Crystals*, Vol. 1, M. O'Keeffe and A. Navrotsky, Eds. (Academic Press, New York, 1981).

[32] Geisinger, K. L., G. V. Gibbs and M. Spackman (In preparation).

[33] Glidewell, C. and D. C. Liles, Acta Crystallogr. B **34**, 124 (1978).

[34] Karle, I. L., J. M. Karle and C. J. Nielsen, Acta Crystallogr. C **42**, 65 (1986).

[35] O'Keeffe, M., M. D. Newton and G. V. Gibbs, Phys. Chem. Minerals **6**, 305 (1980).

[36] Ross, N. L. and E. P. Meagher, Amer. Mineral. **69**, 1145 (1984).

[37] Levien, L. and C. T. Prewitt, Am. Mineral. **66**, 324 (1981).

[38] Warren, B. E. and R. L. Mozzi, J. Appl. Cryst. **2**, 164 (1969).

[39] Geisinger, K. L. and G. V. Gibbs, Phys. Chem. Minerals **7**, 204 (1981).

[40] Gibbs, G. V., E. P. Meagher, M. D. Newton and D. K. Swanson, in *Structure and Bonding in Crystals*, eds. M. O'Keeffe and A. Navrotsky (Academic Press, Ithaca, N.Y., 1981), 195-225.

[41] O'Keeffe, M. and G. V. Gibbs, J. Chem. Phys. 81 **2**, 876 (1984).

[42] Fischer, K., Zeits. fur Krist. **129**, 222 (1969).

[43] O'Keeffe, M. and G. V. Gibbs, J. Phys. Chem. **89**, 4574 (1985).

[44] Phillips, J. C., Solid State Phys. **37**, 93 (1982).

[45] Geisinger, K. L., G. V. Gibbs and A. Navrotsky, Phys. Chem. Minerals, **11**, 266 (1985).

[46] Baur, W. H., in *Structure and Bonding in Crystals*, Vol. II, eds. M. O'Keeffe and A. Navrotsky (Academic, New York, 1981) pp. 31-52.

[47] Navrotsky, A., K. L. Geisinger, P. McMillan and G. V. Gibbs, Phys. Chem. Minerals, **11**, 284 (1985).

[48] Gupta, A. and J. A. Tossell, Am. Mineral., **68**, 989 (1983).

[49] Zhang, Z. G., M. B. Boisen, Jr., L. W. Finger and G. V. Gibbs, Am. Mineral. **70**, 1238 (1985).

[50] Lindsay, C. G. and G. V. Gibbs, Abstract 98th Annual Meeting of the Geological Society of America, 1985.

[51] Bragg, W. L. and J. West, Proc. Roy. Soc. (London), **114A**, 450 (1927).

[52] Bloss, F. D., in *Crystallography and Crystal Chemistry: An Introduction* (Holt, Rinehart and Winston, Inc., New York, 1971).

[53] Slater, J. C., *Symmetry and Energy Bonds in Crystals* (Dover, New York, 1972).

A THEORETICAL STUDY OF SILANOL POLYMERIZATION

LARRY W. BURGGRAF AND LARRY P. DAVIS
Directorate of Chemical and Atmospheric Sciences, Air Force Office of
Scientific Research, Bolling Air Force Base, DC 20332

ABSTRACT

We have applied state-of-the-art semi-empirical molecular orbital
methods to a study of the anionic polymerization of silanols to form silica.
In particular, we have considered nucleophilic attack on silanols and
subsequent reactions of the products. Hydroxide addition proceeds without
activation to form five-coordinate silicate anions. Five-coordinate
structures can also be formed by oligomerization following the attack of
hydroxide on neutral silanols to abstract a proton. These five-coordinate
structures are predicted to play a key role as intermediates in the
polymerization process. Water can be eliminated from these anions, but with
a substantial activation barrier. The activation barrier appears to be lower
for the larger, more complex systems. These predictions are consistent with
a rapid pre-equilibrium to form dimer anions followed by the slower reaction
to form higher oligomers.

INTRODUCTION

Silicon chemistry is increasingly important in the synthesis of many
materials, including catalysts, semiconductors, polymers, ceramics, glasses
and composites. For each of these, an accurate model for predicting silicon
chemistry would be beneficial in tailoring these materials for specific
applications. In particular, a model of silanol polymerization to form
silica would be of great help in preparation of ceramic materials with unique
properties by sol gel methods.

This goal of applying accurate theoretical models to silicon chemistry
in general and silanol polymerization in particular calls for a model which
can handle large molecular systems in a reasonable amount of computation
time. It calls for a theoretical technique which can produce approximate
geometries for reactants, products, and transition states with accurate
energies (to within ten kcal mol-1). The technique of choice which can

perform these fairly accurate calculations in a timely manner is the MNDO (Modified Neglect of Diatomic Overlap) semi-empirical molecular orbital technique developed by Dewar, et. al. [1] This method has been parameterized for silicon [2], as has its earlier version MINDO/3 [3]. Recently, a reparameterization of silicon has resulted in greatly improved results [4].

There have been a number of theoretical studies involving silicon-containing molecules using these and other molecular orbital methods over the last few years [5-12], but all of these have involved fairly small molecules. We have recently compared MNDO calculations with high level ab initio calculations [8], and the conclusion that MNDO should be a useful tool to study larger silicon-containing molecular systems led to our current study of silanol polymerization.

Our ultimate goal in these calculations is to evaluate all possible steps in both anionic and cationic silanol polymerization mechanisms. This paper represents a major portion of our study of the anionic mechanism. In particular, we consider processes beginning with silicic acid ($Si(OH)_4$) and hydroxide ion which can generate a number of complex neutral and anionic structures. We propose a general mechanism for the polymerization based on these calculations.

CALCULATIONS

All calculations were performed with the MNDO method developed by Dewar and co-workers [1]. The MNDO method is a semi-empirical molecular orbital method based on a neglect of diatomic differential overlap (NDDO) scheme. It is parameterized by comparisons with experimental data in the form of heats of formation, molecular geometries, ionization potentials, and dipole moments for a basis group of molecules. The method, encompassed in the form of a computer program called MOPAC [13], is capable of optimizing geometries of stable molecules or transition states, or carrying out reactions along selected reaction coordinates. Options are also available to carry out force constant and thermodynamic calculations on specific geometries, as well as connect a transition state with its reactants and products by means of a calculation of the path of steepest descent. MNDO is now parameterized for all second row elements except lithium and neon, and some third row elements. The silicon parameters used in this study were taken from a recent reparameterization by Dewar's group at the University of Texas [4].

For each reaction studied, the geometries of reactants and products, if known, were completely geometrically optimized. Then, a reaction path calculation was performed by choosing some geometric variable (generally a distance in the molecule) and holding it at one of a series of selected values while the remainder of the geometric parameters were completely optimized. The result was an approximate minimum energy reaction path with the energy at each step along the way. The geometry along the path which appears to have the highest energy was used as the starting point for an optimization of the transition state geometry. The difference in energy between that of the optimized transition state geometry and the reactants' total energy gave the activation energy. Force constant calculations were then used to prove that this optimized point was indeed a transition state by the existence of one, and only one, negative eigenvalue of the Cartesian force constant matrix.

There are two known situations for which MNDO predicts large systematic errors that will affect the interpretation of the results of these calculations. First, MNDO overpredicts heats of formation of very small anions for which most of the charge resides on a single, small (first or second row) atom. This error is common for any molecular orbital method which does not include diffuse functions in the basis set. This type of error must be accounted for when a small, anionic nucleophile (such as hydroxide) is isolated during nucleophilic attack or elimination. Thus, the reactants for any reaction in which hydroxide ion is the nucleophile will be predicted by MNDO to be too unstable. We correct for this error by using the experimental gas-phase heat of formation for hydroxide ion when it is an isolated reactant or product.

The other major systematic error is that MNDO tends to overestimate core-core repulsions between atoms when they are separated by approximately van der Waals distances. The MNDO core-core repulsion function is appropriate for normal bond distances, but it decreases much too slowly as the distance between the atoms is increased. This error accounts for the failure of MNDO to reproduce hydrogen bonds and to accurately calculate heats of formation for strained or crowded molecules. It is also the major cause for gross overestimates of transition-state energies and distortions of transition-state geometries for highly exoergic exchange reactions [8]. This overprediction of activation energies for hydrogen exchange reactions is a nearly-constant 20 kcal mol^{-1} as compared to good ab initio calculations [8]. We must take this overprediction into account for our systems when they involve hydrogen exchange reactions, such as unimolecular eliminations of water from five-coordinate silicon anions.

RESULTS AND DISCUSSION

Because five-coordinate silicon anions appear to be an important class of intermediates in the anionic polymerization of silanols, we will first discuss some general trends in their stabilities. Following this discussion of stabilities, we will propose and give calculational results for the first few steps in anionic silanol polymerization to form silica.

Pentacoordinate Anion Stabilities

We have calculated a number of pentacoordinate silicon anions, and, as Dewar and Healy found earlier [5], we have found that a number of them are predicted to be stable species with respect to decomposition to any of the constituent ligands and the four-coordinate species that would remain after removal of the ligand [8,14,15]. The molecular geometry is almost invariably a trigonal bipyramid, with some slight distortion if the molecule is asymmetric because of different constituent ligands. This distortion can be fairly severe if hydrogen is one of the ligands because of its small size. Indeed, the single exception that we have found to the approximate trigonal bipyramidal structure for pentacoordinate silicon anions is that for H_4SiF^-, which, because of the small size of the hydrogens, may exist as a tetragonal pyramid with the fluorine at the apex of the pyramid. MNDO predicts this tetragonal pyramid to be a stable structure, but high level ab initio calculations predict that the tetragonal pyramid geometry is a transition state for Berry pseudorotation of the trigonal bipyramid [14]. Because of the overall stability of these five-coordinate species, we propose that they may play an important role in silicon reactions involving sufficiently nucleophilic anions, and, in particular, anionic silanol polymerization.

Tables I and II give MNDO stability predictions for adducts of nucleophilic anions with tetramethyl silane and trimethyl silanol, respectively. Note the general trend of decreasing stability of the pentacoordinate adduct with decreasing gas-phase proton affinity of the nucleophile. Also, as the adduct stability gets weaker, the Si-nucleophile bond length gets longer. For nucleophiles with proton affinities less than that of fluoride the pentacoordinate adduct is not predicted to be stable with respect to separated products. Thus, if geometry optimization is attempted in the calculation, the nucleophile separates from the adduct to a distance of several angstroms where a stable charge-dipole complex is formed. The values given for the heats of reaction to form unstable species in Table

Table I. Calculated heats of formation for adducts of anionic nucleophiles (Nu) and tetramethylsilane compared to the proton affinity and softness of the nucleophiles.

NUCLEOPHILE AFFINITIES FOR $Si(CH_3)_4$

REACTION:

$Nu^- + Si(CH_3)_4$

Nu^-	PROTON[1] AFFINITY	SOFTNESS[2]	ΔH°_A (KCAL/MOL)		ΔH°_E (KCAL/MOL)
CH_3^-	417	8.5	-14.8	[3]	-14.8
H^-	400	21	-17.4	[3]	-13.8
OH^-	391	0	-17.6	[3]	-18.9
OCH_3^-	380	3.5	-18.5	[3]	-17.3
F^-	371	-3.0	-8.1	[3]	-10.9
SCH_3^-	359	9.3	UNSTABLE (+3.7 AT 2.2 Å)		
SH^-	352	7.5	UNSTABLE (-3.8 AT 2.2 Å)		
CL^-	333	5.5	UNSTABLE (+4.0 AT 2.2 Å)		
BR^-	324	7.0	UNSTABLE (+7.2 AT 2.3 Å)		
I^-	314	8.3	UNSTABLE (+6.8 AT 2.5 Å)		

1. ΔH° FOR GAS-PHASE REACTION: $NuH_{(G)} \longrightarrow Nu^-_{(G)} + H^+_{(G)}$.

2. BASE "HARDNESS" SCALE OF R.G. PEARSON AND J. SONGSTAD BASED ON K_{EQ} FOR REACTION: $HNu_{(G)} + CH_3OH_{(G)} \rightleftharpoons CH_3Nu_{(G)} + H_2O_{(G)}$.

3. RESULT OBTAINED BY USING EXPERIMENTAL VALUE FOR HEAT OF FORMATION FOR Nu^-.

534

Table II. Calculated heats of formation for adducts of anionic nucleophiles (Nu) and trimethylsilanol compared to the proton affinity and softness of the nucleophiles.

NUCLEOPHILE AFFINITIES FOR $Si(CH_3)_3OH$

REACTION: $Nu^- + Si(CH_3)_3OH \longrightarrow$

Nu^-	PROTON AFFINITY[1] (KCAL/MOL)	SOFTNESS[2]	(1) $\Delta H°_1$[3] (KCAL/MOL)	(2) $\Delta H°_2$[3] (KCAL/MOL)	(3) $\Delta H°_3$[3] (KCAL/MOL)	(4) $\Delta H°_4$[3] (KCAL/MOL)
CH_3^-	417	8.5	-6.7	-5.4	-6.7	-5.4
H^-	400	21	-5.4	-7.7	-4.8	-0.4
OH^-	391	0	-12.1	-6.4	-9.8	-12.1
OCH_3^-	380	3.5	-11.8	-8.7	-7.7	-11.6
F^-	371	-3.0	-2.1	-3.5	-1.1	-5.8
SCH_3^-	359	9.3	——	UNSTABLE	——	——
SH^-	352	7.5	——	UNSTABLE	——	——
Cl^-	333	5.5	——	UNSTABLE	——	——
Br^-	324	7.0	——	UNSTABLE	——	——
I^-	314	8.3	——	UNSTABLE	——	——

1. $\Delta H°$ for gas-phase reaction: $NuH_{(g)} \longrightarrow Nu^-_{(g)} + H^+_{(g)}$.

2. Base "hardness" scale of R.G. Pearson and J. Songstad based on K_{EQ} for reaction: $HNu_{(g)} + CH_3OH_{(g)} \rightleftharpoons CH_3Nu_{(g)} + H_2O_{(g)}$.

3. Results obtained by using experimental values for heat of formation for Nu^-.

I are estimates for a hypothetical species with a typical silicon-nucleophile bond distance.

Both tables give the MNDO results for all possible isomers of the trigonal bipyramidal structures which may be formed, and the most stable one in each case is circled. We should caution, however, that preliminary studies in which we have compared MNDO predictions of isomer stability for silane anions to good ab initio results suggest that MNDO probably does not reliably predict which isomer of a trigonal bipyramidal system will be most stable [14]. The inclusion of the "softness" of the nucleophile (as defined by the logarithm of the equilibrium constant for the gas phase exchange reaction $HNu + CH_3OH \longrightarrow CH_3Nu + H_2O$ [16]) is an attempt to see if any correlation with this parameter and stability was evident. None was obvious.

Pentacoordinate Silicon in Silanol Polymerization

We now turn our attention to an assessment of the importance of pentacoordinate silicon in anionic silanol polymerization. Starting with silicic acid, silica can be prepared by silanol polymerization using either acidic or basic catalysis, although the basic catalysis route predominates at pH's greater than three [17]. In this paper we consider only the first few steps of this polymerization mechanism, and that only for anionic conditions. The model reaction that we consider is the condensation of two silicic acid monomers to form a dehydrated dimer bound by a siloxane bond:

$$2 \; Si(OH)_4 \longrightarrow (HO)_3\text{-}Si\text{-}O\text{-}Si\text{-}(OH)_3 + H_2O \qquad (1)$$

The overall enthalpy of reaction for this process is calculated to be -12 kcal mol^{-1}, a value consistent with aqueous phase estimates [17]. Elimination of water with the minimum expenditure of energy is the key to formation of the siloxane bond. Estimates of the aqueous phase enthalpy of activation for the condensation reaction are approximately 15 kcal mol^{-1} [17].

In order to investigate mechanisms for elimination of water, we first identified the neutral and anionic species which are predicted to be most stable. Figure 1 shows the neutral monomeric and dimeric structures that are predicted to be stable as well as the proton transfer and addition reactions that they may undergo. The heats of formation and reaction given in parentheses are for open-chain versions of the respective anions. Note that silicic acid and its oligomers in basic media can either act as an acid,

Figure 1. Proton transfer and addition reactions possibly involved in first steps of anionic silanol polymerization and their heats of reaction.

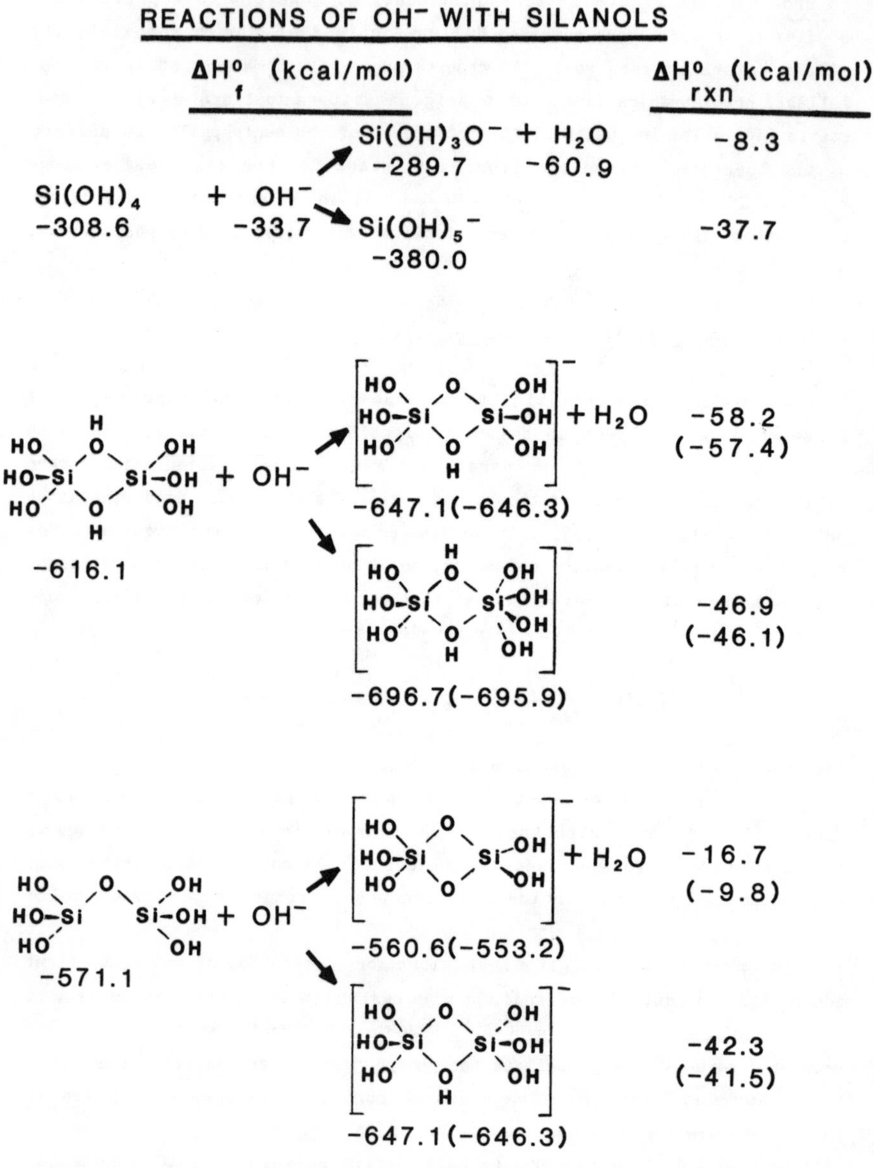

REACTIONS OF OH⁻ WITH SILANOLS

transferring a proton to hydroxide to form water and a silicate anion, or it can accept a hydroxide to form an expanded-coordination silicon anion. The unique chemistry of silicic acid is due to the competition between these reactive sites. Because of the stability of the five-coordinate anion $Si(OH)_5^-$, it appears that it will certainly be an important form of monomeric silicic acid under basic conditions, and it may be present even in mildly acidic conditions. Almost invariably the pentacoordinate anions predicted to be most stable are those with bridging oxygens as opposed to oxygen in a terminal position. We have calculated stable structures of the dimeric anions with terminal oxygens, and they have enthalpies of formation about 40 kcal mol^{-1} higher than the bridged oxide structures. The one exception is the anion product which results from water being removed when hydroxide ion attacks $(HO)_3-Si-O-Si-(OH)_3$ (the dehydrated dimer). In this case, the anion formed with a free oxide end is only about seven kcal mol^{-1} less stable than the anion with the bridging oxide. Thus, in general, the tendency to delocalize the charge over the whole structure rather than a single terminal oxygen strongly disfavors any mechanisms which result in species with terminal oxygen as intermediates.

The silicic acid monomer may dimerize to form a cyclic addition dimer involving two asymmetric bridging OH groups. This neutral pentacoordinate dimer has almost exactly the same stability as that of the two isolated monomers, and it forms essentially without activation. No stable singly-bridged dimer structure was found.

The heat of reaction for proton transfer from the dehydrated dimer is more exothermic than that for the monomer. This calculation agrees with the observation that larger silicic acid oligomers are more acidic than the monomer. The pK_a varies from 9.8 for the aqueous monomer to 6.5 for large oligomers [17].

Addition of hydroxide ion to either the silicic acid monomer or the dehydrated dimer species shown in Figure 1 occurs without activation, as does proton transfer to the hydroxide to form water. Proton transfer, however, could only be made to occur if the O-H--O angle (the last oxygen belonging to the incoming hydroxide) were constrained to remain at 180 degrees throughout the course of the reaction. Otherwise the hydroxyl oxygen falls under the influence of the silicon to form the more stable pentacoordinate anion. For the case of the cyclic dimer, proton transfer to the hydroxide to form water is thermodynamically favored over addition of the hydroxide, although once again both processes occur without activation. In this case addition of the hydroxide forms a hexacoordinate silicon anion, which is still predicted by MNDO to be stable, but not as stable relative to proton transfer products as is formation of the pentacoordinate silicon anions.

Figure 2 shows the outline of the mechanism for forming the first siloxane bond by attack of hydroxide on neutral species. Note that the last step in the mechanism involves regenerating the hydroxide ion catalyst. This step is very unfavorable thermodynamically and, even though the overall process is exothermic, it is not likely to occur in solution where excess reaction energy is likely to be dissipated to the solvent very quickly. The major conclusion of the results thus far is that large cyclic anions containing pentacoordinate silicon are formed quite easily in solutions with even catalytic amounts of hydroxide ions.

An alternative mechanism for forming siloxane bonds by water elimination could involve first forming an expanded-coordination (hydrated silicate) anion by addition of hydroxide and eventually an intramolecular elimination of water. Figure 3 shows the outline for this mechanism. The first two steps in the mechanism occur without activation to first form a monomeric anion and then a dimeric anion. The third step involves the opening of the cyclic dimer anion to form an open-chain dimer anion. This step can occur with only a slight activation energy, and then the open-chain anion can intramolecularly eliminate water to form a smaller dimer anion. This intramolecular elimination of water is predicted by MNDO to have an activation energy greater than 50 kcal mol^{-1} for all five-coordinate anions except the one shown in Figure 3. It is interesting to note, however, that the activation energy for the dimer anions is considerably less (by about 15 kcal mol^{-1}) than that of the monomeric anion. If this trend holds, we expect the activation energy for higher oligomers to be less than that of the dimers. This trend, coupled with the fact that MNDO typically overpredicts activation energies for this type of reaction by about 20 kcal mol^{-1}, indicates that intramolecular water elimination may play a major role in the overall mechanism as the polymerization process proceeds. We are currently investigating further activation energies for water elimination from the higher oligomers.

The process is still incomplete, however. The last step shown in Figure 3 is the same one as that of Figure 2; namely, elimination of hydroxide to regenerate the hydroxide catalyst. As before, this process is unlikely even though the overall reaction is downhill. Therefore it is more likely that the polymerization proceeds via this smaller cyclic anion attacking another monomer with another intramolecular water elimination. Thus the growing polymer continues to carry the negative charge until sometime late in the process. The energy to overcome the activation barrier for each water elimination is obtained from each addition of an anion to another neutral

Figure 2. Silanol dimerization mechanism based on formation of pentacoordinate neutral dimer.

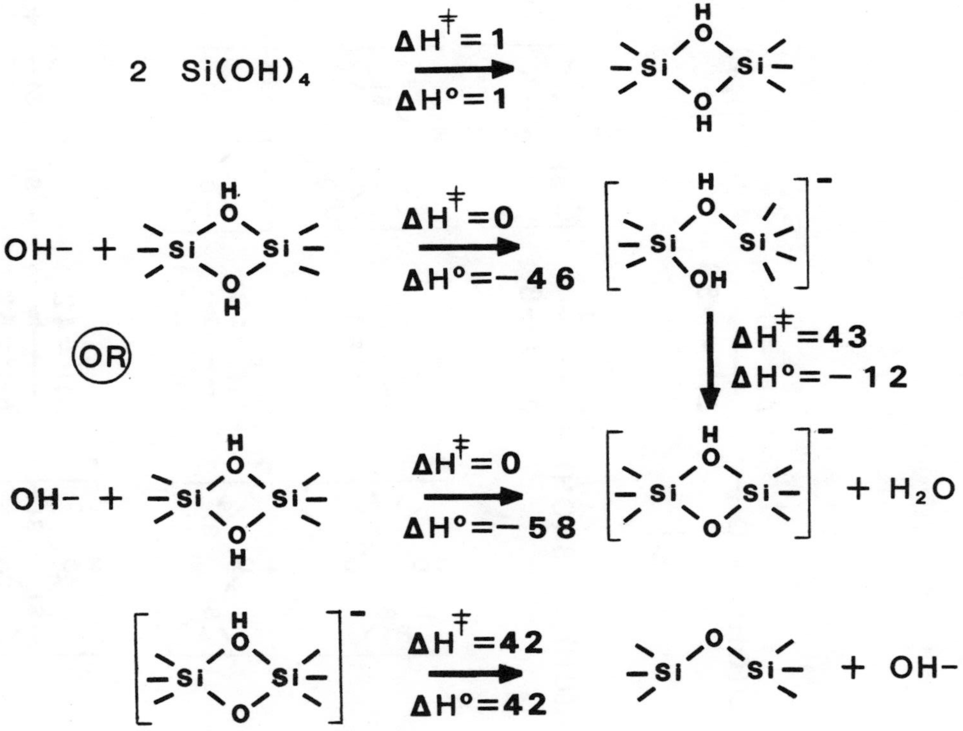

540

Figure 3. Silanol dimerization mechanism based on reactions of the pentacoordinate anion.

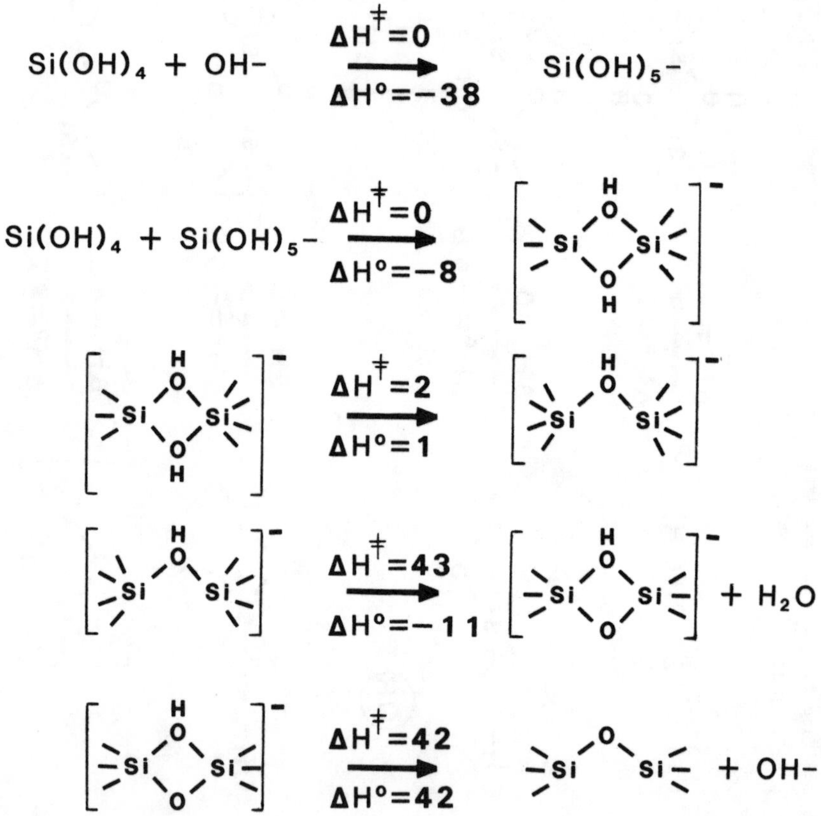

species.

Therefore we consider Figure 3 to represent the mechanism of polymerization in that the open chain pentacoordinate dimer really represents the active part of any of the oligomeric anions. As it eliminates water, it forms a new siloxane bond. Still under investigation is the sequence of steps by which the smaller cyclic dimer attack a silicic acid molecule and forms a new active site for water elimination. We expect the sequence of steps necessary for this process to occur with minimum activation; thus the overall activation energy for the process should be driven by the activation energy for the intramolecular water elimination, which we expect to be on the order of 20 kcal mol^{-1}. The mechanism produces kinetics which are first order in both monomer and anionic oligomer, since these two must add before the rate-determining water elimination can occur. In addition, the hydroxide is serving as a catalyst early in the reaction to form the initial anions. This combination pathway of Figure 2 to form the first siloxane bond with the anionic pathway of Figure 3 also nicely explains the rapid pre-equilibrium (monomer + monomer anion $\longrightarrow$ dimer anion) known to be involved in silanol polymerization.

CONCLUSIONS

MNDO is proving to be an extremely useful tool for studying silicon chemistry in large molecules, particularly if we recognize and deal with its limitations. It appears ideal for studying the first few steps of silanol polymerization, which cannot at present be attacked directly with ab initio methods because of the size of the species required to model the silanol polymerization.

Pentacoordinate silicon anions appear to be quite important in systems which contain small anions because of the ease of addition of these anions to tetrahedral silicon compounds. Thus, our calculations support these pentacoordinate silicon anions as the prime chain-carriers in anionic silanol polymerization. These intermediates are cyclic structures characterized by trivalent bridging oxygen atoms. Reactions of the predicted pentacoordinate neutral dimer are consistent with a rapid pre-equilibrium to form a dimer anion followed by a different and slower reaction mechanism to form higher oligomers.

REFERENCES

1. M.J.S. Dewar and W. Thiel, J. Am. Chem. Soc., 99, 4899 (1977).

2. M.J.S. Dewar, M.L. McKee, and H.S. Rzepa, J. Am. Chem. Soc., 100, 3697 (1977).

3. M.J.S. Dewar, D.H. Lo, and C.A. Ramsden, J. Am. Chem. Soc., 97, 1311 (1975).

4. M.J.S. Dewar, G.L. Grady, E.F. Healy, and J.J.P. Stewart, submitted for publication.

5. M.J.S. Dewar and E.F. Healy, Organometallics, 1, 1705 (1982).

6. W.S. Verwoerd, J. Comput. Chem., 3, 445 (1982).

7. M.J.S. Dewar (private communication).

8. L.P. Davis, L.W. Burggraf, M.S. Gordon, and K.K. Baldridge, J. Am. Chem. Soc., 107, 4415 (1985).

9. M.S. Gordon and C. George, J. Am. Chem. Soc., 106, 609 (1984).

10. M.S. Gordon, J. Am. Chem. Soc., 106, 4054 (1984).

11. M. O'Keeffe and G.V. Gibbs, J. Chem. Phys., 81, 876 (1984).

12. V. Brandemark and P.E.M. Siegbahn, Theoret. Chim. Acta (Berl.), 66, 233 (1984).

13. Quantum Chemistry Program Exchange Program Numbers 455 and 464, Department of Chemistry, Indiana University, Bloomington, Indiana 47405.

14. M.S. Gordon, L.P. Davis, L.W. Burggraf, and R. Damrauer, submitted for publication.

15. L.P. Davis, L.W. Burggraf, and M.S. Gordon, in preparation for publication.

16. R.G. Pearson and J. Songstad, J. Am. Chem. Soc., 89, 1827 (1967).

17. R.K. Iler, The Chemistry of Silica (John Wiley and Sons, New York, 1979), Chapter 3.

PART IX

Poster Session

CERAMIC MICROSTRUCTURE-PROPERTY FRACTAL-DIFFRACTAL CALCULUS :
STATIC SYNERGETICS YIELD OPTIMIZATION DURING PROCESSING VIA REAL-
-TIME Q.A.AND INTERACTIVE Q.C.

EDWARD SIEGEL
Static Synergetics Research Ltd.,183-14[th]Avenue,San Francisco,CA.
94118

ABSTRACT

Ceramic microstructure-property relationships dominate any
and all attempts at"better ceramics through chemistry". Required
is some universal calculus to allow analytic,universal,reversible
scalable computation of one from the other. Static Synergetics
universality-principle provides such a flexible versatile tool,
heretofore not available. It is a reexpression of the very basic
three laws of thermodynamics into $\underline{r}$-,$\underline{k}$-,and w-domains,the equivalence of(symmetry-breaking/defect) Pattern-recognition(the "st-
ructure") to signal-processing(the frequency-dependent FM) prop-
erties/Functions. More basically,it is a manifestation of Noeth-
er's theorem,the basis of mechanics and orgin of the energy continuity equation that is the thermodynamic first law. Such an algorithm,the"software"of yield optimization(quality and quantity)
real-time Q.A. and(in parallel with a specificity dominated process-model)interactive Q.C.requires as "hardware"produced input
the small-angle-scattering(SAS)dominated diffraction-pattern/sta-
tic structure factor $S_{SAS}(\underline{k})$ or Fourier transform $\underline{r}$-domain Patt-
ern/photomicrograph-recognition imaging(of processing-introduced
property-detrimental defects(heterogeneity heirarchy)). Output
are universal FM:1/f flicker(voltage and/or current)noise power
spectrum,signal-to-noise ratio over dynamic range,multi-level system dominated anomalous low temperature/frequency thermal,acous-
tic,...properties,and 1/f relaxation response susceptibility polarization catastrophe derived dielectric,electrical,optical,noi-
se,viscoelastic/mechanical,magnetic,...property Functions. Input
can be the ubiquitous,...universal Mandelbrot <u>fractals</u>,dominating
ceramics;output(and internal)Functions are Berry-Nye-Jakeman w-
domain <u>diffractals</u>,dominating all properties universally.How and
<u>Why</u> it works are detailed exactly;universality,reversibility and
scalability are analytically insured for self-similar(or self-
affine) fractal scaling-relation Pattern-recognition input;appr-
oximate... deviations from universal Functions output obtains
from less than perfect mathematically ideal <u>fractal</u> scaling-relation Pattern-recognition input.Static Synergetics provides a <u>new</u>
<u>practical use for external radiation small-angle-scattering(SAS)</u>
<u>diffraction-pattern/static structure factor measurements in cer-</u>
<u>amic material microstructure-property relationships during proce-</u>
<u>ssing</u>!

INTRODUCTION

Ceramic microstructure-property relationships are desirable
<u>during</u> ceramic processing especially,as well as post-processing.
Large numbers of papers at this"Better Ceramics Through Chemistry"
Symposium deal with measurements of external radiation small-ang-
le-scattering(SAS) diffraction-patterns/static structure factors
$S_{SAS}(\underline{k})$ which are implicitly dependent upon a huge range of processing parameters specific to particular ceramic materials and

processes. Summarizing this processing parameter set(temperature, pressure,...) which itself can vary from process to process and ceramic to ceramic by variable Z, implicitly $S_{SAS}(\underline{k})$ is some(usually unknown) function of processing parameter set $Z; S_{SAS}(\underline{k};Z)$. For a given specific ceramic and its process one might eventually develop the explicit functional dependence upon process parameter set Z of $S_{SAS}(\underline{k};Z)$ by development of a material and process specific process model filled with many details of the chemistry and physics of that specific ceramic being processed by that specific process. While all of this detail is of scientific interest,it does not necessarily present the ceramic processing and ceramic engineering user with a universal flexible versatile tool for evaluating ceramic microstructure-property relationships real-time to provide Q.A.,and perhaps in parallel with such a material -process specific process model,interactive real-time Q.C.to optimize yields. Rather ubiquitous in ceramic materials(among many others) is the concept of _fractal_ structures produced during processing. These may be purposeful or accidental defect fractal aggregates/agglomerates/clumps/clusters/... How may these be used to provide the desired,if not requisite,ceramic microstructure-property relationships?

Kadanoff[1] has recently raised a very closely related question:"Fractals:Where's the Physics?". Given that fractals are seen so ubiquitously in all of Nature,including in ceramic microstructures,what can be done with them? What can they be used for?

Static Synergetics[2] provides the answer in a most startling fashion. It provides the required and desired microstructure-property relationships for ceramics,as well as for many other materials, in a universal manner.

STATIC SYNERGETICS:WHAT IT IS,HOW IT WORKS,WHY IT WORKS

Static Synergetics[2] is a"new" paradigm of Nature:Patterns perform Functions(ie.jobs). If the Patterns are _fractal_,the Functions are universal! Static Synergetics is the equivalence of Pattern-recognition to signal-processing. Pattern-recognition is performed in configuration $\underline{r}$-domain,via a photomicrograph.Alternatively,it is performed in the Fourier transform $\underline{k}$-domain,via an external radiation diffraction-pattern/static structure factor $S(\underline{k})= \int g(\underline{r})e^{i\underline{k}\cdot\underline{r}}d\underline{r}$. If the Patterns in $\underline{r}$-domain they obey scaling relations of self-similarity $g(\% \underline{r})=g(r)/\%^{d-D}=g(r)/\%^A$ (d=geometric embedding Euclidean dimension,D=reduced fractal dimension,A=d-D=dimensional reduction). Their $\underline{k}$-domain Fourier transforms are Berry's[3] _diffractals_ ("waves that have encountered fractals"),obeying self-similar scaling relations $S(\underline{k}/\%)=\%^{d-D}S(k)=\%^A S(k)$. External radiation diffraction-patterns/static structure factors are by definition _diffractals_ if their $\underline{r}$-domain Patterns are _fractals_. _Fractal_ $\underline{r}$-domain $g(\underline{r})$ Patterns will exhibit a heterogeneity heirarchy of agglomeration/aggregation/clumping/clustering/...at large $\underline{r}$. _Diffractal_ $\underline{k}$-domain $S(\underline{k})$ diffraction-patterns will exhibit Fourier transform heirarchal nesting at small $\underline{k}$,so that $S(\underline{k})$ will be dominated by the small-angle-scattering (SAS) contribution $S_{SAS}(\underline{k})$. Static Synergetics 1:1 analytically maps $g(\underline{r})$ or Fourier transform $S(\underline{k})$ Pattern-recognition into w-domain signal-processing Functions(ie.w-dependent properties/jobs). If the $\underline{r}$-domain Patterns are _fractals_,the $\underline{k}$-domain diffraction-patterns $S_{SAS}(\underline{k})$ will be _diffractals_,and the resultant w-domain signal-processing Functions will be _universal_!

How it works is as a synthesis,collapsing to an identity for
fractal symmetry-breaking Patterns,of two classic universality-
principles of physics.

First,input external radiation diffraction-pattern $S(\underline{k})$ was
long ago shown by Brillouin[4],as generalized by Siegel[5],to modul-
ate/filter the collective-mode(phonon,magnon,...)dispersion relation of any and all systems,as a symmetry-breaking,whatever its
orgin,as
$$w(\underline{k}) = \hbar\ \underline{k}^2/\ 2\ m\ S(\underline{k}) \cong \underline{k}^2/S(\underline{k}) \qquad (1)$$
and termed the"generalized-disorder collective-boson(mode)mode-
softening universality-principle". Its meaning is that any symmetry-breaking disorder Pattern,through its $\underline{k}$-domain Fourier transform diffraction-pattern $S(\underline{k})$,produces depressed vibrational
frequencies(mode-softening) with negative-dispersion(negative
group velocity) of the collective-modes,giving rise to stop-bands
in which propagation is impossible. Brillouin's work is the classic basis for all of solid state physics and electrical engineering network design.

Second is the classic Wigner-Dyson,generalized by Handel[6]
and Ngai[7],infra-red divergence universality-principle,predicting
a universal linear density of states for collective-modes
$$N(w) = n\ .\ w \qquad (2)$$
with universal constant n. This is ostensibly a result of random
matrix physics[7]. Its consequence is that if one calculates the
excitation probability of these collective-modes
$$p(w) = \int_0^\infty N(w)\ /\ w^2\ dw \qquad (3)$$
for the linear density of states of the infra-red divergence universality-principle (2),a logarithmic divergence at the w=0 infrared limit obtains
$$p(w) = \int_0^\infty n\ .\ w/w^2\ dw = n\ \int_0^\infty 1/w\ dw \qquad (4)$$
which universally dominates $p(w)$ related Functions(properties)in
the w-domain.

Ostensibly two independent universality-principles (1) and(2)
synthesized together by substitution of (1) into (2) collapse into an identity if the $S(\underline{k})$ symmetry-breaking Pattern Fourier transform diffraction-pattern/static structure factor is off a frac-
tal $g(\underline{r})$,ie.is a diffractal $S_{SAS}(\underline{k})$,the small-angle-scattering
diffraction-pattern. As demonstrated in an accompanying paper in
this Conference in the Symposium on Electronic Packaging Materials Science, the self-similar(or anisotropic self-affine)scaling-
relation for fractal $g(\underline{r})$ and Fourier transform diffractal $S(\underline{k})=$
$S_{SAS}(\underline{k})$ propagates through all of the internal diffractal functions to the output Functions of Static Synergetics,forcing them
be universal in their w-dependence(FM).Static Synergetics obeys
scaling-relations making it either self-similar(isotropic) or
self-affine(anisotropic) with diffractal $S_{SAS}(\underline{k})$ input scattered
off fractal $g(\underline{r})$ symmetry-breaking Pattern,whatever its orgin!
This means that infra-red divergence universality-principle (2)
is a consequence of substitution of diffractal $S_{SAS}(\underline{k})$ into Brillouin universality-principle (1) as its modulating/filtering de-
nominator ;one merely substitutes this diffractal self-similar
(or self-affine) $w(\underline{k})$ into (2) obtaining a universal linear diffractal density of states $N(w)$,guaranteeing universal logarithmic
divergence of excitation probability (3) and (4). This is of critical importance for ceramic material properties/Functions.

Why it works is that it is a reexpression of the very basic three laws of thermodynamics in the $\underline{k}$- and w-domains. Classic Brillouin universality-principle (1) is a result of the continuity equation for energy, the first law of thermodynamics energy conservation. Actual reaching of infinity in the infra-red divergent lower limit of excitation probability (3) and (4) is precluded by the zero point energy(frequency) of the third law of thermodynamics; excitation probabilities and their derived (now universal)w-domain Functions(properties)infra-red diverge toward very large but finite values universally. Second law entropy extremization presumably forms the fractal Patterns universally.

STATIC SYNERGETICS:WHAT IT CAN BE USED FOR

Why should this seemingly abstruse physics be of interest to the ceramic materials processor? Firstly, at this Symposium as well as at many others, $\underline{fractals}$ rear their head ubiquitously, if not totally universally, in a huge variety of materials and processes. As Kadanoff so aptly puts it"Where's the physics?". More to the point of view of the ceramic(and polymer) processing materials scientist:Fractals, where's the (materials science) use?

The use for fractals is the reason for Static Synergetics, It can be used for/as a flexible versatile universal tool to calculate w-domain Functions(properties), $\underline{which\ are\ universal}$, consequences of $\underline{fractal}$ symmetry-breaking Patterns $g(\underline{r})$, as observed via experimental input external radiation ($\underline{diffractal}$) small-angle-scattering(SAS) diffraction-pattern/static structure factor $S_{SAS}(\underline{k})$. It provides a new use for these $S_{SAS}(\underline{k})$ measurements (such as those of the Sandia group) which is very practical. It provides a method for performing real-time quality-assurance(Q.A.) and, when used in parallel with a specific process model, interactive quality-control(Q.C.) $\underline{during}$ processing, provided a method of $S_{SAS}(\underline{k})$ measurement during processing is developed, applied and implimented to provide this real-time input into Static Synergetics, expressed as a mathematical algorithm, a universal $\underline{inexpert}$ $\underline{undedicated}$"software". Wuch is the consequence and power of the universality of w-domain Functions(properties)consequence of $g(\underline{r})$ being $\underline{fractal}$!

Jonscher[8] has shown that dielectric susceptibility, followed by Ngai for dielectric and viscoelastic/mechanical, is a universal Function $\mathcal{X}(w)\cong"1"/w^n$, where n is a constant. Long known as a mystery, recently explained by Handel[6], followed by Ngai[7], is the universal 1/f noise power spectrum $P(w)\cong"1"/w^n$, where n is a constant $n=1\pm10\%$ universally. Both Handel and Ngai have stressed that these two universal Functions, 1/f noise and 1/f polarization catastrophe, are identical, being related by the fluctuation-dissipation theorem of statistical mechanics. More recently, Handel has shown that for wide classes of materials, the "1" numerator of $"1"/w^n$ noise is universal as well (Hooge's constant).

Both Handel and Ngai find that the universal orgin of 1/f noise and 1/f susceptibility is the linear density of states of the infra-red divergence universality principle (2) which produces (3) the 1/w logarithmic divergence of excitation probability (4). However, we have seen that for $\underline{fractal}$ self-similar(or self-affine) symmetry-breaking Pattern $g(\underline{r})$, input as $\underline{diffractal}$ small-angle-scattering diffraction-pattern $S_{SAS}(\underline{k})$, the orgin of the infra-red divergence linear density of states (2) is the Brillouin symmetry-breaking modulation/filtering of dispersion-relation (1). So, universal susceptibility $\mathcal{X}(w)\cong"1"/w^n$ and 1/f noise power spectrum $P(w) \cong "1"/w^n$, and any and all derived Functions(properties)are explicitly computable as explicit functions of input external radiation small-angle-scattering(SAS)diffraction-Pattern $S_{SAS}(\underline{k})$. Static Synergetics predicts a universal <u>fractal</u> Pattern dominated critical exponent n

$$n = S_{SAS}^3(\underline{k};\theta)/\underline{k}^2 \cdot \left[2\underline{k}S_{SAS}(\underline{k};\theta)-\underline{k}^2 \, \partial S_{SAS}(\underline{k};\theta)/ \, \underline{k}\right] \qquad (5)$$

explicitly as a function of Fourier transform <u>diffractal</u> $S_{SAS}(\underline{k};\theta)$ where some implicit processing parameter set θ dependence of $S_{SAS}(\underline{k})$ via either empirical observation or some material-and process-specific process model.

Standard relations[8]between dielectric susceptibility now permit analytic computation of a large range of w-domain frequency-dependent dielectric and electrical Functions as further output of Static Synergetics algorithm,as explicit functions of $S_{SAS}(\underline{k};\theta)$ small-angle-scattering diffraction-pattern,and hence of $\underline{r}$-domain <u>fractal</u> symmetry-breaking Pattern. This Pattern $g(\underline{r})$ may be that of the <u>fractal</u> aggregates/agglomerates/clumps/clusters/... that make up some ceramics, or it may be that of <u>fractal</u> defect disorder,inadvertently processing-introduced and detrimental to desired designer specified Functions(properties),nested in some heterogeneity heirarchy that is self-similar(or self-affine). Since <u>fractal</u> Patterns driven by thermodynamic second law entropy extremization predict n=1.000...for the all important critical exponent, it would require infinite time of ceramic material formation for this to occur; in finite time processing, n will be nearly,but not exactly unity,mirroring the non-perfect nature of the would-be <u>fractal</u> Patterns which were prevented from evolution into perfect self-similar(or self-affine) scaling-relations.But still $n(S_{SAS}(\underline{k};\theta))$ of (5) can be computed,whether <u>diffractal</u> $S_{SAS}(\underline{k};\theta)$ is perfectly self-similar(or self-affine) or not.

Analytically computable as explicit functions of $S_{SAS}(\underline{k};\theta)$, as rapidly and thus as real-time as $S_{SAS}(\underline{k};\theta)$ can be measured are a large number of ceramic material designer specified Functions (properties). Static Synergetics universality-principle,expressed as a mathematical-algorithm"software",utilizing $S_{SAS}(\underline{k};\theta)$ input from diffraction measurement"hardware",functions as an experimental model that is universal rather than specific. It is a true <u>fractal-diffractal calculus</u>. Explicitly computable Functions(properties) of ceramics in the w-domain(frequency-dependent) are:

<u>Dielectric</u>:<u>constant</u>:$\epsilon(w;\theta)=\epsilon_0\{1 +(1+i)/w^n\}$;<u>dispersion</u>:$\epsilon'(w;\theta)=$ $\epsilon_0\{1+1/w^n\}$; <u>dissipation(loss)</u>: $\epsilon''(w;0)=\epsilon_0/w^n$; <u>loss tangent</u>: $\tan\delta(w;\theta)=\{1/w^n\}/\{1+1/w^n\}$; <u>response</u>: $\chi''(w;\theta)/\chi'(w;\theta)=\cot(n\pi/2)$ independent of w and θ; <u>relaxation response susceptibility</u>: $\chi'(w;\theta)$ $= 1/w^n$ and $\chi''(w;\theta)=1/w^n$; <u>A.C.electrical</u>:<u>conductivity</u>:$\sigma_{AC}(w;\theta)=$ $\sigma_{DC}(0;\theta)+\epsilon_0/w^n$; <u>capacitance</u>: $C(w;\theta)=(A/W)\,\epsilon_0\{1+(1+i)/w^n\}$; <u>impedance</u>: $Z(w;\theta)=R +1/iw(A\epsilon_0/W)\,i/w^n +\{1+1/w^n\}$; <u>conductance</u>: $G(w;\theta)$ $=(A/W)\epsilon_0/w^n$; <u>admittance</u>: $Y(w;\theta)=1/Z(w;\theta)$; <u>optical refractive index</u>: $N(w;\theta)+iK(w;\theta)= \epsilon^{1/2}(w;\theta) = \epsilon'(w;\theta)+i\epsilon''(w;\theta)$ (in terms of optical extinction coefficient K);<u>1/f flicker noise power spectrum</u>: $P(w;\theta)=1/w^n$; <u>1/f voltage flicker noise power spectrum</u>: $P_V(w;\theta)-(4k_BT/g)/w^n$;... In the above g=a geometric factor,A=material ceramic cross-sectional area,W=ceramic material thickness. Explicit more detailed expressions have been given twice before and must be omitted here for brevity.From the latter two one can compute as well A.C.(F.M.) dynamic signal-to-noise ratio $(S(w)/P(w;0)$. In all of the above expressions $n=n(S_{SAS}(\underline{k};0)$ is given by (5) as an explicit function of external radiation small-angle-scattering(SAS)diffraction-pattern input,itself an implicit fun-

550

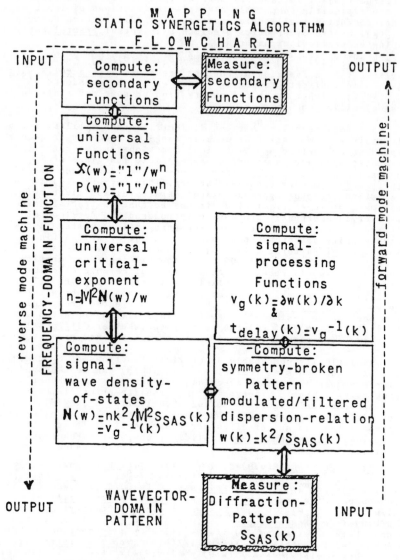

FIG. 1--STATIC SYNERGETICS UNIVERSALITY-PRINCIPLE EXPRESSED
AS A MATHEMATICAL ALGORITHM INEXPERT UNDEDICATED
"SOFTWARE" PACKAGE FOR Q.A./ Q.C. IMPLIMENTATION

ction of processing parameter set θ. Thus,Static Synergetics provides a method of real-time Q.A.;with $S_{SAS}(\underline{k};\theta)$ input as real-time as possible. This is a universal flexible versatile tool independent of the specificity of ceramic material mechanisms and process assumptions,and independent of processing type and details. With suitable process-model specific understanding of the explicit functional dependence of $S_{SAS}(\underline{k};\theta)$ on processing parameter set θ,rather than an implicit one,feedback of output computed Functions (properties),after comparison against some set of designer specified tolerances to ascertain acceptibility(Q.A.),back through the specific process-model can be used for interactive real-time Q.C. with the possibility of yield(quality and quantity)optimization <u>during</u> processing of the ceramic. It should be clearly understood that two distinct uses of Static Synergetics are proposed. In the $S_{SAS}(\underline{k};0)$ input is of the processing introduced <u>defect</u> <u>disorder</u> symmetry-breaking,then the output Functions are the <u>corrections</u> to the ceramic material bulk physical properties Functions for the <u>undefected</u> ceramic. However,if the basic ceramic material itself is composed of fractal aggregates/agglomerates/...,as input $S_{SAS}(\underline{k};\theta)$,the output properties Functions are those of the bulk ceramic material,not defect corrections. Figure 1 summarizes use.

CONCLUSION

Static Synergetics can provide a powerful new universal flex-[9]ible versatile tool in optimizing ceramic material <u>during</u> its processing. Functioning as an <u>in</u>expert <u>un</u>dedicated "software",it is made into an expert dedicated "software" by input of an external radiation small-angle-scattering(SAS) diffraction-pattern $S_{SAS}(\underline{k};\theta)$ which is an implicit function of processing parameter set θ. This permits real-time Q.A.<u>during</u> processing. With use of a process-model,specific feedback permitting interactive real-time Q.C. and with it yield optimization <u>during</u> processing. Application and implimentation depend upon suitable"hardware"application and implimentation of $S_{SAS}(\underline{k};\theta)$measurement.External radiation type should be chosen to avoid radiation damage of the ceramic,and yet of the correct spectral range to overlap the geometric resonance of expected fractal diameters and inter-aggregate distances. This will depend upon the need and ingenuity of the ceramic processor. But it provides a new practical use for the extensive small-angle-scattering measurements now being produced for ceramic materials and in ceramic materials processing,besides physical model development.

REFERENCES

1.L.Kadanoff,Physics Today,p.6(February,1986)
2.E.Siegel,Proc.Electrochem.Soc.83,8,497(1983);Test & Meas.World Expo,San Jose(1985);ASTM Symp.on Semiconductor Materials Processing,San Jose(1986);MRS Spring Mtg.,San Francisco(1985);Palo Alto (Symp.on Electronic Packaging Mtls)(1986);Intl.Conf.on Neutron Scattering,Santa Fe(1985);Intl.Conf.on Phonon Phys.,Budapest(1985)
3.M.V.Berry,J.Phys.A12,6,781(1979);A14,3101(1981)
4.L.Brillouin,<u>Wave Propagation in Periodic Structures</u>,Dover(1953)
5.E.Siegel,J.Noncryst.Sol.40,453(1980);Intl.Conf.on Lattice Dynamics,Paris(1977)-pub.by Flammarion & in J.de Physique(1978)
6.P.Handel,Phys.Rev.A22,2,745(1980)& many other seminal papers!
7.K.-L.Ngai,Comm.on Solid State Phys.,9,4,127(1979);9,5,141(1980); N.R.L.Memo#3917(1979)
8.A.K.Jonscher,Contemp.Phys.24,1,75(1983);in <u>Physics of Dielectric Solids</u>,IOP#58,I.O.P.(1979);Chelsea College Dielectrics Hdbk(1979)
9.E.Siegel,APS March Mtg.(1986);APS April Mtg.(1986)

CERAMICS FROM HYDRIDOPOLYSILAZANE

GARY E. LEGROW, THOMAS F. LIM, J. LIPOWITZ AND RONALD S. REAOCH
Advanced Ceramics Program, Dow Corning Corporation, Midland, MI 48686

INTRODUCTION

The Advanced Ceramics Based on Polymer Processing Program is sponsored by the Defense Advanced Research Projects Agency (Materials Science Division). The objectives of the program include development of a family of Si-C, Si-N, and Si-C-N ceramic fibers which may be used for reinforcement in ceramic, metal and plastic matrices, and development of economical process technology for fabricating ceramic matrix composites with high fracture toughness, high temperature performance and no inherent limitations to forming complex shapes. Attainment of these objectives would circumvent two limitations of existing ceramic material technology - lack of design reliability due largely to catastrophic failure related to brittleness, and inability to fabricate complex shapes [1].

The technical strategy which has been used in this program has involved an integrated multistep process scheme beginning with synthesis of a preceramic polymer. The polymer is converted into a fiber with minimal compositional change by melt spinning. This fiber (hereafter referred to as "uncured") is then crosslinked in the solid state, and pyrolyzed to produce the ceramic fiber. Identification of the relationships between ceramic fiber properties and all important variables in the process scheme is essential for maximization of the ceramic fiber properties. The final step is the use of the ceramic fiber in fabrication of ceramic matrix composites [2].

PRE-CERAMIC POLYMER SYNTHESIS AND CHARACTERIZATION

One pre-ceramic polymer, Hydridopolysilazane (HPZ), [3] is prepared by the mixing of trichlorosilane and hexamethyldisilazane. An exothermic reaction occurs rapidly when these two reagents are mixed wherein silicon-chlorine/silicon-nitrogen redistribution occurs as shown in (1) and (2).

$$HSiCl_3 + Me_3SiNHSiMe_3 \longrightarrow Me_3SiNHSiHCl_2 + Me_3SiCl \qquad (1)$$

$$Me_3SiNHSiHCl_2 + Me_3SiNHSiMe_3 \longrightarrow (Me_3SiNH)_2SiHCl + Me_3SiCl \qquad (2)$$

The mixed chlorine/nitrogen ligand species produced in (1) and (2) are thermodynamically favored species [4]. In addition, as the reaction temperature is raised above the boiling point of trimethylchlorosilane, the latter volatilizes out of the reaction vessel driving (1) and (2) to the right. Continuation of this type of redistribution reaction as shown in (3) results in the formation of a thermodynamically unfavored species [4]. Reaction (3) is both slow and readily reversible.

$$(Me_3SiNH)_2SiHCl + Me_3SiNHSiMe_3 \longleftrightarrow (Me_3SiNH)_3SiH + Me_3SiCl \qquad (3)$$

As the concentration of $(Me_3SiNH)_2SiHCl$ builds in the reaction medium, alternate, more favorable reactions occur. One reaction path leads to the formation of a tetrasilazane and hexamethyldisilazane as shown in (4).

$$2 \ (Me_3SiNH)_2SiHCl \ \longrightarrow \ Me_3SiNHSiHClNHClHSiNHSiMe_3 + Me_3SiNHSiMe_3 \qquad (4)$$

This reaction is favored over the reaction shown in (3) due to formation of a product which retains mixed chlorine/nitrogen ligands on both silicon atoms in the tetrasilazane; reduction in steric strain due to two trimethylsilyl groups per silicon atom in the reagent and only one trimethylsilyl group per silicon atom in the product; and formation of a disilazane which volatilizes from the reaction medium at elevated temperature.

Another reaction which occurs is formation of a trisilyl-substituted nitrogen species with ammonium chloride as a by-product as shown in (5). This reaction appears favorable at low temperatures, ie: below 100 degrees C, and is dependent upon the presence of chlorine in the system. As the temperature is. raised both trimethylchlorosilane and ammonium chloride are volatilized reducing the chlorine content of the system, minimizing this reaction.

$$2 \ (Me_3SiNH)_2SiHCl \ \longrightarrow \ Me_3SiNHSiHClN(Me_3Si)HSi(NHSiMe_3)_2 + HCl \qquad (5a)$$

$$3 \ HCl + Me_3SiNHSiMe_3 \ \longrightarrow \ NH_4Cl + 2 \ Me_3SiCl \qquad (5b)$$

As the polymerization reaction proceeds, the large bulky trimethylsilyl groups along the silazane chain promote cyclization. Thus cyclotetrasilazanes and cyclopentasilazanes should form readily. The multitude of cyclization and branching alternatives prevents identification of individual molecular structures in the growing pre-ceramic polymer. The concentration of bulky monofunctional trimethylsilyl groups relative to $HSiN_3$ branching groups is a primary factor which controls the molecular weight of the pre-ceramic polymer. The 3-dimensional structure of the $HSiN_3$ groups is far more important, controlling most of the physical properties of the pre-ceramic polymer.

The pre-ceramic polymer has been characterized by several techniques including total elemental analysis, proton nuclear magnetic resonance spectroscopy, infrared spectroscopy and gel permeation chromatography. Typical data are shown in Table I, and Figures I and IV.

SPINNING OF BULK POLYMER INTO UNCURED FIBER

Additional techniques which have been used to characterize the bulk polymer include thermomechanical analysis (TMA), thermogravimetric analysis (TGA) and rheology.. All are important properties relevant to spinning of fibers. The softening point of the polymer, Tg, must be well above ambient temperature to permit take up of the fiber on a spool at high speed without fiber coalescence. The rheology of the polymer (Fig. II) must show a steep temperature/viscosity slope which permits spinning at a small incremental temperature over the softening point. If the spinning temperature is excessive, thermal instability of the polymer may lead to irreproducibility in the spinning process. Finally, differences in TGA data in both air and an inert atmosphere (Fig. III) are used to determine the sensitivity of the polymer to oxygen and moisture. The polymer reacts with oxygen and moisture, especially at elevated temperatures. This leads to high oxygen content in the ceramic fiber, thus fiber spinning must be carried out in an inert atmosphere.

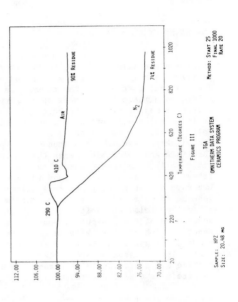

M_n = 3758
M_w = 15105
M_z = 37963
D = 4.01

8.40 7.77 7.13 6.49 5.86 5.22 4.59 3.95 3.31 2.68 2.04

BASELINE SUMMATION SUMMATION BASELINE

Figure 1
GPC of HP2 Polymer

TABLE 1

CHARACTERIZATION OF HYDRIDOPOLYSILAZANE POLYMER

- ELEMENTAL ANALYSIS: Si - 47.2; C - 23.0; N - 22.1; H - 7.8

- ^{1}H NMR SPECTROSCOPY: MeSi - BROAD DOUBLET AT 0.1, 0.2 PPM
  HSi - BROAD DOUBLET AT 4.70, 4.87 PPM
  NH - BROAD SINGLET AT 0.4-1.8 PPM

- GEL PERMEATION CHROMATOGRAPHY: M_n - 3758; M_w - 15105; M_z - 37963
  (SEE FIGURE 1)

- EMPIRICAL FORMULA
  (BASED ON ELEMENTAL
  ANALYSIS, NMR AND GPC): $(SiH)_{39.7}(Me_3Si)_{24.2}(NH)_{37.3}(N)_{22.6}$

- THERMOMECHANICAL ANALYSIS: Tg - 95 C

- THERMOGRAVIMETRIC ANALYSIS: CHAR YIELD (N_2) - 74%
  (SEE FIGURE 111)

- RHEOLOGY: 100 POISE AT 230 C
  (SEE FIGURE 11)

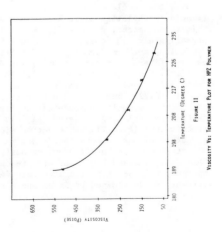

FIGURE II

VISCOSITY VS: TEMPERATURE PLOT FOR HP2 POLYMER

TGA
OMNITHERM DATA SYSTEM
CERAMICS PROGRAM

FIGURE III

SAMPLE: HP2
SIZE: 20.48 MG

METHOD: START 25
FINAL 1000
RATE 20

AIR   90% RESIDUE
N_2 74% RESIDUE
410 C
290 C

With proper draw down, fibers with diameters of 15-20 μm are readily produced.

CURING OF UNCURED FIBER

Up to this point in the process scheme every effort has been made to prepare a specific chemical and structural composition during the pre-ceramic polymer synthesis step and maintain that composition during the fiber spinning step. In the curing step the physical shape of the uncured fiber, on a macro level, must be preserved; however, on a micro level, the molecular weight of the pre-ceramic polymer must be increased to an infinite value via intermolecular crosslinking. To accomplish this in the solid state, narrow limits on process conditions and the nature of the crosslinking reactions are imposed. A properly crosslinked fiber is both infusible and insoluble. When this is accomplished, pyrolysis of the cured fiber to 1200 degrees C can be carried out without deformation of the fiber. HPZ pre-ceramic polymer is particularly suited for facile cure. Exposure of the green fiber to a multifunctional chlorosilane of the general formula $RSiCl_3$ at a temperature above the boiling point of the silane but below the softening point of the polymer permits rapid adsorption of the gaseous silane onto the surface of the fiber and chemical reaction with reactive sites as depicted in (6). If R=H in $RSiCl_3$, removal of Me_3Si groups is quite rapid resulting in formation of a cured fiber containing a lesser amount of carbon. A precursor to a mixed silicon carbide and silicon nitride ceramic is produced. The ceramic is referred to herein as a Si-C-N ceramic.

$$3 \text{ } Me_3SiNHSiH= \text{ } + \text{ } RSiCl_3 \longrightarrow \text{ } RSi(NHSiH=)_3 + 3 \text{ } Me_3SiCl \tag{6}$$

PYROLYSIS

The final step in the process is conversion of the cured fiber into a ceramic fiber. This is accomplished by heating the cured fiber under nitrogen to 1200 degrees C. During this pyrolysis reaction, residual volatiles are removed from the fiber, homolytic scission of some Si-C bonds occurs, and complete loss of hydrogen occurs. Pyrolysis gases include ammonia, methylsilanes, methane and hydrogen. Not only does significant weight loss occur, ca: 30-35%, but an increase in density from approximately 1.0 to 2.3 g/cm^3 occurs, leading to large volume shrinkage. Typically, a 20 μm diameter cured fiber is converted into a 15 μm diameter ceramic fiber. A typical composition of the ceramic fiber is Si = 60.0%, C = 2.3%, N = 32.6%, and O = 2.2%. This corresponds to a rule-of-mixture composition (mole ratio basis) of SiO_2 = 0.12, Si_3N_4 = 1.00, SiC = 0.33, and Si = 0.22. The ceramic composition is not truly a mixture of the above components since it is amorphous by X-ray diffraction. A variety of techniques including infrared (Fig. IV), Raman, ESCA/Auger and ^{29}Si NMR spectroscopy (Fig. V), and electron microscopy, show that the chemical structure of the ceramic approaches random bonding to silicon. That is, silicon atoms are bonded simultaneously to nitrogen, carbon, and oxygen. Only SiN, SiC, and SiO bonds are present. Excess carbon is present in a microcrystalline graphitic structure very similar to that of pyrolytic carbons, as determined by Raman spectroscopy.

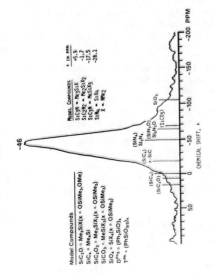

Model Compounds

$SiC_3O = Me_3SiX(x = OSiMe_3, OMe)$
$SiC_4 = Me_4Si$
$SiC_2O_2 = Me_2SiX_2(x = OSiMe_3)$
$SiCO_3 = MeSiX_3(x = OSiMe_3)$
$SiO_4 = SiX_4(x = OSiMe_3)$
$D^{Ph_2} = (Ph_2SiO)_x$
$T^{Ph} = (PhSiO_{3/2})_x$

Model Compounds	δ IN PPM
$SiC_3N = Me_3SiX$	+5.9
$SiC_2N_2 = Me_2SiX_2$	-1.7
$SiCN_3 = MeSiX_3$	-17.5
$SiN_4 = SiX_4$	-28.1
$X = NMe_2$	

FIGURE V
^{29}Si MAS-NMR SPECTRUM OF POWDERED
HP2 CERAMIC FIBER

TABLE II

COMPOSITION OF HP2 POLYMER AND ITS DERIVED CERAMICS

SAMPLE	% Si	% C	% N	% H	% Cl	% O
HP2 BULK POLYMER	46	23	22	6	-	-
$HSiCl_3$ CURED FIBER	35	8	18	4	26	-
UNCURED BULK CERAMIC	59	11	27	0	-	-
$HSiCl_3$ CURED CERAMIC FIBER	60	2.3	32.6	0	0	2.2

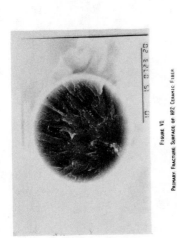

FIGURE VI
PRIMARY FRACTURE SURFACE OF HP2 CERAMIC FIBER

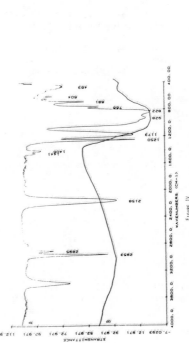

FIGURE IV
INFRARED SPECTRA OF (A) HP2 POLYMER, (B) HP2 DERIVED CERAMIC

Various techniques, including nitrogen desorption and density measurements show that the volume fraction of porosity is approximately 0.3. This porosity results from the rapid loss of a large volume of gases during pyrolysis. Such a high volume fraction of porosity will lead to a lower Young's modulus and a loss in tensile strength as compared to a fully dense ceramic [5]. These fibers exhibit classical tensile fracture behavior of brittle materials (Fig. VI) and follow the Griffith equation for tensile strength [6]. Predominant strength-limiting critical flaws are interior granular defect regions and spinning-induced or mechanical damage-induced surface defects. Ceramic fibers have been produced with up to 450 ksi tensile strength and 30 Msi Young's modulus. Modulus and especially tensile strength of the fibers continue to increase as better control of each of the process steps and their overall integration is achieved [7].

REFERENCES

1. Kenneth J. Wynne and Roy W. Rice, Ann. Rev. Mater. Sci. 14, 297-334 (1984).

2. Ronald H. Baney, Polymer Preprints, 25, 1-3, (1984).

3. John P. Cannady, US Patent No. 4 535 007 (13 August 1985).

4. Donald R. Weyenburg, Louis G. Mahone and William H. Atwell, Ann. NY Acad. Sci. 159, 38-55, (1969).

5. D. C. Larsen and J. W. Adams, "Property Screening and Evaluation of Ceramic Turbine Materials", AF Contract F33615-79-C-5100 (1983).

6. C. T. Li, Dow Corning Corporation, unpublished work.

7. M. I. Haider, Celanese Research Company, unpublished work.

Si-O-N CERAMICS FROM ORGANOSILICON POLYMERS

YUAN-FU YU AND TAI-IL MAH
Universal Energy Systems, Inc., 4401 Dayton-Xenia Road, Dayton, OH 45432

ABSTRACT

The preparation of new, hybrid pre-ceramic organosilicon polymers are described. These hybrid polymers were prepared by reactions of polysilazanes and polysiloxanes. Ceramic materials containing Si-O-N were obtained with high yields ($\sim$ 80%) by pyrolyzing these new polymers at low temperatures (< 800°C). A wide range of chemical compositions and properties can be obtained by using different ratios of polysilazanes and polysiloxanes. Microstructural characterization and densification studies of these ceramic materials are presented.

INTRODUCTION

In recent years, great interest has developed in silicon-containing pre-ceramic polymers whose pyrolysis provides silicon containing ceramics such as silicon carbide, silicon nitride, silicon carbonitride, silicon oxynitride, etc. [1]. The use of these organosilicon polymers provides a unique approach for preparation of numerous ceramic materials in all their useful forms (powders, coatings, fibers, foams, or monoliths) [2].

In the present work, new organosilicon polymers were prepared by reacting easily prepared polysilazane and Si-H containing polysiloxanes. Pyrolysis of these new hybrid organosilicon polymers under flowing NH_3 gas produced high purity silicon oxynitride Si_2ON_2 in good yield. Densification studies on the pyrolyzed powder were performed using the hot-press. The dense monolithic bodies were characterized using x-ray diffraction (XRD) analysis and scanning electron microscopy (SEM).

BACKGROUND

Silicon oxynitride is a good refractory material of highly desirable properties for the fabrication of parts such as nozzles, turbine blades, or other structural elements subjected to high temperatures [3,4]. Si_2ON_2 is stable up to 1550°C in an inert atmosphere. When exposed to air at temperatures in the range of 1400 to 1750°C, Si_2ON_2 has an oxidation resistance superior to that of silicon nitride.

Despite the close structural simulations between Si_3N_4 and Si_2ON_2, ceramics based on Si_2ON_2 have received little attention. The lack of an economical and reproducible method to produce pure Si_2ON_2 is the reason for this. Washburn [5] has shown that the nitridation of a mixture of silica and elemental silicon in a controlled atmosphere containing nitrogen and oxygen with metal oxide (CaO, BaO, MgO, etc.) as the catalyst will yield Si_2ON_2.

$$3Si + SiO_2 + 2N_2 \xrightarrow[1450°C]{Catalyst} 2Si_2ON_2 \qquad (1)$$

However, in addition to the formation of Si_2ON_2 ($\sim$ 75%), a considerable amount of silicon nitride ($\sim$ 15%) was formed and some unreacted SiO_2 was also found.

HYBRID POLYMER SYNTHESIS [6]

Experiments were carried out with polysiloxane $[CH_3Si(H)O]_m$ prepared using conditions under which the yield of the cyclic oligomers (m = 4,5,6) is maximized

$$CH_3SiHCl_2 + H_2O \xrightarrow{CH_2Cl_2} [CH_3Si(H)O]_m + 2HCl \qquad (2)$$

The ammnolysis of CH_3SiHCl_2 in THF produced a mobile oil in higher than 80% yield. Spectroscopic data suggested that this colorless oil had a mostly cyclic $(CH_3SiHNH)_n$ structure [7].

$$CH_3SiHCl_2 + NH_3 \xrightarrow[0°C]{THF} [CH_3SiHNH]_n + NH_4Cl \qquad (3)$$

In one approach, the "graft" method, a reactive polymeric alkali metal silylamide was prepared by using $[CH_3SiHNH]_n$ and catalytic quantities of KH in THF. To this living polymer solution the $[CH_3Si(H)O]_m$ oligomers were added slowly and the active metal species in the solution were quenched with CH_3I. The average molecular weight of the hybrid polymer prepared from 1:1 weight ratio of silazane to siloxane is $\sim$ 1700 g/mol, much higher than its starting materials ($\sim$ 300 g/mol for both silazane and siloxane). This polymer has excellent solubility in hexane, benzene, and THF.

$$(CH_3SiHNH)_n \xrightarrow[THF]{KH} [(CH_3SiHNH)_a(CH_3SiN)_b(CH_3SiHNK)_c]_n,$$
$$\downarrow [CH_3Si(H)O]_m$$
$$\downarrow CH_3I$$

"Graft" Hybrid Polymer   (4)

Using an alternative synthesis method of $[CH_3SiHNH]/[CH_3Si(H)O]$ combined polymers, the polysilylamide was generated "in-situ" in the presence of $[CH_3Si(H)O]_m$.

$$[CH_3SiHNH]_n + [CH_3Si(H)O]_m \xrightarrow[THF]{KH} \xrightarrow{CH_3I} \text{"In-Situ" Polymer} \qquad (5)$$

The combined polymer prepared by the "in-situ" and "graft" polymer methods differed in some ways. The TGA curves of the "graft" polymer and the "in-situ" polymer differ and the ceramic products, when pyrolyzed under a stream of NH_3 gas, are different as well. Properties of some of the hybrid polymers synthesized are summarized in Table I.

CONVERSION OF POLYMER TO CERAMIC

When polymers with organic substituents on the Si atom are used to prepare ceramic materials, the products generally will contain carbon and silicon carbide if they are pyrolyzed in vacuum or inert atmospheres. However, when the pyrolysis is carried out in a stream of ammonia, the ceramic remains usually contains less than 0.5% carbon. One example of this process is pyrolyzed SiO_2-containing polycarbosilane fibers (cured precursor fiber of NICALON) under NH_3 to give silicon oxynitride fibers [8].

TABLE I. Properties of Hybrid Polymers

Hybrid Polymer	Yield(%)	Weight Ratio of Siloxane : Silazane	Synthesis Method	M.W.[a] (g/mol)	Appearance[b]
I	88	1:1	In-Situ	1670	White Solid
II	83	1:1	Graft	1700	White Solid
III	69	5:1	In-Situ	760	White Solid
IV	80	5:1	Graft	2400	White Solid

[a]Cryoscopic in Benzene Solution

[b]All four polymers are very soluble in hexane, benzene, and THF.

Pyrolysis of the hybrid polymers (Table I) to 1000°C under inert atmosphere gave black ceramics with good yields (Table II). More significantly, we have found that pyrolysis of these hybrid polymers, heated to 800°C in gaseous ammonia, provided white solids in high yields. Analysis confirmed that these white solids are silicon oxynitrides. These white ceramics contain little if any, carbon. Some preliminary weight loss versus temperature experiments, indicate the majority of reactions between polymer and NH_3 gas are most likely to occur around 600 to 700°C. It is suggested that at higher temperatures (> 400°C), the NH_3 molecules effect nucleophilic cleavage of the Si-C bonds present in the polymer and the methyl groups are lost as CH_4.

TABLE II. Conversion of Hybrid Polymers to Ceramic

Hybrid Polymers	TGA[a] Yield(%)	Ceramic[b] Yield(%) NH_3	Composition[c]	Carbon[c] (%)	Density[c] (g cm^{-3})
I	84	86	Si_2ON_2	0.15	2.48
II	78	78.5	Si_2ON_2 + β-Si_3N_4	0.48	2.66
III	61	82	Si_2ON_2 + SiO_2	0.13	Porous

[a]In argon, 10°C/min., 1000°C

[b]In NH_3, 100°C/hr., 800°C

[c]Hot-pressed sample, 1700°C

A ceramic pellet was prepared by charging the fine hybrid polymer powder I into a 3/4 in. in diameter stainless-steel die. The sample was pyrolyzed in gaseous ammonia, heated to 800°C, and held at that temperature for one hour. The product was a coherent, round white ceramic pellet which had not cracked or bloated. The sample lost 14% of its original weight and the average diameter reduced to 9/16 in. The SEM micrograph of the fracture surface shows both dense and powdery agglomerated regions. The specimen was then heat-treated in argon at 1300°C for two hours. The SEM micrograph of the fracture surface is shown in Fig. 1. The picture shows that a certain degree of sintering had taken place during the 1300°C heat-treatment, however, it appears that a much higher temperature is needed to achieve appreciable densification. Hot-pressing was used to further heat-treat the material at higher temperatures and to obtain a

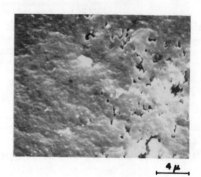

Figure 1. Sem Micrograph of Hybrid Polymer I Derived Ceramic at 1300°C in Argon.

Figure 2. SEM Micrograph of Hybrid Polymer I Derived Ceramic at 1700°C Hot-Pressed.

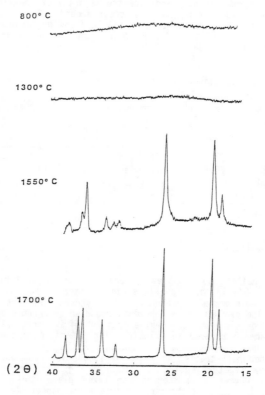

Figure 3. XRD Patterns of Hybrid Polymer Derived Si_2ON_2 at Different Temperatures.

slightly and loaded into a 1-in. in diameter graphite die. The final hot-pressing conditions were 1700°C, 2000 psi for 15 min. in nitrogen atmosphere. Figure 2 is a SEM micrograph of the fracture surface of the hot-pressed dense pellet. The micrographs revealed two microstructural features; (a) a very dense region, and (b) porosity nests which are associated with fine Si_2ON_2 grains and interconnecting pores. The average grain size appeared to be slightly less than one micrometer. The uniformity of the grain size implies homogeneous nucleation and growth process during crystallization from the amorphous state (800°C pyrolysis). The crystallization process of the hybrid polymer I derived silicon oxynitride obtained at 800 to 1700°C in NH_3 gas was examined by the XRD technique with monochromated CuKα radiation (result shown in Fig. 3). At lower temperatures, 800 and 1300°C, XRD patterns of the pyrolyzed samples shows only background diffraction, suggesting a complete amorphorus nature of the samples. This amorphorus character to XRD is not uncommon for ceramic materials derived from organosilicon polymers pyrolysis. The crystallization of Si_2ON_2 from the pyrolyzed specimen took place between 1300 and 1550°C. XRD patterns of the 1550 and 1700°C samples show strong peaks which can be easily assigned to silicon oxynitride diffractions.

Similar ceramic body fabrication experiments were carried out using the fine "graft" hybrid polymer II powder. A slightly brown and very dense ceramic pellet was obtained by hot-pressing the white ceramic powder at 1700°C, 2000 psi for 15 min. in nitrogen atmosphere. The surface XRD pattern of the pellet derived from polymer II, showed Si_2ON_2 diffractions and β-Si_3N_4 diffractions as the minor phase. This result indicates that even though hybrid polymers I and II were prepared from identical starting materials and pyrolyzed under the same condition, the ceramic products are not the same. In terms of Si_2ON_2 preparation, the "In-Situ" hybrid polymer I is a better pre-ceramic polymer than the "graft" hybrid polymer II. The surface XRD pattern of the hot-pressed pellet derive from the siloxane reached hybrid polymer III exhibited Si_2ON_2 diffractions and the expected SiO_2 broad peaks.

CONCLUSION

In this paper we have shown that pure silicon oxynitride can be prepared by pyrolysis of new hybrid organosilicon polymers in gaseous ammonia. The major advantage of this polymer pyrolysis route over conventional methods is that stoichiometric Si_2ON_2 can be obtained at low temperatures with high-purity and the absence of minor phases. Another advantage of this hybrid polymer system is the utilization of readily available and relatively cheap polysiloxane and polysilazane. The third advantage of this system is that the composition of hybrid polymers, as well as ceramic products, can be easily modified by adjusting the ratio of polysiloxane and polysilazane.

564

ACKNOWLEDGEMENTS

This work was sponsored by the Air Force Office of Scientific Research (AFOSR) under Contract No. F49620-85-C-0118. The hybrid polymers used in this research were developed at MIT by Y. F. Yu with Professor D. Seyferth.

REFERENCES

1. D. R. Ulrich and L. L. Hench, Ultrastructure Processing of Ceramics, Glasses and Composites, (Wiley Interscience, NY, 1984) p. 5.

2. R. W. Rice, Bull. J. Am. Ceram. Soc. 62, 889, (1983).

3. "Engineering Property Data on Selected Ceramic," Vol. 1, Nitrides, MCIC-HB-07, Battelle Laboratories, Columbus, Ohio.

4. M. E. Washburn, Bull. J. Am. Ceram. Soc. 46, 667 (1967).

5. M. E. Washburn, U.S. Patent No. 3356513 (5 December 1967), U.S. Patent No. 3639101 (1 February 1972).

6. D. Seyferth and Y. F. Yu, and T. S. Targo, U.S. Patent Pending.

7. D. Seyferth and G. W. Wiseman, U.S. Patent No. 4482669 (13 November 1984).

8. K. Okamura, M. Sato, Y. Hasegawa, and T. Amano, Chem. Lett. 2059 (1984).

SYNTHESIS AND CHARACTERIZATION OF ALUMINUM PROPIONATE SOL-GEL DERIVED Al_2O_3

J. COVINO AND R. A. NISSAN
Research Department, Naval Weapons Center, China Lake, CA 93555-6001

ABSTRACT

Al_2O_3 has been synthesized from aluminum propionate [$Al(CO_2CH_2CH_3)_3$] by Sol-Gel techniques. Methods of characterization include: x-ray powder diffraction, scanning electron microscopy (SEM) and nuclear magnetic resonance (NMR). Solution and solid state ^{13}C and ^{27}Al NMR data is coupled with both x-ray diffraction powder data and SEM in order to understand the transition from sol → gel → amorphous powder → crystalline powder in the sonicated and unsonicated aluminum propionate precursor, and Al_2O_3 powders. Results indicate that the sonicated dried powder is crystalline, having the γAl_2O_3 structure while the unsonicated dried powder is amorphous and crystallizes at 800°C with the αAl_2O_3 structure. These differences in crystallinity are further substantiated by solid state ^{27}Al NMR chemical shifts and line widths at half height.

INTRODUCTION

Sol-gel technology has been extensively utilized for processing nuclear fuel pellets and powders. Currently, the direct firing of gels is being explored to produce ceramics without the use of any intervening powder steps as an extension of Yoldas work on glasses and polycrystalline oxides[1-3].

This is motivated by (1) the high purity and homogeneity available in solids; (2) the potential ability in a viscous liquid to minimize the sources of defects; (3) the ability to visually examine many gel products for defects after drying; and (4) the shaping potential offered by a "plastic" gel. Furthermore, much lower temperatures can be used to fire gels to a fully dense ceramic (e.g., ThO_2[4]) than are required for conventional powder processed bodies.

Key problems in the direct firing of gels are the determination of (1) how to control the large shrinkages involved in the gellation of sols, (2) the size and shape limitations necessary to avoid cracking or distortion, and (3) the range of compositions and materials to which direct gel processing can be applied.

The sol-gel process is now a well-accepted technique for preparing monolithic glass articles without melting. This technique has been applied not only to the preparation of single component oxide glasses, for example, SiO_2[5-7], but also more recently to the preparation of multicomponent oxide[8] and oxynitride[9] glasses. For processes that use metal alkoxides or metal organics as glass precursors, monolithic glass formation consists of (1) growth and linkage of polymer units to form a gel, (2) desiccation of the gel under ambient or hypercritical conditions to form, respectively, a porous xerogel or aerogel, and (3) heat treatment of the porous gel at a temperature sufficiently high to convert it to a dense, glass-like solid.

The purpose of this paper is to describe a technique for producing Al_2O_3 powders via a sol-gel process starting from the aluminum propionate. The mechanistic aspects of the sol → gel → amorphous powder → crystalline powder transitions will allow for the systematic synthesis of particular crystalline forms under reduced temperatures and pressures. Our approach has been to prepare gels of aluminum propionate under acidic conditions and to dehydrate these materials at temperatures of 600°C and 800°C. The effect of pretreatment of the sol phase with sonication was also investigated.

Our methods of characterization included x-ray powder diffraction scanning electron microscopy and nuclear magnetic resonance (NMR). While it was expected that NMR data alone would not be sensitive to long range order (crystallinity) it was hoped that observation of the entire transition from sol → gel → amorphous powder → crystalline powder for the aluminum propionate system by ^{13}C and [27]Al would lead to some understanding of the basic process. This data would be coupled with both x-ray diffraction powder patterns and scanning electron microscopy in order to understand the sol-gel process of Al_2O_3 from aluminum propionate.

EXPERIMENTAL

Sample Preparation

One gram of $Al(CO_2CH_2CH_3)_3$ was mixed with 26 ml of distilled water. To this mixture 20ml of concentrated HCl was added. To this solution 10 ml of methanol was also added and a pH of 0.7 was measured. This solution was heated in a H_2O bath to 40-50°C for 6 hours and left to gel for three days.

Sonication of a second solution prepared identical to the above described method, was performed at 4.4 amp D.C. for 10 minutes. The temperature of this solution ranged from 34-60°C during the sonication process. This solution was also left to gel over the weekend.

The sonicated sample never gelled even after numerous repeated attempts to prepare the solution. Crystallization took place and the product was characterized. On the other hand, the non-sonicated sample did gel after approximately 3 days. This sol → gel transition was also characterized. Following the gel formation, the aluminum gel was dehydrated to a powder at temperatures of 600°C and 800°C. These two powders were also characterized.

Sample Characterization

X-ray analysis. Samples prepared by the above procedures have been analyzed by x-ray powder diffraction. Diffractometer scans were taken on a Phillips diffractometer with a θ-compensating slit, diffracted beam monocromator, scintillator with pulse-height discrimination, and a copper source (CuKα = 1.5418Å, Kα$_1$ = 1.5405Å, Kα$_2$ = 1.5444Å).

Scanning Electron Microscopy (SEM). Scanning electron microscopy was employed in order to measure particle size distribution on the powder samples. Scanning electron micrographs were taken on an Amray 1400 with 40Å lateral resolution electron microscope.

Nuclear Magnetic Resonance (NMR). All NMR spectra were recorded with the use of an NT-200-WB spectrometer operating at 50 MHz for [13]C. NMR spectra of solutions and gels under acidic aqueous conditions were acquired with inverse gated broad band 'H decoupling (to reduce sample heating). 3-(trimethylsilyl)-1-propane sulfonic acid (DSS) in D_2O was used as an external solution reference. Solution state [27]Al spectra were recorded with a one pulse sequence and referenced to $Al(H_2O)_6^{3-}$. Solid samples were combined with 5% by weight KBr (to allow for spinning speed calibration) and ground with a mortar and pestal prior to loading in the sample rotors.

Samples were spun at the magic angle 54.7° with respect to the field at 3-3.5 KHz. Free induction decays were acquired with a one pulse sequence and referenced to the solution state sample of $A\ell (H_2O)_6^{3+}$.

Results and Discussion

A series of ^1H decoupled ^{13}C spectra were recorded for solutions and gels of acidic aluminum propionate. These are displayed in Figure 1. In all cases, three ^{13}C resonances are apparent corresponding to the three carbons of the propionate group, with the resonances centered at 12 ppm (broad, methyl carbons), 31 ppm (methylene carbons), 183 ppm (carboxyl carbons). Spectrum A is representative of a solution of $A\ell(CO_2CH_2CH_3)_3$ with acid and methanol. The broad nature of the methyl resonance is perhaps a consequence of a hindered rotation or possibly due to its being the free end of a "leash" and assuming a number of unequivalent conformations with respect to the growing $A\ell-O-A\ell$ matrix. Presonication of this solution leads to the ^{13}C spectrum shown in Figure 1b. We note that the carboxyl resonance at ≈183 ppm remains sharp with no apparent change in its chemical shift or line shape. The resonance assigned to the methylene carbons centered at α 32 ppm appears to have sharp and broad components in contrast to the unsonicated sample. The methyl resonance centered at 12 ppm displays a similar broadening though, in this case, it is not quite so obvious given the nature of the methyl resonance for the unsonicated sample. There is also other evidence for the breakdown of the $A\ell(CO_2CH_2CH_3)_3$ skeleton, as demonstrated by other broad resonances observed in the aliphatic and carboxylate regions of the ^{13}C NMR spectrum. Our rationalization for these differences requires an apriori knowledge of the effect of sonification on the resultant aluminum oxides formed upon dehydration.

It was found by x-ray powder diffraction that presonication of the $A\ell(CO_2CH_2CH_3)_3$ solutions leads upon dehydration to a crystalline $A\ell_2O_3$ structure. This result indicated to us that presonication of the $A\ell(CO_2CH_2CH_3)_3$ solution had the effect of breaking up $A\ell - CO_2CH_2CH_3$ groups and inducing the polymerization of $A\ell-O$ units prior to the dehydration step. In light of this result, we can understand the ^{13}C NMR spectrum of sonicated $A\ell(CO_2CH_2CH_3)_3$ solution as indicating some degree of aluminum oxide polymerization, and thus many environments for the propionate groups (broad ^{13}C resonances) depending on the extent of polymerization.

The ^{13}C NMR spectrum of the $A\ell (CO_2CH_2CH_3)_3$ gel very clearly displays the three resonances expected for the propionate group. All ^{13}C resonances are broad (≈120 Hz at half height), however, this can be rationalized from the nature of the sample, a viscous gel which did not flow very readily. Thus the motions of the propionate groups are hindered and a line shape approaching the solid state line shape is observed.

Attempts to observe solid state ^{13}C spectra of dehydrated powders and calcined powders were unsuccessful due to dilution of the carbon containing species.

The results of some solution state $^{27}A\ell$ NMR[10-11] studies on $A\ell(CO_2CH_2CH_3)_3$ solutions and gels are shown in Figure 2. All of the solutions gave very similar spectra with one sharp resonance (18 Hz wide at half height) centered at 0 ppm as expected for an octahedral $A\ell$ environment like $A\ell(H_2O)_6^{3+}$ the reference sample. The solution state $^{27}A\ell$ NMR spectrum of an $A\ell(CO_2CH_2CH_3)_3$ gel indicates a mixture of two types of tetrahedral $A\ell$ sites[11] with components centered at 0 ppm and -25 ppm. The line width for these resonances is estimated at 1500 Hz at half height.

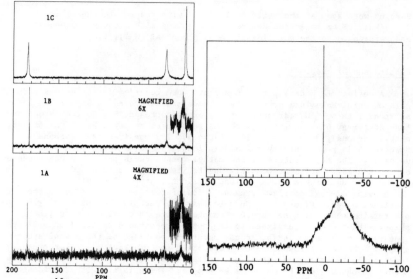

FIG. 1. ^{13}C Solution Spectra of
Aluminum Propionates. a. Nonsonicated solution, b. Sonicated
solution, c. Gelled nonsonicated
solution.

FIG. 2. Solution State ^{27}Al NMR on
$Al(CO_2CH_2CH_3)_3$ Solutions and Gels.

In Figure 3 the x-ray powder spectra for the sonicated and unsonicated
dried powders and the calcined samples (600°C and 800°C) are shown. As can
be seen, the dried alumina sample and the dried sonicated alumina sample
have significantly different x-ray powder patterns. The sonicated sample
forms a crystalline powder which can be indexed as the $\gamma-Al_2O_3$ form while
the unsonicated sample remains poorly crystalline to temperatures as high as
600°C. The energy absorbed by the solution during sonication causes the
organic species to dissociate and form a stable $\gamma-Al_2O_3$ phase faster than
the unsonicated sample.

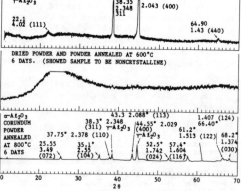

FIG. 3. X-ray Powder Spectra for
the Sonicated, Unsonicated, Dried
Powders and the Calcined Samples
(600°C and 800°C).

Scanning electron micrographs of the dried powder, sonicated powder,
and annealed powder are shown in Figure 4.

(a) Dried powder derived
from Aluminum gel.

(b) Sonicated powder.

(c) αAℓ₂O₃ powder annealed @ 800°C derived from Aluminum gel.
FIG. 4.   SEM Photographs of Aluminum Oxide Powders
Prepared Under Different Conditions.

Solid state ^{27}Aℓ NMR[12-13] of various dehydrated powders derived from
sonicated and unsonicated sol–gel mixtures were studied. A representative
series of spectra are displayed in Figure 5. Powder samples derived from
sonicated and from unsonicated solutions were examined by solid state ^{27}Aℓ
NMR. In addition, two unsonicated samples calcined at 600 and 800°C were
studied. First and second order spinning side bands are observed in each
^{27}Aℓ spectrum at the spinning speed of ≈3500 Hz. The ^{27}Aℓ chemical shifts
and line widths are listed in the Table below along with the x-ray powder
diffraction results.

FIG. 5.   Solid State ^{27}Aℓ NMR of Various
Dehydrated Powders Derived From Sonicated
and Unsonicated Sol–Gel Mixtures.
(a) Sonicated powder, (b) Unsonicated
800°C, (c) Unsonicated 600°C,
(d) Unsonicated powder.

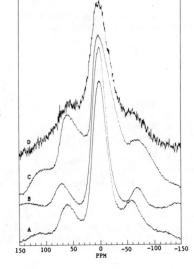

570

The ^{27}Aℓ chemical shifts are relatively constant, however, linewidths appear to yield a test of crystallinity. Amorphous samples, have linewidths of 2000-2100 Hz while crystalline samples have linewidths of 1200-1300 Hz. It would be helpful to run these samples at higher field strengths to ascertain whether it might be possible to observe separate ^{27}Aℓ resonances for different Aℓ environments[12].

Sample	^{27}Aℓ Chemical Shift[1] (PPM)	Line Width at Half Height[2] (HZ)	Powder Diffraction
Sonicated Powder	2.0	1300	Crystalline γAℓ$_2$O$_3$
Unsonicated-800° Calcined	0.6	1200	Crystalline αAℓ$_2$O$_3$ (Corundum Structure)
Unsonicated-600° Calcined	-3.5	2100	Amorphous
Unsonicated Powder	0.1	2000	Amorphous

[1]Chemical shifts were obtained from a line fitting program provided by GE/NMR and are referenced to external Aℓ(H$_2$O)$_6$$^{3+}$ at 0 ppm
[2]Line widths in Hz were obtained from a line fitting program provided by GE/NMR

CONCLUSIONS

The sol-gel derived synthesis of a variety of Aℓ$_2$O$_3$ from aluminum propionate has been discussed. Solution and solid state ^{13}C and ^{27}Aℓ NMR data was coupled with both x-ray powder diffraction data and SEM photographs in order to understand the transition from sol → gel → amorphous powder → crystalline powder in the sonicated and unsonicated aluminum propionate precursor, and Aℓ$_2$O$_3$ powders. Results from these experiments indicate that the sonicated dried powder is crystalline, having the γ-Aℓ$_2$O$_3$ structure while the unsonicated dried powder is amorphous and crystallizes at 800°C with α-Aℓ$_2$O$_3$ structure.

ACKNOWLEDGEMENT

The authors would like to acknowledge the Naval Air Systems Command for their support of this research.

REFERENCES

1. B. E. Yoldas, J. Mater. Sci., 12, 1203-1208 (1977).
2. B. E. Yoldas, J. Mater. Sci., 10, 1856-1860 (1975).
3. B. E. Yoldas, Bull. Am. Ceram. Soc., 54, 286-288 (1975).
4. M. J. Bannister, J. Am. Ceram. Soc., 58, (1-2), 10-14 (1975).
5. N. Nogamic and Y. Moriya, J. Non-Cryst. Solids, 37, 191-201 (1980).
6. M. Yamane, S. Aso, S. Okano, and T. Sakaino, J. Mater. Sci., 14, 607 (1979).
7. J. Zarzycki, M. Prassas, J. Phalippou, J. Mater. Sci., 17, 3371 (1982).
8. C. J. Brinker, J. Am. Ceram. Soc., 65, C4 (1982).
9. S. Sakka and K. Kamiya, J. Non-Cryst. Solids, 48, 31 (1982).
10. "NMR and the Periodic Table," edited by R. K. Harris and B. E. Mann, Academic Press, London, 1978.
11. J. W. Akitt. Annual Reports on NMR Spectroscopy 5A, 465 (1972).
12. "Solid State NMR for Chemists," Colin A. Fyfe, CFC. Press, Guelph (1983).
13. D. Muller, W. Gessner, A. J. Behrens, and G. Scheler, Chem. Phys. Lett., 79, 59 (1981).

POLYMERIC PRECURSOR SYNTHESIS OF CERAMIC MATERIALS

NICHOLAS G. EROR* AND HARLAN U. ANDERSON**
* Oregon Graduate Center, Beaverton, OR   97006
** University of Missouri-Rolla, Rolla, MO   65401

ABSTRACT

For the past twenty years we have successfully synthesized a wide range of ceramic materials by using organic precursors. The organic precursors are formulated into a glass before pyrolysis. Synthesized compounds include, for example, titanates, zirconates, silicates, chromites, niobates, tantalates, ferrites and molybdates.

This preparation technique is based upon having individual cations complexed in separate weak organic acid solutions. The individual solutions are gravimetrically analyzed for the respective cation concentration to a precision of 10-100 ppm. In this way it is possible to precisely control all of the cation concentrations, and to mix the ions on an atomic scale in the liquid state. There is no precipitation in the mixed solution as it is evaporated to the rigid polymeric state in the form of a uniformly colored transparent glass. The glass retains homogeneity on the atomic scale, and may be calcined at a relatively low temperature of only a few hundred degrees Celsius to the homogeneous single phase of precise cationic stoichiometry and particle size of a few hundred Å.

The advantage of having such well characterized, thermodynamically defined compounds is illustrated by the resulting understanding of the solid state chemistry of multicomponent compounds that has occurred.

INTRODUCTION

Over the last few years a great emphasis has been placed on powder preparation processes which yield both homogeneous and fine particulate powders. As a result a number of preparation techniques such as sol-gel [1], freeze drying [2,3] and other organo-metallic [4,5] synthesis have been developed. One of these processes which was developed by Pechini [6] in the 1960's to prepare capacitor oxides such as titanates and niobates has never received much attention as a general preparation technique. It is the intent of this paper to describe the process and to show that it is applicable to the preparation of a much wider range of oxides than Pechini originally suggested.

The Liquid Mix Process (LM Process)

The LM process involves the ability of certain weak acids (alpha-pyroxycarboxylic acids) to form polybasic acid chelates with various cations from elements such as Ti, Zr, Cr, Mn, Ba, La, etc. These chelates can undergo polyesterification when heated in a polyhydroxyl alcohol to form a polymeric glass which has the cations uniformly distributed throughout. Thus the glass retains homogeneity on the atomic scale and may be calcined at low temperatures to yield fine particulate oxides whose chemistry has been precisely controlled.

A typical flowsheet for the preparation of oxides from the LM process is shown in figure 1. The cationic sources which have been successfully used are carbonates, hydroxides, isopropoxides, and nitrates.

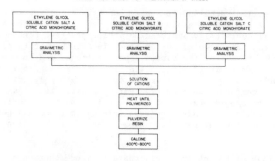

Figure 1. Typical Flow Sheet for Preparation of Oxides

The amount of each raw material necessary to make a particular composition is calculated and then weighed out and mixed with 400 gms of anhydrous citric acid and 600 ml of ethylene glycol in a 2000 ml beaker. The solution is heated at approximately 90°C until all of the cations sources, either carbonates, hydroxides, isopropoxides or nitrates, go into solution. This is the most important step of the process, as complete dissolution of the cations is necessary to insure homogeneity and composition.

The resulting clear solution is evaporated until an amorphous, organic polymer forms. This solid is heated to 400°C in the beaker to burn off as much of the organics as possible. The solid turns into a black, brittle mass. The solid is then ground, screened and transferred to a crucible and calcined at 700-800°C for eight hours. What remains is a homogeneous oxide whose crystallite size is about 50 nm.

Examples of Oxide Preparation

Lanthanum Manganite

Lanthanum and Mn carbonates were quantitatively mixed in the citric
acid and ethylene glycol solution and the system was polymerized.
Figure 2 shows that the initially amorphous x-ray diffraction pattern
changes to the crystalline $LaMnO_3$ pattern as calcination proceeds. The
crystallite size of the powder is shown in figure 3. As can be seen the
crystallite size is less than 0.1 μm. The relative weight loss and
differential thermal analysis of the polymetric precursor are shown in
figure 4. The total weight change was greater than 75%.

Barium Titanate

Barium carbonate and tetraisopropyltitanate were quantitatively
mixed in the citric acid and ethylene glycol solution. After
dissolution was completed, the solution was polymerized, and eventually
calcined to yield powder with a crystallite size of about 0.05 μm. The
DTA and relative weight loss of the polymer and the microstructure of
the resulting oxides are shown in figures 5 and 6, respectively.

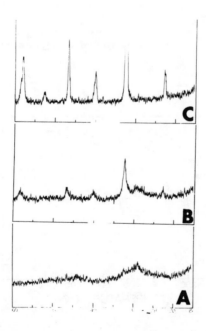

Figure 2.   X-ray Diffraction Pattern of $LaMnO_3$
Preparation: A)   calcined 400 °C - 8 hours
                 B)   calcined 600 °C - 8 hours
                 C)   calcined 700 °C - 8 hours

574

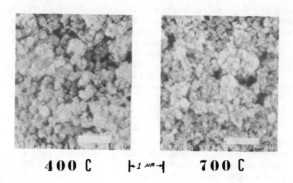

400 C |-1 μm-| 700 C

Figure 3.   Scanning Electron Micrographs of LaMnO₃ from Polymer Process

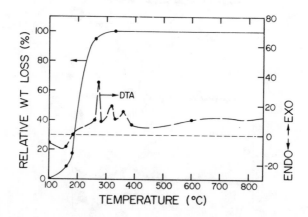

Figure 4.   Differential Thermal Analysis (DTA) and Relative Weight Loss
of Polymetric Precursor for LaMnO₃ Preparation Which Had
Been Heated to 150°C to Polymerize

575

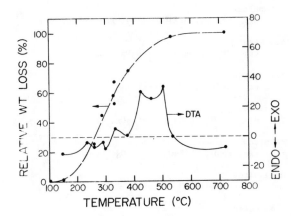

Figure 5. Differential Thermal Analysis (DTA) and Relative Weight Loss
of Polymetric Precursor for BaTiO₃ Preparation Which Had Been
Heated to 150°C to Polymerize.

400 C |–1 μm–| 700 C

Figure 6. Scanning Electron Micrographs of BaTiO₃ from Polymer Process

TABLE I.   OXIDES WHICH HAVE BEEN FORMED BY THE LIQUID MIX PROCESS

Titanates
eg: $BaTiO_3$, $SrTiO_3$, TiO_2, PZT, PLZT, etc.

Niobates
eg: Nb_2O_5, $BaNb_2O_6$, $Pb_3MgNb_2O_9$, etc.

Zirconates
eg: $CaZrO_3$, ZrO_2, etc.

Chromites
eg: Cr_2O_3, $LaCrO_3$, $MgCr_2O_4$, etc.

Ferrites
eg: Fe_2O_3, $LiFeO_2$, $CoFe_2O_4$, etc.

Manganites
eg: $LaMnO_3$, $YMnO_3$, etc.

Aluminates
eg: Al_2O_3, $LaAlO_3$, $MgAl_2O_4$, etc.

Cobaltites
eg: $LaCoO_3$, $YCoO_3$, $PrCoO_3$, etc.

Silicates
eg: Zn_2SiO_4, etc.

DISCUSSION AND CONCLUSION

Over 100 different oxides have been successfully prepared by the LM
process. A few of the oxides prepared are shown in Table I. As can be
seen a wide variety of compounds have been prepared. It has been found
that any oxide which is thermodynamically stable can be prepared by the
LM process as long as precipitation can be prevented during the
polymerization process.

Some advantages of this process are listed as follows:

1)   Well characterized, thermodynamically defined compounds may be
      prepared.

2)   Compounds may be homogeneously doped.

3)   Complicated compounds may be prepared.

4)   Preparation temperatures are less than $800°C$.

5)   The oxides produced are uniform in crystallite size and are
      typically about 50 nm in size.

Some of the disadvantages of the process are:

1)   Procedure is slow and generally takes 1-3 days to complete.

2)   Weight losses are high, typically 70 to 80%

3)   Calcination process is exothermic so that agglomeration of powders occurs.

4)   Powder agglomerates may be hard, so that dispersion of monosized particles may be difficult.

In conclusion, it is obvious that the LM process has a much wider application than was originally outlined by Pechini. The procedure offers a method of making complicated oxides with precisely controlled chemistry, however it is not completely universal - compounds which contain easily reducible cations such as Li, Sn and Zn are difficult to hold in the polymer since very low oxygen activities are encountered as charring proceeds.

REFERENCES

1.   J.M. Fletcher and C.J. Hardy, Chem. Ind. 87 48 (1968).

2.   D.W. Johnson and F.J. Schnettler, J. Amer. Cer. Soc., 53 440 (1970).

3.   F.R. Sale, Metall. Mater. Technol., 9 439 (1978).

4.   C. Marcilly, Ph.D. Dissertation, L. Universite de Grenoble, 1968.

5.   M.S.G. Baythoun and F.R. Sale, J. Mater. Sci., 17 2757 (1982).

6.   M. Pechini, U.S. Patent No. 3,330,697, July 11, 1967.

DISPERSION OF CERAMIC PARTICLES IN ORGANIC LIQUIDS

P.D.CALVERT, R.R.LALANANDHAM, M.V.PARISH, J.FOX, H.LEE, R.L.POBER,
E.S.TORMEY AND H.K.BOWEN.
Ceramics Processing Research Laboratory, Massachusetts Institute of
Technology, Cambridge, MA. 02139.

ABSTRACT

Good dispersion of oxide ceramics in organic solvents can be achieved
using many different dispersants. Several types of dispersants, including
fatty acids, coupling agents, polar aromatic compounds and polymers, are
discussed to illustrate the important phenomena. Many new problems arise
in actual slips during ceramics processing; these are briefly discussed.

INTRODUCTION

During the last few years researchers in this laboratory have studied
a wide range of dispersants for ceramic powders. At its simplest, the
purpose of a dispersant is to suspend particles in a liquid so as to
minimise the strength of the interparticle interactions. The argument can
be made that to achieve maximum strength and perfect homogeneity in a fired
ceramic, it is necessary to start from a green state which is as close to
being perfectly packed as possible. Perfect packing require particles
having a very narrow size distribution allowed to sediment slowly out of
suspension. The slow sedimentation prevents agglomeration; particles act
individually until they form a close packed array.

Such packing is certainly achievable for narrow size distribution
latex particles (1) and has been demonstrated for silica particles under
suitable conditions (2). We have identified a large number of systems that
fulfil the primary goal of separating particles in suitable organic
solvents allowing them to sediment to high packing densities. Real
ceramics processing systems, however, are rather remote from these ideal,
single-solvent/single-dispersant systems, and the problems of producing
dense packing in a ceramic green state within a reasonable time are
complex. This paper outlines recent work on dispersants and discusses how
it might be applied to real processing systems.

FORCES BETWEEN SUSPENDED PARTICLES

Particles dispersed in a solvent attract one another by dispersion
forces arising from the polarisability differences between the particles
and the liquid. In the case of two particles of the same radius separated
by a distance which is small compared to the particle radius, the
interaction energy is given approximately by (3):

$$V_a = -A \, a \, / \, 12 \, h$$

where **a** is the particle radius and **h** is the separation between the two
particles. **A** is the Hamaker constant for the particle-solvent interaction
which is related to the solvent and particle constants by:

$$A = (A_s^{1/2} - A_p^{1/2})^2$$

The practical use of this theory to calculate interparticle forces
depends on reliable methods for calculating Hamaker constants.
Unfortunately, the relevant difference terms cannot usually be calculated

with sufficient precision because the relevant UV absorption data is not available for many materials. The only direct measurements are those for the force between solid surfaces approaching closely in a liquid such as has recently been obtained for mica (4,5) where A is about 10^{-20} J in a hydrocarbon solvent. Only mica and silica have been studied in this way. It has been suggested (1) that systems in which there is a refractive index match between the particles and the solvent should also correspond to a low Hamaker constant since both effects depend on polarisability differences. Although this is an attractive idea, it has not really been tested.

Dispersive forces cause particles to coagulate but the process would be expected to be reversible unless the particles actually form covalent or hydrogen bonds with each other. Since ceramic powders are usually dry and in contact with one another at some time before dispersion, the chemistry of the particle surface is an important factor in the formation of agglomerates which cannot subsequently be broken up.

In order to prevent coagulation, a dispersant must introduce repulsive forces between particles. Charge stabilization and steric stabilization are widely recognised as two ways of doing this (3). Ions attached to particle surfaces may cause coulombic repulsion between particles; these forces can be described by the DLVO theory. This is kinetic stabilization, in that the coulombic repulsion introduces an energy barrier which may be crossed when particles approach with sufficient kinetic energy; thus, in principle, coagulation is retarded rather than eliminated. It is not clear that charge stabilization is important in many organic solvents in which the very low ion concentrations make the DLVO theory untenable.

The alternative approach is steric stabilization, by which polymers attach to particle surfaces forming "clouds" around the particles. When particles approach each other, the overlap of the polymer clouds provides an osmotic pressure which keeps the particles apart. This effect may be treated by the lattice theory of polymer solutions (6). Steric stabilization can also be achieved by attaching short chains to particle surfaces through reactive end groups, forming "hairy" particles. As an approximation it can then be assumed that the particles cannot approach one another closer than the combined lengths of the two layers of attached chains. The interaction energy is therefore reduced to that appropriate to the dispersion energy at the imposed separation. If this energy is on the order of kT, the suspension should be stable against flocculation.

HEAD AND TAIL GROUPS

It is helpful to think of dispersants in terms of a "head" function, responsible for attaching the molecule to the particle, and a "tail" function, responsible for the repulsive interaction.

Carboxylic acids have long been known to provide effective dispersants for oxide ceramics such as titania, zirconia, alumina, and barium titanate, through hydrogen bonding between the surface hydroxyls and the acid. Some esterification of the surface occurs, but the effect is small. We have found that 10% of a carboxylic acid monolayer remains after long Soxhlet extraction with ethyl acetate. For more acidic surfaces such as silica, amines are more effective hydrogen-bonding head groups.

The dispersing tail provides the repulsive force between particles. In most of our systems this is by steric stabilization where the tails should be highly soluble in the surrounding solvent. As the condition at which the tails tend to precipitate is approached (i.e., the point at which self-interactions become stronger than tail-solvent interactions), the particle agglomerate and sediment rapidly. In some cases, as discussed below, repulsion appears to arise via a dipolar repulsion mechanism.

ASSESSMENT OF DISPERSION QUALITY

The main test for dispersion quality is the sediment volume test: a powder-dispersant-solvent mixture is thoroughly dispersed by milling and ultrasonication, then allowed to settle under gravity in a graduated cylinder. The final sediment density is expressed as a fraction of the theoretical density of the solid. A sediment density below 20% and rapid settling are both signs of agglomeration. The best dispersants give sediment densities of up to 50% of theoretical. Centrifugation and colloid pressing both produce higher densities than does simple settling, but the degree of order in the packing is not necessarily improved.

Viscosity measurements on dispersions reveal agglomeration as non-Newtonian viscosity, particularly pseudoplasticity associated with the breakup of agglomerates under shear. Viscosity measurements are not as sensitive to any aggregates which remain unbroken by the initial ultrasonic treatment.

FISH OIL AND GLYCERYL TRIOLEATE

Fish oil is widely used as a dispersant for tape casting ceramics from a variety of organic solvent systems. The oil is extracted from the Atlantic Menhaden fish and is, in principle, largely triglycerides of saturated and unsaturated fatty acids, with chain lengths in the range from 12 to 30 carbon atoms, as shown in figure 1. Tests with glyceryl trioleate showed this model compound to be a very poor dispersant (7).

FIGURE 1
GLYCERYL TRIOLEATE

$$CH_2OOC(CH_2)_7CH=CH(CH_2)_7CH_3$$
$$CHOOC(CH_2)_7CH=CH(CH_2)_7CH_3$$
$$CH_2OOC(CH_2)_7CH=CH(CH_2)_7CH_3$$

FISH OIL

OXIDISED AND CROSS-LINKED PRODUCTS OF POLYUNSATURATED TRIGLYCERIDES

$$CH_2OOC(CH_2)_{18}CH_3$$
$$CHOOC(CH_2)_7CH=CH(CH_2)_7CH_3$$
$$CHOOC(CH_2)_3(CH_2CH=CH)_6CH_3$$

Commercial fish oils are commonly subjected to high-temperature oxidation treatments, which induce cross-linking and probably give rise to some carboxylic acid functions to yield a partly polymerized structure. Fish oil as supplied is a viscous amber syrup; glyceryl trioleate is a nearly clear, light oil. The infrared spectra of fish oil show a much broadened peak in the carbonyl region which would agree with the presence of a range of oxidised species. Adsorption isotherms show that about three times more fish oil than glyceryl trioleate is taken up on alumina from toluene. In glyceryl trioleate, changes in the ester C=O peak are seen on adsorption to alumina, attributable to the dispersant binding to the surface through the

582

ester groups; no such change is seen with fish oil. The coverage for glyceryl trioleate apparently corresponds to the molecules lying flat on the surface (3 nm^2/molecule). Thus, the oleate probably attaches to the surface through the ester function, but the whole molecule lies flat on the surface because the ester-surface interaction is not strong enough to displace the hydrocarbon tails when their entropy has already been reduced by attachment through the ester. In the fish oil, the carboxylic acid binding is much stronger, so neither the ester nor the hydrocarbon sees the surface; unlike in the case of glyceryl trioleate, the bulk of the chains extend into solution to act as steric stabilisers. Glyceryl trioleate becomes a much more effective stabilizer if it is partly oxidised at high temperature (7).

Doroszkowski and Lambourne studied the stabilization of titania dispersions with a variety of carboxylic acids and found the best effect to be produced by large, irregularly branched structures (8). Fish oil seems to fit well into this category, though it has the disadvantage of being irreproducible so that a synthetic analogue would be preferable. To some extent the polyhydroxystearates used for paint dispersants do fit the desirable synthetic category (8).

COVALENTLY BONDED DISPERSANTS

Silane and titanate coupling agents have long been used to aid in the dispersion of reinforcing fillers in polymers. Figure 2 shows the improvement in sedimentation volume of silica particles in hexane on treatment with silanes carrying different hydrocarbon chain lengths. As can be seen, a critical chain length of about 12 carbons is needed to obtain good dispersion, after which there is little improvement. Thus the chain length needed for steric stabilization in this system is quite short.

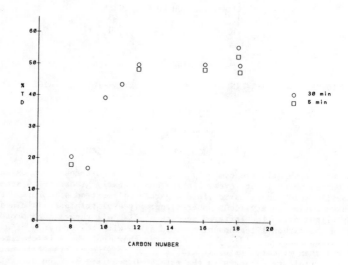

FIGURE 2

EFFECT OF SILANE CHAIN LENGTH, R, ON PACKING DENSITY FROM DISPERSIONS OF STOBER SILICA IN HEXANE. $RSi(OCH_3)_3$.

The titanates form a similar group of coupling agents for which the dialkoxy-dicarboxylates can readily be prepared. The alkoxy groups are then able to react with a particle surface to form a covalent bond while the chains of the carboxylate units can act to give steric stabilization. We have investigated the dispersion of alumina and barium titanate in hexane by the addition of 12-stearoyloxystearic acid or diisopropoxytitaniumdistearoyloxystearate. The titanium coupling agent produced better packing than did the free acid, and the effect withstood prolonged Soxhlet extraction (9). Packing densities above 50% were obtained with gravity sedimentation.

In general, covalently bonded dispersants have the advantage that they can be transferred with the particles between different solutions without dispersant being lost. With the titanates a wide range of different tail groups can be used to suit particular process requirements. On a 0.1 um alumina particle the extra titanium added in the coupling agent amounts to about 0.1 at% titanium content.

AROMATIC DISPERSANTS

In 1961 Lewis described the dispersion of oxide ceramics in water using aromatic acids (10). Recently Parish (11) rediscovered these compounds as good dispersants for oxides in alcohols and in toluene. Both research groups concluded that 1) there is no correlation between the dispersing ability and the ionization properties of an acid, and 2) dispersion is not due to charge stabilization. We believe the effect may be due to the strong dipole moment of the compounds, giving an effective double layer to the surface without ionisation.

POLYMERIC DISPERSANTS

Homopolymers are good dispersants for many particle systems, provided that the molecular weight is sufficiently high for surface bonding to occur despite the polymer being highly soluble. This case has been extensively discussed by Napper (6). The drawback with polymeric dispersants in organic solvents is that the system is vulnerable to small amounts of polar impurities such as water, which may displace the weakly bound polymer from particle surfaces. Copolymers solve this problem in that a few strongly binding groups may be incorporated into a soluble polymer. Many commercially available ceramic dispersants are random copolymers.

We have studied a block copolymer of styrene and methylmethacrylate as a dispersant for barium titanate in benzene where it proved to work well (12). The sediment densities were lower than those which could be achieved with good small dispersants (to a maximum of around 40%) but this probably represents the extra volume of the attached polymer layer. The dispersing action is believed to be due to the methacrylic ester units attaching to the surface while the styrene remains in solution. Adding a small amount of isopropanol caused the powder to agglomerate as the ester was displaced from the surface. Agglomeration could also be induced by taking the polymer close to its precipitation point by adding hexane to the benzene.

A more effective block copolymer would have more strongly binding groups than the ester, but at that point it is no longer necessary to use a block copolymer as opposed to a random- or end-terminated system.

One good end-terminated polymeric dispersant is OLOA 1200 (TM Chevron Corp.), a commercial dispersant for use in mineral oils. A variety of analyses indicate this material to be an amine-terminated polyisobutylene having a chain length of about 60 carbon atoms. Oloa has proved to be a good dispersant for silicon carbide in hexane. Silicon carbide can be

584

considered to be a silica-like powder in that it has many hydroxylated
surface groups. Packing densities 40% of theoretical are obtained.

PROCESSING CONSIDERATIONS

In order to apply these dispersants to ceramics processing, other
interactions must be considered. For tape casting, the slip consists of
powder, polymeric binder, plasticiser, and dispersant. In the slip,
dispersant must remain bonded to the particle surface despite competition
from the polymer and plasticiser. The dispersant tail must also remain
miscible with the polymer-plasticiser combination as the solvent leaves.
Theoretical considerations (3) suggest that slips should flocculate at
intermediate solvent levels even though they may be well dispersed in
solvent or in concentrated polymer; if this is the case ,undesirable
separation may occur during drying. In the dry state, the strength of a
green body must derive from polymer-particle interactions; thus, the
polymer must be able to bind either to the particle surface or to the
dispersant-coated particle. Binding might occur either through adsorption
or through entanglement of the polymer with the dispersant. The ideal
green state consists of regularly packed particles, but such regular
packing is unlikely to occur by settling during the drying stage; to what
extent the drying shrinkage may drive the packing toward perfection is
unclear. Thus in any real ceramic system, many interactions are important
in the choice of a dispersant.

REFERENCES

1) P.N.Pusey and W.van Megen, Nature 320, 340 (1986).
2) B.Vincent, presented at 59th meeting Colloid and Surface Div.
   Amer.Chem.Soc., Potsdam, NY, 1985.
3) R.Buscall and R.H.Ottewill in Polymer Colloids, edited by R.Buscall,
   T.Corner and J.F.Stageman (Elsevier, New York, 1985), pp141-218.
4) Y.Almog and J.Klein, J.Colloid Interf. Sci. 106, 33 (1985).
5) R.G.Horn and J.N.Israelachvili, Chem.Phys.Lett. 71, 204 (1980).
6) D.H.Napper, Polymeric Stabilization of Colloidal Dispersions (Acad.
   Press, New York, 1983).
7) P.D.Calvert, E.S.Tormey and R.L.Pober, Am. Ceram. Soc. Bull. 65, 669
   (1986).
8) A.Doroszkowski and R.Lambourne, Chem.Soc.Farad.Discuss. 65, 252 (1978).
9) R.R.Lalanandham, H.K.Bowen and P.D.Calvert, to be published.
10) A.E.Lewis, J.Amer.Ceram.Soc. 44, 233 (1961).
11) S.Mizuta, M.V.Parish and H.K.Bowen, Ceram. Intl. 10, 43,83 (1984).
12) H.Lee, R.Pober and P.Calvert, J.Colloid Interf. Sci. 110, 144 (1986).

^{29}Si NMR, SEC AND FTIR STUDIES OF THE HYDROLYSIS AND CONDENSATION OF Si(OC$_2$H$_5$)$_4$ AND Si$_2$O(OC$_2$H$_5$)$_6$

CHIA-CHENG LIN AND JOHN D. BASIL
PPG Industries, Inc., Glass Research and Development Center, P.O. Box 11472, Pittsburgh, Pennsylvania 15238 U.S.A.

ABSTRACT

The hydrolysis and initial condensation reactions of Si(OC$_2$H$_5$)$_4$ and low order polyethoxysiloxanes have been studied with high resolution ^{29}Si NMR and size exclusion chromatography/FTIR spectroscopy. The effects of various parameters such as H$_2$O/TEOS mole ratio, catalyst, pH, solvent, temperature and aging have been observed in the different nature of the reaction products. The products include partially hydrolyzed and unhydrolyzed Si(OC$_2$H$_5$)$_4$, Si$_2$O(CH$_2$H$_5$)$_6$, and Si$_3$O$_2$(OC$_2$H$_5$)$_8$. The rate of formation and concentration of each product obtained under different conditions provides information about the reactivity of various silanol-containing species and the factors that affect the overall reaction scheme.

INTRODUCTION

The physical properties of sol-gel materials are determined by the competing hydrolysis, reesterification and condensation reactions [1-3]. Other workers have obtained valuable insight into these relationships by studying the initial hydrolysis and condensation reactions of: silicon tetramethoxide (TMOS) by GC, SAXS [4], ^{29}Si NMR and Raman spectroscopy [5,6]; silicon tetraethoxide (TEOS) [7] by GC [8,9], SAXS [8], IR [10], ^{1}H NMR [8,11] and SEC [12,13]; and substituted trialkoxysilanes by IR [14], UV [15] and SEC/FTIR [16]. Highly aqueous TEOS-derived sols near gelation were studied by ^{29}Si NMR [17].

The use of alkoxides that have more than one Si atom and alkoxide groups of different reactivites should provide additional clues concerning the reactions that are important during the formation of high molecular weight polyalkoxysiloxanol gel precursors. This paper presents the results of experiments in which the acid catalyzed partial hydrolyses of Si(OEt)$_4$, Si$_2$O(OEt)$_6$, and Si$_3$O$_2$(OEt)$_8$ were examined by ^{29}Si NMR and SEC/FTIR.

EXPERIMENTAL

Hexaethoxydisiloxane, Si$_2$O(OEt)$_6$, and octaethoxytrisiloxane, Si$_3$O$_2$(OEt)$_8$, were isolated from commercial ethylsilicate-40 (Union Carbide) and checked for purity by SEC, ^{29}Si NMR (Table I) and refractive index [18]. Titanium tetraethoxide (TET, Alfa) and TEOS (Fisher) were used as received.

The hydrolysis conditions were selected for initial SiO$_2$:H$_2$O:HNO$_3$:EtOH mole ratios of 1:1:0.00056:4.5 at 66°C. Peaks representing silanols were identified by their disappearance after adding 0.2 equiv. TET.

The ^{29}Si NMR spectra were obtained at Case Western Reserve University, Cleveland, OH. Glass sample tube and probe resonances were suppressed by using a depth profiling pulse sequence [19].

Chromatography was performed at ambient temperature with CHCl$_3$ or toluene, Ultrastyragel columns, and a DuPont 850 LC with Waters differential refractometer. Commercial ethylsilicate-40 was used as a five point calibration standard for the molecular weight range 208-744.

FTIR spectra of the column eluant flowstream were obtained with a Perkin Elmer 1800 spectrometer.

Mat. Res. Soc. Symp. Proc. Vol. 73. ©1986 Materials Research Society

Table I. ^{29}Si Chemical shifts of TEOS and ethoxysilane dimer and trimer, in ppm relative to TMS at 0.00.

Structural Unit	Chemical Formula	δ	$\delta - \delta_{TEOS}$
Q^0	$Si^*(OEt)_4$	-81.95	0.00
Q^1_2	$(EtO)_3Si^*OSi^*(OEt)_3$	-88.85	-6.89
Q^1_3	$(EtO)_3Si^*OSi(OEt)_2OSi^*(OEt)_3$	-88.99	-7.04
Q^2_3	$(EtO)_3SiOSi^*(OEt)_2OSi(OEt)_3$	-96.22	-14.27

RESULTS AND DISCUSSION

^{29}Si NMR Data

Examples of the typical ^{29}Si NMR results appear in Figure 1. The ^{29}Si chemical shifts of Q^0 and Q^1 units in identified species produced from the hydrolysis of TEOS and $Si_2O(OEt)_6$ are shown in Tables II and III. Assignment of the resonances is similar to previous methods [5,20] with the subscripts indicating the number of the hydroxyl groups. The chemical shifts of the end group Si atoms (Q^1_2) depend on the number of hydroxyl groups directly attached to the observed Si nucleus, indicated by the B subscripts, and the number of hydroxyl groups in the other Q^1 unit, indicated by x in Table III. Each successive OH group in the observed and adjacent Si unit results in downfield shifts of 2-2.5 and $\sim$0.1 ppm, respectively.

Table II. Chemical shifts of the Q^0 species (R=Et) produced from the hydrolysis of TEOS; in ppm relative to TMS at 0.00.

	Structure	δ	$\delta - \delta_{(TEOS)}$
A_0	$Si(OR)_4$	-81.95	0.00
A_1	$Si(OR)_3(OH)$	-79.07	2.88
A_2	$Si(OR)_2(OH)_2$	-76.58	5.37
A_3	$Si(OR)(OH)_3$	-74.31	7.64

Table III. Chemical shifts of the Q^1_2 silicon nuclei, Si*, in identified species produced from the hydrolysis of $Si(OR)_4$ and $Si_2O(OR)_6$ (R=Et).

	Disiloxane Structure	x=0	x=1	x=2	x=3
B_0	$Si^*(OR)_3OSi(OR)_{3-x}(OH)_x$	-88.85	-88.75	-88.64	
B_1	$Si^*(OR)_2(OH)OSi(OR)_{3-x}(OH)_x$	-86.27	-86.17	-86.06	-85.96
B_2	$Si^*(OR)(OH)_2OSi(OR)_{3-x}(OH)_x$	-83.92	-83.80	-83.72	-83.64
B_3	$Si^*(OH)_3OSi(OR)_{3-x}(OH)_x$	-81.72	-81.62	-81.55	

During the course of TEOS hydrolysis the primary products maintained the order $[TEOS] \cong [(EtO)_3Si(OH)] > [(EtO)_2Si(OH)_2] > [(EtO)Si(OH)_3]$, but the latter two species clearly preferentially undergo condensation. Figure 1a shows that the first disiloxane products formed are predominately $Si(OEt)_2(OH)OSi(OEt)_{3-x}(OH)_x$, (x=1 > x=2), $Si(OEt)(OH)_2OSi(OEt)_{3-x}(OH)_x$, (x=0,1,2), and some $Si(OEt)_2(OH)OSi(OEt)_3$, but not $Si(OEt)_3OSi(OEt)_3$.

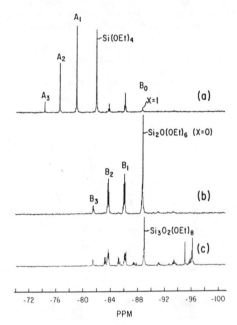

Figure 1. Typical 400 MHz ^{29}Si NMR spectra indicating the species present in hydrolysis solutions of TEOS (a), hexaethoxydisiloxane (b), and octaethoxytrisiloxane (c) after 6 min.

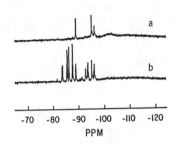

Figure 2. Effect of adding TET (a) to a solution of partially hydrolyzed $Si_3O_2(OEt)_8$ (b) after 120 min.

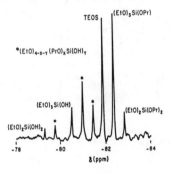

Figure 3. 200 MHz ^{29}Si NMR spectrum (Q^0 region) of a TEOS hydrolysis solution comparable to Figure 1a but with 2-propanol as the solvent.

Table IV presents relative concentration data obtained from integrated intensities of signals due to TEOS, dimer, trimer and their hydrolysis products. By combining the data in the first column with the number of OH groups each species contains, the total number of mmoles of OH groups represented by all of the peaks was calculated to equal the number of mmoles of H_2O (or TEOS) originally present. Only two-thirds of the TEOS was used to consume all of the H_2O by the time spectrum 1a was recorded. By 4 hours the sum of all silanols had decreased by one-third due to condensation while 15% TEOS still remained unreacted.

Andrianov reported that the rate of polyethoxysiloxane hydrolysis decreases significantly with increasing siloxane chain length at constant [H_2O] [18]. In this investigation the H_2O:Si mole ratio was fixed at one and Table IV shows that the amount of unhydrolyzed TEOS (A_0, 33 mole %), dimer end group (Q^1_2, 43 mole %), and trimer (2/3 Q^1_3 + 1/3 Q^2_3 = 42 mole %) after 6 min. are quite similar. The end groups of the trimer (Q^1_3) also hydrolyze more rapidly than the middle (Q^2_3) group.

No $Si(OH)_4$ or $Si_2O(OH)_6$ was detected by NMR during the hydrolysis of TEOS and dimer under these conditions (Figure 1). No branching (Q^3) Si units ($\delta < -97$ ppm) were observed in any of the spectra up to 4 hours, but the addition of TET to the trimer (Figure 2) hydrolysis solutions resulted in the disappearance of all silanols and the formation of products with ^{29}Si chemical shifts characteristic of Q^2 and Q^3 units. The quenching

of silanols by titanium alkoxides is thought to result from the formation
of mixed alkoxides due to the rapid condensation reaction -SiOH + RO-Ti-
$\longrightarrow$ -Si-O-Ti- + ROH [21] or from Ti(OR)$_4$-catalyzed silanol condensation
[22].

Table IV. Evolution of silicon ethoxide hydrolysis products, expressed
in mole %, obtained by ^{29}Si NMR (x=0-3, y=0-3, R=Et).

		Time (min.)			
	Species	6	40	120	240
I.	**TEOS Hydrolysis**				
A_0	Si*(OR)$_4$	33	25	18	15
A_1	Si*(OR)$_3$(OH)	31	30	21	15
A_2	Si*(OR)$_2$(OH)$_2$	16	11	7	3
A_3	Si*(OR)(OH)$_3$	3	2	1	0
B_0	Si*(OR)$_3$OSi(OR)$_{3-x}$(OH)$_x$	2	11	23	30
B_1	Si*(OR)$_2$(OH)OSi(OR)$_{3-x}$(OH)$_x$	11	16	19	19
B_2	Si*(OR)(OH)$_2$OSi(OR)$_{3-x}$(OH)$_x$	4	5	4	3
C&D	Q^2	0	-	8	9
II.	**Dimer Hydrolysis**				
B_0	Si*(OR)$_3$OSi*(OR)$_3$	35		10	
B_0	Si*(OR)$_3$OSi(OR)$_{3-x}$(OH)$_x$	43		23	
III.	**Trimer Hydrolysis**				
Q^1_3	Si*(OR)$_3$OSi(OR)$_{2-x}$(OH)$_x$Si*(OR)$_3$	36	18	14	
Q^2_3	Si(OR)$_{3-x}$(OH)$_x$OSi*(OR)$_2$OSi(OR)$_{3-y}$(OH)$_y$	55	33	26	

Figure 3 presents one type of solvent effect on alkoxide hydrolysis
reactions; the rate of alkoxide exchange is comparable to the rate of hydrolysis.

SEC/FTIR Results

The SEC/FTIR method provides molecular weight and functional group
information about species formed during the hydrolysis reactions (Figures
4 and 5). The spectrum recorded at 6.6 min. is virtually identical to
the spectrum of TEOS in CHCl$_3$, and the identification of the silanols at
higher and lower retention times is based on the similarity of the 1190-880
cm^{-1} region of the spectrum and the appearance of $\tilde{\gamma}$(SiOH) at 3675 cm^{-1}
in the two eluant fractions at 6.3 and 7.3 min. corresponding to the monomeric (Q^0) and disiloxane (Q^1) silanols, respectively (Tables II and III).

All $\tilde{\gamma}$(SiOH) peaks disappeared upon addition of TET. The high molecular
weight products of the TET addition (~1500 mw) exhibit no $\tilde{\gamma}$(SiOH) but
clearly contain Si-OEt.

Figure 6 illustrates the slower formation of monomeric and dimeric
silanols and disiloxanes from TEOS at lower temperature. Figure 7 illustrates
the pH effect on TEOS hydrolysis.

CONCLUSIONS

A comparative study of the partial hydrolysis of silicon ethoxide
monomer, dimer and trimer by high resolution ^{29}Si NMR has resulted in the
detection and peak assignments of three group A silanols and eight disilox-
anols containing up to five OH groups. The initial disiloxanes formed
from the condensation of Q^0 silanols are polyalkoxysilanols. One equivalent of H$_2$O was rapidly consumed by only 2/3 as much TEOS in an initial
hydrolysis stage that was accompanied by some condensation, even at 66°C.

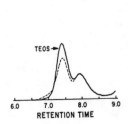

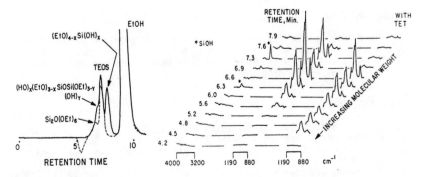

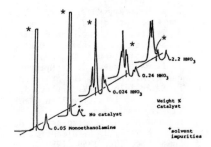

Figure 4. SEC chromatogram of TEOS hydrolysis solution at 60 min. (—) and after adding TET (--).

Figure 5. FTIR data for the 60 min. TEOS hydrolysis solution of Figure 4.

Figure 6. SEC chromatogram illustrating the composition of the TEOS hydrolysis solution after 40 min. at 66° (--) and 24° (—).

Figure 7. SEC illustration of how sol composition varies with pH after 2 hours hydrolysis of TEOS (*). At higher acid concentration the hydrolysis of TEOS is faster.

Silanol reactivity toward condensation followed the order $(EtO)Si(OH)_3 > (EtO)_2Si(OH) > (EtO)_3SiOH$, which is the reverse of the relative rates of formation. The rates of initial hydrolysis of Q^0, Q^1_2 and Q^1_3 structural units are similar, but the terminal ethoxide groups (Q^1_3) of the trimer hydrolyze more rapidly than the middle (Q^2_3) ethoxide groups. Under these reaction conditions one equivalent of H_2O produced detectable amounts of all possible monomer and dimer hydrolysis products except $Si(OH)_4$ and $Si_2O(OH)_6$, but no branching units (Q^3) were detected.

Although SEC did not completely resolve the individual silanol species, they can be identified by FTIR spectroscopy and their reactivity with TET, and the Q^0 silanols are clearly distinguishable from the Q^1 silanols. The addition of TET to the trimer hydrolysis solution results in the formation of Q^2 and Q^3 structural units.

This work has clearly demonstrated that the combination of high resolution ^{29}Si NMR and SEC/FTIR is one of the most useful methods for studying the initial hydrolysis and polycondensation of silicon alkoxides and silicon alkoxide containing multicomponent sol-gel solutions.

ACKNOWLEDGEMENTS

The authors gratefully acknowledge the support provided by PPG Glass R&D and wish to credit J. C. Vanek and E. G. Goralski for obtaining the SEC/FTIR data and Dr. Adrian Valeriu of Case Western Reserve University for his efforts in obtaining the NMR data.

REFERENCES

1. B. E. Yoldas, J. Non-Cryst. Solids 51, 105 (1982).
2. S. Sakka and K. Kamiya, J. Non-Cryst. Solids 48, 31 (1982).
3. C. J. Brinker, D. E. Clark and D. R. Ulrich, Mater. Res. Soc. Symp. 32 (1984).
4. M. Yamane, S. Inoue and A. Yasumori, J. Non-Cryst. Solids 63, 13 (1984).
5. I. Artaki, M. Bradley, T. W. Zerda and J. Jonas, J. Phys. Chem. 89, 4399 (1985).
6. I. Artaki, S. Sinha, A. D. Irwin and J. Jonas, J. Non-Cryst. Solids 72, 391 (1985).
7. M. F. Bechtold, R. D. Vest and L. Plambeck, Jr., J. Am. Chem. Soc. 90, 4590 (1968).
8. C. J. Brinker, K. D. Keefer, D. W. Schaefer, R. A. Assink, B. D. Kay and C. S. Ashley, J. Non-Cryst. Solids 63, 45 (1984).
9. R. Aelion, A. Loebel and F. Eirich, J. Am. Chem. Soc. 72, 5705 (1950).
10. H. Schmidt and A. Kaiser, Glastechn. Ber. 54, 338 (1981).
11. R. A. Assink and B. D. Kay, Mater. Res. Soc. Symp. 32, 301 (1984).
12. Y. I. Rastorguev, E. A. Ryabenko, A. I. Kuznetsov, B. Z. Shalumov and L. A. Zhukova, Zh. Prikladnoi Khimii 50, 2602 (1977).
13. L. M. Antipin, V. V. Krylov, A. I. Borisenko, R. V. Klygina and Y. K. Shaulov, Izvestiya Akad. Nauk USSR, Neorg. Materialy 14, (5), 935 (1978).
14. H. Schmidt, H. Scholze and A. Kaiser, J. Non-Cryst. Solids 63, 1 (1984).
15. K. J. McNeil, J. A. DiCaprio, D. A. Walsh and R. F. Pratt, J. Am. Chem. Soc. 102, 1859 (1980).
16. J. D. Miller and M. Ishida, Anal. Chem. 57, 284 (1985).
17. B. E. Yoldas, J. Polymer Sci. Chem. Ed., to be published.
18. K. A. Andrianov, Organic Silicon Compounds, State Scientific Technical Publishing House for Chemical Literature, Moscow, 1955.
19. M. R. Bendall and R. E. Gordon, J. Mag. Res. 53, 365 (1983).
20 R. K. Harris, C.T.G. Knight and D. N. Smith, J.C.S. Chem. Comm. 421, 726 (1980).
21. B. E. Yoldas, J. Non-Cryst. Solids 38&39, 81 (1980).
22. W. Noll, Chemie und Technologie der Silicone, 2nd Ed., Verlag Chemie, Weinheim, 1968.

THE ACCURATE DETERMINATION OF BULK GLASSES AND GLASS FILM COMPOSITIONS BY INDUCTIVELY COUPLED PLASMA-ATOMIC EMISSION SPECTROMETRY

T. Y. KOMETANI
AT&T Bell Laboratories, 600 Mountain Avenue, Murray Hill, N.J. 07974

ABSTRACT

A sequential inductively coupled plasma-atomic emission spectrometric (ICP-AES) method has been developed for the complete quantitative analyses of silicate glasses. B, P, Ge, and Si have been determined with 2% precision and accuracy in bulk glass and glass films on Si wafer substrates. Two chemical sample processing steps, dissolution in hydrofluoric acid at room temperature and fusion in a sodium carbonate flux, were equally effective for decomposing samples without loss of volatile components. Rapid analysis for controlling glass film composition during IC process development were provided by the ICP-AES method. Where accurate film weights were not obtainable, compositions were established by complete analyses of all glass components except oxygen. The sum of the weights of components converted to their oxides served as the sample weight for calculation purposes.

INTRODUCTION

Chemical vapor deposited silicate glasses are important in communications technology. Glasses used for optical fiber waveguides must meet specifications for refractive index, optical transmission, and viscosity. These properties are dependent upon the concentration of components such as Si, Ge, B, and P.[1,2] Deposited films of phosphosilicate and borophosphosilicate glasses (BPSG) provide a variety of important functions in the fabrication of semiconductor devices.[3,4] Optimization of glass composition for various applications require accurate determination of major and minor components.

Phosphorus and Si in phosphosilicate films have been determined by x-ray fluorescence spectrometry.[5] Methods of determination of B and P in BPSG films by secondary ion mass spectrometry, Rutherford backscatter spectrometry, electron probe technique and wet chemical analysis have been compared.[6] The analyses of BPSG films by ion chromatography and infrared spectroscopy have been reported.[7] Most of the instrumental techniques are dependent on the use of calibration films of known composition and thickness. The classical titration method for B is limited in sensitivity and subject to interferences.[8] DC plasma emission spectrometry has been successfully applied to the determination of major components in optical waveguide materials.[9] A related plasma technique uses an inductively coupled argon plasma as an emission source for rapid sequential determination of most of the elements with detection limits in the parts-per-billion range for Si, B, P, and Ge.[10]

Inductively coupled plasma atomic emission spectrometry (ICP-AES) involves the measurement of intensities of atomic emission lines emanating from an RF coupled argon plasma, into which is introduced an aerosol of the sample solution. A sequential ICP spectrometer diagramed in Fig. 1 locates and measures the peak intensities of characteristic atomic emission lines and directly relates them to concentration of elements in the sample solution. Because of the high temperatures (6000–10,000°K) and stability of the plasma, ICP spectrometry is applicable for quantitative determination of most of the elements over a wide concentration range. It is especially suitable for the rapid analysis of multi-component solutions. Samples are converted to aqueous solutions and compared to calibration solutions. Various dissolution methods were investigated to avoid the loss of volatile components during sample preparation.

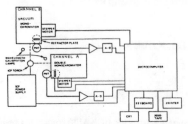

Figure 1. Inductively Coupled Plasma Spectrometer, Allied Analytical Systems PLASMA 200.

Table I. Instrumental Conditions, PLASMA 200.

RF GENERATOR	27.12 MHz
POWER- 4 setting	1.2 kw
HF RESISTANT PLASMA TORCH ASSEMBLY	
NEBULIZER	CROSS FLOW
ARGON FLOW:	
NEBULIZER	1.0 L/min
PLASMA SECONDARY	0.5 L/min
COOLANT FLOW	20 L/min
SAMPLE UPTAKE RATE	1.0 mL/min
INTEGRATION TIME	2 SEC
NUMBER OF READINGS	3
EMISSION WAVELENGTH:	
B	249.77 nm
P	213.62
Si	251.61
Ge	209.43
Al	396.15

EXPERIMENTAL

The Plasma 200 sequential inductively coupled plasma spectrometer (Allied Analytical Systems, Waltham, MA) was used in this study. The instrument consists of an HF-resistant ICP torch assembly source, sample introduction system with a sapphire cross-flow nebulizer, and a two-channel light detection system consisting of an Ebert-Fassie double monochromator (primary-300 mm focal length, secondary-165 mm focal length) and a 1/3 m vacuum monochromator. All components were computer controlled to locate and measure in sequence the peak intensities of characteristic atomic emission lines. The instrumental conditions are summarized in Table I. The RF power setting was 4 (induction coil power 1.2 kW and argon coolant gas flowrate 18 LPM) for all the elements. Three 2-sec. integrations were averaged for intensity measurements at a given wavelength. Background corrections were performed where appropriate by measuring intensities on either side of the line peak. The observation height of the plasma was 20 mm.

Procedure for Dissolution of Bulk Glass

Fusion: 100 mg of pulverized sample were weighed into a 10 ml Pt crucible containing 1g Na_2CO_3 and mixed thoroughly. The crucible was heated in a muffle furnace at 900°C for 1 hour. The cooled crucible was placed in a 100 ml Teflon beaker with 30 ml H_2O and allowed to stand at room temperature until the melt was thoroughly disintegrated. The solution and precipitate were transferred to a 100 ml plastic volumetric flask and diluted to the mark with H_2O. After the precipitate settled the clear supernatant solution was used for analysis.

Dissolution by HF: 1 g of powdered sample (100 mesh) was weighed into a 30 ml Teflon screw-top bottle. 5 ml chilled HF were added and the bottle was cap tightly. The solution was allowed to stand at room temperature until dissolution was complete. The solution was transferred to a 100 ml plastic volumetric flask and diluted to the mark with H_2O. The resultant solution was used for analysis.

Procedure for Dissolution of Glass Films

A weighed quadrant (about 2 g weighed to nearest 0.01 mg) of a 4-inch diameter silicon wafer with deposited glass film was placed in a plastic Fluoroware® dish with cover. Using a plastic pipet, 5.0 ml of 10% HF (prepared by diluting 10 ml of HF reagent to 100 ml) were dispensed onto the wafer so that the solution wetted the film surface. The plastic cover was placed on the wafer dish and allowed to stand 20 minutes. As the film dissolved completely from the Si substrate, the solution beaded up on the polished Si surface and rolled off, causing the wafer to float on top of the solution. The solution was transferred to a 15 ml plastic vial and 5 ml 10% HNO_3 solution containing 0.1% Titron X100 surfactant (ethoxy octylphenols) were added. The stripped wafer was washed, dried, and weighed. The difference in weight before and after stripping represented the weight of the film sample.

Analysis of Small BPSG Sample

A silicon wafer chip with BPSG film was placed in a 10 ml plastic vial containing 2.5 ml of 10% HF. The solution was transferred to another 10 ml vial and 2.5 ml of a solution containing 10% HNO_3 and 0.1% Titron X100 surfactant were added. The concentration of B, P, and Si was determined by ICP-AES.

Preparation of Standard Solution

Na_2CO_3 fusion method: Into each of four 100 ml plastic volumetric flasks 1.0g Na_2CO_3 and 25 ml H_2O were added. Appropriate volumes of 1000 µg/ml B stock standard solution were added to give 0, 10, 50, 150 µg/ml concentrations when diluted to 100 ml.

HF dissolution: Into each of four 100 ml plastic volumetric flasks 5 ml of 48% HF and 5 ml HNO_3 were added. Appropriate volumes of stock standard solutions of B, Si, Ge, and P were added to each of the flasks to give a range of concentration from 0–150 µg/ml. After adding 1 ml of 5% Titron X100 solution the solutions were diluted to the mark with water.

RESULTS AND DISCUSSION

Most high silica materials can be decomposed via attack with 48% HF solution at temperatures up to 100° C.[11] The fluorides of Si and B are sufficiently volatile so that they may be partly or completely lost when aqueous solutions are heated. Silica reacts with HF according to Eq. 1

$$SiO_{2(s)} + 4HF_{(\ell)} \rightarrow SiF_{4(g)} \uparrow + 3H_2O_{(\ell)} \tag{1}$$

to form volatile silicon tetrafluoride, SiF_4. However, in the presence of excess HF the SiF_4 is converted to the complex hydrofluorosilicic acid, H_2SiF_6[12]

$$SiF_4 + 2HF \rightarrow H_2SiF_6. \tag{2}$$

At room temperature this complex was found to be sufficiently non-volatile in agreement with Langmyhr and Graff[13] who reported no significant loss of Si from cold HF solutions. Similarly, B_2O_3 reacts with HF in two steps to yield hydrofluoroboric acid which is not volatile at room temperature.[14,15,16]

$$B_2O_{3(s)} + 6HF_{(\ell)} \rightarrow 2BF_{3(g)} \uparrow + 3H_2O_{(\ell)} \tag{3}$$

$$BF_3 + HF \rightarrow HBF_4.$$

Determination of Silicon

Special precautions were taken to compensate for Si contamination originating from the reagents and the plasma torch. Although critical parts of the sample introduction system consisted of sapphire and alumina to resist HF solutions, appreciable Si blanks resulted from ablation by the plasma of the outer quartz tubes of the torch. The Si 251.61 nm peak for deionized water and air, shown in Fig. 2, indicated that Si contamination was coming from the torch.

594

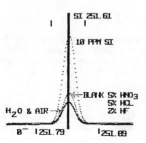

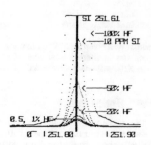

Figure 2. Silicon Contamination from Quartz Plasma Torch.

Figure 3. Effect of HF Concentration on Si Leaching from Quartz Plasma Torch.

The concentration of the HF in solution had a dramatic effect on the amount of Si contamination from the torch as shown ion Fig. 3. The beneficial effect of increasing HNO_3 concentration in a 5% HF solution containing 100 ppm Si is shown in Fig. 4. The precision of measuring the emission intensity of 100 ppm Si solutions with ten 2-second integrations varies from an unacceptable 2.5% RSD at 1% HNO_3 to less than 0.5% RSD for solutions containing >5% HNO_3.

Since the nebulization instability and the Si leaching effect were sensitive to acid concentration it was critical to match the acid content in the blanks, calibration, and sample solutions. The addition of 0.05% Titron X100 surfactant as a wetting agent promoted uniform drainage of the sample spray chamber for stablized intensity measurements.

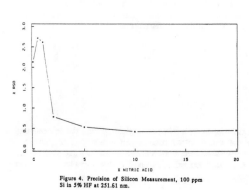

Figure 4. Precision of Silicon Measurement, 100 ppm Si in 5% HF at 251.61 nm.

Table II. Analysis of Optical Waveguide Preform by ICP, HF Dissolution.

	%	RSD
B_2O_5	0.0525 ± 0.0013	2.5
SiO_2	85.92 ± 0.32	0.4
GeO_2	9.38 ± 0.06	0.6
P_2O_5	3.76 ± 0	0
TOTAL OXIDES	99.11	

	%	
NBS 93a	FOUND	CERTIFIED
B_2O_3	12.18	12.53
SiO_2	79.48	80.8

Bulk Glasses

The analyses of optical waveguide preform core glass containing B, P, Ge, and Si and National Bureau of Standards SRM 93a glass by ICP-AES using the HF dissolution method described under EXPERIMENTAL is given in Table II. Replicate determinations on the preform gave good precision and accuracy with excellent recovery of the total oxides at 99.11%. The agreement of ICP-AES data with certified values in the NBS 93a glass indicated that B and Si were not volatilized by HF decomposition.

The HF dissolution method was tested on a wide variety of B-containing glasses. In Table III are listed the ICP-AES results obtained using the Na_2CO_3 fusion method and HF dissolution method. Data obtained using the two procedures are consistent with each other and compare favorably with prompt gamma data[17] The determination of B was shown to be effective over a large concentration range in the presence of widely varying matrices. The simpler HF method was preferred and used only with an HF resistant sample introduction ICP torch system.

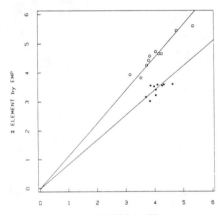

Figure 5. Comparison of B and P Determination, Inductively Coupled Plasma Spectrometry Electron Microprobe Analysis.

Table III. Boron Oxide in Various Glasses
BY ICP AT 249.8 nm

GLASS	NOM.	Na_2CO_3 FUSION	RSD	HF DISS.	RSP	PROMPT GAMMA
NBS 93a	12.53	12.7 ± 0.22	1.7	12.2 ± 0	0	—
7913	3	3.62 ± 0.02	0.6	3.50 ± 0.02	0.6	3.7
1720	5	4.91 ± 0.02	0.5	4.89 ± 0	0	5.1
N51A	9	11.1 ± 0.15	1.4	10.67 ± 0.06	0.6	10.8
7740	13	13.3 ± 0.23	1.7	12.56 ± 0.09	0.7	13.0
7720	15	14.7 ± 0.15	1.0	13.65 ± 0.49	3.6	14.4
7050	24	22.7 ± 0.06	0.3	21.83 ± 0.27	1.2	22.9
7070	26	24.5 ± 0.37	1.5	24.02 ± 0.04	0.2	25.0

Glass Films

The application of the HF method for glass film samples was investigated. Since HF selectively etches glass from Si wafers the film weight was readily determined by difference. The determination of B and P in film samples were performed on quadrants of 4-inch diameter Si wafers typically containing about 10-20 mg of deposit. The appreciable systematic disagreement of results between ICP-AES with electron microprobe (EMP) data on identical film samples as displayed in Fig. 5 was attributed to inherent calibration difficulties of the EMP method.

The results of individual ICP-AEC determination of samples 22 and 23 are given in Table IV. The weights in μg of B, P, and Si were determined in each sample solution and the film weights were calculated as the sum of corresponding oxides B_2O_2, P_2O_5, and SiO_2. The concentrations of B, Si, and P calculated using the sum of oxides as the sample weight were virtually identical to concentrations calculated using film weight measured by difference. Thus, it was not necessary to weigh the samples if total analysis of the components were performed. This was especially useful when the sample was too small for accurate weighing.

Table IV. Determination of B, P, and Si in BPSG Films on Si Wafers by ICP.

FILM	ug OXIDE CALCULATED FROM B, P, AND Si DETERMINED			TOTAL OXIDES	WEIGHT LOSS BY DIFFERENCE
SAMPLE	B_2O_3	P_2O_3	SiO_2	ug	ug
22	68.6	392	3570	4031	3980
23	251	729	6727	7707	7640

FILM	%B CALCULATED FROM		%P CALCULATED FROM		%Si CALCULATED FROM	
SAMPLE	TOTAL OXIDES	WEIGHT LOSS	TOTAL OXIDES	WEIGHT LOSS	TOTAL OXIDES	WEIGHT LOSS
22	.53	.54	4.24	4.13	41.4	42.1
23	1.01	1.02	4.31	4.17	40.8	41.2

Table V. Precision on Replicate BPSG Film Samples on Small Si Wafers, Sample Weights Ranged 50–300 μg Each Determination Made in Triplicate.

Sample	B		P	
	% by wt	% RSD	% by wt	% RSD
10-19-85				
18	4.27±.03	.8	4.51±.22	5.0
19	4.21±.05	1.1	4.49±.28	6.3
20	4.21±.03	.8	4.49±.16	3.6
11-6-85				
18	3.01±.10	3.3	6.06±.32	5.2
19	3.00±.13	4.5	5.92±.45	7.9
20	2.96±.07	2.4	5.97±.28	4.6
12-9-85				
18	5.52±.05	.9	4.01±.12	3.1
19	5.57±.08	1.5	3.72±.12	3.3
20	5.52±.09	1.5	3.78±.17	4.4
AVERAGE		1.9	AVERAGE	4.8

The precision with which B and P can be determined in small BPSG film samples by total analysis is shown in Table V. Samples from three positions of the deposition chamber on different composition runs were analyzed in triplicate on film weights ranging from 50-300 μg with average precisions of 1.9% for B and 4.8% for P. It was evident that adjacent CVD chamber positions produced BPSG films of uniform composition. Excellent agreement of B and P results are shown in Fig. 6 for various film samples determined by the small sample (weight as total oxides) and the large sample (weight by difference) methods. Total component analysis of silicate sample by ICP-AES provided a rapid, quantitative method for the determination of various glasses.

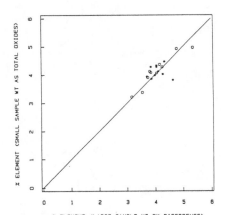

Figure 6. B and P Determination by ICP, Small vs.
Large Samples.

ACKNOWLEDGEMENTS

The author wishes to thank R. A. Levy, J. B. Machesney, and J. W. Mitchell for the various BPSG and bulk glass samples, J. E. Riley, Jr. for the prompt gamma data, and D. R. Wonsidler for the electron microprobe data.

REFERENCES

1. J. B. Mac Chesney, Proc. IEEE **68**, 1181 (1980).
2. S. R. Nagel, J. B. Mac Chesney, and K. L. Walker, IEEE J. of Quant. Elect. **QE-18**, 459 (1982).
3. W. Kern and R. C. Hein, J. Electrochem. Soc. **117** 562 and 568 (1970).
4. G. L. Schnable, W. Kern, and R. B. Comizzoli, J. Electrochem. Soc. **122**, 1092 (1975).
5. S. H. Weissman, Proc. of SPIE — The International Society for Optical Engineering **463**, 130 (1984).
6. P. K. Chu and S. L. Grube, Anal. Chem. **57**, 1071 (1985).
7. J. E. Tong and K. Schertenleib, and R. A. Carpio, Solid State Techn. **161**, (1984).
8. C. L. Wilson and D. W. Wilson, *Comprehensive Analytical Chemistry*, Vol. Ic, (D. Van Norstrond Co., Princeton, NJ, 1962), p. 96.
9. R. A. Burdo, J.of Non-Cryst. Solids **38** and **39**, 171 (1980).
10. V. A. Fassel, Anal. Chem. **51**, 1290A (1979).
11. B. Bernas, Anal. Chem. **40**, 1682 (1968).
12. R. Bock, *A Handbook of Decomposition Methods in Analytical Chemistry*, Wiley, N.Y., 1979, p. 56.
13. F. J. Langmyhr and P. R. Graff, Anal. Chem. Acta. **21**, 334 (1959).
14. R. W. Morrow, Anal. Lett. **5(6)**, 371 (1972).
15. W. P. Kilroy and C. T. Moynihan, Chem. Acta. **83**, 389 (1979).
16. R. A. Burdo and M. L. Snyder, Anal. Chem. **51**, 1502 (1979).
17. J. E. Riley, Jr. and R. M. Lindstrom, J. of Radioanal. Chem., to be published.

ZIRCONIA CHARACTERIZATION BY
PERTURBED ANGULAR CORRELATION SPECTROSCOPY

HERBERT JAEGER,* JOHN A. GARDNER,* JOHN C. HAYGARTH,** AND ROBERT L.
RASERA***
*Dept. of Physics, Oregon State University, Corvallis, OR 97331
**Teledyne Wah Chang Albany, Albany, OR 97321
***Dept. of Physics, University of Maryland Baltimore County, Catonsville,
MD 21228

ABSTRACT

Perturbed Angular Correlation spectroscopy of ^{181}Ta nuclei occupying
cation sites has been used to investigate microscopic structural properties
of several pure and stabilized zirconia materials. The PAC spectra are used
to derive the electric field gradients at the tracer nucleus. Thse are compared with calculations based on a simple point-ion model. Agreement between
experiment and theory is good for stabilized material and qualitatively
reasonable for pure zirconia.

INTRODUCTION

Perturbed Angular Correlation (PAC) spectroscopy is a method of measuring interactions of certain nuclei with the local atomic environment by
detecting gamma rays emitted during a radioactive decay.[1] The technique
is applicable at any temperature. We have recently described the application
of PAC to characterization of zirconia structural polymorphs.[2] In this
paper we report further experimental results and describe initial computer
modeling calculations of PAC spectra for monoclinic and tetragonal zirconia
and for cubic zirconia stabilized by yttria.

EXPERIMENTAL PROCEDURE

The tracer nucleus used in this work is ^{181}Hf, which decays to ^{181}Ta
predominantly by beta emission followed by emission of two gamma rays. The
goal of the experiment is to determine the function $G_2(t)$ that describes the
perturbation of the nucleus by local electric field gradients during the period between the gamma ray emissions. The radioactive tracer nuclei were produced by neutron-irradiation of high-purity zirconia and zirconia/yttria in
which the zirconia contains approximately one mole percent hafnia.

The gamma rays are measured with a four-detector experimental apparatus
having CsF scintillation detectors in a plane at 90° angular intervals. A
time-to-digital converter is started and stopped by timing signals from the
detectors, and an energy analyzer distinguishes between the first (133 keV)
and second (482 keV) gamma ray. A computer accumulates the spectrum $F_{ij}(t)$
of events in which the first gamma ray enters detector i and the second
enters detector j at time t later. By taking an appropriate ratio of the
experimental spectra[2], one determines the product of G_2 with the (known)
effective nuclear angular correlation parameter A_2. If each tracer nucleus
has the same environment, $G_2(t)$ is an oscillating function with three fre-
quencies, ω_1, ω_2, and $\omega_3 = \omega_1 + \omega_2$, characteristic of the nuclear electric
quadrupole interaction. If the electric field gradient is not the same for
all nuclei, $G_2(t)$ is a statistically-weighted average of the perturbation
functions for all nuclei. $G_2(t)$ for samples having two polymorphs present
simultaneously will display six frequencies in general.

EXPERIMENTAL RESULTS: MONOCLINIC AND TETRAGONAL ZrO_2

In Figure 1, the function $A_2G_2(t)$ is shown for pure monoclinic and tet-ragonal zirconia. The spectrum of a mixed-phase sample is also given. The Fourier transform power spectra of these functions show the two lower fre-quencies of each three-frequency group clearly. The highest frequency is barely detectable because of the finite detector time-resolution. No infor-mation is lost however, because ω_3 is the sum of the two lower frequencies. We note that the three frequencies of the tetragonal spectrum are in the ratio 1:2:3, as expected for axially symmetric sites. The corresponding ratio for the monoclinic frequencies is approximately 1:1.6:2.6. For non-axial symmetry, $1 < \omega_2/\omega_1 < 2$ in general.[1]

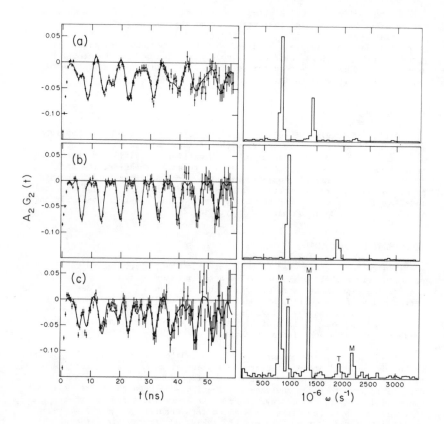

1.   PAC spectrum of ^{181}Ta in pure zirconia at (a) 400°C. (b) 1310°C, and (c) 1160°C. Time spectra are on the left, and the corresponding Four-ier transform power spectra are shown on the right. The solid curves on the time spectra are computer fits discussed in the text. (c) is a sum of spectra from monoclinic (shown by M on the Fourier spectrum) and tetragonal (shown by T) grains. In (c) the amplitude of the third T-frequency is too small to be seen.

The T/(T+M) fraction of the sample in Fig. 1(c) is approximately 0.3. Measurements on this sample were made at several temperatures as T was increased through the M/T transition region and then decreased. The tetragonal fraction is shown in Fig. 2, and the data display the well-known hysteresis characteristic of this martensitic transformation. Zirconia measurements over a wider temperature range were made later with two different samples. Near the transition, the data agreed well with that of the first sample. The frequencies and the resulting values of the electric field gradient magnitude V_{zz} and asymmetry parameter η were determined by computer-fitting these spectra. V_{zz}, η, and the relative frequency width δ are shown in Fig. 3. The parameter δ describes the frequency broadening due to small variations in the field gradients due to random defects, impurities, disorder, etc. δ is unexpectedly large in monoclinic zirconia and is sample-

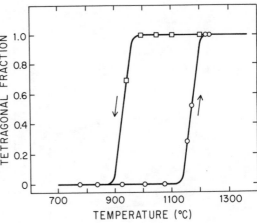

2. Tetragonal fraction determined for pure ZrO_2 from [181]Ta PAC spectra. Circles are data on heating, squares on cooling. The lines are a guide to the eye.

dependent. We note that δ is larger for the monoclinic sample that had been previously transformed to tetragonal than for the sample that had not been heated above the calcining temperature of 800°C. For the latter sample, low temperature spectra were taken before and after the highest temperature PAC measurements, and the values of δ were reproducible. Within experimental uncertainty, δ was zero for all tetragonal spectra. These data indicate that there is significant structural disorder on the scale of only a few lattice constants in the monoclinic zirconia samples.

Fig. 3 also shows model calculations of the electric field gradient using the simplest possible point-ion model. Model assumptions are +4 charge for Zr and -2 for O. The Sternheimer antishielding parameter is taken as -61.[3] The tetragonal structural data used for line T_A, T_B is taken from ref. 4, 5 respectively. The monoclinic structural data for line M are taken from refs. 6 and 7. For M and T_A, the basis is assumed to expand uniformly with temperature, with lattice expansion coefficients from ref. 8. For T_B, the basis temperature dependence is taken as a least square linear fit to structural data of ref. 4. This model is too simple, and it is not surprising that the calculations do not agree quantitatively with experiment. The model does predict the correct asymmetry parameter for the monoclinic phase. V_{zz} is extremely sensitive to the oxygen ion basis position, as indicated by T_A and T_B in Fig. 3(b). The oxygen basis positions differ slightly in the structural data used for T_A and T_B. This sensitivity has been verified by a more accurate theory.[9] Further experimental and

theoretical efforts[9,10] are underway that should provide a better understanding of the microscopic structure and resulting electric field gradients.

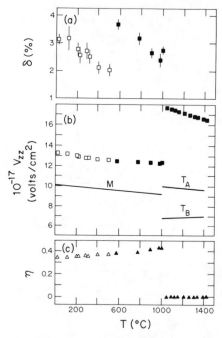

3. (a) Frequency width δ, (b)
electric field gradient magnitude
V_{zz}, and (c) anisotropy parameter
η, for ^{181}Ta PAC spectra in pure
zirconia. Data shown by open
symbols are from samples that
were never heated above the monoclinic-tetragonal transition temperature. Data shown by solid
symbols are from a sample whose
spectra were initially measured
above the monoclinic-tetragonal
transformation temperature. The
solid lines in (b) are model fits
discussed in the text. The calculated (η) values are not shown
but agree well with (c).

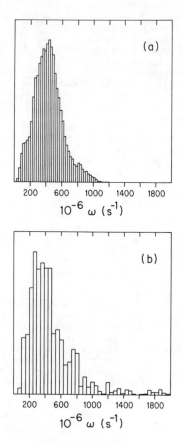

4. (a) Computer simulation discussed in text and (b) experimental
Fourier transform of 1450°C ^{181}Ta
PAC time spectrum of cubic zirconia
stabilized by 18.4 wt. % yttria.

EXPERIMENTAL RESULTS: CUBIC YTTRIA-STABILIZED ZrO_2.

Figure 4 shows the Fourier power transform spectrum for a cubic yttria-
stabilized zirconia sample at very high temperature. Approximately 20% of
the cations in this material are yttrium, and approximately 5% of the oxygen
lattice sites are vacant. The frequency distribution is wide, but the

average frequency is smaller than the frequencies in pure ZrO_2. Low temperature spectra of this sample are qualitatively similar but are characterized by much larger frequencies. We believe that the oxygen vacancies contribute significantly to the electric field gradient only at low temperature. At very high T, the vacancy diffusion rate is much larger than the PAC frequency, and the electric field gradient due to vacancies averages to a negligible value. If this interpretation is correct, the high-T electric field gradient and resulting PAC frequency distribution for Ta tracers is due to variation in the distribution of yttria on surrounding cation sites. The point-ion-model computed frequency distribution due to a random yttria atom distribution is shown in Fig. 4. One expects the point-ion model to be more appropriate for this situation, and agreement between model and experiment is excellent. Experimental measurements and theoretical modeling are still in progress for low and intermediate temperatures.

ACKNOWLEDGEMENTS

This research was supported by Teledyne, Inc. and by the US Department of Energy under contract DE-FG0685ER45191.

REFERENCES

1. H. Frauenfelder and R. M. Steffen, in Alpha-, Beta- and Gamma-Ray Spectroscopy, Vol. 2, ed. K. Siegbahn (North Holland, 1965), p. 997.

2. H. Jaeger, J. A. Gardner, J. C. Haygarth, and R. L. Rasera, J. Am. Ceram. Soc. 69, in press.

3. F. D. Feiock and W. R. Johnson, Phys. Rev. 187, 39 (1969).

4. G. Teufer, Acta Cryst. 15, 1187 (1962).

5. P. Aldebert, J. P. Traverse, J. Am. Ceram. Soc. 68, 34 (1985).

6. D. K. Smith and H. W. Newkirk, Acta Cryst. 18, 983 (1965).

7. J. D. McCullough and K. N. Trueblood, Acta Cryst. 12, 507 (1959).

8. R. N. Patil and E. C. Subbaroao, J. Appl. Cryst. 2, 281 (1969).

9. H. J. F. Jansen and J. A. Gardner, Bull. Am. Phys. Soc. 31, 529 (1986).

10. H. Jaeger, J. A. Gardner, R. L. Rasera, and J. C. Haygarth, Bull. Am. Phys. Soc. 31, 529 (1986).

EPR OF Cu(II) IN SOL-GEL PREPARED SiO$_2$ GLASS

S. DAVE* AND R.K. MACCRONE**
* Perkin Elmer, 761 Main Ave., Norwalk, CT 06859-0420
**Materials Engineering Department and Department of Physics,
  Rensselaer Polytechnic Institute, Troy, NY 12180-3590

ABSTRACT

SiO$_2$ glass has been prepared by the hydrolysis of TEOS in alcohol
using HCl and NH$_4$OH to control the pH. Copper ions were incorporated
from the acetate to act as structural probe ions during the process of
gellation and calcination. The local structure of copper ions was investigated using EPR determined at 300 K and 10 K.

It was found that the pH of the solution had a profound effect on
the EPR spectra of the copper in the gels and glasses, indicating differences
in the local structure.

INTRODUCTION

In the hydrolytic polymerization of TEOS it is now well known that
the pH of the solution effects the mode of chain formation: in acid solutions
long chains are formed while branched structures arise in alkali solutions
[1,2,3,4]. These different polymer morphologies effect the gross microstructure profoundly, giving monolithic structures in one case and spherical
particles in the other. In order to investigate whether these polymeric
differences result in differences in atomic scale structure, it was decided
to incorporate transition metal probe ions into the process, ions whose
optical and magnetic properties are sensitive to the local environment.
In particular we decided to use the EPR (electron paramagnetic resonance)
of copper(II) ions as the major structural probe, this ion being well
studied [5,6], and in particular, in glasses [7,8,9].

EXPERIMENTAL

The glass material was prepared by mixing TEOS and ethyl alcohol
in equal amounts. One half of the stoichiometric amount of water was
added, together with the appropriate catalyst, and the reaction allowed
to proceed for one hour at 30 Celsius. Copper acetate was dissolved
in the remaining water and introduced into the system. Gellation took
place usually within 24 hours. After standing for one week to allow the
solvents to evaporate, the temperature was raised to 60 Celsius and the
gels held for an additional three days. Transformation to glass was
achieved by heating the gel at 1 C/min to 400°C, finally increasing to
600°C and soaking for 12 hours.

The EPR was measured at X-band using a homemade homodyne and a commercial
instrument.

RESULTS

The typical EPR spectra at room temperature of gels formed using
0.01 mole NH$_4$OH catalyst (high pH) and that of the glass resulting from
their calcination are shown in Fig. 1a and 1b, respectively. The corresponding spectra for the material formed using 0.01 mole HCl catalyst (low
pH) are shown in Figs. 2a and 2b.

As can be seen the copper ions in the gel formed under alkali conditions
show well developed hyperfine lines, and the copper in the derived glasses
show a wide distribution of overlapping hyperfine lines. On the other

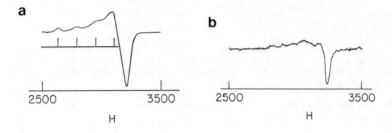

Fig. 1a and 1b.   The EPR spectra at room temperature of
high pH gel and resulting glass.

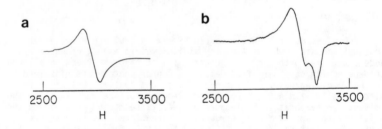

Fig. 2a and 2b.   The EPR spectra at room temperature of low
pH sol and resulting glass.

hand, no hyperfine lines are seen in the acid formed gels and the corresponding derived glass. Thus it is immediately evident that the structural environments of the copper ions differ between the two materials. Indeed this is apparent by eye, where the alkali material is a deep blue color, compared to the pale green-yellow color and light blue color of the acid gel and acid glass, respectively.

The low temperature spectra of the alkali catalysed gel and glass are shown in Figs. 3a and 3b, respectively, and correspondingly for the acid catalysed gel and glass in Figs. 4a and 4b. The spectra of the copper ions in all four materials change dramatically from that observed at room temperature.

Most noticeable is that all material now exhibits well resolved hyperfine lines. The magnitude of these hyperfine lines relative to the magnitude of the lines in the g_p region (high field region) have changed sharply in the alkali catalysed material, and indeed there seems to be no obvious correlation between the magnitude of the hyperfine lines and the high field features in these spectra.

DISCUSSION

The Hamiltonian appropriate to the copper resonance is given by:

$$H = \beta g//(Hz.Sz) + \beta g_p(Hx.Sx + Hy.Sy) + A//(Sz.Iz) + A_p(Sx.Ix + Sy.Iy)$$

where the terms have their usual meaning. Where feasable, the EPR spectra in the above figures have been fitted to the above expression. The values of the fitting parameters are shown in Table I.

Maki and McGarvey [11], and Kinelson and Heyman [10] have calculated the parameters of the spin Hamiltonian using LCAO's of the copper ion and the four oxygen ligands. The explicit values for g and A depend upon the admixture of the ligand-metal wave functions as well as other parameters such as the energy splittings, the spin orbit coupling and the Fermi contact term. In melted glasses, the work of Imagawa [7] and Kawazoe et al. [8] has shown that the EPR results reflect most sensitively on the parameter β_1, which indicates the degree of covalency of the excited B_{2g} hole state, through the equation:

$$\psi(B_{2g}) = \beta_1 d_{xy} - (1-\beta_1{}^2) \frac{P_y{}^{(1)}+P_x{}^{(2)}-P_y{}^{(3)}-P_x{}^{(4)}}{2} ,$$

Qualitatively the values of A// and g// correlate with β_1, the degree of covalency, in the following way: the larger the value of A// and the smaller the value of g//, the smaller the value of β_1. This in turn reflects on a larger covalency. We use this correlation in our further discussion, since we do not have all the required parameters to determine the wave functions quantitatively, (this would require the resolution of the perpendicular hyperfine lines, and independent determination of ΔE_{xy} and ΔE_{xz} by optical measurements, cf. Duran and Fernandez Navarro [9].

The alkali catalysed gel shows the most straightforward result, namely that the copper is contained in a site well defined at both high and low temperatures. The values of the fitting parameters and A// in particular, are not surprisingly perhaps, very different to the values reported for copper ions in melted glasses; A// $\sim$ 200x10^{-4} cms^{-1} compared to A// $\stackrel{\sim}{\sim}$ 150 x 10^{-4} cms^{-1} for melted glasses. The values for g// at low temperature are correspondingly different, 2.248 compared to $\sim$2.35. These results, based on the discussion above, show that the copper ion is apparently in a site where there is appreciably more covalent bonding than in melted glasses. This is not inconsistent with a structure containing many organic components.

On the other hand, the copper ions in the acid catalysed gel at

608

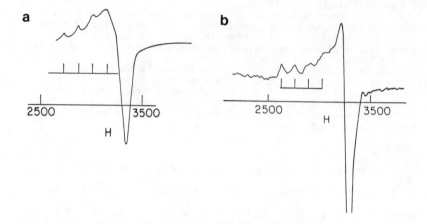

Fig. 3a and 3b. The low temperature EPR spectra of high
pH gels and resulting glass.

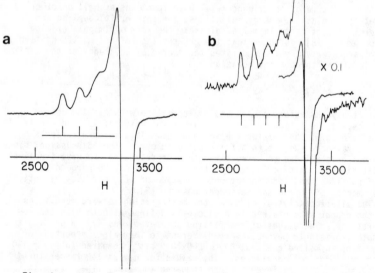

Fig. 4a and 4b. The low temperature EPR spectra of low
pH gels and resulting glass.

TABLE I

T = 300 K				
Catalyst	Material	$g_{//}$	$g_{\rho-}$	$A_{//}(10^{-4}\text{cms}^{-1})$
NH$_4$OH	Gel	2.347	$\sim$2.16	198
	Glass		$\sim$2.14	
HCl	Gel	2.28		
	Glass	2.232	2.14	
T = 10 K				
NH$_4$OH	Gel	2.248	2.044	190
	Glass	2.349	2.064	143
HCl	Gel	2.292	2.075	150
	Glass	2.378	2.054	145

low temperatures display hyperfine and g values much closer to those observed
in melted glasses at room temperature.

At low temperatures, the copper ions in the glasses prepared by both
routes give very similar EPR spectra. Again, we find that the hyperfine
term, $A_{//}$ are $g_{//}$ are close to those of copper in melted glasses. Close
inspection reveals important differences in the melted glass Spin Hamiltonian
parameters and those observed here. For acid catalysed glasses, the parameters
are very close, indicating that the melted and sol-gel glasses are essentially
indistinguishable. For alkali catalysed glasses, the value of $g_{//}$ is
significantly lower. Additional work is needed to understand this observation.

At high temperatures on the other hand, the state of the copper ions
in the glasses prepared by the alkali and acid route are very different.
In the alkali catalysed glass, the distribution of sites has become very
much wider, to such an extent that no distinct hyperfine lines are observed.
In the acid catalysed glass, even the site symmetry has disappeared,
as indicated by the symmetric Lorentzian line. This implies that a re-
structuring of the environment of the copper ions is taking place on times
short compared to $1/\nu$, where ν is the microwave frequency $\sim 10^{10}$ Hz.
Of the possible mechanisms we mention that of extremely soft vibrational
modes associated with copper ion environment, that of diffusion of the
copper ions in the glass and that electron hopping between Cu^{+1} and Cu^{+2}
ions, Griscom [12].

CONCLUSION

The EPR spectra of Cu(II) at room temperature and near liquid helium
temperature show that the local structure in alkali and acid calaysed
gels and glasses are very different from each other.

· The alkali catalyzed gel shows the most well defined copper ion site,
  whose parameters suggest the most covalently bonded site of all the
  systems considered here.

· The alkali and acid catalysed glasses at low temperatures are very
  similar with respect to the copper ion sites, and these sites are
  similar to those found in melted glasses. The alkali catalysed glass
  shows the greatest difference.

· The copper ion sites in the alkali and acid catalysed glasses at
  room temperature are very different: the alkali catalysed glasses

610

shows a tetragonal site for the copper with a wide distribution in hyperfine coupling while the acid catalysed glasses show a symmetric site.

ACKNOWLEDGEMENTS

We appreciate the support of an IBM Fellowship (S.D.) and NSF support under Contract No. DMR-8510617.

REFERENCES

1. K.D. Keefer, in Better Ceramics Through Chemistry, MRS Symposium Vol. 32, Eds. C.J. Brinker, D.E. Clark and D.R. Ulrich, (1984), p. 15.
2. D.W. Schaefer and K.D. Keefer, ibid., p. 1.
3. C.J. Brinker, W.D. Drotning and G.W. Scherer, ibid, p. 25.
4. See also many papers in this symposium, esp. V.V. Mainz et al. and C.A. Salfe.
5. B. Bleaney, Proc. Roy. Soc. 42 (1951) 441.
6. B.J. Hathaway and P.G. Hodgson, J. Inorg. Nucl. Chem., 35 (1983) 4071.
7. H. Imagawa, Phys. Stat. Sol., 30 (1968) 469.
8. H. Kawazoe, H. Hosono, and T. Kanazawa, J. Non-Cryst. Solids, 29 (1978) 173.
9. A. Duran and J.M. Fernandez Navarro, Phys. Chem. Glasses, 26 (1985) 126.
10. D. Kivelson and R. Neiman, J. Chem. Phys. 35 (1961) 149.
11. A.H. Maki and B.R. McGarvey, J. Chem. Phys. 33 (1960) 1074.
12. D. Griscom, Private Communication (1986).

PHYSICO-CHEMICAL CHARACTERIZATION OF ALUMINA SOLS PREPARED FROM ALUMINUM ALCOXIDES

William L. Olson, Signal Research Center, 50 E. Algonquin Rd., Des Plaines, IL 60017.

ABSTRACT

Alumina sols derived from aluminum sec-butoxide (Yoldas) were characterized. The distribution of the polymer sizes within the sol, determined by gel filtration chromatography (GFC), was found to be dramatically affected by small changes in the chemical processing or preparative procedure. Aging the sol at room temperature for two weeks produced no significant change in the GFC elution curves of the alumina sol. Sols with a "milky" appearance were found to exhibit a wider distribution of polymers by GFC than transparent sols. Rotary evaporation of the sol followed by redissolution of the residue was found to change the polymer size distribution described by the gel filtration elution curves. These observations coupled with ^{27}Al NMR spectroscopy and viscometry measurements were used to elucidate the effects of process conditions and aging on the molecular structure of the sol.

INTRODUCTION

Metal oxide powders prepared by the hydrolysis of metal alcoxides possess several technical advantages over conventionally prepared materials and, as such, are currently under intense scrutiny. The potential applications of these materials include such diverse areas as high purity, controlled morphology powder preparation, dielectric and corrosion resistant coatings, optical fibers, sensors and separation membranes[1-2].

In order for the numerous technical advantages of sol-gel derived metal oxide ceramics to be fully realized, methods for characterizing the ceramic precursor prior to powder isolation must be developed [2]. By developing a better understanding of the effects of temperature, aging, atmosphere and precursor choice on the structure of the polymers within the sol, much of the empiricism that exists in sol-gel technology today could be eliminated.

At present, little is known about factors such as the average polymer molecular weight, molecular weight distribution, size, or the molecular structure of sols prepared from metal alcoxides other than silicon [3]. In an effort to quantify and better understand the polymer structure-material property relationships that exist in this area of ceramics, several techniques widely applied for the characterization of organic polymers were used for the characterization of sols prepared from aluminum sec-butoxide. The results of that study are reported herein.

EXPERIMENTAL

The alumina sol was prepared as described by Yoldas [4]. In a typical experiment, deionized water (750mL) was placed in a 2 liter reaction vessel equipped with a Teflon® coated stir bar. Aluminum sec-butoxide (244g) was added to the water in 50g portions. When the addition was complete, the system was allowed to equilibrate at 70-80 C for thirty minutes. Concentrated HNO_3 (3mL, acid/Al mole ratio=0.05) was added and the reaction heated with stirring at 100C for 24 hours. Periodically, the reaction vessel was shaken to aid in redissolving any gelatinous deposits which formed on the reactor walls. The reaction was then cooled to room temperature. Upon standing, the reaction mixture quickly separated into two transparent layers. The lower water layer which contained the sol was collected, filtered (Whatman 41) , and stored in a brown glass bottle.

For the aging study, a transparent sol prepared by the above procedure was divided into two 400mL portions. The first portion was stored at room temperature while the second was stored in a refrigerator at 7°C for the duration of the study. Samples of each sol were periodically removed for characterization.

The gel filtration chromatographic experiments were carried out on a Waters GPC II Liquid Chromatograph using a column of dimensions 12.7mm X 23.5cm. Sephadex G-10 was the support material. The flow rate was maintained at 1mL/min. The viscosity studies were conducted on sols that were gradually reduced in volume by rotary evaporation at room temperature. Periodically during concentration of the sol, 5mL samples were removed for analysis on a Haake Rotovisco viscometer equipped with a cone and plate sensor system. All viscosity measurements were made at 25C. The ^{27}Al NMR spectra were measured on a Nicolet NT-300 spectrometer at 78MHz. The experimental parameters used for data acquisition were a pulse width of 30 microseconds, a pulse delay of 1 second, and a sweep width of 15000 Hz.

RESULTS AND DISCUSSION

Gel filtration chromatography (GFC) is a powerful tool for the characterization of organic polymers. The ability of the method to give a quantitative analysis of the molecular weight distribution of the polymer [5-6], given suitable standards, enables researchers to fully document the effects of processing conditions on the structure and properties of the material. However, GFC has found limited application to inorganic systems [7-13]. This partly arises from the difficulty in obtaining gel supports that are compatible (i.e., do not interact) with the material under study. To date, efforts to characterize the polymer distributions that exist within alumina sols have relied primarily on two techniques, aluminum-27 NMR [7,14-17] and colorimetric titration [18-19]. While useful, these techniques fall short of providing the detailed determination of the polymer size distribution that is available via GFC [7,18].

Gel filtration chromatography achieves this analysis by separating chemical species by size. Separation is achieved by passing the sample mixture over a controlled porosity gel. Small molecules will rapidly diffuse into the open pore structure and are retained while larger molecules are excluded from the internal porosity of the gel. Thus, molecules are eluted from the column in order of decreasing size. This is in marked contrast to all other types of chromatography, where larger molecules are more strongly retained and are sometimes eluted only with difficulty.

Although interpretation of the GFC elution curves appears straightforward, it could be considerably more complicated. While in theory gel filtration chromatography separates polymeric mixtures by size with the largest species being eluted first, in practice this is true only if the solute-support interactions are minimal. Preferential absorption of large molecules onto the column support material can disrupt the elution order of the molecules. In addition, any ionic polymers in the sol which might normally be completely excluded by a particular gel network can be retained because of counterion diffusion into the gel pores.

It is not thought that the support material used in the study, Sephadex G-10, suffered from these limiting factors. The material performed well over the course of the study. There was no increase in column backpressure over dozens of runs, indicating that retention/adsorption effects were minimal. Column performance was reliable and reproducible.

This is in marked contrast to other support materials that were tested. Our efforts to extend this work to include gel supports with higher exclusion (i.e., molecular weight) limits were only partially successful. Introduction of the alumina sol into columns containing Fractogel HW 55F, for example, was found to result in column backpressures exceeding 1000psig. The support material became hard, indicating that some reaction between the sol and support had occured. Other materials that were tested with limited success included Sephacryl S-300 and Fractogel HW40S. As can be appreciated from these results, compatibility of the sol with the gel support is critical for the success this technique. This factor alone may limit the utilization of this method for examining other sol compositions and chemistries.

GFC Results: The unique property of gel filtration chromatography to separate complex polymeric mixtures by molecular size was used to examine the effects of aging on the molecular weight distributions of alcoxide-derived alumina sols. Three systems for detecting the elution of the sol-gel polymers (electrical conductivity, UV absorption (254nm), and index of refraction) were evaluated. Representative elution curves for each of them are presented in Figure 1. Of the three, the electrical conductivity detector was the least useful. Even at the highest level of sensitivity, the detector response was very poor, consistent with the fact that Al sol polymers exist largely as neutral species in solution.

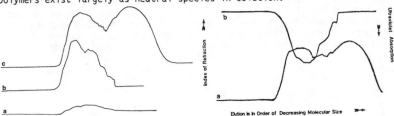

Figure 1. Elution curves of three detector systems used for gel filtration chromatography: a) conductivity b) ultraviolet absorbance (254 nm) and c) index of refraction

Figure 2. Elution curves of an alumina sol obtained utilizing a dual detector system. Curve a) corresponds to the index of refraction detector, b) is the ultraviolet absorption elution curve.

The UV absorption and index of refraction detectors were found to compliment each other and were used in tandem on most of the chromatographic runs. UV absorption spectra of the individual sol components indicated that the absorption at 254nm was specific for the alumina polymers within the sol. Comparison of the UV and index of refraction detector responses shown in Figure 2, reveals that there is a close correspondence of the first absorption in both traces. The second peak of curve (a) is due to the elution of sec-butanol. This component is not detected by the UV detector since the alcohol does not absorb at 254nm. The assignment of the second peak in curve (a) as arising from 2 butanol was verified by comparison with GFC elution curves of pure sec-

614

butanol.

Portions of a sol were stored at 7 and 24C and GFC elution curves of the sols were obtained over a two week period. Since a 10 degree drop in temperature will decrease the rate of a chemical reaction by an order of magnitude (approximately), any "aging" processes involving a redistribution of the polymer sizes should be fairly evident by comparison of the two samples.

Chromatograms obtained on the refrigerated and room temperature sols over a period of several weeks showed no differences between the two samples (see Figure 3). Elution curves obtained early in the aging study were comparable to chromatograms run two weeks later. The features of the peaks, although rather broad and ill-defined, were found to be very reproducible on a day-to-day basis. The consistency of the elution curves indicates that no significant change in the polymer size distribution took place under the conditions of the study. These observations suggest that either the sol is at equilibrium or that the redistribution reactions of the polymers are slow at room temperature.

Figure 3. Gel filtration chromatograms of an aluminum sec-butoxide derived sol stored at a) 24 C and b) 7 C for three days following their preparation. The system utilized an ultraviolet absorption detector.

GFC elution curves obtained on portions of aged sols in which the sec-butanol was removed via rotary evaporation are presented in Figures 4 and 5. In this procedure, alumina sols were concentrated to half volume by rotary evaporation followed by redilution to their original volume with distilled water. This treatment was repeated twice to insure the complete removal of the sec-butanol (sec-butanol/water azeotropic composition 68/32) which makes up approximately 9-15% (by weight) of the original sol. The elution curves for the resultant sols were found to be quite comparable to the original (alcohol containing) sol except for the notable absence of the second peak (lower trace: refractometer response) previously ascribed to sec-butanol. The excellent agreement between the elution curves of the original and alcohol-free sols indicates that rotary evaporation had little effect on the size distribution of the alumina polymers in the sol.

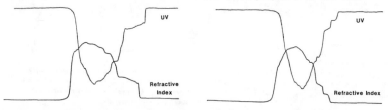

Figure 4. Molecular exclusion chromatogram of a rotary evaporated alumina sol stored at 7 C for 2 weeks

Figure 5. Molecular exclusion chromatogram of a rotary evaporated alumina sol stored at 24 C for 2 weeks.

Changes in the polymer size distributions described by the GFC elution curves were observed however, when alumina sols which had been taken to dryness by rotary evaporation were chromatographed. The glassy boehmite precursor residue which resulted was redissolved in water and analyzed by GFC. The GFC chromatogram, displayed in Figure 6a, shows that the polymer size distribution within the sol has sharpened dramatically relative to the original given above it. From the shift of the chromatogram to shorter retention times, it appears that the average size of the polymer has increased.

Another intriguing result was obtained when a gel filtration chromatogram was obtained on a milky alumina sol. Milky sols are prepared in much the same way as transparent sols except they are heated for a shorter time at 100C (12 hrs) or at a lower temperature. The elution curve, Figure 6(c), reveals that the polymer size distribution is considerably broader relative to that of the transparent sol, Figure 6(b), used in the aging study.

While it is clearly apparent from these results that GFC is a powerful tool for the qualitative characterization of alumina sols, quantitation of the results is considerably more difficult. Ultimately, quantification of these results requires that suitable standards be developed for calibrating the retention time-molecular weight relationships of this system. This is important not only for the retention times but also for calibration of the response of the detector in order to determine the number of molecules at each point in the elution curve. Thus while excellent for qualitative sol analysis, no definitive determination of the molecular weight distribution is possible at this time. Work on this problem is in progress.

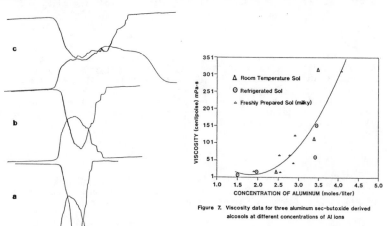

Figure 6. Molecular exclusion elution curves of three alcosols. a) A sol which had been taken to dryness and redissolved. b) Room temperature sol after alcohol removal by rotary evaporation. c) Fresh alumina sol prepared by a similar experimental procedure but which has a "milky" appearance

Figure 7. Viscosity data for three aluminum sec-butoxide derived alcosols at different concentrations of Al ions

Viscosity Study : Portions of a sol aged for two weeks at 7 and 24 C and of a freshly prepared "milky" sol were concentrated at room temperature on a rotary evaporator and samples removed periodically for analysis. The results, given graphically in Figure 7, show that the viscosity of the sol increased slowly until the aluminum concentration reached 3M after which the viscosity of the sol rapidly increased in an exponential-like fashion. Further concentration of the sol past 4M in [Al] was not attempted since a semi-solid, gel-like state had been attained. Several of the gels at this point were thixotropic (their resistance to flow decreasing as a function of the shear rate).

Due to the scatter in the data presented in Figure 7, it is difficult to draw any definitive conclusions concerning the effects of aging or turbidity on the viscometric properties of this system. Overall, the sols appeared to have similiar viscosity characteristics. If there were differences in this regard, the experiment was not precise enough to uncover them.

Aluminum 27 NMR : The ^{27}Al NMR spectra of the Al sols, in agreement with the GFC results, did not indicate any differences between refrigerated and room temperature sols over a period of several weeks. The linewidth and intensity of the broad Al resonance at 5ppm arising from octahedrally coordinated aluminum species [15] was found to be completely independent of aging.

Differences could be resolved however, between sols made by slightly different experimental procedures. While the differences were subtle they were highly reproducible. Under identical experimental conditions, milky alumina sols gave ^{27}Al NMR resonances which were considerably broader (30%) than transparent sols. This difference in line widths can be readily ascribed to the larger distribution of polymers within the milky sol.

CONCLUSIONS

Molecular exclusion chromatography was found to be a powerful tool for the qualititative characterization of alumina sols prepared from aluminum alcoxides. This technique coupled with ^{27}Al NMR and viscometry gave no evidence of aging at room temperature. Overall, these techniques have considerable promise for developing the data base necessary for elucidating the polymer structure/material property relationships of sol-gel materials.

ACKNOWLEDGEMENTS

I would like to thank Bob Swensen and Lorenz Bauer for their assistance in obtaining the GFC and ^{27}Al NMR results and Michael Schoonover for his helpful comments on various aspects of this work.

REFERENCES

1. J. D. Mackenzie, Proceedings of the International Conference on Ultrastructure Processing of Ceramics, Glasses, and Composites, pp. 15-25 (1984).
2. D. R. Uhlmann, B. J. J. Zelinski, and G. E. Wnek, Better Ceramics Through Chemistry, Mat. Res. Soc. Symp. Proc., 32, pp. 59-70 (1984).
3. K. D. Keefer, Better Ceramics Through Chemistry, Mat. Res. Soc. Symp. Proc., Vol. 32, pp. 15-24 (1984).
4. B. E. Yoldas, Am. Ceram. Soc. Bull. 54, 289 (1975).
5. K. A. Granath and B. E. Kvist, J. Chromatog. 28, 69 (1967).
6. J. C. Giddings and K. I. Mallek, Anal. Chem. 38, 997 (1966).
7. J. W. Akitt and A. Farthing, J. Chem. Soc. Dalton 1981, 1606.
8. S. Ohashi, N. Yoza, and Y. Ueno, J. Chromatog. 24, 300 (1966).
9. S. Felter, G. Dirheimer, and J. P. Ebel, J. Chromatog. 35, 207 (1968).
10. P. A. Neddermeyer and L. B. Rogers, Anal. Chem. 41, 94 (1969).
11. C. A. Streuli and L. B. Rogers, Anal. Chem. 40, 653 (1968).
12. T. Tarutani, J. Chromatog. 50, 523 (1970).

617

13. R. A. Henry and L. B. Rogers, Separation Sci. <u>3</u>, 11 (1968).
14. J. W. Akitt and A. Farthing, J. Chem. Soc. Dalton <u>1981</u>, 1617.
15. J. W. Akitt and A. Farthing, J. Chem. Soc. Dalton <u>1981</u>, 1624.
16. J. W. Akitt, N. N. Greenwood, B. L. Khandelwal, and G. D. Lester, J. Chem. Soc. Dalton <u>1972</u>, 604.
17. J. W. Akitt and A. Farthing, J. Mag. Res. <u>32</u>, 345 (1978).
18. V. S. Schonherr and H. P. Frey, Z. Anorg. Allg. Chem. <u>452</u>, 167 (1979).
19. V. W. Gessner, and M. Winzer, Z. Anorg. Allg. Chem. <u>452</u>, 157 (1979).

REACTIVITY OF SILICATES 1. KINETIC STUDIES OF THE HYDROLYSIS OF LINEAR AND CYCLIC SILOXANES AS MODELS FOR DEFECT STRUCTURE IN SILICATES*

CAROL A. BALFE, KENNETH J. WARD, DAVID R. TALLANT, AND SHERYL L. MARTINEZ
Sandia National Laboratories, Albuquerque, NM 87185

ABSTRACT

The kinetics of hydrolysis of hexamethylcyclotrisiloxane and di-t-butyldimesitylcyclodisiloxane in tetrahydrofuran solution have been determined and compared to hydrolysis rates of silica defects. In the presence of sufficient excess water, the first-order rate constant of the cyclotrisiloxane, $k_1 = 3.8 \times 10^{-3}$ min^{-1} is similar to the rate constant, $k = 5.2 \times 10^{-3}$ min^{-1}, of the disappearance of the D2 Raman silica defect band it has been proposed to model. Limited hydrolysis rate data for the cyclodisiloxane suggests that it hydrolyzes at least four times faster than does the cyclotrisiloxane. These data are consistent with rate data available for silica crack growth and support the assignment of highly strained siloxane bonds at the crack tip to cyclodisiloxanes. Infrared spectra determined for the cyclodisiloxanes lend further support to this model.

INTRODUCTION

The relationship between local structure and chemical reactivity is fundamental to understanding the properties of materials. In a number of silica systems, local structures not accounted for by a random network model have been identified by the presence of unusually sharp bands in their Raman spectrum [1-5], or "anomalous" bands in their infrared [6], vacuum ultraviolet [7], or EPR spectra [8]. Typically, the presence of these features is dependent on environmental effects such as temperature [3,10], neutron irradiation [9], and the presence of moisture [3,4]. Consequently, the structures responsible for these anomalies are thought to be defects either in the bulk silica or on the surface [2,3].

Among the known silica defect structures are those responsible for strongly polarized Raman bands [1-5] at 606 cm^{-1} and 495 cm^{-1}. Comparable bands have been reported in sol-gel prepared silicates [13,14], and in leached borosilicate glasses [15]. The band at 495 cm^{-1} is commonly referred to as D1, and that at 606 cm^{-1} as D2. A number of structures such as broken bonds, peroxy bridges, and highly strained cyclodisiloxanes have been proposed to explain D1 and D2 [3,10,12,16-18]. However, comparison of the Raman spectra of model cyclotrisiloxanes and of rock-forming cyclotrisilicates with that of D2 has provided strong evidence that cyclotrisiloxane, $(SiO)_3$, units attached to the random network are responsible for the 606 cm^{-1} band [19]. Similarly, cyclotetrasiloxane, $(SiO)_4$, units are probably responsible for D1 at 495 cm^{-1} [19]. Furthermore, the planar cyclotrisiloxane ring consists of strained siloxane bonds, while the puckered cyclotetrasiloxane ring contains no strained bonds [19-21]. Hence, a cyclotrisiloxane ring is predicted to be more reactive than a cyclotetrasiloxane unit. This is consistent with the

*This work performed at Sandia National Laboratories supported by the U.S. Department of Energy under contract number DE-AC04-76DP00789.

observation that the intensity of D2 is more sensitive to change in temperature and humidity than is the intensity of D1 [3,13,14,32], lending further support to the assignments described above.

Under different experimental conditions, a set of infrared bands appears at 880 and 910 cm^{-1} when silica samples are subjected to vacuum degassing at temperatures above 400°C [6]. The active species thus generated on the surface of silica is a site for chemisorption of Lewis bases such as NH_3 or H_2O [4,22-24]. Because of the observed reactivity and strong sensitivity to moisture of the species responsible for these IR bands, it has been proposed to consist of highly strained siloxane bonds, such as those of the edge-shared tetrahedra reported for silica-W [25] or of the tetra-substituted cyclodisiloxanes synthesized by West and coworkers [26,27].

In assigning structures to local defect sites, comparison of defect spectra with those of structurally characterized model compounds has been used in conjunction with molecular orbital calculations [29,30]. Although the reactivity of silica model compounds has been qualitatively predicted from calculations of bond energies and strain [20,29], little has been done to verify these predictions with quantitative experimental evidence. Some experimental data does exist, however, for the reactivity of the defect systems themselves. The rate of disappearance at 25°C of D2 for a sol-gel derived silicate in an atmosphere of 100% relative humidity has been shown to be first order with a rate constant [40] of 5.2×10^{-3} min^{-1}. In work on stress corrosion cracking of glasses, Bunker and Michalske [31] have proposed that highly strained siloxane bonds exist at the stressed crack tips. They have suggested that the degree of strain is comparable to that found in a cyclodisiloxane structure such as that proposed for the dehydroxylated silica system. Kinetic studies of this system for the rate of hydrolysis at room temperature for cyclodisiloxane rings in an atmosphere of 4 mtorr water vapor gives a rate constant of 10^{-1} min^{-1}[15]. Thus, the rate of hydrolysis of the structure responsible for these highly strained siloxane bonds is much faster than the rate of hydrolysis of the D2 defect.

In an effort further to explore the validity of the various models proposed to exist for defect structures in silica, we have determined the hydrolysis kinetics for a series of organosiloxanes with varying degrees of built-in strain by following changes in the infrared and Raman spectra during exposure to water. By comparison of the hydrolysis rates of the various model compounds, we shall demonstrate that the reactivity, as well as the structure, of the models is consistent with the proposed defect structures described above.

EXPERIMENTAL

Materials: All permethylsiloxanes were purchased from Petrarch Systems, Inc.. The cyclodisiloxanes were provided by Professor Robert West and Gregory Gillette of the University of Wisconsin. Some samples were received as the precursor disilene in which case they were oxidized either in dry air or in an atmosphere of $^{18}O_2$. Using inert atmosphere techniques, each sample was transferred immediately after opening to a Schlenk tube [33] in which it was stored under argon until ready for use. All siloxanes were used without further purification. Tetrahydrofuran (THF), purchased from Burdick and Jackson, was freshly distilled from the sodium salt of benzophenone just before use. Water was pyrolytically distilled in a manner similar to that described by Conway [36] to eliminate organics responsible for fluorescence in the Raman.

Kinetics: THF solutions of known concentration in siloxane were prepared
and their IR and Raman spectra determined. A known quantity of water was
then added to each solution and the IR or Raman spectrum were taken over
time. Spectra were recorded until no further changes were observable. In
the case of siloxanes which appeared to exhibit no reactivity, spectra were
taken up to 14 days after the addition of water.

Instrumental: Fourier transform infrared spectra were acquired at 4 cm^{-1}
resolution using a Nicolet 7199 spectrometer equipped with a wide band
Hg-Cd-Te detector. A Perkin-Elmer attenuated total reflectance (ATR) flow-
cell having an internal volume of 0.5 ml with a trapezoidal KRS-5 window
was used with a beam incidence of 45 degrees.
    The spectroscopic data were analyzed using a two-step classical
multivariate least-squares procedures using all available frequencies from
4000 to 400 cm^{-1} [34]. Four component spectra, THF, siloxane starting
material, water, and product(s) were used to calculate solution
concentrations during hydrolysis for the kinetic rate determinations.
    Raman data were obtained using a computer driven monochromator
equipped with holographic gratings and a photon-counting detection system
[37]. The 514.5 nm argon ion line and the 647.1 nm krypton ion line were
used to illuminate the samples.

RESULTS

    Except for the cyclodisiloxanes, the polysiloxanes studied were chosen
to be the simplest, least sterically hindered representative of the chain
length or ring size under investigation. The cyclodisiloxanes were limited
to those known in the literature [27]. Di-t-butyldimesitylcyclodisiloxane
was chosen for the solution studies because of its solubility in THF. In
every case, the organic side groups and the Si-C bonds were expected to be
unreactive under the experimental conditions. Hence, any observed
hydrolysis reactivity can only be due to that of a siloxane bond, rendering
it a good model for hydrolysis reactions of siloxanes in silica systems.
    Representative infrared and Raman spectra of unreacted and of
partially or fully hydrolyzed samples of hexamethylcyclotrisiloxane are
shown in Figure 1 and Figure 2, respectively. The most significant
infrared spectral changes observed during the hydrolysis of the
cyclotrisiloxane involve the disappearance of the bands at 1016 and
609 cm^{-1} and the concomitant shift of the Si-C asymmetric stretch from 814
to 802 cm^{-1} (Figure 1). The 1016 and 609 cm^{-1} bands represent skeletal
motions unique to the strained, planar trisiloxane ring. [35] The
hydrolysis products are expected to be linear HO(Me$_2$SiO)$_n$OH which do not
have siloxane bond strain. Hence, the reaction products are expected to
have their Si-O-Si asymmetric and symmetric stretches near 1075 and
500 cm^{-1}, respectively [35]. Likewise, the Si-C asymmetric stretch of the
unstrained product is expected [35] to be shifted to ~800 cm^{-1} as is
observed.

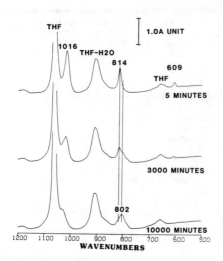

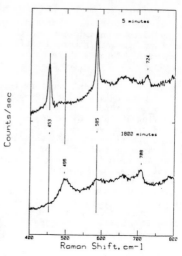

Figure 1. Infrared spectra of the hydrolysis of 0.21 M hexamethyl-cyclotrisiloxane in THF with 5:1 (H_2O:siloxane bonds) at 5, 3000, and 10000 minutes after water addition at 21°C.

Figure 2. Raman spectra of 0.21 M hexamethylcyclotrisiloxane in THF with 10:1 (H_2O:siloxane bonds) at 5 minutes and 1800 minutes after water addition at 25°C.

Raman spectral changes monitored during the course of the hydrolysis reaction include the disappearance of the cyclotrisiloxane ring modes at 453 and 585 cm^{-1} (See Figure 2). A product band at 496 cm^{-1} without a 633 cm^{-1} band, is near the expected Raman frequency for the Si-O-Si mode of a linear siloxane, $HO(Me_2SiO)_nOH$, $n > 3$. [39]

For cyclodisiloxane the spectral changes observed during the course of hydrolysis are consistent with previous assignments. [29,30] In the infrared the strongest decreases in intensity were observed at 824, 802, and 657 cm^{-1} (See Figure 3), while in the Raman, the most significant change was a decrease in intensity of the 868 cm^{-1} mode. Since these bands are due to vibrational modes of the cyclodisiloxane ring, their disappearance is consistent with the rupture of the Si-O-Si ring bonds during hydrolysis. The Raman spectrum of di-t-butyldimesitylcyclodisiloxane had previously been determined and assigned [28a,b]. The Raman frequencies observed in this work for tetramesityldisiloxane are in close agreement with these observations.

The half-lives obtained for the hydrolysis kinetics of hexamethylcyclotrisiloxane and di-t-butyldimesitylcyclodisiloxane in THF solution by both Raman and infrared measurements are listed in Table 1. Discrepancies in the rates observed for the 5:1 hydrolysis of the cyclotrisiloxane by the two vibrational techniques may be accounted for by the difference in temperatures at which the sample solutions were held during the reaction.

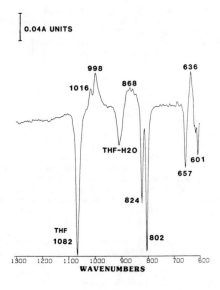

Figure 3. Infrared spectrum of the difference between the spectra at 2400 minutes and 15 minutes after hydrolysis for 0.07 M di-t-butyldimesitylcyclodisiloxane in THF at 21°C. Negative bands indicate modes due to starting material and positive bands indicate modes due to product(s).

TABLE 1

Half Lives for the Hydrolysis of Cyclodi- and Cyclotrisiloxanes in THF. Ratios represent H_2O:Siloxane Bonds.

	Infrared	Raman
$(Me_2SiO)_3$		
5:1	29.9 ± 3 hrs.	20 ± 3 hrs.
10:1	3.1 ± 0.5 hrs.	3.0 ± 0.3 hrs.
$(t - BuMesSiO)_2$		
5:1	7.5 ± 1 hr.	

DISCUSSION

The reactivities of the various siloxanes reported in this work are in good agreement with those expected for the silica defects which they are proposed to model. All linear as well as cyclotetra- and cyclopenta-organosiloxanes which have little or no siloxane bond strain are inert to hydrolysis under neutral conditions. The D1 silica defect structure, proposed to be due to cyclotetrasiloxane [19,28], is also inert to hydrolysis [28,40].

Hydrolytic reactivity was observed to increase in the model compounds with increasing siloxane bond strain. Both IR and Raman data indicate that the hydrolysis at hexamethylcyclotrisiloxane in THF proceeds with first order kinetics with respect to siloxane at a rate of 3.8×10^{-3} min^{-1}. The Raman D2 defect at 608 cm^{-1} associated with cyclotrisiloxane rings [19,28], hydrolyzes at a rate of 5.2×10^{-3} min^{-1}. Although water activity in a nonaqueous solvent will certainly differ from that of a vapor/solid interface, the similarity in this kinetic data at room temperature to that for hexamethylcyclotrisiloxane is further evidence that the D2 Raman band is due to cyclotrisiloxane rings. In both the infrared and Raman experiments, the spectral changes observed during hydrolysis (disappearance of the 608 cm^{-1} band) correspond to the disappearance of the defect on exposure to water. Limited infrared data for the organocyclodisiloxanes suggests that they are hydrolyzed at least four times faster than cyclotrisiloxanes (Table 1).

Further data are required to confirm the assignment of the infrared dehydroxylated silica defect to that of a cyclodisiloxane species. Infrared measurements of an ^{18}O-labeled dehydroxylated silica surface are in progress. Larger samples of the di-t-butyldimesitycyclodisiloxane are being obtained which will allow a more complete set of kinetics data to be obtained for comparison with crack growth rates and surface defect studies.

ACKNOWLEDGMENTS

The authors gratefully acknowledge the assistance of Karen Higgins for obtaining the Raman spectra and to Bruce Bunker, Terry Michalske, Dave Haaland and Jeff Brinker for many helpful discussions.

REFERENCES

1. P.H. Gaskell, Phys. Chem. Glasses 8, 69 (1976).
2. R.H. Stolen, J.T. Krause, C.R. Kurkjian, Disc. Faraday Soc. 50, 103 (1970).
3. R.H. Stolen, G.E. Walrafen, J. Chem. Phys. 64, 2623 (1976).
4. F.L. Galeener, in Lattice Dynamics, M. Bablanski, ed. (Flammarion, Paris) 1978, p. 345.
5. F.L. Galeener, Phys. Rev. B19, 1292 (1979).
6. A) B.A. Morrow and A. Devi, J. Chem. Soc., Faraday Trans. 1, 163, 403 (1972). b) Morrow and I.A. Cody, J. Phys. Chem. 79, 761 (1975).
7. W.D. Compton, G.W. Arnold, Disc. Farady Soc. 31, 130 (1961).
8. D.L. Griscom, Phys. Rev. B22, 4192 (1980).
9. E.J. Friebele, D.L. Griscom, M. Stapelbrock, and R.A. Weeks, Phys. Rev. Lett. 42, 1346 (1979).
10. A.R. Silin, P.J. Bray, J.C. Mikkelsen, Jr., J. Non-Crystalline Sol. 37, 71 (1981).

11. B. Subramanian, L.E. Halliburton, J.J. Martin, J. Phys. Chem. Solids 45, 575 (1984).
12. J.C. Phillips, J. Non-Crystalline Solids 63, 347 (1984).
13. A. Bertoluzza, C. Fagnano, M.A. Marzelli, V. Gottardi, M. Guglielmi, J. Non-Crystalline Solids 48, 117 (1982).
14. D.M. Krol, J.G. van Lierop, J. Non-Crystalline Solids, 63, 131 (1984).
15. B.C. Bunker, Sandia National Laboratories, Albuquerque, NM, unpublished results.
16. R.B. Laughlin, J.D. Joannapoulous, C.A. Murray, K.J. Hartnett, and T.J. Greytak, Phys. Rev. Lett. 40, 461 (1978).
17. G.Lucovsky, Phil. Mag. 39, 513 (1979).
18. C.J. Brinker, E.P. Roth, G.W. Scherer, and D.R. Tallant, J. Non-Crystalline Solids 71, 171 (1985).
19. F.L. Galeener, J. Non-Crystalline Solids, 49, 53 (1982).
20. M. O'Keefe and G.V. Gibbs, J. Chem. Phys., 81, 876 (1984).
21. W.P. Griffith, J. Chem. Soc. (A), 1372 (1969).
22. B.A. Morrow and I.A. Cody, J. Phys. Chem. 80, 1997, (1976).
23. B.A. Morrow and I.A. Cody, J. Phys. Chem. 80, 1998, (1976).
24. B.A. Morrow and I.A. Cody, J. Phys. Chem. 80, 2761, (1976).
25. V.A. Weiss and A. Weiss, Z. Anorg. Allg. Chem. 276, 95 (1954).
26. M.J. Michalczyk, R. West, and J. Michl, J. Chem. Soc., Chem. Commun., 1525 (1984).
27. M.J. Fink, K.J. Haller, R. West, and J. Michl, J. Am. Chem. Soc., 106, 822 (1954).
28. a) C.J. Brinker, D.R. Tallant, E.P. Roth, and C.S. Ashley, J. Non-Crystalline Solids, in press; b) C.J. Brinker, D.R. Tallant, E.P. Roth and C.S. Ashley, Proc. of the Fall 1985 MRS Symposium: Defects in Glass, eds. F.L. Galeener, D.L. Griscom, M. Weber, in press.
29. T. Kudo and S. Nagase, J. Am. Chem. Soc., 107, 2589 (1985).
30. J.S. Binkley and C.F. Melius, Sandia National Laboratories, Livermore, CA, unpublished results.
31. T.A. Michalske and B.C. Bunker, J. Appl. Phys. 56, 2686 (1984).
32. F.L. Galeener, in Proceedings of the Second International Conference on the Structure of Non-Crystalline Materials, eds. P.H. Gaskell, J.M. Parker, E.A. Davis, Taylor and Francis, Ltd., London, 1983.
33. D.F. Shriver, The Manipulation of Air-sensitive Compounds, New York, McGraw-Hill, 1969.
34. D.M. Haaland, R.G. Easterling, and D.A. Vopicka, Appl. Spectrosc., 39, 73 (1985).
35. A.L. Smith and D.R. Anderson, Appl. Spectrosc., 38, 822 (1984).
36. B.E. Conway, H. Angerstein-Kowzlowska, W.B.A. Sharp, and E.E. Criddle, Anal. Chem., 45, 1331 (1973).
37. D.R. Tallant, K.L. Higgins, L.I.A., 42, 12 (ICALEO 1983).
38. G. Fogarasi, H. Hacker, V. Hoffman, S. Dobos, Spectrochim. Acta, 30A, 629 (1974).
39. I.F. Kovalev, I.V. Shevchenko, M.G. Voronkov, and N.V. Kozlova, Dokl. Akad, Nauk. S.S.S.R., 212, 01 (1973).
40. D.R. Tallant, B.C. Bunker, C.J. Brinker and C.A. Balfe, "Raman Spectra of Rings in Silica Materials", to be published in this proceedings.
41. D.R. Tallant, unpublished results.

RAMAN SPECTROSCOPIC STUDIES OF TITANIUM ALKOXIDES USING UV EXCITATION

MARK J. PAYNE AND KRIS A. BERGLUND
Departments of Agricultural and Chemical Engineering, Michigan State
University, East Lansing, MI 48824.

ABSTRACT

The use of Raman spectroscopy can be greatly hindered by the presence
of fluorescing impurities. Even at low concentrations, fluorescence can
completely obscure the Raman signal. In the current study, Raman spectra
were recorded for various titanium alkoxides (ethoxide, isopropoxide,
isobutoxide) as a function of concentration and laser excitation wavelength.
It has been shown that fluorescence can be avoided by using uv-excitation
(363.8 nm). In addition, titanium alkoxides exhibit a preresonance Raman
enhancement as the excitation wavelength appreaches the UV. This result is
confirmed by a uv-visible absorption spectrum of the isopropoxide.

INTRODUCTION

Interest in metal alkoxides has increased in areas such as sol-gel
processing [1] and more specifically, hydrolytic synthesis of ceramic
powders [2]. Very little information has been published concerning Raman
spectra of these compounds.

The Raman effect is a very weak photophysical event. It can be
overwhelmed by fluorescence of the molecule being studied or of impurities,
even at very low concentrations. When fluorescence obscures Raman spectra,
two steps occur. The first step involves absorption of radiation as the
molecule is promoted to an electronically excited state. The second step
involves the release of radiation of slightly lower energy, since some
energy is lost due to thermal degradation. When Raman spectra are taken,
one scans for radiation of slightly lower energy than the incident
radiation. Thus, fluorescence and Raman effects can coincide, and
fluorescence, being the stronger effect, will dominate.

Although the Raman effect occurs at any excitation wavelength, there
can be advantages in using one excitation wavelength over another. A
correct choice of the laser line would take into account the following
considerations. first, light scattering is related to wavelength. The
intensity of the scattered light is inversely proportional to wavelength
raised to the fourth power. Thus, by using excitation of a lower
wavelength, Raman scattering of the same intensity can be obtained at a much
lower laser power. Secondly, if the excitation wavelength corresponds to an
absorbance band for the molecule being studied, resonance enhancement will
occur, and the intensity of certain peaks in the Raman spectrum will be
greatly increased. Finally, fluorescence may be avoided by a correct choice
of excitation wavelength. When the excitation wavelength is changed, the
fluorescing molecule is no longer supplied with the energy needed for the
electronic transition, and the wavelengths being scanned for the Raman
spectrum will move from those in which the fluorescing molecule releases
radiation.

In the current work, the effect of changing concentration and
excitation wavelength were examined. This is a preliminary study in a
research program to evaluate the efficacy of using time-resolved Raman
spectroscopy to study the kinetics of hydrolysis reactions of three
titanium alkoxides (ethoxide, isopropoxide, and isobutoxide).

MATERIALS AND METHODS

Titanium tetraisopropoxide (TiPT) was obtained from both DuPont and Aldrich Chemical Company, Inc. Titanium tetraisobutoxide (TiBT) and tetraethoxide (TET) were obtained from Morton/Thiokol, Inc. Reagent grade isopropanol (iPrOH) was used for the isopropoxide and isobutoxide solutions. Completely dry ethanol was obtained by adding titanium isopropoxide to 200 proof ethanol and stirring overnight. After distillation, ethoxide could be added to the ethanol with no precipitation occurring.

The Raman system used in the current study was a SPEX 1877 Triplemate with an EG&G Princeton Applied Research OMA II detection system equipped with a Coherent Innova 90-5 Ar$^+$ laser. Typical laser powers were 100-150 mW at 363.8 nm and 200 mW at 488 nm and 514.5 nm. A Perkin-Elmer Lambda 3A UV/Vis Spectrophotometer was used for the absorbance spectrum.

RESULTS

Figures 1 and 2 show the Raman spectrum of 25% TiPT (from Aldrich) in isopropanol using 488 nm and 363.8 nm excitation.

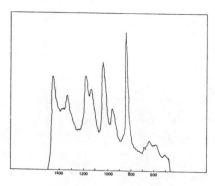

Figure 1. Raman spectrum of 25% TiPT in iPrOH, using 363.8 nm excitation.

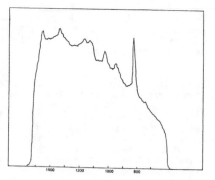

Figure 2. Raman spectrum of 25% TiPT in iPrOH using 488 nm excitation.

The sloping baseline in Figure 1 indicates the presence of fluorescence. The spectrum using 514.5 nm excitation was totally obscured by fluorescence, and no Raman peaks were visible. As can be seen in Figure 2, the fluorescence has been avoided by using 363.8nm excitation. Thus, quality Raman spectra were not attainable using visible excitation due to fluorescence, while changing the excitation wavelength to the uv resulted in good spectra.

The fluorescence observed is not inherent to the alkoxides. As the samples aged, the fluorescence became more pronounced. It is unclear what causes the fluorescence. Possibilities include an absorbed impurity and polymerization of the alkoxides. It was noticed that the alkoxides develop a yellow color as they age, which can be removed via vacuum distillation [2]. However, this failed to reduce the fluorescence in the Raman spectra when visible excitation was used.

Another advantage in using uv excitation for the alkoxides is that a preresonance condition is obtained. Figures 3 and 4

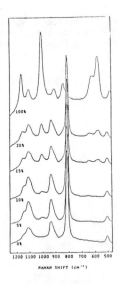 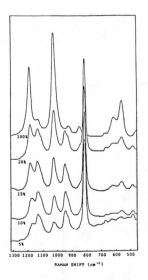

Figure 3. Raman spectra of
varying concentrations of
TiPT in iPrOH using 514.5 nm
excitation.

Figure 4. Raman spectra of
varying concentrations of
TiPT in iPrOH using 363.8 nm
excitation.

show an increase in intensity of the 1025 cm^{-1} and 1180^{-1}peaks when 363.8 nm
excitation is used. Both figures show a range of concentrations of TiPT (from
DuPont) in isopropanol. These spectra were taken before the alkoxides had a
chance to age and develop the fluorescence seen in Figure 1. The absence of
flourescence in the newer samples at 514.5 nm excitation is striking.

The uv absorption spectrum of TiPT was taken to verify the preresonance
condition. As expected, Figure 5 shows a strong absorption in the uv. An
isopropanol reference was used against 2.75 x 10^{-4} molar TiPT (from Aldrich)
in isopropanol, using a 1 cm path length cell. The molar absorptivity was
calculated to be about 7000 M^{-1} cm^{-1}. It would be expected that if TiPT
solutions were excited with laser lines further into uv that the intensity
of the 1025 cm^{-1} and 1180 cm^{-1} peaks would increase further.

Figure 5. Absorption spectrum of .275 mM TiPT in
a 1 cm path cell with an iPrOH reference.

Using uv excitation for TiBT in isopropanol has similar advantages. 363.8 nm excitation results in a reduction in fluorescence and preresonance of the 1023 cm^{-1} and 1176 cm^{-1} peaks, as can be seen in Figures 6 and 7.

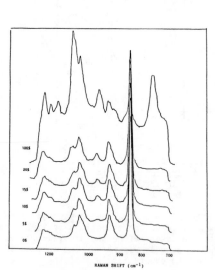

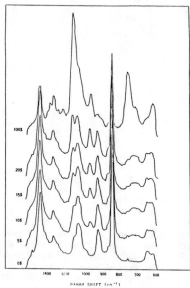

Figure 6. Raman spectra of varying concentrations of TiBT in iPrOH using 514.5 nm excitation.

Figure 7. Raman spectra of varying concentrations of TiBT in iPrOH using 363.8 nm excitation.

Figure 8 is a concentration study of TET in ethanol at 363.8 nm excitation. The TET spectrum at 514.5 nm excitation was completely obscured.

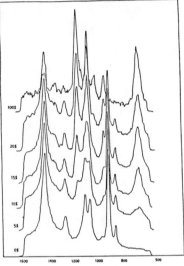

Figure 8. Raman spectra of varying concentrations of TET in ethanol using 363.8 nm excitation.

CONCLUSION

Using uv excitation in the study of titanium alkoxides has been found to have advantages. First, lower laser power was necessary due to the reciprocal dependence of Raman scattering on wavelength. Second, fluorescence which appears in the spectra of older samples when visible excitation is used does not appear in the uv-excited spectra. Third, the preresonance condition intensifies some of the peaks in the spectra. These effects combine to make the uv a good choice of excitation wavelength for studying titanium alkoxides.

ACKNOWLEDGEMENT

This work was supported by Sandia National Laboratories through contract #21-2885.

REFERENCES

1. D.W. Johnson, Jr., Am. Cer. Soc. Bull. 64(12) 1597-1602 (1985).

2. K.A. Berglund, D.R. Tallant, R.G. Dosch, Second International Conference on Ultrastructure Processing of Ceramics, Glasses, and Composites. Palm Coast, Fla. February 25 to March 1, 1985.

Precipitation Kinetics of the Titanium Isopropoxide
Hydrolysis Reaction

R.W. Hartel and K.A. Berglund
Depts. of Agricultural and Chemical Engineering, Michigan State Univ.,
East Lansing, MI   48824

ABSTRACT

The precipitation product of the titanium isopropoxide (TiPT)
hydrolysis reaction was followed using photon correlation spectroscopy.
This technique followed the mean size of the precipitate as the reaction
progressed, as well as giving an indication of the number of particles being
produced. From this analysis, the induction time for the onset of
nucleation, nucleation rate (or the rate of particle production) and mean
growth rate were correlated with reactant concentrations. In the
concentration ranges studied, it was found that the induction time was a
very strong inverse function of the reactant concentrations. Both
nucleation rate and average growth rate were found to increase very rapidly
as reactant concentration increased, with the TiPT concentration being more
important. This reaction was also followed using absorbance
spectrophotometry in the 500 to 320 nm range. After mixing of the
reactants, an absorbance peak was seen between 330 and 335 nm, depending on
the conditions. This peak increased slowly during the induction period and
then increased rapidly after the onset of nucleation due to the increasing
turbidity.

INTRODUCTION

Hydrolysis of metal alkoxides has been demonstrated to be an effective
means of preparing ceramic precursors [1]. However, the kinetic data for
nucleation and growth of the ceramic particles necessary for process design
and optimization are not available. The purpose of the present work is to
investigate experimental techniques for determining this kinetic data for
the hydrolysis of titanium isopropoxide (TiPT).

EXPERIMENTAL

Freshly distilled titanium isopropoxide (TiPT) was diluted in
isopropanol to the desired concentration. A second solution of water in
isopropanol was prepared at the desired molar ratio, R, of water to TiPT in
the final product. These solutions were continuously mixed by pumping (Sage
Syringe Pump) through a four-jet mixing chamber [2] at room temperature.
The mixed product was collected in a cuvet and quickly transferred to the
appropriate measuring device. Final TiPT concentrations ranged from 1.5 to
2.5% (wt.) while R was varied from 7.2 to 14.4. Absorbance at approximately
330 nm and particle size distributions (PSD) were measured at various time
intervals after mixing.

A Coulter Electronics Company Model N4 sub-micron particle analyzer was
used to measure the particle density and size distribution of the
precipitate. A cuvet of the mixed product was inserted into the sample
chamber, from which the auto-correlation functions were generated. Sampling
times varied between 60 and 800 seconds depending on the speed of the
precipitation. The auto-correlation functions were transformed into
particle size distributions through standard estimation techniques, yielding
average size and standard deviations. The average growth rate was found
from the rate of change of the average size with time after mixing. The

induction time for the onset of nucleation was taken to be the time when the particle density, N, as give by the photonic counts per second on the N4, just began to increase above the background count level. After nucleation occurred, increased sharply with time. The initial slope of the rate of change of N with time was taken to be an indication of the nucleation rate for the titanium particles.

The absorbance of the reactant product was monitored as a function of time after mixing using a Perkin-Elmer Lambda 3 spectrophotometer. A TiPT solution of the same final concentration was used as a reference sample for subtraction of TiPT and isopropanol absorbances. The high level of absorbance of TiPT below 320 nm caused saturation of the detector and prevented absorbance measurements extending into the far UV.

RESULTS

Scanning electron micrographs of the titania precipitate showed an aggregate structure of the particles. The aggregate was composed of amorphous units of hydrous titania. The individual units were approximately 0.5 to 1.0 μm in diameter and the aggregates ranged from 2 to 30 μm in size.

Figure 1 shows a typical curve for the change in particle density with time after mixing. In all experiments, there was an induction period where the counts per second determined by the N4 remained at the background level. At some time after mixing, which depended on the reactant concentrations, the number of particles increased rapidly, indicating that the nucleation rate also increased rapidly. The initial slope of N versus t as nucleation occurred was taken to indicate the nucleation rate of particles. Once the nucleation stage had been reached, the PSD of the precipitate was followed.

When particles undergoing Brownian notion are illuminated by a laser light source, the scattered light generates an auto-correlation function. The Coulter Model N4 particle analyzer transforms this auto-correlation function into a size distribution using the method of cumulants. This results in a average size for the PSD as well as the standard deviation of the particle size around this mean.

The average size of the precipitate increased as the particles grew, a typical example of which is shown in Figure 2. This initial linear change in average size with time after mixing was used as the initial average growth rate of particles under the specified conditions. After the particles had grown to a larger size, generally greater than 5 μm, gravitational forces dominated over the Brownian motion causing errors in this type of analysis.

The three parameters, 1) induction time, t_n, 2) average growth rate, G = dL/dt, and 3) nucleation rate as given by N, were correlated with the reactant concentrations after mixing. These results are shown in Figures 3 through 5 for variations in the TiPT concentration and Figures 6 through 8 for variations in R. It can be seen that the induction time was a strong inverse function of both reactant concentrations and that both the average growth and nucleation rates increased rapidly with reactant concentrations. Power-law kinetic expressions were found for these parameters using a linear regression analysis following a logarithmic transformation. These results are summarized in Table I.

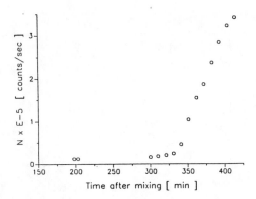

Figure 1. Change in particle density, N, as a function
of time after mixing.

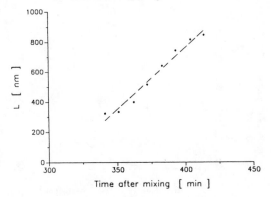

Figure 2. Change in cumulative average size with time
after mixing.

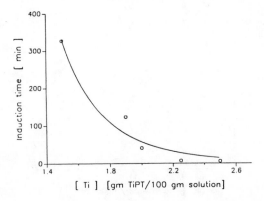

Figure 3. Change in induction time for nucleation with
TiPT concentration at R = 9.6 (mole water/mole TiPT).

636

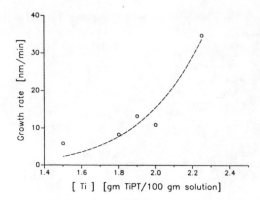

Figure 4. Change in average growth rate with TiPT
concentration at R = 9.6 (mole water/mole TiPT).

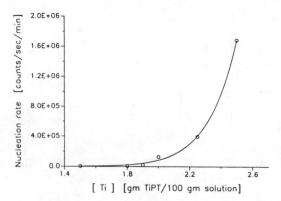

Figure 5. Change in nucleation rate with TiPT concentration
at R = 9.6 (mole water/mole TiPT).

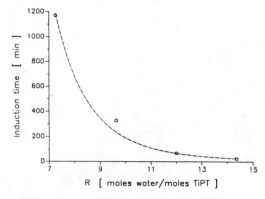

Figure 6. Change in induction time for nucleation
with R at 1.5% (w/w) TiPT.

Table I. Kinetic expressions of the form t_n = a $[Ti]^b$ and t_n = a R^b.

variable	[Ti] [gm TiPT/100gm sol.]			R [mole H_2O/mole TiPT]		
	a	b	r^*	a	b	r^*
t_n [min]	3.0E04	-9.4	0.924	1.6E08	-5.9	0.995
$\bar{G}$ [nm/min]	.90	4.1	0.910	-1.3E03	3.8	0.986
N [counts/s/min]	5.4	13.6	0.951	-8.9E05	7.8	0.996

*Correlation coefficient of linear regression.

An absorbance scan over the wavelength range of 500 to 320 nm of the mixed product prior to nucleation resulted in a single peak at approximately 334 nm. (In the control experiment where the TiPT solution was mixed with pure isopropanol, rather than a water solution, no peak was observed). The absorbance of this peak increased slightly during the induction period. After nucleation had occurred, the absorbance at all wavelengths increased dramatically due to the increased turbidity. The absorbance of the final suspension showed this same peak, but with a long tail extending well into the higher wavelengths. From this, it is evident that this peak represents a reaction product.

The changes in relative absorbance at 334 nm with time after mixing is shown in Figure 9 for reactant concentrations of 1.5% (w/w) TiPT and R = 9.6. During the induction period, the absorbance at 334 nm increased slowly. During this time period, there was no turbidity of the solution and there was no discernible particle density on the N4. After nucleation occurred, the absorbance increased dramatically due to the increased turbidity.

Induction times for nucleation were determined from these absorbance versus time after mixing plots at several reactant concentration levels. These are summarized in Table II along with peak absorbance wavelengths. It can be seen that induction times found in this study generally matched very well with those determined by photon correlation spectroscopy.

Table II. Summary of Absorbance Results.

Final [Ti] %	R [mole H_2O/ mole TiPT]	Peak Absorbance [nm]	t_n [min]
1.5	14.4	331	35.3
1.5	12.0	332.6	63
1.5	9.6	334	319
1.9	9.6	334.7	68
1.9	9.6	334.5	91
1.9	9.6	334.5	65.2

638

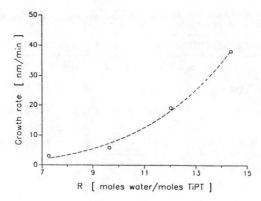

Figure 7.   Change in average growth rate with
             R at 1.5% (w/w) TiPT.

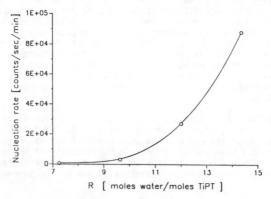

Figure 8.   Change in nucleation rate with R at 1.5%
             (w/w) TiPT.

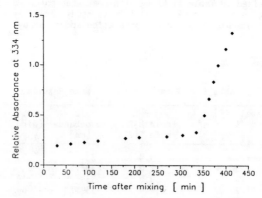

Figure 9.   Change in relative absorbance at 334 nm at
             1.5% (w/w) TiPT and R = 9.6 (mole water/mole TiPT)
             and 1.5% (w/w) TiPT reference.

DISCUSSION

Two steps must occur for the precipitation of the hydrolysis product of titanium isopropoxide. The first step is the chemical reaction between the water and the TiPT. This reaction is presumed to proceed in the following way.

$$Ti(OR)_4 + 4H_2O \rightarrow Ti(OH)_4 + 4ROH$$

The second step is the condensation of the reaction product to form the precipitate due to some spontaneous nucleation mechanism. When the reaction product concentration failed to reach the critical level for nucleation, no precipitation occured. In fact, precipitation did not occur within one day when the TiPT level was below 1.0% (at R = 9.6 moles H_2O/moles TiPT) or when R was below 7 moles H_2O/moles TiPT (at 1.5% TiPT).

Absorbance changes during the induction period also showed a gradual increase in the concentration of the reaction product. As shown in Figure 9, the relative absorbance at 334nm increased gradually up to the point of nucleation. At higher reactant concentrations, this increase was faster.

At the point of nucleation, the concentration of the reaction products was high enough that spontaneous condensation could occur. This was seen by the rapid increase in turbidity after this point. After nucleation had occurred, the number and size of particles continuously increased. The rates of change of particle size and number were found to be very sensitive to reactant concentrations.

The precipitation mechanism described above explains the experimental data very well. However, questions still remain. The problem of reaction by-products and their relative effects on the precipitation rates are unknown. Some evidence of an aging effect of the distilled TiPT on the precipitation rates was observed. That is, freshly distilled TiPT appeared to precipitate more slowly than aged TiPT. This effect was not quantified in this study. These questions can be answered by employing various analytical techniques to study this precipitation. For example, laser Raman spectroscopy is being used to determine the composition of the reaction products at various conditions.

CONCLUSIONS

Photon correlation spectroscopy and UV absorption measurements have been employed to study the hydrolysis reaction of titanium isopropoxide. Preliminary results indicate that precipitation kinetics at moderately low reactant concentrations can be followed in this way. Absorbance spectrophotometry was found to be useful in determining induction time data while photon correlation spectroscopy was required to obtain growth and nucleation data. Future experimental work on the rates of changes of relative absorbance with time might show a correlation with growth and/or nucleation rates.

These preliminary results indicate that the induction time for nucleation depends inversely on the 9.4 and 5.9 powers for TiPT and water, respectively. The mean growth rate of the precipitate was found to depend on approximately the fourth power of the reactant concentrations. In addition, the nucleation rate, as found by the change in particle density, was found to vary as the 13.6 and 7.8 powers of TiPT and water, respectively.

It is hypothesized that the precipitation follows a two-step mechanism. In the initial hydrolysis reaction, a hydrated titanium compound is formed. After the concentration of this reactant has reached a high enough level, spontaneous nucleation occurs and the precipitate is formed. Further analytical work is necessary to verify this mechanism.

ACKNOWLEDGEMENT

This work was supported by Sandia National Laboratories through contract #21-2885 and by the Division of Engineering Research at Michigan State University. The Model N4 was provided by Coulter Electronics Company.

REFERENCES

[1] Mazdiyasni, K.S., Ceramics International, 8 (2), 42-56 (1982).

[2] Caldin, E.F., Fast Reactions in Solution, Blackwell Scientific, Oxford, p.31 (1964).

DRYING BEHAVIOR OF SOL-GEL DERIVED Al_2O_3 AND Al_2O_3-SiC COMPOSITES

R.H. KRABILL AND D.E. CLARK
Department of Materials Science and Engineering, University of Florida,
Gainesville, FL 32611

ABSTRACT

Gel drying is a critical step in the sol-gel synthesis of Al_2O_3 and Al_2O_3-SiC composites. Problems exist during the drying stage that affect the monolithic properties of the sintered products. Classical drying theory was applied to the drying behavior of Al_2O_3 and Al_2O_3-SiC composites in an effort to optimize the drying process and understand the controlling mechanisms.

INTRODUCTION

The Al_2O_3-20 v/o SiC ceramic composite system has been shown to have the strength and toughness necessary for certain structural applications [1]. The production of this composite via sol-gel technology offers several processing advantages. However, problems during the drying stage affect the composite monolithicity. This work encompasses the application of the classical drying theory to sol-gel synthesis in an attempt to understand the prevailing drying mechanisms.

Much of the current interest in sol-gel technology stems from the potential of forming monolithic pieces of glasses, ceramics and composites [1-13]. However, very few sources in the literature explain the processing required for gel monolithicity. The majority of these proposed solutions are essentially empirical and deal mostly with the pure silica system. In this laboratory, research efforts have been concentrated on the fabrication of ceramic/ceramic composites via the sol-gel route [2-4].

The Al_2O_3-20 v/o SiC system is the principal focus of this paper. However, pure Al_2O_3 was also investigated for the purpose of comparison. The long range objective of this work is to understand the mechanisms of drying sufficiently well to minimize processing time and control microstructure of sol-gel derived composites.

DRYING THEORY

One of the most critical stages in the fabrication of these monolithic composites is drying. It is during the drying cycle that surface tension forces created during solvent removal can cause differential stresses leading to warpage and crack propagation. Yoldas [5] reported a monolithic transparent alumina obtained via hydrolysis of an aluminum alkoxide. A critical acid concentration (0.03 to 0.1 mol acid/mol $Al(OR)_3$) was determined necessary to retain monolithicity, but insufficient details were given about the drying stage. Shoup [6] fabricated silica monoliths by adding polar solvents such as formamide or ethyl acetate and gelling at pH > 10. Similarly, Wallace et al. [7] have successfully overcome the drying stresses in silica gels by the incorporation of chemical additives (DCCA) that control the pore size distribution. Yamane et al. [8] and Nogami and Morya [9] obtained monolithic silica discs of small dimensions (1 cm dia) by hydrolysis and polycondensation of tetramethylorthosilicate (TMS). Also working with silica gels, Klein and Garvey [10] report the fabrication of crack-free gels, but drying times were up to 1 month in duration.

642

Zarzycki et al. [11] provide a first look into some of the theoretical aspects of silica gel drying and suggest a drying mechanism that is subdivided into two stages: (1) at the beginning the volume decrease of the material is equal to the volume of evaporated liquid and (2) subsequently the volume is reduced by an amount smaller than the amount of the water lost, causing menisci to be formed and the particles to be pressed together by capillary attraction. The authors suggest a drying rate of 0.03 to 0.08 g/h for 100% monolithicity, but few attempts were made to correlate results with current drying theories.

Gurkovich [12] investigated the preparation of a monolithic lead-titanate gel, but all samples fractured, the largest fracture size being 1.5 cm in diameter. Again, samples were dried at low temperatures for 3 to 4 weeks and solvent evaporation rates were not controlled. Kawaguchi et al. [13] recognized the need for controlling the parameters during the drying stage and suggested that controlling the speed of evaporation of the volatile matter from the gels was very important for sample coherence.

Solids drying behavior has been previously investigated to some depth in the pioneering work of Sherwood [14-16], who studied clay systems. Weight-time relationships were obtained on a laboratory balance and the slopes of the resulting curve were measured to obtain the instantaneous drying rates. Figure 1 represents typical clay drying behavior. The suspension begins drying at point A at a constant rate until B is reached. During this stage there is a uniform volume decrease equal to the water lost so that point B represents the point of particle/particle contact, and no further consolidation will occur. Although the drying rate per unit of wet surface will remain constant during this period, the bulk drying rate will decrease, since dry spots begin to appear on the surface at point B. At point C the original liquid surface will have completely evaporated, forcing the moisture to be drawn from within the pores. The drying rate will undergo a sharp decrease (from C to D) and capillary movement will become the dominant drying mechanism. Point D represents the equilibrium moisture value for the prevailing air humidity. Thus, point D can occur anywhere on the bulk volume-moisture content curve in Fig. 1 given the appropriate temperature/humidity conditions. It should also be pointed out that, in the case of sols, the bulk volume at which point B occurs depends primarily on colloid aggregation in the sol stage. Thus the effectiveness of the peptizing agent [5] can be determined based on point B.

The drying behavior of only a few pure systems has been investigated. To date, no drying behavior studies have been reported with reference to sol-gel derived ceramic composite systems.

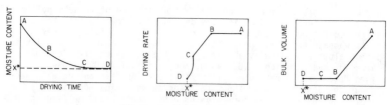

Figure 1. Typical drying curves

EXPERIMENTAL

Alumina sol was made from the hydrolysis of aluminum sec-butoxide using the acid ratio recommended by Yoldas [5] for gel monolithicity (0.07 mole nitric acid per mol $Al(C_4H_9)_3$). A sufficient amount of SiC whiskers[t] to produce a 20 v/o composite were cleaned in 2 M HNO_3 and subsequently ultrasonicated for fifteen minutes in the sol with the aid of a commercial deflocculant to break up the aggregates present in the as-received whiskers. The whisker slurry was then added to the rest of the batch which was then reduced to one-sixth of its original volume. Glycerol was added as a plasticizer to the batch, as recommended in previous work by Lannutti and Clark [2-3]. The Al_2O_3 sol-SiC-whisker slurry was then cast into poly-styrene petri dishes which were precoated with a mold-release agent, R-272[‡].

A control batch of pure Al_2O_3 sol was made for comparison with the Al_2O_3-SiC system and both composite and pure samples were placed in a Blue M steady state constant temperature/humidity cabinet[#] at the desired drying conditions. Weight and size measurements were monitored during the drying cycle.

RESULTS AND DISCUSSION

(a) Temperature/humidity effects on drying behavior

Figure 2 shows the effects of air-drying gels under uncontrolled laboratory conditions. One of the first problems encountered is gel warpage and cracking, largely due to the wide range of humidities (35-80%) prevalent in the laboratory which promoted uncontrolled solvent evaporation and differential drying stresses. The other serious processing problem is the long drying time to reach an equilibrium moisture value. Figure 2 also shows moisture-time data for Al_2O_3 and Al_2O_3-SiC at constant temperature/humidity conditions. No cracking was observed and the drying time for Al_2O_3 was decreased by a factor of 3 as compared to the results for uncontrolled drying conditions. The Al_2O_3-SiC composites behaved in a similar fashion with 100% monolithicity in all cases. It is quite clear that one can manipulate either the temperature, humidity, or both, to achieve the desired results. Increasing the temperature and holding the humidity constant has the effect of increasing the drying rate as observed in Figure 3. However, even though increasing the temperature reduces the drying times at a given humidity, gels having a higher equilibrium moisture content are obtained. If, on the other hand, one holds the temperature constant and reduces the relative humidity, as shown with the 50°C run, higher drying rates can be obtained (Figure 3). However, samples for the 50°C/70% RH run showed some curling, indicating that this humidity was too low for the given temperature. Warpage indicates that drying at these conditions becomes diffusion-controlled as opposed to being primarily evaporation-controlled during the 85% RH runs. It should be noted that a family of curves can be produced and an optimum drying schedule designed to fabricate crack-free, sol-gel-derived ceramics or composites. By comparing the 85% RH curves in Figure 3, it can be seen that higher temperatures result in faster drying in the early stages, but close to the end of the drying cycle (< 2 Kg moisture/Kg solid) the reverse is true. That is, higher temperatures result in slower drying due to a higher absolute humidities for the higher temperatures. In this region the drying rate can be increased by decreasing the temperature and maintaining a constant RH.

A comparison of the results from the composite and pure system drying data indicates that they possess similar drying behaviors. A slightly lower drying rate was found for the composite and this is thought to be due to the presence of the deflocculant used to disperse the SiC whiskers. As shown in Figure 2, both systems reached the same equilibrium moisture content for the

---

†Arco Metals, ‡Union Carbide Corp., #Blue M

644

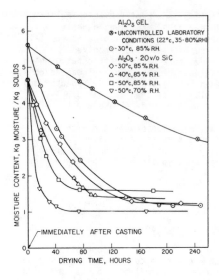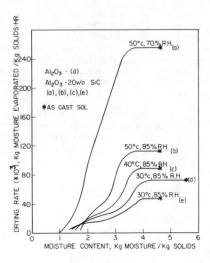

Figure 2. Moisture-time relationships for sol-gel Al₂O₃-SiC

Figure 3. Drying rates of sol-gel Al₂O₃ and Al₂O₃-SiC

same given set of temperature/humidity conditions. Samples in both runs dried uniformly and without cracking.

(b) Effect of humidity on gel structure during drying

Monitoring gel shrinkage as a function of moisture content provides insights into the mechanisms that occur during the drying process. Figure 4 shows shrinkage data for the Al_2O_3-SiC system. Although only 50°C/85% RH and 30°C/85% RH data are plotted for convenience, the remaining data fall on the same normalized curve, indicating that the final structure of the dried gel is fairly insensitive to drying conditions. Based on current drying theories, the point at which volume shrinkage ceases is the point of particle-particle (or aggregate-aggregate) contact. The structure of the gel will determine the position of this point on the curve. Although point B (i.e. structure sensitive) appears to be insensitive to drying conditions, it is expected to vary with the extent of peptization [5]. The gel porosity is about 28%, indicating that the particle packing is closer to close-packed structures than cubic assuming uniform size colloids under the peptizing conditions used in this study.

Figure 5 illustrates a simple schematic of the change in liquid/colloid distribution at various stages of drying. Due to the relative large size of the SiC whiskers compared to the colloids, they have little effect on the drying mechanism for the volume fraction used in this study. Therefore, the whiskers are not illustrated on the figure. Point I represents the beginning of the region in which the fastest drying occurs. A much larger decrease in bulk volume occurs due to the release of moisture and air bubbles produced during the devoluming of the sol. Evaporation of unbound moisture from a surface film occurs during this period. Particles begin to approach each other and area shrinkage is offset by the amount of moisture

and air removed, resulting in a constant drying rate. At point II the air
has evaporated and the surface film becomes unsaturated. Drying follows a
similar pattern to clay drying behavior. Particles approach an initial
equilibrium separation distance and a wet gel is formed. Between stages II
and III capillary movement becomes important, and shrinkage will continue to
occur, but at a slower rate than in stages I and II. Stage III represents
the region in which the liquid bridges between particles begin to disappear,
resulting in particle/particle contact. At stage IV shrinkage will cease
and moisture is removed via a vapor transport mechanism until an equilibrium
moisture content is reached with the surrounding environment. Figure 5
represents the situation in which no warpage or cracking occurs and
equilibrium drying is present at all drying stages as in the 85% RH runs.
Warping is observed when differential drying occurs between the surface
layers and the interior of the gel, causing pore closure and surface
stresses.

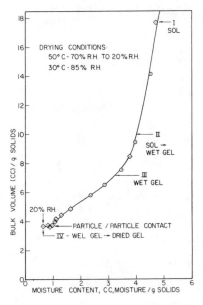

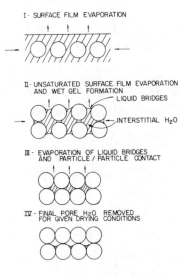

Figure 4. Gel shrinkage as a function of moisture content.

Figure 5. Proposed drying mechanism. Simple cubic packing was used for simplicity.

SUMMARY

(1) Controlled humidity conditions are essential for 100%
monolithicity. (2) There were no observable differences between the drying
mechanism of the pure Al_2O_3 and Al_2O_3-20 v/o SiC systems. (3) The mechanism
of gel composite drying appears to be subdivided into four stages, namely,
(I) surface film evaporation, (II) unsaturated surface evaporation and wet
gel formation, (III) evaporation of liquid bridges and particle/particle
contact, (IV) final pore H_2O removal for given drying conditions. (4)
Careful control must be maintained between stages III and IV, where the rate
is slowed down dramatically and appears to be controlled by capillary
movement. In order to produce a fully dried gel without cracking or

warping, the humidity must be reduced at a rate no greater than the transport of moisture via capillaries.

ACKNOWLEDGMENT

The authors thank the Air Force Office of Scientific Research for their financial support.

REFERENCES

1. Becher, P. F. and G. C. Wei, presentation at the 21st Automotive Technology Development Contract's Coordinators Meeting, Dearborn, Mich., Nov. 14-17, 1983.

2. Lannutti, J. J. and D. E. Clark, Better Ceramics Through Chemistry, C. J. Brinker, D. E. Clark and D. R. Ulrich, eds. (Elsevier Science Publishing Co. 1984), p. 369.

3. Lannutti, J. J. and D. E. Clark, Better Ceramics Through Chemistry, C. J. Brinker, D. E. Clark and D. R. Ulrich, eds. (Elsevier Science Publishing Co. 1984), p. 375.

4. La Torre, G. P., R. A. Stokell, R. H. Krabill and D. E. Clark, in proceedings, 10th Annual Conference on Composites and Advanced Ceramic Materials, Cocoa Beach, Fla., 1986.

5. Yoldas. B. E., Amer. Ceram. Soc. Bull. 54 (1975) 289; 54 (1975) 286.

6. Shoup, R. D., Colloid and Interfacial Science, Vol. 3 (Academic Press, New York, 1976) p. 63.

7. Wallace, S. and L. L. Hench, in Better Ceramics Through Chemistry, C. J. Brinker, D. E. Clark and D. R. Ulrich, eds., J. Wiley & Sons, 1984, p. 47.

8. Yamane, M. et al., J. Mater. Sci. 13 (1978) 865.

9. Nogami, M. and Y. Morya, J. Non-Cryst. Solids 37 (1980)191.

10. Klein, L. C. and G. J. Garvey, J. Non-Cryst. Solids 48 (1982)97.

11. Zarzycki, J. et al., J. Mater. Sci. 17 (1982) 3371.

12. Gurkovich, S. R. and J. B. Blum, in: Ultrastructure Processing of Ceramics, Glasses and Composites, L. L. Hench and D. R. Ulrich, eds., John Wiley & Sons, 1984.

13. Kawaguchi, T. et al., J. Non-Cryst. Solids 63 (1984)61.

14. Sherwood, T. K., Ind. Eng. Chem. 21 (1929) 12; 21 (1929)976; 24 (1932)307.

15. Comings, E. W. and T. K. Sherwood, Ind. Eng. Chem. 26 (1934)1096.

16. Sherwood, T. K. and E. W. Comings, Ind. Eng. Chem 25 (1933)311.

SOL-GEL COATINGS ON CARBON/CARBON COMPOSITES.

S.M. SIM, R.H. KRABILL, W.J. DALZELL JR., P-Y. CHU AND D.E. CLARK
Department of Materials Science and Engineering, University of Florida,
Gainesville, FL 32611

ABSTRACT

The need for structural materials that can withstand severe
environments up to 4000°F has promulgated the investigation of sol-gel
derived ceramic and composite coatings on carbon/carbon composite
materials. Alumina and zirconia sols have been deposited via thermophoresis
on carbon/carbon substrates.

INTRODUCTION

Major efforts are underway to develop and characterize materials that
can withstand the high temperatures required for various systems such as
rocket motors, heat shields, radomes and turbine engines. Ceramic
components for these systems ultimately should be able to withstand severe
oxidative and inert environments with temperatures ranging from 2000°F to
4000°F for extended periods of time and also withstand impact and cyclic
loading at these temperatures. Only a few materials can potentially satisfy
these requirements, namely ZrO_2, HfO_2 and coated carbon/carbon composites
[1,2].

Although much work has been reported in the literature on refractory
coatings [3], very little coating research has been carried out on
carbon/carbon composites. Webb [4] reports on a system under investigation
by AVCO using chemical vapor deposition (CVD) to deposit multilayers of
silicon carbide with silica to seal microcracks created during heat
treatment, and using boron as an oxidation inhibitor. Likewise, LTV
Aerospace and Hitco use similar systems with different processing methods to
manufacture commercially available Advanced Carbon Composites and Ultra
Carbon Composites, respectively [5,6]. As mentioned above, oxides of Zr and
Hf appear most promising for very high temperature applications. Coatings
from these oxides have been produced by Mazdiyasni and Lynch [7] who
pyrolyzed alkoxides by vapor decomposition onto hot graphite substrates.
Coatings of these same oxides are now under investigation in our laboratory
using the sol-gel process.

Dip-coating has been extensively studied and is of wide commercial
interest [8,9]. Through proper control of sol rheology and chemistry,
coatings with a wide range of microstructures can be deposited [10-13].
Thermophoresis, a dip-coating technique developed in our laboratory [14],
has been used to produce sol-gel derived coatings on carbon/carbon composite
substrates, and results from this method are compared with traditional dip-
coating methods. Emphasis in this paper is placed on the processing phase
of the coating deposition process even though a few preliminary results are
given from heat treatment experiments.

EXPERIMENTAL PROCEDURE

Three types of sols were selected for coating carbon/carbon composite
substrates: aluminum alkoxide, zirconium alkoxide and zirconia acetate
sols. The alumina sol was prepared by hydrolysis and peptization of
aluminum sec-butoxide* in excess water at an elevated temperature as

described by Yoldas [15]. An alkoxide colloidal sol was prepared by mixing zirconium n-propoxide* with water, which corresponds to a 115:1 molar ratio of water to alkoxide. Prior to mixing, zirconium n-propoxide and water were diluted separately in excess isopropanol. The hydrolysis product of zirconium alkoxide was stabilized by adding acetic acid with a final pH of 3.5. The equivalent oxide concentration of colloidal sol of zirconium hydrate was ~3% by weight. The hydrolysis reaction of zirconium alkoxide has been investigated previously by Bradley and Mehrotra [16] and Mazdiyasni et al. [17]. A third sol, zirconia acetate† containing 20 wt% ZrO_2, was obtained from a commercial source.

The carbon/carbon composites‡ were supplied by LTV Aerospace as rectangular bars with dimensions of about 190 x 25 x 5 mm. Samples with dimensions of 12.7 x 6.3 x 5 mm were cut using a diamond wafering saw and ultrasonically cleaned in acetone prior to coating.

In order to evaluate wettability, small drops of sols were placed on the substrate with a microsyringe. The contact angle as a function of equivalent oxide concentration was measured from the substrate surface to the tangent of the interface between the substrate and the sol using a microscope. The contact angles were measured immediately after preparation since some aging effects were observed.

For thermophoretic deposition, a temperature gradient was set up between the sols and the substrates by cooling the substrate with liquid nitrogen during immersion. Other details of the coating process are described elsewhere [14]. For the purpose of comparison, substrates were also coated without a thermal gradient. Dipping speed was the same for all samples. All coated samples were then air-dried at room temperature under prevailing humidity conditions within the laboratory. Some of the samples were prepared by multiple dip-coating to produce continuous and thick coatings.

RESULTS AND DISCUSSION

One of the requirements for a good coating is that it wet the substrate. The contact angles of the sols on the carbon/carbon composites depends on sol concentrations as shown in Figs. 1 and 2. Significant differences in wettability (i.e. contact angle) were found between the fabricated surface and cut surface of the substrate. Smaller contact angles corresponding to better wettability were observed on the cut surface. Optimum sol concentration coating ranges are also indicated in Fig. 1 for the alumina sols. Sols with oxide concentrations below this range were too dilute to produce continuous coatings. Above this range, the sols were too viscous to provide uniform and crack-free coatings. Although an optimum range of sol concentrations is indicated in Fig. 2 for the zirconia sol, crack-free coatings were not obtained with any of these.

FT-IR diffuse reflection spectra** of coatings are shown in Fig. 3. A broad and strong reflection peak in the spectra of the uncoated carbon/carbon composite substrates occurs at 503 cm^{-1} and persistently exists in all samples. Diffuse and weak peaks also occur between 1000 and 1300 cm^{-1} [18]. Spectra in Fig. 3(a) were obtained from the samples coated with alumina sol and air-dried. As a result of heating to 1300°C, the aluminum hydroxide coating transformed to α-Al_2O_3 as indicated by peaks at 584 and 444 cm^{-1}. Spectra in Fig. 3(b) and (c) were obtained from the

---

*Alfa Products
†Nyacol Products
‡ACC-4, LTV Aerospace
**Nicolet, MX-1

649

samples coated with zirconia acetate and zirconium alkoxide sols containing
20 and 7.5 wt% of oxide, respectively. Both spectra of air-dried coatings
show the presence of organics and strong response to the carbon/carbon
composite substrate. The broad peak of carbon around 505 cm^{-1} in all
samples is due to the substrate and indicates that these coatings are fairly
thin. Coatings heated to 1300°C do not clearly provide reflection peaks
corresponding to ZrO$_2$. However, the peaks around 814 and 531 cm^{-1} can be
assigned to a monoclinic ZrO$_2$ from the reflection spectra data of pure ZrO$_2$
which occur at 420, 520, 610, 750 and 800 cm^{-1}.

A second requirement for a good coating is that it be continuous and
adhere well to the substrate. Thermophoretic deposition of alumina using

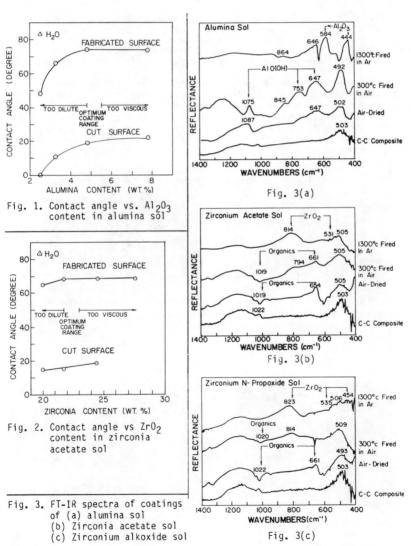

Fig. 1. Contact angle vs. Al$_2$O$_3$
content in alumina sol

Fig. 2. Contact angle vs ZrO$_2$
content in zirconia
acetate sol

Fig. 3. FT-IR spectra of coatings
of (a) alumina sol
(b) Zirconia acetate sol
(c) Zirconium alkoxide sol

Fig. 3(a)

Fig. 3(b)

Fig. 3(c)

sols within the optimum coating range reduces crack formation and provides good adhesion to the substrate. Multiple dip-coating after thermophoretic coating sealed all previous cracks resulting from drying. The SEM micrograph of a dried sample which was thermophoretically coated and repeatedly dip-coated with sols containing 4.8 and 5.3 wt% of alumina, in Fig. 4(b), shows continuous and crack-free coatings. The substrate texture is visible but the coatings are easily distinguishable when compared with the rough surface of the uncoated substrate (Fig. 4(a)). In contrast, Fig. 4(c) shows cracks in the coating along the carbon fiber direction of the sample coated with a sol containing 7.4 wt% of alumina and heated at 300°C for 1 hour in air at a heating rate of 60°C/hr. Drying stresses resulting from too thick coating were responsible for these cracks.

Fig. 4 (d) is a SEM micrograph of an alumina coating consisting of 5.5 wt% of alumina in the sol, prepared by thermophoretic deposition and heated to 1300°C for 30 minutes in an argon atmosphere. The coating exhibits a porous plate-like morphology after heating which is probably a result of large volume shrinkages during the gel $\rightarrow \gamma$-$Al_2O_3 \rightarrow \alpha$-$Al_2O_3$ transformations.

Samples coated without a thermal gradient and dried in air were also examined to evaluate the thermophoretic effect. These coatings were similar in appearance to the thermophoretic coating prepared under the same conditions. However, results from several samples suggest that the thermophoretically deposited coatings are more adherent to the substrates and provide more protection than the coatings produced in the absence of a thermal gradient.

Fig. 4(a). Fabricated surface of uncoated c/c composite

Fig. 4(b). Al_2O_3 sol coating applied via thermophoresis and multiple dip-coating

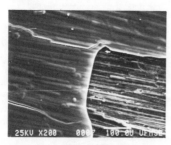

Fig. 4(c). Al_2O_3 sol coating fired to 300°C in air

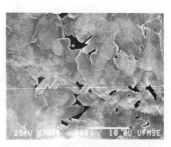

Fig. 4(d). Al_2O_3 sol coating fired to 1300°C in Ar

In contrast to the alumina sols, thermophoresis of zirconia sols did not produce continuous coatings and resulted in cracks along the carbon fiber direction. A SEM micrograph in Figure 5(a) shows a dried thermophoretic coating of zirconia acetate sol containing 21.7 wt% of zirconia. A SEM micrograph in Figure 5(b) shows a dried coating of zirconium alkoxide sol containing 25 wt% of zirconia deposited in the same manner as the coating in Fig. 5(a). Both dried coatings exhibit cracking, but coatings of the zirconia produced from zirconium n-propoxide appear to be more uniform and adherent to the substrate. Figure 5(c) and (d) illustrate the strong dependence of coatings on the carbon fiber array. Fig. 5(c) is a SEM micrograph of zirconia acetate sol coating deposited with 23.5 wt% of zirconia and heated to 300°C in air. In Fig. 5(d), a fairly thick coating of zirconium alkoxide sol is seen from the sample coated thermophoretically and subsequently dip-coated three times with 25 wt% of zirconia. Severe cracking occurred during drying. The cracks were not healed after firing to 1300°C in Ar.

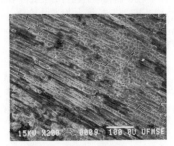

Fig. 5 (a). ZrO_2 acetate
    sol coating

Fig. 5(b). Zr alkoxide
    sol coating

Fig. 5(c). ZrO_2 acetate sol coat-
    ing fired to 300°C in air

Fig. 5(d). Zr alkoxide sol coat-
    ing fired to 300°C in air

The effectiveness of the coating in reducing oxidation of the carbon/carbon composite substrate was evaluated by heating the samples to 1000°C in air. After 10 minutes, the uncoated sample experienced 67 mg/cm² weight loss while the sample dip-coated with alumina lost 61 mg/cm² and the sample thermophoretically coated lost 56 mg/cm². The samples coated with zirconia sols lost between 51-60 mg/cm² and did not show any correlation with the coating process. These results indicate that the coatings did provide some protection to the substrates and thermophoresis provides some advantage in the alumina coating but not in the zirconia coating.

High temperature applications will require the coating to be crack-free and provide a barrier against oxidation. Work is now in progress to produce

652

more effective coatings of Al_2O_3 and ZrO_2 using multiple dipping in combination with thermophoresis.

Large volume changes result when pure ZrO_2 transforms from the tetragonal to monoclinic forms at about 1200°C. These transformations can potentially damage the coating and reduce its effectiveness. In order to increase the useful temperature of the coating, HfO_2 is being investigated. It experiences similar transformations but at much higher temperatures and the accompanying volume changes are slightly less.

SUMMARY

Carbon/carbon substrates have been coated with alumina and zirconia sols using thermophoresis. Results to date indicate that: (1) an optimum sol concentration range exists for producing uniform and crack-free coatings of alumina; (2) the use of a thermal gradient (~200°C) results in better coatings with the alumina sols, but has little influence on the coatings produced with zirconia sols; (3) multiple dippings will most likely be required to produce oxidation-resistant coatings, and (4) tests at 1000°C in air indicated that the thermophoresis provides advantages for the alumina coating.

ACKNOWLEDGMENTS

The authors thank the Air Force Office of Scientific Research for their financial support.

REFERENCES

1. Subbarao, E. C., Advances in Ceramics, Vol. 3, A. H. Heuer and L. W. Hobbs, ed., The American Ceramic Society, Columbus, OH, 1981, p. 1.
2. Lynch, C. T., Refractory Materials: High Temperature Oxides, Vol. 5-II, A. M. Alper, ed., Academic Press, NY, 1960, p. 193.
3. Fisher, G. , Am. Cer. Soc. Bull., 65 (2) 283 (1986).
4. Webb, R. D., NASA Conference Publication 2406, NASA, 1985, p. 149.
5. Ohlhorst, C. W. and P. O. Ransone, ibid., p. 277.
6. Johnson, A. C. and J. W. Finley, ibid., p. 175.
7. Mazdiyasni, K. S. and C. T. Lynch, Report No. ASD TDR 63-322, May 1963.
8. Mackenzie, J. D., Ultrastructure Processing of Ceramics, Glasses and Composites, L. L. Hench and D. R. Ulrich, ed., John Wiley & Sons, 1984, p. 15.
9. Dislich, H., J. Non-Cryst. Solids, 63, 237 (1984).
10. Dislich, H. and E. Hussmann, Thin Solid Films, 77, 129 (1981).
11. Sakka, S., K. Kamiya, K. Makita and Y. Yamamoto, J. Non-Cryst. Solids, 63, 223 (1984).
12. Yoldas, B. E., Appl. Opt. 19, 1425 (1980).
13. Martinsen, J., R. A. Figat and M. W. Shafer, Better Ceramics Through Chemistry, C. J. Brinker, D. E. Clark and D. R. Ulrich, ed., North-Holland, 1984, p. 145.
14. Dalzell, W. J. and D. E. Clark, to be published in Ceram. Eng. Sci. Proc. (1986).
15. Yoldas, B. E., Am. Cer. Soc. Bull., 54 (3) 289 (1975).
16. Bradley, D. C., Mehrotra and D. P. Gaur, Metal Alkoxides, Academic Press, NY, 1978.
17. Mazdiyasni, K. C., C. T. Lynch and J. S. Smith, J. Am. Ceram. Soc., 48 (7) 372 (1965).
18. Gadsden, J. A., Infrared Spectra of Minerals and Related Inorganic Compounds, Butterworth, Reading, Mass., 1975.

THE GEL ROUTE TO TiO$_2$ PHOTOANODES

S. DOEUFF, M. HENRY and C. SANCHEZ
Spectrochimie du Solide, Université Paris VI, 4 place Jussieu, Tour 44,
2è étage, 75230 Paris Cedex 05, France.

ABSTRACT

Cr^{3+} and Al^{3+} doped TiO$_2$ can be easily made via the sol-gel process. Mixing the metal-organic solutions give rise to a random dispersion of the doping ions into the TiO$_2$ network. Amorphous and crystalline phases have been characterized all the way from the gel to the crystalline products by thermal analysis, X-ray diffraction, vibrational spectroscopy and E.S.R. Organic groups appear to be involved in the formation of the gel network and could lead to a better control of the morphology of the xerogel. A thermal stabilization of the anatase phase, up to 900°C, is observed when Cr^{3+} is introduced as a dopant. This would lead to the preparation of anatase TiO$_2$ photoanodes.

I. INTRODUCTION

Electrode materials for the photoelectrolysis of water has received an increasing interest during the last decade. The first semiconductor used for that purpose was TiO$_2$ (1). It still remains the most investigated material because of its chemical stability in aqueous electrolytes. Although single crystals were first studied, it was later shown that poly-crystalline TiO$_2$ also exhibits efficiencies close to that of single crystal electrodes (2)(3). A further improvement was obtained by doping TiO$_2$. Cr^{3+} for instance is known to extend optical absorption in the visible range (4-6) while Al^{3+}, improve the quantum efficiency of carrier generation (5).

TiO$_2$ photoanodes can be obtained through a large variety of methods : thermal oxidation of titanium (7), thermal decomposition of salts (8), anodic oxidation (9), chemical vapor deposition (10)(11) or thick film technology (12). One of the main problems to be solved is to obtain a random dispersion of the doping ions throughout the TiO$_2$ lattice. This usually requires high temperatures and therefore leads to the rutile TiO$_2$ phase. However, it was reported that anatase could exhibit a higher photocatalytic activity (11)(13). It would therefore be interesting to develop new methods allowing both, a homogeneous doping and lower temperatures.

These requirements suggest that the sol-gel process could be particularly suitable for making TiO$_2$ photoanodes. Doping can be easily done, at a molecular level, by mixing the precursor solutions. Coatings or pellets can be obtained at temperatures low enough to stabilize the anatase phase. This paper reports on the sol-gel synthesis and characterization of Cr^{3+}-Al^{3+} doped TiO$_2$. The electrochemical properties of the photoanodes will be described elsewhere.

II. RESULTS AND DISCUSSION

2.1 Sample preparation

TiO$_2$ gels or colloidal solutions, were made through the hydrolysis of Ti(OBu^n)$_4$ in the presence of acetic acid according to a procedure previously described (14). The molecular precursors solution is obtained by mixing 65.4 g of TiOBu^n)$_4$ with 85.7 g of a n-butanol solution of Al(OBu^s)$_3$ (1 wt%)

654

and 63.1 g of a n-butanol solution of Cr(acac)$_3$ (1 wt%). All metal-organics were purchased from Ventron and used without further purification. After mixing, the solution is refluxed for 2 hours at 120°C, then transfered into a rotating evaporator in order to reduce the total volume to about one third. 17.2 g of glacial acetic acid are then added prior to hydrolysis which is performed with 128.4 g of a water n-butanol solution (10 wt%). Gelation occurs within a few hours giving rise to a green monolithic gel. Drying is performed in air at 80°C.

E.S.R. spectra were recorded at 77K on a Varian E109 spectrometer.

IR Spectra were performed on a Perkin Elmer 580 in the 4000 cm^{-1}-200 cm^{-1} frequency range while Raman spectra were carried out by using a DILOR Spectrometer.

2.2 Thermal analysis of the xerogel

The xerogel appears to be amorphous by X-ray diffraction. The D.T.A. curves are shown in fig.(1), together with the weight loss.

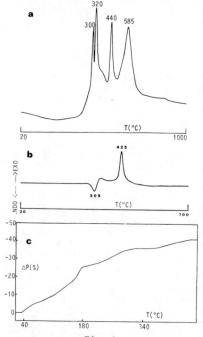

Fig. 1
Thermal analysis of a Cr^{3+}-Al^{3+} doped TiO$_2$ xerogel.

a : D.T.A performed in air
b : D.T.A performed under nitrogen
c : weight loss in air.

A large endothermic peak, corresponding to a continuous weight loss, is observed below 200°C. It corresponds to the removal of physically adsorbed solvent molecules. Heating in air then results in a series of exothermic peaks between 300°C and 600°C (fig. 1a). An accurate assignement of each of them would be difficult, but they must correspond to organic ligands removed by combustion. A comparison with D.T.A. curves of pure titanium acetate and previous publications (15), suggests that the first peaks, between 300°C and 360°C, should correspond to acetate groups while the following ones should be due to butoxy groups associated with alkoxides. The D.T.A. curve is much simpler when the experiment is performed under nitrogen in order to avoid combustion. Two sharp peaks can be seen (fig. 1b). An endothermic one at 305°C that should correspond to the departure of organic ligands (acetates, alkoxides) and an exothermic one at 425°C that corresponds to the crystallization of the TiO$_2$ anatase phase as confirmed by X-ray diffraction. A chemical analysis of the xerogel, together with the observed weight loss shows that we have roughly one organic ligand for two titanium atoms and about two acetates for one butoxy.

Thermal X-ray diffraction experiments were performed on samples annealed for 1 hour at different temperatures. They show that a badly crystallized anatase phase appears at 400°C. It remains stable up to 900°C while the rutile phase begins to be visible around 700°C. A mixture of both phases is obtained between 700°C and 900°C, and pure rutile above 900°C. Similar experiments performed on gels containing differents amounts of Cr^{3+} show that the anatase phase is stabilized in the presence of doping impurities. A pure TiO_2 xerogel gives rise to 30% of anatase and 70% of rutile when heated at 800°C for 1 hour. In the same conditions 90% of anatase are still observed when TiO_2 is doped with 5% of chromium.

2.3 Infra-red and Raman analysis

The infra-red spectrum of the xerogel is shown in fig.(2). The absorption bands of the organic ligands can be clearly seen.

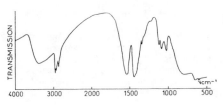

Fig. 2
Infra-red spectrum of xerogel

The butoxy groups, linked to the titanium, give rise to a series of bands between 1100 cm^{-1} and 1000 cm^{-1}, showing that all OR ligands have not been removed by hydrolysis. Acetate ligands are characterized by a doublet around 1500 cm^{-1}. It could be assigned to the $\nu_a(COO) = 1550$ cm^{-1} and $\nu_s(COO) = 1445$ cm^{-1} stretching vibrations (16). The small frequency separation ($\Delta\nu = 105$ cm^{-1}) suggests that acetate behaves as a bidentate chelating ligand (17). A broad absorption band is observed at low frequencies, below 900 cm^{-1}. It corresponds to the envelope of the phonon bands observed in crystalline titanium oxides (18). It is too poorly resolved to allow any accurate assignement.

Upon heating, organic groups are removed and the infra-red spectrum does not give anymore information. The Raman spectrum however is more sensitive toward the Ti-O-Ti vibrations. According to the literature, the Raman spectra of crystalline anatase and rutile are quite different (18). We thus see that the anatase phase appears around 300°C (fig. 3a). The Raman peaks are still very broad, with a half-width of about 50 cm^{-1}. This suggests a badly crystallized phase with organized domains of about 100 Å in diameter. Crystallization improves upon heating and well crystallized anatase is observed around 600°C (fig. 3b). The Raman spectrum of the rutile phase begins to be clearly visible around 800°C and above 900°C, only rutile is observed (fig. 3c).

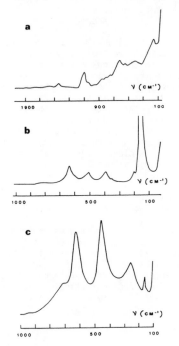

Fig. 3
Raman spectra of xerogel after annealing at different temperatures :

a : 300°C
b : 600°C
c : 900°C

2.4 E.S.R. analysis

The band E.S.R. spectrum of the xerogel is shown in fig. (4a). It is typical of Cr^{3+} ions ($S = 3/2$) in a strongly distorted crystal field. This spectrum can be described with the usual spin Hamiltonian :

$$\mathcal{H} = g\beta H.S + D(S_z^2 - \frac{1}{3} S^2) + E(S_x^2 - S_y^2)$$

where $g = 1.98$, $D = 0.15$ cm^{-1} and $E = 0.05$ cm^{-1}.

These parameters are quite different from those reported in the literature for Cr^{3+} doped TiO_2 (anatase or rutile). This shows that the chemical environment of chromium should not be the same. The strong distorsion ($\lambda = E/D = 0.33$) even suggests that organic ligands rather than only oxygen may be linked to the Cr^{3+} ion.

No significant modification of the ESR spectrum is observed when the xerogel is heated up to 240°C. Around 300°C, a sharp signal, centered around the free electron g value, then appears (fig. 4b). It can be assigned to free radicals arising from the combustion of organic ligands (19). Some slight modification of the ESR parameters are then observed when the xerogel is heated to higher temperatures. The D value decreases suggesting that the crystal field around Cr^{3+} changes when the organic groups are removed.

Two ESR signals are observed at 400°C (fig. 4c). They are both centered around $g = 1.98$ and can be assigned to Cr^{3+} ions. One signal is rather broad ($\Delta Hpp = 170$ G) while the other one is much sharper ($\Delta Hpp = 30$ G). According to previous results reported for Cr^{3+} impregnated polycrystalline TiO_2 (20), we may think that the broad signal corresponds to Cr^{3+} ions dispersed in the poorly crystallized anatase phase while the sharp one would be due to surface chromium species (21).

The intensity of the sharp signal progressively decreases upon heating while the broad one sharpens. The corresponding fine structure becomes therefore clearly visible at 600°C (fig. 3d), allowing an accurate determination of the ESR parameters : $g = 1.98$ and $D = 0.0374$ cm^{-1}. These values are in close agreement with those reported for Cr^{3+} doped TiO_2 anatase (22).

A new signal progressively appears at higher temperatures while the previous one decreases in intensity. Above 900°C the ESR signal observed (fig. 4d) corresponds to Cr^{3+} ions in a rutile lattice characterized by $g = 1.98$, $D = 0,68$ cm^{-1} and $E = 0.14$ cm^{-1} (20)(23).

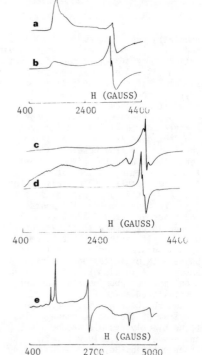

Fig. 4
X-band ESR spectra of Cr^{3+} doped xerogels heated at different temperatures

a : 80°C
b : 240°C
c : 400°C
d : 600°C
e : 900°C

III. CONCLUSION

Our results show that the sol-gel process can be used for making doped TiO_2 semiconducting photoanodes. Three main advantages can be pointed out.:

. A homogeneous doping can be easily obtained by mixing the appropriate solutions of metal-organic precursors, before hydrolysis. The sharpness of the ESR lines in both crystalline TiO_2 phases shows that magnetic inter-actions between Cr^{3+} neighbours are negligeable. No aggregation is observed and a random dispersion of the doping impurities is easily obtained.

. The sol-gel process leads to a thermal stabilization of the anatase phase by Cr^{3+} ions. It has therefore been possible to make anatase photoanodes in order to study their photoelectrochemical properties.

. We have shown that the organic additives, such as acetates, are directly bonded to the metal ions. Acetic acid changes the precursor at a molecular level and slows down the hydrolysis process. Organic ligands are not all removed after hydrolysis and polycondensation. They are therefore directly involved in the formation of the network and can be removed only upon hea-ting above 300°C. We may then think that chemical additives would allow a better control of the hydrolysis-condensation process, as already shown by L. Hench (24)(25). It would then be possible to make "tailor-made" TiO_2 particles, giving monodisperse powders for making pellets or polymeric gels for coating films.

Acknowlegments :

We are particularly grateful to Dr Paul DUMAS for valuable advice and discussion in the Raman study.

REFERENCES

1. K. Honda and A. Fujishima, Bull. Chem. Soc. Jap., 44, 1148 (1971).

2. K.L. Hardee and A.J. Bard, J. Electrochem. Soc., 122, 739 (1975).

3. S.N. Subbarao, Y.H. Yun, R. Kershaw, K. Dwight and A. Wold, Mat. Res. Bull., 13, 1461 (1978).

4. J. Augustynki, J. Hinden and C. Stalder, J. Electrochem. Soc., 124, 1063 (1977).

5. A.K. Ghosh, H.P. Maruska, J. Electrochem. Soc., 124, 1516 (1977).

6. G. Campet, J. Verniolle, J.P. Doumerc and J. Claverie, Mater. Res. Bull., 15, 1135 (1980).

7. D. Haneman and P. Holmes, Solar Energy Mater., 1, 233 (1979).

8. Y. Matsumoto, J.I Kurimoto, T. Shimizu and E.I. Sato, J. Electrochem. Soc., 128, 1040 (1981).

9. Y. Matsumoto, T. Shimizu, A. Tuyoda and E.I. Sato, J. Phys. Chem., 86, 3581 (1982).

10. Y. Takahashi, K. Tsuda, K. Sugiyama, H. Minoura, P. Makino and M. Tsuiki, J. Chem. Soc. Faraday Trans. I, 77, 105 (1981).

658

11. H. Minoura, M. Nasu and Y. Takahashi, Ber. Busenges Phys. Chem., $\underline{89}$, 1064 (1985).

12. K.D. Kochev, Solar Energy Mater., $\underline{12}$, 249 (1985).

13. M.D. Ward, J.R. White, A.J. Bard, J. Am. Chem. Soc., $\underline{105}$, 27 (1983).

14. S. Doeuff, M. Henry, C. Sanchez and J. Livage, J. Non-Crystalline Solids (submitted).

15. C.J. Brinker and J.P. Mukherjee, H. of Materials Sci., $\underline{16}$, 1980 (1981).

16. K. Nakamoto, Infra-red and Raman spectra of inorganic and coordination compounds (1978) 3thd ed. John Wiley & Sons, N.Y.

17. K.H. Von Thiele, M. Panse, Z. Anorg. Allg. Chem., $\underline{441}$, 23 (1978).

18. N.T. Mc. Devitt, W.L. Baun, Spectrochimica Acta, $\underline{20}$, 799 (1964).

19. A.A. Wolf, E.J. Friebele, D.C. Toran, J. Non-Crystalline Solids, $\underline{71}$, 345 (1985).

20. J.C. Evans, C.P. Relf, C.C. Rowlands, J.A. Egerton and A.J. Pearman, J. Mater. Sci. Letters, $\underline{3}$, 695 (1984).

21. J.C. Evans, C.P. Relf, C.C. Rowlands, T.A. Egerton and A.J. Pearman, J. Mater. Sci. Letters, $\underline{4}$, 809 (1985).

22. T.I. Barry, Solid State Comm., $\underline{4}$, 123 (1986).

23. H.J. Gerritsen, S.E. Harrison, H.R. Lewis and J.P. Wittke, Phys. Rev. Lett., $\underline{2}$ (4), 153 (1959).

24. L.L. Hench, in "Ultra structure processing of ceramics, glasses and composites" ed. L.L. Hench and D.R. Ulrich (Wiley-Interscience) (1984).

25. D.R. Ulrich, Cer. Bull., 64 (II), 1444 (1985).

MICROPOROUS LAYERS FROM SOL-GEL TECHNIQUES

A. LARBOT, J.A. ALARY, J.P. FABRE, C. GUIZARD and L. COT
Laboratoire de Physico-Chimie des Matériaux (UA 407)
Ecole Nationale Supérieure de Chimie - 8, Rue de l'Ecole Normale -
34075 - MONTPELLIER Cédex - FRANCE -

In separative process, the use of mineral membranes instead of organic ones is more interesting for it offers numerous advantages as follows :
- High temperature and pressure resistant (no compression of the membrane).
- Corrosion and abrasion resistant.
- Not sensitive to bacterial action.
- Steam sterilisable.
- Longer life time.

The sol-gel process is specially adequate to realize (at a lower temperature than for a normal sintering) porous ceramics in thin layers with a very narrow porous distribution (1-2) By means of a sol-gel process we have obtained a new generation of membranes of Alumina and Titanium dioxide.

The steps of the process can be summarized as follows :

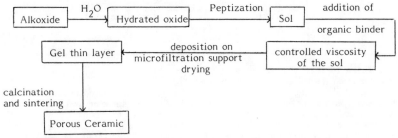

I - ALUMINA MEMBRANE

1) Experimental date

The choice of boehmite powder as hydrated oxide has been driven by the porosity properties of this particular variety (3-4-5-6).

Because of many works on alumina gels, (7-8) we have choosen nitric acid as "electrolyte" for its stability to peptize the boehmite particles and at the same time to modify their superficial charge.

Boehmite powder has been added to different pH acid solutions. These experiments are described fig. 1. It shows that we only obtain boehmite gels for solutions in which pH is inferior to 1.1.

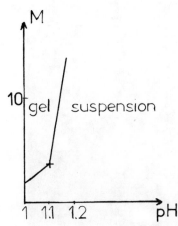

Fig. 1 : Peptization of boehmite as a function of pH.
M : Weight of boehmite added to 25 gr of acid solution.

660

In the second phase of the preparation, we mix the sol with an organic binder.This solution can then be applied homogeneously on a porous ceramic support. That is to say that this sol must have such a viscosity that it can form a layer of regular thickness before being absorbed through the support.We studied the parameters such as acid and organic binder quantity and concentration, and boehmite powder quantity.

2) Drying and thermal treatment

The sol so applied is dried under 4° C during 48 h. This treatment improves considerably the thinness and homogeneity of the layer. The thermogravimetric analysis of the gel allows us to determine a calcination program taking into account departure of water and organic products.

We have studied the influence of the final calcination temperature on the crystalline structure of the membrane and on the value of the pore diameter.Table I summarises the results of the X ray diffraction study on the gel kept 1 h at the final temperature.

Table I : Phases obtained after thermal treatment.

Temperature °C	25	400	900	1100	1200
Structure		γ Al$_2$O$_3$	θ Al$_2$O$_3$	α Al$_2$O$_3$	
State	boehmite gel	amorphous solid	crystalline solid		

These transformations are in accordance with those announced in the literature when the sol-gel technique is used (9).

Fig. 2 shows the variations of the pore diameter (measured by mercury porosimetry) according to the temperature. The pore diameter evolution is linear for calcination temperatures between 500° C to 1000° C, which corresponds to the existence field of the γ phase. Between 400° C and 500° C the X ray diffraction diagrams shows a structure badly crystallized. Above 1000° C the appearance of phases θ and α of alumina gives a more radical increase of the pore diameter. So adjusting the final calcination temperature allows us to obtain from the same sol, mineral membranes with pore diameters from 4 nm to 100 nm. Let us remember that the ultrafiltration field covers 1 nm to 100 nm.

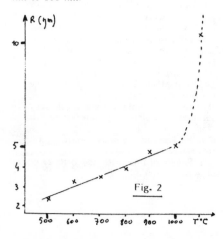

Fig. 2

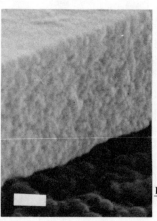

Fig. 3

As an example, fig. 3 shows the morphology of a calcined membrane under 500° C. We can notice the homogeneïty of the particles and their low granulometry.

On fig. 4 we have shown the graphic of the pore volume of the same membrane, showing the pore diameter as well as the distribution. The latter proves that the pore spectrum is very narrow, which is desirable to obtain selective separation in good conditions.

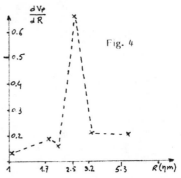

Fig. 4

II - TITANIUM DIOXIDE MEMBRANE

1) Experimental data

The main difference between the experimental process for an alumina coating and for a titanium dioxide coating consists in the raw materials.Here, the first chemical is not a commercial hydrated oxide powder but a suspension of titanium hydrated oxide produced by the hydrolysis of an alkoxide. The latter can be Titanium IV isopropoxide, or the Titanium IV butoxide.In both cases the complete hydrolysis is done at room temperature, under stirring, and yields a hydrated oxide (10-11-12-13). The precipitate is peptized with a dilute solution of nitric or hydrochloric acid.

The kinetic of peptisation can be accelerated by increasing temperature and acidity of the solution. The addition of organic binders such as cellulosic compounds, is an important step to get the best viscosity for the thickness and the homogeneïty of the final layer. The coating is deposited on macroporous ceramic substrate by pouring, then dried and at last sintered up to 500° C.

The ceramic membranes are observed by SEM and characterized by mercury porosimetry.

2) Description of the TiO_2 layer

Thanks to this technique we get reproducible Titanium dioxide layers. Their thickness varies in the same way as the viscosity of the sol. So, we can adjust the viscosity from 20 cp to 50 cp in order to control the thickness to a fixed value between 0.4 and 2.2 micrometers.

In every case, the pore diameter is below 50 Å after sintering at 400° C. Here are 2 views of such a layer : Fig. 1 : Surface of membrane, fig. 2 : layer on its substrate.

3) Structural evolution of a TiO$_2$ Gel

The results of the study presented in this paper concern gels obtained from sols peptized with nitric acid, without organic binders.

3.1. - Decomposition of TiO$_2$ gels as a function of temperature - Infrared study

T° C Groups	25° C	160°C/15h	180°C/15h	200°C	270°C	350°C ≤
-C-O-Ti	0	0	0	0	0	0
-OH	S	VW	VW	VW	0	0
-NO$_3$	S	S	S	S	VW	0

(0 = not ; S = strong ; VW = very weak)

These groups are observed through their absorption at, respectively :

Ti-O-C : 1005 cm^{-1}, 950 cm^{-1}, 619 cm^{-1} (14)
-(O-H) : 3400 cm^{-1}, 1625 cm^{-1} (13)
-NO$_3$:   1390 cm^{-1} (15)

- Gels obtained from sols peptized by nitric acid contain NO$_3$ groups which are strongly linked to the oxide structure.
- A thermogravimetric study shows the successive departures of chemicals and confirm the results of the previous table.
The same experiments with a gel ready to be turned into a ceramic layer permitted to define an appropriate rate of heating avoiding cracks.

4.2. Crystallization of TiO$_2$ Gels

- Differential Scanning Calorimetry

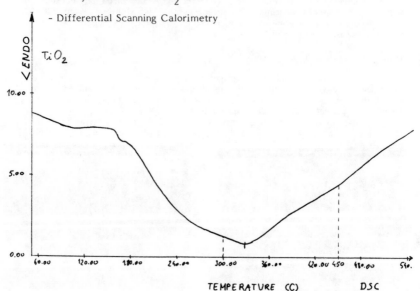

TEMPERATURE (C)   DSC

- The first endothermic band indicates departure of water linked to the Titanium oxide.
- From 300 to 450° C, an exothermic phenomenon takes place, gradually. As shown by XR. diffraction, that means a gradual crystallization of the material.
- No important variations happened up to 1500° C.

. XR Diffraction Study

T° C Order	25°C	160°C/15h	180°C/15h	200°C	270°C	350°C	550°C
Amorphous	S	S	S	S	S	V W	0
Crystallized	0	0	V W	V W	S	anat. + ruti	anat. + ruti

Crystallization from the amorphous state is gradual from 350° C to 450° C, yielding a mixture of anatase and rutile structures, both tetragonal (16).Comparison of all these results shows a good agreement.

IV - CONCLUSION

This study has permitted the realization of a mineral membrane, using the sol-gel process.

The experimental conditions to prepare the sol and to obtain the gel cast on the porous support have been given.

For alumina membrane, variations in the thermal treatment have allowed graduations in the pore diameter from 4 nm to 100 nm ; with this technique the membrane can cover the whole field of ultrafiltration. The purification of polluted water has been realized.(17)

We also made a TiO_2 membrane without cracks, up to 2.2 micrometers thick, $\leqslant$ 5 nm of pore diameter at 400° C, crystallized in anatase plus rutile.

REFERENCES

(1) - S. Sakka, K. Kamiya, K. Marita and Y. Yamamoto -
Journal of Non-crystalline Solids 63 (1984) 223-225.

(2) - H . Dislich and E. Hussman
Thin solid films, 77 (1981) 129-139.

(3) - European patent application n° 0034889 Al.

(4) - J. Bugosh, R. L. Brown - A novel fine alumina powder fibrillar
boehmite. Res. and Dev. - Vol. 1 n° 3 Sept. 1982.

(5) - Pascal - Vol VI, 574 - 581.

(6) - W. H. Gitzen
The american ceramic society - Inc - Special publication n° 4.

(7) - M. Vital-Mathieu - Etude des gels d'alumine désorganisée.
Thèse Lyon 1955.

(8) - B. E. Yoldas
American ceramic society bulletin - Vol 54 n° 3 51975) 285-290.

(9) - J. Wiley and sons
American ceramic society book service.

(10) - Heitner, Winguin and Albu-Yakon
J. Inorg. Nucl. Chem., 1966, 28, 2379.

(11) - I. N. Belyaev and S. A. Artamonova
Russ. J. Inorg. Chem. 1966, 11, 253.

(12) - G. V. Jere and C. C. Patel
J. Sci. Indust. Research 1961, 20 B, 292.

(13) - Vivien, Livage, Mazières.
J. Chim. Phys. 67 (1970), 199.

(14) - Bradley - " Metal Alkoxides " (Ac. Press) 1978, 118.

(15) - Erokhina and Prozorovskaya
Zhur. Neorg Khim, 24, 899, 1979.

(16) - Vallet Regi and Veiga Blanco
Ann. Quim 76 B, 172 (1980).

(17) - C. Guizard, J. A. Alary, A. Larbot, L. Cot, M. Rumeau,
B.Castelas, J. Gillot - Le lait (1984), 64, 276.

EFFECTS OF HYDROLYSIS ON METALLO-ORGANIC
SOLUTION DEPOSITION OF PZT FILMS†

RUSSELL A. LIPELES, DIANNE J. COLEMAN, AND MARTIN. S. LEUNG
The Aerospace Corporation, Chemistry and Physics Laboratory,
P.O. Box 92957, Los Angeles, CA 90245

ABSTRACT

The effects of hydrolysis on the degree of polymerization during metallo-organic solution deposition of lead zirconate titanate (PZT) films have been investigated. The reaction of lead 2-ethylhexanoate, zirconium n-tetrapropoxide, and titanium tetrabutoxide in isopropanol with water were studied using thermogravimetry, specular reflectance Fourier transform infrared spectroscopy (FTIR) and optical and electron microscopy.

Films prepared from coating solutions having varying amounts of water exhibited dramatic differences in morphology. The films were spin-coated on platinum coated fused silica substrates and annealed at 525°C for 30 minutes. Unhydrolyzed coating solutions and solutions with a mole ratio of water to total metal of 0.5 yielded perovskite films with 0.5–5μm grains. A mole ratio of 1.5 (the amount of water required to completely hydrolyze the metallo-organics in the solution) formed amorphous, porous films. The stability of the prepolymerized films inhibits crystallization and densification at moderate temperatures.

INTRODUCTION

Ferroelectric lead zirconate titanate (PZT) and doped PZT have been used for electrooptical light modulators [1], high speed total internal reflection switches [2], and surface acoustic wave devices [3]. Improved control of film morphology would enhance the performance of these devices by increasing the transparency of PZT films. Recently, PZT films have been prepared primarily by physical deposition methods such as radio frequency diode and planar magnetron sputtering [4-7]. Preparation of PZT films by a chemical method has also been reported [8,9]. Wet chemical methods using metal alkoxide starting materials offers the potential of improving control over grain size and morphology.

Recent progress has been made in the processing of silica glasses prepared by sol-gel methods [10]. However, much less work has been reported on ferroelectric materials. In contrast to the formation of glasses, the processing of PZT films is complicated by the need for crystallinity and transparency for use in optical devices. Thus, the grain size, crystal structure, and surface morphology have to be tailored for specific optical applications.

In this paper, we report a study of the effects of hydrolysis and prepolymerization of metal alkoxides on the structure of PZT films. Our results show that the amount of water added to the coating solution can be used to vary the crystallinity and surface morphology.

EXPERIMENTAL

The starting mixture consisted of lead 2-ethylhexanoate (Pb=22.1 weight percent), zirconium n-tetrapropoxide (Zr=24.5 weight percent), and titanium tetrabutoxide (Ti=14.0 weight percent) to obtain a final composition of $Pb_{1.0}Zr_{0.55}Ti_{0.45}O_3$. All starting materials were obtained from Alfa Products, Thiokol/Ventron Division. The mixture was diluted with anhydrous isopropanol to form a coating solution that contained between 30 to 50 weight per cent starting materials. Small amounts of water were added to hydrolyze and prepolymerize the starting materials. The viscosity of the solution was kept below 500 cpoise so that the solution could be used for spin coating.

Films were formed by spin coating the solution on platinum coated fused silica substrates. The film was dried in air first at 110°C for 10 minutes and then at 300°C for 10 minutes to remove most of the organics. The drying temperature was deliberately kept low to avoid pyrolysis of the organics and to inhibit crystallization of the film. This coating cycle was repeated until the desired film thickness was obtained. The film was then annealed at 525°C for 30 minutes to remove the remaining organics and to crystallize the film into the perovskite structure required for ferroelectric behavior. This thermal treatment was used for all of the samples reported here. A typical thickness of PZT with 6 coatings was 0.5μm thick. Films up to 1.5μm thick have been prepared by this method.

The structure of the films was characterized by x-ray diffraction and spectral reflectance Fourier transform infrared spectroscopy (FTIR). FTIR spectra were obtained by reflection from the PZT films through a mask with a 3mm diameter hole using a Nicolet Model MX1 FTIR. Scannning electron microscopy (SEM) and optical microscopy were used to examine the grain structure and surface morphology of the films.

RESULTS AND DISCUSSION

The mixture of starting materials is expected to undergo some exchange of ligands with the solvent [11]. However, polymerization requires the addition of water. Hydrolysis reactions are summarized below:

$$Pb-(OOCC_7H_{15})_2 + H_2O \longrightarrow$$
$$H--O--Pb-(-OOCC_7H_{15}) + C_7H_{15}COOH \qquad (1)$$

$$Zr-(-OC_3H_7)_4 + H_2O \longrightarrow$$
$$H--O--Zr-(-OC_3H_7)_3 + C_3H_7OH \qquad (2)$$

$$Ti-(-OC_4H_9)_4 + H_2O \longrightarrow$$
$$H--O--Ti-(-OC_4H_9)_3 + C_4H_9OH \qquad (3)$$

Hydrolysis of more than one alkoxide ligand probably occurs in the zirconium and titanium compounds [11]. Condensation forms metal-oxygen-metal linkages [11], e.g.,

$$H--O--Zr-(-OC_3H_7)_3 + H--O--Ti-(-OC_4H_9)_3 \longrightarrow$$
$$(OC_4H_9)_3--Ti--O--Zr-(-OC_3H_7)_3 + H_2O \qquad (4)$$

Hydrolysis and condensation can be repeated to form long linear and branched metal-oxygen-metal polymers. To identify the amounts of water added to each coating solution, the water concentration is expressed as a mole ratio (h) of water to the total number of mole of metal:

$$h = [H_2O]/([Pb] + [Zr] + [Ti]). \quad (5)$$

For $Pb_{1.0}Zr_{0.55}Ti_{0.45}O_3$, h=1.5 corresponds to the stoichiometric amount of water needed for complete hydrolysis.

Hydrolysis of the coating solution forms volatile alcohols according to the reactions in Eqns 2 and 3. The temperature dependence of weight loss due to the evolution of organic by-products was studied using thermogravimetry of air dried powders prepared from 30 weight per cent solutions of PZT starting materials with h=0 and h=1.5. The sample with h=0 in Figure 1 shows that higher temperature is required to remove the

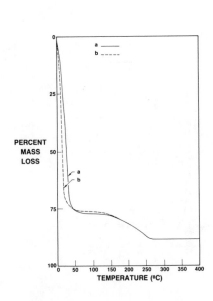

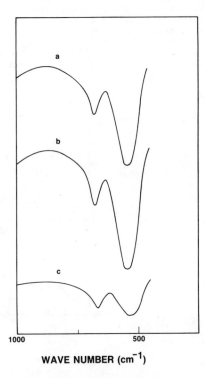

Figure 1. Thermogravimetry of PZT Powders with (a) h-0 and (b) h-1.5.

Figure 2. Spectral reflectance FTIR of 0.8μmthick PZT films prepared on platinum coated 525°C for 30 minutes with (a) h-0, (b) h-0.5, (c) h-1.5.

668

solvent and by-product alcohols than for the h=1.5 material. These results are consistent with the formation of low boiling n-propanol (bp=97°C) and butanol (bp=82°C) during hydrolysis. Based on these results, a preanneal temperature of 300°C was used to evaporate the organic by-products without pyrolysis.

Condensation of hydrolyzed alkoxides results in the formation of metal-oxygen-metal bonds in the film. Spectral reflectance Fourier transform infrared spectroscopy (FTIR) was used to measure the spectrum of 0.8μm thick films deposited on platinum coated substrates and annealed at 525°C. The absorption band centered at about 540 cm^{-1} in Figure 2 is larger for the h=0 and h=0.5 films, than for fully hydrolyzed (h=1.5) film. This band has been assigned to a TiO_6 octahedra vibrational mode by Last [12] and Spitzer et. al [13]. Based on standard reference spectra for lead zirconate and lead titanate, zirconium will also have a band at the same position [14]. The growth of this band shows that limited hydrolysis (h=0.5) increases local order compared to the dry (h=0) samples. The lower intensity of this band in the h=1.5 film indicates a lack of the perovskite phase with its octahedral coordinated titanium and zirconium.

The presence of the perovskite phase in the h=0 and h=0.5 films after annealing at 525°C is confirmed by x-ray analysis. Identical x-ray diffraction spectra were obtained for the h=0 and h=0.5 films. The x-ray diffraction spectra in Figure 3 for h=0 and h=0.5 agree with other spectra for crystalline, perovskite PZT films reported in the literature[4,7,8]. However, the spectrum in Figure 3 for h=1.5 films is fairly amorphous. This result confirms that the intensity of the infrared band at 540 cm^{-1} can be used to characterize the crystallinity of PZT films.

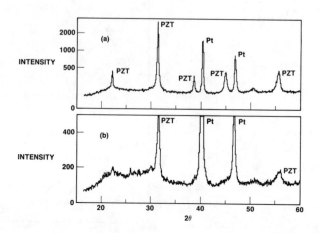

Figure 3. X-ray diffraction spectra of 0.8um thick PZT films for (a) h=0 and h=0.5, and (b) h=1.5.

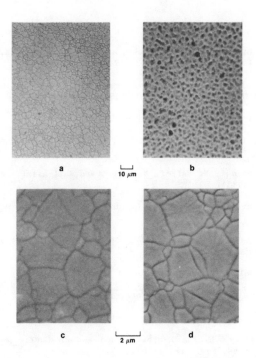

Figure 4. Optical micrographs of PZT films
annealed at 525°C with (a) h=0 and (b) h= 1.5
SEM of PZT films (c) with h=0 and (d) h=0.5.

The effect of water on surface morphology is shown in the
micrographs   in Figure 4.   For h=0 films, a polycrystalline film was
obtained with a distribution of grain size from 0.8 to 4µm . In the h=1.5
film, a very porous structure was observed that was similar to that
obtained in silica [10].   The porosity results from trapped solvent in a
rigid prepolymerized PZT matrix.   Annealing at 525°C only removes the
solvent and does not densify the film.   A comparison of the scanning
electron micrographs   in the Figure of h=0 and h=0.5 films show a
difference in structure due to prepolymerization.

CONCLUSIONS

Water hydrolyzes the alkoxides which can condense to form a metal-
oxygen-metal   polymer.   Our   results   indicate   that   the   degree   of
prepolymerization controls the structure of the dried and annealed PZT
films.   Extensive hydrolysis of alkoxides in the coating solution results
in amorphous, porous films that do not densify at temperatures up to
575°C.   In contrast, polycrystalline, perovskite PZT films can be grown
from coating solutions containing relatively little water.

References

†.   Supported in part by Aerospace Sponsored Research

1.   M. Ishida, H. Matsunami, and T. Tanaka, Appl. Phys. Lett., 31, 433-434

670

(1977).

2. K. Wasa, O. Yamazaki, H. Adachi, T. Kawaguchi, and K. Setsune, J. Lightwave Technol., LT-2, 710-13 (1984).

3. S. B Krupanidhi, M. Sayer, K. El-Assal, C. K. Jen, and G. W. Farnell, J Canadian Ceram. Soc., 53, 28-33 (1984).q

4. A. Okada, J Appl. Phys., 48, 2905-2909 (1977).

5. H. Adachi, T. Kawaguchi, K. Setsume, K. Ohji, and K. Wasa, Appl. Phys. Lett., 42, 867-8 (1983).

6. S. B. Krupanidhi, N. Maffei, M. Sayer, and K El-Assal, Ferroelectrics, 51, 93-98 (1983).

7. H. Matsunami, M. Suzuki, M. Ishida, and T. Tanaka, Japan. J. Appl. Phys., 15, 1163-4 (1976).

8. J. Fukushima, K. Kodaira, and T. Matsushita, J. Material Sci., 19, 595-598 (1984).

9. R. A. Lipeles, N. A. Ives, and M. S. Leung, in Ultrastructure Processing of Ceramics, Glasses and Composites, edited by L. L Hench and D. R. Ulrich (Wiley, New York, 1986).

10. D. R. Ulrich, Ceramic Bull., 64, 1444-8 (1985).

11. D. C. Bradley, R. C. Mehrotra, and D. P. Gaur, Metal Alkoxides (Academic Press, New York, 1978).

12. J. T. Last, Phys. Rev., 105, 1740-50 (1957).

13. W. G. Spitzer, R. C. Miller, D. A. Kleinman, and L. E. Howarth, Phys. Rev., 126, 1710-21 (1962).

14. C. D. Craver, The Coblentz Society Desk Book of Infrared Spectra, (The Coblentz Society, Inc., Kirkwood, MO, 1977) pp 100,106

SOL-GEL AR FILMS FOR SOLAR APPLICATIONS*

CAROL S. ASHLEY AND SCOTT T. REED
Sandia National Laboratories, Albuquerque, New Mexico, 87185

ABSTRACT

Sol-gel derived antireflective films have been prepared for a variety of solar applications. The optical properties of the films are optimized by microstructure tailoring in solution by aging and/or in the film by heating and etching. The resulting film provides a quarter-wave, single layer interference surface with a reflectance minimum of <1% at 600 nm. We have applied sol-gel derived AR films to glass and plastics for solar thermal and photovoltaic applications, e.g., parabolic trough collector envelopes, plastic Fresnel lenses, and glazing materials for flat plate collectors.

INTRODUCTION

Single layer antireflection (SLAR) coatings deposited from multicomponent sol-gel compositions are being utilized in various solar energy related programs at Sandia National Laboratories, including solar thermal and photovoltaic systems. In solar thermal systems, antireflection (AR) coatings have been used in parabolic trough collectors to increase the transmittance of protective glass envelopes that are positioned around the central receiving tube to minimize convective losses. By antireflecting both surfaces of the envelope with sol-gel derived AR films, the solar energy averaged transmittance can be increased from 0.92 to 0.96-0.97, an improvement of ~5%. The solar photon averaged transmittance of polycarbonate, a candidate material for use in flat plate collectors, can be increased by ~9% with the application of an AR film to both surfaces. In photovoltaic concentrator systems, acrylic Fresnel lenses are used to focus sunlight on small area solar cells. The solar photon averaged transmittance of the lens can be increased by as much as 5% with the application of sol-gel AR films onto both the smooth and faceted surfaces.

Both glass and plastics exhibit reflectance losses at the material/air interface (due to differences in refractive index) that average ~7% for two surfaces. This reflectance loss can be minimized over a wide range of the spectrum by the application of a graded index coating with a refractive index (n) that varies smoothly from the value for the substrate (n_2) to the value for air (n_1=1.0). Alternatively, reflectance at a single wavelength can be reduced to a value near zero by the use of a film of the correct thickness and index of refraction. This condition is met by a film with an index of refraction, n_f, given by

$$n_f = (n_2 n_1)^{1/2} \qquad (1)$$

where n_1 and n_2 are the refractive indices of air and the substrate, respectively. The film thickness, d, required to give minimum reflectance at the specified wavelength, λ, is defined by the equation

$$\lambda = 4 n_f d . \qquad (2)$$

*This work performed at Sandia National Laboratories supported by the U.S. Department of Energy under contract number DE-AC04-76DP00789.

672

If the requirements of equations 1 and 2 are met, the reflected energy from
the air/film interface will be 180° out of phase with the reflectance from
the film/substrate interface. Under these conditions, the reflections will
destructively interfere, thereby reducing the reflectance at the desired
wavelength to a value near zero. For example, a SLAR film on PYREX
(n_2=1.47) with minimum reflectance at 600 nm, must have a thickness (d) of
123 nm and an index (n_f) of 1.21.

Antireflection is typically achieved by surface modification
techniques (e.g., selective leaching of a phase separable glass layer) or
by the deposition of either single or multiple, discrete dielectric films.
Porous, single layer films, often prepared using sol-gel techniques, may
result in low reflectance at a single wavelength or exhibit the low
broadband reflectance commonly observed with graded refractive index films
[1,2] and surfaces [3]. Sol-gel deposited films contain open,
interconnected pores with dimensions controlled by the composition, size,
and morphology of polymers grown in solution prior to deposition. The
inherent porosity (and thus the index of refraction) of the film can be
further modified by subsequent processing, e.g., etching and/or heat
treatments. As discussed in greater detail elsewhere [4,5], etching causes
an increase in the average pore diameter (reduction in n), whereas thermal
treatments result in a uniform reduction in pore size (increased n) and
eventually complete densification.

This paper reports the development of a process to apply a SLAR film
onto PYREX for parabolic trough applications. Low reflectance at 600 nm is
achieved by thickness and refractive index tailoring of the sol-gel film.
For application to PYREX, this tailoring was accomplished through a
combination of heating and etching. For use on plastic substrates for
solar photovoltaic applications, the process was modified to eliminate the
heating and etching steps.

EXPERIMENTAL

Solutions of the composition 71 SiO_2, 18 B_2O_3, 7 Al_2O_3, 4 BaO (wt%)
were prepared as described previously [6]. Sol-gel films were deposited
onto PYREX by dip or drain coating from solutions aged for 1 week at 50°C.
Dip coating involves controlled rate withdrawal of a substrate from the
solution. With drain coating, the sol is drained away from the stationary
substrate at a controlled rate. Similar films were produced by both
processes for equal coating rates.

PYREX envelopes, 10 feet long x 2.4 inches diameter, were coated using
20 liters of the aged sol-gel AR composition, according to the process
described previously [6]. Briefly, the tubes were lowered into a
cylindrical PVC coating tank and the solution was pumped out of the tank at
10 in/min. To standardize drying conditions, the tubes were coated and
dried in flowing dry nitrogen. After drying, the coating was partially
densified by heating the tubes at 500°C in a conveyor belt furnace. Heat
treatment was followed by etching for 3-5 minutes in 0.015% NH_4HF_2/0.26N
H_2SiF_6 to optimize the thickness and refractive index of the film for
maximum AR effect.

Optical Measurements

The total amount of incident energy on a surface is the sum of the
reflected energy (R), the energy transmitted through the substrate (T), and
the energy absorbed by the substrate material (A), as shown by the
relationship:

$$1 = R + T + A .$$

(3)

Assumptions for the value A are: 1) absorptance is constant for materials
of the same composition and thickness, and 2) contributions to absorptance
by the sol-gel films are insignificant. As a result, any reduction in
reflectance results in a corresponding increase in transmittance. For
PYREX, the reflectance for two surfaces averages 0.07, fixing the maximum
increase in transmittance to 0.07. For solar thermal systems, the
appropriate optical quantities are solar energy averaged values (R_E, T_E,
A_E). These values are calculated using spectral data obtained over the
wavelength range 265 nm-2400 nm and averaged over a solar energy spectral
distribution [7,8]. For photovoltaic systems, the appropriate spectral
quantities (R_p, T_p, A_p) are calculated by averaging over a solar photon
spectrum that is limited to the active region of the photovoltaic device
(e.g., 265 nm-1060 nm for silicon).

Optical measurements of reflectance or transmittance were made as
described earlier [6] using either a Beckman 5270 spectrophotometer
equipped with an integrating sphere, or a portable solar reflectometer
(Model SSR, Devices and Services Co., Dallas, TX). Thickness and
refractive index measurements were made, using a Gaertner L119X research
ellipsometer equipped with a 6328Å He/Ne laser, on films applied to silicon
wafers.

AR FILMS ON PYREX

As discussed in the previous section, the microstructure of this sol-
gel film is primarily determined by the polymer growth in solution prior to
film deposition. Films deposited from unaged (freshly prepared) solutions
exhibited no AR effect after heating and etching. However, films prepared
from solutions aged at 50°C for up to 2 weeks showed progressively better
AR properties with increased solution age [4]. For a sol-gel film
deposited on PYREX from an aged solution, the spectral reflectance
decreases after heating and etching as shown in Figure 1. After coating,
the film was rapidly heated to 500°C
for a few minutes to allow partial
densification, thus forming a durable,
porous layer. The heated film exhibits
a minimum reflectance of ~0.038 at 800
nm. Etching for short times in

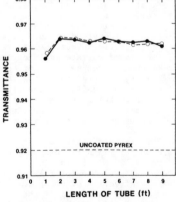

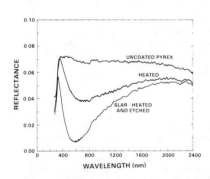

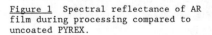

Figure 1 Spectral reflectance of AR
film during processing compared to
uncoated PYREX.

Figure 2 Solar averaged transmittance (average of 9 tubes) of
as-processed tubes (o), and after
16 weeks outdoor exposure (●).

NH_4HF_2/H_2SiF_6 reduces both the thickness and refractive index resulting in the discrete minimum reflectance (<0.01 at 600 nm) characteristic of a single layer interference film.

To determine the suitability of these films for use as AR coatings on parabolic trough collector envelopes, a full-scale process was developed. A previous Sandia AR process [3] increased the solar transmittance (over the wavelength range 500 nm to 2000 nm) of the PYREX envelopes from 0.92 to >0.97 by using phase separation techniques. Phase separation of the PYREX was achieved by extended heat treatment (575°C, 24 hrs) near the softening temperature of the glass (often resulting in deformation), followed by selective acid etching to form the porous surface. In contrast, sol-gel preparation methods permit lower temperature processing (eliminating tube deformation) while allowing substantial control over final film porosity by microstructure tailoring in solution, i.e., via composition or aging; or by porosity tailoring of the film using combinations of heating and etching.

Twelve parabolic trough collector envelopes were processed as described in the previous section, and installed in a modular line-focus solar test facility at Sandia National Laboratories. After processing, the solar averaged transmittance values of the coated substrates ranged from 0.96 to 0.97, compared with ~0.92 for uncoated PYREX. Maximum end-to-end variation in transmittance along the 10 foot length of a single tube was <0.01 transmittance units. As noted in Figure 2, optical measurements made on 9 of the tubes after 16 weeks operation in the test facility showed no decrease in transmittance within the reproducibility (±0.002) of the instrumentation. Thus, this process was successful in depositing a sol-gel AR coating onto receiver envelopes that showed no significant decrease in transmittance after environmental exposure in a parabolic trough collector system.

AR FILMS ON PLASTICS

It was necessary to modify the process developed for PYREX for use on the thermoplastics of interest to the solar industry: acrylic and polycarbonate. These materials are light weight, low cost alternatives to glass for many solar applications, and are more easily formed into the complex shapes required for some concentrator lens designs. Typical solar applications for plastics include Fresnel lenses in both photovoltaic concentrator systems and thermal receivers, as well as glazings for flat plate collectors.

The transmittance of plastics, as with glass, is decreased by surface reflectance; therefore methods of antireflection are required. Techniques for antireflecting plastic surfaces include vapor deposition of films, reactive plasma surface modification, and fluorination processes [9-12]. Most of these processes are not only expensive and subject to sample size and geometry constraints, but also may cause deformation of the plastic surface due to localized temperature increase. In contrast, sol-gel AR processing is relatively inexpensive, and allows simple deposition techniques that permit the coating of large, complex substrates, e.g., the simultaneous coating of the inner and outer surfaces of a 10 foot long tube.

The greatest restriction to the adaptation of the sol-gel AR process developed for PYREX for use on plastics is the upper temperature limit (generally 150°C) of plastics. Additionally, acid etching according to the procedures developed for films on PYREX was ineffective in decreasing the index of refraction of unheated films. However, through careful microstructure tailoring of the sol-gel solution to achieve the required low index of refraction, and precise thickness control (via coating rate), an AR film was deposited on plastics that required no heating or etching.

Reflectance spectra for sol-gel AR films applied to acrylic and polycarbonate substrates are shown in Figures 3 and 4, respectively. These films, which are dried under a heat lamp to ~90°C, produce spectra which are nearly identical to that of a heated (500°C) and etched SLAR film on PYREX (see Figure 1). Excellent AR properties are obtained on both materials using this process. The reflectance of acrylic at 600 nm is reduced from 0.074 for the uncoated substrate to 0.005 for the SLAR coated material. Polycarbonate exhibits a reflectance value of <0.005 at 600 nm after SLAR processing, compared to 0.092 for the uncoated substrate. In both cases, reflectance becomes irregular in the near-infrared wavelengths beyond 1000 nm due to absorption by the plastic substrates.

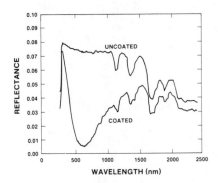 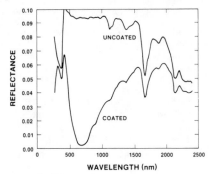

Figure 3 Spectral reflectance of an uncoated and AR coated acrylic substrate.

Figure 4 Spectral reflectance of an uncoated and AR coated poly-carbonate substrate.

The solar energy averaged transmittance (T_E) and solar photon averaged transmittance (T_p) of uncoated and SLAR coated PYREX, acrylic and polycarbonate are shown in Table I. By using sol-gel AR films, transmittance values comparable to or better than values for AR films on PYREX can be obtained for plastics. Depending on the application (and thus the method of solar averaging used), AR films can increase the transmittance of plastics from 5-9%. Differences in T_E and T_p for the same material result from differences in absorption over the wavelength range used to calculate these values.

TABLE I. SOLAR ENERGY AVERAGED (T_E) AND SOLAR PHOTON AVERAGED (T_p) TRANSMITTANCE OF UNCOATED AND SINGLE-LAYER AR COATED MATERIALS.

SUBSTRATE	UNCOATED		SLAR		% IMPROVEMENT	
	T_E	T_p	T_E	T_p	T_E	T_p
PYREX	0.92	0.92	0.97	0.97	5.4	5.4
Acrylic	0.84	0.91	0.88	0.96	4.8	5.5
Polycarbonate	0.82	0.87	0.88	0.95	7.3	9.2

AR films were applied to Fresnel lenses using the modified process. Acrylic lenses (6" x 6"), each with a concentric pattern of angled facets molded into one side, were dip-coated using a pulley and counter-balance system. A uniform film could be applied to both the smooth and faceted lens surfaces by controlling the coating rate (4-8 in/min depending on

solution viscosity), and by compensating for the volume change and buoyancy
introduced by the lens. Due to their size and geometry, the optical
properties of the coated lenses could not be measured using standard
techniques; however, the lenses exhibited the uniform blue color
characteristic of these AR films. A control sample of flat acrylic,
processed in an identical manner, demonstrated the desired optical
properties, i.e., <1% reflectance at 600 nm.

Slight decreases in the transmittance of AR coated plastics were
observed after exposure for up to three months to outdoor environmental
conditions. However, as noted in Table II, nearly equal decreases were
observed for the uncoated materials. Transmittance values of both coated
and uncoated samples could be restored to near original values by rinsing,
suggesting that one mechanism of transmittance loss is contamination of the
surface. In concurrence with the findings of other workers [1], it appears
that although these films may not be extremely abrasion resistant, their
weatherability may be acceptable for many applications.

TABLE II. WEATHERABILITY OF UNCOATED AND SLAR COATED ACRYLIC AND
POLYCARBONATE.

CONDITIONS	UNCOATED (T_E)	SLAR (T_E)
Acrylic		
Initial	0.85	0.90
11 wks outdoors	0.83	0.88
Rinsed	-	0.89
Polycarbonate		
Initial	0.83	0.89
18 days outdoors	0.82	0.88
Rinsed	0.82	0.88

SUMMARY

Sol-gel derived antireflective films were prepared for both solar
photovoltaic and solar thermal applications. The process for AR film
formation on PYREX involves film deposition from an aged, multicomponent
solution, followed by heat treatment (500°C) and acid etching to tailor
film porosity and thickness for maximum AR effect. This process was used
to apply AR coatings to parabolic trough collector envelopes that increased
the transmittance of the tubes by ~5%. The coated tubes exhibited no
significant decrease in transmittance after 16 weeks outdoor exposure.

The process for PYREX was adapted for use on acrylic and
polycarbonate. A sol-gel AR film, requiring no heat treatment or etching
could be deposited on these substrates through microstructure tailoring in
solution and precise thickness control. The solar photon averaged
transmittance of acrylic and polycarbonate were improved by 5-9% after
processing. Finally, the suitability of sol-gel AR processing for
commercial solar applications was successfully demonstrated by the
application of these films onto full-scale parabolic trough collector
envelopes and Fresnal concentrating lenses.

ACKNOWLEDGMENTS

The authors thank Rod Mahoney for assistance with optical
measurements, and Jeff Brinker and Dick Pettit for continued support.

REFERENCES

1. B. E. Yoldas and D. P. Partlow, Applied Optics 23, 1418 (1984).
2. H. L. McCollister and N. L. Boling, U.S. Patent No. 4 273 826 (June 1981).
3. H. L. McCollister and R. B. Pettit, Solar Energy Eng. 105, 425 (1983).
4. C. J. Brinker and R. B. Pettit, Sandia National Labs Report SAND 83-0137 (1983), pp. 68-80.*
5. R. B. Pettit, C. S. Ashley, S. T. Reed and C. J. Brinker, in Sol-Gel Technology, edited by Lisa Klein (Noyes Publications, Park Ridge, NJ, 1986), to be published.
6. C. S. Ashley and S. T. Reed, Sandia National Labs Report SAND 84-0662 (1984).*
7. M. A. Lind, R. B. Pettit and K. D. Masterson, J. of Solar Energy Eng. 102, 34 (1980).
8. R. B. Pettit, Solar Energy Mat. 1, 125 (1979).
9. D. Milam, Lawrence Livermore National Lab Report Energy and Technology Review, p. 9 (March 1982).
10. E. H. Land, H. G. Rogers and S. M. Bloom, U.S. Patent No. 1 444 152 (July 1976).
11. L. F. Urry, U.S. Patent No. 3 996 067 (December 1976).
12. NASA, U.S. Patent No. 1 530 833 (June 1976).

_____

* Available from: National Technical Information Service
                   U.S. Department of Commerce
                   5285 Port Royal Road
                   Springfield, VA 22161 USA

IRON-OXIDE AND YTTRIUM-IRON-OXIDE THIN FILMS
A TEST OF THE FEASIBILITY OF PRODUCING AND MEASURING
MICRO- AND MILLIMETER-WAVE MATERIALS

RALPH W. BRUCE* AND GEORGE KORDAS**
* Electrical and Biomedical Engineering Dept.,
** Mechanical and Materials Engineering Dept., Vanderbilt University, Nash-
ville, TN 37235

ABSTRACT

Tin-oxide/iron-oxide and yttrium-iron-oxide thin-films have been produced
with the sol-gel method. Film thicknesses from 50 to 155 nm were deposited by
the sol-to-gel transformation onto borosilicate substrates. The dielectric
losses of the films were deduced by the complex reflection coefficient of
these materials in the range from 2 GHz to 18GHz using an HP network analyzer.
Samples of 5 cm x 5 cm were centered on the waveguide flange and a shorting plate (12.2 cm x 15.2 cm) of brass centered over the sample. Measurements were automatically made using the HP software provided.
In the X-Band region, three absorption bands at 8.3, 9.4, and 11.5 GHz
were detected for the substrate, with losses (absorption peaks) ranging from
13.5, 11.9, and 9.9, respectively. For the tin-oxide/iron-oxide sol-gel
coated substrate, three bands at 8.4, 9.5, and 11.6 GHz were observed having
losses of 16.2 to 13.4, 12.9 to 11.4, and 11.3 to 9.7, respectively, depending
upon thickness and orientation. For the yttrium-iron-oxide sol-gel coated
substrate, three bands at 8.4, 9.5, and 11.6, with losses depending upon sol-
gel thickness and orientation.

INTRODUCTION

In order to use materials at micro- and millimeter-wave frequencies, the
electrical properties must be well known. Current techniques for the measurement of these properties are based upon the use of samples that are significantly thick or bulky in relation to the wavelength of the measurement frequency. When the sample is very thin, these techniques are not as useful.
This is especially the case with sol-gel thin films. A technique based upon a
modified (or lossy) cavity resonator has been developed that holds the promise
of providing the information needed.
Von Hippel's method and its derivatives require a machined sample that
fits into a section of waveguide terminated in a short-circuit [1]. Based
upon the change of the VSWR (or reflection coefficient) from the unloaded to
the loaded condition, the dielectric properties can be determined. But this
requires: 1). A sample thickness which is a significant portion of a wavelength; 2). A sample that can be machined; and 3). A sample that can only be
used for the measurement, i.e., destructive testing. For sol-gel thin films,
only the machining can be achieved but the sample is no longer useable. Additionally, measurements are accomplished at single frequencies and are liminted
also by the resolution of the scale used on the probe-carriage.
Recently, a technique has been developed that does not require the
machining of the sample [2]. In this method, the sample is placed on the
waveguide flange and measurements of the reflected power made. Based upon
these values, the dielectric properties as a function of frequency are obtained. But one of the conditions of this technique is that the reflected
signal is not influenced by the presence of any object in the vicinity of the

sample. This is especially the case if a shorting plate is placed on the sample. This implies that once again the material must be substantially thick in order to make the measurement. Yet, a modification of this last technique (in which the effects of the shorting plate are used) holds the promise of resolving the problem of measuring very thin films. In the method described, the waveguide-sample-shorting plate act as a leaky cavity resonator. Initial results show changes in the resonance of the cavity resonator both as a function of the sol-gel material and of the material thickness.

EXPERIMENTAL

Tin-oxide/iron-oxide and yttrium-iron-oxide thin films were prepared using the sol-gel method [3]. These were then tested in the X-Band (8.2 to 12.4 GHz) using the HP 8409C Automatic Network Analyzer (ANA) (Figure 1). A detailed view of the waveguide flange shows the arrangement of the sample and the shorting plate (Figure 2). The ANA was calibrated using a modified HP 11863F Accuracy Enhancement Software Package. Subsequent to this, the calibration was verified using a 50 ohm load, and short and open circuits. Since there was no transmission path in the current set-up, only the reflected signal was used to obtain material characteristics. The reflected signal (S_{11} in the S-parameter description) was described by both magnitude and phase. The Return Loss ($= -20\log_{10} /S_{11}/$) was then plotted as a function of frequency.

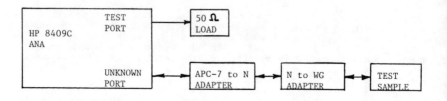

Figure 1. Modified HP network analyser used for the measurements.

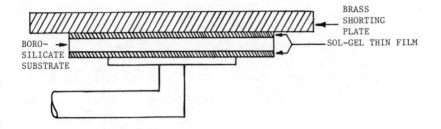

FIGURE 2. Detailed view of the waveguide flange, sample and brass shorting plate.

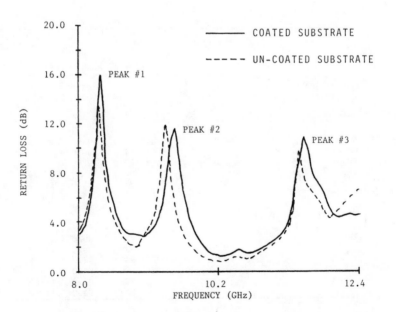

FIGURE 3. Absorption peaks of the substrate with and without coatings.

RESULTS AND DISCUSSION

Figure 3 shows typical absorption peaks of the borosilicate substrate with and without coatings. Three distinctive absorption peaks were detected in the substrate (fig. 3) the position and return loss of which is significantly altered by the composition and thickness of the coatings. The peak positions as well as return losses are not significantly affected by the ninety degree rotation of the borosilicate glass with respect to the waveguide flange. When the substrate was coated, changes of these values were obtained indicating surface textures. This suggests that this technique is appropriate to qualitatively assess the quality of the films.

The absorption loss of the second peak is independent of the composition as well as the thickness of the films. However, the resonance frequency of this peak shifts to lower values while the resonance positions of the first and third peak remain practically unchanged. The peak position and absorption loss depend on both the real and imaginary parts of the dielectric constant. Since the return losses of the second peak are about the same with and without coatings, we conclude that the imaginary part of the dielectric constant of the substrate determines these losses. The lower frequency of resonance would indicate that real part of the effective dielectric constant is predominant and changes after coating. The power absorption of the first peak increases with the thickness of the yttrium-iron-oxide film while it is the same within the errors of measurements for different thicknesses of the tin-oxide/iron-oxide film. The above discussion is summarized in Tables I and II.

TABLE I

	ORIENTATION	FREQUENCIES OF ABSORPTION PEAKS MHz		
		PEAK 1	PEAK 2	PEAK 3
BOROSILICATE	0˙	8330	9390	11480
SUBSTRATE	90˙	8340	9390	11530
YTTRIUM-IRON-OXIDE				
50nm	0˙	8310	9350	11430
80nm	0˙	8400	9510	11600
80nm**	0˙	8640	9750	11850
120nm	0˙	8350	9520	11550
	90˙	8350	9520	11550
120nm**	0˙	8350	9470	11530
TIN-OXIDE/IRON-OXIDE				
145nm	0˙	8380	9530	11600
	90˙	8360	9470	11550
150nm	0˙	8360	9370	11600
	90˙	8290	9430	11490
155nm	0˙	8360	9450	11550
	90˙	8290	9400	11490

TABLE II

	ORIENTATION	POWER ABORPTION AT PEAKS dB		
		PEAK 1	PEAK 2	PEAK 3
BOROSILICATE	0˙	13.5	11.9	9.6
SUBSTRATE	90˙	13.5	11.9	10.0
YTTRIUM-IRON-OXIDE				
50nm	0˙	12.3	11.6	10.8
80nm	0˙	14.3	11.5	10.8
80nm**	0˙	13.9	10.2	7.1
120nm	0˙	15.2	11.4	10.7
	90˙	14.9	12.0	11.3
120**	0˙	15.4	10.6	9.3
TIN-OXIDE/IRON-OXIDE				
145nm	0˙	16.2	11.9	9.7
	90˙	15.0	12.8	10.2
150nm	0˙	13.7	12.9	10.5
	90˙	13.5	11.8	11.0
155nm	0˙	15.1	12.1	10.1
	90˙	13.6	11.4	11.3

* 90˙ is 90˙ CCW from the 0˙ position
** These substrates were made at a different time than the others. These values are added to maintain completeness of the study.

CONCLUSIONS

Our data support the following points:

1. For the yttrium-iron-oxide film, there is a general increase in the frequency of the peak as the thickness increases.

2. There is also a general increase in the power absorption as the thickness increases for the yttrium-iron-oxide. But this is not so for the tin-oxide/iron-oxide film.

3. Orientation of the substrate can change either the frequency of the peak or the power absortion.

We have shown that this method can distinguish between various materials and material thicknesses of thin-films derived from the sol-gel process. Further work is needed to determine the relationship between the frequency and power absorption and the dielectric values that give rise to these phenomena.

684

REFERENCES

1. A.R. Von Hippel,<u>Dielectric</u> <u>Materials</u> <u>and</u> <u>Applications</u>, (MIT Technology Press,New York,1954).

2. Viron Teodoridis, Thomas Sphicopoulos and Fred E. Gradiol,IEEE Trans. MTT,<u>MTT-33</u>,359 (1985).

3. H. Dislich and E. Hussmann, Thin solid Films,<u>77</u>,129 (1981).

NEW MAGNETIC MATERIALS PRODUCED BY THE SOL-GEL PROCESS: A POSSIBILITY OF PRODUCING EXOTIC DEVICES

G. Kordas, Vanderbilt University, Nashville, TN 37235, USA.

ABSTRACT

Ferromagnetic resonance (FMR) spectra of an iron-oxide thin film with a thickness of 70 A were recorded between 100 and 410 K at 9.5 GHz after heat treatment at about 500 ^{o}C in hydrogen atmosphere. The FMR-signal of this film consisted of two components (A and B) when the film plane was oriented parallel to the external field. The intensity of these components is not proportional to $T^{3/2}$ in the range from 100 to 410 K. The line width of the A-component is determined by inhomogeneities in the magnetic structure. The line width of the B-component may be influenced by the spin-spin relaxation mechanism and skin effect. The temperature behavior of the resonance field of the A- and B-components was tentatively attributed to variation of the local fields with the temperature of measurements.

INTRODUCTION

As it becomes more apparent that communication will increasingly go via light guide fibers rather than wires, more and more interest arises in developing means to switch and modulate the light. Devices accomplishing such switching often involve films of materials in which interaction between optical and magnetic properties exist. Magneto-optic materials can be incorporated into a planar or circular waveguide to produce a miniature circulator or isolator. Such materials are also used for display, computer memory, sensor, and printing devices.

Yttrium Iron Garnet (YIG) films doped with Bi, Sm, Ga, etc. are the basic materials for magneto-optic applications [1]. Traditionally, these materials have been manufactured by Liquid Phase Epitaxy (LPE), sputtering, hot pressing, and Chemical Vapor Deposition (CVD) techniques [1-4]. YIG films can also be produced with the sol-gel method. Very little research, however, has been carried out on the physical properties of the sol-gel thin films.

The present paper deals with the magnetic properties of an iron-oxide sol-gel thin film determined by the Ferromagnetic Resonance (FMR) technique. This composition was chosen because it is the basic composition for many applications.

Experimental

The iron-oxide thin film was prepared using the dip-coating technique described previously [5,6]. The thin-film thickness was about 70 A. The FMR-measurements were conducted on circular samples with a diameter of about 2 mm. These samples were heat treated at 500 ^{o}C in hydrogen atmosphere for about five minutes prior to the measurements. The work was carried out using an IBM-Bruker ER 200 D-SCR spectrometer operating at 9.5 GHz. The temperature of measurements was varied from 100 to 410 K with an IBM-Bruker ER 4111 VT temperature controller. The FMR spectra were recorded and analyzed using an IBM-9001 computer.

Results

Figure 1 shows a typical FMR spectrum of the film. This spectrum was recorded with the plane oriented perpendicular to the external magnetic field. Two signals are evident from this measurement that will be referred to as A - and B - components in the following. The resonance field of the A - and B - components depends on the orientation of the film with respect to the applied external field.

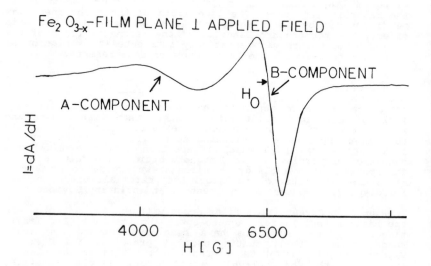

Figure 1. FMR spectrum of an iron oxide film heat treated at about 500 °C recorded at 9.5 GHz and 300 K. The measurements were made with the film plane perpendicular to the external magnetic field.

Figure 2 shows the temperature dependence of the A- and B-component intensity. The intensity of these components was displayed in figure 2 as a function of $T^{3/2}$. A $\log(I*\Delta Hpp^2)$ vs. $\log(T)$ plot is shown in figure 2.

Figure 3 displays the temperature dependence of the peak-to-peak line width and the resonance field of the A- and B- components. One can notice that the resonance field of the A- and B- components depends linearly on the temperature of measurements. The peak-to-peak linewidth of the A-component increases with the decrease of the temperature of measurements while of the B-component increases up to 300 K, remains constant from

300 to 340 K, and then decreases with further increase of the temperature of measurements.

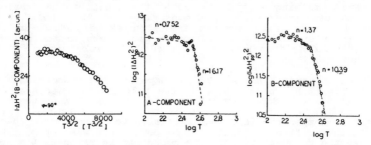

Figure 2. Temperature dependence of the A- and B- components.

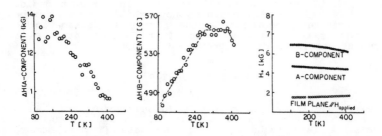

Figure 3. Peak-to-peak line width and resonance field of the A- and B- components as a function of the temperature of measurements.

Discussion

The resonance condition for an infinitely thin disc with the film plane perpendicular to the external magnetic field is given by the equation [7]:

$$h\omega = g\mu_B(H - 4\pi M_s)$$

Scanning Electron Microscopy (SEM) of an iron oxide film revealed particles having small distribution about their means [8]. The ferromagnetic resonance condition includes the resonance field, H_o, and the local field, H_1, in the particle model [9]:

$$H_{eff} = H_o + H_1$$

Phases with appreciably different H_1 or M_s exhibit distinctive different FMR-signals. For example, the spectrum of a Co-Cr film produced by sputtering process consists of two FRM-lines that can be compared to the A- and B- component (figure 1) [10]. One of the FMR-resonances in the Co-Cr thin film was explained assuming a layer formed at the onset of the fabrication process interfacing the substrate and the "bulk" thin film. The similarity between the iron oxide and Cr-Co FMR spectra may suggest the existence of two layers in the sol-gel thin film. Additional work is needed to substantiate this hypothesis.

The area underneath the A- and B- components is proportional to $4pM_s(A,B)*V(A,B)$ where $V(A,B)$ is the relative volume. Assuming that $V(A,B)$ is independent of the temperature of measurements, then the temperature dependence of the double integral of the A- and B- components is directly related to changes of $M_s(A,B)$. Direct integration of the A- and B- components was not possible because their integration limits could not be defined. A simple calculation of the areas was made by multiplying the vertical separation, I, by the square of the peak-to-peak line width, ΔH, of the first derivative of the absorption peak. One can perceive from figure 2 that the intensity of the A- and B- components is not proportional to $T^{3/2}$. This result indicates that the A- and B- components do not exhibit normal ferromagnetic behavior. According to the Stoner ferromagnetic model [11], the temperature dependence of the magnetization of some amorphous materials can be described by the equation:

$$M^2(T) = M^2(0)[1-(T/T_c)^n]$$

For a-$Y_{1-x}Ni_x$ alloys, n is equal 2 (Stoner model) for x between 0.93 and 0.97. The n-coefficient is 1.5 for x=0.872 and 1.0 for x=0.833. A least square fit of the data (fig. 2) revealed values for n that cannot be expected by the Stoner ferromagnetic model. Since the magnetic film was produced after heat treatment at about 500 $^{\circ}$C in hydrogen atmosphere a distribution of chemical states can be expected, for example Fe^+, Fe^{2+}, and Fe^{3+}. This distribution may also result in the formation of several iron oxide magnetic phases. Consequently, the temperature dependence of the magnetization might not be described by a single model.

The line width of the A- and B- components exhibits different dependence on T (figure 2) indicating that different relaxation mechanisms determine their magnitudes. The line width of the A-component decreases with increasing temperature of measurements (fig. 2). Inhomogeneities in the magnetic structure cause the line width of a FMR-signal to increase as the temperature drops approximately proportional to $M_s(T)^{1/2}$ [11]. Inhomogeneities in the sol-gel film can be due to surface or interface defects (macro-inhomogeneities) or random distribution of iron(II)/iron(III) ions at different sites (micro-inhomogeneities). The temperature dependence of the line width of the B-component (fig. 2) is similar to those reported for spinel ferrites, in which there are iron(II) and iron(III) ions in octahedral sites [12]. The spin-spin relaxation mechanism produces a line width that is proportional to the temperature of measurements [12]. The treatment of the iron oxide sol-gel film in hydrogen atmosphere causes the conductivity to increase considerably [13]. The skin effect may also contribute to the line width of the B-component [11]. In this case, the line width is proportional to the square root of the conductivity that increases with the increase of the temperature of measure-

The ferromagnetic resonance field in $Gd_{37}Al_{63}$ decreases with
the increase of the temperature of measurements due to the
decrease of the local field, H_1 [9]. The local field is determined by the reduced magnetization, $m(T)$, in $Gd_{37}Al_{63}$. Distributions in Curie temperatures and cluster sizes cause the
net $m(T)$ to become a linear function with the temperature of
measurements [9]. Since the ferromagnetic resonance field of
the A- and B-components decreases about linearly with respect
to the temperature of measurements, we tentatively attribute
this behavior to the temperature variation of H_1.

ERENCES

1. J. Daval, B. Ferrand, J. Geynet, D. Challeton, J.C. Peuzin,
   A. Leclert, and M. Monerie, IEEE Trans. Magn.,
   Mag-11,5(1975)1115.
2. D.H. Harris, R.J. Janowiecki, C.E. Semler, M.C. Wilson,
   and J.T. Cheng, J. Appl. Phys., 41,3(1970)161.
3. G.R. Blair, A.C.D. Chaklader, and N.M.P. Low,
   Mat. Res. Soc., 8(1973)161.
4. R.L. Gentilman, J. Amer. Cer. Soc., 56(1973)12,623.
5. H. Dislich and E. Hussmann, Thin Solid Films, 77(1981)129.
6. H. Dislich and P. Hinz, J. Non-Cryst. Solids, 48(1982)11.
7. C. Kittel, Phys. Rev., 73(1948)155.
8. G. Kordas, R.A. Weeks, and G. Kordas, J. Appl. Phys.,
   57,8(1985)3812.
9. S.C. Hart, P.E. Wigen, and A.P. Malozemoff, J. Appl. Phys.,
   50,3(1979)1620.
10. P.V. Mitchell, A. Layadi, N.S. VanderVen, and J.O. Artman,
    J. Appl. Phys., 57,8(1985)3976.
11. K. Moorjani and J.M.D. Coey, Magnetic Glasses, Elsevier,
    New York, 1984, p137.
12. S.V. Vonsovskii, Ferromagnetic Resonance, Pergamon Press,
    New York, 1966.
13. G. Kordas and E. Sonder, unpublished data.

PART X

Materials for Electronic Packaging

NON-CONVENTIONAL ROUTE TO GLASS-CERAMICS FOR ELECTRONIC PACKAGING

C. GENSSE AND U. CHOWDHRY
E. I. du Pont de Nemours & Company, Inc., Central Research & Development
Department, Experimental Station, Wilmington, DE 19898

ABSTRACT

This paper describes a non-conventional route to formation of dense
glass-ceramics at relatively low temperatures. Organometallic precursors
were used to synthesize powders with controlled purity and stoichiometry in
the $MgO/Al_2O_3/SiO_2$ system. These amorphous powders sintered by viscous
flow to form a dense glass below 1000°C. Further heating resulted in the
formation of very fine grained glass-ceramics. Beta-quartz, cordierite,
and cordierite-mullite glass-ceramics were obtained with attractive
properties for electronic packaging.

INTRODUCTION

Ceramic powders derived from low temperature chemistry offer the
potential for control over purity, homogeneity, stoichiometry, as well as
physical characteristics such as particle size, size distribution, and
morphology. Impressive reductions in sintering and densification
temperatures have been achieved by using submicron particles, narrowly
sized, and uniformly packed. Low temperature processing is particularly
attractive to the electronic packaging industry wherein the trend is toward
multilayered substrates that can be cofired with good electrical conductors
such as Cu, Au, and Pd-Ag. The goal for these systems is to achieve a
hermetic dielectric below 1000°C in an inert atmosphere to allow use of Cu
metallization. Low temperature chemistry shows promise for bringing us
closer to that goal.

The choice of a cordierite ($2MgO.2Al_2O_3.5SiO_2$) based glass-ceramic
system stems from its attractiveness as an alternative substrate material
to conventionally used alumina. Dielectric constants for materials in this
system range from 5-6 at 1 MHz [1], while alumina has a dielectric constant
of 8-9 at 1 MHz. The thermal expansion coefficient of a pure synthetic
cordierite is anisotropic, being $3.27 \times 10^{-6}/°C$ along the "a" direction and
$0.28 \times 10^{-6}/°C$ along the "c" direction [2]. It has been demonstrated that
a composite containing 36 wt % cordierite and 64 wt % mullite
($3Al_2O_3.2SiO_2$) has a thermal expansion coefficient that closely matches
that of silicon in the temperature range, 25°C-300°C [1]. Lastly, the
cordierite and cordierite-mullite systems offer the potential for lower
sintering and densification temperatures than α-Al_2O_3 which is
conventionally sintered at 1600°C thereby restricting the choice of
cosinterable conductors to refractory metals such as tungsten and
molybdenum.

The fabrication of multiphase silicate based ceramic systems is
commonly practiced with the aid of liquid phase sintering. Densification
generally does not occur until a eutectic temperature or the melting point
of one of the components is exceeded. In the cordierite system the amount
of liquid formed in the system can be large, and this can result in
unacceptably high porosity due to gas entrapment [3].

For certain silicate compositions, glass-ceramic technology is
practiced. This involves melting the oxide constituents and cooling to
form a glass, followed by crystallization of the glass by a controlled heat
treatment to form a polycrystalline body. This paper describes a low
temperature route to a glass-ceramic through use of chemically derived
powders.

EXPERIMENTAL PROCEDURE

Powders of various compositions in the $MgO/Al_2O_3/SiO_2$ system were
obtained by hydrolysis and condensation of an appropriate alkoxide mixture.
Two sets of factors were considered to be important:

(1) nature of precursors, order of mixing, aging of solutions before
     and after mixing.
(2) precipitation conditions such as temperature, pH, and amount of
     water available for hydrolysis.

The following 0.1 molar solutions were used as precursors:

(a) magnesium sec-butoxide in amyl alcohol (+1%wt acetic acid)
(b) tetraethylorthosilicate in methanol (+.2 mole $H_2O/.1$ mole TEOS)
(c) aluminum sec-butoxide in amyl alcohol

Each solution was aged for 48 hours prior to mixing to allow
alcoholysis to be completed. Magnesium alkoxide and TEOS solutions were
initially mixed and the aluminum alkoxide solution was subsequently added.
NMR studies showed that the environment of the silicon remained unchanged
after mixing; therefore aging of the mixture was not necessary. The
resulting mixture was a clear solution. The hydrolysis-condensation
reaction was carried out by adding the alkoxide mixture at a slow rate (2
drops per second) into a stirred solution of ammonia-saturated water at
70°C. The reaction was carried out for 24 hours. During this time,
ammonia was bubbled through the reaction mixture. A reaction yield of 100%
was obtained.

The resultant white powder was washed thoroughly in isopropyl alcohol,
filtered, and dried in air at room temperature. The surface area of these
powders was 300 m^2/g, and they were x-ray amorphous.

PHASE EVOLUTION

1. X-ray Studies

The phase evolution for a powder with the composition $2MgO.2Al_2O_3.5SiO_2$
observed in this study is described below:
The powder remained amorphous up to 900°C. At 920°C, crystalline peaks
of a stuffed β-quartz hexagonal crystal structure appeared in the x-ray
diffraction pattern (Fig 1). Stuffed β-quartz has the crystal structure of
high quartz in which some of the silicon has been substituted by aluminum.
To compensate for the charge deficit, alkali or alkaline earth cations
enter the structure (stuffing) along the helicoidal six-fold axis (Fig. 2).
The stoichiometry of this crystalline phase can be determined, even though
an amorphous phase is most often associated with it, because the cell
parameters "a" and "c" are proportional to the ratio $R=SiO_2/SiO_2 + Al_2O_3 + MgO$ [4] (Fig. 3). Cell dimensions were obtained by least squares
refinement of the x-ray data (Table 1) and the graph in Fig. 3 was used to
deduce the value of R. As the sintering temperature increased, the value
of R increased to about 95% before the structure transformed to the low
temperature hexagonal form of cordierite also known as μ-cordierite. From
950°C upward, there was a gradual and continuous increase in the amount of
amorphous phase present and at 1150°C, the crystallized hexagonal form of
cordierite was present as observed in Fig. 1. This cordierite phase is
metastable, and above 1350°C it transforms to the high temperature, low
symmetry form of cordierite known as α-cordierite.

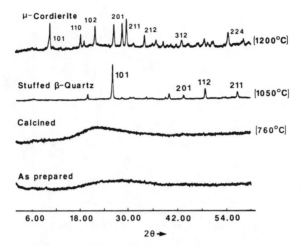

Fig. 1. X-ray diffraction pattern evolution with temperature for a powder with the composition: $2MgO.2Al_2O_3.5SiO_2$.

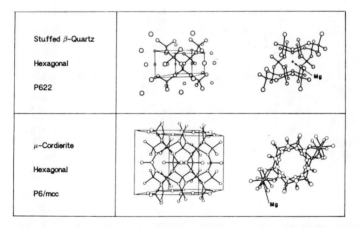

Fig. 2. Unit cells and projections along c-axis for stuffed β-quartz and hexagonal cordierite.

The sluggish transformation involves slow diffusion of silicon and aluminum ions and their redistribution over the tetrahedral sites T_1 and T_2 [5] (Fig. 4). The observed d-spacings and cell parameters of the various phases investigated in this study are shown in Table 1.

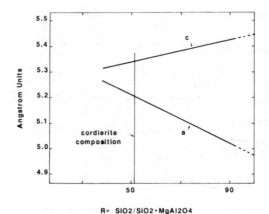

R= SiO2/SiO2+MgAl2O4

Fig. 3. Variation of cell parameters of stuffed β-quartz as a function of $R = SiO_2/SiO_2+Al_2O_3+MgO$.

Table 1: Observed d-spacings and cell parameters of the different phases in the cordierite system.

Stuffed β-Quartz*		μ-Cordierite	α-Cordierite**
950°C	1050°C		
4.48	4.322	8.425	4.878
3.43	3.373	4.854	4.642
2.592	2.290	4.633	4.071
2.295	1.832	4.076	3.363
2.247	1.685	3.368	3.126
2.069		3.126	3.038
1.859	a=5.122	3.015	3.003
1.719	c=5.443	2.632	2.836
1.697		2.451	2.641
1.654		2.349	2.544
1.618		2.202	2.453
1.618		2.091	2.421
		1.941	2.336
a=5.183		1.895	2.278
c=5.341		1.868	2.231
			2.171
		a=9.770	2.107
		c=9.352	
			a=9.721***
			b=17.062
			c=9.339

*temperatures given are the sintering temperatures
**observed after 14 days at 1400°C
***reported

In this study, complete transformation to orthorhombic cordierite was not achieved even after maintaining the sample at 1400°C for 14 days. (See Table 1).

The rate of phase transformation in the system was dependent on the composition of the material. The presence of excess aluminum, for instance, resulted in a larger temperature range of stability for the stuffed β-quartz structure.

697

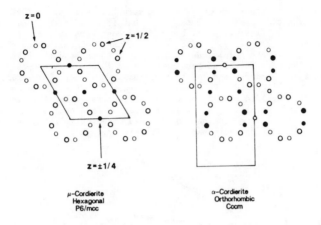

z=0
z=1/2
z=±1/4

μ-Cordierite
Hexagonal
P6/mcc

α-Cordierite
Orthorhombic
Cccm

● Aluminum-rich Tetrahedra

Fig. 4. Structural rearrangement during transformation of hexagonal to orthorhombic cordierite.

2. FTIR Studies

Infrared spectroscopy was used to characterize the powder at different stages of phase transformation. The evolution of the IR spectra with temperature is shown in Fig. 5.

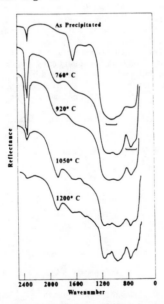

Fig. 5. FTIR spectra as a function of temperature for a powder with the composition: $R = 2MgO.2Al_2O_3.5SiO_2$.

The as-prepared amorphous powders showed a structure similar to that observed in hydrated aluminosilicate gels. Detailed infrared studies of such gels have been provided by A. Leonard and coworkers [6], and by J. Fripiat and coworkers [7], [8]. Two bands are used to monitor the changes which take place in the gels: one centered at 1040cm^{-1} and the other centered at 750cm^{-1}. The broad band centered at 1040cm^{-1} is composed of several peaks which correspond to the four modes ν_1 to ν_4 of the stretching vibration of the Si-O bond. We found the frequency of the ν_3 component (maximum reflectance) to be influenced by two factors: the stoichiometry and the degree of hydration. A linear decrease of frequency of the ν_3 component with increasing aluminum content was observed in the as prepared powders as well as in the calcined powders (Fig 6).

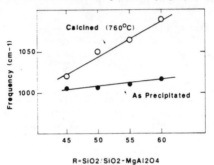

R=SiO2:SiO2-MgAl2O4

Fig. 6. Variation of Si-O stretching frequency (ν_3) with composition.

Such a change in frequency is related to the loss of cohesion between silicon tetrahedra as a result of the substitution. This tendency was offset by the structural rearrangement following increase of temperature and subsequent dehydration. During dehydration, the aluminum cations change from six-fold to four-fold coordination by loss of two hydroxyl groups [8]. The resulting increase of frequency suggests that an ordering of the silicon tetrahedra network takes place during the dehydroxylation process. Figure 7 shows that for samples calcined in air, the rate of dehydroxylation was linearly dependent on temperature.

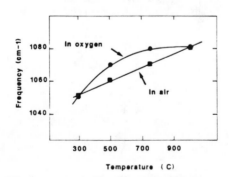

Fig. 7. Variation of Si-O stretching frequency (ν_3) as a function of temperature and calcination atmosphere.

When samples were calcined in oxygen however, dehydroxylation was achieved at lower temperatures. The ability to dehydrate the powders at temperatures lower than the onset of densification (800°C) enabled us to achieve complete densification of the sample by avoiding the formation of bubble structures due to the entrapment of remaining water during pore closure.

Structural rearrangement of the amorphous powder during calcination could also be seen with the appearance of a band at 750 cm^{-1}. This band corresponds to the ν_6 deformation vibration of the Si-O-Si bands. Its existence has been related by Matossi [9] to the formation of a ring structure, and marks the difference between hydrated and dehydrated amorphous powder. At 920°C, the spectrum was similar to that of a glass of the same composition, and no significant change was observed after crystallization to form stuffed β-quartz. These data are in agreement with studies [10],[11],[12] which have shown that the short range order in an aluminosilicate glass is identical to the short range order found in open quartz.

SINTERING/MICROSTRUCTURE EVOLUTION

The as-prepared powders consisted of equiaxed, 100Å particles (Fig. 8a) which agglomerated into 0.1 micron equiaxed microagglomerates (Fig. 8b). The surface area of these powders was 300 m^2/g, and did not change significantly after calcination. Calcination was carried out in an oxygen atmosphere, at 760°C. A slow heating rate of 1°C per minute was used during calcination and sintering. After calcination, the powders were pressed into pellets to 40% green density; no binder was used.

Upon heating, the pellets densified by a viscous flow mechanism to form a dense glass at 850°C (Fig. 9b). Increase in temperature beyond 920°C led to crystallization of the glass Fig. 10. As already stated, the first phase to form was stuffed β-quartz followed by a progressive increase in the value of R. The reconstructive transformation to the metastable hexagonal cordierite phase between 1100 and 1150°C was accompanied by the formation of cracks. Formation of such cracks was avoided by adding excess aluminum sec-butoxide (2% wt) to the original alkoxide mixture.

Several samples were prepared with compositions along the cordierite-mullite tie line in the ternary phase diagram. A dense glass sample of composition: 60 wt% cordierite and 40 wt% mullite was obtained at 920°C. The mullite phase appeared between 950°C and 1000°C. Crystallization to form a cordierite-mullite composite was complete at 1200°C. It was heated to 1300°C to reveal the micron sized grains of mullite in a fine grained cordierite matrix (Fig 11).

Previously reported studies on cordierite and cordierite-mullite composites [1] indicate that the bodies do not densify until the melting point of cordierite is exceeded. Porosity in the form of spherical 15-30μm pores is a persistent problem which has been attributed to either outgassing of the starting materials during firing or to the high wetting angle that exists between mullite and molten cordierite. The high wetting angle is thought to encourage formation of an interconnected network of mullite grains which reduces the densification rate.

The work reported in this study circumvents the above mentioned problems presumably because the alkoxide derived powders are highly reactive and densify by viscous flow well below the melting point of any constituent.

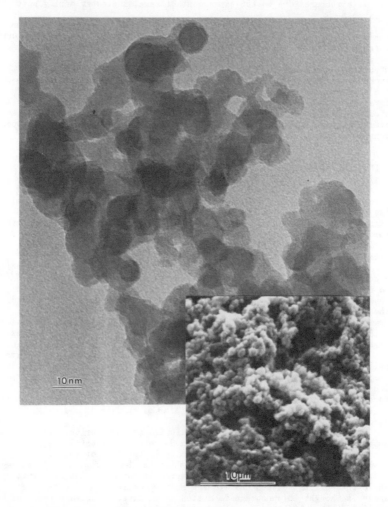

Fig. 8. Transmission electron (a), and scanning electron (b) micrographs of as-prepared powder.

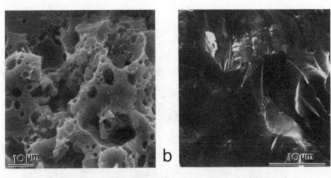

Fig. 9. Influence of calcination atmosphere on microstructure of samples
sintered at 900°C.
   (a) calcination atmosphere: air
   (b) calcination atmosphere: oxygen

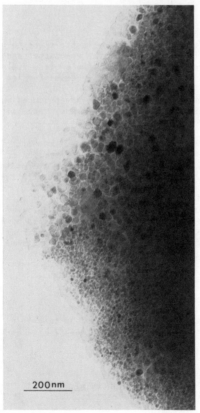

Fig 10. Transmission electron
micrograph of a stuffed
β-quartz glass-ceramic.

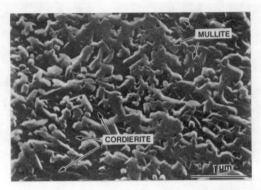

Fig. 11.   Scanning electron micrograph of a cordierite (60%) - mullite
(40%) sample sintered at 1800°C. (Etched fracture surface).

CONCLUSIONS

   Organometallic precursors offer an attractive synthetic route to
formation of very fine, reactive ceramic powders. The reactants tend to be
dilute solutions, and when prepared in a batch mode, the process yield is
low. The powders are, therefore, expensive but offer promise for low
temperature processing.
   Using organometallic precursors, amorphous powders are produced which
densify at low temperatures by viscous flow. A dense glassy material with
the cordierite composition can be obtained at 850°C, prior to
crystallization.
   A glass-ceramic containing predominantly a stuffed β-quartz phase can
be obtained without resorting to the conventional melting, cooling and
recrystallization. The pure stuffed β-quartz material has a temperature
range of stability of 920-1050°C, can be densified in this temperature
range, and has a dielectric constant of 5.5 at 1 MHz. Its thermal
expansion coefficient is $3 \times 10^{-6}/°C$ close to that of Si.
   Multiphase composites, such as cordierite-mullite can be prepared by
adjusting the ratio of precursors. A dense glass can be obtained at 920°C,
and dense crystalline samples can be obtained in the temperature range
1150-1200°C. They are very fine grained, and are mechanically strong. The
sinterability of these materials in an inert atmosphere and their
compatibiliy with copper, however, remains to be explored.

ACKNOWLEDGEMENTS

   The authors gratefully acknowledge helpful discussions with A. W.
Sleight and microscopy assistance from K. J. Morrissey and M. Van Kevalaar.

REFERENCES

[1]   B.H. Mussler and M.W. Shafer, Ceramic Bulletin 63, 5 (1984).
[2]   J.S. Presnall, J.J. Fitzpatrick, and P. Predecki, Advances in X-ray
Analysis 27 (1978).

[3]   J.D. Hodge presented at the fall meeting of the Electronics
        Division, American Ceramic Society; Orlando, Florida (1985).
[4]   W. Schreyer and J.F. Schairer, Z. Krist 116, 60 (1961).
[5]   Y.H. Kim, D. Mercurio, J.P. Mercurio, and B. Frit, Materials
        Research Bulletin 19, 209 (1984).
[6]   A. Leonard, S. Suzuki, J.J. Fripiat, and C. De Kimpe, J. Phys. Chem.
        68, 2608 (1964).
[7]   J.J. Fripiat, A. Leonard, and J.B. Uytterhoevon, J. Phys. Chem. 69,
        3274 (1965).
[8]   J.J. Fripiat, A. Leonard, and N. Barake, Bulletin Soc. Chem. France
        122, (1963).
[9]   F. Matossi, J. Chem. Phys. 17, 247 (1949).
[10]   M. Hass, J. Phys. Chem. Solids 31, 415 (1970).
[11]   J.B. Bates and A.S. Quist, J. Chem. Phys. 56, 1528 (1972).
[12]   J.B. Bates, J. Chem. Phys. 56, 1910 (1972).

This paper also appears in Mat. Res. Soc. Symp. Proc. Vol. 72.

DIELECTRIC PROPERTIES OF SOL - GEL SILICA GLASSES

G. V. CHANDRASHEKHAR and M. W. SHAFER
IBM Thomas J. Watson Research Center, P. O. Box 218, Yorktown Heights, New York
10598

ABSTRACT

Dielectric properties have been measured for a series of porous and fully densified silica glasses, prepared by the sol-gel technique starting from Si-methoxide or Si-fume. The results for the partially densified glasses do not show any preferred orientation for porosity. When fully densified (~2.25 gms/cc) without any prior treatment of the gels, they have dielectric constants of ≥ 6.5 and loss factors of 0.002 at 1 MHz, compared to values of 3.8 and <0.001 for commercial fused silica. There is no corresponding anomaly in the d.c. resistivity. Elemental carbon present to the extent of 400-500 ppm is likely to be the main cause for this enhanced dielectric constant. Extensive cleaning of the gels prior to densification to remove this carbon were not completely successful pointing to the difficulty in preparing high purity, low dielectric constant glasses via the organic sol-gel route at least in the bulk form.

INTRODUCTION

Sol-gel processing of glasses and ceramics has attracted enormous research interest in the past several years, due in part to apparent advantages over more conventional processing in special areas of glass formation [1]. The high surface area of the gels and xerogels leads to high reactivity and in some cases to low temperature processing and even formation of non-equilibrium phases. The products can be made in thin film or bulk form and can be cast into convenient shapes at low temperatures. Greater purity and homogeneity are also possible.

Most of the interest has been focussed on preparation of single and multi-component glasses, making fine powders for further processing, study of the solution chemistry involved and microstructural investigations. There is little information in the literature [2] about the electrical properties of the gel derived materials, despite the fact that the high porosity of the xerogels offers the possibility of forming materials with low dielectric constants and other unique properties. In this paper we present the results of our electrical measurements on sol-gel derived silica glasses, their comparison to conventionally made silica glass and attempts to correlate with porosity.

EXPERIMENTAL

The silica glass samples used in the present study were prepared by various methods:

1. Sols were prepared starting from tetra-methoxy silane (TMOS), methanol and water corresponding to a 10:1 (sample a) or 4:1 (sample b) ratio of water to silicon. A small amount of nitric acid was added to catalyze the hydrolysis. The mixture was refluxed for several hours, cooled and allowed to gel for 24 hours at room temperature in closed

containers. Then they were further aged in sealed jars for 16 hours at 50°C and finally heated at 80°C for 24 hours in 100% humidity. In some cases a drying control agent like formamide was added during the preparation of the sol. The gels thus obtained were heated to various temperatures from 500 to 1050°C to give a series of glasses with varying porosities. Beyond 1050°C x-ray diffractometer patterns showed evidence of crystallization.

2. A fully aged gel made as in (1) was heated at various temperatures in air to get rid of most of the organics and water and then flame melted on a quartz rod to a clear bubble free glass (sample c).

3. Cabosil M-5 silica fume (Cabot Corporation, Boston, MA) of nominal purity 99.8%+ was dispersed in ethanol [3], the resulting gel aged as in (a) and then sintered at 1300°C (sample d).

4. An organic free gel was made following the double dispersion method of Rabinovich et al.[4] by dispersing Cabosil M-5 silica fume in water, the resulting gel aged and then sintered at 1420°C (sample e).

An impurity analysis of representative samples of the various glasses was done by atomic absorption spectrometry. Carbon analysis was done by heating ground samples in air and measuring the amount of CO_2 evolved (LeCo Corporation, St. Joseph, MI). Densities were measured by weighing geometrically shaped samples (6x6x1 mm). Infrared spectra were recorded using a Perkin-Elmer 1430 ratio recording infrared spectrophotometer. Porosity and surface areas were measured by mercury porosimetry and in some cases by B.E.T. absorption isotherms.

For dielectric measurements, square samples (8x8x1 mm) had flat faces ground and polished and on which 500Å thick gold electrodes of area 6x6 mm were sputtered were used. Both two and three electrode configurations were used. Dielectric constant and loss factor in the 1KHz to 10MHz range were measured with a Hewlett-Packard 4192-A low frequency impedance analyzer. D.C. resistivity measurements were made on the same samples using a Hewlett-Packard 4392-A high resistance meter and cell assembly.

RESULTS AND DISCUSSION

All the samples used in the measurements were transparent. Thermogravimetric analysis results showed that most of the weight loss occurred by 500°C, so that samples heated above this temperature are essentially porous silica bodies. Figure 1 shows a plot of the measured densities as a function of the sintering temperature. The results show that for the alkoxide gel glasses, most of the sintering takes place between 850 to 1050°C. Samples heated to 1050°C are fully densified with densities identical to that of fused silica glass (2.25 ± 0.05 gms/cc). The silica fume-ethanol and the silica fume-water based glasses also densify to this value but the temperatures for final densification are higher (1300 and 1420°C respectively).

Electrical Properties of Fully Densified Glasses

In Table 1 are listed the electrical properties of the various fully densified glasses measured. For comparison a commercial silica glass sample (sample f) has also been included. It can be seen that the alkoxide based glasses and the Si-fume-ethanol glass have consider-

707

TABLE 1
Properties of Fully Densified Glasses

Sample	Density gms/c.c	ε_{1MHz}	tan δ 1MHz	ρd.c. Ωcms
10:1 Silica gel (a)	2.275	7.05	0.002	$\geq 10^{15}$
4:1 Silica gel (b)	2.270	6.95	0.002	$\geq 10^{15}$
10:1 gel flame melted (c)	2.260	6.4	0.0015	$\geq 10^{15}$
Silica fume and ethanol (d)	2.250	6.40	0.0015	$\geq 10^{15}$
Silica fume and water (e)	2.250	4.4	0.0015	$\geq 10^{15}$
Commercial fused silica (f)	2.260	3.80	<0.001	$\geq 10^{15}$
10:1 gel purified and melted (g)	2.240	5.1	0.0015	$\geq 10^{15}$
10:1 gel purified and melted (h)	2.245	4.6	0.0012	$\geq 10^{15}$

ably higher values of dielectric constant (~6.5 - 7.0) than the commercial fused silica glass (3.8) at 1MHz while the dielectric loss factor varies between 0.0015 and 0.002 compared to less than 0.001 for fused silica. The D.C. resistivities are about the same for all the glasses studied ($>10^{15}$). Further the dielectric constants show almost no frequency dependence in the 10^3-10^7 Hz range, very similar to fused silica glass (Fig. 2).

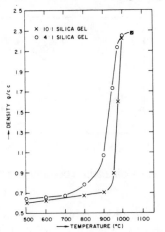

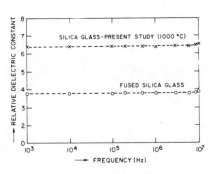

Fig. 1.　　Density as a function of sintering temperature for alkoxide gel glasses.

Fig. 2.　　Frequency dependence of the relative dielectric constant for a typical gel glass.

Thus it is seen that there is a considerable anomaly in the dielectric constants of the organic gel derived glasses without any corresponding changes in the dielectric loss factor or D.C. resistivities when compared to fused silica glass. We now consider the origins of these high values in terms of water or OH⁻ groups, carbon and other impurities.

708

Atomic absorption spectrometry results showed that the organic gel derived glasses have no major impurities except 100-150 ppm of Na and 10 to 40 ppm of K. A survey of literature values of the 1MHz dielectric constants against the Na-content of alkali silicate glasses [5] show that about 20 mole% of Na_2O is required to get dielectric constants of 6.5 or more and hence the 200 ppm or so of total alkali content in our glasses cannot be the reason for the high dielectric constants.

Let us now look at the effect of water or OH⁻ on the dielectric constant. Andeen and coworkers have reported [6] that the dielectric constant of fused silica increased linearly from 3.78 to 3.83 when the OH⁻ concentration increased from 0.01 to 0.1%. The water content of our samples as determined by infrared spectroscopy (Fig. 3) are listed in Table 2 along with the corresponding dielectric constants. It can be seen that there is no correlation between the dielectric constants and the water or OH⁻ content. The 4:1 and 10:1 alkoxide gel glasses and silica fume-ethanol glasses have different amounts of water but the same dielectric constant while the silica fume-water derived glass having the maximum amount of water has a dielectric constant of 4.4, only slightly higher than fused silica. Also, a (10:1) alkoxide based glass when flame melted has as little water as the fused silica glass but a dielectric constant of 6.4. Thus we can conclude that water or hydroxyl ions alone are not responsible for the high dielectric constants.

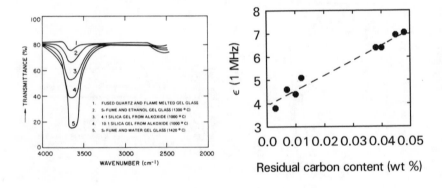

Fig. 3. Comparative infrared spectra of the various glasses showing water content.

Fig. 4. Relative dielectric constant as a function of residual carbon content for the various glasses.

Table 2 shows that the organic gel based glasses with the high dielectric constant have carbon in the 350 to 500 ppm range while the silica fume-water derived glass with the low dielectric constant has about 100 ppm of carbon. Carbon can be incorporated into these glasses as a result of the combustion of organic matter held in micropores still present even after most of the densification has taken place. It can be present either as elemental carbon or as silicon carbide. Because of the very high surface area and surface free energy of the gels, the organic materials in the micropores can react with SiO_2 to form SiC [7]. From our Raman spectral data we can say that the concentration of SiC, if present at all, is less than 1%. Silicon carbide has a dielectric constant of 40 (though numbers as low as 15 have been quoted). At less than 1% concentration, it is unlikely, by any known mechanism, to increase the dielectric constant of SiO_2 to 6.5 or more, even if it is present as filaments or pore

TABLE 2
Water and Carbon Contents of Fully Densified Glasses

Sample	H_2O, wt %	C, ppm	ε_{1MHz}
10:1 Silica gel (a)	0.12	480	7.05
4:1 Silica gel (b)	0.08	450	6.95
10:1 gel flame melted (c)	0.04	400	6.4
Silica fume and ethanol (d)	0.05	380	6.4
Silica fume and water (e)	0.30	100	4.4
Commercial fused silica (f)	0.025	30	3.8
10:1 gel purified and melted (g)	0.035	120	5.1
10:1 gel purified and melted (h)	0.035	70	4.6

surface coatings with connectivity.

That leaves us with elemental carbon as the most likely cause for the enhanced dielectric constant. Carbon in substantial amounts (about 0.1% is thought to be the percolation limit) is known to increase the dielectric constant and loss factor of cordierite ceramics [8] while at the same time drastically reducing the D.C. resistivity to $\sim 10^5 \Omega cms$ from values of $\sim 10^{15} \Omega cms$ for the pure cordierite. Our samples have a maximum of about 500 ppm of carbon (below the percolation threshold) and none of our samples had a resistivity of less than $10^{15} \Omega$ cms. So the mechanism by which carbon enhances the dielectric constants without any corresponding decrease in resistivity (i.e., without any connectivity of the carbon) has to be different.

To further establish the role of carbon, we carried out two experiments to reduce the amount of carbon in our samples and see the effect on the dielectric constant. A (10:1) silica gel of about 400 m^2/gm surface area aged at 100°C as before was ground, heated slowly in O_2 to 350°C and held there for several hours. Then it was treated in hot 1 molar nitric acid for 24 hours, cleaned with distilled water, treated with warm hydrogen peroxide, rinsed, dried and then heated slowly in flowing O_2 to 800°C. The resulting material was melted on a quartz rod in an oxy-hydrogen flame (to avoid any carbon uptake). The glass thus obtained (sample g) had about the same amount of water as fused silica glass but less carbon (120ppm) and a dielectric constant of 5.1 at 1 MHz. A similar procedure with a silica gel with much higher surface area ($\sim 800 m^2/gm$) - to reduce the carbon still further by making the cleaning process more effective - gave SiO_2 glass (sample h) with a dielectric constant of 4.6. In Fig. 4 are plotted the carbon content of these various glasses against the corresponding dielectric constants and the results show that the dielectric constant increases linearly with the residual carbon content.

So it appears that residual carbon even in small amounts can increase the dielectric constant of sol-gel silica glasses and that there is a relation between the dielectric constant and carbon content. Further work in establishing this relationship and understanding the mechanism of this enhancement are necessary. But as can be seen, extensive purification of the gel before densification was not very successful in completely removing the carbon. This raises another question - about the feasibility of making low dielectric constant glasses via the organic sol-gel route at least in bulk form.

710

Dielectric Data for Porous Glasses

While we are unable as of now to provide a mechanism by which carbon can increase the dielectric constant of the fully densified organic gel derived glasses, we can on the basis of this high value try to explain the dielectric data for the partially densified glasses. These porous glasses can be treated as mixtures of ideal dielectrics - in this case silica and empty pores with the porosity being either along or normal to the applied field. When the porosity is along the capacitor plates, the structure corresponds to capacitive elements in series and the inverse capacitances are additive. Then $1/\varepsilon^1 = v_1/\varepsilon_1^1 + v_2/\varepsilon_2^1$ where v_1 and v_2 are the volume fractions of the two phases with relative dielectric constants ε_1 and ε_2. In contrast, when the porosity is normal to the capacitor plates, the applied field is similar to each of the elements so that the capacitances are additive: $\varepsilon^1 = v_1\varepsilon_1^1 + v_2\varepsilon_2^1$. In Fig. 5, are plotted the measured dielectric constants at 1MHz for the partially densified glasses as a function of their porosities. Also plotted are two curves, calculated for the two extreme cases of porosity being either parallel or perpendicular to the applied field and assuming the high dielectric constant for the silica phase and a value of one for the pores (air). The results show that the experimental data lie between the two calculated curves indicating that there is no preferred orientation for the porosity.

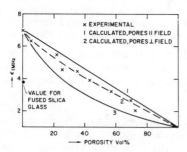

Fig. 5. Experimental and calculated dielectric constants as a function of porosity for alkoxide glasses.

Acknowledgement: Thanks are due to R. A. Figat for preparing some of the samples and to B. E. Olson for impurity analysis.

REFERENCES

1. *Better Ceramics through Chemistry*, Materials Research Society Symposia Proceedings **32**, 1984.
   Also S. Sakka, Am. Ceram. Soc. Bull. **64 (11)**, 1985, 1463 and D. W. Johnson, Jr., Am. Ceram. Soc. Bull. **64 (12)**, 1985, 1597.

2. T. Kawaguchi, H. Hishikura, J. Iura and Y. Kokubu, J. Non. Cryst. Solids **63**, 1984, 61.

3. F. J. Bonner, G. Kordas and D. L. Kinser, J. Non. Cryst. Solids **71**, 1985, 361.

4. E. M. Rabinovitch, D. W. Johnson, Jr., J. B. MacChesney and E. M. Vogel, J. Am. Ceram. Soc. **66 (10)**, 1983, 683.

5. *Dielectric Materials and Applications*, ed. A. R. VonHippel, John Wiley, 1954, 311.

6. C. Andeen and D. Schuele, J. Appl. Phys. **45 (3)**, 1974, 311.

7. S. Yajima, J. Hayashi, M. Omori and K. Okamura, Nature **261**, 1976, 683.

8. S. M. Zalar, private communication.

This paper also appears in Mat. Res. Soc. Symp. Proc. Vol. 72.

THE EFFECT OF HYDROLYSIS CONDITIONS ON THE CHARACTERISTICS OF PbTiO$_3$ GELS AND THIN FILMS

K. D. BUDD, S. K. DEY, AND D. A. PAYNE
Department of Ceramic Engineering and Materials Research Laboratory
University of Illinois at Urbana-Champaign, Urbana, IL 61801

ABSTRACT

Sol-gel processing represents a promising method of fabrication for thin films of electronic ceramics which are useful in a number of packaging and device applications. In this study, the influence of acid and base catalysts on the structure of PbTiO$_3$ gels and films (0.1-1.0 μm) was investigated, for the purpose of inducing and identifying gel structures which were the most suitable as precursors for thin dielectric layers. Continuous, crack-free films, with dielectric strengths in excess of 10^6 V/cm were developed. Basic solutions gelled rapidly, phase separated, and were probably more crosslinked than acidic gels. Acidic gels seemed more capable of polymeric rearrangement during drying, yielding denser amorphous structures with microcrystalline regions. High-field dielectric constants (1 MV/m ac) in the range K=30-40, and K=160-170, were determined for amorphous and crystalline films, respectively.

INTRODUCTION

Sol-gel processing has generated considerable interest in recent years as a chemical method of ceramic preparation. This is partly because of the unique forming methods which are possible. For example, fine powders, bulk monolithic parts, fibers, composites, and thin films have all been prepared. With the exception of powders, most of the previous work has involved glass-forming systems [1]. Recently, potential applications in microelectronic packaging and thin film devices have created interest in extending sol-gel thin film technology to electronic ceramics, including titanates and other materials. Sol-gel derived BaTiO$_3$ thin films were among the first to be reported [2], followed by Pb(Zr,Ti)O$_3$ [3], PLZT [4], and others.

This paper reports on the influence of hydrolysis conditions on structure development in PbTiO$_3$ gels and films. Specifically, the effects of acid and base catalyst on gellation rate, gel structures, densification, and crystallization behavior were investigated.

Purpose

The purpose of the investigation was to induce structural variations in PbTiO$_3$ gels and films, characterize the evolution of structure, and determine which structures (and therefore processing conditions) were most suitable for the fabrication of thin dielectric layers. By contrast with the lack of information for complex titanate systems, numerous basic studies have been reported on the polymerization of silicon alkoxides. It has generally been found that the growth of silicate polymers could be biased to give markedly different gel structures, by establishing hydrolysis conditions which resulted in preferential hydrolysis, or growth, on specific types of sites. This bias stemmed from differences in the reaction mechanisms which prevailed under acidic versus basic conditions [5,6]. For example, base-catalyzed structures tended to be much more crosslinked than acid-catalyzed gels, which subsequently influenced the physical characteristics and densification behaviour [7].

Mat. Res. Soc. Symp. Proc. Vol. 73. ⁻ 1986 Materials Research Society

EXPERIMENTAL

Precursor Synthesis

Precursor solutions of a complex Pb-Ti alkoxide were prepared by reacting lead acetate with titanium isopropoxide in methoxyethanol, similar to a method reported previously by Gurkovich and Blum [8]. Gas chromatography indicated an exchange reaction occurred, which was previously unaccounted for, namely:

$$Ti(OR')_4 + 4(ROH) \rightarrow Ti(OR)_4 + 4R'OH$$
$$R' = iC_3H_7, \quad R = CH_3OCH_2CH_2. \tag{1}$$

Consequently, the experimental conditions were modified, to identify and control the synthesis reactions.

Lead acetate was dissolved in methoxyethanol, and dehydrated by three successive distillations, and redilutions, with "dry" methoxyethanol. In a separate vessel, methoxyethanol and titanium isopropoxide were mixed in a molar ratio of 8:1. Isopropanol from the exchange reaction, and two-thirds of the remaining methoxyethanol, were removed by distillation. Gas chromatography and distillate density indicated that the exchange reaction had gone to completion. The two solutions were then recombined, boiled, and concentrated by vacuum distillation. Three distillations were carried out to ensure complete reaction, and removal of methoxyethyl acetate, according to

$$Pb(Ac)_2 + Ti(OR)_4 \rightarrow PbTiO_2(OR)_2 + 2 \, RAc \quad , \tag{2}$$

where "Ac" represents an acetate group. The oxo-alkoxide formed was partially polymerized, with a number average molecular weight of approximately 1050 (i.e., ~2.5 formula units). A third reaction,

$$Pb(Ac)_2 + 4ROH \rightarrow Pb(OR)_2 + 2RAc + 2H_2O \quad , \tag{3}$$

could also occur, either during the lead acetate dehydration, or upon mixing of solutions, if catalyzed by titanium alkoxide. In fact, a small amount of methoxyethyl acetate, representing 6% of the total acetate added, was recovered in the dehydration distillate. With or without reaction 3, an overall reaction equivalent to reaction 2 would occur, provided that all of the OR groups generated in reaction 3 were hydrolyzed, and then fully condensed. However, reaction 3 does provide a mechanism for the formation of -Pb-O-Pb- bonds, and for additional OR and/or OH groups.

Gel and Film Formation

All of the acetate originally added was recovered in the form of methoxyethyl acetate. The alkoxide solution was adjusted to 1 molar concentration, and stored in a dry box. 0.5 molar gels, and thin film precursor solutions, were formed by combining equal volumes of stock solution and methoxyethanol-catalyst-water solutions. HNO_3 and NH_4OH were used as the catalysts. Thin films were spin-cast at 2000 rpm on silicon or platinum substrates, using a photoresist spinner. An ellipsometer was used to determine both refractive index and thickness of the thin films.

RESULTS AND DISCUSSION

Solutions

Characteristics of the hydrolyzed precursor solutions were studied using Karl Fischer titrations, viscometry, and vapor phase osmometry. Hydrolysis was found to be rapid in all cases; and the influence of catalyst on structure development was thought to be mainly through the condensation reactions. Stable 0.5 molar solutions, with unchanging viscosities, were obtained for water additions up to 1.5 moles H_2O per mole of $PbTiO_3$ (Fig. 1), suggesting, either a threshold amount of hydrolysis water was necessary for polymerization, or, the formation of stable oxo-alkoxide species. The latter explanation was favored because an increase in molecular weight was observed.

As illustrated in Fig. 1, acid and base catalysts had a significant effect on gellation rate. Solutions containing 10^{-1} molar HNO_3 required 1 to 2 orders of magnitude longer time to gel than those catalyzed with 10^{-1} molar NH_4OH. Figure 2 illustrates the effect of catalyst additions on phase separation for the wet gels. Very basic gels exsolved a large amount of liquid phase, which was found to be nearly pure solvent. Phase separation in gels has been related to chain stiffness, backbone charge, and the amount of crosslinking in the structure [5]. Greater amounts of phase separation are to be expected for more highly crosslinked structures. These results indicate that structural differences were indeed induced in $PbTiO_3$ gels, and were in fact analogous to behaviour previously reported for silicate systems [6].

Dried Gels

Evidence of structural differences were directly observed for dried gels. Very interesting differences were apparent between the $PbTiO_3$ system and silicates. Figure 3 illustrates TEM photomicrographs of acid and base catalyzed gels (for 10^{-1} molar catalyst additions), formed by dip-coating on copper grids, and drying at 150°C. Acid gels diffracted electrons (Fig. 4), indicative of the presence of microcrystalline regions. Strand-like morphologies were also observed, although they were on a relatively large scale (hundreds of Angstroms). Basic gels were more textured, and had amorphous diffraction patterns.

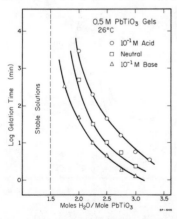

Fig. 1   Effect of Hydrolysis Water on Gelation Rate.

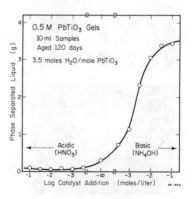

Fig. 2   Effect of Catalyst Addition on Phase Separation in Gels.

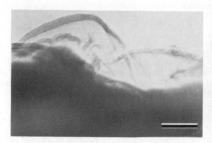

Fig. 3 TEM Photomicrographs of Acid-Catalyzed (Left) and Base-Catalyzed
       Gels. Bar = 500 Angstroms.

Figure 5 illustrates DSC results for gels pre-heated in air at 350°C for
20 minutes. The exotherms were essentially crystallization peaks, because
little change in weight or surface area occurred in the relevant temperature
range. Basic gels had significantly higher heats of crystallization than
acidic gels. This contrasted with what might be expected for gel-derived
glasses, and represents an interesting difference between the $PbTiO_3$ system
and silicates; and possibly between traditional non-glass forming systems and
glass formers. Crystallization enthalpies should increase with increasing
free volume, which in turn should be higher for less crosslinked structures.

For the $PbTiO_3$ system, the structure of dried gels appeared to be
profoundly influenced by polymeric rearrangement during drying. Acidic
structures were found to order themselves into microcrystalline regions, with
correspondingly low heats of crystallization. This phenomenon is probably
enhanced in non-glass forming systems, and might also be related to the
relatively dilute gel concentrations used (0.5 moles of $PbTiO_3$ per liter).
The behaviour was somewhat analagous to organic polymer melts in that linear

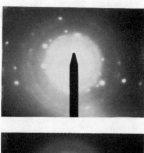

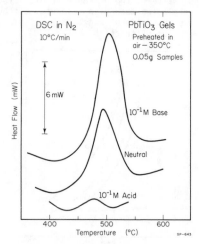

Fig. 4 Electron Diffraction Patterns
       of Acidic (Top) and Basic Gels.

Fig. 5 Crystallization Behavior
       of $PbTiO_3$ Gels.

polymers often crystallize upon solidification, while crosslinked polymers do not. It was also observed that gels which dried slowly had low crystallization enthalpies, while more rapidly dried gels had correspondingly larger enthalpies.

Gel to Ceramic Conversion

Figure 6 illustrates the variations in refractive index observed for $PbTiO_3$ thin films as a function of processing conditions. Acidic films heated at 350°C had higher refractive indices than corresponding basic films, consistant with the concept that polymeric rearrangement led to denser structures. Air-dried or mildly dried films (70°C, 5 min.) exhibited the opposite trend. The more acidic films also had a higher weight loss (18 wt.% versus 14 wt.%), indicative of the tendency to trap solvent or other organic matter in the fine pore structure of acidic gels. Apparently, the collapse and shrinkage for this system was limited not only be the stiffness of the polymeric structure, but also by the organic matter present. In fact, densities of bulk gels, as determined by helium gas pychnometry, were only slightly greater than densities obtained from bulk measurements, throughout the entire heat-treatment range.

On heating, excess organic material was removed from the acidic gels, and a crossover in the refractive index relationship, with heat treatment temperature, for acidic and basic films was consistantly observed (Fig. 7).

Microstructures and Properties

Structural differences induced in gels were found to significantly affect the evolution of microstructures in $PbTiO_3$ thin films. As illustrated in Fig. 8, acid-catalyzed films developed dense regions, which were surrounded by extensive intergranular porosity. These features, which were absent from base-catalyzed films, were possibly remnants of denser regions which formed during polymeric rearrangement, and which became surrounded by continuous pore channels as the microstructure evolved [9]. Interesting differences in

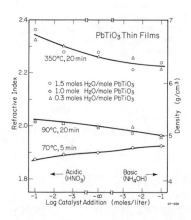

Fig. 6   Refractive Index for Amorphous
         Thin Films.

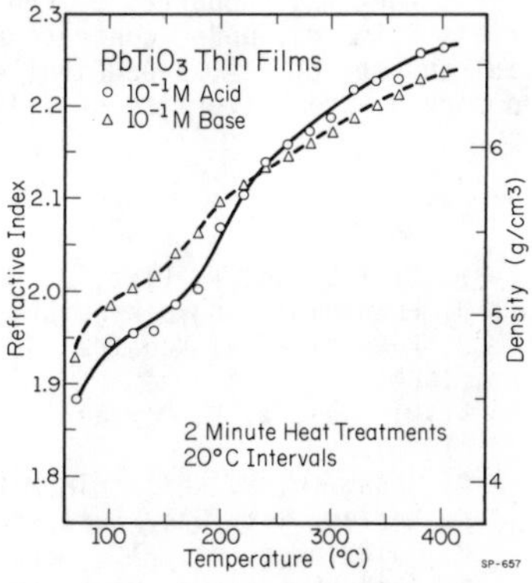

Fig. 7   Densification of Acidic
         and Basic Thin Films.

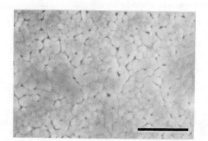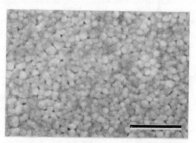

Fig. 8 Microstructures of Acid-Catalyzed (left) and Base-Catalyzed Thin
Films of PbTiO$_3$ Fired at 700°C. Bar = 1 micron.

electrical properties were also observed between acid- and base-catalyzed
gels. The more dense and partially ordered amorphous, acid-catalyzed films
had higher dielectric constants (K = 35-40) than base-catalyzed films (K = 28-
32). Crystalline acid-catalyzed films had slightly smaller dielectric
constants and strengths, presumably because of the aforementioned defects.

CONCLUSIONS

Structural variations in PbTiO$_3$ gels and thin films were induced by
control of the hydrolysis conditions. Analagous behaviour to silicate gels
was observed, in that basic solutions gelled more rapidly, with phase
separation, and were probably more crosslinked than acidic gels. However,
acidic gels appeared to be more capable of polymeric rearrangement during
drying, and formed denser amorphous structures with microcrystalline regions
and low heats of crystallization. This resulted in intergranular porosity in
polycrystalline acid-catalyzed films, causing somewhat lower dielectric
constants and strengths than those for base-catalyzed films.

ACKNOWLEDGEMENTS

This work was supported by the U.S. Department of Energy, Division of
Materials Sciences, under contract DE-AC02-76ER01198. KDB acknowledges
partial support of an ONR-ASEE Fellowship. The use of the Center of Electron
Microscopy at the University of Illinois is gratefully acknowledged.

REFERENCES

[1]   H. Dislich and P. Hinz, J. Non. Cryst. Sol., 48, 11-16 (1982).
[2]   J. Fukushima, Yogyo Kyoaishi, 83, 204 (1975).
[3]   J. Fukushima, K. Kodaira, T. Marsushita, J. Mater. Sci., 19, 595-598
       (1984).
[4]   K. D. Budd, S. K. Dey and D. A. Payne, Brit. Cer. Proc. 36, 107-121
       (1985).
[5]   D. Schaefer, K. Keefer in Mat. Res. Soc. Proc. 32 ,1-14 (1984).
[6]   K. Keefer, Mat. Res. Soc. Proc. 32, 15-24 (1984).
[7]   C. J. Brinker, et. al., Mat. Res. Soc. Proc. 32, 25-32 (1984).
[8]   S. Gurkovich and J. Blum, in: Ultrastructure Processing of Ceramics,
       Glasses and Composites, Ed. by L. Hench and D. Ulrich, John Wiley &
       Sons, 152-160 (1984).
[9]   K. D. Budd, Ph.D. Thesis, University of Illinois, (1986).
  This paper also appears in Mat. Res. Soc. Symp. Proc. Vol. 72.

SYNTHESIS, STRUCTURE AND APPLICATIONS OF TiO$_2$ GELS

J. LIVAGE,
Spectrochimie du Solide, Université Pierre et Marie Curie, 4 place Jussieu,
75230 Paris Cedex 05, France.

ABSTRACT

TiO$_2$ gels are usually obtained through hydrolysis of titanium alkoxides. Chemical additives can however react with the precursor at a molecular level and therefore modify the hydrolysis-condensation reactions. Several examples will be described : acetic acid, acetylacetone or Cr(acac)$_3$. The whole sol-gel process is followed all the way from the precursors to the gel and each step is characterized by spectroscopic experiments (Infra-red, N.M.R, E.S.R.). Some electronic properties of TiO$_2$ gels are then described. Chemical additives allow an optimization of the sol-gel process according to each specific applications : electrochromic display devices, photoanodes or photochemical reactions.

I. INTRODUCTION

Sol-gel processing of ceramics has become an area of intense research interest. This method involves the use of molecular precursors, mainly metal alkoxides, as starting materials. A macromolecular network is then obtained as a result of condensation reactions and the microstructure of the ceramic strongly depends on the experimental procedure. Major advances in ultrastructure processing will require an emphasis which relates chemical process to gel formation and powder morphology.

Hydrolysis of the molecular precursors is usually performed in the presence of an acid or a base catalyst that allows experimental control of the rate and extent of the hydrolysis reaction (1). These catalysts are supposed to favor either electrophilic or nucleophilic substitutions, but the chemical role of the counter ion is usually neglected (2).

Multicomponent ceramics can be easily obtained by mixing the solutions of metal-organic precursors (3), but whether a mixed-alkoxide or only an intimate mixing is obtained often remains a matter of controversy.

The addition of Drying Control Chemical Agents (DCCA) such as formamide or oxalic acid, allowed L. Hench to produce optically transparent dried gel monoliths (4). These chemical additives affect the rates of both hydrolysis and polycondensation reactions and therefore control the size and shape of the pore distribution.

Chemical reactions are involved all along the sol-gel process. A better understanding of their mechanisms then appears to be the key point for further developments (5). This paper will report on spectroscopic experiments (Infra-red, N.M.R, E.S.R) performed at each stage of the sol-gel process, from the molecular precursor to the resulting ceramic. A characterization of the different species will be given, allowing a better understanding of the chemical role of each reagent. Since TiO$_2$ is a major constituent in many electronic devices, our study will be focussed on TiO$_2$ gels obtained through hydrolysis of titanium alkoxides. Some electronic properties of these TiO$_2$ ceramics will then be reported and their potential application as colloidal dispersions, thin films or pellets will be briefly described.

II. SPECTROSCOPIC STUDY OF THE SOL-GEL PROCESS

2.1 Infra-red analysis of the role of acetic acid

TiO_2 gels are usually obtained through hydrolysis of titanium alkoxides in the presence of an acid catalyst (6)(7). Our experiments have shown that monolithic transparent gels can be reproducibly obtained when acetic acid is added prior to water (8). We therefore choose the following procedure : glacial acetic acid is first added to pure $Ti(OBu^n)_4$. A strongly exothermic reaction takes place leading to a clear solution. The infra-red spectrum of such a solution is shown in fig.(1a). It exhibits the absorption bands corresponding to the titanium alkoxide together with free acetic acid and the butyl-acetate ester. More interesting however would be to notice a set of two bands around 1500 cm^{-1} that can be assigned to the $\nu_s(CO_2^-)= 1440$ cm^{-1} and $\nu_a(CO_2^-) = 1570$ cm^{-1} vibrations (9). Their frequency separation ($\Delta\nu = 130$ cm^{-1}) is typical of an acetate ion acting as a bidentate ligand (10). It would be difficult to distinguish between chelating or bridging groups, but the doublet clearly shows that acetates are bonded to the titanium. The intensity of the doublet increases with the amount of acetic acid, showing that the exothermic reaction corresponds to the substitution of OR groups by bidentate CH_3COO groups. Acetic acid then changes the metal-organic precursor at a molecular level giving rise to new soluble species, $Ti(OR)_x(Ac)_y$.

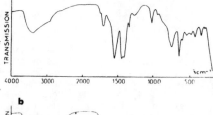

After acidification, hydrolysis is performed by adding a water-butanol solution under vigorous stirring. Gelation then occurs after a period of time that strongly depends on the respective concentrations of titanium, acetic acid and water. The time of gelation decreases when the concentration of titanium or water increases while the reverse effect is observed with acetic acid. This acid appears to slow down the hydrolysis-condensation reactions. The infra-red spectrum of the gel, obtained after hydrolysis, is shown in fig.(1b). The most interesting feature comes from a modification of the previous doublet around 1500 cm^{-1}. It's intensity decreases while it becomes sharper. The frequency separation also decreases ($\Delta\nu = 105$ cm^{-1}) suggesting that only chelating acetates remain. Acetate bridges seem to be broken during hydrolysis, while the increase in intensity of the two bands at 1715 cm^{-1} and 1270 cm^{-1} suggest that some monodentate acetate ligands are formed.

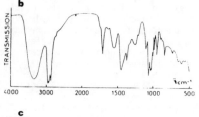

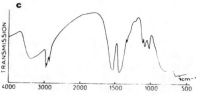

The infra-red spectrum of a xerogel obtained after drying the previous gel for 1 hour at 80°C is shown in fig.(1c). Most bands due to (OR) or (OH) groups have disappeared but the doublet around 1500 cm^{-1} is still clearly visible. This shows that chelating acetates are not easily

Fig. 1
Infra-red spectra recorded at different steps of the sol-gel process.
a. $Ti(OBu^n)_n$ + acetic acid
b. Gel obtained after hydrolysis
c. Xerogel dried for 1 hour at 80°C.

removed by water so that most of them still remain in the xerogel network, slowing down the hydrolysis process. A thermal analysis shows that these acetates are strongly bonded and can only be removed upon heating above 300°C. A broad absorption band is observed on the low energy side of the spectrum. It corresponds to the envelope of the phonon spectrum of the Ti-0-Ti bonds of a titanium oxide network (11).

2.2 N.M.R analysis of chelating (acac) ligands

Colloidal TiO_2 particles can be prepared through hydrolysis of titanium tetraisopropoxide in acidic aqueous solutions (HCl, pH 1.5) (12). A transparent TiO_2 sol is obtained which is stable only at pH < 3. Above this pH, precipitation occurs and it is necessary to employ a stabilizing agent such as polyvinyl-alcohol. It would be interesting however to investigate whether strong chelating ligands could stabilize colloidal dispersions over a wider range of pH. Therefore, acetylacetone is mixed to an equal amount of $Ti(OPr^i)_4$. An exothermic reaction takes place while the solution turns to yellow. This coloration arises from optical absorption around 330 nm, presumably due to charge transfers from the acac ligand to the Ti(IV) ion. As for acetic acid, infra-red spectra show that acac ligands are directly bonded to the titanium fig.(2a). This is evidenced by the splitting of the (C-C) and (C-0) stretching frequencies (9). A broad absorption band is usually observed in free acacH, around 1620 cm^{-1}. It gives rise to a doublet at 1600 cm^{-1} and 1520 cm^{-1} in the presence of $Ti(OR)_4$.

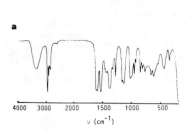

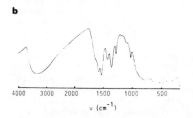

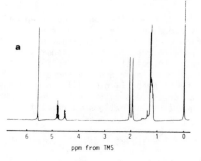

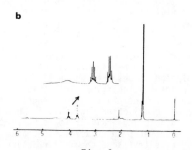

Fig. 2
Infra-red spectra of
a. solution of $Ti(OPr^i)_4$ + acac H
b. Powder obtained after hydrolysis and evaporation of the solvents.

Fig. 3
Proton NMR of
a. A solution of $Ti(OPr^i)_4$ + acac H
b. Colloidal solution after hydrolysis

The proton NMR spectrum of the previous solution is shown in fig.(3a).
It exhibits several peaks that can be assigned to both kinds of ligands :

- (OPr^i) gives rise to the following chemical shifts
  δ = 1.15-1.25 for the (CH_2) protons
  δ = 4.48 and 4.76 for the (CH) protons

- (Acac) gives rise to the following chemical shifts
  δ = 1.92 and 2.03 for the (CH_3) protons
  δ = 5.48 for the (CH) protons

Integration of the NMR spectrum leads to a ratio of $3(OPr^i)$ for 1(Acac) suggesting that the molecular precursor should correspond to $Ti(OPr^i)_3$ (Acac).
Two kinds of (OPr^i) groups are observed in a ratio of 2:1. They could correspond to end groups and bridging groups suggesting that some polymerization occurs. Moreover, the chemical shifts observed for (acac) ligands
are somewhat different from those of pure (acac H) : δ = 2.02 for (CH) and
δ = 5.46 for (CH_3). This reinforces the assumption that (acac) is now bonded
to the titanium.

Hydrolysis of the new titanium precursor is performed by adding a water-
ethanol solution. An NMR study of the first stages of the hydrolysis reaction
shows that the peaks corresponding to the (OPr^i) ligands are deeply modified
while those corresponding to (acac) are not. Beyond a ratio (H_2O / Ti) of 3,
NMR peaks due to (OPr^i) ligands disappear. All these groups seem to have
been removed while the NMR peaks due to (acac) ligands are still visible
fig.(3b).

No precipitation is observed and the solution remains clear when water
is added, even in a large excess, and over a wide range of pH, up to pH 10.
Slow evaporation of the solvents, performed under vacuum at room temperature,
leads to more and more viscous solutions and finally to a powder. An infrared spectrum of this powder fig.(2b) shows that all bonded (OPr^i) groups
have been removed while the doublet due to (acac) ligands is still visible.
A broad absorption band is seen on the low energy side of the spectrum.
It corresponds to the envelope of the phonon spectrum due to Ti-O-Ti bonds
in a titanium oxide network (11).

2.3 E.S.R analysis of Cr^{3+} doped TiO_2 gels

E.S.R has not yet been widely used in the study of the sol-gel process
(13)(14). This spectroscopy of course can detect only paramagnetic species.
Therefore TiO_2 gels have to be doped with some transition metal ions such as
Cr^{3+}. This paramagnetic ion was choosen as an E.S.R probe because it
exhibits a large zero field splitting (S = 3/2), very sensitive to small
variations of the local crystal field. Chromium was introduced as $Cr(acac)_3$,
dissolved in n-butanol, and mixed to a $Ti(OBu^n)_4$-butanol solution. A violet
solution is obtained whose ESR spectrum is typical of isolated $Cr(acac)_3$
species fig.(4a) (15). This solution however turns green when heated at
120°C and the ESR spectrum progressively changes. This ESR spectrum is
rather complicated fig.(4b). It must be due to several chemical species
and a detailed analysis would be rather difficult. However, no modification
is observed when $Cr(acac)_3$ is heated in n-butanol without $Ti(OBu^n)_4$. This
suggests that some chemical reaction occurs between both metal-organic
precursors leading to a mixed Ti-Cr compound in which Cr^{3+} should be
surrounded by (OBu^n) and (acac) ligands.

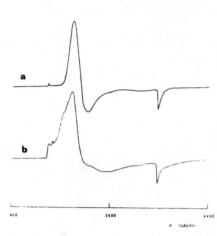

Fig. 4
E.S.R spectra of Cr^{3+} doped TiO_3 recorded at different steps of the sol-gel process.

a. $Ti(OBu^n)_4$ + Cr(acac) solution
b. After heating 120°C
c. Previous solution + CH_3COOH
d. Gel obtained after hydrolysis

Another important modification is observed when acetic acid is added to the solution. The ESR spectrum fig.(4c) shows a larger orthorhombic distortion of the ligand field around Cr^{3+}. This suggests some change of the chromium chemical environment that presumably involves acetate ligands as previously shown by infra-red for $Ti(OBu^n)_4$.

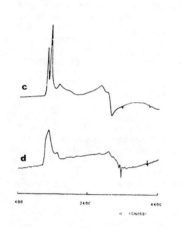

Hydrolysis of the solution leads to a green gel that again exhibits a completely different ESR spectrum fig.(4d). This spectrum however is not modified when the gel is dried at 80°C for 2 hours. The measured ESR parameters (g = 1.98, D = 0.15 cm^{-1}) are quite different from those reported in the literature for Cr^{3+} in the anatase or rutile TiO_2 crystalline phases (16) (17). The orthorhombic distortion (λ = 0.33) is much larger suggesting that Cr^{3+} is not surrounded by oxide ions only. Organic groups must be still bonded to the metal ion. These organic ligands are removed upon heating, leading to the anatase phase between 400°C and 900°C, then pure rutile above 900°C. It has to be pointed out that the anatase phase seems to be stabilized up to higher temperatures in the presence of Cr^{3+}.

III. ELECTRONIC PROPERTIES OF TiO_2 GELS AND COLLOIDS

3.1 Electrochromic display devices

The electrochromic properties of transition metal oxides such as WO_3, MoO_3 or V_2O_5 have been extensively studied during the last decade. Electrochromic displays based on TiO_2 have also been reported (18). They were made from a dispersion of TiO_2 fine powder having the anatase structure deposited onto glass electrodes using polyvinylalcohol as a binder. It is well known that coatings can very conveniently be deposited from gels or colloidal solutions(19). Therefore we tried to make a TiO_2 display device via the gel route. The colloidal solution was obtained through hydrolysis of a $Ti(OBu^n)_4$-butanol solution. It was then deposited onto an I.T.O coated glass electrode by spin-coating. A transparent uniform film was obtained upon drying at room temperature. It appeared to be amorphous by X-ray diffraction. This working electrode was then placed into an electrochemical cell containing $LiClO_4$ in propylene carbonate as an electrolyte and a platinium

counter electrode. The transparent TiO_2 film turns blue upon applying a negative voltage of -2 volts. Bleaching can be reversibly obtained upon applying
a reverse voltage of +2 volts. The whole cycle takes about 1 second. Cyclic
voltammetry experiments indicate an energy consumption of about 20 mC per
cm^2. ESR spectra of the blue oxide exhibit a strong signal around g = 1.93
typical of Ti^{3+} ions. A broad absorption band centered at 12.500 cm^{-1} is
observed by optical spectroscopy. It corresponds to intervalence charge
transfer between Ti^{3+}-Ti^{4+} ions. The electrochromic mechanism should therefore involve a double injection process of electrons and Li^+ ions.

3.2 TiO_2 photoanodes

Pure and stoichiometric TiO_2 is an insulator with a wide band-gap
Eg = 3.2 eV. It is therefore of little practical use in photovoltaic or
photoelectrolysis devices unless means are found to enhance both its electrical conductivity and light response into the visible region. Heating
TiO_2 under hydrogen or in a vacuum around 700°C-900°C leads to a non-
stoichiometric oxide that exhibits an n-type semiconducting behavior (20).
Doping TiO_2 with Cr^{3+} ions extends the spectral response into the visible
while the quantum efficiency of carrier generation can be improved with
aluminium doping (21). All these operations require high temperature
processes and therefore the rutile TiO_2 crystalline phase is always obtained.
The sol-gel process could provide an easier method for the preparation
of TiO_2 photoanodes. Doping can be made, at a molecular level, by mixing
butanol solutions of $Ti(OBu^n)_4$, $Al(OBu^n)_3$ and $Cr(acac)_3$. The mixture is then
heated for 2 hours at 120°C so that the molecular precursors can react.
Hydrolysis is performed by adding a water-butanol solution. It leads to a
green colloidal solution that can be easily deposited, by spin-coating, onto
a titanium plate. A rather hard coating is obtained after heating at 400°C
under vacuum. It exhibits the anatase TiO_2 structure. This photoanode is then
dipped into a photoelectrochemical cell containing an acid solution
(H_2SO_4 0.1 N) together with a platinum counter electrode and a S.C.E reference electrode. The current-potential characteristics of the TiO_2 electrode,
in the range (+1 V, -0,5 V) is typical of a n-type semiconductor. It is
completely flat in the black but exhibits a photocurrent of about 1 mA/cm^2
when irradiated with visible light. The sol-gel process then appears to be
very promising for making TiO_2 photoanodes. Moreover the metastable anatase
phase is obtained, which is supposed to exhibit better photoelectrochemical
properties than the rutile phase (22).

3.3 Photochemical properties of colloidal TiO_2

Illumination of semiconductor colloids generates electrons and positive
holes which may react with the aqueous solvent or dissolved species (23)(24).
On the standard electrochemical energy scale, the upper edge of the valence
band of crystalline TiO_2 is located at +3.1 V while the lower edge of the
conduction band is at -0.1 V. Positive holes created upon U.V. irradiation
of TiO_2 are therefore very strong oxidizing agents and can be used for generating photochemical reactions (24). Interfacial electron-transfer processes
are very important and colloidal dispersions of semiconducting oxides in
which the particles are small enough to avoid scattering of light seem to be
good candidates for such applications (12). TiO_2 sols are usually made
through hydrolysis of titanium alkoxides. However, the colloidal solution
can only absord U.V. light and are not stable above pH 3 (12).

We have shown that strong chelating ligands such as (acac) could lead
to transparent sols that remain stable up to pH 10. Moreover, charge transfers from the (acac) ligand to the titanium ion give rise to a strong absorption in the visible region. They could therefore improve the photochemical

efficiency of TiO_2 sols. These sols can be readily reduced upon irradiation with visible light. They turn to blue when hydrolysis is performed with an excess of water at a pH>2. They turn to pink when hydrolysis is performed at pH<2. The ESR spectra of both photochemically colored solutions are typical of Ti^{3+} ions, but the g values are completely different. The blue solution exhibits the same ESR spectrum as $Ti(acac)_3$ with $g_{/\!/}$ = 1.99 and $g_\perp$ = 1.93 while the pink solution exhibits an orthorhombic spectrum with g_1 = 1.979, g_2 = 1.973 and g_3 = 1.907. This suggests that some (acac) ligands are removed upon acid hydrolysis.

It has to be pointed out that the redox potential of the blue species is especially low (-1V vs.NHE). They then exhibit a stronger reducing power than other TiO_2 colloids. They can reduce many chemical species in the solution or even give rise to photocurrents at a collecting electrode. These reducing properties have been evidenced with methyl viologen, a compound often used as an electron relay in photosensitive colloidal systems (MV^{2+}/MV^+, -0.44 V vs. NHE). Under irradiation with visible light, the reduced MV^+ species can be readily seen by ESR and optical spectroscopy. No Ti^{3+} can be detected suggesting that the electron transfer $Ti^{3+} + MV^{2+} \longrightarrow Ti^{4+} + MV^+$ is very fast.

TiO_2 colloids also exhibit some oxidizing properties. These have been studied using the spin -trapping technique (25). ESR experiments performed in the presence of a spin-trap (POBN) show that under irradiation with visible light CH_3-CHOH radicals can be detected showing that the solvent is oxidized.

Colloidal species, obtained through hydrolysis of $Ti(OPri)_4$ in the presence of acetylacetone, behave as microphotoelectrodes exhibiting both a high reducing power and a high oxidizing power.

IV. CONCLUSION

The sol-gel process appear to be a very versatile way of making TiO_2 based devices. Depending on the specific applications, different chemical additives can be used. They change the precursor at a molecular level and may improve the properties of the resulting titanium oxide. Acetic acid slows down the hydrolysis process leading to monolithic gels that can be easily deposited for making amorphous TiO_2 electrochromic electrodes. Strong chelating (acac) ligands prevents a complete hydrolysis and stabilize colloidal solutions over a wide range of pH. Doping TiO_2 gels with $Cr(acac)_3$ improves the optical response of TiO_2 photoanodes in the visible region.

REFERENCES

1. L.C. Klein, Ann. Rev. Mater. Sci., 15, 227 (1985).

2. R.K. Iler, the Chemistry of Silica (Wiley, New-York 1979).

3. K.S. Mazdiyasni, Ceramics International, 8, 42 (1982).

4. S. Wallace and L.L. Hench, in Better Ceramics through Chemistry, edited by C.J. Brinker, D.E. Clark, D.R. Ulrich (North-Holland 1984) p. 47.

5. D.R. Ulrich, Ceramic Bulletin, 64, 1444 (1985).

6. E.A. Barringer and H.K. Bowen, Langmuir, 1, 414 (1985).

7. M.F. Yan and W.W. Rhodes, Mater. Sci. and Eng., 61, 59 (1983).

8. S. Doeuff, M. Henry, C. Sanchez and J. Livage, J. Non-Cryst. Solids, (submitted).

9. K. Nakamoto, in Infra-red and Raman spectra of Inorganic and Coordination Compounds, 3rd Edition (John Wiley, New-York, 1978).

10. Von K.H. Thiele and M. Panse, Z. Anorg. Allg. Chem., 441, 23 (1978).

11. N.T. Mc Devitt and W.L. Baun, Spectrochimica Acta, 20, 799 (1964).

12. D. Duonghong, E. Borgarello and M. Grätzel, J. Am. Chem. Soc., 103, 4685 (1981).

13. G. Kordas, R.A. Weeks and L.C. Klein, J. of Non-Cryst. Solids, 71, 327 (1985).

14. A.A. Wolf, E.J. Friebele and D.C. Tran, J. of Non-Cryst. Solids, 71, 345 (1985).

15. S. Doeuff, M. Henry, C. Sanchez and J. Livage, J. of Non-Cryst. Solids (submitted).

16. T.I. Barry, Solid State Comm., 4, 123 (1966).

17. H.J. Gerritsen, S.E. Harrison, H.R. Lewis and J.P. Wittke, Phys. Rev. Letters, 2, 153 (1959).

18. T. Ohzuku and T. Hirai, Electrochemica Acta, 27, 1263 (1982).

19. H. Dislich and P. Hinz, J. Non-Cryst. Solids, 48, 11 (1982).

20. K.D. Kochev, Solar Energy Materials, 12, 249 (1985).

21. A.K. Ghosh and H.P. Maruska, J. Electrochem. Soc., 124, 1516 (1977).

22. H. Minoura, M. Nasu and Y. Takashi, Ber. Bunsenges Phys. Chem., 89, 1064 (1985).

23. M. Grätzel, Acc. Chem. Res., 14, 376 (1981).

24. A. Henglein, Pure and Appl. Chem., 56, 1215 (1984).

25. C.D. Jaeger and A. Bard, J. Phys. Chem., 83, 3146 (1979).

This paper also appears in Mat. Res. Soc. Symp. Proc. Vol. 72.

CHARACTERIZATION OF SOL-GEL DERIVED TANTALUM OXIDE FILMS

L.A. Silverman, G. Teowee and D. R. Uhlmann, Department of Materials
Science and Engineering, M.I.T., Cambridge, Mass. 02139

ABSTRACT

We have studied the densification and dielectric properties of sol-gel
derived tantalum oxide thin films as the insulators in MIS capacitors.
Hydrolysis of tantalum ethoxide is extremely rapid and goes to completion in
ethanol. Condensation is also rapid, and goes to completion in toluene.
Multiple layers were applied by spin-coating up to thicknesses of 3000 Å
before cracking of the coating during drying ensued. Densification occurs
from room temperature to 450 C, with the original film thickness decreasing
by about half in one hour at 450 C. No additional densification occurs upon
heating to 750 C. The dielectric constant decreases from unfired samples to
those fired at 450 C, and then increases on firing from 600 to 750 C. The
value of the dielectric constant at 1 MHz for samples fired at 750 C for one
hour is 20, similar to that of anodically grown Ta_2O_5. Leakage currents as
low as 2×10^{-7} amp cm^{-2} have been measured for applied fields of 200,000
V cm^{-1}.

INTRODUCTION

Tantalum oxide (Ta_2O_5) is used extensively as a dielectric in precision
capacitors. Its use is not due to an especially high dielectric constant
(20-40) [1-3], but rather to the uniform and controllable thickness of the
oxide layer produced by anodization of tantalum metal. Also, the potential
use of Ta_2O_5 as a possible replacement for SiO_2 in integrated circuits has
recently been considered because its higher dielectric constant will allow
for a large reduction in the area of the MIS capacitors used. Methods of
fabrication of Ta_2O_5 films have included oxidation of sputtered metal films
[4,5], reactive sputtering [6] and CVD starting with metal-organic [7, 8] or
chloride [1] precursors among others. These films have good dielectric
constants and low leakage currents, but the methods of manufacture are complicated and time consuming.

This paper reports progress in our synthesis and characterization of sol-
gel derived Ta_2O_5. With this process, it is possible to make oxide thin films
easily, quickly and inexpensively.

The sol-gel approach to making oxide ceramics is conceptually quite
simple. Water is added to metal-organic precursor solutions, resulting in
their hydrolysis (Eq. 1). The partially hydrolyzed species

$$Ta(OR)_5 + H_2O \longrightarrow (RO)_4TaOH + ROH \qquad (1)$$

condense with OR or OH groups on other molecules, leading to metal-oxygen-
metal bridges (Eq. 2, e.g.). This leads to the formation of oligomers in
solution, and upon further condensation, to the formation of gels. The gels

$$Ta(OR)_5 + (RO)_4TaOH \longrightarrow (RO)_4Ta-O-Ta(OR)_4 + ROH \qquad (2)$$

are loosely packed porous structures containing many residual organic groups,
surrounded and penetrated by solvent. As the solvent evaporates, the gels
shrink and crosslink; and if done with care, monolithic aerogels can be
produced. The aerogels are then heat treated to remove residual water and
organics, producing ceramic bodies.

This simple description understates the complexity of the process [9].
In solution, the alkoxide molecules can form complexes with the solvent as

well as with each other. Reaction rates may differ greatly between different
alkoxides, leading to the formation of inhomogeneous gels. Upon removing
the solvent, stresses resulting from solvent concentration gradients and
capillarity forces can lead to fracture. Heat treatments done under the wrong
conditions can also lead to fracture or to trapping residual organics which
can cause partial reduction and bloating at high temperatures.

A thin film geometry lends itself well to sol-gel processing. Adhesion
to the substrate restricts shrinkage in the plane of the film, so that drying
stresses are reduced and the films remain intact. Diffusion distances perpendicular to the substrate are short, allowing fast transport of reaction
products and reducing the trapping of residual organics. By using a single
cation system in a thin film geometry, crack-free homogeneous oxide layers
can routinely be produced.

The present paper reports on the densification and evolution of the
dielectric constant in sol-gel derived Ta_2O_5 films as functions of the
deposition conditions and heat treatment temperatures and times.

EXPERIMENTAL PROCEDURES

The coating solutions were made from optical grade tantalum ethoxide
(Alfa Products) dissolved in ethanol which was freshly distilled from magnesium ethoxide. To this, deionized water dissolved in ethanol was added to
bring the water/alkoxide molar ratio to 1 to 1.3. All solutions contained
5 weight percent oxide. Infrared spectroscopy to follow hydrolysis was done
in CaF_2 cell, 0.1 mm thick, on a Perkin Elmer 297 spectrometer. Proton NMR
run on a Brucker 300 MHz machine was used to examine the structure of the
oligomers formed in deuterated toluene.

The solutions were spin-coated in air onto n-doped [111] silicon wafers,
2.25 inches in diameter, at 4000 or 5000 rpm for 30 sec. Great care was used
to minimize the time that the solution was exposed to the atmosphere before
spinning. During an additional spin, the coated wafers were twice exposed
to N_2 saturated with H_2O at about 40 C. The slightly visible haze which
resulted was allowed to evaporate between and after exposures and subsequent
layers of coating were spun on following the same procedure.

The coated wafers were scribed and broken into 1 cm squares. A random
sample of wafers was masked and etched in HF so that the average film thickness prior to heat treatment could be measured on a Sloan Dektak II. Of the
remaining squares, each was heat treated as desired. All heat treatments
above 300 C were done in quartz-line furnaces under N_2 to prevent exposure
to alkali ions and the formation of a thermal oxide on the Si. Lower temperature treatments were done in ovens in Pyrex containers.

Al was evaporated onto the densified oxide; and samples fired above
300 C were run through a standard positive photoresist process to develop
1 mm square electrodes. Unfired sample as well as those fired at 150 C were
masked with wax to develop the electrodes. The areas of the irregularly
shaped electrodes were measured using a planimeter on enlarged photographs
of the samples. Al was evaporated onto the backs of the wafers to provide
contacts; and capacitance-voltage measurements on the MIS capacitors at MHz
were made on a Hewlett Packard 4061 A system linked to an HP9236 personal
computer. Dielectric constants at 1 MHz were calculated based on the
maximum capacitance measured under a positive bias [4]. Current-voltage
measurements were made on an HP4145 system. Finally, the wafer was masked
again and the oxide film was etched using HF. After unmasking, thickness
of the densified film was measured.

RESULTS AND DISCUSSION

Examining the hydrolysis rate of $Ta(OEt)_5$ in ethanol using IR spectroscopy at 1650 cm^{-1} water band, we learned that the reaction occurs so rapidly at room temperature that the free water is consumed before the cell can be loaded into the instrument. Condensation is also rapid, as evidenced by the immediate precipitation of powder when the water/alkoxide ratio exceeds 2. As this ratio falls toward 1.6, the rate of formation of visible powder is slowed and eventually stops. This confirms the data of Bradley [10] as to the degree of hydrolysis at which precipitation occurs.

Proton NMR on the hydrolyzed and condensed oligomers in deuterated toluene shows only protons associated with ethoxy groups, none due to residual uncondensed hydroxy groups. Hence condensation in toluene is complete. It should be noted that condensation in ethanol may not occur to such a large extent due to a reverse condensation reaction. Two coalescence temperatures were present, with the ratio of integrated peak intensities below the lower T_c remaining constant before and after polymerization. Before reaction, the ethoxide exists almost entirely as a dimer in toluene [11] because the solvent is not very nucleophilic. This leads to a 2:2:1 ratio in peak intensities attributed to the two types of terminal and one bridging ethoxy groups in the unreacted species [12, 13]. The observation that the peak intensities do not change after reaction suggests that the oligomers are randomly polymerized, and have not preferentially reacted at certain sites of the dimer.

During and after spin coating, interference colors were visible. Wafers coated at less than 4000 rpm showed large color gradients from the edge of the wafer to the center, indicating that the films were not of uniform thickness. Those coated at 4000 rpm and above were uniform in color. Multiple coatings were only possible after the previous layer had been exposed to moisture. Failure to do this after coating under dry N_2 resulted in no additional thickness after the first layer was applied. Hence it seems that the oligomers in solution must react with surface hydroxyl groups in order to become immobilized on that surface.

Upon exposure to humidity after the gels had dried, the colors of the films always changed toward the blue end of the spectrum. During storage in air, the color continued to change until an equilibrium color was attained some 16-24 hours after the coating was applied. The shift toward blue indicates that the optical thickness is decreasing. This can happen either by the loss of ethoxy groups due to the continuing hydrolysis from atmospheric humidity leading to a decrease in overall density and refractive index of the film, or by additional condensation and densification leading to a decrease in thickness. Films comprised of multiple layers whose final thickness was greater than 3600 Å cracked due to drying stresses immediately after coating; while those between 3000 and 3600 Å in thickness cracked upon heating to 150 C. Films thinner than 3000 Å could be processed without cracking.

Adhesion of the heated films to the substrates is quite good. Unfired films were easily scratched and could be removed from the substrates by scraping. Films heated to a temperature as low as 150 C were not removed or damaged by limited physical abrasion. Hence it seems that many of the primary covalent bonds to the surface are formed due to condensation on surface hydroxyl groups at temperatures between 25 C and 150 C.

During firing, all coatings with initial thicknesses below 3000 Å remained intact. The interference colors changed dramatically within the first seconds of being heated and then remained constant through the longest heating time of 24 hours. In addition, the thickness of the film does not change significantly with time after the first hour of heating at temperatures above 300 C nor with temperature for temperatures above 450 C (Fig. 1). This indicates that the densification of the films is quite rapid at these temperatures. This is not surprising in light of densification being purported to occur only through condensation at temperatures where the viscosity of the gel is quite high [14], and condensation in solution for this system has been shown to be rapid. There is little densification between 450 C and 750 C. This correlates well with the TGA data, which show no weight loss above 450 C.

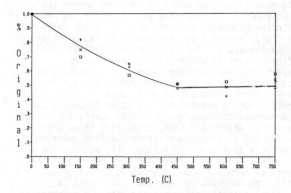

Fig. 1 - Shrinkage vs. temperature for films fired for (+) 1 hour, (x) 5 hours and (□) 24 hours.

Increased resistance to etching was observed as the sample was heated above 300 C. The most notable change occurred between samples fired at 600 C and those treated at 750 C. From this we can infer that some structural changes occur between these two temperatures. However, from the thickness data we see that this change does not result in densification.

The dielectric constant (ϵ) of these materials depends strongly on their thermal histories (Fig. 2). All unfired samples, as well as those fired at 150 C, exhibited dielectric constants above 25. These data do not appear in the figure because the C-V curves were not of classical form. Specifically, the capacitance at positive bias voltages did not level out before large current leakage occurred (e.g., Fig. 3a).

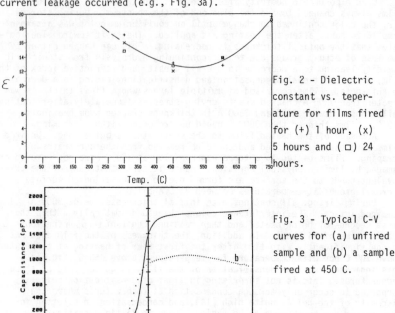

Fig. 2 - Dielectric constant vs. teperature for films fired for (+) 1 hour, (x) 5 hours and (□) 24 hours.

Fig. 3 - Typical C-V curves for (a) unfired sample and (b) a sample fired at 450 C.

This did not give a reliable value of the maximum capacitance for calculating a dielectric constant. A possible explanation for the premature leakage is that the gel at temperatures below 150 C contains a continuous network of interconnected pores. These channels may act as pipes for current along their organic-containing hydrated surfaces. Samples fired at higher temperatures, where the extent of densification was higher, produced curves of classical shape (e.g., Fig. 3b).

As the sample is heated from room temperature to 450 C, the dielectric constant decreases considerably (from 25 to 12). It changes little between 450 C and 600 C; and between 600 C and 750 C, increases dramatically to 20. The decrease in ε corresponds closely with the weight loss seen in the TGA. The dielectric constants of water and ethanol at 1 MHz are 78.2 and 24.5 respectively [15]. The loss of these adsorbed or chemically bound groups would lead to a drop in dielectric constant if the remainder of the material had lower ε. The significant increase in ε above 600 C must be due to a structural change that is separate from densification.

The crystallization exotherm peak and Tg as measured by DTA on bulk powders precipitated from the coating solutions occur at 740 C and 605 C respectively. However, X-ray diffraction on the same powder heated at 750 C for 1 hour showed only small amounts of crystallinity. A recent study on crystallization above 900 C of thermally oxidized Ta films on Si indicate that ε depends upon crystallization temperature, time and atmosphere [16]. Final values of ε for crystallized Ta_2O_5 in that study ranged from 20 to 28. At present, we have no evidence on the crystallinity of the film; and if crystallization is occurring at 750 C, it is possible that the transformation is incomplete. This could explain why our data on the films heated to 750 C reflect lower values of ε than Ref. 16. Another possibility is that relaxation around and above Tg is responsible for the increase in ε between 600 C to 750 C. It should be recognized that the temperatures and rates of processes in bulk powders may be quite different from those in films due to constraints imposed by the substrate in the latter case. This means that both the glass transition and crystallization temperatures may be shifted higher in the films compared with those measured in the powder. It is likely therefore that the films still may be amorphous after treatment at 750 C.

Preliminary data on leakage current suggest that the films become more resistive as the heat treatment temperature is raised. Films treated at 750 C result in leakage current of 2×10^{-7} amps cm^{-2} for fields of 2×10^{-5} volts cm^{-1} across a 1300 Å film.

CONCLUSIONS

Sol-gel derived tantalum oxide films exhibit dielectric constants that are comparable to anodically grown on thermally oxidized Ta_2O_5 when fired under certain conditions. Densification at temperatures above 300 C is rapid, with the final density of films fired between 450 C and 750 C being similar. The dielectric constant decreases from 25 for unfired films to 12 for samples treated at 450 C; and above 600 C, the dielectric costant increases sharply to 20 with no change in film thickness. The decrease in dielectric constant is attributed to loss of residual water and organics. The increase above 600 C is attributed to either slight relaxations in the amorphous solid, or partial crystallization of the film.

ACKNOWLEDGEMENTS

Financial support for the present work was provided by the Air Force Office of Scientific Research. This support is gratefully acknowledged, as is the help of B.J.J. Zelinski, B.D. Fabes, B.D. Doyle, C.S. Parkhurst, M.P. Andersen and G. Ditner. Thanks to IBM for providing the silicon wafers.

REFERNCES

1. W.H. Knausengerger, and R.N. Tauber, J. Electrochem. Soc., 126, 927 (1973).
2. A.S. Povlovic, J. Chem. Phys., 40, 951 (1964).
3. D. Gerstenberg, J. Electrochem. Soc., 113, 1542 (1966).
4. G.S. Oehrlein, and A. Reisman, J. Appl. Phys., 54 6502 (1983).
5. H. Rizkalla, and S.T. Wellinghoff, J. Mater. Sci., 19, 3895 (1984).
6. S. Kimura, Y. Nishioka, A. Shintani and K. Mukai, J. Electrochem. Soc., 130, 2414 (1983).
7. E. Kaplan, M. Balog, and D. Bentchkowsky-Frohman, J. Electrochem. Soc., 123, 1570 (1976).
8. C.C. Wang, K.H. Zaininger, and M.T. Duffy, RCA Review, Dec. 1980.
9. B.J.J. Zelinski, and D.R. Uhlmann, J. Phys. Chem. Solids, 45, 1069 (1985).
10. D.C. Bradley and H. Holloway, Can. J. Chem., 39, 1818 (1961).
11. D.C. Bradley, W. Wardlaw and A. Whiteley, J. Chem. Soc., 5, 726(1956).
12. C.E. Holloway, J. Coord. Chem., 1, 253 (1971).
13. L.G. Pfalzgraf, and J.G. Reiss, Bull. Soc. Chim. France, 11, 4348 (1968).
14. C.J. Brinker, and G.W. Scherer, J. Am. Cer. Soc., 69, C-12 (1986).
15. A. Von Hippel, Dielectric Materials and Applications, M.I.T. Press, Cambridge, Massachusetts (1954).
16. G.S. Oehrlein, F.M. d'Heurle and A. Reisman, J. Appl. Phys., 55, 3715 (1984).

CRYSTALLIZATION OF OXIDE FILMS DERIVED FROM METALLO-ORGANIC PRECURSORS

K.C. CHEN, A. JANAH AND J.D. MACKENZIE
Department of Materials Science and Engineering,
University of California, Los Angeles, CA 90024

ABSTRACT

Ferroelectric lead zirconate titanate and ferrimagnetic nickel ferrite films were prepared by dip-coating of metal organic solutions. In the amorphous state, both films were porous, with specific porous surface areas on the order of tens of square meters. Crystallization temperatures were in the range of 400 - 500°C. The crystallization kinetics of the porous films and powders of both compounds have been studied by using non-isothermal DSC and isothermal XRD. The kinetic parameters were empirically described by the Johnson-Mehl-Avrami transformation equation $\alpha = 1 - \exp(-Kt^n)$. The order of reaction (n) of PZT film is approximately 1 but that of the powder is 2, and their activation energies are 80 and 70 kcal/mole, respectively. It was found that the PZT film crystallizes more easily than the powder at the same temperature. The nickel ferrite film and powder, on the other hand, have similar n values ranging from 0.6 to 0.7, with activation energies of 20 and 35 kcal/mole, respectively. The possible mechanisms were also briefly discussed.

INTRODUCTION

There is widespread current interest to prepare glasses and ceramics from metal alkoxides precursors via sol-gel and solution techniques. These processes have the advantages of better homogeneity of mixing and lower processing temperatures. The materials obtained are usually amorphous and highly porous at low temperatures. To make crystalline ceramics, these amorphous and highly porous materials have to be both crystallized and densified. The scientific understanding of the processing parameters and phases crystallized from gels have been advancing rapidly [1,2,3,4], and many investigators have reported on the crystallization of powders [5,6,7]. However, there is relatively little information on the kinetics and mechanisms of crystallization of amorphous porous thin films [8].

Since a study of the crystallization mechanism may lead to better control of microstructures and properties of films, the present work is aimed towards a more detailed understanding of the crystallization kinetics, limiting parameters and mechanisms of transformation from amorphous to crystalline films of two electronic materials, ferroelectric lead zirconate titanate and ferrimagnetic nickel ferrite. The results of these compounds are compared and possible crystallization mechanisms are discussed.

EXPERIMENTAL

The PZT with Zr/Ti ratio equal to 52/48 and $NiFe_2O_4$ solutions were prepared as shown in Figures 1 and 2. All chemicals were obtained from Alfa Products. The solutions were mixed in an ultrasonic bath for 30 minutes to ensure homogeneity. To obtain films, fused silica glass slides were dipped into the solutions and were slowly withdrawn at 2.5cm/min. The

Mat. Res. Soc. Symp. Proc. Vol. 73. ©1986 Materials Research Society

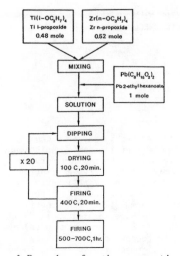

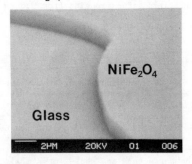

Fig. 1 Procedure for the preparation of PZT film.

Fig. 2 Procedure for the preparation of $NiFe_2O_4$ film.

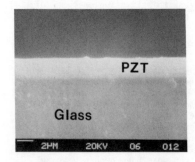

Fig. 3 SEM micrographs of the PZT and $NiFe_2O_4$ films on glass substrates.

chemical composition of the PZT was confirmed by X-ray fluorescence and spectrographic methods were used for the nickel ferrite. SEM micrographs of the PZT and $NiFe_2O_4$ films are shown in Figure 3.

Surface area was measured using a nitrogen adsorption surface area analyzer (Micromeritics Flowsorb II 2300). TGA and DSC studies were conducted on Perkin Elmer TGS2 and DSC thermal analyzer. A fixed quantity of 10 mg was used for all DSC studies. X-ray diffraction data was obtained on a Norelco diffractometer using monochromatic Cu Kα radiation.

RESULTS AND DISCUSSION

TG and DTG analysis

For crystallization kinetic studies, organic free films and powders are essential. Thermogravimetric analysis was employed to identify the organic burn-out temperatures of both solutions as shown in Figure 4. For PZT solution, the total weight loss (in dynamic oxygen) was about 72%. There were two major stages of organic burn-out. The first stage, occurring at 100°C, was attributed to alcohol evaporation while the second at 300°C, corresponded to burn off of the 2-ethylhexanoate group and other organics. These weight loss results are consistent with the calculated ones, which are indicated by arrows in Figure 4. The removal of organics between 250 - 300°C consist of three steps which have not been yet identified. The total weight loss of the nickel ferrite solution is about 90%, which also agrees with theoretical calculation. Similar burn off stages were also observed for this solution.

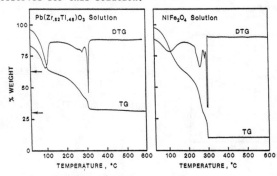

Fig. 4. TGA and DTG of the organo - metallic solutions of PZT and $NiFe_2O_4$.

Surface area measurements

Surface area studies on the PZT film on fused silica substrates, show that the specific surface areas are on the order of square meters per gram (Figure 5). This indicates the presence of a large number of fine pores in the film. Further, the large surface area may create a geometric constraint on the crystallization kinetics. The decrease in surface area with heat treatment related to densification, is seen to be more dependent on the temperature than on the time of heat treatment. The surface area of the film at 700°C was found to be below the detectable limit of the surface area analyzer. It should be noted that at 700°C for 1 hr, the film still has about 7 m^2/g specific surface area. Therefore the film can only be densified through a sintering process. The specific surface area of the nickel ferrite film was larger than that of the PZT, ranging from 90 to 150 m^2/g, for the same heat treatment schedule.

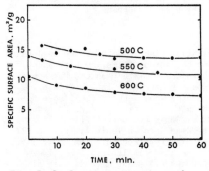

Fig. 5. Surface area vs. time and temperature for the PZT film.

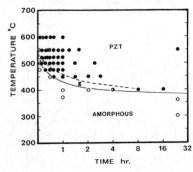

Fig. 6. Time temperature transformation diagram for the PZT film.

X ray diffractometry

Figure 6 shows the time-temperature transformation diagram obtained by XRD. Below 400°C, only the amorphous phase was observed. An unknown phase was observed after short heat treatment time at 500°C and it was still detectable up to 550°C.

Figure 7 shows the fraction transformed of the PZT film, determined by the peak intensity of the (111) peak of perovskite. Similar curves were obtained on the $NiFe_2O_4$ film using the intensity of the (311) peak. The peak intensity of the PZT film heat treated at 550°C for 80 hrs and that of the $NiFe_2O_4$ film heated to 600°C for 168 hrs, were used as the 100% crystallized intensity values.

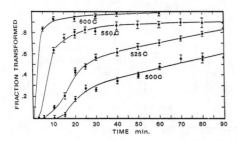

Fig. 7. Isothermal curves for the crystallization of PZT film obtained from XRD using the intensity of the (111) peak.

As the transformation is "allotropic", the Johnson-Mehl-Avrami transformation kinetic equation [9] as shown below, was used to evaluate the XRD data.

$$\alpha = 1 - \exp(-Kt^n) \qquad (1)$$

Here, α is the volume fraction transformed, K is the effective overall reaction rate and the value of the exponent n depends on the mechanism of nucleation and growth and the growth geometry. The value of K is assigned an Arrhenius temperature dependence with apparent activation energy E and frequency factor ν, as follows :

$$K = \nu \exp(-E/RT) \qquad (2)$$

Values of n were determined by plotting $\ln(-\ln(1-\alpha))$ vs $\ln$ time and are given in Table 1. The crystallite size of both crystalline films were approximately 300 Å as determined by peak broadening.

Non-isothermal DSC

It was thought that the crystallization kinetic data of the powders should be comparable to the kinetics of the porous films, as both had equivalent surface areas. Figure 8 shows the DSC study on the PZT powder. By increasing the heating rate β, the peak temperature T_p shifted to higher temperatures. The heat of transformations were 1.1 kcal/mole for the PZT powder and 305 kcal/mol for the nickel ferrite powder. Mathematical analyses were performed by the methods of Augis and Bennett [10], Kissinger [11] and Ozawa [12]. The results were summarised in Table 1, for the PZT and Nickel ferrite powders. Similar studies were conducted on the PZT film on SiO_2 substrate and the results are also given in Table 1.

The values of E, ν and n obtained using different methods vary slightly. The PZT powder has ν values ranging from $10^{17} - 10^{18}$, an E value of 70 kcal/mole and an n value of 2. The same experiments on the PZT film gave higher values of ν and E of $10^{20} - 10^{22} s^1$ and 80 kcal/mole, respectively, but

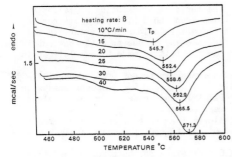

Fig. 8. Non isothermal DSC plots at different heating rates, for the amorphous PZT powder.

a lower n value, close to 1. A similar value of n was obtained from XRD. The reaction rate constants calculated from (2) showed that the film crystallized more easily than the powder at the same temperature.

The apparent activation energy of crystallization may consist of activation energies of the diffusion of ionic species, structural relaxation and nucleation and growth. The activation energy values for the diffusion of Pb^{2+} in $PbTiO_3$ [13] and that of O^{2-} ion in TiO_2 [14] and ZrO_2 [15] ranges from 45-66 kcal/mole. The difference in activation energies may arise from structural relaxation and nucleation and growth.

As the PZT crystallizes in a cubic structure, n may be more related to the nucleation and growth mechanism, than the growth geometry. It is known that the crystallization of sol-gel derived amorphous phase is catalyzed by adding dispersed 'seeds' [16]. Therefore the partially crystallized powder will behave as if seeds were present, with subsequent

Table I. Comparison of the kinetic parameters for crystallization of amorphous PZT and $NiFe_2O_4$ calculated from non-isothermal DSC and isothermal XRD studies, using different mathematical analyses.

	Methods	Powder			Film			Ref.
		ν s^{-1}	E kcal/mol	n	ν s^{-1}	E kcal/mol	n	
PZT		Non-isothermal DSC						
	$\ln\beta$ vs $1/T_p$		74			84		[12]
	$\ln(-\ln(1-\alpha))\mid_T$ vs $\ln\beta$			2.1			0.9	[12]
	$\ln(\beta/(T_p-T_i))$ vs $1/T_p$	1×10^{18}	61	2.0	6×10^{21}	74	1.3	[10]
	$\ln(T_p^2/\beta)$ vs $1/T_p$	6×10^{16}	71		3×10^{20}	79		[11]
		Isothermal XRD						
	$\ln(-\ln(1-\alpha))$ vs $\ln t$	–	–				0.8- 1.0	[17]
$NiFe_2O_4$		Non-isothermal DSC						
	$\ln\beta$ vs $1/T_p$		40			–		[12]
	$\ln(-\ln(1-\alpha))\mid_T$ vs $\ln\beta$			0.7			–	[12]
	$\ln(\beta/(T_p-T_i))$ vs $1/T_p$	3×10^{11}	32	0.6	–	–	–	[10]
	$\ln(T_p^2/\beta)$ vs $1/T_p$	3×10^9	37		–	–		[11]
		Isothermal XRD						
	$\ln(-\ln(1-\alpha))$ vs $\ln t$	–	–		19		0.6- 0.7	[17]

736

nucleation and growth occuring on the surfaces of existing nuclei. This
process can give rise to an n of value equal to 2 [17]. The discrepancy
in n of powder and film does not necessarily indicate different crystallization mechanisms [18]. This difference may be due to different submicron
structures of the powder and film as the film is made by a multi-dipping
technique.

The nickel ferrite powder had a frequency factor ν of around $10^9 - 10^{11}$
sec^{-1}, an E of $\simeq 35$ kcal/mole and an n of around 0.6. As contrasted with
the values of n differing for the PZT film and powder, similar n values
were obtained for the $NiFe_2O_4$ film and powder.

The activation energy is much lower than that of solid state reactions
[19] and the diffusion of Ni^{2+} and O^{2-} in $NiFe_2O_4$ [20,21] and close to the
diffusion of Fe^{3+} in $\alpha-Fe_2O_3$ [22].

The n value of approximately 0.6 obtained for both powder and film
cannot be explained at present without electron microscopy and other
structural studies.

ACKNOWLEDGEMENT

The authors are grateful to the Air Force Office of Scientific
Research, Directorate of Chemical and Atmospheric Sciences, for support of
this study. Thanks are also due to Dr. A. Osaka, Mr. T.J. Yuen and K.
Thorne.

REFERENCES

1. R.A. Lipeles, N.A. Ives and M.S. Leung, private communication.
2. J. Fukushima, K. Kodaira and T. Matsushita, J. Mater. Sci. 19 595-598
   (1984).
3. K.D. Budd, S.K. Dey and D.A. Payne, Am. Ceram. Soc. Annual Meeting,
   May 1985, Cincinnati, Ohio.
4. K. Suwa, S. Hirano, K. Itozawa and S. Naka, Proc. International Conf.
   on Ferrite, 1980, Japan, 23-26.
5. V.V. Sakharov, V.G. Savenko, A.A. Nurgalieva and K.I. Petrov, Russian
   Journal of Inorg. Chem., 26 (9) 1251-1255 (1981).
6. S.R. Gurkovich and J.B. Blum, Ferroelectrics, Vol. 62 189-194 (1985).
7. V.S. Rao, S. Rajendran and H.S. Maiti, J. Mat. Sci., 19 3593 (1984).
8. A.S. Shaikh and G.M. Vest, Am. Ceram. Soc. Annual Meeting, May 1985,
   Cincinnati, Ohio.
9. M.E. Fine, Introduction to Phase Transformation in Condensed Systems,
   McMillan Co., 542, New York (1963).
10. J.A. Augis and J.D. Bennett, J. Thermal Anal., 13 283 (1978).
11. H.E. Kissinger, J. Res. Nat. Bur. Stds., 57 (1956) 217; Anal. Chem. 29
    (1957) 1702.
12. T. Ozawa, J. Thermal Anal., 2 (1970) 301 ; Polymer 12 150 (1971).
13. A.P. Lyubimov, A.A. Kalashnikov and B. Nuriddinov, Dokl. Akad. Nauk
    Uzb SSSR, 29 5 (1972) 24-26.
14. E. Eguchi and K. Kajima, Trans. Japan Inst. Met., 13 (1972) 45.
15. F.J. Kenesha and D.L. Douglass, Oxid. Metals, 3 (1971) 1.
16. M. Kumagai and G.L. Messing, J. Am. Ceram. Soc., 68 [9] 500 (1985).
17. J.W. Christian, Transformation in Metals and Alloys, Part I, 2nd Ed.,
    Pergamon Press, N.Y. (1975) 542.
18. M.C. Weinberg, J. Non-cryst. Solids, 72 (1985) 301-314.
19. N.A. Eissa and A.A. Bhagat, J. Am. Ceram. Soc., 59 (1976) 7-8.
20. G. Maxima and M.L. Craus, Rev. Roum. Phys., 16 [6] (1971) 655-657.
21. H.M. O'Brien and F.V. DiMarcello, J. Am. Ceram. Soc., 53 [7] (1970)
    413.
22. R.H. Chang and J.B. Wagner, J. Am. Ceram. Soc. 55 [4] (1972) 211.

This paper also appears in Mat. Res. Soc. Symp. Proc. Vol. 72.

New Initiatives/Novel Materials

CHEMISTRY AND APPLICATIONS OF INORGANIC-ORGANIC POLYMERS
(ORGANICALLY MODIFIED SILICATES)

H. SCHMIDT AND B. SEIFERLING
Fraunhofer-Institut für Silicatforschung, Neunerplatz 2, D-8700 Würzburg,
Federal Republic of Germany

ABSTRACT

The combination of inorganic polymeric networks with organic components
leads to inorganic-organic polymers. A convenient method for the introduction of organic radials into an inorganic backbone is the use of organosubstituted silico esters in a polycondensation process. This leads to
≡Si-O-Si≡ network containing materials, so-called organically modified silicates (ORMOSILs). For the synthesis of the inorganic backbone, in opposition to the high temperature preparation of non-metallic inorganic materials
like ceramics, "soft chemistry" methods have to be applied in order to preserve organic groupings to be incorporated. Therefore, the sol-gel process
is a suitable technique [1-5]. A review over basic synthesis principles and
chemical methods, their effect on special material properties and the application potential will be given.

1. INTRODUCTION AND GENERAL CONSIDERATIONS

The desire to combine properties of very different materials in one
and the same product has led to composite materials on a macroscopic scale.
Examples therefore may be filled polymers, laminates like automotive
screens, fibre reinforced epoxides or even coating all type of materials
with paints. Another possibility is opened by the idea to combine properties of different components on a molecular scale, as it is shown on large
scale with organic copolymers. The idea to combine inorganic with organic
components on a microscopic scale requires a formation process of an inorganic network which is compatible to the thermal stability of organic components. For example it is not possible to add organic monomers or polymers
to a glass melt in order to get hybrid materials. A "soft" chemical synthesis of inorganic polymers can be provided by the sol-gel process. In most
cases, sol-gel synthesis of inorganic polymers leads to porous gels as intermediates, which in general are more or less amorphous. For densification
a heat treatment has to be applied. The temperature depends on the system,
but with few exceptions the temperatures to be applied are above the level
of thermal stability of organic groupings. That means that in the case of
ORMOSILs high temperature moulding procedures will be restricted to very
special systems. The question arises, whether it is possible to change the
properties of the inorganic network by the incorporation of organic groupings in a way, that the described disadvantages can be avoided, e.g. by
gaining network flexibility in order to receive dense products whithout
high temperatur treatments. The opposite problem may arise, if porous materials have to be prepared.
An introduction of an organic group into an inorganic network may act
in two basically different ways [6, 7]. It may act as a network modifier or
as a network former. Both functions can be realized in ORMOSILs. A suitable
way to achieve this is the use of organosubstituted silicic acid esters of
the general formula $R'_nSi(OR)_{4-n}$, where R' can be any organofunctional grouping [8-10]. If R' is a non-reactive group, it will have a network modifying effect; if it can react with itself or additional components, it acts
as network former. For the synthesis of the described hybrid materials, the
chemistry plays the key role. Reaction condition have to be developed for

each system which take into consideration the sol-gel requirements as well as those defined by the organic groups.

2. REACTION PRINCIPLES

The reaction principles of the sol-gel process are well known and not described here. Examples for the introduction of organic components using sol-gel techniques are schematically given in eq. (1-5):

$$-\overset{|}{\text{M}}\text{-OH} + \text{HO-}\overset{\overset{\text{R}}{|}}{\underset{\underset{\text{R}}{|}}{\text{Si}}}\text{-} \longrightarrow -\overset{|}{\text{M}}\text{-O-}\overset{\overset{\text{R}}{|}}{\underset{\underset{\text{R}}{|}}{\text{Si}}}\text{-} \tag{1}$$

$$-\overset{|}{\text{M}}\text{-O-}\overset{|}{\text{Si}}\sim\!\!\!\sqrt{\!\!\!/} + \sqrt{\!\!\!\sim}\overset{|}{\text{Si}}\text{-O-}\overset{|}{\text{M}}\text{-} \longrightarrow -\overset{|}{\text{M}}\text{-O-}\overset{|}{\text{Si}}\sim\!\!\!\sim\!\!\!\sim\!\!\!\sim\overset{|}{\text{Si}}\text{-O-}\overset{|}{\text{M}}\text{-} \tag{2}$$

$$-\overset{|}{\text{M}}\text{-O-}\overset{|}{\text{Si}}\sim\!\!\!\sqrt{\!\!\!/} + /\!\!/ + /\!\!/ + /\!\!\sim\overset{|}{\text{Si}}\text{-O-}\overset{|}{\text{M}}\text{-} \longrightarrow -\overset{|}{\text{M}}\text{-O-}\overset{|}{\text{Si}}\sim\!\!\!\sim\!\!\!\sim\!\!\!\sim\overset{|}{\text{Si}}\text{-O-}\overset{|}{\text{M}}\text{-} \tag{3}$$

$$-\overset{|}{\text{M}}\text{-O-}\overset{|}{\text{Si}}\sim\!\!\!\sim\!\!\!\sim\!\!\!\underset{\text{O}}{\overset{\text{O}}{\triangleleft}}\!\!\!\sim\!\!\!\sim\overset{|}{\text{Si}}\text{-O-}\overset{|}{\text{M}}\text{-} \longrightarrow$$

$$-\overset{|}{\text{M}}\text{-O-}\overset{|}{\text{Si}}\sim\!\!\!\sim\!\!\!\sim\underset{\underset{\text{O}}{\overset{\text{O}}{\diagup}}}{\overset{\text{O}}{\diagdown}}\!\!\!\sim\overset{|}{\text{Si}}\text{-O-}\overset{|}{\text{M}}\text{-} \tag{4}$$

$$-\overset{|}{\text{M}}\text{-OH} + \text{HO-}\overset{|}{\text{Si}}\text{-OH} + \text{HO-}\overset{|}{\text{M}}\text{-OH} + \text{OH-}\overset{|}{\text{Si}}\text{-OH} + /\!\!/ + /\!\!/ + /\!\!/ + /\!\!/ \text{---}$$

$$\tag{5}$$

R = organofunctional group, e.g. amino, carboxy,
M = network forming metal, e.g. Ti, Al, Zr

Eq. (1) represents the introduction of network modifying units. Eqs. (2-4) show examples for building up an organic polymeric network in addition to the sol-gel derived inorganic one. In eq. (5) the formation of two independent interpenetrating networks is indicated.

3. CHEMISTRY OF SYNTHESIS AND TAILORING OF SPECIAL MATERIAL PROPERTIES

3.1. General considerations

Synthesizing inorganic polymers by the sol-gel route requires soluble reactive monomers. An easy way to do this, is to use soluble alkoxides and carry out a hydrolysis and condensation reaction. As known from investigations from different authors, the reactions and the polymeric structure formation processes are very sensitive to reaction condition, composition and starting monomers [11-14]. The knowledge of detailed mechanisms of hydrolysis and condensation is very poor, partially due to the complexity of these processes, partially due to the lack of interest based on an underestimation of their importance. In spite of this, basic features for the reaction of simple systems, e.g. Si-, Al- and Ti-alkoxides, could be developed and the influence of reaction condition on the derived materials could be evaluated. The introduction of organic components into the sol-gel process in general means a drastic change of the system: Firstly substituted esters $R_nSi(OR)_4$ behave quite different than other esters and other alkoxides in hydrolysis and condensation as shown in [15] and secondly the organic groups or the addition of organic monomers can influence the structure of the inorganic backbone. That means, that one has to expect a complex set of influencing parameters caused by organics, which presumably will affect more then a "simple" additional behavior. In the following, examples will be given to illustrate how special properties can be achieved or manipulated in ORMOSILs.

3.2. Influence of synthesis on material properties

As pointed out above the step of formation of polymeric structures plays a key role for the properties of the derived material. Two main different processes have to be distinguished. The first one can be described as a "monomer connection" step. With three dimensional crosslinking monomers a large variety of structures and, based on them, of properties is possible. Thus, even in one component systems very different materials can be prepared as shown by the variety of silica gels. In multicomponent systems a second step, which can be described as a "component distribution" process, in addition has to be considered. The latter one is related to different reactivities in hydrolysis and condensation. Eq. (6-8) illustrate the influence of the reaction kinetics on homogenity of a two components system (A, B: starting alkoxides; A', B': hydrolysed monomers; A", B": condensates; k_i: rate constants)

$$mA \xrightarrow{k_1} mA'; \quad nB \xrightarrow{k_2} nB'$$

$$mA' \xrightarrow{k_1'} A_m''; \quad nB' \xrightarrow{k_2'} B_n'' \quad (k_1', k_2' \gg k_3) \tag{6}$$

$$mA' + nB' \xrightarrow{k_3} (m+n)(A''\cdot B'') \quad (k_3 \gg k_1', k_2') \tag{7}$$

$$nA_m'' + vB_n'' \xrightarrow{k_4} (u+v)(A_m''\cdot B_n'') \quad \begin{array}{l} I \equiv \text{homogeneous} \\ II \equiv \text{inhomogeneous} \end{array} \tag{8}$$

It is clear, that the component distribution influences properties drastically. If homogeneous materials are required, techniques have to be developed to overcome the effect of different reactivities.

Therefore the chemically controlled condensation ("CCC") method was developed which allows a precise control of hydrolysis and condensation rate, by chemical water generation within the system. The water producing reaction is an ester formation, which uses the solvent alcohol as one and the catalytic effective acid as the other reaction partner (eq. 9). By pro-

$$ROH + RCOOH \xrightarrow{H^+} RCOOR + H_2O; \quad \equiv MOR + H_2O \xrightarrow{H^+} MOH + HOR \tag{9}$$

per choice of the type of alcohol, concentration of acid and temperature, very different water formation rates can be established. Another important advantage of the method is the homogeneous water formation within the reac-

tion mixture, avoiding concentration gradients which necessarily occur if
pure water or water containing phases are mixed with water free ones. It is
supposed that in a mixture of two alkoxides with very different hydrolysis
rates, like $Ti(OEt)_4$ and $Si(OEt)_4$, the CCC-method leads preferably to the
partial hydrolysis of $Ti(OEt)_4$. After only one sixteenth to one tenth of
the CCC-formation of water necessary for hydrolysis of the total amount of
OR-groups, pure water can be added to the system without causing precipitation of TiO_2. One has to conclude, that the $(TiO_4)^{4-}$ units are immobilized
in an polymeric structure, probably including $(SiO_4)^{4-}$ units, since pure
$(TiO_4)^{4-}$ based oligo- or polymeric units should not be able to prevent completely TiO_2 precipitation with excess water. Thus, it is possible to prepare easily water based laquers from systems of the type $R'Si(OR)_3/M(OR)_n$
(with n = 3, 4, M e.g. = Al, Zr, Ti and R' e.g. = epoxy) by use of the CCC-
method (10).

$$Ti(OR)_4 + \text{epoxy-}Si(OR)_3 \xrightarrow[H^+]{CCC} \left[Ti-O-Si-\text{epoxy}\right]_{\text{oligomer}}$$

$$\text{oligomer} \xrightarrow[\text{water}]{\text{excess}} \left[Ti-O-Si-\text{epoxy}\right]_{\text{polymer}} \tag{10}$$

The polymer is a viscous liquid which reacts within some days to a monolithic transparent solid. The viscous liquid can be used in coating techniques and cured by heat within minutes to scratch resistant films. IR-spec-
troscopy shows, that the curing step from the viscous to the solid state is
connected with loss of OH groups, which indicates a condensation based curing. After a heat treatment for serval hours at 130 °C, there is still a remarkable amount of OH groups left. This phenomena was investigated more in
detail in the system 65 $MeViSiO/32.5 (C_6H_5)_2SiO/2.5 SiO_2$ (Me = methyl, Vi =
vinyl). The material is prepared by reacting diphenylsilandiol, $Si(OEt)_4$
and $MeViSi(OEt)_2$ in a mixture of toluene and ethanol and hydrolysing the
whole mixture with excess water under H^+ catalysis by refluxing for several
hours (11). After solvent removal a highly viscous product remains, which is

$$-\underset{C_6H_5}{\overset{C_6H_5}{Si}}OR + HO-\underset{Vi}{\overset{Me}{Si}}-OH + RO-\underset{Vi}{\overset{Me}{Si}}-OR \xrightarrow[H^+]{H_2O} -\underset{C_6H_5}{\overset{C_6H_5}{Si}}-O-\underset{Vi}{\overset{Me}{Si}}-O-Si- + 2\ HOR \tag{11}$$

cured thermally. Fig. 1 shows the IR-spectra after different temperature
treatments. It is shown, that at least two different types of OH groups appear, which can be assigned to bridged (lower frequency) and unbridged (higher frequency) species. The bridged species can condense more easily during
the curing procedure than the unbridged ones. At T > 100°C, the samples are
liquid and remain lowly viscous during the whole heat treatment. In spite of
this low viscosity, the condensation process is very slow. After cooling
down, the products show a thermoplastic behavior. The presence of diphenyl
silane seems to avoid a higher degree of three dimensional crosslinking by
steric hindrance. If KOH or NaOH is added to the mixtures, even very low concentrations (0.01 mmole NaOH/l) affect a rapid condensation. In fig. 2, the
spectra before and after the base addition demonstrate the effect. As known
from other investigations, too, bases are more efficient condensation catalysts than acids. Similar tendencies are observed within the system epoxysi-
lane/TiO_2 from spectroscopic data too. Condensation experiments with different catalysts and IR analysis indicate a remarkable difference between HCl,
NH_3 and KOH. The KOH catalysis shows the far lowest, HCl catalysis the highest OH-group content. All three type of reactions leads to viscous liquids

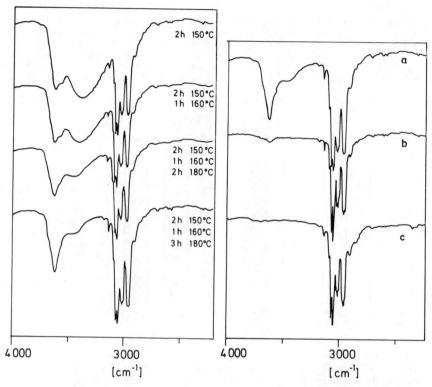

Fig. 1. Effect of thermal curing on ORMOSILs of the system MeViSiO/Ph$_2$SiO/SiO$_2$

Fig. 2. Effect of catlyst on ORMOSILs of the system MeViSiO/Ph$_2$SiO/SiO$_2$
a) thermally cured resin prepared with HCl catalysis
b) resin heated up with 0.4% NaOH to 150°C for 30 min
c) thermally cured resin prepared with KOH catlysis

which can be cured to monoliths. Underwater storage of the monoliths reveals quite different behavior depending on catalyst type, too: Whereas the HCl and NH$_3$ catalysed species show a swelling of about 5 to 10 % with crack formation after redrying, the KOH catalyzed species show no detectable swelling and are extremely stable against any water treatment. The latter behavior is assigned to a low water take up due to low OH group content. This results turned out to be extremely important for all type of coatings to be used under wet conditions. Furthermore, in these experiments it could be proved that dense monoliths can be prepared at low temperatures due to the higher network flexibility caused by organic ligands.

For a sufficient understanding of inorganic-organic polymers, not only the inorganic process, but also the organic network forming should be known quite well. It has been shown elsewhere that epoxy groups are responsible for interesting mechanical properties like surface hardness [16]. The effect of different alkoxides on the epoxide polymerization has been described in [17]. It turned out that Zr(OR)$_4$ was most effective in epoxide polymerization. Further investigations were carried out to optimize reaction conditions. It could be shown, that even $Zr(OR)_4$ contents of about 1 mole-% affect rapid polymerization. Increasing $Zr(OR)_4$ contents increase the polymerization rate, whereas the HCl concentration does not affect the rate.

In this connection the question of the influence of acids with respect to the ring opening reaction (12) is of high interest and was investigated with different acids. Fig. 3 shows the results. $HClO_4$, H_2SO_4 and partially

$$\equiv Si \sim\sim\sim \overset{\triangle}{\underset{O}{}} + H_2O \longrightarrow \equiv Si \sim\sim\sim CHOHCH_2OH \qquad (12)$$

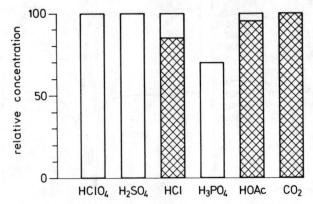

Fig. 3. Effect of catalysts on the relative concentration (based on starting material) of glycol and epoxy (hatched) groups in completely hydrolized epoxysilane

H_3PO_4 causes a complete ring opening under the applied experimental conditions ($25°C$ for 17 hours, 1 mole H^+/l) whereas acetic acid and CO_2 hardly attack the epoxy group. All the applied acids act as good catalysts in hydrolysis and condensation. Especially CO_2 is very convenient to promote the condensation process without ring opening. This process is important, if the epoxy ring has to be preserved during the inorganic backbone forming reaction for a subsequent epoxy polymerization or for preventing an internal reesterification reaction (13).

$$\equiv Si \sim\sim\sim CHOH-CH_2OH + RO-Si\equiv \longrightarrow \equiv Si \sim\sim\sim CHOH-CH_2-O-Si\equiv + HOR \qquad (13)$$

As mentioned above dense materials can be prepared within the system epoxysilane/TiO_2 with the CCC-method up to at least 40 mole-% of TiO_2. Substitution of TiO_2 by 25-30 mole-% SiO_2 leads to brittle and porous species. This may be attributed to the inorganic network densifying effect of TiO_2, known from the silicon chemistry [18]. In analogy to TiO_2, ZrO_2 can be incorporated into the network (14):

$$\equiv Zr-OR + RO-Si\sim\sim\sim R \longrightarrow \equiv Zr-O-Si\sim\sim\sim R \qquad (14)$$

Thereby, in case of epoxy groups the polyethylene oxide formation is favored even at room temperature due to the excellent catalytic activity of $Zr(OR)_4$ [17]. Dense monoliths can be prepared easily with this system, too. The absolute densities of these products are surprisingly low, even at higher metal oxide contents, indicating "open" structures compared to the "inorganic" corresponding systems TiO_2/SiO_2 and ZrO_2/SiO_2. In fig. 4 the

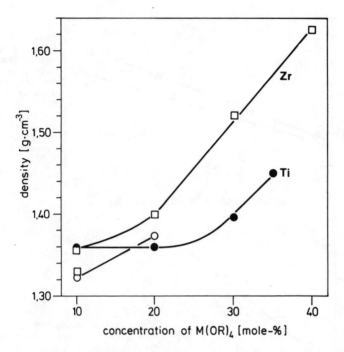

Fig. 4. Influence of M(OR)$_4$ concentration on the density of thermally cured
ORMOSIL condensates (epoxysilane/M(OR)$_4$); squares: Zr; circles: Ti;
full symbols: one step hydrolysis of the whole mixture; open symbols: two step hydrolysis, reacting the epoxysilane before M(OR)$_4$

dependence of density on the metal oxide content is shown. One step reaction TiO$_2$ containing species with higher Ti contents exhibit low densities
compared to ZrO$_2$ containing materials. At lower contents of M the opposite
behavior is shown. Up to 20 mole-% there is almost no change in density in
the TiO$_2$ case. Similar tendencies are shown by the ZrO$_2$ system. There is a
big gap between the actual and the theoretical density, computed from the
increments (polyethylenoxide: $\rho \approx 1.2$, SiO$_2$ glass: $\rho \approx 2.2$ and crystalline
ZrO$_2$: $\rho \approx 6$. The resulting density with 40 mole-% ZrO$_2$ would be about 3,
the measured one is about 1.7. This demonstrates that the ORMOSIL structure
does not follow an additional behavior based on the single components and
that this is an influence of the reaction conditions. The more open structure of the ORMOSIL may result from the organic residue, but no plausible
models could be developed up to now.

The refractive index shown in fig. 5 increases with increasing metal-
oxide content, but seems to be surprisingly low with respect to the corresponding inorganic system. Considering the low densities, the measured
values are in a good agreement with Lorentz-Lorenz calculations. Compared
to this, phenyl group containing ORMOSILs show higher refractive indices,
due to the contribution of the phenyl group (fig. 6). The addition of epoxy
groups decreases the refractive index remarkably in this system.

Epoxy ORMOSILs have been developed as scratch resistant coating and
contact lens materials. It could be shown, that the scratch resistance in
the TiO$_2$/epoxy system mainly depends on curing and preparation conditions
[16, 17]. Since the TiO$_2$ systems shows some general disadvantages with

746

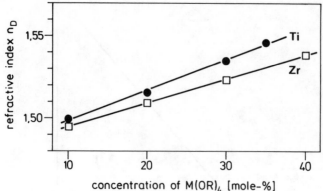

Fig. 5.
Relation between refractive index n_D and concentration of $M(OR)_4$ in thermally cured ORMOSIL condensates (epoxysilane/ $M(OR)_4$)

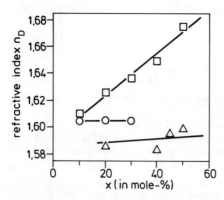

Fig. 6.
The effect of concentration of $M(OR)_4$ (M = Ti, Zr) in different ORMOSIL systems

□: $(100-x)$ $Ph_2Si(OH)_2/x$ $Ti(OEt)_4$
o: $(100-x)$ $Ph_2Si(OH)_2/x$ $Zr(OPr)_4$
△: $(80-x)$ $Ph_2Si(OH)_2/x$ $Zr(OPr)_4/$ 20 epoxysilane

respect to UV stability (if not protected by UV absorbers), the effect of other metal oxides with respect to their scratch resistance was studied. The scratch resistance is measured by a modified Erichsen test, where a Vickers diamond is scratched over the surface and where the load causing the first microscopically visible scratch is determined. Table I shows the comparison of different compositions.

Table I. Scratch behavior of the compositions (mole-%) 50 epoxysilane/ 30 SiO_2/20 MO_x (curing temperature 90 °C)

M	diamond load
Si	< 1
Zr	5
Al	10-20
Ti	10

Standard UV tests with the unprotected systems show, that the Al_2O_3 and ZrO_2 systems are about 10 times more resistant than the TiO_2 system. The poor results from the TiO_2 system can be attributed to its photocatalytic activity [17]: UV light irradiation leads to Ti^{3+} formation which could be proved by spectroscopy.

With SiO_2 only instead of TiO_2 it is very difficult to receive compact materials. The samples are very brittle and are cracking during curing, that means they show a very poor mechanical stability. TiO_2 containing monoliths show a reasonable mechanical stability. The effect of composition was studied on this system. The results are shown in fig. 7. There seems to be a maximum at about 5 mole-% TiO_2, due to the network densifying effect of TiO_2. Higher contents lead to a higher modulus of elasticity with a corresponding increase of brittleness and decreasing tensile strength. The system approaches to a "glass like" behavior. The incorporation of a polymethylmethacrylate (PMMA) network improves tensile strength remarkable (15).

$$\equiv Ti-OR + RO-Si-OR + 2ROSi\sim\sim OOC-\underset{\underset{CH_3}{|}}{C}=CH_2 + nCH_2=\underset{\underset{CH_3}{|}}{C}-COOR \longrightarrow$$

$$-Ti-O-Si-\text{polymethylmethacrylate chain} - Si - \tag{15}$$

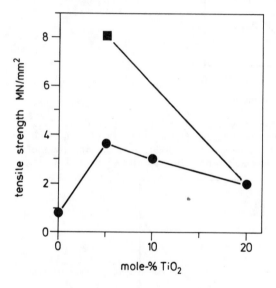

Fig. 7.
Effect of composition on tensile strength:
■ epoxysilane/TiO_2;
● 20 mole-% methylmethacrylat and 5 mole-% of methacryloxysilane added

The results show that the polymeric network can be used to improve mechanical properties of ORMOSILs. It should be mentioned that due to the possibility of a radical polymerization mechanism, photo catalysis can be used and photo curing procedures can be applied.

Beside mechanical properties, chemical surface properties of coatings are of interest, if these properties can be used for special reactions, e.g. for sensor purposes: Systematic investigations have been carried out in order to synthesize chemical reactive coatings for the interaction with gaseous components. The idea was, to transduce the change in electronic state of a reacting surface molecule directly to a microelectronic device, e.g. a field effect transistor (FET). As model systems, the adsorption of CO_2 and SO_2 were chosen. Fig. 8 shows the CO_2 load of porous ORMOSILs ($SiO_2/NH_2(CH)_2)_3$, $SiO_{3/2}$) as a function of composition and BET surface. It is possible to receive remarkable loads. In spite of this no remarkable change of electric properties of the loaded coatings takes place, so that no detectable signals e.g. in capacitance tests can be monitored.

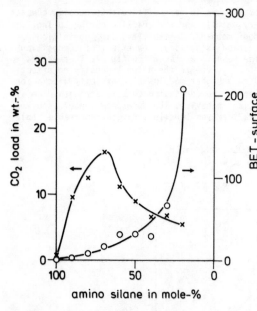

Fig. 8.
CO_2 adsorption at
$40^{\circ}C/10^3$ hPa CO_2
pressure and BET sur-
face depending on
composition of the
system SiO_2/amino-
silane

In opposition to this coatings from the system $SiO_2/(C_2H_5)_2N(CH_2)_3SiO_{3/2}$ exhibit reversible change of capacitance with SO_2 [19]. First results on FETs confirm the capacitance experiments. In order to reduce influence of water vapor, hydrophobic components can be incorporated. Fig. 9 shows the

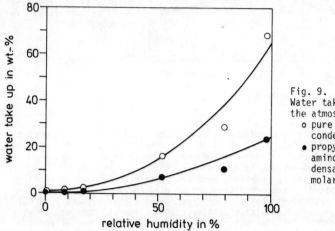

Fig. 9.
Water take up from
the atmosphere.
o pure aminosilane
   condensate
● propylsilane/
   aminosilane con-
   densate (1:1
   molar)

effect of H_2O adsorption on two different systems. It clearly points out the reduction in H_2O adsorption by the introduction of the propyl group. These results open the possiblity of synthesizing a large variety of sen- sitve coatings for very different purposes.

Phenyl group containing adsorbents can play an important role for solvent stripping from air. From capacity reasons high specific surface areas
are required. In former experiments systems from SiO_2 and $(C_6H_5)_2SiO$ have
been synthesized. It was found, that even small contents of the diphenylsilane reduce the surface area drastically, if "normal" liquid reaction conditions (refluxing, ethanol as solvent, aqueous HCl addition) are applied.
Bubbling HCl gas through the reaction mixture improves the specific surface
area even of diphenylsilane contents up to 20 mole-% (fig. 10). Up to now
it is not quite clear, why a minimum at pH 4 appears, but here the maximum
gelation time is observed. Since short condensation times shorten the relaxation period for the system, "open" structures should be more preferably
obtained in this case.

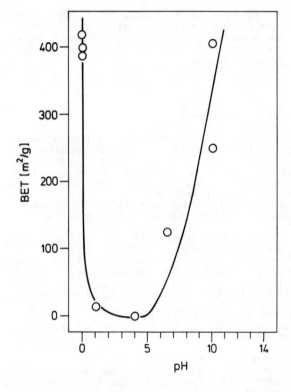

Fig. 10.
BET surface of a
SiO_2: $(C_6H_5)_2SiO$ =
80:20 (molar ratio)
condensate depending
on pH during hydro-
lysis

4. CONCLUSIONS

The investigations on ORMOSILs have shown, that the sol-gel process
can be advantageously used for the preparation of inorganic-organic polymers. Furthermore one can see, that the use of the rules of chemistry allow
the tailoring of special material properties. Up to now, different groups
of materials with common properties have been developed. There is the
scratch resistent group which is mainly based on the epoxysilane with inorganic network formers. This group can be used for the preparation of coatings and bulk materials as well. The thermoplastic group is based on di-

750

phenylsilane contents of more than 40 mole-%. The photocurable group is based on photocurable ligends like methacryl, vinyl or allyl groups and can easily be combined with a big variety of polymerizable monomers. The thermoplastic and photocurable group can be used as coatings, adhesive films and bulk materials. The porous group is characterized by a high content of inorganic network formers and mainly monofunctional organosilanes and can be used as porous coatings, carriers for catalysts and membranes. A variation of this type are functional coatings with different surface groups to be used in the sensor field. Based on the wide modification, applications seem to be possible in very different branches, which opens a high innovative potential of these materials.

5. ACKNOWLEDGEMENTS

The authors thank the Bundesminister für Forschung und Technologie of the Bundesrepublik Deutschland and many industrial plants for the financial support of the work. They appreciate gratefully the help of Dr. A. Kaiser, Dr. F. Hutter and Dr. K. H. Haas for the experimental work and Prof. Dr. H. Scholze for his helpful discussions.

6. REFERENCES

1. R.Roy, J.Amer.Ceram.Soc. 52 (1969) 344.
2. H.Dislich, Angew.Chem. 83 (1971) 428.
3. J.Zarzycki, J.Non-Cryst.Solids 63 (1984) 105.
4. S.Sakka in: MRS Symp. Proc., Vol. 32, Better Ceramics Through Chemistry, C.J.Brinker ed. (North-Holland, New York, Amsterdam, Oxford, 1984). p. 91.
5. J.Wenzel in: Glass ... Current Issues, A.F.Wright and J.Dupuy, ed. (Martinus Nijhoff Publishers, Dordrecht, Boston, Lancaster, 1985), p. 224.
6. H.Scholze, J.Non-Cryst.Solids 73 (1985) 669.
7. H.Schmidt, G.Philipp, H.Patzelt and H.Scholze, Hot melt adhesives for glass containers by the sol-gel process, to be published in J.Non-Cryst.Solids.
8. H.C.Gulledge, US.Pat. 2,512,058, 20. Jun. 1950.
9. K.A.Andrianov and A.A.Zhdanov, J.Polym.Sci. 32 (1958) 513.
10. H.Schmidt in: MRS Symp. Proc., Vol. 32, Better Ceramics Through Chemistry, C.J.Brinker ed. (North-Holland, New York, Amsterdam, Oxford, 1984), p. 327.
11. C.J.Brinker, K.D.Keefer, D.W.Schaefer and C.S.Ashley, J.Non-Cryst. Solids 48 (1982) 47.
12. B.E.Yoldas, J.Mater.Sci 12 (1977) 1203.
13. H.Schmidt and H.Scholze in: Glass ... Current Issues, A.F.Wright and J.Dupuy, ed., (Martinus Nijhoff Publishers, Dordrecht, Boston, Lancaster 1984), p. 263.
14. S.Sakka, J.Non-Cryst.Solids 73 (1985) 651.
15. H.Schmidt, A.Kaiser, M.Rudolph and A.Lentz, Contribution to the kinetics of glass formation from solution II, to be published in Proceedings of Ultrastructure Processing of Ceramics, Glasses and Composites, Florida 1985.
16. G.Philipp and H.Schmidt, J.Non-Cryst.Solids 63 (1984) 283.
17. G.Philipp and H.Schmidt, The reactivity of TiO_2 and ZrO_2 in organically modified silicates, to be published in J.Non-Cryst.Solids.
18. W.Noll, Chemie und Technologie der Silicone, 2nd ed. (Verlag Chemie, Weinheim 1968).
19. F.Hutter, K.H.Haas and H.Schmidt, ORMOSILs - a new class of materials for sensitive layers in the development of gas sensors, to published in Proceedings of the 2nd International Meeting on Chemical Sensors, Bordeaux 1986.

INNOVATIVE CHEMICAL/CERAMIC DIRECTIONS

P.E.D. Morgan
Rockwell International Science Center, Thousand Oaks, CA   91360

ABSTRACT

I have attempted herein to show that relatively simple chemistry is available for much needed application to ceramic technology. The literature, as exemplified in the second section, already contains much more usable information than generally supposed. We need, too, a revival in simple inorganic chemistry.

Increasing Use of Sol-Gel

As many workers are now interested in sol-gel, a count of publications (Fig. 1), primarily with this main topic, will illustrate the growing awareness of the importance of this particular chemical method in ceramics. The number of patents issued, which has been small for many years, is also increasing, perhaps indicating more serious commercial interest. The refocusing of efforts on the uniquely fine microstructures that can be achieved, which have been implicit for many years in work on catalysts, and their morphological control by seeding can only increase the interest.

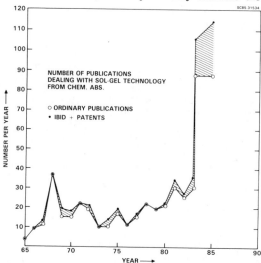

Fig. 1   Number of publications whose main topic is sol-gel (not including publications with gel as a more minor concept).

Earlier "Nanostructural" Work

Catalyst chemists have long used Scherrer line-broadening to assess the size of pore-enclosed particles and were fully aware that they could reach the 1-10 nm regime. Later zeolites, with known, controllable pore sizes, were impregnated and used. Meanwhile, cluster chemists synthesized clusters at the 0.5-10 nm level with known geometries and controllable chemistry.

Some aspects of nanostructures, now being so widely talked of, are really not new, let me quote chapter and verse. For many years, catalyst chemists (among others) have understood the need for extremely finely divided particles in multiphase systems. An area of earlier great interest was the Fischer-Tropsch synthesis (the production of hydrocarbons from carbon monoxide and oxygen), "carried out using a catalyst such as a hydrogenating metal of the eighth group of the periodic system in combination with a promoter, usually a difficultly reducible metal oxide in promotional amounts, and this mixture of metal and metal oxide is supported on a suitable carrier such as Kieselguhr." The patent [1] whence this is quoted issued in 1950, but was applied for in 1944, then goes on to describe a new improved method of making such a triphasic material:

"A silica hydrogel, which is sometimes referred to as a hydrous oxide of silicon in gelatinous or jelly form, is prepared by the reaction of an aqueous sodium silicate solution with sulfuric, hydrochloric or nitric acid. The resultant mixture is allowed to set to the gelatinous state. It is then washed to remove the water-soluble salts and is impregnated by soaking thereinto a mixed solution of the water-soluble salts of the hydrogenation type metals of the eighth group of the periodic system and the water-soluble salts of difficultly reducible metal oxides. Thus, for example, the iron, cobalt or nickel nitrates, the iron, cobalt or nickel acetates or chlorides, etc., and the thorium or magnesium, uranium, manganese, aluminum nitrates, acetates, chlorides, etc., are mixed in aqueous solution and taken up in the hydrous oxide of silicon in jelly form. The excess solutions are then drained therefrom and the precipitate or hydrogel is dried and heated at the lowest temperature necessary to effect decomposition of the metal salts. The material may then be ground to the desired particle size, that is of the approximate order of between 20 and 200 μm in diameter, and reduced by passing a stream of hydrogen thereover under temperatures between about 600° and about 800°F, using space velocities of between about 4000 and about 6000 v/v/h."

The need to carefully control the firing to achieve high activity by controlling particle size, taken as read by catalyst chemists is covered:

"Care in the conversion of the nitrates to the corresponding oxides is desirable, and in fact necessary if high activity of the catalyst is to be obtained. Generally speaking, the catalyst should not be heated during the conversion to a temperature much above 475°F. In general, a temperature between about 400°F and about 450°F is most suitable. The novel catalyst was found to be much more active than was the case when merely impregnating the dry silica gel with aqueous solutions of the nitrates followed by their conversion to oxides at temperatures of between about 800°F and about 1000°F and subsequent reduction. *The impregnation of the hydrous oxide of silicon in jelly form produced a highly active catalyst of better quality than the catalyst prepared directly from dry silica gel as the initial carrier.*" (Italics mine)

The use of mixed diphasic alumina-silicas is also covered:

"Other methods of impregnating and forming the close association between the catalytic components of the catalyst and the silica gel carrier may be employed. Thus, for example, instead of using the hydrous oxide of silicon in jelly form, a mixture of hydrous oxides of silicon and aluminum in jelly form may be employed in one of two forms: either the jelly form of the hydrous oxides of silicon and aluminum may be separately prepared and admixed; or they may be cosetting gels to give the plural gels of hydrous silica and hydrous alumina. A further combination of the silica-alumina hydrous oxides in jelly form is prepared by first preparing the jelly form of hydrous oxide of silicon, washing the same and then impregnating or soaking it in a solution of a suitable aluminum salt, for example, an aluminum nitrate or sulfate or chloride, followed by the precipitation of the aluminum with ammonia solution and washing the same free of resulting salts."

Many more patents teach the production of nanostructural particles, including the recent work of Boudart [2,3] on nm particles of iron, gold, platinum, etc. In their turn, catalyst chemists more recent use of TEM is based on much earlier work where decompositions were used to achieve ultra-fine particles, rarely are the seminal early works cited.

The Polymer Alternative to Sol-Gel

An area that is, I believe, innovative (and now I must be very careful to quote precedents) was suggested by earlier work [4] in which mixed alkoxides were refluxed in solvents to produce an undistillable oil, which on air hydrolysis turns to a gel for subsequent conversion to a glass. We inferred that neat alkoxides might polymerize (pyrolyze) under reflux in argon to produce oxy-alkoxides as oils, resins, gums or glasses, and so it was [5]. It transpired that Bradley (father of alkoxide chemistry) had briefly looked at the pyrolysis of alkoxides [6]; he had used tertiary-alkoxides and concluded that intermediate oxy-alkoxides were unstable. The instability in his case probably is a more direct result of the metal he was using (zirconium) than of the fact that he was using the t-case, which is additionally, however, destabilizing. We find [5] with primary and secondary alkoxides of Al, Ti and Zr that the intermediates, which are typically about 1/2:1/2 oxide:alkoxide, i.e., $AlO(O-sC_4H_9)$ are stable and isolatable. These new polymers, which can contain > 40 w/o oxide, appear to have utility for making films, fibers and as extrusion aids, etc.

Just as has been seen with the crystallization of gels, spray-dry-roasted-solution material and other glassy precursors, so also the thermo-lyzed polymer glasses showed interesting coupled grain-growth-inhibition effects [7]. Thus, a copolymerized oxy-alkoxide of Al + Zr, designed to produce $Al_2O_3/10$ v/o ZrO_2, is air-oxidized at ~ 500°C and then hot-pressed at ~ 1700°C to give the backscatter microstructure in Fig. 2. Large α-alumina grains > 10 μm appear to have grown via γ-alumina and included ultra-fine ZrO_2 within them. The ZrO_2, identified by EDS, remains tetragonal to room temperature even though "unstabilized".

Earlier work used only commercially available simple alkoxides such as $Al(O-sBu)_3$. Another simple class of "alkoxides" are the acetylacetonates (acac) which have much greater stability due to the chelating nature of the acac grouping. Indeed, heating $Al-(C_6H_9O_3)_3$, aluminum triethylacetoacetic ester, under reflux up to ~ 300°C produced only minor amounts of acetone as a pyrolysis product with very little degradation of the $Al(acac)_3$, although the product was reddish in color. However, when $Al(acac)(OC_3H_7)_2$ was heated under reflux from 200-300°C, all the propyl groups could be thermo-lyzed off as propanol and propene, leaving a thick oil at 300°C which cools to a transparent orange/red glass at room temperature. This analyzes as $AlO_{1.16}(C_6H_9O_3)_{0.69}$ containing 32 w/o of Al_2O_3; the color probably indicates an interesting slight change in the chelating acac group to a more aromatic nature. This is the first report of oxy-acac polymers.

754

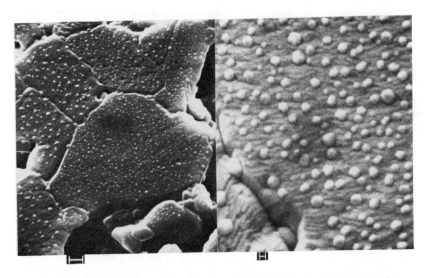

Fig. 2 Al₂O₃/10 v/o ZrO₂: ultra-fine crystals of ZrO₂ in alumina matrix.
Left marker 1 μm , right marker 0.1 μm .

Alkoxides can also be reacted with other simple reagents, possibly of industrial utility. For example, if $SiCl_4$ under argon is dripped into $Al(OBu)_3$ at room temperature, an exothermic reaction ensues. With slow reflux heating, acid gases are evolved; by ~ 130°C, the reaction tends to quicken with faster evolution and frothing. If done slowly, the reaction is manageable, the frothing subsides or does not occur. The thick gummy/glassy product produces white mullite when heated in air to 1000°C (if the initial Si/Al ratio is appropriate). A large, potentially useful, but not much explored, organic/inorganic chemistry exists which can be used for ceramic synthesis.

Other Chemical Mixing

Although the use of gels (or spray-dry-roasting of polymer methods, etc.) may give nearly optimal mixing for monophasic or polyphasic ceramics, with much less effort it is still feasible to achieve mixing so much better than with mixed oxides that excellent results are achieved for studies of phase diagrams and for synthesis of complex crystals, etc. These methods generally follow the infiltration of ions in solution into the base AlO(OH) or similar material, in the manner so familiar to catalyst chemists, followed by controlled firing schedules. This author is particularly interested in alumina systems and has used Disperal®, bohemite, to produce pure $CaTi_3Al_8O_{19}$ (and other members of its new family) [8], $CdAl_2Ti_2O_8$ [9], two types of $Na_{1/2}La_{1/2}Al_{12}O_{19}$,[10] β''' and β'''' aluminas that are phase pure [11], and has shown that probably $CaAl_{12}O_{19}$ and "$NaAl_{11}O_{17}$" have only a very limited solid solution range [12]. Many magnetoplumbites [12], some of which have laser, and other interesting new optical promise, have been easily synthesized by this technique (Table I). The method to produce pure β and β'' alumina powders [11] has been applied [13] to produce phase pure $β''-K_2MgAl_{10}O_{17}$ with the use in all these cases of flash heating or very

Table I

X-Ray Diffraction Analysis of Magnetoplumbite-Related Structures

Compounds	a	c	c/a	Comments
Case 1 Types				
$NdMgAl_{11}O_{19}$	5.585	21.880	3.92	
$NdFeAl_{11}O_{19}$	5.593	22.059	3.96	
$NdZnAl_{11}O_{19}$	5.583	21.882	3.92	All stabilized by divalent ion; form by 1200°C
$NdNiAl_{11}O_{19}$	5.580	21.916	3.93	
$SmMgAl_{11}O_{19}$	5.582	21.851	3.92	
$GdNiAl_{11}O_{19}$	5.577	21.853	3.92	Forms at 1600°C only; at 1200°C, $NiAl_2O_4$ + $Gd_3Al_5O_8$ garnet/perovskite type
"$YMgAl_{11}O_{19}$", "$InMgAl_{11}O_{19}$"				Do not form
Case 2 Types				
$NdFeZn_2Al_8TiO_{19}$	5.624	22.092	3.93	2 Zn^{2+} balanced by Ti^{4+}
$Nd_{1/2}Na_{1/2}Al_{12}O_{19}$	5.563	21.910	3.94	At 1500°C
$Sm_{1/2}Na_{1/2}Al_{12}O_{19}$	5.561	22.136	3.98	Slow firing up to 1200°C. At 1600°C has superlattice
"$Bi_{1/2}Na_{1/2}Al_{12}O_{19}$","$Y_{1/2}Na_{1/2}Al_{12}O_{19}$"				Do not form
$Nd_{1/2}Cs_{1/2}Al_{12}O_{19}$	Poorly crystalline			Forms at < 1200°C No superlattice
$Na_{1/2}Ba_{1/2}Al_{11-1/2}Ti_{1/2}O_{19}$	Poorly crystalline			Forms at 1100°C decomposes at higher temperatures
"$Ba_{1/2}Ca_{1/2}Al_{12}O_{19}$	~ 5.58	~ 22.7	~ 4.07	Poorly crystalline β-type-Phase I like
$La_{3/4}Li_{1/2}Al_{11-3/4}O_{19}$	5.566	21.961	3.95	
Miscellaneous Other Types				
$SrFeAl_2Cr_9O_{19}$	5.769	22.484	3.90	All octahedral sites can be filled with Cr^{3+}?
"$SrAl_3Cr_9O_{19}$","$NdMgAl_2Cr_9O_{19}$"				Do not form under conditions used.
$BaMg_6Ti_6O_{19}$	5.937	23.379	3.94	Normal MP
$PbAl_{12}O_{19}$	5.573	22.041	3.96	Normal MP
"$SrMg_6Ti_6O_{19}$","$PbMg_6Ti_6O_{19}$"				Do not form
"$BaMg_6Zr_6O_{19}$","$BaSc_{12}O_{19}$"				Do not form
"$BaLi_6Nb_6O_{19}$","$BaLi_4Ti_8O_{19}$"				Do not form
"$BaTi_3Mg_3Al_6O_{19}$"				Complex mixtures, no MP or β
$Cs_{1+x}Fe_{11}O_{17+x/2}$	5.923	24.158	4.08	β type

slow heating rates [14]. In spite of this, ceramists and others persist
with atavistic methods that do not easily achieve equilibrium and therefore
publish sometimes doubtful results for phase diagrams. It becomes clear,
using chemical methods, that many more compounds exist, especially at low
temperature, than have been discovered with ceramic methods. The point
seems often to be missed also that dense ceramics, e.g., grinding grain,
are being made directly from hydroxides, rather than oxides, the term colloidal alumina for colloidal boehmite or suchlike, is extremely unfortu-
nate, misleading and should cease. The seeming predjudice against materials such as hydroxides has definitely deterred progress, although they do
have their own particular problems.

Fused Salt Powder Synthesis for Chalcogenides

We have an interest in chalcogenides for IR windows in the 8-14 μm
region, one of the two atmospheric IR windows. This requires weak bonds
between heavy atoms (e.g., oxides are no good, chalcogenides potentially
suitable). Strength and thermal shock resistance are also required, and
for this the best situation is strong bonds between light atoms.

Two types of compromise have been attempted. In one, low coordina-
tions with relatively strong covalent bonding such as for ZnS and $ZnSiP_2$
[15] do not quite give the desired long wavelength transmittance, but do
have better thermal shock resistance, while compounds such as ZnSe have
good transmittance, but are too soft. The other, typified by the prescient
choice of $CaLa_2S_4$ [16-19], is based on the use of many weaker ionic bonds
with high coordinate cations and anions. In the latter case, thermal shock
resistance has been unsatisfactory (at least with the large grain size
material), although the long wavelength properties are extremely promising,
while still unoptimized.

For suitable ceramic fabrication, we recognize the need to produce
fine-grained ceramics with final grain size of less than ~ 5 μm, whose
toughness and thermal shock resistance should be superior without degrading
the transmittance. To achieve this, we need < 1 μm uniform particles of
the material well compacted into a dense green body, and we probably must
hot-isopress (HIP) at temperatures preferably < ~ 1000°C.

With this in mind, we used $CaLa_2S_4$ as a prototype to test fused salt
powder techniques [20-23] (Ref. 20 is an earlier example of this technique
than is usually cited) to attain useful results quickly.

A series of runs was initiated to determine conditions for precipitation and recrystallizing $CaLa_2S_4$ from a Na-Ca-La-S-Cl eutectic melt. Na_2S
itself melts at 1175°C and NaCl at 801°C; the eutectic in this system has
not been determined as far as we know, but the similar case of Na_2S-
Na_2CO_3(MP 856°C) has a eutectic at 755°C. It is reasonable to assume a
eutectic at < ~ 750°C for the Na_2S-NaCl system. The reaction:

$$CaCl_2 + 2LaCl_3 + 4Na_2S + \text{excess } Na_2S \xrightarrow[900°C]{H_2S} CaLa_2S_4 + 8NaCl + \text{excess } Na_2S$$

was tried with varying amounts of Ca, La and Na salts weighed out, mixed in
a dry box, and fired in a 100% H_2S atmosphere in quartz glass boats in an
alumina tube muffle furnace to 800-900°C. Slight melting was obvious at

~ 800°C and was pronounced at 900°C. The melts were initially extracted with cold water in which the basic Na_2S-NaCl eutectic is easily soluble and the remnant powder rinsed with acetone and air-dried. XRD analysis showed the presence of no $CaLa_2S_4$, but the obvious presence of recently discovered rock salt structure $NaLaS_2$ [24] and La_2O_2S contaminant. XRD analysis on the melts, before water extraction, indicated that La_2O_2S was already present and was not a decomposition product of water reactions. Many runs, changing the relative amounts of Ca, La, Na, showed no sign of giving more than traces of $CaLa_2S_4$, but always $NaLaS_2$ was a major component. The by-product La_2O_2S, a sign of contaminant oxygen, was greatly reduced, but not eliminated, by long slow heatups with excess H_2S.

The greater ease of formation of $NaLaS_2$ and the fact that it survived the water washes (and even boiling water for 30 min) suggested it to be a more stable compound (i.e., more negative ΔG) than $CaLa_2S_4$ and might be a worthy candidate for testing out the approach. Having a somewhat ionic 6:6:6 coordination, it is a compromise between the types discussed earlier.

For the next tests the almost irreducibly simple reaction:

$$4Na_2S + LaCl_3 \xrightarrow[900^0C]{H_2S} NaLaS_2 + \underbrace{2Na_2S + 3NaCl}_{eutectic\ liquid}$$

was tried. Approximately 90% pure $NaLaS_2$ with ~ 10% La_2O_2S was always produced; changes in the ratio of Na:La or extra times in H_2S did not seem to improve the purity. With large excess of Na_2S, an unknown water-sensitive compound, which we suspect could be something like Na_5LaS_4, appeared also.

The nagging problem of byproduct La_2O_2S was solved by replacing the H_2S stream with argon plus CS_2 vapor; argon was bubbled through CS_2 liquid in a glass jar before passing over the mixtures throughout the heating and cooling. It was reasoned that if carbon was formed, then it would form a removable layer on top of the melt. In fact, carbon was not seen and the La_2O_2S contaminant was removed. Substantially pure $NaLaS_2$ was easily produced. The eutectic melt is now dissolved by soxhleting with (m)ethanol or THF. SEM pictures (Fig. 3) of the product indicate that, indeed, we can produce narrow size distributions of $NaLaS_2$, resulting from the Ostwald ripening phenomenon in the melt. The particles may be showing some incipient cubic faceting and perhaps some superficial decomposition which should be eliminated by organic extractions.

Synthesis in the Transformation-Toughening Debate

It is not well recognized in the materials community that chemical synthesis not only makes things, but, by elucidating reaction mechanisms, kinetics, thermodynamics, etc., can lead to judgments about other physical phenomena. Such a case, which is of interest because of the focus on transformation toughening, is the synthesis of 6 nm monoclinic ZrO_2 [25]. This suggested that the surface energy/size effect, suggested by Garvie [26,27] for the tetragonal → monoclinic transformation at low temperatures, was at least incomplete in some way. ' It was acknowledged [25] that surface energy could have been affected by the method of preparation, although in reply Garvie seemed to suggest that he alone noted this. However, later results of heating the 6 nm particles in air are illuminating: 140°C overnight, 7.4 nm monoclinic, 302°C for 1 h, 9 nm monoclinic with

758

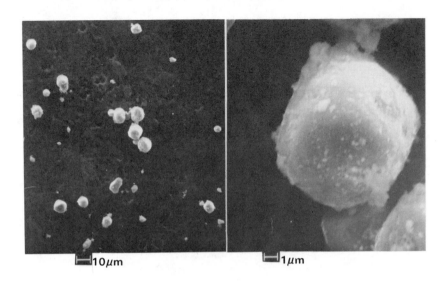

10μm   1μm

Fig. 3   Particles of NaLaS₂ from NaCl/Na₂S fused salt synthesis.

~ 1% tetragonal, 500°C/h, 9.5 nm monoclinic with ~ 5% tetragonal. It does
not appear that the monoclinic is converting to tetragonal even though, at
these temperatures, surface water and nitrate groups should have been
easily removed. I believe the traces of tetragonal are coming from traces
of amorphous material in the original monoclinic and not seen in the XRD
patterns. Contrary to Garvie's assertion [27], in the hypothetico-deduc-
tive scheme of science (promulgated by Karl Popper and accepted, with some
modifications, by most philosophers and aware scientists), it is not necessary for the agent of falsification to suggest a new hypothesis.

Chemistry of Liquid Phases at Boundaries

There has been growing awareness that chemistry plays a big role in
grain boundaries of ceramics (how could it be otherwise!) and indeed that
liquid phases are more commonly present in impure oxides than formerly
imagined. There is an old technique [28-30] to detect low levels of liquid
in high resistivity ceramic materials using electrical conductance. Details are elsewhere [31], but in Fig. 4 is shown the electrical conductivity and temperature vs 1/T for pure alumina (from Disperal®) and for Al_2O_3
doped with 1 m/o TiO_2 and 0.5 m/o $Na_{1/2}O$. The inflection in the doped case
exactly corresponds to where a liquid is now known to form [32] and suggests, as the rise is by an order of magnitude, that levels of liquid at
~ 0.1% could be detected. Methods of microwave loss should see this also.
The mechanism of TiO_2 enhancement of sintering in commercial aluminas
(which always have significant soda) is believed now to be by a liquid

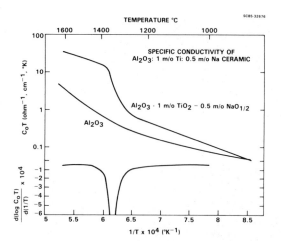

Fig. 4 Electrical conductivity and temperature vs reciprocal temperature for Al_2O_3:1 m/o TiO_2:0.5 m/o NaO_2.

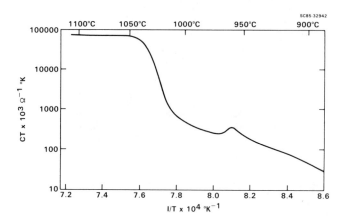

Fig. 5 Electrical conductivity and temperature vs reciprocal temperature for Al_2O_3:2 m/o TiO_2:2 m/o CuO.

phase mechanism. Figure 5 illustrates the same result type for Al_2O_3-2 m/o CuO-2 m/o TiO_2; liquid is again indicated for former sintering studies. The technique is equally applicable to natural rock materials and has been applied to basalt [31].

Si-S-Cl-N Chemistry

The reaction of cheap $SiCl_4$ and NH_3 is industrially important [33]. It has been known for a long time that in this reaction, $\rightarrow$ Si(NH)$_2$ + NH_4Cl, it is difficult to remove chloride from the Si(NH)$_2$ [34]. This has been done industrially by the use of the liquid ammonia method [35]. Another potentially interesting technique, also using irreducibly cheap chemicals, is to use the gas phase reaction:

$$3SiCl_4 + 2N_2 + 6H_2 \rightarrow Si_3N_4 + 12HCl \uparrow$$

At ~ 1500°C, thermodynamics indicate the anticipated reaction will proceed in the forward direction (as long as unexpected intermediates to not intercede). As SiCl4 reacts readily with Al_2O_3 and mullite at these temperatures, the reaction was carried out in a nonporous graphite tube (nonporous SiC and Si_3N_4 tubes are unavailable). The use of graphite limited the top temperature of reaction to ~ 1400°C; at higher temperatures, thermodynamically favored SiC formed. α-Si_3N_4 was formed in small yields, Fig. 6, below 1400°C with a flexible fiber morphology, but "seeding" might greatly increase the yields. The Cl content was low by EDS, as anticipated, where

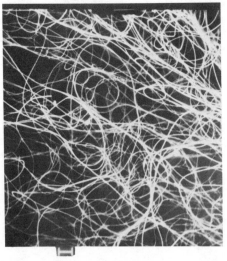

Fig. 6 Fine fibrous α-Si_3N_4 mass by reaction of $SiCl_4 + N_2 + H_2$ at 1400°C.

the crystalline product forms at such high temperatures. The fibers are polycrystalline by XRD. It appears that some gaseous intermediates form, perhaps such as Cl_3Si-$SiCl_3$, which do not further react sufficiently with $N_2 + H_2$. Also implied by the results is that above ~ 1400°C, the reaction:

$$SiCl_4 + C + 2H_2 \rightarrow SiC + 4HCl \uparrow$$

could be used to produce SiC.

We first reported the use of silicon sulfide to produce Si_3N_4 at the first "Better Ceramics Through Chemistry" [36]. Sulfur too is extremely cheap, and evidence suggests that (related to the much weaker Si-S bond) S is much more easily driven off from product Si_3N_4 [37].

The work has been extended to the reactions of $SiCl_4$ and SiS_2 with hydrazine. In each case, rapid reaction occurs at room temperature with gaseous or liquid hydrazine to produce amorphous intermediate materials. In the $SiCl_4/N_2H_4$ case, after gentle heating under vacuum to drive off ex-

cess hydrazine and hydrazinium hydrochloride (warning: hydrazine can deto-
nate on heating), chloride was retained. With the SiS_2/N_2H_4 material, at
1200°C amorphous, and then at 1400°C α-Si_3N_4, was formed when the product
was heated in argon.

After the reaction of SiS_2 + N_2H_4 at room temperature, unusual effects
were seen on heating the product in argon to high temperatures. Between
900°C and 1400°C, only amorphous product was seen, but at > 1200°C, a product distilled in the tube to form whiskers which subsequently on removal
were amorphous. Sulfur was detectable by EDS (~ 10%) in the products.
Examination of the 1200°C needles by SEM shows evidence of VLS-type mechanism in "mushroom" outgrowths, Fig. 7. Evidently, interesting liquid
phases are transiently available in the Si-S-N system, which could be used
for new fabrication possibilities.

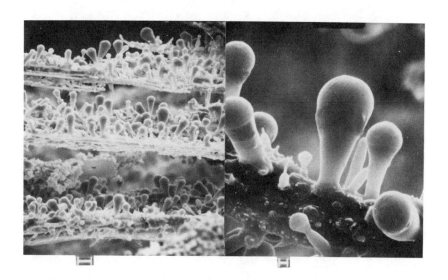

Fig. 7 SiS_2/N_2H_4 derived material heated in argon to 1200°C.   Left marker
10 μm, right marker 1 μm.

This idea also led to attempts to hot-press Si_3N_4 using Si-S-N glasses
instead of the usual intergranular oxynitride glass approach. With up to
5% SiS_2 incorporated into Si_3N_4 (Sumitomo brand) and hot-pressing up to
~ 1700°C, very little densification in fact was achieved. Useful liquids
did not seem to form here and the samples showed signs of large evaporation
condensation effects with acicular α-Si_3N_4 in the porous body. Sulfur
(like chloride) seems to stabilize the α-form of Si_3N_4, and only above
1700°C was the β form observed quite suddenly to form.

CONCLUSION

I hope with a few examples to have shown that much unexplored chemistry is available for use in ceramic processing. Perhaps we need University Departments of Applied Chemistry, as in Japan, to stimulate efforts.

REFERENCES

1. J.P. Bilisoly, U.S. Patent 2,496,265 (February 7, 1950).
2. K. Fujimoto and M. Boudart, J. Phys. Colloq. 40, C2, Pt. 3 (1979); Proc. Int. Conf. Mossbauer Effect, Kyoto Japan (1978).
3. M. Boudart and Y.L. Lam, U.S. Patent 4,136,062 (January 23, 1979).
4. G. Carturan, V. Gottardi and M. Graziani, J. Noncryst. Solids 29, 41 (1978).
5. P.E.D. Morgan, H.A. Bump, E.A. Pugar and J.J. Ratto, in press, Ultrastructure Processing of Ceramics, Glasses and Composites II, Proc. Conf., Palm Beach, FL, March 1985.
6. D.C. Bradley and M.M. Faktor, Trans. Farad. Soc. 55, 2117 (1959).
7. E.A. Pugar and P.E.D. Morgan, "Coupled Grain Growth Phenomena in Al_2O_3-ZrO_2 Diphasic Ceramics," accepted, Comm. Am. Ceram. Soc.
8. P.E.D. Morgan and M.S. Koutsoutis, J. Matl. Sci. Lett. 4, 321 (1985).
9. P.E.D. Morgan and M.S. Koutsoutis, J. Matl. Sci. Lett. 4, 1230 (1985).
10. P.E.D. Morgan, Matl. Res. Bull. 18, 231 (1983).
11. P.E.D. Morgan, U.S. Patent No. 4,339,511 (July 13, 1982).
12. P.E.D. Morgan, accepted, J. Am. Ceram. Soc.
13. N.R. Osborne, Am. Ceram. Soc. Annual Meeting, Chicago, IL, April 29, 1986; Bull. Am. Ceram. Soc. 65 (33), 518 (1986).
14. P.E.D. Morgan, Mat. Res. Bull. 19, 369-76 (1984).
15. G.Q Yao, H.S. Shen, R. Kershaw, K. Dwight and A. Wold, "Preparation and Characterization of $ZnSiP_2$ and $ZnGeP_2$ Single Crystals," Tech. Report No. 1, Office of Naval Research, Contract No. N00014-85-K-0177 (1986).
16. P.O. Provenzano, S.I. Boldish and W.B. White, Mat. Res. Bull. 12, 936 (1977).
17. W.B. White, D. Chess, C.A. Chess and J.V. Biggers, SPIE 297, 38 (1981).
18. D.L. Chess, C.A. Chess, J.V. Biggers and W.B. White, J. Am. Ceram. Soc. 66, 18 (1983).
19. O. Schevciw and W.B. White, Mat. Res. Bull. 18, 1059 (1983).
20. D.G. Wickham, Ferrites, 105-7, Proc. Int. Conf., Kyoto, Japan (1970).
21. P.J. Walker and R.C.C. Ward, Mat. Res. Bull. 19, 717 (1984).
22. C.C. Hsu, J. Electrochem. Soc. 131, 1632 (1984).
23. H.J. Scheel, J. Crystal Growth 24/25, 669 (1974).
24. M. Sato, G. Adachi and J. Shiokawa, Mat. Res. Bull. 19, 215 (1984).
25. P.E.D. Morgan, Comm. Am. Ceram. Soc. 67, C-204 (1984).
26. R.C. Garvie, J. Phys. Chem. 69, 1238 (1965).
27. R.C. Garvie and M.F. Goss, J. Matl. Sci. 21, 1253 (1986).
28. J.O.M. Bockris, J.A. Kitchener, S. Igatowicz and J.W. Tomlinson, Trans. Farad. Soc. 48, 75 (1952).
29. R. Cypres and B. Van Ommeslaghe, Bull. Soc. Franc. Ceram. 54, 65 (1962).
30. G.C. Nicholson, J. Am. Ceram. Soc. 48 (10), 525 (1965).
31. P.E.D. Morgan and M.S. Koutsoutis, "Electrical Measurements to Detect Suspected Liquid Phase in the Al_2O_3-1 m/o TiO_2-0.5 m/o $NaO_{1/2}$ and Other systems," submitted to Com. Am. Ceram. Soc.
32. P.E.D. Morgan and M.S. Koutsoutis, Comm. Am. Ceram. Soc. 68, C156 (1985).

33. P.E.D. Morgan and E.A. Pugar, J. Am. Ceram. Soc. 68 (12), 699 (1985).
34. M. Billy, M. Brossard, J. Desmaison, D. Giraud and P. Goursat, J. Am. Ceram. Soc. 58 (5-6), 254 (1975).
35. T. Iwai and T. Kawahito, U.S. Patent No. 4,196,178 (April 1, 1980).
36. P.E.D. Morgan, Proc. MRS Meeting, "Better Ceramics Through Chemistry," Albuquerque, Feb. 1984; MRS Symp. Proc. 32, 213, eds., C.J. Brinker, D.E. Clark and D.R. Ulrich, North-Holland (1984).
37. P.E.D. Morgan and E.A. Pugar, U.S. Patent No. 4,552,740 (November 12 1985).

PREPARATION OF GRADIENT-INDEX GLASS RODS BY THE SOL-GEL PROCESS

Masayuki Yamane,* J. Brian Caldwell** and Duncan T. Moore**
*Yamane Department of Inorganic Materials, Tokyo Institute of Technology,
2-12-1 Ookayama, Meguro-ku, Tokyo 152 Japan
**Caldwell and Moore, The Institute of Optics, University of Rochester,
Rochester, NY 14627

ABSTRACT

The fabrication of gradient-index material via the sol-gel technique
is reported. The material is based on the $PbO-B_2O_3-SiO_2$ system and a K_2O-
PbO liquid phase interdiffusion is used to produce the necessary composition gradient.

INTRODUCTION

The usual goal of sol-gel processing is to produce optically clear
homogeneous glass. However, the sol-gel process also lends itself in several unique ways to the production of glass in which the refractive index
varies continuously. This type of glass is called gradient-index (GRIN)
glass and is currently used for optical fiber, compact photocopier lens
systems, endoscopes, pick-up lenses for compact audio/video discs, et
cetera.[1-3]
The manufacture of GRIN materials is currently based on one of the
following methods: (a) the exchange of ions in an optical glass for ions
in a molten salt bath; (b) stuffing and partial unstuffing of a porous
glass with a water soluble salt followed by subsequent drying and sintering, and (c) chemical vapor deposition (CVD) of the soot produced by flame
hydrolysis of $SiCl_4$ and $GeCl_4$ onto a silica glass rod followed by densification of the soot.[4-6]
Although the material produced by these techniques is suitable for the
applications listed above, there exists a need for GRIN materials which are
large in size, have a large variation of refractive index, and can be economically produced. Many design studies have been done which show that the
use of gradient-index lens elements can significantly reduce the required
number of elements in complicated lens systems for cameras, binoculars,
microscopes, et cetera.[7-8] The ion exchange technique is limited to a
distance less than 6-7 mm or so due to small diffusion coefficients of ions
in a solid glass. This technique is also generally limited to the use of
single valence ions which severely restricts the variety of GRIN materials
which can be made. The stuffing and partial unstuffing technique can produce large size GRIN materials with a very broad range of composition, but
it cannot easily produce large index changes. The CVD technique is capable
of producing large size GRIN materials with large index changes but it is a
very expensive process.
This report is on the preparation of gradient-index glass rods via the
sol-gel technique. This method involves first making a multi-component gel
containing one or more index modifying ions, then soaking the gel, while it
is still wet and porous, in a solution containing different index modifying
ions, allowing the ions in the gel to interdiffuse with the ions in the solution. After interdiffusion, the gel is dried and then sintered into gradient-index glass. The advantages of this technique are as follows: (a)
a large amount of index modifying materials, such as lead, can be incorporated into the gel, thus making large index changes possible; (b) deep diffusion can be done quickly because the ions being interdiffused are in acqueous solution; (c) almost any two ions can be interdiffused regardless of
valence, and (d) the processing equipment and procedures are fairly simple

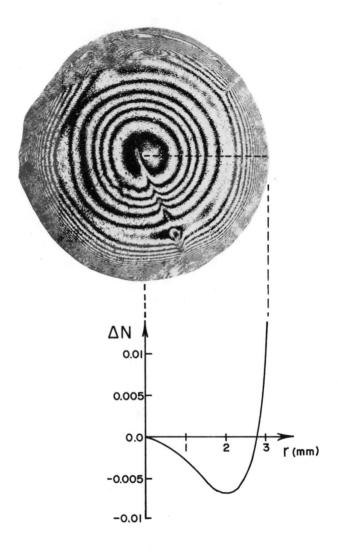

Figure 1. Interferogram and refractive index profile of a radial gradient--
index material made by the sol-gel process at the University of
Rochester. The sample is 0.50 mm thick and was measured in a
Mach-Zehnder interferometer using an argon-ion laser operating at
a wavelength of 5145 angstroms.

and inexpensive.

PREPARATION

First, a sol was prepared by stirring together 20 ml of tetramethoxy-silane (TMOS), 20 ml of tetraethoxysilane (TEOS), and 40 ml of an acqueous solution containing 0.8 moles/liter boric acid (H_3BO_3) and 5 ml/liter hydro-chloric acid (HCl). By the time this mixture became homogeneous, it was rather warm due to the heat of reaction, so it was cooled to room tempera-ture before the rest of the procedure was carried out.

Next, 15 ml of this sol was slowly added with stirring to 5 ml of a 0.5 moles/liter acqueous lead nitrate solution followed by the addition of 30 ml of a 0.25 moles/liter acqueous lead acetate solution. The lead-containing sol was then poured into a 23 mm inside diameter polyethylene vial and allow-ed to gel at room temperature. The vial was tightly covered during the gel-ling stage.

After aging for about 12 hours, the gel, which had shrunk slightly to 20 mm in diameter and 40 mm in length, was removed from the vial and placed in a 0.2 moles/liter acqueous solution of potassium nitrate for 30 minutes. During this time, Pb^{2+} ions in the gel were able to exchange with K^+ ions in the surrounding solution. The ion-exchanged gel was placed in a covered polystyrene weighing dish for 4 hours and was then placed in a drying oven at 60°C for about 72 hours.

The dried gel (xerogel) was about 10 mm in diameter and 20 mm in length. It was placed in a fused quartz reaction tube which was, in turn, placed into a horizontal electric furnace. The heat treatment consisted of a 5°C/hour ramp in temperature from room temperature up to 600°C with 24 hour isother-mal treatments at 200°C, 360°C, and 460°C, and a 48 hour isothermal treat-ment at 600°C. Oxygen gas was fed into the reaction tube at a rate of about 50 ml/minute up to a temperature slightly above 460°C, at which point helium was substituted for oxygen for the remainder of the heating cycle. At the end of the 600°C treatment, the temperature was reduced to room temperature at a rate of 60°C/hour.

CONCLUSION

At the end of heat treatment, the sample was observed to have several small internal cracks and a rather rough surface. However, the cracks were not extensive and the sample had clearly sintered into a glass. A thin wafer was cut perpendicular to the axis of the cylindrically shaped sample and was polished plane parallel to a thickness of 0.5 mm. The wafer was then analyzed in a Mach-Zehnder interferometer using an argon/ion laser operating at $\lambda = 0.5145$ microns. The interferogram is shown in Figure 1. Each successive fringe corresponds to an index change of λ/t, where λ is the wavelength used and t is the thickness of the wafer.

The fact that the refractive index increases over the outer millimeter indicates that lead ions migrated toward the surface during drying. Clearly, this is a problem which needs to be corrected. Nevertheless, the results indicate that the sol-gel technique has potential for producing large GRIN rods with large index changes at a reasonable cost.

REFERENCES

1. W. J. Tomlinson, Appl. Opt. 19, 1127 (1980).

2. M. Kawazu, Y. Ogura, Appl. Opt. 19, 1105 (1980).

3. T. Yamagishi, K. Fujii, I. Kitano, Appl. Opt. 22, 400 (1983).

4. H. Kita, I. Kitano, T. Uchida, M. Furukawa, J. Am. Ceram. Soc. 54, 321 (1971).

5. J. H. Simmons, R. K. Mohr, D. C. Tran, P. B. Macedo, J. A. Litovitz, Appl. Opt. 18, 2732 (1979).

6. S. Sudo, M. Kawachi, T. Edahio, T. Izawa, T. Shioda, I. Gotoh, Electron. Lett. 14, 534 (1978).

7. L. G. Atkinson, S. N. Houde-Walter, D. T. Moore, D. P. Ryan, J. M. Stagaman, Appl. Opt. 21, 993 (1982).

8. J. D. Forer, S. N. Houde-Walter, J. J. Miceli, D. T. Moore, M. J. Nadeau, D. P. Ryan, J. M. Stagaman, N. J. Sullo, Appl. Opt. 22, 407 (1983).

SILOXANE MODIFIED SiO_2-TiO_2 GLASSES VIA SOL-GEL

C.S. Parkhurst, W.F. Doyle, L.A. Silverman, S. Singh, M.P. Andersen,
D. McClurg, G.E. Wnek and D.R. Uhlmann, Department of Materials Science and
Engineering, M.I.T., Cambridge, Mass. 02139

ABSTRACT

Polydimethylsiloxane [PDMS]-modified SiO_2-TiO_2 glasses have been prepared
via the sol-gel route. Polymer compositions varied between 17 and 67 wt %
PDMS, using PDMS of molecular weights 1,700 and 36,000. Also varied was the
Si/Ti ratio for a given polymer content and the nature of the Ti alkoxide.
A general synthetic procedure was found which made optically clear samples.
Dense monolithic structures were obtained at room temperature for all compositions. The room temperature densification is attributed to relaxation
and flow in the sample due to the presence of the polymer. The effects on
properties of the overall composition and molecular weight of the polymer are
reported, and implications in terms of structural models are considered.

INTRODUCTION

Organically modified sol-gel derived glasses are of considerable
interst [1,2, e.g.] because of their potential for providing unique combinations of properties. These materials are synthesized by chemically
incorporating organic polymers into the forming inorganic network. The properties of the resulting materials are anticipated to be highly dependent
upon the composition, molecular weight and chemistry of the polymer as well
as the process history (and hence the microstructure). Many of the materials
exhibit unusual densification behavior at low temperatures, reflecting their
exceptional relaxation and flow characteristics.

The present paper describes a general synthetic approach for characterizations of the resulting materials.

SYNTHESIS AND CHARACTERIZATION PROCEDURES

PDMS has been incorporated into SiO_2-TiO_2 glasses by reacting OH-
terminated PDMS (petrarch) of either 36,000 or 1700 molecular weight with
tetraethylorthosilicate (TEOS) and titanium(IV)n-butoxide or titanium(IV)-
ethoxide(Alfa) employing p-toluenesulfonic acid (Aldrich) as a catalyst
(see Fig. 1).

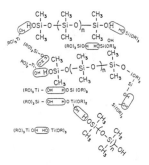

Fig. 1 Chemical Incorporation
of PDMS into SiO_2-TiO_2
Glass

The chemistry of the terminal SiOH on the polymer is analogous to that of the SiOH from the hydrolyzed Si alkoxide. Condensation of the polymer with the alkoxide affords Si-O-Si and Si-O-Ti bonds which are thermodynamically stable [3]. PDMS is a rubber at room temperature and is responsible for imparting properties such as flexibility to the resulting material.

Ti alkoxides are known to condense rapidly with monomeric silanols [4, e.g.], to react readily with OH-terminated PDMS [5, e.g.], and to serve as crosslinking agent for OH-terminated PDMS [6, e.g.]. The use of Ti alkoxides is expected to enhance the incorporation of PDMS by condensing with its terminal OH. Inorganic alkoxides are expected to increase the thermal stability of the PDMS, since it is reported that Ti alkoxides increase the thermal stability of silicones [7].

Successful preparation of PDMS-modified SiO_2-TiO_2 glasses is neither straightforward nor trivial. Examination of the respective chemical species indicates gross differences in their reactivities, as:

$$R_{Ti(OR)_4} \ggg R_{Si(OR)_4} > R_{HO-PDMS-OH}$$

Moreover, the polymer is hydrophobic; and in high molecular weight form, it separates from aqueous solutions. These difficulties have been substantially overcome by using a stepwise batch polymerization process which yields transparent dried products.

The transparent samples were obtained by first pre-equilibrating the Si alkoxide with water(Si alkoxide/water mole ratio, 1/2) in refluxing tetrahydrofuran. The calyst (0.1 wt %) was p-toluenesulfonic acid. The polymer was then added; and the mixture reacted long enough for all free water to be consumed. The reagent concentration was 10 wt %. The Ti alkoxide was then added and allowed to react (if the Ti alkoxide is added too soon, while free water is present, a TiO/OH material precipitates immediately upon addition of the Ti alkoxide). The resulting clear solutions were placed in an aluminum dish loosely covered with parafilm in air. After partial evaporation of the solvent, they gelled; and upon further air drying, dense bodies were formed. The order of addition of reagents is critical. If the order is reversed by adding Ti alkoxide before pre-equilibrated Si alkoxide to the polymer, the resulting dried sample is white even though the solution is clear.

One potential side reaction which may occur during synthesis involves the degradation or equilibration of the polymer molecular weight under the influence of catalysts such as strong acids. This equilibration involves rearrangements of the Si-O-Si backbone of the polymer which alter the molecular weight under the influence of catalysts such as strong acids. This equilibration involves rearrangements of the Si-O-Si backbone of the polymer which alter the molecular weight. The mechanism depends upon the SiOH concentration [8]. If the molecular weight equilibrates before reaction of the polymer SiOH with a metal alkoxide, there may be little effect of the initial molecular weight.

A worst-case situation was investigated, wherein the polymer, catalyst and water were reacted for the total reaction time of 48 hours, and the molecular weight distributions of the polymer before and after treatment were compared using size exclusion chromatography [9].

Differential scanning calorimetry was performed on a duPont 1090B Thermal Analysis System equipped with a duPont 910 Differential Scanning Calorimetric module at a heating rate of 10 C min^{-1} under N_2. Glass transition temperatures (Tg's) were measured at the midpoint of the discontinuity of the specific heat.

RESULTS AND DISCUSSION

The samples which were synthesized are listed in Table I. All samples produced using the normal sequence of addition were transparent except for the 36,000 molecular weight PDMS/Si(OR)$_4$/Ti(OR)$_4$, 4/1/1 material which was opaque-white. The opacity of this sample reflects a solubility problem, associated with the smaller enthalpy of mixing of the high molecular weight PDMS, compared with the optically clear analogous sample synthesized with 1,700 molecular weight PDMS.

TABLE I
Compositions in the PDMS-$\overline{\text{Si(OR)}_4}$-Ti(OR)$_4$ System (parts by weight)

Polymer	Si(OR)$_4$	Ti(OR)$_4$	Procedure	Appearance of bulk sample
A. Polymer of 36,000 molecular weight				
4	1	1	A	white, opaque, crack-free
1	1	1	A	clear, crack-free
1	1	1	A*	white, opaque, crack-free
0.4	1	1	A	clear but cracked
1	1	4	A	clear but cracked
1	1	0.4	A	clear, crack-free
1	1	1	B	white, opaque, crack-free
4	1	1	B	white, opaque, crack-free
B. Polymer of 1,700 molecular weight				
4	1	1	A	clear, crack-free
1	1	1	A	clear-crack-free
0.4	1	1	A	clear but cracked
1	1	4	A	clear but cracked
1	1	0.4	A	clear, crack-free
1	1	1	A*	white, opaque, crack-free

A - normal mode of addition, use of Ti(OBu)$_4$
A*- normal model of addition, use of Ti(OEt)$_4$
B - reverse addition

Samples made from Ti(OEt)$_4$ formed opaque monoliths from a clear reaction solution, even for formulations which produced transparent monoliths when Ti(OBu)$_4$ was used. This likely reflects the increased reactivity of Ti(OEt)$_4$. Unreacted Ti(OEt)$_4$ hydrolyzes faster with atmospheric moisture compared with Ti(OBu)$_4$.

The 1/1/1 samples prepared using the reverse sequence addition were also phase separated. This suggests that differences in sample appearance occur as a result of which alkoxide is allowed to react first with the polymer terminal function group.

Polymer molecular weights were compared before and after treatment with the catalyst to study whether the polymer chain degrades in the presence of the acid. The results, reported in polystyrene equivalent molecular weight (PSEQ), of the size exclusion chromatography(SEC) study are given below.

Sample	Number Average Molecular Weight (PSEQ)
36,000 PDMS before treatment	6.8×10^4
36,000 PDMS after treatment	3.6×10^4

The results of the SEC study indicate a decrease in molecular weight. The decrease is not extreme, even for the worst-case conditions investigated.

772

It seems likely, therefore, that the PDMS is degrading somewhat as the synthesis proceeds. The resulting molecular weight of the 36,000 material is still significantly higher than that of the 1,700 material; and a useful comparison between the two may still be made.

The results from the DSC studies are summarized in Table II. The pure polymers have a Tg about -120 C; and the 36,000 molecular weight polymer melts at -40 C. All transparent samples show thermal transitions indicative of Tg's. In a few samples, these transitions occur at the same temperature as the Tg of the pure polymer, which indicates domains of polymer-rich material in the samples. In most samples, the transitions are higher in temperature than the Tg of the pure polymer. These transitions are neither as intense nor distinct as the Tg transitions in the pure polymer.

TABLE II

Thermal Analysis Results

Sample*	PDMS mol. wt	Procedure	Tg(C)	Tm(C)
PDMS	36,000	-	-122	-38
PDMS	1,700	-	-108	-
4:1:1	36,000	A	-121	-35
4:1:1	1,700	A	-110, -90,0	-
1:1:1	36,000	A	- 85, 8	-
1:1:1	1,700	A	- 83, -10	-
0.4:1:1	36,000	A	-121	-
0.4:1:1	1,700	A	- 88, -3	-
1:1:1	36,000	A*	- 95	-45

A - normal addition using $Ti(OBu)_4$
A*- normal addition using $Ti(OEt)_4$
* - weight ratios of polymer:silicon alkoxide:titanium alkoxide

It has previously been observed that Tg increases with increasing restriction of the ends of the PDMS by reaction with Ti alkoxide [10], as shown by the data below $Ti[O (DMS)_n OH]_4$ in which one end of the PDMS is reacted:

n	Tg (degrees C)
5	-69
9	-110
13	-120

On this basis, the increase in the polymer Tg seen in most samples is consistent with incorporation of the polymer into the glass on a chemical level. The PDMS in our samples is more restricted than in $Ti[O (DMS)_n OH]_4$ with both ends potentially reacted. The environment along the chain backbone, which should have significant glass character, is also confining and contributes to the increase in Tg. Some samples have two transitions, which indicate two types of environment for the incorporated polymer. The material giving the highest Tg is the most incorporated oxide-modified or restricted; while the material showing the lower Tg is more polymer-rich, less incorporated/modified/restricted. These results suggest some degree of phase separation. Such phase separation in the transparent samples must occur on a small size, as the indices of refraction of PDMS on SiO_2-TiO_2 are quite different, or the separated phases must be appreciably modified.

The loss of the PDMS melting point in the optically clear samples is also consistent with a high degree of chemical incorporation and confinement of the polymer. A study of crystallization kinetics indicates that incorporation of Ti alkoxide, reacted with OH-terminated PDMS, slows the crystallization rate [11]. In contrast, fillers increase the rate of crystallization [11].

Dilute solutions of 36,000 molecular weight PDMS/$Si(OEt)_4$/$Ti(OBu)_4$, 4/1/1 and 0.4/1/1 were prepared and gelled. Samples of these gels were dried at room temperature overnight. During this process, the gelled samples lost 85% and 90% of their initial weights, respectively. The 4/1/1 sample formed a dense, uncracked, translucent monolith. Its appearance is significantly different from the white, opaque 4/1/1 sample which was gelled at a higher solids content. The difference reflects still another processing variable of consequence. The 9.4/1/1 sample cracked on densification.

More generally, the densification behavior of the PDMS-modified SiO_2/TiO_2 glasses seems notable. Incorporation of PDMS at concentrations as small as 33 wt % (0.36 mole %) gave rise to the formation of transparent, dense, crack-free monoliths even at room temperature. Such densification must reflect exceptional relaxation and flow characteristics imparted by the polymer-characteristics which are presently under investigation in our laboratory.

CONCLUSIONS

Samples of PDMS-modified SiO_2-TiO_2 glasses covering a range of composition have been successfully prepared. The properties and appearance of the samples could be drastically altered by varying the synthetic procedure, composition, starting material and gelation treatment. All samples containing sizable concentrations of polymer could be formed as monolithic structures at room temperature. The low temperature densification is attributed to the exceptional relaxation and flow characteristics imparted by the polymer. The appearance of the samples varied from transparent to opaque-white depending on the Ti alkoxide used, the order of addition and whether the 4/1/1 36,000 molecular weight polymer sample was gelled prior to or after air drying. The transparent samples showed higher thermal transitions than the pure polymer and a lack of crystallization, both of which support chemical incorporation of the DMS into the glass by reaction of the OH-PDMS end groups with the alkoxides and subsequent glass formation about the polymer chain.

ACKNOWLEDGEMENTS

Financial support for the present work was provided by the Air Force Office of Scientific Research, International Partners in Glass Research, and Rogers Corp. This support is gratefully acknowledged, as is the help of B. Fabes, B. Zelinski, and Prof. Y.M. Chiang.

REFERENCES

1. H. Schmidt, U.S. Pat. 4,374,696 (1983).
2. G. Wilkes, et al., Polymer Preprints 26, 300 (1985).
3. K. Andrianov, et al., Vysokomol. Soed. 1, 743 (1959).
4. I. Shiihara, et al., Chem. Reviews, 61, 1 (1961).
5. A. Zhadanov, Doklady Akad. Nauk USSR, 138, 361 (1961).
6. D.R. Weyenberg, U.S. Pat. 3,013,992 (1961).
7. J.B. Rust, et al., U.S. Pat. 3,013,992 (1961).
8. D.T. Hurd, J. Am. Chem. Soc. 77, 2998 (1955).
9. J. Janca, Steric Exclusion Liquid Chromatography of Polymers (Dekker, N.Y., 1984).
10. K.A. Andrianov, N.A. Kurasheva, Bull. Acad. Sci. USSR, 1011 (1962).
11. K. Andranov, J. Polymer Sci. A-1, 23 (1972).

THERMAL, ACOUSTICAL AND STRUCTURAL PROPERTIES OF SILICA AEROGELS

J. FRICKE and G. REICHENAUER
Physikalisches Institut der Universität, Am Hubland, D-8700 Würzburg,
Federal Republic of Germany

ABSTRACT

Silica aerogels either in monolithic or in granular form provide excellent thermal insulation and thus may be used as superinsulating spacer
in all kinds of window systems. Highly porous aerogels also are exciting
acoustic materials with sound velocities in the order of 100 m/s and
acoustic impedances between 10^4 and 10^5 kg/(m$^2 \cdot$ s). Silica aerogels produced from TMOS seem to consist of massive primary particles ($\phi \cong$
1 nm, $\rho \cong$ 2000 kg/m^3) which form secondary particles ($\phi \cong$ 5 to 6 nm,
$\rho \cong$ 800 kg/m^3) displaying fractal properties (D $\cong$ 2). The further build-
up creates the highly porous low-density structure which is responsible
for the special thermal, acoustical and optical properties of aerogels.
Above about 100 nm, transparent aerogels should be homogeneous.

THERMAL TRANSPORT

Evacuated silica aerogels either in monolithic or in granular form
provide excellent thermal insulation at ambient temperature: Thermal
loss coefficients in the order of 0.5 W/(m$^2 \cdot$ K) have been measured for
15 to 20 mm thin layers under external atmospheric pressure load /1/,
/2/. Considering the clear transparency of tiles and the translucency of
pellet fillings, a vast potential for the reduction of heat losses in all
kinds of window systems (Fig. 1) can be recognized. In addition, aerogel
layers may serve as superinsulating covers for passive use (Fig. 2) and
long-term storage of solar energy /3/, /4/.

In order to improve this fascinating material with respect to its
thermal insulation and its optical properties, a detailed understanding of
the heat loss channels and their coupling as well as of its structural properties is necessary. We want to summarize the most recent findings
/1/, /2/, /5/, /6/:

• The radiative heat transfer in thin layers of SiO$_2$ aerogel depends

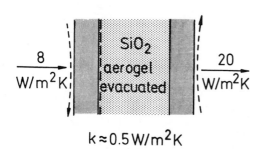

Fig. 1. Superinsulated
window system; an evacuated SiO$_2$-aerogel layer of
15 to 20 mm provides a
thermal loss coefficient of
about 0.5 W/(m$^2 \cdot$ K) which
corresponds to an R-value
of 11.

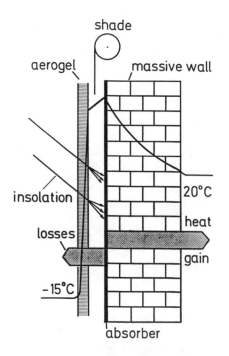

Fig. 2. Translucent wall system; the absorbed solar radiation is partly available for room heating /3/.

on the spectral infrared absorption and the boundary emissivity. A large absorption coefficient in the wavelengths region above 7 μm provides effective blockage of the thermal infrared spectrum at ambient temperatures. Leakage of ir radiation increases for rising temperatures, with the ir spectrum being shifted towards the region of low absorption (wavelengths $\Lambda \cong 3$ to 5 μm). Thus an important task for the future is the integration of an effective ir opazifier into the aerogel skeleton, which then could be used in high temperature systems, too.

● The solid conduction λ_{sc} in monolithic aerogel layers strongly depends on the density ρ. A variation $\lambda_{sc} \propto \rho^\alpha$ with $\alpha \cong 1.6$ can be found (Fig. 3). For granular aerogel layers, the solid conduction of the filling changes more slowly with the pellet density ρ (e.g. $\alpha \cong 0.6$). This can be derived from a simple contact model of touching spheres /2/. Still unsolved is the optimization problem of ρ with respect to the minimization of total thermal losses.

● Thermal conduction due to the presence of a gas within the monolithic porous system is appreciable only above 100 mbar (Fig. 4). For granular fillings, the onset of gas conduction can be observed (Fig. 5) already at pressures as low as 10^{-2} mbar /2/. These fillings thus need a glass-metal rim seal in order to maintain the vacuum over a sufficiently long time.

● The non-linear coupling between solid conduction and radiative transport creates a very unique temperature distribution. Striking is a radiative boundary layer /5/, which is especially pronounced for low-emissivity walls. In this layer, a steep temperature gradient allows for thermal losses mainly due to solid conduction, while deep inside the aerogel

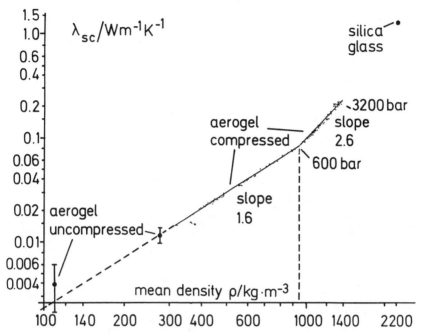

Fig. 3. Scaling of solid conduction with aerogel density; the full line represents the compressed aerogel /6/, the separate points the uncom-
pressed samples of different densities /1/.

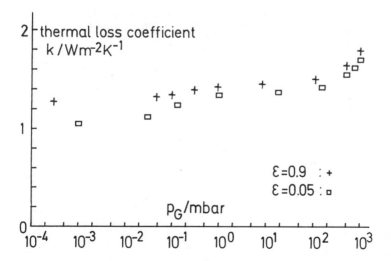

Fig. 4. Variation of thermal loss coefficient with internal gas (N_2)
pressure for a tile system (provided by the Lund group); $\rho \cong 270$ kg/m^3,
$T \cong 300$ K, outer load 1 bar , emissivity $\varepsilon = 0.05$ or 0.9 /1/.

778

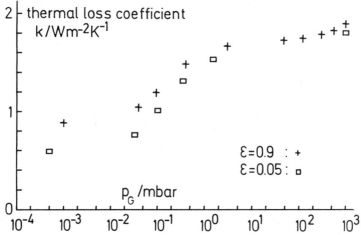

Fig. 5. Variation of thermal loss coefficient with internal gas (N_2) pressure for a granular system (granules provided by BASF/Ludwigshafen); density of the granules 230 kg/m^3, $\phi \cong 3$ mm, T $\cong$ 314 K, outer load 1 bar /2/.

also radiative transport occurs. Though approximations for the complex heat transport in SiO$_2$ aerogel are available /5/, more exact coupled multi-band calculations still need to be performed.

ACOUSTIC PROPERTIES

As was discovered recently, silica aerogels are exceptional acoustic materials, too /7/. The sound velocity can be as low as 100 m/s, corresponding to extremely small values of the Young's modulus EM $\cong 10^6$ N/m^2. As expected for percolating systems, the Young's modulus of porous silica scales with density: EM $\propto \rho^{\beta}$, with $\beta \cong 3.7$ (see Fig. 6).

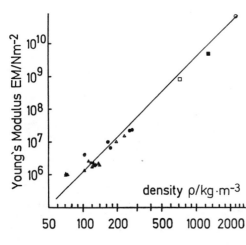

Fig. 6. Scaling of Young's modulus for silica systems; the highest value is for massive silica glass, the two next points are for silica gels, and all the lower remaining points are for silica aerogels /7/.

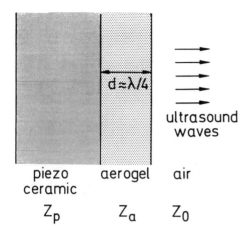

ultrasound
waves

piezo   aerogel   air
ceramic

Z_p Z_a Z_0

Fig. 7. Arrangement to increase ultra-sonic intensities in air. Z_p, Z_a and Z_0 are the acoustic impedances of piezo ceramic, aerogel and air, respectively. Impedance matching requires $Z_a = (Z_p \cdot Z_0)^{1/2}$ /8/.

Due to the low sound velocities and the small acoustic impedances $Z \cong 10^4$ to 10^5 kg/(m$^2 \cdot$ s), SiO_2 aerogels may be used as acoustic $\lambda/4$ antireflection layers (Fig. 7). Thus it might be possible to considerably increase the ultra-sonic intensity radiated off a piece of piezo ceramic into air /8/.

STRUCTURAL ASPECTS

Up to now, no consistent and plausible picture of aerogel structures from the nm to the μm range has been presented. We thus tried to extract relevant information from a variety of structural investigations on aerogels, performed by several groups in the world. We derived the following picture:

The Range $\phi \cong 1$ nm

Small angle X-ray scattering (SAXS) reveals /9/, /10/ that aerogels (made from TMOS) - like silica gels - consist of massive colloidal particles with smooth boundaries up to the 1 nm-range (see Fig. 8). This can be concluded from the slopes in the Porod plot which are -3 for a slit geometry /9/ and -4 for pin hole collimation /10/. The density of these primary particles ought to be $\cong 2000$ kg/m^3 as measurements with helium pycnometry demonstrated /11/.

The 1 to 5 nm Range

TEM bright field images of silica aerogels show features a few nm in diameter /9/. We shall call these entities secondary particles. In this range aerogels are considered mass fractals /10/ with fractal dimension $D \cong 2$, which can be deduced from Fig. 8 b, where the Porod slope is -2 in the range 0.02 Å^{-1} < K < 0.1 Å^{-1}. The existence of particles with diameters of 5 to 6 nm is supported by the derived probability distribution (Fig. 9) for the radius of gyration $\overline{R}$ /9/. Such data can be extracted from the Guinier plot (log intensity versus square of

780

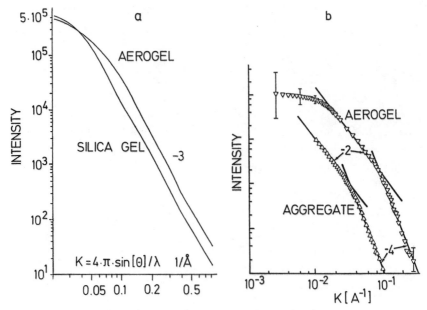

Fig. 8. Porod plots from SAXS measurements; a) aerogel (Lund, ρ ≅ 145 kg/m³), slit geometry in comparison with silica gel /9/, b) aerogel (Lund, ρ ≅ 88 kg /m³) and silica gel, pin hole collimation /10/. The postulated uniformity for lengths ≳ 10 nm /10/ is contradicted by other observations (see next page).

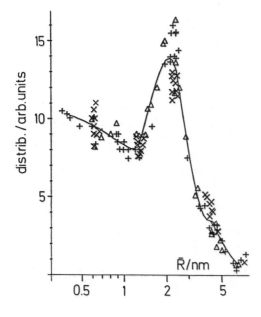

Fig. 9. Probability distribution for the radius of gyration $\bar{R}$ from SAXS /9/; three different aerogel densities have been investigated (+: 275 kg/m³, x: 145 kg/m³, △ : 105 kg/m³, Lund group)

scattering angle) for SAXS data. For spherical particles, one gets
$\phi \cong 2.6 \cdot \bar{R}$.

BET surface measurements and the determination of the differential
pore volume support the above picture: Baked hydrophilic SiO_2 aerogels
which had been exposed to water show a distinct peak for the pore diameter at 5 nm (Fig.10). Structures above $\phi \cong 6$ nm have been destroyed by
the water surface tension /9/. Only the rather stable secondary particles seem to have survived. It is noteworthy that the high specific
N_2 BET surface area of about 800 m^2/g is reduced by only 10 % if
SiO_2-aerogels are soaked in water. We may conclude that the large surface area is predominantly located within the (surviving) secondary particles and is provided by the surfaces of the great number of 1 nm primary particles.

The 5 to 100 nm Range

Hints on the structure in this range can be gained from a number
of observations:
- The clear transparency of monolithic aerogel gives an upper limit
for the pore sizes of about 100 nm.
- Rayleigh scattering generally emerges from structures which are
small compared to the wavelengths of visible light; one thus would conclude that the largest inhomogeneities measure less than about 100 nm.
- The thermal conduction of silica aerogel tiles rises significantly
only at gas pressures above 100 mbar; using the Knudsen formula, this
indicates voids in the order of 50 - 100 nm /12/.
- Gas flow resistance is extremely large even for thin aerogel layers; from leak rates mean channel widths in the order of 100 nm can
be estimated.
- Adsorption/desorption techniques are probably not reliable to probe
the pore volume of aerogels above 10 nm; firstly one expects a partial
destruction of the delicate structure. This has been demonstrated with
water adsorption (see Fig. 10), however, may also be true to some extent for N_2. Secondly the Kelvin equation generally used to derive a
pore size distribution links the meniscus directly to the pore width;
this may not be allowed for aerogels, where the secondary particles
possibly form branching chains which enclose 50 nm voids - thus creating the highly porous body.
- Hg-porosimetry is also considered unreliable. As we have demonstrated, aerogels are easily compressed, i. e. their Young's modulus is
very small /7/. If one starts to increase the Hg pressure, the aerogel
samples will first be compressed and only later be penetrated by the
liquid. The separation of compression and penetration effects in the
data reduction process is rather difficult /11/.
- SEM pictures show structures in the 10 to 50 nm range.

OUTLOOK

Though a lot of applications for aerogels have been proposed /13/
and many properties have been thoroughly investigated, up to now a detailed picture of their structural appearances is not on hand. For future
activities it seems to be important to employ non-destructive testing

782

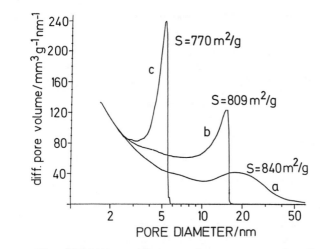

Fig. 10. Differential pore volume versus pore diameter for silica aerogel (Lund, 145 kg/m^3); a) original sample; b) after storage in saturated water vapor; c) after soaking in water, pore volume 0.86 cm^3/g; S = BET-surface /9/.

methods like X-ray or light scattering. Furthermore one should compare aerogels of different densities, coming from different sol-gel processes, made with acid or base catalysis, dried in different supercritical environments etc..

Of central importance is the understanding of the chemical and physical processes which lead to the formation of surfaces and interfaces on a microscopic scale. The next step would be the control and manipulation of these processes, generally called ultrastructural processing /14/.

References

1. D. Büttner, R. Caps, U. Heinemann, E. Hümmer, A. Kadur, and J. Fricke, in Aerogels, Springer Proceedings in Physics, 6, edited by J. Fricke (Springer Verlag, Berlin, Heidelberg, New York, Tokyo 1986), pp. 104-109.

2. D. Büttner, E. Hümmer, and J. Fricke, ibid., pp. 116-120.

3. A. Goetzberger and V. Wittwer, ibid., pp. 84-93.

4. J. Fricke, ibid., pp. 94-103.

5. R. Caps and J. Fricke, ibid., pp. 110-115.

6. O. Nilsson, Å. Fransson, and O. Sandberg, ibid., 121-126.

7. M. Gronauer, A. Kadur, and J. Fricke, ibid., 167-173.

8. M. Gronauer and J. Fricke, Acustica 59, 177 (1986).

9. G. Schuck, W. Dietrich, and J. Fricke, in Aerogels, Springer Proceedings in Physics, 6, edited by J. Fricke (Springer Verlag, Berlin, Heidelberg, New York, Tokyo 1986), pp. 148-153.

10. D. W. Schaefer, J. E. Martin, A. J. Hurd, and K. D. Keefer, in Physics of Finely Divided Matter, Springer Proc. in Phys., 5, edited by N. Boccara and M. Daoud (Springer Verlag Heidelberg, Berlin, New York, Tokyo, 1985), pp. 31-37.

11. F. J. Broecker, W. Heckmann, F. Fischer, M. Mielke, J. Schröder, and A. Stange, in Aerogels, Springer Proceedings in Physics, 6 edited by J. Fricke (Springer Verlag, Berlin, Heidelberg, New York, Tokyo 1986), pp. 160-166.

12. A. Kadur, Diplom Thesis, Universität Würzburg, 1986 (unpublished)

13. J. Fricke, in Aerogels, Springer Proceedings in Physics, 6 edited by J. Fricke (Springer Verlag, Berlin, Heidelberg, New York, Tokyo 1986), pp. 2-19.

14. D. R. Ulrich, in Ultrastructure Processing of Ceramics, Glasses and Composites, edited by L. L. Hench and D. R. Ulrich (John Wiley & Sons, New York 1984), pp. 6-11.

Acknowledgement

This work was partially supported by the Deutsche Forschungsgemein - schaft.

SYNTHESIS OF METAL SULFIDE POWDERS FROM ORGANOMETALLICS

CURTIS E. JOHNSON, DEBORAH K. HICKEY AND DANIEL C. HARRIS
Chemistry Division, Research Department, Naval Weapons Center, China Lake, CA
93555-6001

ABSTRACT

Organometallic reagents have been examined for the low-temperature
preparation of metal sulfide powders which are desired as precursors to optical ceramics. The general approach is typified by the reaction of diethylzinc with hydrogen sulfide in toluene solution at room temperature. Electron
microscopy shows that the white ZnS product consists of aggregates of
$\leq 0.1 \ \mu m$ particles. The product is further characterized as predominantly
cubic zinc sulfide (X-ray diffraction) which contains some residual hydrocarbon. Various organozinc reagents and experimental procedures have been
tried in order to optimize the reaction. The reaction has also been extended
to organometallic complexes of Al and Mg, and also to a ternary system,
$ZnAl_2S_4$.

INTRODUCTION

New low-temperature routes to powder precursors for 8-12 μm IR-
transmitting ceramics are being explored. Organometallic reagents of the
type MR_n (R = alkyl) are examined in room temperature reactions in organic
solvents. These organometallic reagents can often be obtained in high
purity. In solution they can be mixed on the molecular level (bimetallic
complexes are also possible), which is an important consideration for making
ternary materials. Low-temperature reactions in solution are expected to
yield small uniform particles which can be processed under relatively mild
conditions into fine-grained ceramics. These ceramics should be low in
defects and have improved optical and mechanical properties. If the grain
size is substantially smaller than the wavelength of light (8-12 μm), light
scattering between grains in noncubic materials does not occur and a much
wider range of materials would be available.

SYNTHESIS AND CHARACTERIZATION OF ZnS

The reaction of diethylzinc with hydrogen sulfide in solution produces a
fine white precipitate immediately upon mixing (Eq. 1) [1]. Our initial

$$Et_2Zn + H_2S \xrightarrow{\text{solvent}} ZnS + 2 \ EtH \qquad (1)$$

procedure involved purging a Et_2Zn/heptane solution with H_2S through a
syringe needle or small diameter glass tube. This method suffers from plug-
ging of the H_2S outlet by the ZnS product. A better procedure involves the
addition of a Et_2Zn/toluene solution to a toluene solution that is saturated
with H_2S. The ZnS precipitate is then washed with toluene and pentane and
dried at 100°C under vacuum.

The product is characterized as ZnS by elemental analysis and X-ray
powder diffraction. The diffraction pattern in Figure 1 exhibits three broad
lines attributed to β-ZnS [2].

Scanning electron micrographs reveal that the powder consists of aggregates of particles 0.1 μm or smaller in size (Figure 2a).

The major impurity in the ZnS powder is due to residual hydrocarbon.
Elemental analyses show up to 2% carbon and diffuse reflectance FTIR spectra

786

show peaks due to hydrocarbon groups as well as peaks at 2534 and 2446 cm^{-1} tentatively assigned to S-H groups. These results indicate that the reaction of Et_2Zn and H_2S has not proceeded to completion. In order to quantify the amount of residual Zn-Et groups, ZnS is dissolved in acid to liberate ethane which is analyzed by gas chromatography (Eq. 2). These results have been used to optimize the synthesis procedure for ZnS. The data are presented in Table I.

$$ZnS \text{ (residual Zn-Et)} \xrightarrow{H^+} Zn^{2+}(aq) + H_2S + EtH \qquad (2)$$

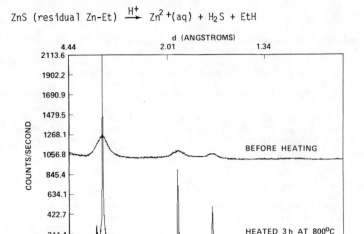

FIGURE 1. X-Ray Powder Diffraction Patterns of ZnS Before and After Heating.

TABLE I. Acid Hydrolysis of ZnS.

Method of ZnS preparation	Alkane, ppm
H_2S purge of 2 M Et_2Zn/heptane	30,000
Standard addition, 0.25 M Et_2Zn/toluene	90
Reverse addition	520
0.25 M Et_2Zn/heptane	260
0.025 M Et_2Zn/toluene	60
0.025 M Et_2Zn, -40°C	35
1.1 M Et_2Zn, liquid H_2S at -78°C	35
Me_2Zn, gas phase	1000 (CH_4)
0.25 M Me_2Zn/toluene	400 (CH_4)
t-Bu_2Zn/toluene	7 (C_4H_{10})

The procedure involving H_2S purge of Et_2Zn/heptane is clearly inferior to the "standard addition" procedure which involves the slow addition of 40 mL of Et_2Zn/toluene solution to 40 mL of H_2S-saturated toluene (30,000 ppm vs. 90 ppm ethane). Three contributing factors have been identified. First, the reaction proceeds further to completion if H_2S is present in excess during the reaction (520 ppm ethane for the "reverse addition" procedure involving the addition of H_2S-saturated toluene to the Et_2Zn/toluene solution).

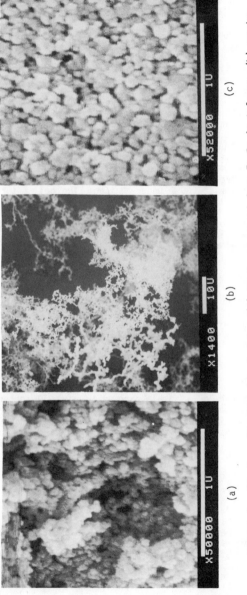

(a)   (b)   (c)

FIGURE 2. Electron Micrographs of ZnS. (a) Powder prepared from Et_2Zn in toluene, (b) powder prepared from Me_2Zn in the gas phase, and (c) pellet pressed at 800 MPa and then heated 3 hours at 800°C.

Second, heptane is not as good a solvent as toluene for the reaction (compare entries 2 and 4 in Table I). Third, slightly better results are obtained for dilute Et_2Zn solutions (this raises the ratio of H_2S to Et_2Zn, compare entries 2 and 5 in Table I). Note also that lower temperatures lead to better results. This is most likely due to the increased solubility of H_2S in toluene at lower temperatures.

Table I also lists results for two other alkyl groups, methyl and tertiary butyl. Dimethylzinc reacts less completely than Et_2Zn while t-Bu_2Zn gives the least amount of residual hydrocarbon. These results parallel the order of stability of the dialkylzinc compounds which is $Me_2Zn > Et_2Zn > t$-Bu_2Zn.

When the reaction of Me_2Zn with H_2S is carried out in the gas phase at room temperature, quite different results are obtained. Mixing H_2S gas with Me_2Zn vapor (argon carrier gas) leads to rapid formation of a highly flocculent precipitate. Electron micrographs show that the solid consists of a fibrous network with chains of tenth-micron sized particles (Figure 2b). The X-ray powder diffraction pattern has fairly sharp lines due to approximately equal amounts of α- and β-ZnS. The average crystallite size calculated from the Scherrer equation is ~250 Å which compares to crystallite sizes ranging from ~30 to 100 Å calculated for ZnS preparations in solution.

The effect of heating on the ZnS has been briefly examined. A pellet of ZnS (prepared by the "standard addition" method) was pressed at 800 MPa and then heated 3 hours at 800°C under vacuum. The pellet turned gray due to residual carbon in the sample. Scanning electron micrographs show that the particle size has increased about five-fold from ~0.02 to ~0.1 μm (Figure 2c). Similar conclusions can be drawn from X-ray diffraction data (Figure 1). The calculated average crystallite size has increased from 45 to 320 Å in the heated material.

SYNTHESIS OF Al_2S_3

The reaction of triethylaluminum with H_2S (Eq. 3) has been performed by the "standard addition" procedure discussed earlier. A fine white powder precipitates rapidly. The powder is extremely reactive in air producing

$$\begin{array}{c} Et \diagdown \quad Et \diagdown \quad Et \\ \quad \diagup Al \diagdown \quad \diagup Al \diagdown \\ Et \diagup \quad \diagdown Et \diagup \quad \diagdown Et \end{array} + H_2S \xrightarrow{\text{toluene}} \text{"}Al_2S_3\text{"} + EtH \qquad (3)$$

H_2S. X-ray analysis of a sample exposed to air ~10 seconds shows only a small percent sulfur. Electron micrographs of this mostly hydrolyzed material show that the particle size is well under 1 μm. Acid hydrolysis of the powder liberates large amounts of ethane indicating that the reaction in Eq. 3 is far from completion. The product contains 6% carbon by elemental analysis, equivalent to 0.25 residual ethyl groups per aluminum.

SYNTHESIS OF $ZnAl_2S_4$

When a 2:1 mixture of triethylaluminum and diethylzinc in toluene is reacted with H_2S, a fine white powder is again obtained. As for aluminum sulfide in the previous section, acid hydrolysis liberates large amounts of ethane. The amount of ethane produced is proportionately the same as that found from the "Al_2S_3" sample, indicating that the residual ethyl groups are due to the aluminum and also that the presence of Et_2Zn has not significantly altered the reaction of Et_3Al with H_2S. The reaction does produce intimately mixed Zn and Al in the product, as X-ray images for Zn, Al, and S look identical at a resolution of ~0.3 μm.

SYNTHESIS OF MgS

The reaction of Et_2Mg with H_2S in diethylether solvent produces a white precipitate which was washed with diethylether and pentane. The X-ray powder diffraction pattern contains weak broad lines for MgS plus some weak unassigned lines. From elemental analysis the solid contains 7.8% carbon. Acid hydrolysis, however, produces only ~25 ppm ethane. Thus, the source of the carbon is not Et-Mg groups, but most likely residual diethylether solvent. Electron micrographs show that the solid consists of highly agglomerated tenth-micron size particles.

CONCLUSIONS

Metal sulfide powders with very small particle sizes have been prepared by the reaction of metal alkyls with H_2S in solution at room temperature. The principal limitation of the method is the presence of hydrocarbon impurities. These impurities can be minimized by adjusting experimental parameters including the choice of metal alkyl. Future directions include processing studies on ZnS and extension of the synthetic method to other metals such as Ti, Y, and Mo.

ACKNOWLEDGMENTS

The authors gratefully acknowledge the assistance of Dr. C. K. Lowe-Ma (X-ray diffraction), R. W. Woolever (SEM), Dr. R. A. Nissan (NMR), Dr. M. P. Nadler (IR), and J. H. Johnson (GC). We also thank the Office of Naval Research for support of this work.

REFERENCES

1. During the course of our work, a brief report of ZnS synthesis from Et_2Zn and H_2S in diethyl ether appeared. Takakazu Yamamoto and A. Taniguchi, Inorg. Chim. Acta <u>97</u>, L11 (1985).
2. JCPDS powder diffraction file 5-0566 (β-ZnS) and 5-492 (α-ZnS).

PREPARATION OF Al_2O_3-ZrO_2 COMPOSITES BY ADJUSTMENT OF SURFACE CHEMICAL BEHAVIOR[*]

S. BAIK, A. BLEIER, AND P. F. BECHER
Structural Ceramics Group, Metals and Ceramics Division, Oak Ridge
National Laboratory, P. O. Box X, Oak Ridge, Tennessee 37831

ABSTRACT

Aqueous colloidal routes for processing binary suspensions containing Al_2O_3 and ZrO_2 were designed and tested in order to achieve homogeneous microstructures. Effects of particle size and size ratio of each component, pH, and electrolyte concentration of composite suspensions on sedimentation, green density, and ZrO_2 distribution in sintered microstructures were examined. The pH conditions for inhibiting differential sedimentation without impairing green density were optimized. Overall suspension and coagulation behavior for these composite systems were explained using the DLVO approach. Optimum balance of colloidal and gravitational forces occurred when the secondary minimum heterocoagulation was maximized.

INTRODUCTION

Zirconia toughened composite ceramics can have unique mechanical properties associated with the phase transformation of dispersed zirconia particles in ceramic matrices [1-3]. The increase in fracture toughness of such systems (e.g., Al_2O_3-ZrO_2 composites) depends on the morphology and distribution of ZrO_2 particles — their size, shape, concentration, and location in Al_2O_3 matrix [2,4,5], which are critically influenced by the processing methodology (e.g., mixing, consolidation and firing).

Sol-gel processing [1,6] is effective in dispersing ZrO_2 in either alumina powders or sols. But, it is not always possible to vary microstructural parameters in a systematic way. Mechanical mixing [2,7], on the other hand, is inadequate due to the inherent inhomogeneity in the original powders and the contamination during vigorous mechanical stirring or milling. In comparison, the colloidal processing routes have several advantages for producing homogeneous composite systems and tailoring the microstructures [8,9]. They include (1) removal of the inhomogeneities in the original powders, (2) dispersion and mixing in colloidal states, and (3) consolidation into green pieces which maintain the uniformity of colloidal state, a favorable packing arrangement for subsequent densification.

However, when two or more powders with different size, density, shape, or surface charge characteristics are mixed and consolidated in dispersed states, one severe problem will be inevitably encountered: demixing or phase separation due to differential settling during consolidation stage. Larger or denser particles settle faster than the small or lighter particles. Two approaches have been taken to overcome such problems: (1) to use highly concentrated slips containing a specific polymeric stabilizer, or (2) to consolidate in the flocculated states. While the former approach has been adopted successfully in various single component ceramic systems, the application to the multicomponent composite ceramics has not been systematically studied. The latter approach has been

*Research sponsored by the Division of Materials Sciences, U.S. Department of Energy, under contract DE-AC05-840R21400 with Martin Marietta Energy Systems, Inc.

used in various binary systems, e.g., Al_2O_3-ZrO_2 [10], Al_2O_3-SiO_2 [11]. Spontaneous heteroflocculation can preserve the mixing state during consolidation, but produce "open" flocculate structure [12] which is undesirable for subsequent densification.

This paper reports an alternative colloidal processing concept that produces homogeneous microstructures in the Al_2O_3-ZrO_2 composites for a wide variety of morphological characteristics of ZrO_2 particles. The electrostatic interactions of each particle are predicted by the DLVO model and tested by sedimentation. The optimum pH routes to avoid differential settling while maintaining favorable particle packing states are identified.

PRIMARY AND SECONDARY MINIMUM COAGULATION IN BINARY SUSPENSIONS

According to the classical DLVO treatment [13] of the stability of lyophobic colloids, the sum of the electrical double layer repulsion and van der Waals attraction leads to a relationship between the interaction energy and the interparticle distance, which exhibits two minima in energy separated by a maximum as shown schematically in Fig. 1. The coagulation can occur in either of two low energy states. One that occurs at the deep minimum near to the contact point is defined as the primary minimum coagulation (PMC) and the other with a shallow minimum at further distance is the secondary minimum coagulation (SMC) [14,15]. The PMC is irreversible due to its deep energy well (>100 kT), whereas the SMC is believed to be reversible and can be redispersed with little external force because the energy well is shallow (approximately a few kT) and has no potential barrier.

ORNL-DWG 86C-9980

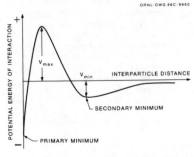

Fig. 1. Schematic diagram of potential curve as a function of the distance between two particles.

The phase separation due to differential settling in binary suspensions can be prevented by heterocoagulation in either modes — PMC or SMC — by regulating the relative surface charge characteristics. For instance, adjusting pH of the mixed colloidal suspension to the region between the points of zero charge (p.z.c.) of each component, the potential barrier due to the double layer repulsion can be removed and the mixed states can be preserved by rapid PMC. However, it would be advantageous

Table I. Particle sizes after sizing

	Sizing conditions* (g·h)	Size range (μm)	Average size (μm)
Al_2O_3, HPS-40	24-100	0.2-1.0	0.5
ZrO_2, SC105	1-24	0.3-1.5	0.71
ZrO_2, TZ-0	48-200	0.1-0.5	0.20

*Dispersed in water at pH 2.5; g·h represents gravity times hours.

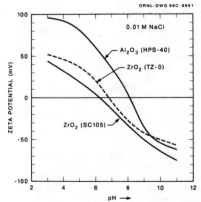

Fig. 2.   Zeta potential versus pH
at 0.01 M ionic strength.

to utilize the heterocoagulation in SMC mode because irreversible physical contact can be avoided and favorably packing states can be realized during consolidation as demonstrated recently for homocoagulation of mono-dispersed latex [16] and possibly TiO_2 [17]. The purpose of this study is to answer whether we can identify the pH range where the secondary minimum heterocoagulation (SMHC) is fast and stable enough to be used in Al_2O_3-ZrO_2 composite system.

MATERIALS

Two commercial grade zirconia* and one alumina** powders were selected due to their large differences in particle size (see Table I). As-received powders were highly agglomerated but ball milling in water at pH 2.5 broke the agglomerates. The powder size distribution was further tightened by sedimentation and centrifugation. Table I shows the final particle size, size range and sizing conditions. Final slurries were stored at pH 2.5 for subsequent experiments.

ZETA POTENTIAL

The electrophoretic mobilities were measured for each powder, using a Pen Kem 3000 electrokinetic analyzer as a function of pH, and used to calculate zeta potentials according to Henry's equation [18],

$$\xi = 6 \ \pi \eta \mu / \epsilon [1 + 0.5 \ f(\kappa a)] \ , \qquad (1)$$

where μ is the electrophoretic mobility of the particle in a solution of viscosity η and dielectric constant ϵ. The correction factor, $[1 + 0.5 \ f(\kappa a)]$, is a function of particle radius, a, and double layer thickness, κ^{-1}. It varies between 1.0 (low κa) and 1.5 (high κa) [19]. Before the measurements, the powders were cleaned by repeated washing with doubly-distilled water and aged for 72 h at desired pH values. The ionic strength of the solution was 0.01 M NaCl. The pH adjustment was done with either dilute HCl or NaOH.

Figure 2 shows the zeta potential versus pH for the powders used. The p.z.c. of alumina (HPS-40) is approximately 8.2. The values obtained for zirconia are 6.2 for SC105 and 6.8 for TZ-0, which agree closely with the values reported recently [9].

---

*Magnesium Elektron, SC105 and Toyo Soda Manufacturing Co., TZO.
**Sumitomo Chemical Co., HPS-40.

PAIRWISE POTENTIAL CALCULATION — THE DLVO THEORY

Using the zeta potentials measured as a function of pH, the pairwise interaction energies were calculated based on the DLVO theory and the treatment by Hogg et al. [20] for the repulsive energy, V_R. The surface potentials were assumed to be constant, i.e., the surface potential of each particle is taken to be independent of the distance of separation [21]. The attractive potential energy, V_A, due to van der Waals forces was calculated using the equation developed by Hamaker [22]. Hamaker constants used in the calculation are $A_{131} = 4.5 \times 10^{-20}$ J, $A_{232} = 13.95 \times 10^{-20}$ J, and $A_{132} = 7.9 \times 10^{20}$ J where 1, 2, and 3 represent alumina, zirconia, and water, respectively. They are determined using the equation discussed by Israelachvili [23] on the basis of the Lifshitz theory. The total interactive potential, V_T, was taken as the sum of V_R and V_A after taking into account a displacement of 1-nm arising from the Stern layer. Then the maximum and minimum (secondary) energies and corresponding interparticle distances were determined in the pH range 3-11 for each pair of homo- and heterointeractions.

Figure 3 shows the potential barriers, $V_{max} - V_{min}$, versus pH for the case of Al_2O_3 (HPS-40) - ZrO_2 (TZ-0) pairs. Due to the large Hamaker constant for zirconia compared to that of alumina, the potential barrier for zirconia was removed over a wide range of pH (5.0 ~ 8.5). The potential barrier for heterointeraction between alumina and zirconia was found to be zero between pH 6.5 and 8.5, which implies that in this pH range irreversible heterocoagulation contributes significantly to overall stability of the binary suspension. Therefore, in order to avoid primary minimum heterocoagulation (PMHC) the pH values have to be either lower than 6.5 or higher than 8.5. In the case of SC105, the corresponding pH range is lower than 5.5 or higher than 9.0. An increased contribution of van der Waals attraction arising from the larger particle size of ZrO_2 contributes to lowering the energy barrier.

On the other hand, as shown in Fig. 4, in order to optimize SMHC it is desirable to stay at a pH value close to 6.0-6.5 for TZ-0 and 5.0—5.5 for SC105. Under these conditions, the secondary minimum energy can be as deep as ~10 kT for TZ-0 and ~20 kT for SC105, which may be sufficient for SMHC to contribute significantly to the overall stability of the mixed suspension.

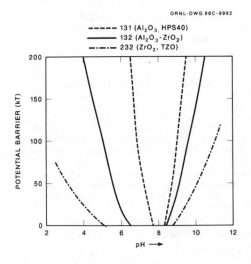

Fig. 3.
The potential barrier, $V_{max} - V_{min}$, versus pH for pairwise interactions in the binary suspension of Al_2O_3-ZrO_2 (TZ-0).

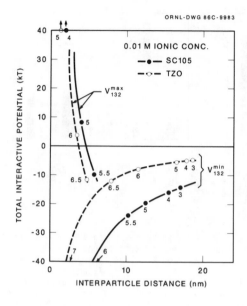

ORNL-DWG 86C-9983

0.01 M IONIC CONC.

●— SC105
○— TZO

Fig. 4.
The maximum and minimum
potential energy for
heterointeraction as a
function of separation.
The numbers in the figure
represent the pH value.

SEDIMENTATION

For the purpose of studying experimentally the overall stability
of the binary mixture, as a function of pH, the sedimentation behavior
was observed with 5 vol % of 1:1 mixture of Al_2O_3-ZrO_2 and compared with
those of single component suspensions in the pH range of 4-8. Mixed
suspensions were initially prepared at pH 2.5. After sonication, the pH
was adjusted by adding 0.1 M NH_4OH solution. The ionic strength of the
suspensions after pH adjustment is estimated to be between 0.006 and 0.01.
Preiodically, Periodically, the settling heights were measured, and the pH
was readjusted, if necessary.

The bar diagrams in Fig. 5 compare the settling heights after two
weeks. The stability behavior of alumina and zirconia was consistent
with the DLVO model. The binary suspensions showed some interesting
behavior especially at pH 5 and 6, the conditions with which we are
especially concerned. All the suspensions at pH 4 were found to be highly
stable, as evidenced by the diffused boundaries between the solid and
water and low settled volume. Nonetheless, the evidence of differential
settling was observed clearly in the case of binary mixture with SC105.
On the other hand, at pH 7 and 8, rapid flocculation took place and large
settled volumes were established in less than 2 to 3 h indicating that
irreversible heterocoagulation occurred rapidly and formed the "open"
flocs. At pH 5 and 6, binary suspensions indicate some degree of
unstability while the Al_2O_3 suspension remains highly stable. However,
ZrO_2 suspensions clearly indicate much less stability in comparison to the
binary suspension. Moreover, the settling volumes of binary suspension at
pH 5 and 6 are less than those of destabilized single component, ZrO_2 in
this instance. Such observations clearly demonstrate the presence of
heterointeraction in the binary suspension in this pH region. However,
the nature of interaction is markedly different from those of the
heterointeraction observed at pH 7 and 8.

796

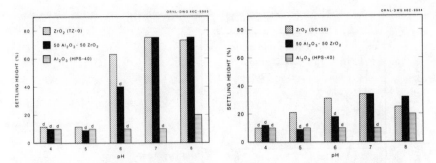

Fig. 5. Settling heights measured after two weeks for Al_2O_3, ZrO_2 and binary suspensions (a) for TZ-0 (b) for SC105. Diffused boundaries are indicated by "d".

GREEN AND SINTERED DENSITY

In order to examine the effect of pH on the zirconia particle distribution and the packing density, several samples were prepared using the composition of Al_2O_3-10 vol % ZrO_2. A 5 vol % solids binary mixture was prepared at pH 2.5 and the pH was adjusted to 4, 5, 6, 7, and 8. Then green compacts were prepared by the filter-pressing technique at 15 psi of Ar gas. After drying in a vacuum oven, green densities were measured and the sintered densities were determined after firing in air at 1475°C/1 h. The distribution of zirconia particles in alumina matrices was examined using sintered pieces after polishing and thermal etching (1400°C/air).

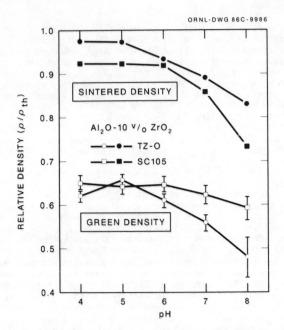

Fig. 6. Green and sintered density versus the pH for consolidation.

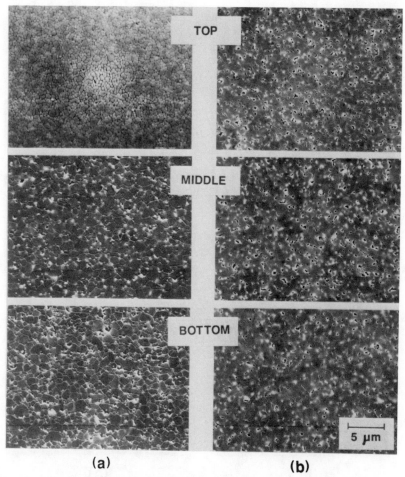

Fig. 7. Microstructures at the top, middle, and bottom of the specimen cross section. The Al_2O_3-10 vol % ZrO_2 composites with (a) TZ-0, pH = 4, (b) TZ-0, pH = 5. (Continued on next page.)

As shown in Fig. 6, relatively high green density as well as sintered density were maintained up to pH 5 for TZ-0 and up to pH 6 for SC105. Above these critical pH points, a monotonic decrease in density was observed. These results clearly demonstrate the advantage of colloidal processing with stable suspensions in terms of particle packing and subsequent densification. However, phase separation due to differential settling was confirmed for specimens prepared at pH 4. The zirconia concentration near the top surface of the composite made with TZ-0 is much higher than the average concentration while the bottom portion shows the depletion of zirconia (Fig. 7a). The situation is reversed with the composite made with SC105. The top layer is almost entirely alumina while the bottom portion is made of entirely zirconia (Fig. 7c). On the contrary, homogeneous distributions of zirconia were observed with the specimens prepared at pH 5 for TZ-0 (Fig. 7b) and at pH 6 for SC105 (Fig. 7d).

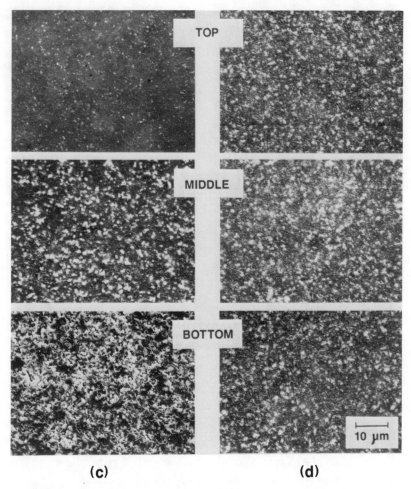

(c)   **(d)**

Fig. 7 (Cont.). (c) SC105, pH = 4, (d) SC105, pH = 6.

DISCUSSION

Table II summarizes the optimum pH values determined from the DLVO calculation and experimental observations. Fair agreements could be found among them. The criteria used for the model calculation — the maximization of secondary minimum heterocoagulation — predict the optimum pH values close to the p.z.c. of one component. However, they can vary slightly depending on the relative particle size and total number ratio of the binary mixture, as observed through experiments. In the case of 10 vol % SC105 the particle number ratio between alumina and zirconia is approximately 20:1, which implies that a pH value even nearer to the p.z.c can be used, because the probability of heterocollision is much greater than the homocollision of ZrO_2 and the heterocoagulation is somewhat controlled.

Table II. Optimum pH values determined by various methods
for processing Al_2O_3-ZrO_2 composites

ZrO_2 (p.z.c)	DLVO theory	Sedimentation	Density and microstructure
TZ-0 (6.8)	6—6.5	5—6	5
SC105 (6.2)	5.5	5—6	6

On the contrary, the number ratio becomes approximately 4:6 in the case
of 10 vol % TZ-0, implying that heterocollisions are as likely as homo-
collisions. Thus, heterocoagulation and the homocoagulation of zirconia
have to be carefully balanced in order to maintain high packing density in
this case.

It is also possible to use pH ranges higher than the p.z.c of alumina
(~8.2). Mixing can be performed at pH 11 ~ 12 where both components are
highly stable with negative surface changes. The pH is then adjusted to
pH ~ 9 to destabilize the alumina particles and optimize the secondary
minimum heterocoagulation. This particular pH route may be preferable if
the number of particles of zirconia is much higher than those of alumina,
such as the composites made of TZ-0. One of the disadvantages is,
however, the narrow pH range associated with the stability-instability
transition for alumina.

Another important variable that can change the optimum conditions
significantly is the ionic strength of the solution. The double layer
repulsive energy is also strongly dependent on the ionic strength. It
can be controlled either by adding salt or by selecting the proper pH
during initial mixing. In general, high ionic strength increases the
depth of secondary minimum energy by tightening the double layer thickness.
However, if ionic strength is too high, according to the DLVO model, PMC
will ensue.

CONCLUSION

A new colloidal processing concept for producing the Al_2O_3-ZrO_2
composite ceramics was investigated. To prevent the differential
settling, after mixing at the low pH, the pH of the solution is readjusted
before consolidation to the range where the secondary minimum heterocoagulation is maximized. The advantage of this particular colloidal processing
route is to obtain favorable green states for densification while maintaining homogeneous mixing. We believe that this processing concept can
be applied in principle to all the binary composite systems involving two
physically and chemically different powders.

ACKNOWLEDGEMENTS

The authors appreciate S. B. Waters and C. G. Westmoreland for
experimental assistance, M. A. Janney and R. J. Lauf for reviewing the
manuscript, and F. Stooksbury for its preparation. The helpful discussions
with Drs. R. M. Cannon and R. Raj are also greatly appreciated.

REFERENCES

1. N. Claussen and M. Rühle, in Advances in Ceramics, Vol. 3, edited by A. H. Heuer and L. W. Hobbs (American Ceramic Society, Columbus, OH, 1981), pp. 137-63.
2. F. F. Lange, J. Mat. Sci. 17, 247-54 (1982).
3. P. F. Becher, Acta Metall., in press.
4. A. H. Heuer, N. Claussen, W. M. Kriven, M. Rühle, J. Amer. Ceram. Soc. 65, 642-650 (1982).
5. A. G. Evans, N. Burlingame, M. Drory, W. M. Kriven, Acta Metall. 29, 447-456 (1981).
6. P. F. Becher and V. J. Tennery, in Fracture Mechanics in Ceramics, Vol. 6, edited by R. C. Bradt et al. (Plenum Press, New York, 1983), pp. 383-99.
7. F. F. Lange, J. Am. Ceram. Soc. 66, 396-98 (1983).
8. I. A. Aksay, F. F. Lange, B. I. Davis, ibid., 66, C-190 (1983).
9. E. M. DeLiso, W. R. Cannon, A. S. Rao, presented at the 1985 MRS Annual Meeting, Boston, MA, 1985.
10. E. Carlström and F. F. Lange, J. Am. Ceram. Soc. 67, C-169 (1984).
11. P. E. Debély, E. A. Barringer, H. K. Bowen, ibid., 68, C-76 (1985).
12. J.Th.G. Overbeek, in Emergent Process Methods for High-Technology Ceramics, Materials Science Research, Vol. 17, edited by R. F. Davis et al. (Plenum Press, New York, 1984), pp. 25-43.
13. B. V. Derjaguin and L. D. Landau, Acta Physicochim. URSS 14, 631 (1941); E. J. Verwey and J.Th.G. Overbeek, Theory of the Stability of Lyophobic Colloids (Elsevier, Amsterdam, 1949).
14. J.Th.G. Overbeek, Chapter III in Colloid Science, Vol. 1, edited by H. R. Kruyt (Elsevier Publishing Company, 1952).
15. R. Hogg and K. C. Yang, J. Colloid Interface Sci. 62, 407 (1977).
16. G. Y. Onoda, Phys. Rev. Letters 55, 226 (1985).
17. E. A. Barringer and H. K. Bowen, J. Am. Ceram. Soc. 65, C-199 (1982).
18. D. C. Henry, Proc. Roy. Soc. A133, 106 (1931).
19. A. L. Smith, Chapter 2 in Dispersion of Powders in Liquid, edited by G. D. Parfitt (Elsevier, New York, 1969).
20. R. Hogg, T. W. Healy, D. W. Fuerstenau, Trans. Faraday Soc. 62, 1638 (1966).
21. A. Bleier and E. Matijevic, J. Colloid Int. Sci. 55, 510 (1976).
22. H. L. Hamaker, Physica 4, 1058 (1937).
23. J. N. Israelachvili, Intermolecular and Surface Forces (Academic Press, London, 1985), p. 144-145.

DIRECT PROCESS METAL ALKOXIDES
AS CERAMIC PRECURSORS

R. J. AYEN AND J. H. BURK
Stauffer Chemical Company, Dobbs Ferry, NY 10522

ABSTRACT

Metal alkoxides have great promise as oxide ceramic precursors. Direct manufacture from the metal and the alcohol permits the production of high purity materials and of those alkoxides which cannot be easily obtained by other routes. Examples which are of great interest to ceramics are the alkoxides of silicon, aluminum, magnesium and yttrium.

Currently, only tetraethyl silicate is produced in large volumes by the direct route, employing a catalytic continuously fed process, yielding a distilled product. The reactivity of silicon, aluminum, magnesium and yttrium is governed by the metal surface state and activation and varies with different catalysts and alcohol chain length and branching.

The major use of tetraethyl silicate is in partially hydrolyzed form as binder for precision casting molds and zinc-rich primer coatings. Judicious choice of alkoxide ligands allows the manufacture of soluble or liquid Si, Al, Mg, and Y derivatives. These are especially suitable for molecular doping and mixing in the manufacture of yttria-stabilized zirconia, mullite, spinel, cordierite and other oxide ceramic precursors.

INTRODUCTION

Metal alkoxides are gaining industrial importance as catalysts, binders, crosslinkers, and coupling agents. The expectations and promise of metal alkoxide use in ceramics, glass, fibers and coatings have increased over the years [1]. Progress in this area is evidenced by application of alkoxide/oxide powder and sol-gel technology [2,3].

Currently, the most widely used alkoxides in industry are those of silicon, titanium, aluminum, magnesium and sodium. Their methods of manufacture are governed by the reactivity of the metal or the metal derivatives and of the alcohol. Generally, the electropositive metals, such as the alkali and alkaline earth metals, or aluminum, or yttrium, can react with an alcohol directly to form the metal alkoxide and hydrogen (Eqn. 1).

$$M + n\ ROH \longrightarrow M(OR)_n + n/2\ H_2 \qquad (1)$$

The rate of this reaction increases with the electropositive character of the metal and decreases with increasing ROH alkyl chainlength and branching. In contrast, titanium and zirconium alkoxides are most conveniently synthesized from the metal halides and the alcohol in the presence of a base (Eqn. 2).

$$MCl_4 + 4\ ROH \xrightarrow[\text{Solvent}]{:B} M(OR)_4 + 4\ HB^+Cl^- \qquad (2)$$

General reviews of metal alkoxide chemistry, including synthetic routes, physical and chemical properties, and uses can be found in "Metal Alkoxides" by Bradley, Mehrotra and Gaur [4], and in Kirk-Othmer "Encyclopedia of Chemical Technology" [5].

Except for the most electropositive alkali and alkaline earth metals, direct process metal alkoxide technology necessitates physical or chemical activation of the metal to render the surface reactive and to improve rate and yield of alkoxide formation. Activation has included grinding or milling of magnesium, aluminum or yttrium under nitrogen with or without a catalyst [6,7]. Use of silicon alloys with calcium, magnesium, copper or iron has been described for production of alkyl silicates [8,9,10,11]. Most common has been the reaction of the metal with iodine or other reactive metal halides to etch the metal surface [4,12,13]. Etching and/or amalgamation with mercury compounds has been employed [14]. Amine surfactants and alkoxy alcohols and their salts have been used in catalyst systems and also as ligands [15, 16, 17]. Also very common is the activation of the metal surface with preformed alkali metal alkoxide or the product alkoxide [11,18].

Catalysts and activation of the metal are needed for the direct process to counteract surface passivation by oxides, decomposition products, etc. Thus, very clean freshly generated "active silicon" was found to react as vigorously as an alkali metal with methanol at 20-60°C [19]. Production of methoxysilanes and tetramethoxysilicate from Si/MeOH reactions support a Rochow reaction mechanism for metal/alcohol reactions, with metal insertion into the H-OR bond [10].

DIRECT PROCESS METAL ALKOXIDES

Conventional Ethyl Silicate Process

For over a century, the preferred method of production, laboratory scale or commercial, was von Ebelmen's original 1846 synthesis [20] from silicon tetrachloride; i.e.,

$$SiCl_4 + 4\ C_2H_5OH \longrightarrow Si(OC_2H_5)_4 + 4\ HCl \qquad (3)$$

The process can be run in a batch or continuous mode. The reaction is exothermic (64.8 kj/mol), and the heat of reaction can be used to drive off the by-product hydrogen chloride. The process yields a bottoms-out product which contains at least 90% tetraethyl orthosilicate. This material is then neutralized, using an alkali such as sodium hydroxide or sodium carbonate. At this stage, the product is referred to as "condensed" ethyl silicate. As will be seen below, most of this material is converted to other products. In preparing some of these products, by-product ethanol is formed, which is recovered and recycled to the $SiCl_4$/ethanol reaction step. Obviously, recovery and disposal of the by-product HCl is a major consideration in this process.

Production of Ethyl Silicate by Direct Process

Ethyl silicate is produced commercially in large volumes. Considerable effort has therefore been expended on developing a direct process, with the goals of eliminating the formation of HCl, reducing raw material costs, and producing a higher quality product. Numerous patents have been issued in this area, with most emphasizing the use of novel catalyst systems [11,15,16,17]. In 1985, full scale commercial production of tetraethyl orthosilicate was started using a direct, semi-continuous process [21].

In this process, an alkali metal salt of a high boiling hydroxy compound, such as an alkoxyalcohol or an alkoxy-alkoxyalcohol, is used to catalyze the reaction between silicon and ethanol [17]. The catalyst is added to the reactor as a liquid solution, and powdered silicon metal is charged to provide the excess needed for the reaction. The production of ethyl silicate is then carried out <u>via</u> a continuous reaction (Eqn. 4).

$$4 \ C_2H_5OH + Si \longrightarrow Si(OC_2H_5)_4 + 2 \ H_2 \qquad (4)$$

The reactor is heated to 160-180°C, and the ethanol feed is started. Once the reaction is proceeding at a steady rate, silicon metal is fed at a rate just sufficient to maintain the initial excess amount. The reaction rate is monitored by measuring the hydrogen evolution. Ethyl silicate product and unreacted alcohol are removed continuously from the reactor by distillation and are fed to a fractionation column. Alcohol is recovered from this column and is recycled to the reactor. The product, ethyl silicate "condensed", is obtained from the column reboiler. A per-pass ethanol conversion of about 50% is typical.

After the production of 40-50 kg of ethyl silicate per kg of catalyst solution, the catalyst activity begins to decline. When the reaction rate reaches an unacceptable level, the silicon and ethanol feeds are stopped. The material remaining in the reactor, which contains spent catalyst solution and unreacted silicon, is cooled and removed for disposal. The reactor is then cleaned to prepare it for the next production cycle. A flow chart for the direct process is shown in Figure 1.

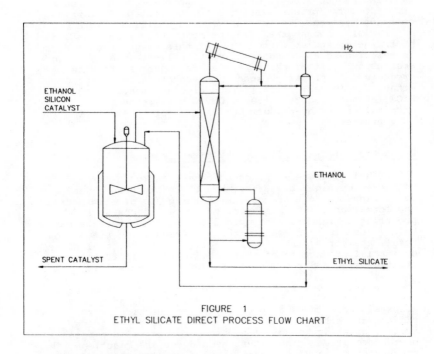

FIGURE   1
ETHYL SILICATE DIRECT PROCESS FLOW CHART

Prehydrolyzed Ethyl Silicate

Ethyl silicate from the direct process has a typical assay of 96% and is sold as "condensed" grade. A small fraction of the "condensed" is redistilled and is sold as a "pure" grade, with a typical assay of 99.5%. However, most ethyl silicate produced is not sold as tetraethyl ortho-silicate, but it is instead converted to a 40% hydrolyzed product or further hydrolyzed and formulated and sold as "binder". The reaction to produce the 40% hydrolyzed product (Eqn. 5).

$$Si(OC_2H_5)_4 + 0.8\ H_2O \xrightarrow{H^+} Si(OC_2H_5)_{2.4}\ O_{0.8} + 1.6\ C_2H_5OH \quad (5)$$

utilizes a strong mineral acid, such as hydrochloric or sulfuric, as a catalyst. The hydrolysis proceeds in steps with ethoxy groups being replaced by hydroxy groups, followed by condensation.

Ethyl silicate 40 contains ethanol, ethyl silicate monomer, and various oligomeric products. These range from dimer and trimer to more highly condensed linear, branched, and cyclic structures (Table I).

TABLE I

TYPICAL DISTRIBUTION OF LOW MOLECULAR WEIGHT
MATERIALS IN ETHYL SILICATE 40*

	Area %
C_2H_5OH	2.1
Monomer	25.4
Dimer	19.8
Trimer	14.6
Tetramers	11.0
Pentamers	10.6
Hexamers	5.3
Heptamers	4.5
Octamers	3.2
Unknown	3.5

* Gas chromatography

Commercial binders are formed by further hydrolysis, up to 90%, usually in the presence of another alcohol (other than ethanol) or an alkoxyalcohol. The most common binders are hydrolyzed to 70-90% levels, again using strong mineral acid catalysts. Another series of binders are hydrolyzed to levels below 70% using alkaline catalysts such as organic amines.

Ethyl Silicate Uses

Approximately 80% of the ethyl silicate produced is sold in prehydrolyzed forms for major applications as binders for zinc-rich coatings and precision investment casting [5]. In zinc-rich coatings the binders are used to bond zinc dust to a metal surface. The polymeric silicates in the binder are soluble in the coating formulation. After being coated onto the metal surface, these polymers further condense and deposit out of solution because of reaction with atmospheric moisture and evaporation of solvents. Some free, amorphous silica is formed which reacts with the zinc dust to form zinc silicate. Zinc is above iron in the electromotive force series of elements. Its greater tendency to oxidize than iron suppresses the oxidation, or corrosion, of iron with which it is in contact. This proceeds until the zinc has been consumed. Thus, the zinc "sacrificially" protects the iron. Coatings of this type are said to provide cathodic protection.

Investment casting, also known as the "lost wax process", is a technique for producing very detailed, precisely dimensioned cast metal objects. The process is begun by preparing a detailed wax pattern. Ethyl silicate binders are then mixed with refractory flours such as powdered zirconia and powdered fused silica to form a slurry. The wax pattern is dipped into this slurry, the slurry-coated part is coated with a fine, dry ceramic sand, and the resulting coating is cured in hot air or ammonia. This process is repeated several times, using progressively coarser grades of the dry ceramic sand, until a self-supporting shell, 3/16 to 3/8" thick, is formed. The coated part is then heated to melt the wax pattern, and the wax is drained from the mold. The mold is fired to burn out the last traces of wax, and the hot, molten metal is poured into the mold.

Aluminum, Magnesium and Yttrium Alkoxides

The production levels and use of aluminum and magnesium alkoxides lag far behind those of the silicates or titanates. The aluminum isopropoxide and butoxides are items of commerce [5,22]. The alkoxides are produced directly from aluminum metal either without a catalyst or using catalysts such as iodine, mercuric chloride, aluminum chloride, tin chloride, alkoxy alcohols, and preformed aluminum alkoxides [23,24]. The straightforward batch reaction of the metal and alcohol is difficult to control, and many of the patents in this area are directed towards controlling the reaction rate and, therefore, the exotherm and the generation of hydrogen.

Techniques for producing magnesium alkoxides directly from the metal are similar to those used for aluminum alkoxides, although less emphasis is placed on controlling the reaction rate. Catalysts cited include mercuric chloride, iodine, and the preformed product alkoxide [25]. Use of a solvent with a boiling point in the 100-150°C range, combined with operation at 5-20 atm pressure results in reaction times of a few hours for complete conversion. Adding a catalyst then further reduces the reaction time. The C_1 through C_4 alkoxides can be produced in this manner.

Yttrium alkoxide as an oxide precursor is of great potential interest in a variety of structural ceramics, fluxing, coating and optoelectronic applications. Use and preparative scale are still small. Yttrium isopropoxide has been prepared by Brown and Mazdiyasni [26] by reacting yttrium metal powder or turnings with isopropanol using mercuric chloride as a catalyst (Eqn. 6).

$$Y + 3 \text{ i-}C_3H_7OH \xrightarrow{\text{HgCl}_2} Y(\text{Oi-}C_3H_7)_3 + 3/2 \text{ H}_2 \qquad (6)$$

The isopropanol was dried before use, and water and oxygen were carefully excluded from the reactor. Isopropanol, in large excess, plus yttrium metal and trace amounts of catalyst were charged to a reaction flask, and the reaction was carried out under reflux conditions (82°C) for 24 hours. Yields of up to 75% based on the metal were reported. Milling of the yttrium in the presence of a catalyst improves yields [7]. Application of yttrium isopropoxide in the preparation of yttria-doped zirconia particles from the controlled hydrolysis of the metal alkoxides was recently reported [27].

Complex metal alkoxide salts can be employed as molecular precursors for multi-metal oxide ceramics [1,28,29]. Early reports describe the formation of a large number of metal alkoxide complexes [12]. Thus, Meerwein et al. synthesized the ethyl- and propyl-derivatives of $MgAl_2(OR)_8$ directly from the metals with $HgCl_2$ and I_2 catalysts. Other preparative routes to spinel alkoxide precursors include use of alkoxyalcohols [30] and mixing of the preformed magnesium and aluminum alkoxides [1]. Similarly, polycrystalline mullite ($3Al_2O_3 \cdot 2SiO_2$) has been synthesized by hydrolysis of stoichiometric mixtures of the aluminum and silicon isopropoxides [29].

CONCLUSION

The technology for continuous manufacture of tetraethyl orthosilicate directly from the metal and alcohol has been developed and commercialized. This allows cost-effective and environmentally clean production of this alkoxide. Additional work is needed for the development of catalysts and processes for aluminum, magnesium and other metal alkoxides of special interest for ceramic and glass precursor applications.

REFERENCES

1. H. Dislich, Angew. Chem., Int'l. Edn., 10, 363(1971)
2. H. Dislich, J. Non-Crystalline Solids, 57, 371(1983)
3. Electronic Chemicals News, 1/6/86, pp 7-8
4. D. C. Bradley, R. C. Mehrotra and D. P. Gaur, Metal Alkoxides, Academic Press Inc. (London) LTD., 1978.
5. Kirk-Othmer, Encyclopedia of Chemical Technology, 3rd Edn., Volume 2, pp 1-17; Volume 20, pp 912-921.
6. John M. Gaines, U. S. Patent No. 2 927 937 (1960)
7. C. C. Greco and K. B. Triplett, Stauffer Chemical Co., Patent application filed.
8. F. Speer and E. Wiberg, U. S. Patent No. 2 909 550 (1955)
9. C. P. Haber, U. S. Patent No. 2 445 576 (1948)
10. W. E. Newton and E. G. Rochow, Inorg. Chem. 9, 1071 (1970)
11. Otto Bleh, U. S. Patent No. 3 627 807 (1971)
12. H. Meerwein and T. Bersin, Ann. Chem., 476, 113-150 (1929)
13. W. A. Rex, U. S. Patent No. 2 666 076 (1954)
14. K. Mazdiyasni, L. M. Brown and C. T. Lynch, U. S. Patent No. 3 757 412 (1973)

808

15. G. Kreuzburg, A. Lenz and W. Rogler, U. S. Patent No. 4 113 761 (1978)

16. A. R. Anderson and T. H. Porter, U. S. Patent No. 3 803 197 (1974)

17. W. L. Magee and J. E. Telschow, U. S. Patent No. 4 288 604 (1981)

18. W. Joch, A. Lenz and W. Rogler, U. S. Patent No. 4 197 252 (1980)

19. E. Bonitz, Angew, Chem., Int'l Edn., 5, 462(1966)

20. J. J. von Ebelmen, Ann. Chem., 57, 319-355(1846)

21. Stauffer plants in Weston, MI and Paulinia, Brazil with multi-million pound capacity.

22. G. C. Whitaker, Aluminum Alcoholates and the Commercial Preparation and Uses of Aluminum Isopropylate, Advances in Chemistry Series No. 23, pp 184-9 (1959)

23. P. Kobetz, H. Shapiro and F. J. Impastate, U. S. Patent No. 3 717 666 (1973)

24. A. J. Buzas and R. T. E. Schenk, U. S. Patent No. 3 446 828 (1969)

25. H. Feichtinger, H. Noeske and H. W. Birnkraut, U. S. Patent No. 3 920 713 (1975)

26. L. M. Brown and K. S. Mazdiyasni, Inorg. Chem., 9, 2783 (1970)

27. B. Fegley, Jr., P. White, and H. K. Bowen, Ceram. Bull., 64, 1115(1985)

28. M. Natsui and T. Takahashi, U.S. Patent No. 4 543 346 (1985)

29. K. S. Mazdiyasni and L. M. Brown, J. Amer. Ceram. Soc., 55, 548(1972)

30. I. M. Thomas, U.S. Patent No. 3 903 122 (1975)

OXIDE - NON-OXIDE COMPOSITES BY SOL-GEL

EDWARD J. A. POPE AND J. D. MACKENZIE
Department of Materials Science and Engineering
University of California, Los Angeles, CA 90024

ABSTRACT

In this paper, three types of composites prepared using the sol-gel process are examined. Porous composites, based upon the dispersion of solid, oxide and non-oxide ceramic particles in a sol-gel derived matrix are described. A transparent composite, prepared by the impregnation of a polymer phase into a porous gel matrix is presented. Light-weight, "triphasic" composites, possessing good abrasion resistance and high fracture ductility, can also be fabricated by the polymer impregnation of a porous gel composite.

INTRODUCTION

Recently, the sol-gel route to glass and ceramics has received widespread interest for the preparation of bulk glasses, porous solids, and thin films[1]. Unlike traditional processing techniques, in which powders are reacted at high temperatures, the sol-gel process relies upon polymerization reactions in liquid solution at temperatures near ambient. This liquid solution route to ceramics results in the added advantage of high homogeneity of the reaction product.

This paper describes three types of composites prepared via the sol-gel processing route. These are:1)organic polymer-oxide composites, prepared by the polymer impregnation of a sol-gel matrix; 2) porous composites, based upon the dispersion of solid particles in a sol-gel derived matrix, and; 3) triphasic composites, obtained via the polymer impregnation of porous composites. The processing routes that result in these composites are presented in figure 1. Other composites have also been made utilizing the sol-gel route[2-6].

Some of the advantages of the sol-gel route for the fabrication of composites include the ability to prepare compositions and microstructures that are presently impossible or exceptionally difficult by non-sol-gel techniques. Wide variations in porosity, from 2 to 80 percent, are readily achievable[7]. Small pore diameters ranging from 10 to 1000 angstroms are common. Moreover, continuous "dual matrix" materials are possible by polymer impregnation.

ORGANIC POLYMER - OXIDE COMPOSITES

Properties of the Porous Gel

Transparent organic polymer - oxide composites have been prepared by impregnating a porous oxide gel with a monomer and polymerizing it *in situ*. The silica gel is initially transparent and 65 percent porous prior to impregnation. Using transmission electron microscopy, it has been observed that the gel network consists of approximately 50 A spherical particles. By nitrogen adsorption BET, the average pore diameter was determined to be approximately 100A. It has been shown that properties such as strength and elastic modulus are both strongly dependent upon porosity[9,10]. As with strength and elastic modulus for porous bodies, Vicker's hardness of porous silica gels depends inverse exponentially with increasing porosity, as shown in figure 2, in which the porosity was controlled by the catalyst[7,11].

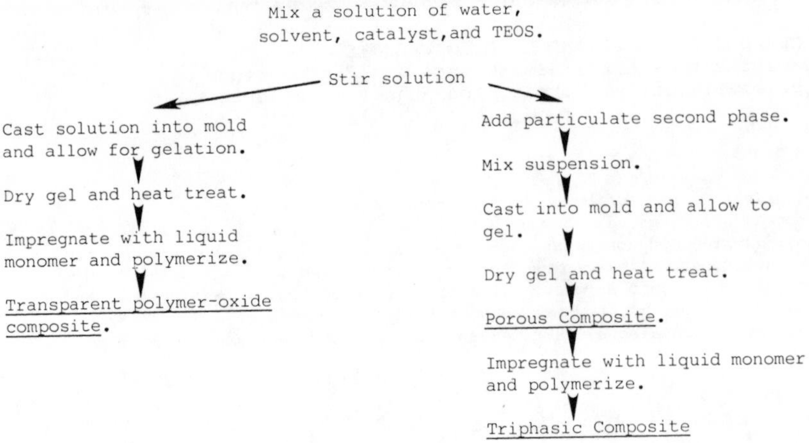

Mix a solution of water,
solvent, catalyst, and TEOS.
Stir solution
Cast solution into mold and allow for gelation.
Dry gel and heat treat.
Impregnate with liquid monomer and polymerize.
Transparent polymer-oxide composite.
Add particulate second phase.
Mix suspension.
Cast into mold and allow to gel.
Dry gel and heat treat.
Porous Composite.
Impregnate with liquid monomer and polymerize.
Triphasic Composite

Figure 1: Preparation of Sol-Gel Derived Composites.

Optical Properties

Many polymers, including PMMA, silicone, and copolymers of PMMA--butyl acrylate (BA), styrene-BA, PMMA-dimethyl butadiene, and silicone--styrene have been impregnated into 65 percent porous silica gel. Some examples of transparent composites of this type are presented in figure 3. The porous silica gels utilized in this study are transparent in air, water, carbon disulfide, and when impregnated with a variety of polymers, all of which have widely differing indices of refraction. The composite index of refraction seems to obey a simple rule of mixtures between the two phases present. The ultrafine perturbations in refractive index are much smaller than the wavelength of visible light.

VICKERS HARDNESS NUMBER
PERCENT POROSITY
0
100
200
300
400
500
600
700
800
10
20
30
40
50
60
70
80

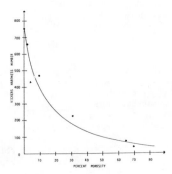

Figure 2: Vicker's Hardness of Silica Gel vs. Porosity.

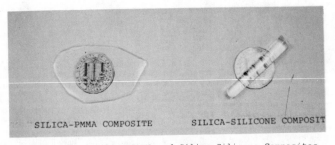

SILICA-PMMA COMPOSITE
SILICA-SILICONE COMPOSIT

Figure 3: Photograph of Silica-PMMA and Silica-Silicone Composites.

Comparison of Bulk vs. Impregnated Polymer

It is well known that as the size of a particle decreases, the surface area to volume ratio increases, such that a 100A particle has a surface area to volume ratio 10,000 times that of 100 micron particle, for example. The thermodynamics of surfaces differ significantly from that of the interior of a material, raising the question as to whether differences can be observed between the bulk polymer and the polymer impregnated into the 100A diameter pores of the silica gel. In table I, data for PMMA prepared under identical conditions, except that the second sample polymerized inside the silica gel, is presented. A significant depression of the glass transition and curing temperatures was observed. The density of the polymer phase was measured to be the same in both cases (1.16 gm/cc).

Table I: Glass Transition and Curing Temperature for PMMA

Sample	Glass Transition Temp.	Curing Temp.
PMMA (bulk)	97.2°C	157.1°C
PMMA (in silica gel)	81.9°C	120.5°C

Elastic Modulus

These transparent composites have relatively high strengths and low densities. Strength data has been published previously[8]. The elastic modulus for 65 percent PMMA-Silica composite is 1.5×10^6 psi. In figure 4, modulus values are plotted for a 35 percent PMMA-silica glass composite made from vycor porous glass and a 65 percent PMMA-silica gel composite. The elastic modulus of both composites fall within the region defined by the Hashin-Shtrikman model.

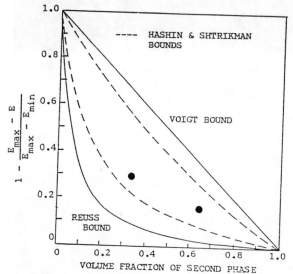

Figure 4: Observed and Predicted Variations of Young's Modulus with Volume Fraction PMMA.

POROUS COMPOSITES

A porous composite can be formed if an "inert" filler is dispersed in a sol-gel solution. Composites have been prepared based upon the following systems: SiO_2-SiC(particles), SiO_2-SiC(whiskers), SiO_2-Al_2O_3, SiO_2-Si_3N_4, SiO_2-TiC, SiO_2-Al(metal dust), SiO_2-SiO_2(microspheres), and SiO_2-SiO_2(fumed powder).

Large, fully dried, crack free, monolithic samples are typically diffi-cult to obtain by sol-gel. The pronounced reduction of drying shrinkage, however, due to the addition of a solid dispersant allows the fabrication of large, crack free composites[8]. In figure 5, a 2.5 inch diameter, one inch thick fully dried composite disc of $33SiC$-$67SiO_2$(by weight percent) is shown. The effect of heat-treatment on the physical properties of the SiO_2--SiC composite is presented in table II. Past 600°C, at which temperature the matrix has reached theoretical density, the porosity and surface area decrease with increasing heat treatment temperature due to pore collapse. The average pore diameter increases, indicating that the smaller pores tend to collapse first.

Figure 5: Photograph of Porous SiC-SiO_2 Composite Gel (scale in inches).

Table II: Porosity, Surface Area, Average Pore Diameter, Apparent Density, and Volume Shrinkage as a Function of Heat Treatment Temperature for 33w/o SiC-SiO_2 Composite Gel.

Heat Treatment Temperature (°C)	Percent Porosity	Average Pore Diameter (anstroms)	Surface Area (m^2/gm)	Apparent Density (gm/cc)	Volume Shrinkage (percent)
25	68	92	397	2.19	68
400	68	90	---	2.20	68
600	72	89	418	2.40	68
800	64	96	193	2.40	--
950	55	--	---	2.40	80
1100	35	262	18	2.40	85

TRIPHASIC COMPOSITES

A new family of "triphasic" composites can be made by impregnating porous composites with a polymer, as shown in figure 1. One system currently under study is SiC-SiO$_2$-PMMA, some properties of which have been published previously[8]. In figure 6, compressive stress is plotted against strain for this system. A remarkable 13.5 percent strain prior to fracture is observed. This is surprising in that the ceramic phase is continuous. Looking closely at this figure, three regions can be discerned, an elastic region, a possible slow crack growth region, and a brief rapid crack growth region. This extensive plastic strain behavior has not been observed for polymer impregnated composites that do not contain a dispersed second phase. This suggests that the SiC phase may act to deflect and/or retard crack propogation.

This composite can be prepared in a variety of configurations with a maximum processing temperature of 60°C. Heat treatment of the gel matrix prior to impregnation, however, can be utilized to modify the properties of the composite.

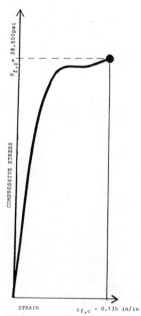

Figure 6: Compression Stress vs. Strain Curve for Fracture of SiC-SiO$_2$-PMMA Composite.

Abrasion Rate

For many applications, the abrasion rate of a material is of great importance. In table III, relative abrasion rates are compared under standardized conditions for 8 materials, including dense silica gel, pure PMMA, and 3 composites prepared in this study. Despite the fact that it contains 68 percent PMMA, which is very soft, the triphasic composite has an abrasion rate comparable to melt cast soda-lime-silicate glass manufactered commercially.

Table III: Relative Abrasion Rates for Selected Materials.

Material	Relative Abrasion Rate (mm^3/cycle)
Commercial Silica Glass	0.012
Diamonite 99% Dense Alumina	0.004
Soda-Lime-Silicate Glass	0.023
Dense Silica Gel	0.014
PMMA	0.349
Silica-63% PMMA Composite	0.092
SiC-SiO$_2$-PMMA Composite	0.027
Silicone-Silica Composite	0.501

220 Grit SiC coated abrasive / 750 gm load

CONCLUSIONS

1. Bulk composites can be prepared via the dispersion of an oxide or non-oxide second phase into a sol-gel matrix.

2. The pore stability and shrinkage behavior of these porous composites can be greatly affected by the quantity, chemistry, and geometry of the second phase dispersant.

3. A wide variety of transparent, polymer-oxide composites can be prepared by the polymer impregnation of porous gels. Moreover, these materials are as strong as melt cast glass and allow for the tailoring of refractive index by a simple rule of mixtures approach.

4. New triphasic composites consisting of a continuous oxide matrix, a ceramic dispersant, and a continuous polymer third phase can be prepared. This low density, non-porous material possesses high fracture ductility, good strength and abrasion resistance, and can be prepared at temperatures slightly above ambient.

ACKNOWLEDGEMENTS

This work was supported by the Directorate of Chemical and Atmospheric Sciences, Air Force Office of Scientific Research. Special thanks to Ms. Doris Plenert and Mr. Tim Joseph.

REFERENCES

1. J. D. Mackenzie, Ultrastructure Processing of Ceramics, Glasses, and Composites, Ed. by L. L. Hench and D. R. Ulrich (J. Wiley & Sons, N.Y.:1984).

2. S. P. Mukherjee and J. Zarzycki, J. Am. Ceram. Soc., 62 (1979).

3. D. W. Hoffman, R. Roy, and S. Komarneni, ibid. 67 (1984) 468.

4. R. A. Roy and R. Roy, Mats. Res. Bull.,19 (1984) 169.

5. J. J. Lannutti and D. E. Clark, papers B4.6 & B4.7, 1984 Spring Meeting Materials Research Society.

6. G. S. Moore, N. Toghe, and J. D. Mackenzie, presented at the 36th Annual Pacific Coast Meeting, American Ceramic Society, 1983.

7. E. J. A. Pope and J. D. Mackenzie, J. Non-Cryst. Sol., in press.

8. E. J. A. Pope and J. D. Mackenzie, "Porous and Dense Composites from Sol--Gel", proceedings of the 21st University Conference on Ceramic Science, July, 1985.

9. R. M. Spriggs, J. Am. Ceram. Soc., 45 (1962) 454.

10. E. Ryshkewitch, ibid, 36 (1953) 65.

11. E. J. A. Pope, Masters Thesis, Univ. of Calif.,Los Angeles, 1985.

SILICON CARBIDE FROM ORGANOSILANES AND
APPLICATION IN SILICA GEL GLASS COMPOSITES

BUTRAND I. LEE* AND L.L. HENCH
Advanced Materials Research Center, College of Engineering, One Progress Blvd., #14, Alachua FL 32615

ABSTRACT

Several organosilanes were crosslinked and pyrolyzed to produce silicon carbide (SiC). Use of chemical crosslinking agents required lower temperatures and shorter times for curing and increased SiC char yields. Impregnating the silanes into sol-gel derived silica monoliths followed by crosslinking and pyrolysis resulted in hard and tough SiC/SiO_2 composite bodies.

INTRODUCTION

Since Yajima and coworkers [1-4] produced silicon carbide (SiC) fibers from a polymer precursor starting from polydimethylsilane, several potentially superior precursors to Yajima's have been reported [5-10]. In the Yajima process, see below, the starting precursor polymer is converted to polycarbosilane by thermal rearrangement followed by crosslinking via oxygen.

Lee and Hench [11] using polydimethyl methylphenylsilane (polysilastyrene:PSS),** vinylic silanes*** and allyl modified PSS showed that the silanes can be converted to SiC without the intermediate step of the thermal rearrangement of polydimethylsilane to polycarbosilane and without the oxygen crosslinking.

Further work has shown that various SiC precursor polymers [12] can be used with nonoxygen crosslinking and pyrolysis to obtain SiC. SiC/SiO_2 molecular composites can be made using this process as described in ref. [13].

---

*Currently at Department of Ceramic Engineering, Clemson University, Clemson, SC 29634
**In collaboration with R. Sinclair (3M Co.) and R. West (Univ. of Wisconsin).
***Courtesy of Dr. Curt Shilling of Union Carbide, Terrytown, NY.

EXPERIMENTAL

PSS was synthesized by copolymerizing dichloromethylphenyl silane and dichlorodimethyl silane in the presence of metallic sodium.

$$Me_2SiCl_2 \atop PhMeSiCl_2 \quad \xrightarrow[\text{Toluene, } 110^\bullet\text{C}]{Na} \quad \left[Me_2Si - \underset{\underset{Ph}{|}}{\overset{\overset{Me}{|}}{Si}} \right]_n \qquad (1)$$

Allyl modified PSS was prepared by substituting 10 mole % dichloro-allylmethylsilane in place of dimethyldichlorosilane. Details on the reactions and fractionation are given by Lee [12]. The structure formulas of the polysilanes are:

PSS-o (oligomer)
PSS-p (polymer)

Allilic PSS
(A-PSS)

Vinylic Silane
(ViSO: oligomer)
(ViSP: polymer)

Crosslinking of the silanes was carried out by thermal treatment, addition of a chemical free radical initiator (CFRI: benzoyl peroxide, azobisbutyronitrile and dicumyl peroxide: DCP), chloroplatinic acid and - irradiation. Pyrolysis of the polymers was done in a furnace with a flowing inert gas and also in a DuPont TGA 951 Thermogravimetric Analyzer with a heating rate of 10°C/min. Differential scanning calorimetry (DSC) using the same DuPont Model 951 was carried out with a heating rate of 5°C/min. Crosslinking reaction products of PSS via DCP were identified by gas chromatographs (GC) in order to establish the reaction mechanism. These silanes were used to impregnate silica gel matrices, see ref. 12. Microhardness of the pyrolyzed polysilane/SiO$_2$ gel composites was determined with a Kentron Microhardness Tester with 1 kg load and 5-10 measurements per sample.

RESULTS

Among the CFRI's investigated, only DCP in the range of 2-15 wt% showed a positive crosslinking reaction. In the crosslinking of ViSP without DCP under the same conditions as with DCP (110°C), no solidification was observed within 20 hrs. However, curing at 30°C higher temperature, 140°C, resulted in solidification of the liquid ViSP, signifying crosslinking. The difference in the chemical structure of the ViSP samples crosslinked thermally compared with ViSP samples with DCP is shown by FTIR spectra in Fig. 1.

Differential scanning calorimetry thermograms (DSC) are given in Figs. 2 and 3 to compare crosslinking mechanisms. In Fig. 3, DSC's of oligomer, polymer PSS, and allylic PSS are compared.

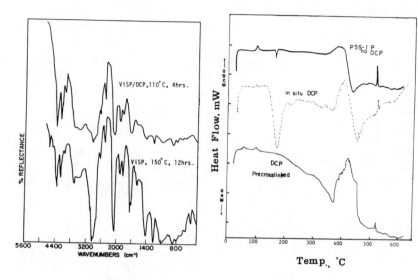

Fig. 1. FTIR spectra of crosslinked vinylic silane showing the effect of DCP, temperature and time.

Fig. 2. DSC thermograms of PSS and PSS/DCP before and after the cross-linking reaction

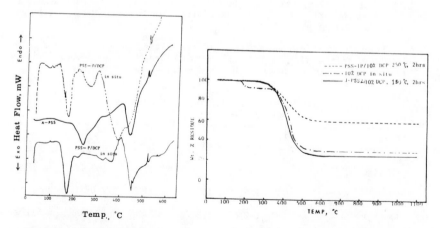

Fig. 3. DSC thermograms of PSS-oligomer and PSS-polymer in the presence of DCP and A-PSS as synthesized.

Fig. 4. TGA char yields of SiC for PSS showing the effect of time and temperature of the crosslinking reaction.

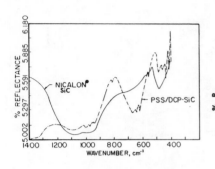

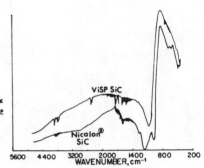

Fig. 5. FTIR spectra of Nicalon®
and PSS/DCP-SiC after pyrolysis
at 1050°C.

Fig. 6. FTIR spectrum of SiC
obtained from ViSP/DCP compared with
the spectrum of Nicalon®. Note the
absence of Si-O band in ViSP/SiC.

TGA char yields of PSS and the effect of DCP are shown in Fig. 4. The TGA yields of ViSP/6% DCP and ViSO/6% DCP are 72% and 25% respectively. FT-IR spectra of the pyrolyzed products are given in Fig. 5 and 6, along with Nicalon® SiC made by the Yajima process. The spectra show the absorption band characteristic of a Si-C stretching vibration at 793 cm^{-1} along with a small SiO$_2$ band at ~1040 cm^{-1}.

The char yield of PSS without crosslinking was <20 wt%, which is close to the char yield of ViSO, while the char yield of the PSS/DCP systems showed 52-61 wt% SiC. Results of a diamond point microhardness (DPN) test for SiO$_2$ gel impregnated with silanes pyrolyzed at various temperature are given in Table I and Fig. 7.

Fracture toughness values (K_{IC}) of the silane impregnated silica gels calculated from a Vickers indentation crack lengths [14] showed 80-100% greater K_{IC} values as compared with commercial fused silica glass.

Table I. Microhardness of SiC/SiO$_2$ Composites as a Function
of Pyrolysis Temperatures.

Composites & Control	DPN T →	520°C	600°C	700°C	800°C	900°C
OA-ViSO		139±1	233±15	239±14	669±173	814±200
OA-gel		146±58	157±1	174±25	210±51	305±103
FA-ViSP		192±66	342±75	351±87	403±46	550±180
FA-gel		129±64	171±24	210±51	291±67	

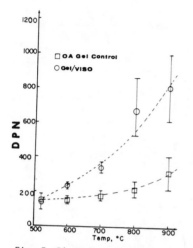

Fig. 7. Diamond pyramid hardness
of silica matrix gel and poly-
silane impregnated silica matrix
gel monolith as a function of
pyrolysis temperature.

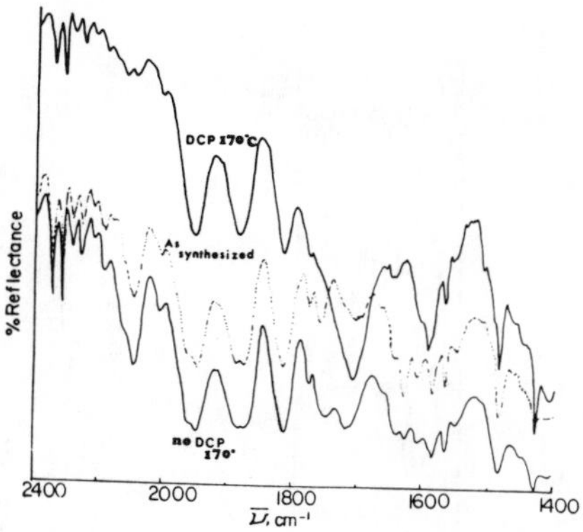

Fig. 8. FTIR specta of allylic PSS
before and after crosslinking.

DISCUSSION

All polymerization reactions carried out for PSS produced ~3X larger
oligomer than polymer fractions. The oligomer could be repolymerized and
crosslinked by DCP, however. This may be important in the impregnation of
porous ceramic bodies in order to strengthen them. Oligomers of PSS are
believed to be in cyclic form so that the crosslinking mechanism is
expected to be different from the polymer fraction by opening up the ring.
This is shown by a higher TGA yield of the oligomer (67%) than that of a
polymer [11]. The DSC data (Fig. 3) show that oligomers require a higher
temperature for crosslinking than polymers and the decomposition begins
50°C lower than the polymers. The high char yield of the oligomers should
be caused by partial decomposition of the precursor at the crosslinking
temperature. Among the several CRFIs studied, only DCP yielded an insoluble and infusible solid of polysilanes. This is probably due to the
active methyl radical, which was not present in any other CFRI used.

An allyl group on PSS is advantageous in crosslinking. As shown in
Fig. 3 and 8, A-PSS can be crosslinked thermally or via DCP. A-PSS is
also crosslinkable by a H_2PtCl_6 catalyst. Greater than 8×10^{-7} mol Pt^{4+}
per 1 g A-PSS was required for an effective crosslinking [12]. Crosslinking of A-PSS by Pt^{4+} must be between Si-H and C=C, Eq. (2) [15].

$$\text{SiH} + \text{C=C} \xrightarrow{\text{catalyst}} \text{Si-C-C-H} \qquad (2)$$

Fig. 9. FTIR spectra of a region showing the effect of crosslinking of Si-H band intensities at -2080 cm^{-1}.

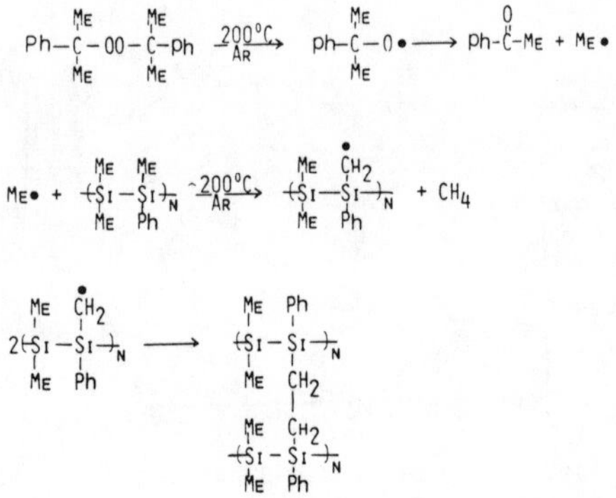

Fig. 10. The crosslinking mechanism of PSS by DCP.

Vinylic silanes can be crosslinked both thermally and via the CFRI, DCP. With DCP, crosslinking is faster and required a lower temperature: 110°C for 4 hours as compared to 150°C for 12 hours without DCP. Without DCP, 150°C for 12 hours treatment still did not produce complete crosslinking, as shown by the large Si-H IR peak in Fig. 1. In Fig. 9, the as-received PSS shows a sharp and stong absorption band for Si-H at 2100 cm^{-1}, a medium sized band for γ-ray irradiated PSS, and a small band for 10 wt% DCP treated PSS. This means that PSS crosslinking can occur between Si-H's as well as by methyl free radicals from DCP (Fig. 10) at higher temperatures. The bond energy of Si-H (314 KJ/mole) [16] is much smaller than that of ≡Si-CH$_2$-H (418KJ/mol) [17] hence the crosslinking of PSS by DCP proceeds with Si-H bonds breaking at ∿150°C followed by Si-C-C-Si linkages via a methyl radical at ∿250°C. In Fig. 2, as-received PSS shows a small exothermic peak at ∿160°C, which should correspond to the crosslinking reaction via Si-H. Rearrangement of the polymer chain is thought to occur at ∿400°C. The small spikes at 100°C correspond to water evaporation. The reason for sharp endothermic peaks at ∿520°C is not known.

A GC study of PSS [12] with DCP showed a large amount of methane and acetophenone, which are some of the products from the proposed crosslinking reaction given in Fig. 10. The amount of methane is too much to come from the Si-H coupling alone. Thus, the methane must be formed by the methyl radicals of DCP after abstracting methyl hydrogen from Si-CH$_3$. The possibility of crosslinkage via Si-Ph-Ph-Si is doubtful because of the greater bond energy for ⊗—H (112 Kcal/mol) than for CH$_2$-H (104 Kcal/mol) [18]. The minimum at ∿200°C for PSS with in-situ DCP crosslinking must be due to the decomposition of DCP. Under the heating rate of DSC (5°C/min), the slow DCP crosslinking reaction was not able to keep up with the heat-ing rate. The exothermic reaction was incomplete until the temperature was ~220°C. This supports the previous observation of incomplete cross-linking with DCP at temperatures below 200°C and the low TGA char yield of the in situ DCP crosslinked PSS (Fig. 4) In the case of more reactive ViSP, temperatures above 110°C with DCP and above 150°C without DCP are required for complete crosslinking (Fig. 1).

It has been demonstrated that incorporating a small amount of SiC phase into a pure SiO_2 glass matrix in a form of organosilane increases the microhardness of the glass material. The acid catalyzed gels used in this work produced cracks and/or foaming at temperatures ~850°C because of the small mean pore radius in the range of ~15 Å [12]. The pores are very small for gases to leave the gel structure before pore closure at the surface.

CONCLUSION

The exploratory organosilane (OS) precursors used in this work have the potential to be formed into desired shapes using conventional low temperature plastic processing, then pyrolyzed to obtain SiC material with a yield that is nearly the theoretical limit. The highest char yield is given by the polymer fraction of vinylic silane (72%). An allyl group on the PSS chain improves crosslinkability. A complete crosslinking of OS precursors increased the char yield of SiC. For complete crosslinking, DCP with temperatures greater than 250°C for PSS and greater than 130°C for vinylic silanes are required. Other combinations of the functional groups such as Si-H, vinyl, allyl, etc. should further improve crosslink-ability and ceramic yield. Char yield of SiC is roughly shown to be a functon of concentration of the crosslinking agent DCP, and a function of temperature and time. Sol-gel derived monolithic silica glasses can be reinforced by impregnating the SiC phase by way of an organosilane. The reinforcing effect measured by microhardness is nearly three times greater for the SiC infiltrated composite glass after heating to 900°C, than the matrix under the same condition.

ACKNOWLEDGEMENT

Authors are grateful for the financial support of AFOSR Contract #F49620-83-C-0072 for this support of this research.

REFERENCES

1. S. Yajima, J. Hayashi, and M. Omori, Chem. Lett., 931 (1975).
2. S. Yajima, K. Okamura, and J. Hayashi, Chem. Lett., 1209 (1975).
3. S. Yajima, K. Okamura, J. Hayashi, and M. Omori, J. Am. Ceram. Soc., 59, 324 (1976).
4. S. Yajima, T. Shishido, and H. Kayano, Nature, 264, 237 (1976).
5. C. Schilling, J. Wesson, and T. Williams, Am. Ceram. Soc. Bull., 62, 912 (1983).
6. C. Schilling, J. Wesson, and T. Williams, U.S. Patent No. 4,414,403 (1983).
7. C. Schilling, T. Williams, ACS Polymer Reprints, 25, 1 (1984).
8. R. West, L. David, P. Djurovich, H. Yu, and R. Sinclair, Am. Ceram. Soc. Bull., 62 899 (1983).
9. R. Baney and J. Gaul, U.S. Patent No. 4,310,651 (1982).
10. R. Baney and J. Gaul, U.S. Patent No. 4,298,558 (1981).

11. B.I. Lee and L.L. Hench, "Crosslinking and Pyrolysis of Polysilanes," in Ultrastructure Processing of Ceramics, Glasses, and Composites, L.L. Hench and D.R. Ulrich, eds., John Wiley and Sons, New York, 1986, in press.

12. B.I. Lee, Chemically Derived Ceramic Composites, Ph.D. Dissertation, University of Florida, Gainesville, FL, 1986.

13. B.I. Lee and L.L. Hench, presented at the 10th Annual Conference on Composites and Advanced Ceramic Materials, Jan. 19-24, 1986, Cocoa Beach, FL.

14. G.R. Antis, P. Chantikul, B.R. Lawn, and D.B. Marshall, J. Am. Ceram. Soc., 64, 532 (1981).

15. J. Speïer, J. Webster, and G. Barnes, J. Am. Chem. Soc., 79, 974 (1957).

16. S. Yajima, Y. Hasegawa, J. Hayashi, and M. Iimura, J. Mat. Sci., 13, 2569 (1978).

17. J. Roberts and M. Caserio, Basic Principles of Organic Chemistry, 2nd edition, W.A. Benzamin, Inc., Menlo Park, California, 1977, p. 92.

18. CRC Handbook of Chemistry and Physics, 52nd edition, p. F-186.

Author Index

824

Subject Index